Ecological Connectivity among Tropical Coastal Ecosystems

Ivan Nagelkerken

Editor

Ecological Connectivity among Tropical Coastal Ecosystems

Springer

Editor
Ivan Nagelkerken
Radboud University Nijmegen
Faculty of Science
Department of Animal Ecology
and Ecophysiology
P.O. Box 9010
6500 GL Nijmegen
the Netherlands
i.nagelkerken@science.ru.nl

ISBN 978-90-481-2405-3   e-ISBN 978-90-481-2406-0
DOI 10.1007/978-90-481-2406-0
Springer Dordrecht Heidelberg London New York

Library of Congress Control Number: 2009926883

Cover photo: Coastline of Bawi Island (Zanzibar, Tanzania) with exposed reef at low tide. Photo by
Martijn Dorenbosch

Printed on acid-free paper

Springer is part of Springer Science+Business Media (www.springer.com)

Foreword

The scale of effects of human activities on our ecosystem services in the past half century has increased to the level where we are now compelled to consider interactions among complex systems for responsible management of our resources. Human activities have been causing global effects on climate, the abundance and distribution of nutrients, and the sea level and chemistry of the oceans. There have been a number of books in the past half century on the ecology and management of coral reefs, of mangroves, and of seagrass meadows as separate systems. This book on 'Ecological connectivity among tropical coastal ecosystems' is timely because it is focused on providing understanding of the higher level of interactions between these systems. Ivan Nagelkerken has spent his career determining the extent and complexities of population connectivities of fishes among tropical coastal habitats. He now takes on the role of editor to pull together biogeochemical, ecological, and population linkages among coastal habitats and guiding us to conclusions for management policies and socioeconomic implications.

The capacity of systems for self-sustainability can increase with diversity at all levels. A more diverse genotype provides a greater potential capacity for a species to adapt to climate change and other large-scale effects of human activities. A greater species diversity of primary producers, framework constructing species, herbivores, and predators provide potential capacity for a habitat or ecosystem to accommodate eutrophication and other effects of human activities. We must now include consideration of the diversity of interactions among habitats. Coral reefs protect inshore habitats from wave action while mangroves can buffer coral reefs from terrestrial input of sediment and other pollutants, and so while the coastal habitats can exist in isolation, they are probably more resilient to large-scale changes from human activities when they constitute a diverse interacting seascape. This book addresses not only these interactions of coastal habitats among themselves, but also considers interconnectivity and relationships with surrounding terrestrial uplands, rivers, and offshore marine systems.

Corals and mangroves are 'foundation species' in that they actually construct and expand the coastline and provide the physical structure of much of the tropical coastal ecosystem. The popular term 'reclamation' demonstrates the lack of understanding of the general public of the fundamental importance of these systems. It is inappropriate to assume the right to 'take back' land originally created by mangroves and corals. The habitats are not just 'foundations' by themselves, but they also serve as parts of an interacting system. Many fish and crustacean species of commercial importance spend different stages in their life cycles in different habitats and for some, the neighboring habitat is required. Some fishes and crustaceans move between habitats on a daily basis, providing a daily interconnection of biomass, nutrients, and effects of predation. This book is needed to summarize and clarify the complex interactions that lead to the ecosystem services provided by these coastal habitats. By providing the latest information on the ecological interactions among the coastal habitats in terms of physical processes, nutrients, organic matter, living organisms, and effects of predation, shelter, and substrata, and by providing the latest techniques in studying these processes, this book addresses the fundamental importance of dealing with the needs and perspectives of local human populations.

Although coral reefs, mangroves, and seagrass meadows have among the highest gross primary productivity of terrestrial or marine ecosystems, they are also in especially vulnerable situations. Unfortunately, the best habitats for productivity, diversity, and coastal formation are also the most beneficial and logistically efficient for human settlement and activity. Sixty percent of the human population lives within 50 miles of the ocean coasts. Anthropogenic and natural disturbances such as sea level rise, sedimentation, and cyclones are especially focused at the boundaries of the three coastal ecosystems. With human population growth and with the increased technological abilities of humans to harvest and remove resources at a greater per capita rate, degradation of coastal habitats and resources are increasing with positive feedback from the increasing demands of the growing human population. The need for increased understanding of the interactions among these essential coastal habitats is becoming more critical as the demands of growing human populations for organic resources and ecological services are increasing. I hope this book is distributed broadly and rapidly so that the decision-makers and managers of tropical coastal resources and development are brought into awareness of the need to not just protect habitat and species, but to sustain ecosystem services and resources by maintaining the higher level interactions among coastal systems.

Professor in coral reef ecology and management,   Dr. Charles Birkeland
Editor 'Life and Death of Coral Reefs'

Preface

The idea to edit this book started with an e-mail from Suzanne Mekking of Springer Science and Business Media who wanted to make an appointment to talk about current needs for new books in the field of aquatic sciences. During that meeting, she attempted to persuade me into writing a book about my field of research – ecological interactions among coral reefs, mangroves, and seagrass beds by reef fishes. At first, I was not interested due to the large amount of work this would encompass, and my already overloaded work schedule. After giving it some thought over the following month or so, I quickly realized that many advances on this topic had been made in the last decade, and that this would be the perfect time to put together the scattered knowledge on this topic, for the first time, in the form of an edited book. The fast demise and degradation of coral reefs, mangroves, and seagrass beds worldwide also was an important consideration to edit this book, hoping that it would increase the appreciation for these tropical coastal habitats, and provide insights that could contribute to their conservation. Within a month, I had made a list of urgent topics needing review, and had contacted various specialists from around the world requesting their contribution to the book. I was delighted by the fast and enthusiastic response from the majority of the people that I approached. Aside from a few individuals not keeping their promise to contribute a chapter, I have been exempt of various frustrations that are known to occur when editing a book. In the following two years, 28 authors from Australia, USA, and various European countries, worked hard to bring together this book. I thank them for this great effort, and for responding to my requests for improvements, changes, and help in a timely manner. The quality of the book could not have been improved without the help of many peer reviewers. I am extremely grateful to the following people who have provided fast and critical reviews of the various book chapters: Aaron Adams, Charles Birkeland, Steve Blaber, Dave Booth, Steven Bouillon, Paul Chittaro, Patrick Collin, Stephen Davis, Thorsten Dittmar, Ashton Drew, Dave Eggleston, Craig Faunce, Bronwyn Gillanders, William Gladstone, Mick Haywood, Alan Jones, Rob Kenyon, Craig Layman, Jeff Leis, Christian Lévêque, Ivan Mateo, Bob McDowall, Jan-Olaf Meynecke, Rick Nemeth, Heather Patterson, Simon Pittman, Yvonne Sadovy, Joe Serafy, Steve Simpson, and Marieke Verweij. I am also indebted to Charles Birkeland for taking the time to write a foreword for the book, and to Martijn Dorenbosch for providing the front cover picture for the book. Lastly, I thank my wife Shauna Slingsby and my son Diego Nagelkerken for their support and understanding, during the many days, nights, weekends, and holidays that I was working on this book instead of being with them. Now that the book is finished, I hope it will prove valuable for ecosystem managers, fisheries ecologists, graduate students, and other researchers in the field.

Ivan Nagelkerken

Contents

Contributors

Aaron J. Adams Center for Fisheries Enhancement, Habitat Ecology Program, Mote Marine Laboratory, Charlotte Harbor Field Station, P.O. Box 2197, Pineland, FL 33945, USA, aadams@mote.org

Michael Arvedlund Reef Consultants, Rådmand Steins Allé 16A, 2-208, 2000 Frederiksberg, Denmark, arvedlund@speedpost.net

Stephen J.M. Blaber CSIRO Marine and Atmospheric Research, P.O. Box 120, Cleveland, Queensland 4163, Australia, steve.blaber@csiro.au

Steven Bouillon Katholieke Universiteit Leuven, Department of Earth and Environmental Sciences, Kasteelpark Arenberg 20, B-3001 Leuven, Belgium; and Vrije Universiteit Brussel, Department of Analytical and Environmental Chemistry, Pleinlaan 2, B-1050 Brussels, Belgium, steven.bouillon@ees.kuleuven.be

Chris Caldow NOAA/NOS/NCCOS/CCMA Biogeography Branch N/SCI-1, 1305 East-West Highway, Silver Spring, MD 20910, USA, chris.caldow@noaa.gov

Rod M. Connolly Australian Rivers Institute – Coasts and Estuaries, and School of Environment, Griffith University Gold Coast campus, Queensland 4222, Australia, r.connolly@griffith.edu.au

Stephen E. Davis III Department of Wildlife and Fisheries Sciences, Texas A&M University, College Station, TX, USA 77843-2258, sedavis@tamu.edu

Thorsten Dittmar Max Planck Research Group for Marine Geochemistry, Carl von Ossietzky University, Institute for Chemistry and Biology of the Marine Environment, 26111 Oldenburg, Germany, tdittmar@mpi-bremen.de

John P. Ebersole Biology Department, University of Massachusetts Boston, 100 Morrissey Boulevard, Boston, MA 02125, USA, john.ebersole@umb.edu

Craig H. Faunce National Marine Fisheries Service, Alaska Fisheries Science Center, 7600 Sand Point Way NE, Seattle, Washington 98115, USA, Craig.Faunce@noaa.gov

Thomas K. Frazer University of Florida, Institute of Food and Agricultural Sciences, School of Forest Resources and Conservation, Program in Fisheries and Aquatic Sciences, Gainesville, FL 32653, USA, frazer@ufl.edu

Bronwyn M. Gillanders Southern Seas Ecology Laboratories, DX 650 418, School of Earth and Environmental Sciences, University of Adelaide, SA 5005, Australia, bronwyn.gillanders@adelaide.edu.au

William Gladstone School of Environmental and Life Sciences, University of Newcastle Central Coast, P.O. Box 127, Ourimbah NSW 2258, Australia, William.Gladstone@newcastle.edu.au

Rikki Grober-Dunsmore Institute of Applied Sciences, Private Bag, Laucala Campus, University of South Pacific, Suva, Fiji Islands, dunsmore_l@usp.ac.fj, rikkidunsmore@gmail.com

Michael D.E. Haywood CSIRO Division of Marine and Atmospheric Research, P.O. Box 120, Cleveland, 4163, Queensland, Australia, mick.haywood@csiro.au

Rudolf Jaffé Southeast Environmental Research Center and Department of Chemistry, Florida International University, Miami, Florida 33199, USA, jaffer@fiu.edu

Kathryn Kavanagh School of Marine and Atmospheric Sciences, Stony Brook University, Stony Brook, NY 11794, USA, kathryn_kavanagh@yahoo.com

Matthew S. Kendall NOAA/NOS/NCCOS/CCMA Biogeography Branch N/SCI-1, 1305 East-West Highway, Silver Spring, MD 20910, USA, Matt.Kendall@noaa.gov

Robert A. Kenyon CSIRO Division of Marine and Atmospheric Research, P.O. Box 120, Cleveland, 4163, Queensland, Australia, Rob.Kenyon@csiro.au

Boris Koch Alfred Wegener Institute for Polar and Marine Research, Department of Ecological Chemistry, Am Handelshafen 12, D-27570 Bremerhaven, Germany, boris.koch@awi.de

Uwe Krumme Leibniz-Center for Tropical Marine Ecology (ZMT), Fahrenheit-strasse 6, 28359 Bremen, Germany, uwe.krumme@zmt-bremen.de

Craig A. Layman Marine Sciences Program, Department of Biological Sciences, Florida International University, 3000 NE 151st Street, North Miami, Florida 33181, USA, cal1634@yahoo.com

Diego Lirman Rosenstiel School of Marine and Atmospheric Science, University of Miami, 4600 Rickenbacker Causeway, Miami, FL, USA 33149-4000, dlirman@rsmas.miami.edu

David A. Milton Wealth from Oceans Flagship, CSIRO Marine and Atmospheric Research, P.O. Box 120, Cleveland, Queensland 4163, Australia, david.milton@csiro.au

Ivan Nagelkerken Department of Animal Ecology and Ecophysiology, Institute for Water and Wetland Research, Faculty of Science, Radboud University Nijmegen, Heyendaalseweg 135, P.O. Box 9010, 6500 GL Nijmegen, the Netherlands, i.nagelkerken@science.ru.nl

Richard S. Nemeth Center for Marine and Environmental Studies, University of the Virgin Islands, MacLean Marine Science Center, 2 John Brewer's Bay, St. Thomas, US Virgin Islands, 00802, rnemeth@uvi.edu

Simon J. Pittman NOAA/NOS/NCCOS/CCMA Biogeography Branch N/SCI-1, 1305 East-West Highway, Silver Spring, MD 20910, USA; and Marine Science Center, University of the Virgin Islands, St. Thomas, United States Virgin Islands, 00802, USA, Simon.Pittman@noaa.gov

Jeffrey R. Wozniak Department of Wildlife and Fisheries Sciences, Texas A&M University, College Station, TX, USA 77843-2258, wozniak@tamu.edu

Chapter 1
Introduction

Ivan Nagelkerken

Coral reefs, mangrove forests, and seagrass beds are dominant features of tropical coastlines. These tropical coastal ecosystems have long been known for their high productivity, rich biodiversity, and various ecosystem services (Harborne et al. 2006). For example, coral reefs have important economic, biological, and aesthetic values; they generate about $30 billion per year in fishing, tourism, and coastal protection from storms (Stone 2007). The extent of mangroves has frequently been linked to a high productivity in adjacent coastal fisheries (Manson et al. 2005, Meynecke et al. 2008, Aburto-Oropeza et al. 2008) which can approach economic values of up to US$ 16,500 per hectare of mangrove (UNEP 2006). Nutrient cycling of raw materials by seagrass beds has been estimated to value US$ 19,000 ha^{-1}. yr^{-1} (Constanza et al. 1997).

In the last few decades, these ecosystems have suffered from serious degradation due to human and natural impacts, such as pollution, eutrophication, sedimentation, overexploitation, habitat destruction, diseases, and hurricanes (Short and Wyllie-Echeverria 1996, Alongi 2002, Hughes et al. 2003). It has been estimated that 20% of the world's coral reefs have been destroyed, while 50% are under direct or long-term risk of collapse (Wilkinson 2004). Mangroves and seagrass beds have declined up to 35% worldwide in their surface area (Shepherd et al. 1989, Valiela et al. 2001, Hogarth 2007). Of the island coral reef fisheries, 55% is currently unsustainable (Newton et al. 2007). Overfishing is one of the principal threats to coral reef health and functioning, and has led to detrimental trophic cascades and phase shifts from coral reefs to macroalgal reefs (Jackson et al. 2001, Hughes et al. 2007).

The need for the protection of these ecosystems is clear, but from a management perspective their connectivity has hardly been taken into consideration (Pittman and McAlpine 2003). Earlier research and management efforts have typically focused on single ecosystems. Although these coastal ecosystems can thrive in isolation (Birkeland and Amesbury 1988, Parrish 1989), it is clear that where

I. Nagelkerken (✉)
Department of Animal Ecology and Ecophysiology, Institute for Water and Wetland Research, Faculty of Science, Radboud University Nijmegen, Heyendaalseweg 135, P.O. Box 9010, 6500 GL Nijmegen, the Netherlands
e-mail: i.nagelkerken@science.ru.nl

I. Nagelkerken (ed.), *Ecological Connectivity among Tropical Coastal Ecosystems*,
DOI 10.1007/978-90-481-2406-0_1, © Springer Science+Business Media B.V. 2009

they occur together considerable interactions may occur (Ogden and Zieman 1977, Sheaves 2005, Valentine et al. 2008, Mumby and Hastings 2008). We are just beginning to understand their ecological linkages, but for optimal management an ecosystem-approach is needed where cross-ecosystem linkages are also considered (Friedlander et al. 2003, Adams et al. 2006, Aguilar-Perera and Appeldoorn 2007, Mumby and Hastings 2008).

Cross-ecosystem interactions can largely be subdivided into biological, chemical, and physical interactions (Ogden 1997). Examples of interactions are exchange of fish, shrimp, nutrients, detritus, water bodies, sediment, and plankton among systems. The type of ecosystem connectivity that is covered in this book refers to ecological interactions among ecosystems. The term 'ecological connectivity' is used here as the book is focused on interactions among ecosystems by movement of animals, and by exchange of nutrients and organic matter which form part of the ecological processes in these systems. In the last decade or so, an increase in knowledge has been gained on cross-ecosystem interactions in the tropical seascape warranting a comprehensive review of this topic, as presented in this book for the first time. The major focus is on the coral reef, mangrove, and seagrass ecosystems, and on interactions that result from the mutual exchange of nutrients, organic matter, fish, and crustaceans. Bringing together the existing knowledge on this topic will hopefully contribute to a better appreciation for these systems, provide insights into the mechanisms that underlie their ecological linkages, and provide tools and information for more effective management.

Early studies investigating cross-ecosystem ecological linkages in the tropical seascape focused, amongst other things, on the concept of mangrove outwelling which postulated that detritus from mangrove ecosystems fuels adjacent food webs (Odum 1968). Other early connectivity research focused more on feeding migrations and degree of overlap in fish faunas among ecosystems (Randall 1963, Ogden and Buckman 1973, Ogden and Ehrlich 1977, Ogden and Zieman 1977, McFarland et al. 1979, Weinstein and Heck 1979), or migration by decapods from nearshore to offshore areas (Iversen and Idyll 1960, Costello and Allen 1966, Lucas 1974, Kanciruk and Herrnkind 1978). These studies were predominantly done in the Caribbean region, particularly on grunt species (Haemulidae) and penaeid shrimp, and as a result our understanding of the patterns and mechanisms playing a role in the much larger Indo-Pacific region remains hampered and is still debated (Nagelkerken 2007).

This book is not exhaustive for all existing interactions among tropical ecosystems, as this is too much to review within a single book. Hydrological connectivity, i.e., resulting from exchange of water bodies and sediment, is an important type of physical interaction. A very recent and comprehensive book entitled 'Estuarine ecohydrology' by Wolanski (2007) is recommended for further reading. Another important omission in the current book is that of ecosystem linkages by pelagic larvae of marine fauna. The buzzword 'connectivity' has mainly been used for this type of oceanographic connectivity, i.e., how reefs and different geographic areas are connected by flow of larvae due to oceanic currents and swimming capabilities of fish larvae. Recent reviews include those by Cowen (2006), Cowen et al. (2006), and

Leis (2006). Another topic that is not covered in detail in this book is how climate change and the resulting increase in seawater levels and/or outflow from rivers will affect the interactions among and functioning of tropical ecosystems (e.g., Roessig et al. 2004, Day et al. 2008, Gilman et al. 2008, but see Chapters 3, 9, and 16).

The present book consists of four parts, each covering a different topic: biogeochemical linkages, ecological linkages, tools to study these linkages, and management and socio-economic implications. Part 1 starts with the biogeochemical linkages among tropical ecosystems. Chapter 2 reviews the exchange of nitrogen and phosphorus among coastal systems, while Chapter 3 focuses on the exchange of organic and inorganic carbon. Various pathways of exchange are discussed in these two chapters, such as water-mediated fluxes, biogeochemical cycles, and movement by marine fauna. Anthropogenic and terrestrial inputs into tropical coastal systems are examined, including the effects of human perturbations and climate change. The importance of carbon exchange among systems for faunal and microbial communities is evaluated.

In Part 2, eight chapters review the ecological linkages among tropical coastal ecosystems. Chapter 4 starts with examining how reefs are connected through spawning migrations of fish and decapods, and the effects of these migrations on local food webs. Reference is also made to species that link shallow estuarine habitats with offshore marine areas through spawning migrations. Many demersal animals living in tropical coastal habitats have a pelagic larval stage before starting their benthic life phase. Chapter 5 reviews the senses and cues used by these pelagic larvae to find their respective settlement habitats in the tropical seascape. The life stage around settlement is characterized by heavy mortality and thus has important demographic implications. Chapter 6 reviews various mechanisms during the early life phase of fish and decapods that affect their distribution and abundance. After settlement, animals may use multiple tropical coastal habitats at one time, or shift between them through ontogeny. Chapter 7 evaluates the various types of ontogenetic habitat shifts for decapods and discusses several underlying mechanisms. During their residency in coastal habitats, animals also connect habitats on a short time scale, through diel and tidal migrations. This is often based on connecting resting and feeding sites, and is reviewed in Chapter 8. Rivers form corridors for migrating animals between inland freshwater areas, coastal estuaries, and offshore marine habitats. The ways in which these ecosystems are connected by diadromous fishes is discussed in Chapter 9. As freshwater flow is the main physical driver for this connectivity, changes in flow due to global warming and construction of dams is also assessed. Shallow coastal areas are assumed to function as important nurseries for juveniles of a variety of fish and decapod species that live on coral reefs or offshore areas as adults. The existing evidence for this concept is reviewed in Chapter 10, with reference to the underlying mechanisms. The nursery role of tropical habitats is affected by many sources of variability. Chapter 11 evaluates these sources and how they have caused different conclusions on the nursery function of these habitats.

Our understanding of the ecological connectivity among tropical coastal ecosystems has been partly impeded by the lack of (advanced) techniques to measure connectivity. Only quite recently have modern techniques become available due to technological advancements. Part 3 reviews various advanced and modern techniques that can be used to measure biogeochemical (Chapter 12) and biological (Chapter 13) linkages among tropical ecosystems. In addition, these two chapters discuss traditionally used techniques. Ecosystem linkages operate at different spatial scales and connect a mosaic of habitats. The way in which terrestrial landscape ecology concepts and approaches can be used to address questions regarding the influence of spatial patterning on ecological processes in the tropical seascape is evaluated in Chapter 14.

Shallow-water tropical ecosystems provide many ecosystem services for humans, but they are heavily impacted through anthropogenic effects. In Part 4, Chapter 15 evaluates the importance of coastal habitats for offshore fishery stocks, while Chapter 16 discusses in detail how these systems can be conserved and managed.

References

Aburto-Oropeza O, Ezcurra E, Danemann G et al (2008) Mangroves in the Gulf of California increase fishery yields. Proc Natl Acad Sci USA 105:10456–10459

Adams AJ, Dahlgren CP, Kellison GT et al (2006) Nursery function of tropical back-reef systems. Mar Ecol Prog Ser 318:287–301

Alongi DM (2002) Present state and future of the world's mangrove forests. Environ Conserv 29:331–349

Aguilar-Perera A, Appeldoorn RS (2007) Variation in juvenile fish density along the mangrove-seagrass-coral reef continuum in SW Puerto Rico. Mar Ecol Prog Ser 348:139–148

Birkeland C, Amesbury SS (1988) Fish-transect surveys to determine the influence of neighboring habitats on fish community structure in the tropical Pacific. In: Co-operation for environmental protection in the Pacific. UNEP Regional Seas Reports and Studies No. 97, pp. 195–202. United Nations Environment Programme, Nairobi

Costello TJ, Allen DM (1966) Migrations and geographic distribution of pink shrimp, *Penaeus duorarum*, of the Tortugas and Sanibel grounds, Florida. Fish Bull 65:449–459

Costanza R, d'Arge R, de Groot R et al (1997) The value of the world's ecosystem services and natural capital. Nature 387:253–260

Cowen RK (2006) Larval dispersal and retention and consequences for population connectivity. In: Sale PF (ed) Coral reef fishes. Dynamics and diversity in a complex ecosystem, pp. 149–170. Academic Press, U.S.A.

Cowen RK, Paris CB, Srinivasan A (2006) Scaling of connectivity in marine populations. Science 311:522–527

Day JW, Christian RR, Boesch DM et al (2008) Consequences of climate change on the ecogeomorphology of coastal wetlands. Estuar Coast 31:477–491

Friedlander A, Nowlis JS, Sanchez JA et al (2003) Designing effective marine protected areas in Seaflower Biosphere Reserve, Colombia, based on biological and sociological information. Conserv Biol 17:1769–1784

Gilman EL, Ellison J, Duke NC et al (2008) Threats to mangroves from climate change and adaptation options: a review. Aquat Bot 89:237–250

Harborne AR, Mumby PJ, Micheli F et al (2006) The functional value of Caribbean coral reef, seagrass and mangrove habitats to ecosystem processes. Adv Mar Biol 50:57–189

Hogarth PJ (2007) The biology of mangroves and seagrasses. The biology of habitats series. Oxford University Press, Oxford

Hughes TP, Baird AH, Bellwood DR et al (2003) Climate change, human impacts, and the resilience of coral reefs. Science 301:929–933

Hughes TP, Rodrigues MJ, Bellwood DR et al (2007) Phase shifts, herbivory, and the resilience of coral reefs to climate change. Curr Biol 17:360–365

Iversen ES, Idyll CP (1960) Aspects of the biology of the Tortugas pink shrimp, *Penaeus duorarum*. Trans Am Fish Soc 89:1–8

Jackson JBC, Kirby MX, Berger WH et al (2001) Historical overfishing and the recent collapse of coastal ecosystems. Science 293:629–638

Kanciruk P, Herrnkind W (1978) Mass migration of spiny lobster, *Panulirus argus* (Crustacea: Palinuridae): behavior and environmental correlates. Bull Mar Sci 28:601–623

Leis JM (2006) Are larvae of demersal fishes, plankton, or nekton? Adv Mar Biol 51:57–141

Lucas C (1974) Preliminary estimates of stocks of king prawn, *Penaeus plebejus*, in south-east Queensland. Aust J Mar Freshwat Res 25:35–47

Manson FJ, Loneragan NR, Skilleter GA et al (2005) An evaluation of the evidence for linkages between mangroves and fisheries: a synthesis of the literature and identification of research directions. Oceanogr Mar Biol Annu Rev 43:483–513

McFarland WN, Ogden JC, Lythgoe JN (1979) The influence of light on the twilight migrations of grunts. Env Biol Fish 4:9–22

Meynecke JO, Lee SY, Duke NC (2008) Linking spatial metrics and fish catch reveals the importance of coastal wetland connectivity to inshore fisheries in Queensland, Australia. Biol Conserv 141:981–996

Mumby PJ, Hastings A (2008) The impact of ecosystem connectivity on coral reef resilience. J Appl Ecol 45:854–862

Nagelkerken I (2007) Are non-estuarine mangroves connected to coral reefs through fish migration? Bull Mar Sci 80:595–607

Newton K, Cote IM, Pilling GM et al (2007) Current and future sustainability of island coral reef fisheries. Curr Biol 17:655–658

Odum EP (1968) Evaluating the productivity of coastal and estuarine water. Proceedings of the Second Sea Grant Conference, University of Rhode Island, pp. 63–64

Ogden JC (1997) Ecosystem interactions in the tropical coastal seascape. In: Birkeland C (ed) Life and death of coral reefs, pp. 288–297. Chapman & Hall, U.S.A.

Ogden JC, Buckman NS (1973) Movements, foraging groups, and diurnal migrations of the striped parrotfish *Scarus croicensis* Bloch (Scaridae). Ecology 54:589–596

Ogden JC, Ehrlich PR (1977) The behavior of heterotypic resting schools of juvenile grunts (Pomadasyidae). Mar Biol 42:273–280

Ogden JC, Zieman JC (1977) Ecological aspects of coral reef-seagrass bed contacts in the Caribbean. Proc 3rd Int Coral Reef Symp 1:377–382

Parrish JD (1989) Fish communities of interacting shallow-water habitats in tropical oceanic regions. Mar Ecol Prog Ser 58:143–160

Pittman SJ, McAlpine CA (2003) Movements of marine fish and decapod crustaceans: process, theory and application. Adv Mar Biol 44:205–294

Randall JE (1963) An analysis of the fish populations of artificial and natural reefs in the Virgin Islands. Caribb J Sci 3:31–46

Roessig JM, Woodley CM, Cech JJ et al (2004) Effects of global climate change on marine and estuarine fishes and fisheries. Rev Fish Biol Fisheries 14:251–275

Sheaves M (2005) Nature and consequences of biological connectivity in mangrove systems. Mar Ecol Prog Ser 302:293–305

Shepherd SA, McComb AJ, Bulthuis DA et al (1989) Decline of seagrasses. In: Larkum AWD, McComb JA, Shepherd SA (eds) Biology of seagrasses, pp. 346–393. Elsevier, Amsterdam

Short FT, Wyllie-Echeverria S (1996) Natural and human-induced disturbance of seagrasses. Environ Conserv 23:17–27

Stone R (2007) A world without corals? Science 316:678–681

UNEP (2006) Marine and coastal ecosystems and human well-being: a synthesis report based on the findings of the Millennium Ecosystem Assessment. United Nations Environment Programme, Nairobi

Valentine JF, Heck KL, Blackmon D et al (2008) Exploited species impacts on trophic linkages along reef-seagrass interfaces in the Florida keys. Ecol Appl 18:1501–1515

Valiela I, Bowen JL, York JK (2001) Mangrove forests: one of the world's threatened major tropical environments. BioScience 51:807–815

Weinstein MP, Heck KL (1979) Ichtyofauna of seagrass meadows along the Caribbean coast of Panamá and in the gulf of Mexico: composition, structure and community ecology. Mar Biol 50:97–107

Wilkinson C (2004) Status of coral reefs of the World. Australian Institute of Marine Science, Townsville

Wolanski E (2007) Estuarine ecohydrology. Elsevier, Amsterdam

Part I
Biogeochemical Linkages

Chapter 2
Nitrogen and Phosphorus Exchange Among Tropical Coastal Ecosystems

Stephen E. Davis III, Diego Lirman and Jeffrey R. Wozniak

Abstract   The concentration and flux of nitrogen (N) and phosphorus (P) through mangrove wetlands, seagrass meadows, and coral reef habitats are mediated by a wide range of hydrodynamic and chemical pathways determined by both natural and anthropogenic drivers. The direct proximity of these coastal habitats to burgeoning urban centers makes them quite susceptible to excessive nutrient loading, subsequent land-use impacts, the related effects of eutrophication and of course the associated loss of ecosystem services. For this reason mangrove, seagrass, and coral reef ecosystems are among the most threatened ecosystems in the tropics. While quantifying the exchange of materials between coastal wetlands and nearshore waters has been the focus of estuarine research for nearly half a century, a concerted effort to understand the net exchange of N and P across these habitats has only begun in the last 20 years. Furthermore, attempts to better understand the interplay of N and P cycles specifically between each of these three habitats has been all but nonexistent. The role mangrove and seagrass ecosystems play in buffering nearshore coral habitats from land-based influences remains a topic of great debate. Critical to understanding the nutrient dynamics between these ecosystems is defining the frequency and magnitude of connectivity events that link these systems together both physically and biogeochemically. In this chapter we attempt to address both N and P water column concentrations and system-level exchanges (i.e., water-mediated fluxes and nutrient loading). We consider how the interactions of N and P between these systems vary with geomorphology, hydrography, seasonal programming, and human influences.

Keywords   Mangrove · Seagrass · Coral reef · Nutrient · Flux

S.E. Davis (✉)
Department of Wildlife and Fisheries Sciences, Texas A&M University, College Station, TX, USA,
e-mail: sedavis@tamu.edu

I. Nagelkerken (ed.), *Ecological Connectivity among Tropical Coastal Ecosystems*,
DOI 10.1007/978-90-481-2406-0_2, © Springer Science+Business Media B.V. 2009

2.1 Introduction

Mangrove, seagrass, and coral reef ecosystems are among the most threatened ecosystems in the tropics due primarily to human impacts such as overfishing, land conversions and subsequent land-use impacts, and climate change (Jackson et al. 2001, Valiela et al. 2001, Hughes et al. 2003, Pandolfi et al. 2003, Short et al. 2006). These ecosystems—especially seagrass and coral reefs—are often oligotrophic with clear water conditions and can be susceptible to excessive nutrient loading and the effects of eutrophication (Szmant 2002, Short et al. 2006, Twilley 1995). Based on evidence from the literature, the impact of nutrient loading on coral reefs and seagrass beds is more localized and diminished with distance offshore, as dilution and flushing minimize impacts (Bell 1992, Szmant 2002, Atkinson and Falter 2003, Rivera-Monroy et al. 2004). However, mangrove wetlands have been shown to effectively reduce nutrient loading from wastewater and agricultural effluent to seagrass and coral reef ecosystems (Tam and Wong 1999, Lin and Dushoff 2004). Despite this functional attribute of mangroves, there have been documented effects of large-scale storm events resulting in significant runoff and nutrient loading impacts to these offshore ecosystems (Tilmant et al. 1994, Short et al. 2006). Further, seagrass-dominated areas adjacent to highly developed shorelines and within restricted lagoonal systems (with increased water residence times) also seem to be susceptible to chronic nutrient loading (Hutchings and Haynes 2005, Short et al. 2006). In a meta-analysis, Valiela and Cole (2002) concluded that in estuaries with well-developed fringing coastal wetlands (mangrove and saltmarsh), seagrass production was oftentimes higher and loss of seagrass habitat was lower as these fringing transitional/wetland ecosystems buffer loads of upland-derived nutrients (particularly nitrogen) to sensitive, subtidal seagrass beds. Seagrass and mangrove ecosystems may in turn serve as an upland nutrient buffer for coral reefs.

2.1.1 Background on Coastal Flux Studies

Quantifying exchanges of materials between coastal wetlands and nearshore waters has been the focus of estuarine research for nearly half a century (Teal 1962, Nixon 1980, Childers et al. 2000). Much of this work was inspired by the 'outwelling hypothesis' that was formulated through research and observations conducted in saltmarsh-dominated estuaries of the southeast Atlantic coast of the USA (Teal 1962, Odum and de la Cruz 1967, see also description in Chapter 3). Although studies testing this concept have not actually proven its universality, they have led to a better understanding of the patterns and range of variability of wetland–estuarine and estuarine–nearshore exchanges of nitrogen (N) and phosphorus (P). Seminal among this body of work is Nixon's (1980) review of the literature on N and P fluxes between saltmarshes and adjacent estuaries where he concluded a general trend of nitrate (NO_3^-) and nitrite (NO_2^-) uptake by the marshes and an export of dissolved organic nitrogen (DON) and phosphate (PO_4^{3-}) from the marshes to estuarine waters. Until that time, little was known about the fate and transport of these important macronutrients in analogous tropical and subtropical coastal wetlands (i.e., mangrove swamps) and nearshore waters supporting seagrass and coral reef ecosystems. Despite the body of work reviewed by Nixon (1980) and subsequent reviews that incorporated tropical coastal ecosystems (Boto 1982, Alongi et al. 1992, Lee 1995, Childers et al. 2000), little research has been done to track the net exchange of N and P across mangrove, seagrass, and coral reef ecosystems.

At the coastal margin, upland-derived sources of inorganic and organic nutrients are often intermittent as a function of seasonal patterns in rainfall and runoff, producing intra-annual patterns of water source (river vs. marine), nutrient concentrations, and nutrient flux (Twilley 1985, Rivera-Monroy et al. 1995, Ohowa et al. 1997, Davis et al. 2003a). Furthermore, in many estuarine ecosystems, the direction and magnitude of nutrient flux has been shown to correspond to nutrient concentrations in the water column, highlighting an important link between water quality and the direction and magnitude of nutrient exchange (Wolaver and Spurrier 1988, Whiting and Childers 1989, Childers et al. 1993, Davis et al. 2003a). Natural disturbances such as tropical storms, frontal passages, and hurricanes not only affect the structure of these tropical coastal ecosystems but can also account for a significant spike in the exchanges of N, P, and sediment within and among them (Tilmant et al. 1994, Sutula et al. 2003, Davis et al. 2004).

From a mass balance standpoint, mangrove wetlands are generally considered to be net exporters of organic materials (Lee 1995), suggesting they may also represent a source of organically bound nutrients to seagrass beds and, possibly, coral reefs. The influence of mangrove and upland sources of materials naturally becomes more diminished with distance offshore and is replaced by marine-dominated (mainly upwelling) or *in situ* processes governing nutrient exchange (Monbet et al. 2007). However, there is little consensus regarding the magnitude of the contribution of this exported material on seagrass and coral reef nutrient cycles and food web dynamics (Odum and Heald 1975, Robertson et al. 1988, Alongi 1990, Fleming et al. 1990, Lin et. al. 1991, Hemminga et al. 1994, see Chapter 3).

Given the lack of consensus regarding the magnitude of mangrove contributions to these offshore tropical ecosystems, as well as the variability in nutrient sources across both spatial (mangrove ←→ seagrass ←→ coral reef) and temporal (e.g., diurnal, seasonal, inter-annual, etc.) scales, an understanding of the factors that regulate nutrient concentrations in each of these tropical coastal ecosystems may yield valuable insight into how these systems transform and exchange materials such as nutrients. Such information can also provide us with better approaches to management, particularly in response to anthropogenic alterations in the quality and quantity of freshwater flows to the coastal zone. Therefore, the primary goal of this chapter is to summarize the current state of our understanding with respect to patterns of N and P concentration and exchange across tropical coastal margins.

In this chapter, we seek to summarize published water column concentrations of N and P as well as fluxes of these elements between different ecosystem components (sediment, vegetation, water, detritus, and biota) in mangrove, seagrass, and coral reef areas. In order to understand the degree of connectivity among these threatened coastal ecosystems, our next goal is to summarize available literature on system-level exchanges (i.e., loads or water-mediated fluxes) of N and P. Given such a limited body of literature addressing the latter, we will focus on within-ecosystem exchanges and speculate on the latter by considering the different factors affecting flux dynamics and the spatial and temporal extent of biogeochemical connectivity among these tropical coastal ecosystems. Specifically, we will consider the roles of hydrologic flushing/water residence time, spatial connectivity, proximity to sources of nutrients (i.e., rivers and zones of upwelling), and human impacts in driving patterns of concentration and flux of nitrogen and phosphorus.

2.1.2 Conceptual Model of N and P Exchange Among Tropical Coastal Ecosystems

The conceptual model presented in Fig. 2.1 is intended to reflect the potential paths of water-mediated exchange of N and P among tropical coastal ecosystems and will

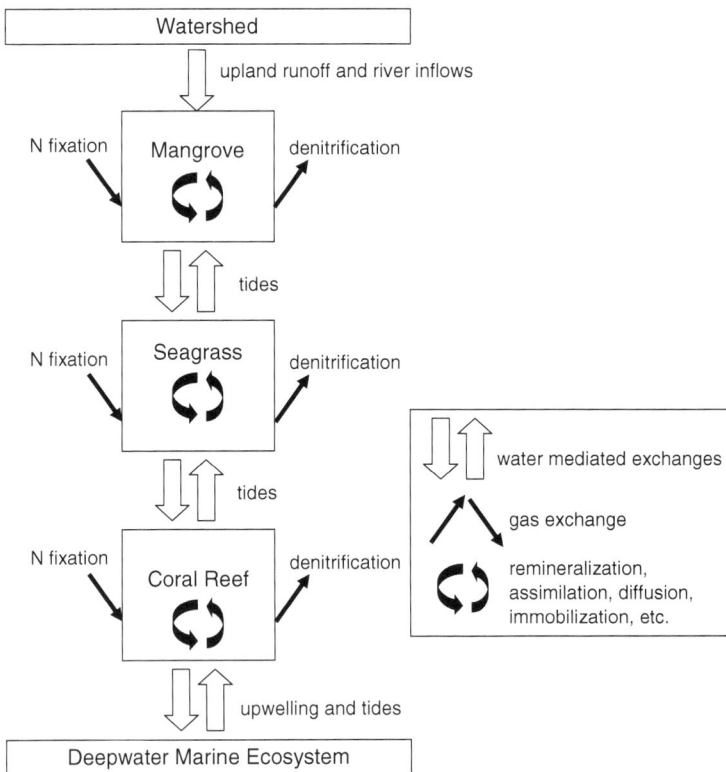

Fig. 2.1 Conceptual diagram showing pathways of lateral (i.e., water-mediated) and vertical (plant-water column or sediment-water column) fluxes of nitrogen and phosphorus between mangrove, seagrass, and coral reef ecosystems

guide discussion of our synthesis of concentration and flux data from the litera-ture. It is comparable to Fig. 3.1 in Chapter 3. Because of tidal influences, flow-mediated ecosystem exchanges of materials are presented as bi-directional paths of equivalent magnitude. However, episodic pulses in river inflow to the coastal margin and upwelling events can temporarily shift the balance of these bi-directional flows either seaward or landward, respectively. This basic model also reflects the contribu-tion of these end-member sources such as deepwater marine and upland ecosystems and acknowledges the active internal recycling (assimilation and remineralization) of N and P within each ecosystem type.

For the sake of simplicity and due to the constraints of available information for each ecosystem, we have limited this conceptual model to surface water-borne trans-port and exchange of N and P (Fig. 2.1). Obviously, atmospheric deposition, ground-water discharge, and biological processes such as nitrogen fixation and denitrifica-tion contribute greatly to coastal N and P cycling and will be discussed throughout this chapter (Zimmerman et al. 1985, Mazda et al. 1990, Sutula et al. 2003, Lee and Joye 2006). Evidence even suggests that coral reefs may receive some N that is fixed in these other shallow water environments (France et al. 1998). However, we will not focus on these types of processes at the level of ecosystem exchange, as the contribution of these processes would naturally be imbedded in empirical measurements of N or P within or between these settings.

2.2 N and P in Tropical Coastal Ecosystems

Given the growing impact of nutrient enrichment and the potential for eutrophi-cation, as well as the ubiquitous influence of tides and river inflows linking these ecosystems, understanding the surface water exchanges of ecologically impor-tant elements such as nitrogen (N) and phosphorus (P) within and among these ecosystems is needed. Phosphorus and nitrogen are of great importance in biologi-cal systems, as these elements are required for structural (N and P), electrochemical (P), and mechanical functions (P) of biological organisms (Sterner and Elser 2002). Aside from biological uptake, different forms of these two elements can also be effectively removed from a system via abiotic processes such as volatilization and loss to the atmosphere, adsorption onto particles, or bound in mineral forms. As a result of the limited availability of N and P relative to other biologically required elements, primary producers in tropical coastal marine ecosystems often display a limitation by either one of these elements (Fourqurean et al. 1992, Lapointe and Clark 1992, Amador and Jones 1993, Agawin et al. 1996, Feller et al. 2002), thus increasing the need to understand N and P dynamics.

The concept of nutrient limitation—as conceptualized by Justus von Liebig in the 1840s and considered from a stoichiometric perspective by Sterner and Elser (2002)—predicts that organisms will be limited by the resource that is in lowest supply (i.e., availability) relative to the needs of that organism. However, a recent meta-analysis by Elser et al. (2007) suggests that, at the level of an ecosystem, the concept of a single limiting nutrient may not be the rule and that tropical coastal ecosystem such as mangroves, seagrasses, and coral reefs are going to respond to changes in both N and P. Recent experimental evidence in mangrove and seagrass ecosystems in the neo-tropics supports this notion (e.g., Feller 1995, Ferdie and Fourqurean 2004).

Nitrogen and phosphorus may enter mangrove, seagrass, and coral reef ecosystems via a number of different pathways (Boto 1982, Liebezeit 1985, D'Elia and Wiebe 1990, Hemminga et al. 1991, Leichter et al. 2003). These nutrients are transmitted in organic or inorganic forms to coastal ecosystems via surface water, groundwater, and atmospheric deposition (both wet and dry). Relative to the water column, the sediment/soil and biomass in these ecosystems represent the largest reservoirs of N and P. However, freshwater inputs from rivers and coastal upwelling are often the primary source of natural loads of N and P to mangrove wetlands and coral reefs, respectively (D'Elia and Wiebe 1990, Nixon et al. 1996, Monbet et al. 2007), and changes in the quantity and quality of river inflows are often implicated for enhanced loading of these elements to the coastal zone (Nixon et al. 1996, Valiela and Cole 2002). Once N and P are immobilized within mangrove, seagrass, or coral reef ecosystems, the different forms of these elements are susceptible to transformation via an array of biogeochemical pathways, depending on conditions such as sediment type (terrigenous vs. biogenic), redox, pH, light, temperature, and availability of labile organic substrate (Nixon 1981, D'Elia and Wiebe 1990, Bianchi 2007).

Lastly, an important caveat for understanding nutrient dynamics within these ecosystems is that nutrient concentration does not necessarily translate directly into nutrient availability, as nutrients may remain within a system but become temporarily unavailable for utilization by primary producers. An example of this is the case of nutrients (e.g., ammonium, phosphate) adsorbed to sediment particles or bound in refractory organic matter.

2.2.1 *N and P Concentration in Mangrove Ecosystems*

The interaction of tides, wind, precipitation, and upland runoff plays an important role in determining the hydrodynamics and chemistry of mangrove waterways (Lara and Dittmar 1999, Davis et al. 2001a, Childers et al. 2006, Rivera-Monroy et al. 2007). However, human-associated impacts to coastal mangroves can overwhelm any of these natural drivers of water quality oftentimes resulting in excessively high concentrations of N and P (Nedwell 1975, Nixon et al. 1984, Rivera-Monroy et al. 1999). In tidally-dominated systems with little upland influence, inorganic N and P concentrations can be quite low (Boto and Wellington 1988), although groundwater inputs can enhance concentrations of these elements (Ovalle et al. 1990). Microtidal systems with a seasonal upland influence have surface water salinity patterns that are noticeably lower during the wet season and highest during the dry season, reflecting the contribution of end-member sources of water. Water column concentrations of N and P typically reflect this changing source water signature (Davis et al. 2003a). On the other hand, mangrove waterways that are strongly river-dominated typically show a year-round upland influence on surface water quality patterns (Nixon et al. 1984).

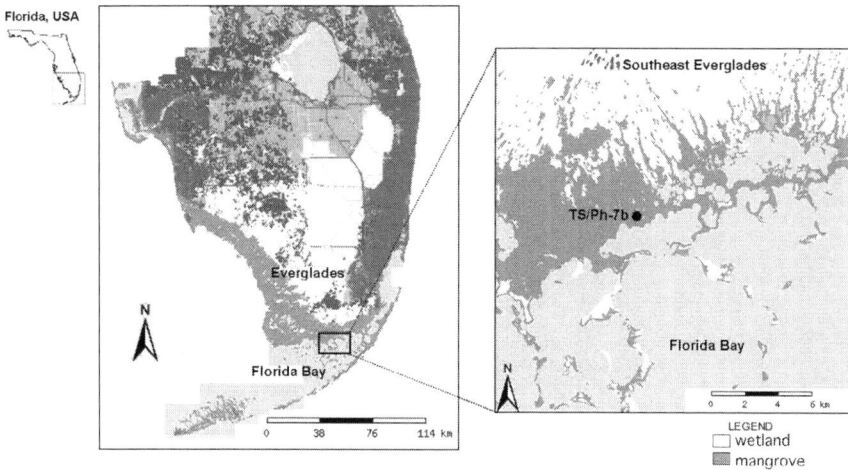

Fig. 2.2 Image of south Florida showing Florida Bay, which is situated between the Everglades (to the north) and Florida Keys (to the south). The expanded image on the right highlights the location of the mangrove ecosystem that lies between the freshwater Everglades marshes and seagrass-dominated Florida Bay. Map was generated using Florida Coastal Everglades LTER Mapserver project (http://fcelter.fiu.edu/gis/everglades-map)

Davis et al. (2001a,b) showed that total nitrogen (TN) concentrations in lower Taylor River (Florida, USA, site TS/Ph 7b in Fig. 2.2) could approach 90 μM and were significantly higher during the wet season compared with the dry season, sometimes by more than 40 μM (Table 2.1). The pattern shown by TN reflected that of dissolved organic carbon (DOC), indicating that much of the TN in this seasonal mangrove system fed by Everglades runoff may be organic in character. Rivera-Monroy et al. (1995) found a similar seasonal trend for dissolved organic nitrogen (DON) in a fringe mangrove wetland of Estero Pargo (Mexico), a tidal mangrove system with little upland influence. Still, their highest wet season value for TN (approximately 65 μM, estimated by summing reported concentrations for DON, particulate nitrogen (PN), NH_4^+, NO_2^-, and NO_3^-; Rivera-Monroy et al. 1995) was lower than the wet season average reported for Taylor River (77 μM) and the upper Sangga River (Malaysia, 60–80 μM), a tidal mangrove river with a strong upland influence (Nixon et al. 1984). Boto and Wellington (1988) found considerably lower levels of DON, reflecting the weak upland connection in the Coral Creek system near Hinchinbrook Island, Australia.

Despite the similarity in TN concentrations, concentrations of total phosphorus (TP) in the Sangga system were more than an order of magnitude higher than the values measured in Taylor River (Table 2.1), with molar ratios of TN:TP ranging from 20 to 40 (Nixon et al. 1984). In Taylor River, TP was usually <0.5 μM and TN:TP often exceeds 100, reflecting the oligotrophic and P-limited status of this region (Davis et al. 2001a, b). Such low concentrations of surface water TP are not limited to the southern Everglades mangrove transition zone. They are typical

Table 2.1 Water column concentration ranges of nitrogen (N) and phosphorus (P) from various mangrove ecosystems. All units are in µM. NH_4^+ = ammonium, NO_x^- = NO_3^- and NO_2^-, DON = dissolved organic N, TN = total N, SRP = soluble reactive P, TP = total P

Reference	Location	NH_4^+	NO_x^-	DON	TN	SRP	TP
Boto and Wellington (1988)	Fringe mangrove along Coral Creek, Hinchinbrook Island, Australia	0.1–1.6	<0.1–0.3	2.0–8.0	–	<0.01–0.22	–
Davis et al. (2001a)	Dwarf mangrove along Taylor River, Florida, USA	0.1–5.2	0.1–5.8	–	46–94	<0.01–0.15[a]	0.2–0.7[a]
Davis et al. (2001b)	Fringe mangrove along Taylor River, Florida, USA	0.1–6.3	0.2–5.8	–	41–89	<0.01–0.24	0.18–0.67
Lara and Dittmar (1999)	Riverine/fringe mangrove along Furo de Chato, Brazil	<5–30	~2–4	–	–	~1–5	–
Nixon et al. (1984)[b]	Matang mangrove forest along Sangga River, Malaysia	<5–24	<0.2	20–50	40–85	0.2–1	~1–3
Ohowa et al. (1997)	Mangrove rivers (Mkurmuji and Kidogoweni) in Gazi Bay, Kenya	0.3–3	0.2–8	–	–	0.5–3.9	–
Ovalle et al. (1990)	Fringe mangrove along tidal creek in Sepetiba Bay, Brazil	0.9–7	0.8–5	–	–	0.5–1.8	–
Rivera-Monroy et al. (1995)	Fringe mangrove in Terminos Lagoon, Mexico	1.1–51.7	0.2–4.9	7.8–42.9	~65[c]	–	–
Rivera-Monroy et al. (2007)	Riverine mangrove along Shark River, Florida, USA	<0.1–4.8	<0.1–3.5	–	~12–40	<0.01–0.8	0.2–2.9
Robertson et al. (1993)[d]	Riverine mangrove along Fly River, Papua New Guinea	0.1–1.4	1.8–11.8	–	–	0.5–5.3	–

[a] SRP and TP data were as high as 0.46 and 1.32 µM, respectively, during a single sampling in May 1998
[b] Data from a single cruise in 1979. Only mangrove sites along estuarine transect considered here
[c] Wet season TN estimated by sum of DON, PN, and DIN
[d] Data from two cruises along salinity gradient in 1989 and 1990

of both the freshwater southern Everglades (Noe et al. 2001, Childers et al. 2006) and eastern Florida Bay estuary (Boyer et al. 1999), as well as other carbonate-dominated mangrove settings, such as Coral Creek, Australia (Boto and Wellington 1988).

Concentrations of dissolved inorganic nitrogen (DIN) and phosphorus (DIP) are similar across many of the mangrove systems reviewed and suggest an oligotrophic nature with regard to the water column pool of 'available' (i.e., dissolved inorganic) nutrients (Table 2.1). Mangrove waters, in general, have relatively low levels of DIP or (soluble reactive P—SRP) and DIN ($NH_4^+ + NO_x^-$; Alongi et al. 1992). In some cases, the extent of human impact controls inorganic nutrient profiles (Nedwell 1975, Nixon et al. 1984), while in others the degree of upland and groundwater influence on the system appear to be of greater importance (Boto and Wellington 1988, Ovalle et al. 1990).

In South Florida, where systems tend to be oligotrophic and limited by the availability of phosphorus, SRP concentrations are extremely low. SRP concentrations in Taylor River (Fig. 2.2) are typically 0.01–0.05 μM, and sometimes below the limits of analytical detection (<0.01 μM). These concentrations are much lower, in some cases more than two orders of magnitude lower, than SRP values from other mangrove systems (Table 2.1; see also Alongi et al. 1992). Alternatively, NH_4^+ and NO_x^- numbers were comparable across all systems. Despite the similarity in DIN concentration ranges between these systems, molar ratios of DIN:DIP in Taylor River are much higher (sometimes exceeding 300) than the others, reflecting the low availability of inorganic P in this 'upside-down' estuary (Chiders et al. 2006). These high ratios of N:P in the environment are also reflected in mangrove leaf detritus ratios of N:P reported by Davis et al. (2003b) with N:P of yellow, nearly senesced *Rhizophora* leaves averaging 75.

2.2.2 N and P Concentration in Seagrass Ecosystems

Under normal oligotrophic conditions, levels of dissolved inorganic nutrients in the surface water of seagrass beds are generally low (Table 2.2). A review of seagrass studies conducted by Touchette and Burkholder (2000) showed that SRP levels are commonly 0.1–<2 μM, NH_4^+ ranges from 0 to 3.2 μM, and NO_x^- levels range from 0.05 to 8 μM in seagrass habitats. Data from few of the studies reviewed by Touchette and Burkholder (2000) as well as several others are provided in Table 2.2. In contrast to surface waters, levels of inorganic nutrients within sediment pore water pools are often two orders of magnitude larger, with SRP levels of up to 20 μM, NH_4^+ concentrations up to 180 μM, and NO_x^- concentrations up to 10 μM. Of course, the concentration values for inorganic nutrients reported for normal oligotrophic conditions can be exceeded significantly under eutrophic conditions, especially where anthropogenic sources of discharges are observed. In a recent review by Lee et al. (2007) water-column concentrations of NH_4^+ exceeded 50 μM and pore water concentrations exceeded 400 μM under these types of human-influenced eutrophic conditions.

Table 2.2 Water column concentration ranges of N and P from various seagrass ecosystems. All units are in µM. For abbreviations see Table 2.1

Reference	Location	NH_4^+	NO_x^-	TN	SRP	TP
Agawin et al. (1996)	Two sites near Cape Bolinao, West Philippines	1.6–1.9	0.4–0.8	–	0.1–0.2	–
Boon (1986)	Moreton Bay, Australia	0.7	<0.1	–	0.53	–
Boyer et al. (1999)	Florida Bay, USA	1.05–3.41	0.2–0.9	30.5–80.8[a]	0.03–0.05	0.25–0.65
Carruthers et al. (2005)	Three sites near Bocas del Toro, Panama	0.2–0.26	0.20–0.27	–	0.07–0.14	–
Erftemeijer and Herman (1994)	Two sites in Indonesia	2.2–3.2	0.9–1.4	–	0.8–1.4	–
Fourqurean et al. (2003)[b]	Florida Bay, USA	<0.1–120.04	–	~53[c]	<0.01–1.57	0.02–4.21
Hemminga et al. (1995)	Three sites in Gazi Bay, Kenya	0.4–0.8	0.3–0.45	–	–	–
Stapel et al. (1996)	Indonesia	1–2.6	–	–	0.1–0.6	–
Tomasko and Lapointe (1991)	Carrie Bow Cay, Belize	0.28	0.09	–	–	–
Tomasko and Lapointe (1991)	Twin Keys, Belize	1.05	0.05	–	–	–
Uku and Bjork (2005)	Sites near Vipingo and Nyali Beach, Kenya	1.5–2	2.4–7.6	–	0.7–0.8	–

[a] Total organic nitrogen (TON) comprised the bulk of TN fraction
[b] Values represent min/max of multiple samplings
[c] Estimated from sum of median values for DIN and TON

Spatial and temporal patterns in water column concentrations of N and P in seagrass beds can be attributed to many other non-human factors such as variation in water residence time (affecting the contribution of internal recycling), proximity to inflow sources, storm and wind events, and patterns of vegetation turnover and decomposition. Long-term data from multiple sites across Florida Bay indicate that surface water NH_4^+ concentrations can exceed 100 μM, especially in the hydrographically isolated, central region of this oligotrophic bay (Boyer et al. 1999, Fourqurean et al. 2003; Table 2.2). Further, Boyer et al. (1999) showed that increased inflow to Florida Bay between 1989 and 1997 may have resulted in reduced bay-wide TP concentrations, as freshwater derived from the Everglades watershed is depleted in P. This trend is also supported by the well documented gradient of N:P that decreases from east to west Florida Bay as a result of higher P availability near the interface with the Gulf of Mexico and reduced P availability towards the eastern (i.e., interior), Everglades-influenced region of the bay (Fourqurean et al. 1993, Childers et al. 2006; Fig. 2.2).

The dominant inorganic form of N within pore water pools is NH_4^+, with a relatively lower contribution of NO_x^-. In the water column, NO_x^- tends to be the dominant form of inorganic N, but NH_4^+ can be locally dominant (Touchette and Burkholder 2000, Lee et al. 2007). In addition to these inorganic sources, organic compounds such as amino acids, urea, dissolved organic phosphorus (DOP), and particulate organic phosphorus (POP) can provide significant sources of P and N within seagrass habitats (Bird et al. 1998, Perez and Romero 1993). In fact, Hansell and Carlson (2002) suggest that the pool of dissolved organic N and P can be several times larger than the concentration of inorganic nutrients, but may not be immediately available for uptake and utilization.

As with mangroves, the concentration of macronutrients in seagrass tissue is commonly used as an indicator of nutrient status and the ratio of C:N:P is commonly used to evaluate spatial and temporal patterns of nutrient availability (Duarte 1990, Fourqurean et al. 1992, 1997). Fourqurean et al. (1992) also showed that spatial variability in these ratios also reflected patterns in seagrass abundance and productivity in Florida Bay. In general, seagrasses are considered N limited in most environments, with P limitation being prevalent in carbonate-dominated settings (Short 1987, Short et al. 1990, Fourqurean et al. 1992, Burkholder et al. 2007). Nevertheless, these general patterns of nutrient limitation are influenced locally by species and the dominant sources of nutrient input.

2.2.3 N and P Concentration in Coral Reef Ecosystems

The development, condition, and long-term survivorship of coral reefs are closely tied to nutrient fluxes and nutrient dynamics within each system. From some of the earliest research, coral reefs have been commonly referred as the 'oases' of the ocean and their ability to thrive and achieve high rates of productivity in oligotrophic environments prompted significant research on the role of nutrients such as N and P. Symbiotic zooxanthellae play a significant role in coral nutrition by providing coral hosts with organic photosynthetic products (e.g., glycerol, glucose) that are rapidly incorporated into animal tissue as well as enhancing calcification rates. Within this symbiotic relationship, coral hosts provide zooxanthellae with sources of inorganic nutrients through their metabolic wastes as well as physical habitat and an enhanced light environment to sustain photosynthesis (Muller-Parker and D'Elia 1997, Anthony et al. 2005). The ability of coral reefs to import dissolved N and P from the water column and the tight recycling of nutrients by the coral-zooxanthellae relationship enables these systems to sustain high rates of productivity even under the seemingly low nutrient availability commonly observed over coral reefs (e.g., Johannes et al. 1972, Atkinson 1992).

Coral reefs were initially considered as systems that can only thrive within a narrow set of physical parameters that include light, temperature, and nutrients. The view of coral reefs as fragile systems with narrow environmental optima has been challenged by more recent observations of reef development and growth in environments previously described as 'marginal' for coral survivorship (Perry and Larcombe 2003). The documentation of coral growth and coral reef development in areas influenced by upwelling that introduces both high levels of nutrient levels and cold temperatures (e.g., Glynn 1977), and in nearshore habitats with elevated nutrients, sedimentation, and reduced light levels (Fabricius 2005, Lirman and Fong 2007), suggests an ecological niche for coral reefs that may be much wider than previously expected as well as a potential beneficial role of moderate levels of nutrients (Anthony 2000, Anthony and Fabricius 2000). The survivorship and growth of corals in these marginal environments may be directly tied to the ability of corals to supplement autotrophic sources of nutrition with heterotrophic ones. For example, Fabricius (2005) reports that the intake of moderate levels of Particulate Organic Matter (POM) can enhance coral growth and compensate for the negative impacts caused by increased DIN, light reduction, and sedimentation. Similarly, Edinger et al. (2000) indicated that corals can supplement their energy supplies by feeding on particulate or dissolved organic matter. Finally, increased availability of heterotrophic energy and nutrient sources in nearshore coastal habitats has been linked to higher coral growth, increased energy storage, and increased resilience to disturbances such as coral bleaching (Edinger et al. 2000, Anthony 2006, Grottoli et al. 2006).

Relative to mangrove and seagrass ecosystems, the body of literature containing surface water concentration data (not to mention pore water and sediment N and P content) for different forms N and P is relatively small. A review of various reef ecosystems by Szmant (2002) showed that surface water dissolved inorganic N and P concentrations are typically low (usually around 1 μM or less for DIN and <0.5 μM for SRP) on the landward side of the reef with increasing concentrations towards the reef crest adjacent to areas of high flushing and upwelling. Other studies by Szmant and Forrester (1996) and a review by Costa et al. (2006) show this same spatial variation in water column concentrations in addition to potentially significant seasonal variability within each of these locations. More work is needed to improve our understanding of the forces that drive this variability.

2.2.4 N and P Flux in Mangrove Ecosystems

Studies of materials exchange in tidal mangrove wetlands are becoming more prevalent in the estuarine literature (see also Chapter 3). Aside from the pioneering works of Golley et al. (1962) and Odum and Heald (1972), it has been only in the last 15–20 years that tropical, mangrove-dominated estuaries have been the setting for this type of ecosystem-level research. Adapting many of the techniques developed in temperate saltmarsh systems, investigators of recent mangrove studies have shown that tidally driven mangrove wetlands can effectively serve as sinks for total suspended solids (Rivera-Monroy et al. 1995) and dissolved inorganic nitrogen (Kristensen et al. 1988, Rivera-Monroy et al. 1995). As a result of these studies, we are gaining a better understanding of the environmental factors that regulate the exchanges of mangrove-derived matter in these tropical and subtropical estuarine systems.

Leaf litter fall and decomposition is an important recycling pathway for nutrients and fixed carbon in all forested aquatic ecosystems (Fisher and Likens 1973, Brinson 1977, Tam et al. 1990). Although biological processes are important in governing the ultimate fate of leaf litter, evidence from numerous field and lab studies indicates that physical leaching is largely responsible for initial losses of these materials (e.g., Brinson 1977, Rice and Tenore 1981, Middleton and McKee 2001). Rates of leaf litter leaching are sensitive to environmental factors such as temperature, sunlight, water availability, and salinity (Nykvist 1959, 1961, Parsons et al. 1990, Chale 1993, Steinke et al. 1993). Some researchers have suggested that the biotic contributions in this early stage of decomposition are minimal and most often limited to microbial conditioning of the litter (Nykvist 1959, Cundell et al. 1979, France et al. 1997). Other studies, however, have shown a significant microbial response within 24 hrs to fixed carbon and nutrients leached from mangrove leaves (Benner et al. 1986, Davis and Childers 2007).

In tropical mangrove ecosystems, leaf litter leaching rates decline dramatically after a few days of immersion in water, yet this process is responsible for substantial losses of N and P to the water column and soil (Rice and Tenore 1981, Chale 1993, Steinke et al. 1993, Davis et al. 2003b; Fig. 2.3). When in low supply, these leached nutrients can then be utilized by epiphytic bacteria that are decaying the more refractory leaf tissue, resulting in a gradual enrichment of the tissue through time (Davis et al. 2003b, Davis and Childers 2007). On a regional scale, the coupled process of mangrove leaf litterfall and leaching contributes to intra-annual patterns in water quality and materials flux unique to these coastal wetlands (Twilley 1985, Maie et al. 2005). This may be particularly important in nutrient-poor, dwarf mangrove wetlands where water residence times are often high and herbivory rates are very low (Twilley 1995, Feller and Mathis 1997). This combination of ecosystem properties naturally leads to more reliance on internal recycling (i.e., detrital pathways) as a means of controlling nutrient availability and productivity.

Water temperature and salinity are two of the most important factors controlling the global and local distributions of mangrove ecosystems along the world's shorelines (Kuenzler 1974, Odum et al. 1982, Duke 1992). Fluctuations in either of

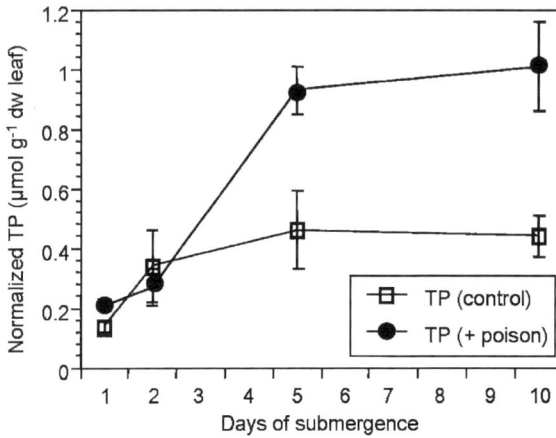

Fig. 2.3 Cell plot showing the total amount of phosphorus (TP) leached from red mangrove (*Rhizophora mangle* L.) leaves over a 10-day incubation. Incubations with poison had no biological activity, and losses to the water column indicated the amount of leachable P. The difference between this and control incubations reflected the contribution of epiphytic bacteria that removed P from the water column and relocated leached P to the leaf surface. All values are normalized to the initial dry mass (dw) of each leaf

these factors can have profound effects on aboveground and benthic productivity in mangrove wetlands (Alongi 1988, Clough 1992). In tropical and subtropical areas, water temperatures are generally a function of season or time of year, reflecting changes in air temperature, light intensity, or precipitation/cloud cover. Salinity is usually an indicator of a combination of season (wet or dry), water source (upland runoff or marine/tidal), and physical position within an estuary. Changes in salinity and temperature can affect the availability, uptake, or release of a given constituent in a mangrove wetland. Relationships of this type have already been documented in temperate saltmarsh systems (e.g., Wolaver and Spurrier 1988).

Total nitrogen dynamics in mangrove wetlands, like organic carbon dynamics, appear to vary according to the relative contributions of tide, season, and upland influence in a given system. However, Alongi et al. (1992) showed that fluxes of nitrogen in mangrove wetlands could be low and erratic, showing little effect of season or location within an estuary. Rivera-Monroy et al. (1995) measured significant exports of the bulk TN components (dissolved organic and particulate N) from a fringe mangrove near Estero Pargo, a tidal mangrove creek along the gulf coast of Mexico. Dissolved organic nitrogen export was consistent across most samplings in their system, while particulate nitrogen (PN) exports were seasonal with the highest exports measured after precipitation events (Rivera-Monroy et al. 1995). On the other hand, a tidal mangrove wetland along Coral Creek (Australia), a system with little upland influence, was found to export PN and import DON at considerably higher rates (Boto and Bunt 1981, Boto and Wellington 1988). Quarterly flux data from Taylor River indicated consistent import of TN at rates similar in magnitude to the DON and PN flux measurements from Mexico (Rivera-Monroy et al. 1995,

Davis et al. 2001a). The phenomenon of TN uptake in these dwarf mangrove sites was also evident in long-term water quality data from the Everglades and Florida Bay, which showed consistently higher concentrations in the mangrove ecosystem than at the downstream Florida Bay sites (Childers et al. 2006).

Mangrove flux studies from across the tropics have also shown discrepancies in the relative exchanges of NH_4^+ and NO_x^-. For instance, Boto and Wellington (1988) measured rather low DIN fluxes in Coral Creek, with uptake of NH_4^+ and export of NO_x^- (Table 2.3). Even though the majority of the tides measured in this study yielded significant fluxes of these constituents, the authors concluded that the system was near-equilibrium in terms of dissolved inorganic nutrient exchange (Boto and Wellington 1988). Dittmar and Lara (2001) also found a non significant flux of NO_x^- in a Brazilian mangrove forest in the Caeté Estuary. Still, additional studies have shown import of both DIN constituents into tidal mangrove wetlands. Using benthic chambers to measure sediment–water column fluxes at Ao Nam Bor mangrove swamp in Thailand, Kristensen et al. (1988) observed consistent and similar uptakes of both DIN constituents in light and dark chambers (Table 2.3). They later determined that the rates of nitrification and denitrification in this system were roughly equal (Kristensen et al. 1995). Similarly, Rivera-Monroy et al. (1995) noted consistent DIN uptake in Estero Pargo (Table 2.3). However, NH_4^+ uptake was roughly an order of magnitude higher than NO_x^- uptake, due to sediment retention of NH_4^+, plant uptake, or high rates of nitrification (Rivera-Monroy et al. 1995). An assay of coupled nitrification–denitrification at this site later revealed that NO_x^- uptake was not necessarily associated with denitrification, but instead with sediment uptake and retention (Rivera-Monroy and Twilley 1996). Using bell jar incubations, Alongi (1996) measured consistent sediment uptake of inorganic nitrogen and phosphorus along the mid-intertidal zone of Coral Creek (Table 2.3). DIN flux data from Taylor River showed that NH_4^+ was consistently imported by the dwarf wetland while NO_x^- was consistently released into the water column (Davis et al. 2001a). These NH_4^+ fluxes into the soil (i.e., an indication of nitrification) could not have been predicted from pore water concentrations, as they often exceed 50 μM NH_4^+ in this area of the Everglades (Koch 1997). This trend is in contrast to what has been shown in many estuarine saltmarshes and even some mangrove sediments, where NH_4^+ is generally exported and NO_x^- imported (Nixon 1980, Childers et al. 1999, K Liu and SE Davis unpubl. data). Lara and Dittmar (1999) also showed that ammonium dynamics in mangrove wetlands could be influenced by diurnal patterns of production and respiration, with NH_4^+ concentrations approximately 44% higher in a Brazilian mangrove forest during the night than during the day.

The results from the dwarf mangroves in Taylor River suggest that nitrification of NH_4^+ in the dwarf mangrove wetlands may provide a considerable source of oxidized inorganic nitrogen to the water column. Ovalle et al. (1990) arrived at a similar conclusion in their study of the factors controlling the chemistry of a tidal mangrove creek of Sepetiba Bay, Brazil. They determined that net nitrification exceeded net denitrification thereby resulting in the observed increase of nitrate in the water column during ebb tide (Ovalle et al. 1990). Depending on hydrology and substrate availability, this source of oxidized inorganic nitrogen can fuel denitrification either

Table 2.3 Flux estimates of N and P in different mangrove ecosystems. Annual estimates and hourly estimates are noted. Positive values indicate mangrove uptake. Negative values indicate export or release to water column. For abbreviations see Table 2.1

Reference	Site	Method	NH$_4^+$ flux	NO$_x^-$ flux	SRP flux	TN flux	TP flux
Alongi (1996)[a]	Coral Creek, Australia	Core incubation[b]	0.07	0.04	0.03	–	–
Boto and Wellington (1988)	Coral Creek, Australia	Creek flux[c]	0.15	-0.03	–	–	0.5
Davis et al. (2001a)	Taylor River, USA	Mangrove enclosures[c]	2.55	-4.32	–	26.1	0.41
Kristensen et al. (1988)	Ao Nam Bor, Thailand	Benthic flux[c]	6.83	4.51	–	–	–
Liu and Davis unpubl. data	Taylor River, USA	Core incubation[b]	-1.74	0.07	0.08	–	–
Rivera-Monroy et al. (1995)	Estero Pargo, Mexico	Flume[c]	0.53	0.08	–	0.06	–
Rivera-Monroy unpubl. data[d]	Boca Chica, Mexico	Core incubation[b]	-2.51	0.95	–	–	–
Rivera-Monroy unpubl. data[d]	Shark River, USA	Core incubation[b]	0.65	0.29	-0.08	–	–
DT Rudnick et al., South Florida Water Management District, Florida, unpubl. data	Taylor River, USA	Benthic flux[c]	3.28	1.2	–	–	–

[a] Mid-intertidal average across multiple seasonal samplings
[b] Season-averaged fluxes in mg. m^{-2}. hr^{-1} for batch core incubation studies
[c] Annual flux estimate based on multiple, samplings. Fluxes in units of g. m^{-2}. yr^{-1} for each
[d] Unpublished data from Rivera-Monroy later appeared in Childers et al. (1999)

in situ or in adjacent systems, resulting in a substantial loss of N from an estuary (Jenkins and Kemp 1984, Henriksen and Kemp 1988, Seitzinger 1988). However, studies indicate that denitrification may not be a significant sink for nitrogen in unpolluted mangrove systems, as losses of nitrate appear to be directed more towards the sediments rather than the atmosphere (Alongi et al. 1992, Kristensen et al. 1995, Rivera-Monroy and Twilley 1996).

Only a modest number of studies have measured significant wetland–water column exchanges of phosphorus in mangrove systems. Most of these indicate net import of phosphorus by the mangrove wetland, as these systems promote deposition of sediment-associated forms of P or the sediments can effectively scavenge P from the water column (Nixon et al. 1984, Boto and Wellington 1988; Table 2.3). In Brazil, Dittmar and Lara (2001) measured significant export of SRP during the dry season, but a small uptake of SRP by the mangrove forest in the rainy season. Since phosphorus concentrations are so low in Taylor River, Davis et al. (2001a, b) were unable to detect significant flux of SRP between the wetland soil, vegetation, and water column. However, they were able to detect significant exchange of TP in the dwarf mangrove wetland during both wet and dry season samplings (Davis et al. 2001a; Table 2.3).

2.2.5 N and P Flux in Seagrass Ecosystems

Along undeveloped tropical coastlines, upland runoff typically passes through mangrove forests before being discharged into nearshore seagrass beds. Flux patterns described above for mangroves are often important in affecting the concentrations and forms of N and P entering these seagrass ecosystems. However, the bulk exchange of materials, particularly in mangroves with a weak tidal signature and a strong wet–dry season pattern of discharge, is overwhelmingly driven by patterns of surface water discharge. For example, in the southern Everglades ecosystem, wet season outflow accounts for approximately 99% of the surface water-borne net export of N and P to Florida Bay (Sutula et al. 2003). Further, storm events such as hurricanes and tropical storms can also account for much of the annual exchanges of N and P between mangrove wetlands and seagrass beds (Davis et al. 2004). Using long-term TN and TP-concentration data from the Florida Coastal Everglades Long-Term Ecological Research program (FCE-LTER, D Childers unpubl. data) and US Geological Survey (USGS) gauging station from the mouth of Taylor River (FCE-LTER site TS/Ph 7 and USGS station # 251127080382100) we illustrate the strong seasonal signature of surface water-mediated exchange between mangrove and seagrass ecosystems in eastern Florida Bay (Figs. 2.2, 2.4). These patterns directly reflect the seasonal discharge pattern of these types of mangrove creeks that have strong positive discharge patterns throughout the wet season and little net exchange of water during the dry season (Fig. 2.4). Further, given the proximity of these systems to one another (Fig. 2.5), upland or mangrove-derived nutrients can have immediate, direct impacts on nearby seagrass beds. The reverse is also true, particularly during storm events such as hurricanes where storm-related surges can result in

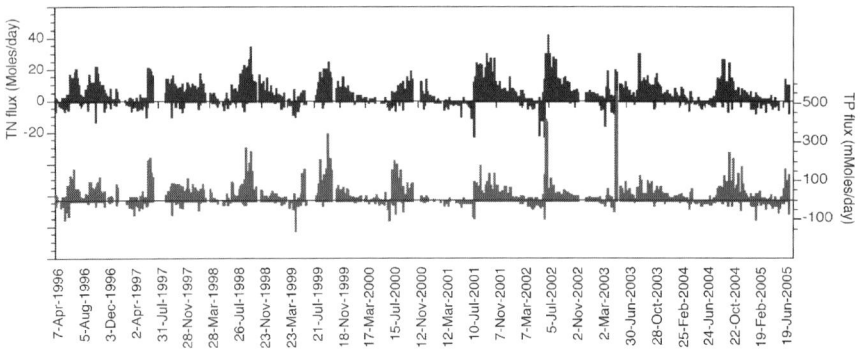

Fig. 2.4 Estimated fluxes of total nitrogen (TN) and total phosphorus (TP) from Taylor River, Florida (USA)—a mangrove creek that empties into NE Florida Bay. These long-term data from 1996 to 2005 show the strong seasonal nature of discharge and water-mediated exchange of N and P between the mangrove ecosystem and seagrass-dominated waters in Florida Bay

significant resuspension of subtidal (i.e., seagrass) sediment and subsequent deposition in nearby mangrove forests (Davis et al. 2004; Fig. 2.5).

Within seagrass beds, significant nutrient uptake can be achieved through leaves and the root-rhizome system. Similarly, leaves and rhizomes can act as nutrient reservoirs, especially for N that can be stored in amino acids and other soluble and non-soluble compounds (e.g., Udy et al. 1999). Moreover, as with other marine and terrestrial plants, seagrasses are able to take up nutrients in excess of their metabolic needs and store these for periods of low availability (Gobert et al. 2006, Romero et al. 2006). Finally, while limited research has been conducted on the transport of nutrients within and between above and belowground tissues in seagrasses, there is evidence that seagrasses are able to translocate nutrients within shoots (Lepoint et al. 2002) and among clonal ramets (Marbà et al. 2002) to sustain growth of new tissue.

The relative uptake of N and P through the leaves and rhizomes can vary significantly among species and habitats. While the higher concentration of nutrients within pore water pools would suggest a benefit for nutrient uptake through belowground structures, leaf uptake of water column nutrients can still supply a considerable proportion of seagrass N (reviewed by Romero et al. 2006). In fact, several experimental studies have documented the high nutrient uptake capacity of leaf tissue, especially at low nutrient concentrations. For example, Lee and Dunton (1999) showed that 50% of N uptake can take place through the leaves in *Thalassia testudinum*. Similarly, high rates of N uptake through the leaves were recorded for *Zostera* (Short and McRoy 1984), *Phyllospadix* (Terrados and Williams 1997), and *Posidonia* (Gobert et al. 2006). Seagrass leaves have a higher affinity for NO_3^- than for NH_4^+. However, NH_4^+ is the dominant form of DIN taken through the rhizomes (Touchette and Burkholder 2000).

Seagrass nutrient pools can be replenished through three processes: sedimentation to the soil from the overlying water column, nitrogen fixation, and nutrient uptake by leaves (Hemminga et al. 1991). Sources of dissolved nutrients to

Fig. 2.5 Moving clockwise from upper left, photos showing (**a**) mangrove leaf detritus in a sea-grass bed, (**b**) carbonate sediment that had been deposited in a South Florida mangrove forest as a result of hurricane Wilma in 2005, (**c**) small core showing the thickness of carbonate sediment deposit in (b), (**d**) seagrass growing immediately adjacent to mangrove, and (**e**) seagrass growing amongst coral heads

seagrasses include those available from the water column as well as those released from decaying organic matter through remineralization. A major source of nutrients to seagrass meadows is derived from the sedimentation of sestonic particles that include organic and inorganic components (Romero et al. 2006). Seagrass mead-ows play a major role in the retention of particles and the accumulation of sedi-ments and organic matter is central to the development and subsistence of seagrass habitats. The buffering activities of seagrass canopies facilitate sedimentation and particle accumulation that increases nutrient availability to these important habitats (Hemminga et al. 1991, Koch et al. 2006).

Total sediment organic pools in seagrass beds depend on litter production by seagrasses and other organisms (e.g., macroalgae, epiphytes, microalgae), organic matter derived from external sources, and the utilization and degradation of these inputs. It is estimated that the input of N (up to 60 g N m^{-2}. yr^{-1}; Romero et al. 2006) and P (up to g P m^{-2} . yr^{-1}; Gacia et al. 2002) from sediment sources can potentially provide most of the annual nutrient requirement for seagrass growth. A flux study in seagrass beds of Laguna Madre (Texas, USA), found significant regeneration of NH$_4^+$ in the water column as well as release from the sediments (Ziegler and Benner 1999). Ziegler and Benner (1999) believed this sediment release was associated with NH$_4^+$ regeneration in the benthos during daylight hours. Similarly, benthic flux studies by Holmer and Olsen (2002) and Mwashote and Jumba (2002) showed mostly a release of DIN from the sediment of seagrass beds in both Phuket Island (Thailand) and Gazi Bay (Kenya), respectively. Finally, seagrass habitats can exhibit high levels of N$_2$ fixation through the activities of bacteria associated with seagrass rhizomes (Welsh 2000) as well as cyanobacteria associated with seagrass leaves, which can supply up to 38% of the N requirements of *T. testudinum* (Capone and Taylor 1977).

An important mechanism for the conservation and recycling of nutrients within seagrass ecosystems and in mangrove forests is the ability of the plants to resorb nutrients from older or senescent tissue (Feller et al. 2003, Romero et al. 2006). In fact, it has been reported that on average >20% of the annual N and P requirements can be obtained from nutrient resorption (Hemminga et al. 1999). However, even if seagrasses are able to reclaim a considerable portion of the nutrients stored in mature leaves, detached leaves that contain >75% of their original nutrient content can represent a significant nutrient drain from the system if they are removed prior to entering the detrital pool.

Loss of N and P from seagrass beds can occur through the process of leaching/exudation from living and dead plant material, diffusion from sediment, denitrification, nutrient transfer by foraging animals, and export of sloughed leaves and leaf fragments (Hemminga et al. 1991). The main source of nutrient losses to seagrass meadows is the removal of leaf material by waves, tides, and currents (Romero et al. 2006). The export of leaf litter and macroalgae provides a link between seagrass meadows and adjacent habitats such as mangroves, hardbottom habitats, and coral reefs but can also represent a major nutrient drain for the source habitats. In a recent review, Mateo et al. (2006) report that up to 100% of production can be exported out of a seagrass habitat due to hydrodynamic forcing and that nutrient losses of >40% of N and >20% of P assimilated can be exported. The export of detached seagrass leaves can be especially significant during storm events (Davis et al. 2004), and marine sediment and seagrass leaf litter is commonly seen along fringe mangrove habitats (see Fig. 3.5 of Chapter 3). Similar accumulations of macroalgae and seagrass litter can be seen around patch reefs habitats of the Florida Reef Tract where seagrass beds composed mainly of *Thalassia testudinum* are abundant on sandy substratum (D Lirman pers. observ.).

Another mechanism resulting in the removal of nutrients from seagrass habitats is via the direct consumption of plant matter by grazers and the detachment of leaves through grazing activity. While herbivores that reside within seagrass beds can release remineralized N and P back into the system through their feces, herbivorous guilds that feed on seagrass beds but reside part or most of the day away from these habitats can lead to a net N and P export from the system. Such is the case of juvenile fishes that reside in mangrove habitats during the day but migrate into adjacent seagrass beds to feed at night (Nagelkerken et al. 2000, Verweij et al. 2006).

Lastly, N and P can remain within the system but become unavailable for seagrass use. This is especially true for P that can adsorb to organic and inorganic particles and become relatively unavailable for plant uptake. In carbonate sediments, P is often bound to calcium and therefore can limit seagrass growth due to its reduced availability. Similarly, both N and P can be bound to refractory organic compounds that can be buried in the sediments and no longer available for uptake (Koch et al. 2001).

2.2.6 N and P Flux in Coral Reef Ecosystems

The flux of N and P between the water column and coral reef communities has been commonly estimated by measuring the changes in the concentration of nutrients over time and as water moves over the reef (e.g., Johannes et al. 1983, Atkinson 1987). The uptake of dissolved inorganic N and P by reefs can be rapid, highly variable in space and time, and is directly dependent on the biological and structural characteristics of the reef community (e.g., productivity, abundance, taxonomic structure, topography; Baird and Atkinson 1997, Koop et al. 2001), hydrodynamics (e.g., water residence time, mixing, velocity; Hearn et al. 2001, Falter et al. 2004), temperature and light (Johannes et al. 1983), and nutrient concentrations (Pilson and Betzer 1973, Smith et al. 1981) in a given system. An example of rapid uptake of nutrients was observed during the 'Elevated Nutrient on Coral Reefs Experiment' (ENCORE) conducted in the Great Barrier Reef (GBR), where levels of NH_4^+ and SRP returned to background levels 2–3 hrs after nutrient additions that increased ambient concentrations to >11 μM NH_4^+ and >2 μM SRP (Koop et al. 2001)—levels considerably greater than typical ranges exhibited in most coral reefs.

Both particulate and dissolved forms of organic and inorganic N and P discharged from land provide significant nutrient inputs into coral reef ecosystems (Furnas et al. 1997). The majority of nutrients discharged enter the coastal environment in particulate form (Furnas 2003). Nutrients remineralized from bacteria, plankton, and detrital matter in suspended particulate matter can be quite high for areas with high sedimentation rates and can be made readily available to coral reef organisms (Fabricius 2005). In fact, the consumption of phytoplankton by benthic feeders that include corals, sponges, tunicates, bivalves, bryozoans, and polychaetes provides one of the main benthic-pelagic coupling mechanisms in reef habitats and a major source of nutrients (Yahel et al. 1998). In addition to oceanic (e.g., bacteria, phytoplankton, zooplankton) and land-based sources of allochthonous nutrients (e.g., overland flow, riverine discharge, groundwater, sewage discharge), nutrients can enter reef systems through upwelling, atmospheric deposition, rainfall, and N_2 fixation (D'Elia and Wiebe 1990).

Coral reefs are commonly surrounded by seagrass beds and macroalgal communities and are in close proximity to mangrove habitats that can provide sources of nutrients bound in detrital matter. Plant detritus can be transported into reef systems through wave action and currents and the nutrients contained in this detrital pool can become available to reef organisms through remineralization. Finally, herbivores that utilize the reef structure for refuge but conduct daily grazing and foraging migrations into adjacent habitats such as seagrass beds can add to the reef's nutrient pool through the deposition of feces. A clear example of the activities of herbivorous guilds that inhabit coral reefs but graze on surrounding habitats either during the night or the day is manifested in the appearance of grazing 'halos' surrounding many patch reefs in the Caribbean and elsewhere. These halos of bare substrate are commonly created by the grazing activities of sea urchins and fishes as they forage away from the reef structure (Randall 1963, Ogden et al. 1973). Another example of grazing activities that can introduce new nutrients into reef habitats was described by Meyer et al. (1983) who reported that the feces of schools of juvenile grunts, that grazed on nearly seagrass beds during the day but aggregated during the day around coral colonies, can be significant sources of N and P that can be readily incorporated into coral tissue, enhancing coral growth.

The relative contribution of these nutrient sources can vary significantly among coral reefs both within and among geographic regions. In the central GBR, Furnas et al. (1997) determined that there are four main sources of allochthonous nutrients to this reef area: (1) rivers account for the greatest potential source with up to 21.3 Kmol N m^{-1} (i.e., linear m of shelf) and up to 2.0 Kmol P m^{-1}, (2) upwelling can contribute up to 5.0 Kmol N m^{-1} and up to 0.4 Kmol P m^{-1}, (3) rainfall accounts for up to 0.84 Kmol N m^{-1} and up to 0.02 Kmol P m^{-1}, and (4) sewage can contribute up to 0.14 Kmol N m^{-1} and up to 0.02 Kmol P m^{-1}. Much of the river-derived N and P enter the reef ecosystem during episodic floods caused by storms or seasonal high-rain events, again highlighting the influence of acute events. Generally speaking, P input is equally divided into dissolved inorganic or soluble reactive P (DIP or SRP), dissolved organic P, and particulate P, while >75% of N input is as dissolved organic N, 18% as particulate N, with a smaller contribution (<5%) from dissolved inorganic N, mainly as NH_4^+ and NO_3^- (Furnas et al. 1997). Lastly, another potential important source of N into the GBR is through N_2 fixation by both benthic and pelagic (mainly *Trichodesmium*) cyanobacteria and contributing up to 72 Kmol N m^{-1} (Furnas et al. 1997).

Coral reefs have been shown to be generally effective at removing dissolved nutrients from the overlying water masses (Koop et al. 2001) as well as retaining internally recycled nutrients. The high rates of gross productivity and biomass accumulation recorded in coral reef habitats compared to the open ocean under low nutrient conditions led researchers to believe that nutrient supply to coral communities is maintained mainly through tight nutrient retention and regeneration (e.g., Pomeroy

1970, Johannes et al. 1972). More recent studies have indicated that direct nutrient uptake from the water column can also be a major source of new nutrients (Falter et al. 2004). While the relative contribution of recycled and allochthonous nutrients to reef productivity can be debated, it is clear that coral reefs possess mechanisms for nutrient retention and regeneration that contribute to productivity.

The exchange of photosynthetic and waste products between coral hosts and their endosymbiotic zooxanthellae provides one clear example of nutrient recycling within reef systems that enhances coral growth and reef accretion. The remineralization of organic detritus by bacterial communities within sediments and numerous cryptic organisms found within the interstitial spaces of reef frameworks provides another major source of recycled nutrients to reef communities (Szmant-Froelich 1983, Szmant 2002). In the GBR, the remineralization of nutrients from benthic organic pools contributed up to 14 Kmol N m^{-1} and up to 1.8 Kmol P m^{-1} to the reef communities (Furnas et al. 1997).

Coral reefs have also been shown to export nutrients, demonstrating their potential to serve as nutrient sources for adjacent seagrass or mangrove ecosystems (Webb et al. 1975, Delesalle et al. 1998, Hata et al. 1998). Nutrients are typically exported as dissolved and particulate forms via water-mediated transport and in gaseous form through denitrification. For example, the removal of macroalgal biomass from reef habitats during storms may benefit corals by removing aggressive competitors, but it can represent an important export of nutrient-rich plant biomass that would no longer be available for *in situ* consumption or remineralization (Lapointe et al. 2006). As is the case in seagrass habitats, nutrients can also remain within reefs but become unavailable for use by reef organisms. Both N and P can be bound to refractory organic compounds that can be buried into sediments or deep into the reef framework and no longer available for uptake.

2.3 Human Impacts on N and P Concentration and Flux

More than 20 years ago, it was suggested that local hydrologic and geomorphologic factors largely governed the flux of organic matter from coastal wetlands to nearshore waters (Odum et al. 1979). This was later supported by data from outwelling studies demonstrating that mangrove wetlands exported organic matter in relation to tidal energy (Twilley 1985, review by Lee 1995). The direct connection to land and rivers puts these coastal ecosystems in a particularly vulnerable situation not only when it comes to land conversions but also eutrophication (Valiela et al. 2001). Aside from the direct hydrologic connections, the physical proximity of these systems to one another (see Fig. 2.5) can also make them susceptible to these influences. The number of nutrient flux studies in mangrove wetlands has increased dramatically over the past 15 yrs. This trend has stemmed from recent deterioration of water quality in many tropical and subtropical coastal areas as a result of deforestation, coastal development, oil spills, and freshwater diversion (Twilley 1998).

By and large, the focus of these recent flux studies has been on the exchange of organic matter between the mangrove and nearshore environment (Boto and Bunt 1981, Twilley 1985, Woodroffe 1985, Flores-Verdugo et al. 1987, Robertson 1986, Lee 1995). However, there have been few studies that have quantified the exchanges of inorganic N and P within mangrove wetlands or between mangroves and nearshore systems (Boto and Wellington 1988, Kristensen et al. 1988, Nedwell et al. 1994, Rivera-Monroy et al. 1995, Davis 1999). Although it is generally held that mangroves export organic matter in relation to tidal energy (Odum et al. 1979, Twilley 1985, Lee 1995), the fate of inorganic nutrients in estuarine mangrove systems is still poorly understood. This is of concern, because these ecosystems exhibit structural/morphological variability in response to variations in nutrient availability and are also susceptible to anthropogenic nutrient loading. Moreover, they also serve as a buffer against nutrient loading to adjacent seagrass beds (Valiela and Cole 2002).

Seagrass communities worldwide have experienced drastic declines in abundance and spatial extent due to a combination of factors that are common sources of stress to all coastal ecosystems. These factors include natural disturbances such as elevated temperatures, sea level changes, changes in water chemistry, diseases, competition (e.g., grazing, epiphytism, invasive species), and storm events, as well as human disturbances commonly associated with increasing population numbers and coastal development such as increases in sedimentation and nutrients, pollution, dredging and boating impacts, and overfishing (reviewed by Orth et al. 2006). Human-induced changes in the quality and quantity of freshwater inputs to tropical coastlines have also been linked to changes in nearshore water quality and seagrass communities (Robblee et al. 1991, McIvor et al. 1994).

The negative influence of eutrophication is often cited as one of the main sources of disturbance to seagrass habitats, especially those found adjacent to urban centers and in shallow, poorly flushed coastal systems (Touchette and Burkholder 2000, Ralph et al. 2006). The mechanisms that mediate the negative impacts of eutrophication on seagrass condition include: (1) direct physiological impacts caused by high, toxic levels of inorganic N forms and increased C demand under high nutrient conditions, and (2) indirect impacts of increased growth of epiphytes, macroalgae, and phytoplankton that can outcompete seagrasses for nutrients, reduce light availability, and create anoxic conditions deleterious for seagrass growth and survivorship (Ralph et al. 2006, Burkholder et al. 2007).

The direct deleterious effects of increased N on seagrass growth and survivorship have been demonstrated experimentally for several seagrass species (reviewed by Touchette and Burkholder 2000). The high energetic cost of N assimilation and the accumulation of toxic levels of nitrate, nitrite, and ammonium were identified as responsible for the documented reduction in growth and mortality in *Zostera* (Burkholder et al. 1992, van Katwijk et al. 1997, Peralta et al. 2003), *Halodule* (Burkholder et al. 1994), and *Ruppia* (Santamaría et al. 1994). The most common impact of eutrophication on seagrass habitats is the rapid growth of primary producers such as epiphytes, macroalgae, and phytoplankton that can outcompete seagrasses for substrate, light, and nutrients. Overgrowth, reduced light levels, and limited nutrient availability can yield negative impacts that range from reduced growth to massive die-offs (reviewed by Burkholder et al. 2007). Moreover, rapid increases in the biomass of seagrass competitors can result in anoxic conditions and promote increased sulfide concentration in the sediments, which can further influence seagrass metabolism, growth, and survivorship (Calleja et al. 2007, Koch et al. 2007). The end result of this would be a conversion of seagrass to an ephemeral algal or phytoplankton-dominated system with a reduced capacity for ecosystem C storage, enhanced mass transport of previously plant-bound nutrients, a reduced importance of denitrification, and, in carbonate-dominated systems, enhanced P release from sediments (McGlathery et al. 2007).

Coral reef ecosystems have experienced a drastic decline in condition, diversity, and extent in the recent past (Gardner et al. 2003, Pandolfi et al. 2003). The causes of this worldwide decline are varied, but the competition of corals with macroalgae is often cited as one of the most significant factors influencing coral persistence (Hughes et al. 2007, Kleypas and Eakin 2007). The role of increased nutrients, mainly from human sources, has been highlighted as one of the main determinants of the outcome of the competition between corals and macroalgae for space and the shift from coral-dominated systems to algal-dominated reef states (Lapointe 1997, Littler and Littler 2007). The fast turnover rates of macroalgae and cyanobacteria compared to corals can facilitate the rapid accumulation of biomass of these taxa that can cause, in the absence of grazing, coral mortality through shading, sediment trapping, abrasion, and allelopathy (reviewed by McCook et al. 2001). A classic example of nutrient-mediated overgrowth of corals by macroalgae was documented in Kaneohe Bay, Hawaii, where the influx of human sewage resulted in the rapid growth of the green macroalga *Dictyosphaeria*. Increases in the abundance of *Dictyosphaeria* and increases in phytoplankton in the water column caused significant coral mortality that continued until the nutrient source was removed by the relocation of the sewage outfall further away from the reefs (Smith et al. 1981).

Increased nutrients have also been shown to influence coral communities by increasing rates of reef bioerosion (Chazottes et al. 2002) and, under high levels of P, decrease whole-reef calcification (Kinsey and Davies 1979). Based on the review of studies of nutrient impacts, Bell (1992) suggested an eutrophication threshold level for reefs of 1 μM DIN and 0.1–0.2 μM SRP. Further research by Lapointe (1997) also suggested that macroalgal overgrowth of corals can be expected at nutrient threshold levels exceeding 1 μM DIN and 0.1 μM SRP. In addition to the community-level impacts described, increased nutrients can have significant impacts at the coral-colony level. Increased concentrations of inorganic N and P, individually and in combination, have been shown to disrupt the coral symbiosis and result in reduced coral calcification and growth and reduced fecundity (see reviews by Szmant 2002, and Fabricius 2005).

2.4 Conclusions

Mangrove wetlands, seagrass meadows, and coral reefs are some of the most threatened ecosystems on the planet and are also among the most susceptible to the effects of nutrient loading and loss of ecosystem services. Empirical data from numerous studies (and common sense) tells us that hydrologic flushing dictates the relative importance of internal recycling versus imports/exports from adjacent aquatic ecosystems in driving the N and P budget of unimpacted mangrove swamps, seagrass beds and coral reefs. Despite the importance of flushing, research also tells us that surface water nutrient concentrations (natural or human-influenced) become increasingly important with distance offshore, as direct uptake can represent an important source of nutrients in coral and seagrass ecosystems. However, ecosystem states can shift to more plankton-dominated situations—reducing light availability to these subtidal producers—if concentrations become excessive.

Little is presently known about the fate of upland or mangrove-derived nutrients seaward of coastal wetlands. Even less is known about the fate of upwelled nutrients landward of coral reefs. As a result of methodological and resource constraints, our tendency has been to focus on the direct linkages between the ecosystems illustrated in the simple conceptual diagram in Fig. 2.1. Simulations and mass-balance studies can provide some insight into net ecosystem exchanges of materials, but there is a pressing need for studies quantifying the exchange of N and P across the land—ocean interface along tropical coastlines. Chapter 12 describes several recent methods that can be used to identify sources of organic materials in the water column, sediments, and live biomass of each of these ecosystems. That chapter also makes a case for combining these source characterization approaches with direct flux measurements as a means for better understanding the biogeochemical connectivity among mangrove, seagrass, and coral reef ecosystems.

These are also 'open' ecosystems from a materials exchange standpoint, as water and materials freely exchange across their ecological bounds. Intuition and research tell us these ecosystems are connected biogeochemically; however, little information exists on the extent of connectivity in space and time as well as the factors driving connectivity. Nutrient dynamics within these coastal habitats are influenced by rates of nutrient import, uptake, recycling, and export. Influx of N and P into each is caused by enrichment from the aforementioned external sources (atmospheric deposition, surface water, and groundwater) and N_2 fixation. Losses of N and P losses are caused mainly by the export of dissolved and particulate sources out of the system via the same routes and denitrification.

As for documentation of large-scale exchange across ecosystem boundaries, Hemminga et al. (1994) showed that outwelled carbon from a Kenyan mangrove creek was balanced by fluxes of seagrass-derived carbon into the mangrove zone, indicating a potentially tight coupling between these ecosystems. Kitheka et al. (1996) further demonstrated the importance of seasonal variations in river inflow on patterns of water quality and water column productivity across a mangrove—coral reef transect, suggesting that the oligotrophic nature of the mangrove ecosystem (i.e., low inorganic nutrient release) and short residence time of Gazi Bay facilitated the flushing of river-borne nutrients during the wet season. In a study encompassing 36 tidal cycles, Dittmar and Lara (2001) showed both seasonal and strong diurnal patterns in N and P flux from a Brazilian mangrove forest, suggesting that flux estimates should consider temporal variability at these scales. Such large-scale studies attempting to link the hydrology and biogeochemistry of these adjacent coastal ecosystems are needed and will continue to provide more clarity with regard to the factors affecting the spatial and temporal extent of their connectivity.

Acknowledgments Data and support for this chapter were provided by the U.S. NOAA Center for Sponsored Coastal Ocean Research, the South Florida Water Management District, the U.S. Geological Survey, and the Florida Coastal Everglades Long-Term Ecological Research (FCE LTER) Program, which is supported by the National Science Foundation under Grant No. DBI-0620409 and Grant No. DEB-9910514.

References

Agawin NS, Duarte CM, Fortes MD (1996) Nutrient limitation of Philippine seagrasses (Cape Bolinao, NW Philippines): in situ experimental evidence. Mar Ecol Prog Ser 138: 233–243

Alongi DM (1988) Bacterial productivity and microbial biomass in tropical mangrove sediments. Microb Ecol 15:59–79

Alongi DM (1990) Effect of mangrove detrital outwelling on nutrient fluxes in coastal sediments of the central Great Barrier Reef Lagoon. Estuar Coast Shelf Sci 31:581–598

Alongi DM, Boto KG, Robertson AI (1992) Nitrogen and phosphorus cycles. In: Robertson AI, Alongi DM (eds) Tropical mangrove ecosystems. American Geophysical Union, Washington DC

Alongi DM (1996) The dynamics of benthic nutrient pools and fluxes in tropical mangrove forests. J Sea Res 54:123-148

Amador JA, Jones RD (1993) Nutrient limitations on microbial respiration in peat soils with different total phosphorus content. Soil Biol Biochem 25:793–801

Anthony KRN (2000) Enhanced particle-feeding capacity of corals on turbid reefs (Great Barrier Reef, Australia). Coral Reefs 19:59–67

Anthony KRN (2006) Enhanced energy status of corals on coastal, high-turbidity reefs. Mar Ecol Prog Ser 319:111–116

Anthony KRN, Fabricius KE (2000) Shifting roles of heterotrophy and autotrophy in coral energetics under varying turbidity. J Exp Mar Biol Ecol 252:221–253

Atkinson MJ (1987) Rates of phosphate uptake by coral reef flat communities. J Exp Mar Biol Ecol 32:426–435

Atkinson MJ (1992) Productivity of Enewetak Atoll reef flats predicted from mass transfer relationships. Cont Shelf Res 12:799–807

Atkinson MJ, Falter JL (2003) Coral reefs. In: Black K, Shimmield G (eds) Biogeochemistry of marine systems. Blackwell Publishing, Oxford

Baird ME, Atkinson MJ (1997) Measurement and prediction of mass transfer to experimental coral reef communities. J Exp Mar Biol Ecol 42:1685–1693

Bell PRF (1992) Eutrophication and coral reefs – some examples from the Great Barrier Reef. Lagoon Water Res 26:553–568

Benner R, Peele ER, Hodson RE (1986) Microbial utilization of dissolved organic matter from leaves of the red mangrove, *Rhizophora mangle*, in the Fresh Creek Estuary, Bahamas. Estuar Coast Shelf Sci 23:607–619

Bianchi TS (2007) Biogeochemistry of estuaries. Oxford University Press, Oxford

Bird KT, Johnson JR, Jewett-Smith J (1998) In vitro culture of the seagrass *Halophila decipiens*. Aquat Bot 60:377–387

Boto KG, Bunt JS (1981) Tidal export of particulate organic matter from a northern Australian mangrove forest. Estuar Coast Shelf Sci 13:247–255

Boto KG (1982) Nutrient and organic fluxes in mangroves. In: Clough BF (ed) Mangrove ecosystems in Australia: structure, function and management. Australian National University Press Canberra, Australia

Boto KG, Wellington JT (1988) Seasonal variations in concentrations and fluxes of dissolved organic materials in a tropical, tidally-dominated, mangrove waterway. Mar Ecol Prog Ser 50: 151–160

Boyer JN, Fourqrean JW, Jones R (1999) Seasonal and long-term trends in the water quality of Florida Bay (1989–1997). Estuaries 22:412–430

Brinson MM (1977) Decomposition and nutrient exchange of litter in an alluvial swamp forest. Ecology 58:601–609

Burkholder JM, Mason KM, Glasgow HB (1992) Water-column nitrate enrichment promotes decline of eelgrass (*Zostera marina* L): evidence from seasonal mesocosm experiments. Mar Ecol Prog Ser 81:163–178

Burkholder JM, Glasgow HB, Cooke JE (1994) Comparative effects of water-column nitrate enrichment on eelgrass *Zostera marina*, shoalgrass *Halodule wrightii*, and widgeongrass *Ruppia maritime*. Mar Ecol Prog Ser 105:121–138

Burkholder JM, Tomasko DA, Touchette BW (2007) Seagrasses and eutrophication. J Exp Mar Biol Ecol 350:46–72

Capone DG, Taylor BF (1977) Nitrogen fixation (acetylene reduction) in the phyllosphere of *Thalassia testudinum*. Mar Biol 40:19–28

Calleja ML, Marbà N, Duarte CM (2007) The relationship between seagrass (*Posidonia oceanica*) decline and sulfide porewater concentration in carbonate sediments. Estuar Coast Shelf Sci 733:583–588

Chale FMM (1993) Degradation of mangrove leaf litter under aerobic conditions. Hydrobiologia 257:177–183

Chazottes V, Le Campion-Alsumard T, Peyrot-Clausade M et al (2002) The effects of eutrophication-related alterations to coral reef communities on agents and rates of bioerosion (Reunion Island, Indian Ocean). Coral Reefs 21:375–390

Childers DL, Cofer-Shabica S, Nakashima L (1993) Spatial and temporal variability in marsh-water column interactions in a southeastern USA salt marsh estuary. Mar Ecol Prog Ser 95: 25–38

Childers DL, Davis SE, Twilley R et al (1999) Wetland-water column interactions and the biogeochemistry of estuary-watershed coupling around the Gulf of Mexico. In: Bianchi T, Pennock J, Twilley R (eds) Biogeochemistry of Gulf of Mexico estuaries. John Wiley & Sons, New York

Childers DL, Day JW, McKellar HN (2000) Twenty more years of marsh and estuarine flux studies: revisiting Nixon (1980). In: Weinstein MP, Kreeger DQ (eds) Concepts and controversies in tidal marsh ecology. Springer, The Netherlands

Childers DL, Boyer JN, Davis SE et al (2006) Nutrient concentration patterns in the oligotrophic 'upside-down' estuaries of the Florida Everglades. J Exp Mar Biol Ecol 51:602–616

Clough BF (1992) Primary productivity and growth of mangrove forests. In: Robertson AI, Alongi DM (eds) Tropical mangrove ecosystems. American Geophysical Union, Washington DC

Costa OS, Attrill M, Nimmo M (2006) Seasonal and spatial controls on the delivery of excess nutrients to nearshore and offshore coral reefs of Brazil. J Mar Syst 60:63–74

Cundell AM, Brown MS, Stanford R (1979) Microbial degradation of *Rhizophora mangle* leaves immersed in the sea. Estuar Coast Mar Sci 9:281–286

Davis SE (1999) The exchange of carbon, nitrogen, and phosphorus in dwarf and fringe mangroves of the oligotrophic southern Everglades. PhD Dissertation, Florida International University, Florida

Davis SE, Childers DL, Day JW et al (2001a) Wetland-water column exchanges of carbon, nitrogen, and phosphorus in a southern Everglades dwarf mangrove. Estuaries 24:610–622

Davis SE, Childers DL, Day JW et al (2001b) Nutrient dynamics in vegetated and non-vegetated areas of a southern Everglades mangrove creek. Estuar Coast Shelf Sci 52:753–765

Davis SE, Childers DL, Day JW et al (2003a) Factors affecting the concentration and flux of materials in two southern Everglades mangrove wetlands. Mar Ecol Prog Ser 253:85–96

Davis SE, Coronado-Molina C, Childers DL et al (2003b) Temporal variability in C, N, and P dynamics associated with red mangrove (*Rhizophora mangle* L) leaf decomposition. Aquat Bot 75:199–215

Davis SE, Cable J, Childers DL et al (2004) Importance of episodic storm events in controlling ecosystem structure and function in a Gulf Coast estuary. J Coast Res 20:1198–1208

Davis SE, Childers DL (2007) Importance of water source in controlling leaf leaching losses in a dwarf red mangrove (*Rhizophora mangle* L) wetland. Estuar Coast Shelf Sci 71:194–201

Delesalle B, Buscail R, Carbonne J et al (1998) Direct measurements of carbon and carbonate export from a coral reef ecosystem (Moorea Island, French Polynesia). Coral Reefs 17: 121–132

Dittmar T, Lara, R (2001) Do mangroves rather than rivers provide nutrients to coastal environments south of the Amazon River? Evidence from long-term flux measurements. Mar Ecol Prog Ser 213:67–77

Duarte CM (1990) Seagrass nutrient content. Mar Ecol Prog Ser 67:201–207

Duke NC (1992) Mangrove floristics and biogeography. In: Robertson AI, Alongi DM (eds) Tropical mangrove ecosystems. American Geophysical Union, Washington DC

D'Elia CF, Wiebe WJ (1990) Biogeochemical nutrient cycles in coral reef ecosystems. In: Dubinsky Z (ed) Coral reefs: ecosystems of the World series. Elsevier Science Publishers, Amsterdam

Edinger EN, Limmon GV, Jompa J et al (2000) Normal coral growth rates on dying reefs: are coral growth rates good indicators of reef health? Mar Pollut Bull 40:404–425

Elser JJ, Bracken M, Cleland E et al (2007) Global analysis of nitrogen and phosphorus limitation of primary producers in freshwater, marine and terrestrial ecosystems. Ecol Lett 10: 1135–1142

Fabricius KE (2005) Effects of terrestrial runoff on the ecology of corals and coral reefs: review and synthesis. Mar Pollut Bull 50:125–146

Falter JL, Atkinson MJ, Merrifield MA (2004) Mass-transfer limitation of nutrient uptake by a wave-dominated reef flat community. J Exp Mar Biol Ecol 49:1820–1831

Feller, IC (1995) Effects of nutrient enrichment on growth and herbivory of dwarf red mangrove (*Rhizophora mangle*). Ecol Monogr 65:477–505

Feller IC, McKee K, Whigham D et al (2003) Nitrogen vs phosphorus limitation across an ecotonal gradient in a mangrove forest. Biogeochemistry 62:145–175

Feller IC, Mathis WN (1997) Primary herbivory by wood-boring insects along an architectural gradient of *Rhizophora mangle*. Biotropica 29:440–451

Ferdie M, Fourqurean JW, (2004) Responses of seagrass communities to fertilization along a gradient of relative availability of nitrogen and phosphorus in a carbonate environment. J Exp Mar Biol Ecol 49:2082–2094

Fisher SG, Likens GE (1973) Energy flow in Bear Brook, New Hampshire: an integrative approach to stream ecosystem metabolism. Ecol Monogr 43:421–439

Fleming M, Lin G, Sternberg LSL (1990) Influence of mangrove detritus in an estuarine ecosystem. Bull Mar Sci 47:663–669

Flores-Verdugo FJ, Day JW Jr, Briseno-Duenas R (1987) Structure, litter fall, decomposition, and detritus dynamics of mangroves in a Mexican coastal lagoon with an ephemeral inlet. Mar Ecol Prog Ser 35:83–90

Fourqurean JW, Zieman JC, Powell GVN (1992) Phosphorus limitation of primary production in Florida Bay: evidence from C:N:P ratios of the dominant seagrass *Thalassia testudinum*. J Exp Mar Biol Ecol 37:162–171

Fourqurean JW, Jones RD, Zieman JC (1993) Processes influencing water column nutrient characteristics and phosphorus limitation of phytoplankton biomass in Florida Bay, FL, USA: inferences from spatial distributions. Estuar Coast Shelf Sci 36:295–314

Fourqurean JW, Moore TO, Fry B et al (1997) Spatial and temporal variation in C:N:P ratios, $\delta^{15}N$, and $\delta^{13}C$ of eelgrass (*Zostera marina* L) as indicators of ecosystem processes, Tomales Bay, CA, USA. Mar Ecol Prog Ser 157:147–157

Fourqurean JW, Boyer JN, Durako MJ et al (2003) Forecasting responses of seagrass distributions to changing water quality using monitoring data. Ecol Appl 13:474–489

France R, Holmquist J, Chandler M et al (1998) $\delta^{15}N$ evidence for nitrogen fixation associated with macroalgae from a seagrass-mangrove-coral reef ecosystem. Mar Ecol Prog Ser 167: 297–299

France R, Culbert H, Freeborough C et al (1997) Leaching and early mass loss of boreal leaves and wood in oligotrophic water. Hydrobiologia 345:209–214

Furnas MJ, Mitchell A, Skuza M (1997) Shelf-scale nitrogen and phosphorous budgets for the central Great Barrier Reef (16–19°S). Proc 8th Int Coral Reef Symp, Panama 1:809–814

Furnas MJ (2003) Catchments and corals: terrestrial runoff to the Great Barrier Reef. Australian Institute of Marine Science and CRC Reef, Townsville, Australia

Gacia E, Duarte CM, Middelburg JJ (2002) Carbon and nutrient deposition in a Mediterranean seagrass (*Posidonia oceanica*) meadow. J Exp Mar Biol Ecol 47:23–32

Gardner TA, Coté IM, Gill JA et al (2003) Long-term region-wide declines in Caribbean corals. Science 301:958–960

Glynn PW (1977) Coral upgrowth in upwelling and non-upwelling areas off the Pacific coast of Panama. J Mar Res 35:567–585

Gobert S, Cambridge ML, Velimirov B et al (2006) Biology of *Posidonia*. In: Larkum AWD, Orth RJ, Duarte CM (eds) Seagrasses: biology, ecology and conservation. Springer, The Netherlands

Golley F, Odum HT, Wilson RF (1962) The structure and metabolism of a Puerto Rican red mangrove forest in May. Ecology 43:9–19

Grottoli AG, Rodrigues LJ, Palardy JE (2006) Heterotrophic plasticity and resilience in bleached corals. Nature 440:1186–1189

Hansell DA, Carlson CA (2002) Biogeochemistry of marine dissolved organic matter. Academic Press, New York

Hata H, Suzuki A, Maruyama T et al (1998) Carbon flux by suspended and sinking particles around the barrier reef of Palau, western Pacific. J Exp Mar Biol Ecol 43:1883–1893

Hearn CJ, Atkinson MJ, Falter JL (2001) A physical derivation of nutrient-uptake rates in coral reefs: effects of roughness and waves. Coral Reefs 20:347–356

Hemminga MA, Harrison PG, van Lent F (1991) The balance of nutrient losses and gains in seagrass meadows. Mar Ecol Prog Ser 71:85–96

Hemminga MA, Slim FJ, Kazungu J et al (1994) Carbon outwelling from a mangrove forest with adjacent seagrass beds and coral reefs (Gazi Bay, Kenya). Mar Ecol Prog Ser 106: 291–301

Hemminga MA, Marbà N, Stapel J (1999) Leaf nutrient resorption, leaf lifespan and the retention of nutrients in seagrass systems. Aquat Bot 65:141–158

Henriksen K, Kemp WM (1988) Nitrification in estuarine and coastal marine sediments. In: Blackburn TH, Sorensen J (eds) Nitrogen cycling in coastal marine environments. John Wiley and Sons, New York

Holmer M, Olsen AB (2002) Role of decomposition of mangrove and seagrass detritus in sediment carbon and nitrogen cycling in a tropical mangrove forest. Mar Ecol Prog Ser 230:87–101

Hughes TP, Baird AH, Bellwood DR et al (2003) Climate change, human impacts, and the resilience of coral reefs. Science 301:929–933

Hughes TP, Rodrigues MJ, Bellwood DR et al (2007) Phase shifts, herbivory, and the resilience of coral reefs to climate change. Curr Biol 17:360–365

Hutchings P, Haynes D (2005) Marine pollution bulletin special edition editorial. Mar Pollut Bull 51:1–2

Jackson JBC, Kirby MX, Berger WH et al (2001) Historical overfishing and the recent collapse of coastal ecosystems. Science 293:629–638

Jenkins MC, Kemp WM (1984) The coupling of nitrification and denitrification in tow estuarine sediments. J Exp Mar Biol Ecol 29:609–619

Johannes RE, Alberts J, D'Elia CF et al (1972) The metabolism of some coral reef communities: a team study of nutrient and energy flux at Eniwetok. BioScience 22:541–543

Johannes RE, Wiebe WJ, Crossland CJ (1983) Three patterns of nutrient flux in a coral reef community. Mar Ecol Prog Ser 12:131–136

Kinsey DW, Davies PJ (1979) Effects of elevated nitrogen and phosphorus on coral reef growth. J Exp Mar Biol Ecol 24:935–940

Kleypas JA, Eakin CM (2007) Scientists perceptions of threats to coral reefs: results of a survey of coral reef researchers. Bull Mar Sci 80:419–436

Koch MS (1997) *Rhizophora mangle*: seedling development into the sapling stage across resource and stress gradients in subtropical Florida. Biotropica 29:427–439

Koch EW, Benz RE, Rudnick DT (2001) Solid-phase phosphorus pool sin highly organic carbonate sediments in North-eastern Florida Bay. Estuar Coast Shelf Sci 52:279–291

Koch MS, Schopmeyer SA, Nielsen OI et al (2007) Conceptual model of seagrass die-off in Florida Bay: links to biogeochemical processes. J Exp Mar Biol Ecol 350:73–88

Koch EW, Ackerman JD, Verduin J et al (2006) Fluid dynamics in seagrass ecology – from molecules to ecosystems. In: Larkum AWD, Orth RJ, Duarte CM (eds) Seagrasses: biology, ecology and conservation. Springer, The Netherlands

Koop K, Booth D, Broadbent A et al (2001) ENCORE: The effect of nutrient enrichment on coral reefs synthesis of results and conclusions Mar Pollut Bull 42:91–120

Kitheka JU, Ohowa B, Mwashote B et al (1996) Water circulation dynamics, water column nutrients and plankton productivity in a well-flushed tropical bay in Kenya. J Sea Res 35(4): 257–268

Kristensen E, Andersen FO, Kofoed LH (1988) Preliminary assessment of benthic community metabolism in a south-east Asian mangrove swamp. Mar Ecol Prog Ser 48: 137–145

Kristensen E, Holmer M, Banta G et al (1995) Carbon, nitrogen, and sulfur cycling in sediments of the Ao Nam Bor mangrove forest, Phuket, Thailand: a review. Phuket Mar Biol Cent Res Bull 60:37–64

Kuenzler EJ (1974) Mangrove swamp systems. In: Odum HT, Copeland BJ, McMahon EA (eds) Coastal ecological systems of the United States I. The Conservation Foundation, Washington DC

Lapointe BE, Clark MW (1992) Nutrient inputs from the watershed and coastal eutrophication in the Florida Keys. Estuaries 15:465–476

Lapointe BE (1997) Nutrient thresholds for bottom-up control of macroalgal blooms on coral reefs in Jamaica and southeast Florida. J Exp Mar Biol Ecol 42:1119–1131

Lapointe BE, Bedford BJ, Baumberger R (2006) Hurricanes Frances and Jeanne remove blooms of the invasive green alga *Caulerpa brachypus* forma *parvifolia* (Harvey) Cribb from coral reefs off northern Palm Beach County, Florida. Estuar Coast 29:966–971

Lara RJ, Dittmar T (1999) Nutrient dynamics in a mangrove creek (North Brazil) during the dry season. Mang Salt Marsh 3:185–195

Lee SY (1995) Mangrove outwelling: a review. Hydrobiologia 295:203–212

Lee KS, Park SR, Kim YK (2007) Effects of irradiance, temperature, and nutrients on growth dynamics of seagrasses: a review. J Exp Mar Biol Ecol 350:144–175

Lee KS, Dunton KH (1999) Inorganic nitrogen acquisition in the seagrass *Thalassia testudinum*: development of a whole-plant nitrogen budget. J Exp Mar Biol Ecol 44:1204–1215

Lee R, Joye S (2006) Seasonal patterns of nitrogen fixation and denitrification in oceanic mangrove habitats. Mar Ecol Prog Ser 307:127–1441

Leichter JJ, Stewart H, Miller S (2003) Episodic nutrient transport to Florida coral reefs. J Exp Mar Biol Ecol 48:1394–1407

Lepoint G, Defawe G, Gobert S et al (2002) Experimental evidence for N recycling in the leaves of seagrass *Posidonia oceanica*. J Sea Res 48:173–179

Liebezeit G (1985) Sources and sinks of organic and inorganic nutrients in mangrove ecosystems. In: Cragg S, Polunin N (eds) Workshop on mangrove ecosystems dynamics. UNDP/UNESCO, Port Moresby, Papua New Guinea

Lin B, Dushoff J (2004) Mangrove filtration of anthropogenic nutrients in the Rio Coco Solo, Panama. Manage Environ Qual 15:131–142

Lin G, Banks T, Sternberg L (1991) Variation in ^{13}C values for the seagrass *Thalassia testudinum* and its relations to mangrove carbon. Aquat Bot 40:333–341

Lirman D, Fong P (2007) Is proximity to land-based sources of coral stressors an appropriate measure of risk to coral reefs? An example from the Florida Reef Tract. Mar Pollut Bull 54: 779–791

Littler MM, Littler DS (2007) Assessment of coral reefs using herbivory/nutrient assays and indicator groups of benthic primary producers: a critical synthesis, proposed protocols, and critique of management strategies. Aquat Conserv: Mar Freshwat Ecosyst 17:195–215

Maie N, Yang C, Miyoshi T et al (2005) Chemical characteristics of dissolved organic matter in an oligotrophic subtropical wetland/estuarine ecosystem. J Exp Mar Biol Ecol 50:23–35

Marbà N, Hemminga MA, Mateo MA et al (2002) Carbon and nitrogen translocation between seagrass ramets. Mar Ecol Prog Ser 226:287–300

Mateo MA, Cebrián J, Dunton K et al (2006) Carbon flux in seagrass ecosystems. In: Larkum AWD, Orth RJ, Duarte CM (eds) Seagrasses: biology, ecology and conservation. Springer, The Netherlands

Mazda YH, Yokochi, Sato Y (1990) Groundwater flow in the Bashita-Minato mangrove area, and its influence on water and bottom mud properties. Estuar Coast Shelf Sci 31:621–638

McCook LJ, Jompa J, Diaz-Pulido G (2001) Competition between corals and algae on coral reefs: a review of evidence and mechanisms. Coral Reefs 19:400–417

McIvor CC, Ley JA, Bjork RD (1994) Changes in freshwater inflow from the Everglades to Florida Bay including effects on biota and biotic processes: a review. In: Davis SM, Ogden JC (eds) Everglades: the ecosystem and its restoration. St Lucie Press, Delray Beach, Florida

McGlathery KJ, Sundback K, Anderson IC (2007) Eutrophication in shallow coastal bays and lagoons: the role of plants in the coastal filter. Mar Ecol Prog Ser 348:1–18

Meyer JL, Schultz ET, Helfman GS (1983) Fish schools: an asset to corals. Science 220:1047–1049

Middleton BA, McKee KL (2001) Degradation of mangrove tissues and implications for peat formation in Belizean island forests. Ecology 89:818–828

Monbet P, Brunskill G, Zagorskis I et al (2007) Phosphorus speciation in the sediment and mass balance for the central region of the Great Barrier Reef continental shelf (Australia). Geochim Cosmochim Acta 71:2762–2779

Mwashote BM, Jumba IO (2002) Quantitative aspects of inorganic nutrient fluxes in the (Kenya): implications for coastal ecosystems. Mar Pollut Bull 44:1194–1205

Nedwell DB (1975) Inorganic nitrogen metabolism in a eutrophicated tropical mangrove estuary. Water Res 9:221–231

Nedwell DB, Blackburn TH, Wiebe W (1994) Dynamic nature of the turnover of organic carbon, nitrogen and sulphur in the sediments of a Jamaican mangrove forest. Mar Ecol Prog Ser 110:223–231

Nixon SW (1980) Between coastal marshes and coastal waters: a review of 20 years of speculation and research on the role of saltmarshes in estuarine productivity and water chemistry. In: Hamilton P, McDowell KB (eds) Estuarine and wetland processes. Plenum Press, New York

Nixon SW (1981) Remineralization and nutrient cycling in coastal marine ecosystems. In: Neilson BJ, Cronin LE (eds) Estuaries and nutrients. The Humana Press, Clifton, New Jersey

Nixon SW, Furnas BN, Lee V et al (1984) The role of mangroves in the carbon and nutrient dynamics of Malaysia estuaries. In: Soepadmo E, Rao AN, Macintosh DJ (eds) Proceedings of the Asian symposium on mangrove environment: research and management. University of Malaya, Kuala Lumpur

Nixon SW, Ammerman J, Atkinson L et al (1996) The fate of nitrogen and phosphorus at the land-sea margin of the north Atlantic Ocean. Biogeochemistry 35:141–180

Noe GB, Childers DL, Jones RD (2001) Phosphorus biogeochemistry and the impact of phosphorus enrichment: why is the Everglades so unique? Ecosystems 4:603–624

Nykvist N (1959) Leaching and decomposition of litter. I. Experiments on leaf litter of *Fraxinus excelsior*. Oikos 10:190–211

Nykvist N 1961 Leaching and decomposition of litter. III. Experiments on the leaf litter of *Betula verrucosa*. Oikos 12:249–263

Odum E, de la Cruz A (1967) Particulate organic detritus in a Georgia salt marsh-ecosystem. Am Assoc Adv Sci Pub 83:383–388

Odum WE, Heald EJ (1972) Trophic analyses of an estuarine mangrove community. Bull Mar Sci 22:671–738

Odum WE, Heald EJ (1975) The detritus-based food web of an estuarine mangrove community. In: Cronin LE (ed) Estuarine research, Vol. 1. Academic Press, New York

Odum WE, Fisher JS, Pickral JC (1979) Factors controlling the flux of particulate organic carbon from estuarine wetlands. In: Livingstone RJ (ed) Ecological processes in coastal and marine systems. Plenum Press, New York

Odum WE, McIvor CC, Smith TJ III (1982) The ecology of mangroves of south Florida: a community profile. US Fish and Wildlife Service FWS/OBS-87/17, Washington DC

Ogden JC, Brown RA, Salesky N (1973) Grazing by the echinoid *Diadema antillarum* Philippi: formation of halos around West Indian patch reefs. Science 182:715–717

Ohowa BO, Mwashote BM, Shimbira WS (1997) Dissolved inorganic nutrient fluxes from two seasonal rivers into Gazi Bay, Kenya. Estuar Coast Shelf Sci 45:189–195

Orth RJ, Carruthers TJB, Dennison WC et al (2006) A global crisis for seagrass ecosystems. BioScience 56:987–996

Ovalle ARC, Rezende CE, Lacerda LD et al (1990) Factors affecting the hydrochemistry of a mangrove tidal creek, Sepetiba Bay, Brazil. Estuar Coast Shelf Sci 31:639–650

Pandolfi JM, Bradbury RH, Sala E et al (2003) Global trajectories of the long-term decline of coral reef ecosystems. Science 301:955–958

Perez M, Romero J (1993) Preliminary data on alkaline phosphatase activity associated with Mediterranean seagrasses. Bot Mar 36:499–502

Peralta G, Bouma TJ, van Soelen J et al (2003) On the use of sediment fertilization for seagrass restoration: a mesocosm study on *Zostera marina* L. Aquat Bot 75: 95–110

Pilson ME, Betzer FB (1973) Phosphorus flux across a coral reef. Ecology 54:581–588

Pomeroy LR (1970) The strategy of mineral cycling. Annu Rev Ecol Systemat 1:171–190

Ralph PJ, Tomasko D, Moore K et al (2006) Human impacts on seagrasses: eutrophication, sedimentation and contamination. In: Larkum AWD, Orth RJ, Duarte CM (eds) Seagrasses: biology, ecology and conservation. Springer, The Netherlands

Randall JE (1963) An analysis of the fish populations of artificial and natural reefs in the Virgin Islands. Caribb J Sci 3:31–47

Rice DL, Tenore KR (1981) Dynamics of carbon and nitrogen during the decomposition of detritus derived from estuarine macrophytes. Estuar Coast Shelf Sci 13:681–690

Rivera-Monroy VH, Day JW, Twilley RR et al (1995) Flux of nitrogen and sediments in Terminos Lagoon Mexico. Estuar Coast Shelf Sci 40:139–160

Rivera-Monroy VH, Twilley RR (1996) The relative role of denitrification and immobilization in the fate of inorganic nitrogen in mangrove sediments (Terminos Lagoon, Mexico). J Exp Mar Biol Ecol 41:284–296.

Rivera-Monroy VH, Torres LA, Bohamon N et al (1999) The potential use of mangrove forests as nitrogen sinks of shrimp aquaculture pond effluents: the role of denitrification. J World Aquaculture Soc 30:12–25

Rivera-Monroy VH, Twilley R, Bone D et al (2004) A conceptual framework to develop long-term ecological research and management objectives in the wider Caribbean region. BioScience 54:843–856

Rivera-Monroy V, de Mustert K, Twilley R et al (2007) Patterns of nutrient exchange in a riverine mangrove forest in the Shark River Estuary, Florida, USA. Hidrobiològica 17:169–178

Robertson AI (1986) Leaf-burying crabs: their influence on energy flow and export from mixed mangrove forests (*Rhizophora* spp) in northeastern Australia. J Exp Mar Biol Ecol 102:237–248

Robertson AI, Alongi DM, Daniel PA et al (1988) How much mangrove detritus enters the Great Barrier Reef Lagoon? Proc 6th Intern Coral Reef Symp 2:601–606

Romero J, Kun-Seop L, Pérez M et al (2006) Nutrient dynamics in seagrass ecosystems. In: Larkum AWD, Orth RJ, Duarte CM (eds) Seagrasses: biology, ecology and conservation. Springer, The Netherlands

Santamaría G, Dias C, Hootsmans MJM (1994) The influence of ammonia on the growth and photosynthesis of *Ruppia drepanensis* Tineo from Doñana National Park (SW Spain). Hydrobiologia 275/276:219–231

Seitzinger SP (1988) Denitrification in freshwater and coastal marine ecosystems: ecological and geochemical significance. J Exp Mar Biol Ecol 33:702–724

Short FT (1987) Effects of sediment nutrients on segrasses: literature review and mesocosm experiment. Aquat Bot 27:41–57

Short FT, McRoy CP (1984) Nitrogen uptake by leaves and roots of the seagrass *Zostera marina* L. Bot Mar 17:547–555

Short FT, Dennison WC, Capone DG (1990) Phosphorus-limited growth of the tropical seagrass *Syringodium filiforme* in carbonate sediments. Mar Ecol Prog Ser 62:169–174

Short FT, Koch EW, Creed JC (2006) Seagrass net monitoring across the Americas: case studies of seagrass decline. Mar Ecol 27: 277–289

Smith SV, Kimmere WJ, Laws EA et al (1981) Kaneohe Bay sewage diversion experiment: perspectives on ecosystem responses to nutritional perturbation. Pac Sci 35:279–397

Steinke TD, Holland AJ, Singh Y (1993) Leaching losses during decomposition of mangrove leaf litter. S Afr J Bot 59(1):21–25

Sterner RW, Elser JJ (2002) Ecological stoichiometry: the biology of elements from molecules to the biosphere. Princeton University Press, Princeton, New Jersey

Sutula MA, Perez BP, Reyes E et al (2003) Factors affecting spatial and temporal variability in material exchange between the Southeastern Everglades wetlands and Florida Bay (USA). Estuar Coast Shelf Sci 57:757–781

Szmant-Froelich A (1983) Functional aspects of nutrient cycling on coral reefs. In: Reaka ML (ed) The ecology of deep and shallow coral reefs. Symposium series for undersea research, NOAA Undersea Research Program 1:133–139

Szmant AM (2002) Nutrient enrichment in coral reefs: is it a major cause of coral reef decline? Estuaries 25:743–766

Szmant AM, Forrester A (1996) Water column and sediment nitrogen and phosphorus distribution patterns in the Florida Keys. Coral Reefs 15:21–41

Tam NFY, Vrijmoed LLP, Wong YS (1990) Nutrient dynamics associated with leaf decomposition in a small subtropical mangrove community in Hong Kong. Bull Mar Sci 47:68–78

Tam NFY, Wong YS (1999) Mangrove soils in removing pollutants from municipal wastewater of different salinities. J Env Qual 28:556–564

Teal JM (1962) Energy flow in the salt marsh ecosystem of Georgia. Ecology 43:614–624

Terrados J, Williams SL (1997) Leaf versus root nitrogen uptake by the surfgrass *Phyllospadix torreyi*. Mar Ecol Prog Ser 149:267–277

Tilmant J, Curry R, Jones R et al (1994) Hurricane Andrew's effects on marine resources. BioScience 44:230–237

Touchette BW, Burkholder JM (2000) Review of nitrogen and phosphorus metabolism in seagrasses. J Exp Mar Biol Ecol 250:133–167

Twilley RR (1985) The exchange of organic carbon in basin mangrove forests in a southwest Florida estuary. Estuar Coast Shelf Sci 20:543–557

Twilley RR (1988) Coupling of mangroves to the productivity of estuarine and coastal waters. In: Jansson BO (ed) Coastal-offshore ecosystem interactions. Springer-Verlag, Berlin

Twilley RR (1995) Properties of mangrove ecosystems in relation to the energy signature of coastal environments. In: Hall CAS (ed) Maximum power, University Press of Colorado, Niwot, Colorado pp.43–62.

Udy JW, Dennison WC, Lee Long WJ et al (1999) Responses of seagrasses to nutrients in the Great Barrier Reef, Australia. Mar Ecol Prog Ser 185:257–271

Valiela I, Bowen JL, York JK (2001) Mangrove forests: one of the world's threatened major tropical environments. BioScience 51:807–815

Valiela I, Cole L (2002) Comparative evidence that salt marshes and mangroves may protect seagrass meadowes from land-derived nitrogen loads. Ecosystems 5:92–102

van Katwijk MM, Vergeer LHT, Schmitz GHW et al (1997) Ammonium toxicity in eelgrass *Zostera marina*. Mar Ecol Prog Ser 157:159–173

Webb KL, Dupaul WD, Wiebe W et al (1975) Enewetak (Eniwetok) Atoll: aspects of the nitrogen cycle on a coral reef. J Exp Mar Biol Ecol 20:198–210

Welsh DT (2000) Nitrogen fixation in seagrass meadows: regulation, plant-bacterialinteraction and significance to primary productivity. Ecol Lett 3:58–71

Whiting GJ, Childers DL (1989) Subtidal advective water flux as a potentially important nutrient import to southeastern USA saltmarsh estuaries. Estuar Coast Shelf Sci 28:417–431

Wolaver TG, Spurrier JD (1988) The exchange of phosphorus between a euhaline vegetated marsh and the adjacent tidal creek. Estuar Coast Shelf Sci 26:203–214

Woodroffe CD (1985) Studies of a mangrove basin, Tuff Crater, New Zealand. III. The flux of organic and inorganic particulate matter. Estuar Coast Shelf Sci 20:447–461

Yahel G, Post AF, Fabricius K et al (1998) Phytoplankton distribution and grazing near coral reefs. J Exp Mar Biol Ecol 43:551–563

Ziegler S, Benner R (1999) Nutrient cycling in the water column of a subtropical seagrass meadow. Mar Ecol Prog Ser 188:51–62

Zimmerman CF, Montgomery JR, Carlson P (1985) Variability of dissolved reactive phosphate flux rates in nearshore estuarine sediments: effects of groundwater flow. Estuaries 8:228–236

Chapter 3
Carbon Exchange Among Tropical Coastal Ecosystems

Steven Bouillon and Rod M. Connolly

Abstract Tropical rivers provide about 60% of the global transport of organic and inorganic carbon from continents to the coastal zone. These inputs combine with organic material from productive mangrove forests, seagrass beds, and coral reefs to make tropical coastal ecosystems important components in the global carbon cycle. Carbon exchange has been measured over multiple spatial scales, ranging from the transport and fate of terrestrial organic matter to the coastal zone, export of organic matter to the open ocean, exchange of leaf litter between mangroves and adjacent seagrass beds, to movement of carbon (at a scale of meters) between adjacent saltmarsh and mangrove habitats. Carbon is exchanged directly as particulate or dissolved material, or through migration of animals or through a series of predator-prey interactions known as trophic relay. This chapter first examines riverine carbon inputs to the tropical coastal zone, and how this material is processed in estuaries. The mechanisms and extent of carbon exchange among tropical coastal ecosystems are then discussed, showing their importance in ecosystem carbon budgets, and the implications for faunal and microbial communities.

Keywords Organic carbon · Mangroves · Seagrasses · Coral reefs · Tropical rivers

3.1 Introduction

Tropical coastal ecosystems are often highly productive, and can receive organic matter from a variety of sources, such as riverine inputs, local production by phytoplankton, or vegetated systems (mangroves, seagrasses). Tropical rivers have a disproportionately high importance in the global delivery of organic and inorganic

S. Bouillon (✉)
Katholieke Universiteit Leuven, Department of Earth and Environmental Sciences,
Kasteelpark Arenberg 20, B-3000 Leuven, Belgium; and Vrije Universiteit Brussel,
Department of Analytical and Environmental Chemistry, Pleinlaan 2, B-1050 Brussels, Belgium
e-mail: steven.bouillon@ees.kuleuven.be

I. Nagelkerken (ed.), *Ecological Connectivity among Tropical Coastal Ecosystems*,
DOI 10.1007/978-90-481-2406-0_3, © Springer Science+Business Media B.V. 2009

carbon to the coastal zone (Ludwig et al. 1996a), but biogeochemical processing and local inputs from primary production in the coastal zone can greatly modify the quantity and composition of carbon. In-depth knowledge of carbon fluxes and transformations in the tropical coastal zone is therefore important for a finer constraining of global carbon budgets. Moreover, considering the rapid and global changes occurring in river flows and associated sediment and organic matter transport, coastal eutrophication and destruction of coastal ecosystems such as mangroves, seagrass beds, and coral reefs, understanding the functioning of these systems and their interactions is important to be able to correctly assess the health of estuaries and coastal systems and predict the impact of climate change or anthropogenic disturbance.

Organic matter differs substantially in biochemical composition and availability to consumers, depending on whether it is imported by rivers or produced locally by various primary producers (plankton, seagrasses, macroalgae, and mangroves). Exchange of organic matter across ecosystem boundaries thus has important consequences for the availability of organic matter and the relative importance of burial, mineralization, and consumption by fauna. It has often been proposed that organic matter exported from tidal wetlands such as mangroves and saltmarshes enhances secondary production in the coastal zone, thus contributing to fisheries production. The mechanisms involved now appear to be much more complex, however, and there is as yet little evidence for a direct trophic link between land-derived organic matter inputs and coastal zone fisheries in the tropics (e.g., see Lee 1995). Exchange of carbon has been studied over multiple spatial scales, ranging from the transport and fate of terrestrial organic matter to the coastal zone, export of organic matter to the open ocean, exchange of litter between mangroves and adjacent seagrass beds, to movement of carbon (at a scale of meters) between adjacent saltmarsh and mangrove habitats (Fig. 3.1).

This chapter attempts to summarize the available information on patterns of carbon movement and exchange, and to discuss the underlying mechanisms and consequences. We focus first on the riverine inputs of organic and inorganic carbon to the tropical coastal zone, synthesize available data on how this material is processed in estuaries, and explain how this differs from temperate estuaries. The second part of this chapter discusses the exchange of carbon among various tropical coastal ecosystems, its importance in understanding ecosystem carbon budgets, and the implications of carbon exchange for faunal and microbial communities.

3.2 Riverine Carbon Transport to the Tropical Coastal Zone

3.2.1 Fluxes, Composition, and Fate of Riverine Organic Matter

The global delivery of organic carbon (C) to the world's oceans is estimated to be in the order of 0.3–0.5 Pg C y^{-1} (1 Pg = 10^{12} g) (e.g., Ludwig et al. 1996a, 1996b, Schlünz and Schneider 2000), partitioned almost equally between dissolved and particulate organic carbon (DOC and POC). Riverine transport of inorganic carbon is globally estimated at approximately 0.3–0.4 Pg C y^{-1} (Ludwig et al.

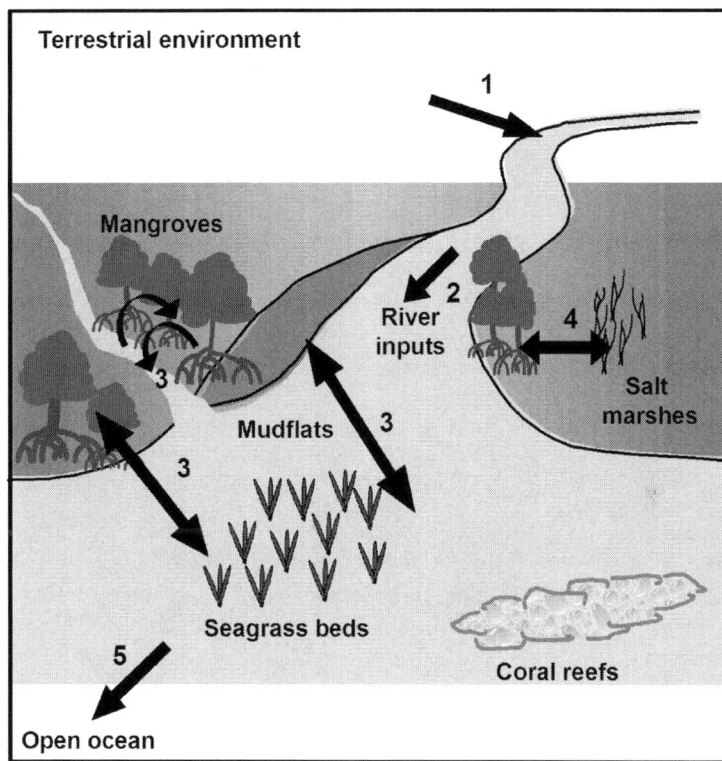

Fig. 3.1 Overview of some of the carbon exchange pathways considered in this chapter: (**1**) terrestrial inputs to rivers, (**2**) river inputs to the coastal zone, (**3**) exchange between intertidal and subtidal aquatic systems, (**4**) exchange between intertidal ecosystems, and (**5**) export towards the open ocean

1996a). The tropics are thought to be responsible for about 60% of these fluxes (Ludwig et al. 1996a, Table 3.1), and are therefore disproportionately important in the global terrestrial—marine carbon transport when considering their areal C fluxes. We can thus estimate that for both inorganic and organic carbon, about 0.2–0.25 Pg C is delivered annually to the tropical coastal zone. It should be stressed, however, that relatively few empirical datasets exist on carbon export in tropical rivers, and that these estimates are to a large extent based on extrapolations of data from a limited number of catchments themselves based on empirical models relating catchment characteristics to carbon export. Hence, errors in either underlying data on catchment characteristics or the relationship with carbon export can bias these estimates.

Considering that certain geographical areas are responsible for a major part of these C fluxes, they are particularly important in determining the overall estimates of riverine C transport. Milliman et al. (1999), for example, estimated that six islands in the Indo-Pacific were responsible for about 20% of the global riverine sediment flux, whereas they represent only about 2% of the terrestrial area draining to the

Table 3.1 Estimates of global riverine carbon transport, as dissolved organic carbon (DOC), particulate organic carbon (POC) and dissolved inorganic carbon (DIC), and the importance of the tropical zone in global carbon transport. These estimates are calculated from the data presented in Ludwig et al. (1996a). Note that DIC export is assumed to be equal to alkalinity export

	Flux (10^{12} g C y^{-1})			Flux ratios	
Region	POC	DOC	DIC	POC/DOC	DIC/(DOC+POC)
Tropical, Atlantic Ocean	45.3	59.1	74.6	0.77	0.71
Tropical, Indian Ocean	34.8	21.4	45.1	1.63	0.80
Tropical, Pacific Ocean	33.9	26.8	60.3	1.26	0.99
Σ Tropical zone (24°S–24°N)	114.0	107.3	180.0	1.06	0.81
Σ World	178.6	179.8	291.8	0.99	0.81
% of global transport in the tropics	63.8	59.7	61.7	–	–

global ocean, and this may suggest that such areas may be similarly important in terms of carbon delivery. In line with this, Baum et al. (2007) recently estimated that DOC export from Indonesia could be equivalent to as much as 10% of the global riverine DOC flux to the coastal zone. Data from a number of east African estuaries (Bouillon et al. 2007a, b, Ralison et al. 2008), indicate that the relative export of inorganic carbon (as compared to organic carbon) is more than 10 times higher than that predicted by the empirical model of Ludwig et al. (1996a). On a regional scale, it appears there is a substantial gap in data on the quantities and partitioning of carbon export. A further problem particularly relevant to quantifying carbon transport in tropical rivers is that (with some exceptions such as the Congo basin; Coynel et al. 2005), river discharge and associated carbon transport are often highly seasonal, with the majority of discharge often occurring in a very narrow time frame (e.g., Eyre 1998, Hung and Huang 2005).

Compositionally, organic carbon in rivers generally shows a strong link to the catchment vegetation and land use. In particular in turbid rivers where primary production is light-limited, organic matter from fringing vegetation, floodplains, and terrestrial soils (through runoff) dominates the river-borne organic matter pool. Nevertheless, not all vegetation types within a catchment contribute equally to riverine organic carbon inputs. In the Congo basin, for example, Coynel et al. (2005) found that forested sub-basins showed about three times higher area-specific fluxes of organic carbon than savannah-dominated basins. Similarly, a number of studies found a smaller contribution of C4-derived material (i.e., derived from tropical grasslands) in riverine organic matter than would be expected based on their relative cover in the rivers' catchment (Martinelli et al. 1999, Ralison et al. 2008). Inputs of organic carbon from (C4) grasslands appear to be more important during flood events or high flow periods (Martinelli et al. 1999) when there is sufficiently strong runoff to mobilize soils and organic matter.

The estimates of carbon transport above (see Table 3.1) refer to what is delivered by rivers to the tropical coastal zone, but do not take into account possible changes occurring within estuaries, bays, and lagoons, and thus do not necessarily reflect what is actually delivered to the open ocean. In these coastal systems, a range of changes can take place which greatly modify the quantity and composition of organic matter pools. Organic matter can be removed through burial, consumption, or mineralization, and new inputs of organic matter can arise, in particular from the often very productive vegetated systems such as mangroves or seagrass beds. This will result in deviation from conservative behavior along the estuarine mixing gradient. Conservative mixing implies no loss or inputs along the estuarine gradient, and hence, a linear concentration gradient between freshwater and marine end-members (see also Chapter 12). Mixing scenarios can also be evaluated using δ^{13}C signatures, whereby conservative mixing follows the general equation (described here for dissolved organic carbon):

$$\delta^{13}C$$
$$= \frac{Sal(DOC_F\delta^{13}C_F - DOC_M\delta^{13}C_M) + Sal_F DOC_M\delta^{13}C_M - Sal_M DOC_F\delta^{13}C_F}{Sal(DOC_F - DOC_M) + Sal_F DOC_M - Sal_M DOC_F}$$

whereby: Sal = the sample salinity, $DOC_F\delta^{13}C_F$ = the DOC concentration and stable isotope composition at the freshwater or least saline end-member, $DOC_M\delta^{13}C_M$ = the DOC concentration and stable isotope composition at the marine end-member.

Examples of such non-conservative behavior are shown in Fig. 3.2, where DOC and $\delta^{13}C_{DOC}$ profiles are shown for two contrasting estuaries. The DOC profile for Mtoni Estuary (Tanzania) shows clear net inputs of DOC along the estuarine gradient, i.e., with DOC data points above the conservative mixing line (Fig. 3.2a). The corresponding $\delta^{13}C_{DOC}$ profile (Fig. 3.2b) indicates that the inputs of DOC in this estuary have a ^{13}C-depleted signature, consistent with the expected DOC inputs from mangroves, which occur along the length of the salinity profile measured (see also Machiwa 1999). The DOC profile from the Tien River estuary (Mekong Delta, Vietnam), where no mangrove vegetation is present, shows a contrasting pattern, with net losses of DOC along the salinity gradient, i.e., most DOC data points below the conservative mixing line (Fig. 3.2c). The $\delta^{13}C_{DOC}$ profile for this site (Fig. 3.2d) is similar in shape to the one from Mtoni and indicates that this loss of DOC coincides with a depletion in ^{13}C of the remaining DOC pool, most likely suggesting selective degradation of a more ^{13}C-enriched fraction of DOC.

The behavior of DOC in estuaries may also change seasonally: Dittmar and Lara (2001a), for example, reported DOC profiles for the Caeté Estuary (Brazil) which show both conservative characteristics and non-conservative behavior during different parts of the year. Similarly, Young et al. (2005) report DOC profiles from a tropical seagrass-covered and mangrove-fringed lagoon which suggest both net losses of DOC or net inputs during the mixing process, depending on the season.

River flows are often highly seasonal in tropical regions (e.g., Vance et al. 1998), with the exception of systems with large catchment areas along the equator such as the Congo River basin (see Coynel et al. 2005), and the composition and degradation status of organic matter can thus be distinctly seasonal (e.g., Ford et al. 2005,

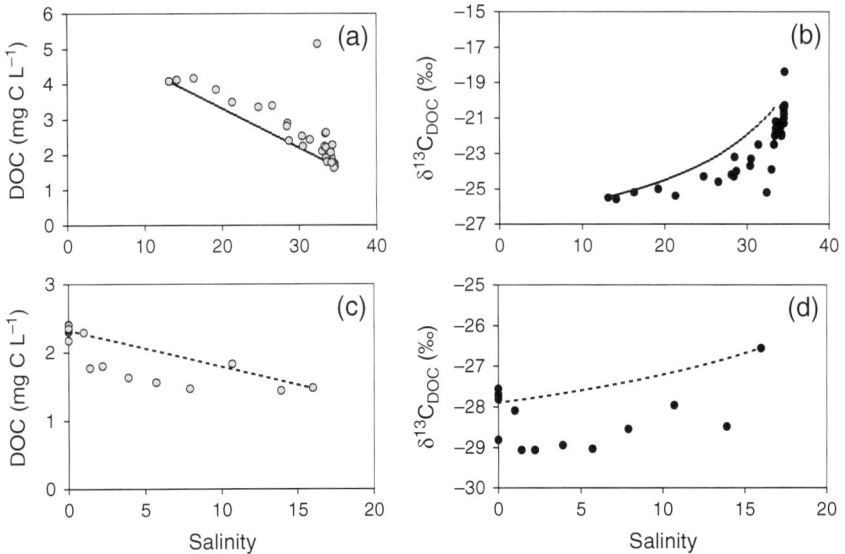

Fig. 3.2 Examples of non-conservative behavior of dissolved organic carbon (DOC) in tropical estuaries showing profiles of DOC (**a**) and $\delta^{13}C_{DOC}$ (**b**) from Mtoni Estuary (Tanzania) and from Tien Estuary (Mekong Delta, Vietnam) (**c, d**). Source: S Bouillon and AV Borges, unpubl. data. Dotted lines show the patterns expected for conservative mixing between the least saline and most saline end-members

Dai and Sun 2007). During low flow periods, estuaries have much longer residence times, with consequently much higher potential for biogeochemical processes to modify the quantity and composition of organic matter and nutrients (Eyre 1998). Conversely, during periods of high flow, large estuarine plumes may develop which allow for riverine material to be transported further offshore and with less processing of organic carbon within the estuary (e.g., Ford et al. 2005). In coastal bays, in contrast, the relative contribution of terrestrial material to the overall organic carbon pool may be more substantial during the dry season. Xu and Jaffé (2007) reported such a pattern for Florida Bay, which was ascribed to reduced primary production within the bay during the dry season. The fate of riverine organic matter is thus likely to differ substantially during high and low flow periods, although few studies have actually documented such patterns.

The delivery of terrestrial organic matter to offshore waters is important in at least some circumstances. Extensive offshore delivery has been demonstrated for a number of large river systems such as the Congo River (e.g., Schefuß et al. 2004), the Fly River in Papua New Guinea (Goñi et al. 2006) and the Ganges–Brahmaputra River system (Galy et al. 2007), and from tidal wetlands such as the extensive mangroves along the coast of Brazil (Dittmar et al. 2006). Carbon delivery from smaller rivers discharging to open coasts is probably less important. In Australia, for example, where small estuarine plumes punctuate long stretches of sandy coastline, a conservative tracer showed that estuarine particulates were distributed over only a small area at the estuarine mouth, and terrestrial carbon contributions had little impact on background coastal sources (Gaston et al. 2006).

The contribution of terrestrial carbon to estuarine metabolism and local food-webs has received relatively little attention so far, in part due to the difficulty of detecting its incorporation. Stable isotope signatures of terrestrial C3 plants overlap with those of often-present local lateral inputs (e.g., mangroves) and may also over-lap with those of *in situ* aquatic producers. In systems where organic matter derived from catchment C4 vegetation contributes significantly to the riverine carbon load, however, it becomes much more feasible to calculate terrestrial carbon contribution. Surprisingly, data from such systems suggest that terrestrial organic matter can be a major source of carbon even in intertidal mangrove sediments, and contributes equally to sedimentary bacterial communities (up to 40–50%; see Bouillon et al. 2007b, Ralison et al. 2008). The extent to which communities of higher organisms such as invertebrates and fish rely on terrestrial organic matter has recently become a topic of study in temperate waters (e.g., Darnaude et al. 2004), but in tropical systems this has to our knowledge not been studied in detail, although this may be a promising line of future work.

3.2.2 Effects of Human Perturbations

3.2.2.1 Changing River Flows and Catchment Land-uses

Freshwater flows from rivers into estuaries and ultimately into coastal waters are fundamentally important to carbon transfer. Dissolved and particulate carbon is transported directly in these waters. Freshwater flows also affect carbon movement indirectly, through their effects on salinity in estuaries that alter distributions of coastal pants and the migratory movements of aquatic animals. Freshwater surface and groundwater flow is an important factor, for example, in the distribution of man-grove (Hutchings and Saenger 1987) and saltmarsh plants (Pennings and Bertness 2001). Anthropogenic changes to freshwater flows from rivers therefore alter car-bon transfer within and among systems via several different mechanisms. Freshwa-ter is now in such short supply that a global shortage is looming (Postel 2000) and increased harvesting is a certainty. There is thus a need for strong, science-based decisions about water releases from dams to maintain ecosystem health (environ-mental flows) under pressing political realities (Arthington et al. 2006).

River flows discharging to the sea generally stimulate productivity (Gillanders and Kingsford 2002). In tropical systems, very clear correlations have been found between river flow and fisheries harvests. Flow in two different river systems on the east coast of Australia, for example, match annual fisheries catches, either with or without a time lag. Flows in the Fitzroy River are correlated with increased sur-vival and growth of cohorts of barramundi (*Lates calcarifer*), and catches of this species are higher several years later (Staunton-Smith et al. 2004). Summer flows in the Logan River are positively correlated with catches of fish, crabs and prawns (Loneragan and Bunn 1999). This effect is detected in the same year, and might sim-ply be a result of increased harvesting of recruits into fishing zones, as is probably the case for banana prawns in the Gulf of Carpentaria, Australia (Vance et al. 1998). Another mechanism which has been suggested is that increased terrestrial organic matter loads to coastal waters increases the abundance of meiofauna and macrofauna, the main prey of the fisheries species (Loneragan and Bunn 1999).

Changing land-use in coastal catchments also affects the amount and nature of organic matter arriving in estuaries and coastal waters. In China, carbon loads from urban and agricultural areas are so now so prevalent that inputs from local mangrove forests have become unimportant in food webs (Lee 2000). The change from forest to agriculture over the last 200 years in catchments adjacent to the Great Barrier Reef, for example, is thought to have increased sediment delivery to the reef about four-fold (see Furnas 2003, cited in Ford et al. 2005), and presumably organic loads along with it.

3.2.2.2 Effects of Climate Change on Carbon Exchange

Patterns in carbon exchange among tropical systems sit within an overarching position of global carbon cycles (Cloern 2001). Carbon is central to the topical issue of climate change. Although the effects of climate change on marine systems has been considered (Poloczanska et al. 2007), we have been unable to find any studies of how climate change might affect carbon exchanges at the land–sea interface. The most certain effect of recent and predicted acceleration in changes to climate on carbon exchange will be through altered rainfall patterns and therefore river flows (Table 3.1). Where rainfall is reduced, the overall delivery of organic matter to estuaries and the coast will be lower. Conversely, where increased rainfall is predicted, we can also predict a greater contribution of terrestrial organic matter to the coastal zone. Overlaying those effects will be the increased variability in rainfall (Poloczanska et al. 2007), with more severe weather events leading to rainfall peaks of greater magnitude and frequency than currently occur. Extreme flow events will likely lead to large pulses of input of terrestrial matter, and as discussed above, probably to an increased importance of C4 material from agriculture.

The ramifications of climate change will, however, be much broader than this. The extent and type of land-use in coastal catchments will presumably be altered, through changes in agricultural activities and urbanization (Cloern 2001). Ultimately this will alter carbon inputs to estuaries and coastal waters (as discussed above).

3.3 Exchange of Carbon Between Vegetated Tropical Systems and Adjacent Systems

3.3.1 Transfer of Carbon from Intertidal to Subtidal — Outwelling

Concepts about carbon transfer among nearshore systems are dominated by the theory of net transfer of carbon from shallow, estuarine habitats to deeper, adjacent waters. This 'outwelling hypothesis' is based on observations, from saltmarsh systems on the Atlantic coast of North America, that secondary production in adjacent waters could only be sustained if the marsh exports energy (Odum 1968). For tropical systems, this translates as a potential dependency on mangrove-derived organic matter for secondary production in adjacent systems. The export of particulate and dissolved organic carbon from mangroves has received considerable attention, even though the number of quantitative studies is rather limited to allow for an accurate assessment of organic carbon export on a global scale (e.g., Bouillon et al. 2008b), and the assessment of export rates is hampered by methodological difficulties (e.g., Ayukai et al. 1998, see also Section 12.4 in Chapter 12). Some stable isotope studies show that invertebrates and fish in habitats within hundreds of meters of mangroves obtain carbon from the mangrove forest (Harrigan et al. 1989, Lugendo et al. 2007), although others have not found this (e.g., Connolly et al. 2005, see also Section 3.3.1.5). Lack of influence has been definitively demonstrated at sites further away (Lee 1995). Mangroves may serve both as exporters of organic and inorganic carbon, but also import organic matter during tidal inundation, and assessing the net balance of these processes is not straightforward.

Organic matter from vegetated, intertidal habitats such as mangroves in tropical waters might be exported via three main avenues (Fig. 3.3): (1) dissolved or particulate matter, (2) through migration of animals from intertidal to subtidal waters, and (3) through a series of predator–prey interactions known as trophic relay (Kneib 1997). Each of these pathways is discussed in more detail below. Pathways of DOC, POC, and macro-litter export are likely to differ substantially. For DOC, a number of studies have stressed the importance of sediment–water exchange and pore water flow as vectors for DOC exchange with estuarine or tidal creek waters (e.g., Ovalle et al. 1990, Dittmar and Lara 2001b, Schwendenmann et al. 2006, Bouillon et al. 2007c). Particulate organic carbon, in contrast, appears to be influenced more by water current velocities and runoff (e.g., Twilley 1985). The importance of tidal dynamics was also suggested by Twilley (1985) who compared organic carbon export in different types of mangrove forests and found that the cumulative tidal amplitude is a main driver of the magnitude of total organic carbon export.

3.3.1.1 Exchange of DOC and POC

Our understanding of organic carbon exchange in mangroves comes from a relatively small number of studies: a recent review documents only six and seven estimates for DOC and POC export, respectively, and 11 estimates for total organic carbon export (Bouillon et al. 2008b). It should also be kept in mind that these estimates have been derived using a variety of approaches, including tidal measurements of organic carbon combined with water current measurements, and flux estimates using flow-through flumes (see also Chapter 12). Global estimates of organic carbon export (POC+DOC) from mangroves are in the order of about 250 g C $m^{-2}y^{-1}$, with DOC and POC each representing about half of this flux. Together, this would amount to approximately 20% of the net primary production by mangroves, although it must be stressed that our current understanding of carbon cycling in

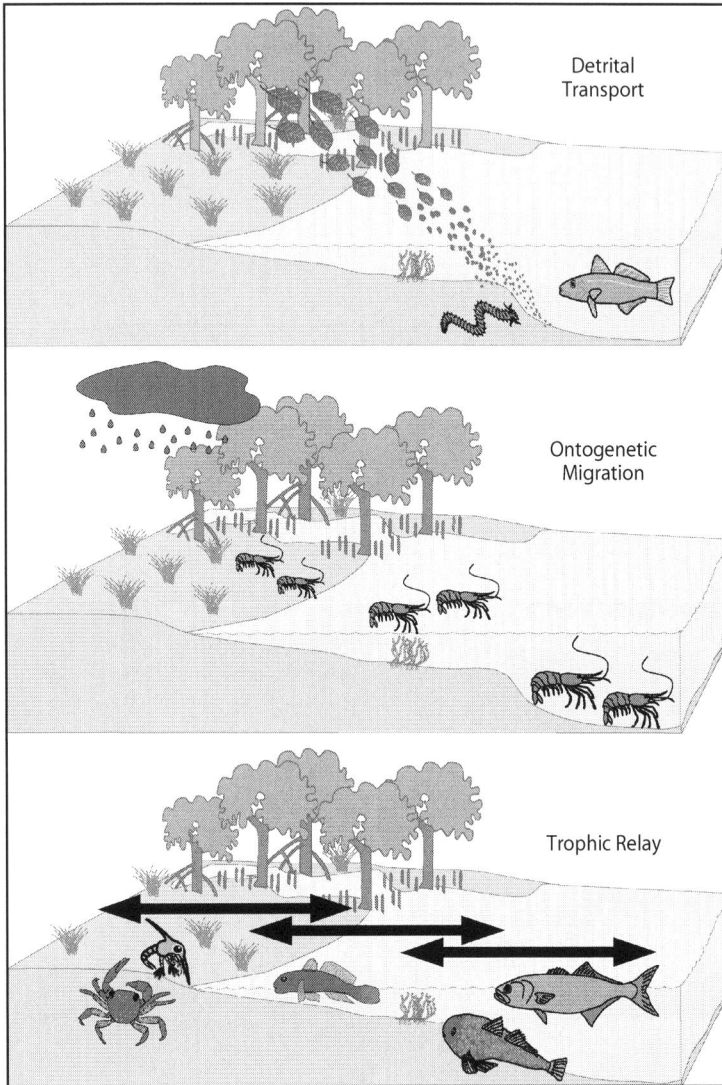

Fig. 3.3 Three mechanisms of transfer of organic matter from mangroves to food webs in deeper water (from Connolly and Lee 2007). Detrital transport includes movement of particulate and dissolved organic matter. Ontogenetic migration is movement in migrating animals such as banana prawns (*Fenneropenaeus merguiensis*). Trophic relay (Kneib 1997) involves a series of predator-prey interactions; in Australia, for example, crabs and their larvae which are high on the shore are eaten by fish such as glassfish (Ambassidae), gobies (Gobiidae) and juvenile mullet (Mugilidae), which in turn are preyed upon by fish such as flathead (Platycephalidae) and tailor (Pomatomidae)

mangroves leaves a large uncertainty in assessing an overall C budget for these systems (see Bouillon et al. 2008b). One major shortcoming of most current flux estimates (apart from the limited number of data used to extrapolate to a global level) is the fact that they rarely include a source characterization, and thus do not differentiate between organic carbon fluxes of mangrove origin and those of other potential carbon sources. In that respect, current flux data could be inherently biased and provide an overestimate of actual organic carbon fluxes from mangroves. Source characterization could be particularly important to integrated POC flux measurements, since it is known that mangroves (and other intertidal systems) can trap significant amounts of particulate material, including organic carbon often of non-mangrove origin, during tidal inundation (e.g., Middelburg et al. 1996, Bouillon et al. 2003). Moreover, import and export fluxes of POC (of different origin) can be closely balanced (e.g., Ayukai et al. 1998), leaving only a small residual net flux. For dissolved organic carbon, there are indeed studies which report a net influx, rather than efflux of organic carbon in certain mangrove systems. The flume experiments by Davis et al. (2001) in fringing mangroves along the Taylor River, for example, show that despite seasonal variations, DOC was generally imported from the water column, while TOC showed only small net fluxes, which ranged from import to export. Boto and Wellington (1988) also noted net DOC uptake in mangrove sediments in Coral Creek in northern Australia. One aspect of the study by Davis et al. (2001) is that their flumes were in continuously submerged mangroves, along the flow-path of a channel, in a non-tidal system, and any influence of the tidal pumping phenomenon cannot be ascertained. Since tidal pumping is likely an important mechanism for solute export (e.g., Dittmar and Lara 2001a, Schwendenmann et al. 2006, Bouillon et al. 2007c), DOC export in non-tidal systems may be significantly lower than in most other mangrove settings. In general, tidal hydrology and cumulative tidal amplitude would appear to be important determinants of the degree of organic carbon export. Subsequently, carbon export has been suggested to be higher in riverine forests than in fringe and basin forests (see Twilley 1985), and higher during periods of higher freshwater runoff in estuarine systems (Sutula et al. 2003). Similarly, Romigh et al. (2006) reported a seasonal pattern in DOC fluxes (i.e., periods with net export as well as periods with net import of DOC) consistent with a strong influence of freshwater discharge and tidal amplitude on DOC fluxes.

3.3.1.2 Exchange of Dissolved Inorganic Carbon

The focus on carbon exchange in tropical coastal ecosystems has so far been directed to organic carbon species, but to our knowledge no studies have attempted to directly quantify exchange of dissolved inorganic carbon (DIC). Nevertheless, tropical coastal ecosystems are sites with intense cycling of inorganic carbon, in particular the classical mangrove-seagrass-coral reef sequence. Mangroves are known for their intense mineralization and high CO_2 exchange (e.g., Borges et al. 2003). Tropical seagrass beds can attain very high primary production rates (e.g., Hemminga et al. 1994) resulting in significant lowering of pCO_2 levels in the water column (Bouillon et al. 2007a). Coral reefs, on the other hand, are a major contributor to overall oceanic $CaCO_3$ production (Gattuso et al. 1998). A recent comparison of DIC and DOC profiles from a number of tidal mangrove creeks and estuaries indicated that the lateral inputs of DIC from the mangroves was on average about eight times higher than for DOC (Bouillon et al. 2008b). If this is confirmed in other systems and/or through direct quantitative estimates of DIC exchange, this would imply that the mineralization of mangrove carbon and its subsequent export as DIC is substantially higher than the export of mangrove-derived material as organic carbon.

3.3.1.3 Migration and Trophic Relay

The transfer of energy from nearshore to offshore waters in migrating animals is an often overlooked but potentially important mechanism (Kneib 2000). Many important fisheries species, including crustaceans such as crabs and prawns, arrive in estuarine waters as larvae or post-larvae, then grow in the upper estuary, before migrating as larger animals (with their carbon) to the sea. In tropical waters, migration of key species such as banana prawns (*Fenneropenaeus merguiensis*) is often strongly seasonal. In the Gulf of Carpentaria, northern Australia, this results from strong freshwater flows through estuaries (Vance et al. 1998). In peninsular Malaysia, where the seasonality of migration for this same species is less pronounced because of more evenly distributed rainfall, the transfer of carbon in the body of animals is still important, because there is the same pronounced net migration out of estuaries (Ahmad Adnan et al. 2002). In southern USA, carbon transfer has been inferred from stable isotopes studies showing the movement of substantial numbers of pink shrimp (*Farfantepenaeus duorarum*) from seagrass meadows to unvegetated fisheries areas (Fry et al. 1999). The total carbon load transferred in this way has not been estimated, and it might ultimately prove to be small relative to particulate and dissolved transfer. This energy source is, however, probably important in coastal food webs because the animals that migrate are highly likely to be predated, and the link with food webs is therefore much more direct than for DOC and POC exported from estuaries.

The phenomenon of trophic relay was first described from temperate saltmarshes, which have small, resident fish and crustacean species that are preyed upon by somewhat larger fish visiting the marsh as transients at high tide. These predators are themselves potentially preyed upon by larger piscivorous fish, thus producing the effect of a relay system that transfers energy from shallow to deeper waters (Kneib 1997). There is preliminary evidence that this concept also applies in tropical systems. For example, glassfish (*Ambassis jacksoniensis*) have been shown to feed on huge quantities of shore crab larvae on a subtropical marsh in Queensland (Hollingsworth and Connolly 2006). Such marshes are inundated only on spring tides, and inundation has an extraordinary effect on the pattern of feeding by fish. Glassfish visiting the marsh on the first night of a tidal cycle feed only lightly, eating a small number of a range of prey types. This inundation apparently acts as a cue for shore crabs to release larvae, and on subsequent nights, glassfish eat an average of 100–200 crab larvae per fish (Fig. 3.4). Glassfish are a small, extremely abundant

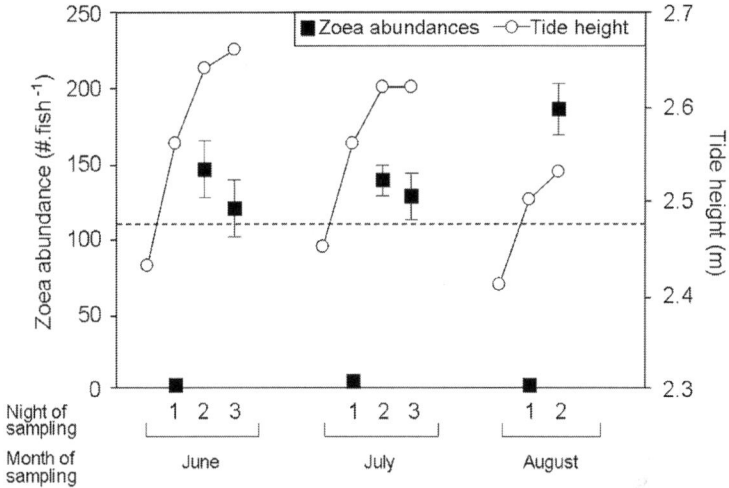

Fig. 3.4 Crab zoea abundances (mean ± SE) in glassfish (*Ambassis jacksoniensis*) stomachs after feeding on subtropical saltmarsh (data from Hollingsworth and Connolly 2006). In each monthly cycle, fish do not feed on zoea on the first night a marsh is flooded but do so on subsequent nights. Tidal height is shown for each night of sampling and the night before sampling. Tidal height at which marsh is inundated (2.48 m) is shown by dotted line

schooling species, and would be preyed upon by many of the larger fish in channels adjacent to these subtropical marshes (Baker and Sheaves 2005). Substantial effort has recently been aimed at understanding the trophic structure of fish communities in tropical systems (Nagelkerken and van der Velde 2004). Although the extent of piscivory generally remains to be demonstrated (Sheaves and Molony 2000), the first studies combining food web and movement analysis show that trophic relay is potentially very important (Kruitwagen et al. 2007, Lugendo et al. 2007).

3.3.1.4 Effects on Organic Matter Transfer on Food Webs and Ecological Structure

The transport and accumulation of macro-litter in adjacent systems has a number of impacts on the sedimentary environment and faunal communities, but few studies have documented such effects. Daniel and Robertson (1990) suggested that the presence and abundance of exported mangrove detritus had a positive influence on certain groups of macrobenthos such as penaeid shrimps, e.g., by serving as a shelter from predation. For benthic microfauna, in contrast, Alongi (1990) did not find convincing evidence that exported mangrove detritus enhanced the densities of flagellates, ciliates or protozoa. In a long-term experiment in which mangrove litter was added to a microcosm with a sandy substrate, Lee (1999) found no marked influence of litter addition to macrofaunal biomass, but species richness and diversity decreased with increasing litter inputs. The latter could be due to the negative effects of tannins leaching from litter (Alongi 1987, Lee 1999). Organic carbon in sediments can obviously serve as an important food source for organisms, but the oxygen depletion and accumulation of toxic by-products occurring when high loads of organic matter are delivered to sediments has been shown to result in potential decreases in the abundance and diversity of benthic fauna (Hyland et al. 2005). Although initial reports suggested an important direct trophic role for mangrove organic matter in adjacent aquatic foodwebs (Odum 1968), most later studies found little or no unambiguous evidence for such a role and suggest that the contribution of mangrove-derived carbon to nearshore foodwebs is minimal (see Bouillon et al. 2008a, and Section 3.3.1.5). Considering the importance of dissolved organic matter exchange, the lack of data on the fate of DOC and DON (dissolved organic nitrogen) is striking, and presents an important area for future work. The experiments by Dittmar et al. (2006) indicated that DOC from mangrove pore waters is partly photo-degraded and chemically modified, but that a major part remains after several weeks of incubation in the presence of a natural bacterial community. This suggests that part of the mangrove-derived DOC pool is sufficiently refractory to be dispersed over large areas when hydrodynamic conditions allow.

The quality of organic matter is also important as a determinant of consequences of carbon transport for fauna. For example, excessive labile organic matter (e.g., from shrimp farming) can lead to extensive hypoxia zones (Chua 1992), whereas more refractory organic matter can accumulate in marine sediments (POC) or can be transported in dissolved form (DOC) offshore (Alongi and Christoffersen 1992).

3.3.1.5 Detecting 'False Positives' in Outwelling Studies—Avoiding Pitfalls in Stable Isotope Gradient Analysis

A large number of studies have used stable isotope ratios to infer the relative contribution of carbon from intertidal vegetation, particularly mangroves, and other potential sources to the sedimentary or suspended organic matter pool (e.g., Machiwa 2000, Kuramoto and Minagawa 2001, Thimdee et al. 2003). A common strategy has been to relate variations in $\delta^{13}C$ values of POC to the admixture of mangrove-derived carbon and 'marine' phytoplankton, where the latter is characterized by typical $\delta^{13}C$ values of about -20 to $-18‰$ (e.g., Rezende et al. 1990, Chong et al. 2001). This oversimplified approach has a major shortcoming because it is based on an assumption that phytoplankton within estuaries or mangrove creeks has a $\delta^{13}C$ signature similar to that of marine phytoplankton, which is unlikely since mangrove creeks and estuaries typically have $\delta^{13}C$ signatures for DIC which are distinctly depleted in ^{13}C by $6–8‰$ (Bouillon et al. 2008a). Primary producers in the water column are therefore expected to show a similar depletion relative to producers from open marine systems, and the same holds for benthic microalgae (Guest et al. 2004).

The depleted $\delta^{13}C$ values of DIC near mangroves also affects values of benthic macrophytes such as seagrasses. Seagrass $\delta^{13}C$ values usually range between -16 and $-12‰$ (Hemminga and Mateo 1996), but $\delta^{13}C$ values of seagrasses adjacent to mangrove forests typically show a gradient of more depleted values close to the mangroves, becoming more enriched with increasing distance towards the sea (e.g., a range of almost 10‰ over <4 km distance found by Hemminga et al. (1994) and Marguillier et al. (1997)). Studies that overlook the DIC isotope gradient with increasing distance offshore from mangroves therefore also overlook a probable gradient in isotope ratios of primary producers with distance offshore. Any gradient in isotope ratios of particulate carbon or even in animal tissues described in such studies might therefore provide 'false positives' in their test of the importance of mangrove carbon.

The isotopic depletion of the DIC pool near mangroves, and its effect on other local autotrophs, means that reliance on any autotroph will look like a mangrove contribution to food webs adjacent to mangroves in isotope gradient studies. It is important, therefore, to adopt specialized strategies to overcome this challenge in studies of potential outwelling. First, isotope values of potential alternative sources should be measured intensively and at a fine spatial scale. If plankton cannot be properly collected, spatially intensive DIC sampling provides a realistic alternative. Second, because carbon isotope measurements alone often cannot resolve the contribution of various sources to the POC pool, a combination of isotopes with other tracers should be considered (such as POC/PN ratios, e.g., Gonneea et al. 2004; POC/Chl. *a* ratios, e.g., Cifuentes et al. 1996; or other biochemical tracers such as lignin-derived phenols, Dittmar et al. 2001, see also Chapter 12).

3.3.2 Transfer of Carbon to Intertidal Habitats—Inwelling

3.3.2.1 Seagrass to Mangroves

The role of macrolitter in material exchange has been poorly studied in tropical coastal systems, and represents an important gap in our knowledge, since the few available studies suggest that the quantities of floating or suspended macrolitter can be high in comparison to the normal POC or DOC concentrations.

Slim et al. (1996) documented tidal transport of seagrass, mangrove, and macroalgal litter in a Kenyan bay. They found clear evidence for bidirectional transport of macrolitter, with a dominance of seagrass litter during both ebb and flood periods, but mangrove litter being more important during ebb than during flood periods. The accumulation of mangrove-derived material in this system has also been demonstrated based on organic carbon and stable isotope evidence (Hemminga et al. 1994, Bouillon et al. 2004). The deposition of litter in intertidal mangroves can be highly conspicuous in sites close to seagrass beds or where macroalgae are abundant (Fig. 3.5), and is also evident based on stable isotope data in bulk sediments which are often distinctly different from that of the dominant local vegetation (Middelburg et al. 1996, Bouillon et al. 2004; Fig. 3.6) and in the distribution pattern of n-alkanes in mangrove sediments close to the seagrass beds (P.V. Khoi and S. Bouillon, unpubl. data). Wooller et al. (2003) found the sediment organic matter in *Laguncularia* mangroves in Twin Cays (Belize) often to be dominated by non-mangrove sources, including seagrass material, as evidenced by some sites having high $\delta^{13}C$ signatures combined with high C/N ratios, consistent with those of

Fig. 3.5 Deposition of
seagrass litter in intertidal
Avicennia marina forests in
Gazi Bay (Kenya) (**a**),
deposition of the macroalgae
Ulva spp. in *Sonneratia alba*
mangroves in Mtoni Estuary,
Dar es Salaam (Tanzania) (**b**)

Thalassia sp. from adjacent seagrass systems. Massive deposits of seagrass material
have also been reported on tropical sandy beaches (Hemminga and Nieuwenhuize
1991) and in intertidal flats (de Boer 2000), but little is known on the fate of this
material and its potential trophic importance in these unvegetated systems.

3.3.2.2 Seagrass to Mudflats

For shallow sand and mud flats, recent experimental work in temperate waters has
resulted in a new conceptualization of food webs. Deliberate ^{13}C tracer experiments
on the intertidal flats of northern Europe clearly show that benthic microalgae in
the sediment are a major contributor to food webs (Middelburg et al. 2000). This
has formed part of the more general realization that benthic microalgae are highly
productive and easily assimilated in a food web context (MacIntyre et al. 1996).

In tropical Australian systems, there is evidence from fatty acid studies that
benthic microalgae make at least some contribution to the nutrition of invertebrates (Meziane et al. 2006). On the other hand, carbon isotope evidence from
the same mudflats shows a strong reliance on allochthonous carbon from adjacent

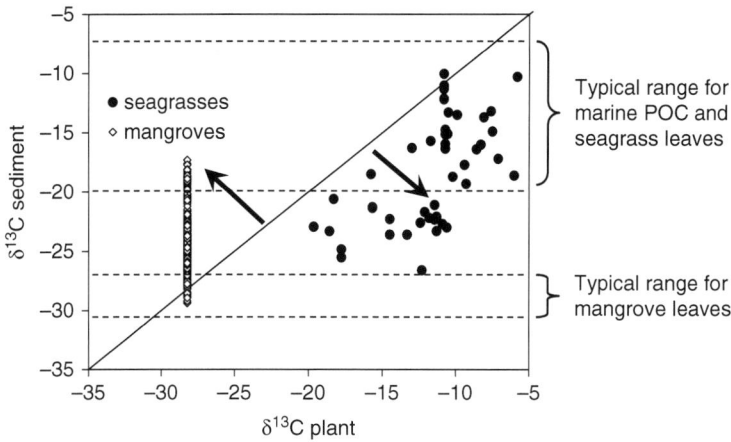

Fig. 3.6 Comparison of sediment organic carbon $\delta^{13}C$ signatures with those of the dominant seagrass species (data for tropical and subtropical systems only) or mangroves. For mangroves, we used a global average $\delta^{13}C$ value for plant material of -28.2% (see Bouillon et al. 2008a). For data sources for seagrass systems, see Bouillon et al. (2004); data for mangrove systems are also presented in Kristensen et al. (2008). Sediment $\delta^{13}C$ signatures in mangroves are often distinctly more ^{13}C-enriched to those of mangrove litter inputs, and conversely, sediment $\delta^{13}C$ data from seagrass beds are consistently ^{13}C-depleted relative to the dominant seagrass vegetation. POC = particulate organic matter

seagrass meadows (Melville and Connolly 2005). This transfer of organic material from seagrass meadows to mud flats is further supported by recent results for the commercially-important portunid mud crab, *Scylla serrata*. Mud crabs generally have relatively enriched carbon isotope ratios, showing reliance on organic matter from either seagrass meadows or saltmarsh grass. Mud crab ratios, however, show very strong spatial variation. A survey of mud crabs at different distances from key habitats found that distance to seagrass, and not distance to saltmarsh (or mangroves), explained much of the variation (Fig. 3.7). This isotope evidence suggests that, where seagrass is present in shallow tropical waters, carbon from the meadows will have a disproportionately high contribution to animal nutrition, whereas further from meadows and where no seagrass exists, animals rely on a generalized carbon pool from a variety of sources.

3.3.3 Scales of Carbon Transfer Among Systems

The source of energy to consumers and its movement among habitats has been a key focus in ecology. Carbon is expected to move more in aquatic than terrestrial systems because water acts as an efficient transport medium (Polis et al. 1997). In practice, however, the degree to which carbon is transported and utilized in food webs varies among systems.

Fig. 3.7 Relationships between carbon stable isotope ratios ($\delta^{13}C$) of mud crabs *Scylla serrata* and the distance crabs were caught from the nearest patch of three habitats (seagrass, saltmarsh, mangroves). The strongest relationship is with seagrass distance, and for this habitat exploded views of small distances show the tight relationship. No relationship exists for saltmarsh or mangroves. Data from Waltham and Connolly (unpubl.)

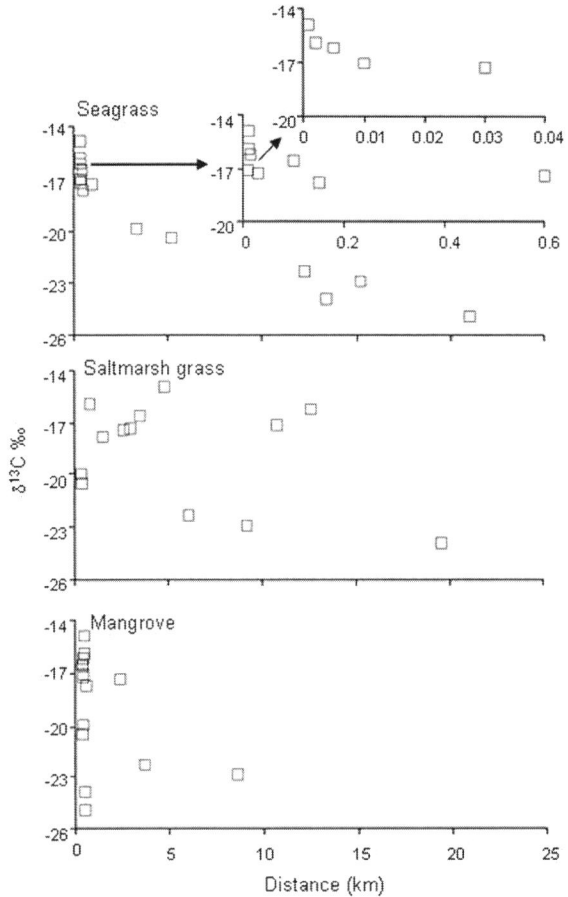

First, the extent of carbon dispersal from river plumes to coastal ecosystems depends on flow rates. Major rivers such as the Amazon River affect pelagic and benthic processes over tens of kilometers (Smith and Demaster 1996). The discharge from smaller rivers, however, can be retained in small, distinct plumes that remain close to the coastline, over an area less than 1 km^2 (e.g., Gaston et al. 2006).

Within estuaries themselves, carbon is potentially moved on tidal currents. The amount of carbon available to move has been difficult to quantify, because of high rates of allochthonous input from riverine sources and autochthonous production from often extensive fringing vegetation, and high secondary productivity and, therefore, consumption of carbon. Depending on season and location, mangrove carbon has been detected as detritus in sediment at between 2 and 4 km from mangrove forests, using both stable isotope (Rodelli et al. 1984) and fatty acid (Meziane et al. 2006) techniques. For some estuaries, however, the large scale movement of carbon expected from the outwelling theory has not been substantiated (e.g.,

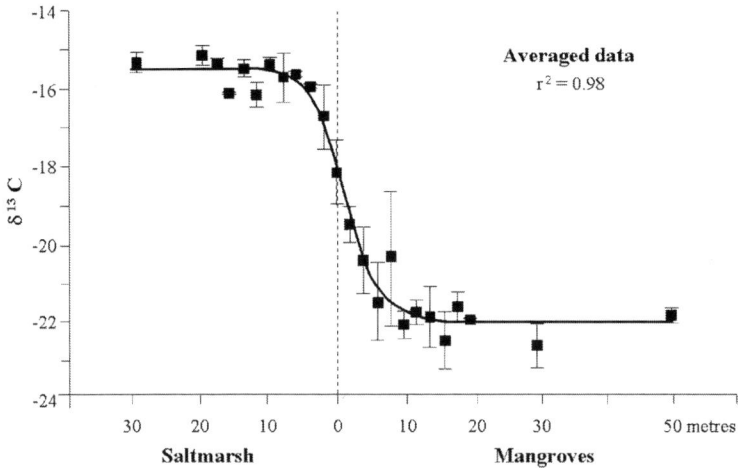

Fig. 3.8 Carbon stable isotope values of the grapsid crab *Parasesarma erythrodactyla* across the mangrove/saltmarsh interface (mean ± SE) from three sites; after Guest and Connolly 2004). The rapid change at the interface indicates that crabs utilize different carbon (energy) sources in the two habitats

Loneragan et al. 1997). More recent evidence suggests that the movement of carbon in estuarine habitats can occur at a finer scale than has previously been considered. For example, in a tropical study examining the carbon isotopes of shore crabs at sites separated by hundreds of meters, Hsieh et al. (2002) found that the crabs derive their carbon from the sites in which they reside rather than from further afield. A subsequent study of the movement and assimilation of carbon by shore crabs in a subtropical estuary showed that crabs obtain their nutrition from sources within the surrounding few meters (Fig. 3.8). Detailed measurements of crab and POM movement suggests that the short distance carbon is transported occurs through movement of POM rather than crabs, which have a very small foraging area (Guest et al. 2004, 2006).

3.3.4 Carbon Exchange in Coral Reefs

Compared to other tropical coastal ecosystems, very few studies have addressed the exchange of carbon between coral reefs and adjacent waters (Gattuso et al. 1998).

Delesalle et al. (1998) estimated for a French Polynesian coral reef system that 47% of organic matter production and 21% of carbonate production was exported, the latter being in agreement with previous estimates by Smith et al. (1978; 25%). These numbers were considered minimum estimates, since they did not consider DOC exchange, exchange of macro-debris, and since their measurements were carried out during relatively calm weather conditions and only considered export towards the ocean on the outer part of this fringing reef. The extensive sampling setup by Delesalle et al. (1998) also revealed that horizontal and downslope advection of particles was the dominant pathway of export, rather than vertical transport offshore. The latter is also supported by the data in Hata et al. (1998, 2002), who estimated that only about 20–35% of the POC exported from reef flats was recovered in sediment traps at 40–50 m depth at some distance outside the reefs.

Hata et al. (1998) estimated that the net export rate of organic carbon from a coral reef in Palau represented about 4% of the gross primary production, but this study considered only export of particulate organic carbon and not DOC. Furthermore, since the majority of gross primary production is generally respired in such systems (up to 94%, estimated by Hata et al. 2002), this may still represent a significant part of the (relatively small) net organic carbon production. Hata et al. (2002), for example, estimated that the majority of net community production (80–100%) was exported as DOC or POC, with DOC fluxes being 5–6 times higher than POC fluxes. Considering the open character of coral reefs, such a high degree of export, in particular of the DOC produced within these systems, is not entirely surprising. Coral cays can act, however, to trap and store carbon. Pile (2005) showed on the Great Barrier Reef, for example, that almost all of the ultra-nanoplankton in ambient water is removed by filter feeding organisms on the coral reefs in one pass of the water over the reef. This powerful filtering role suggests net uptake of carbon on reefs, at least over short periods, once the activities of all sessile organisms are taken into account. Similarly, de Goeij and van Duyl (2008) found that the surface of coral reef cavities (including the associated biota) can act as net sinks of DOC.

Export of the excess organic carbon production in the form of living biomass (e.g., juvenile fish migrating to other environments to complete their life cycle) has been proposed to be a major component of the organic matter export in coral reefs (50–75%, see Gattuso et al. 1998), as well as export of drifting algae in systems where algae are an important component of the reef (Kilar and Norris 1988, see also Hata et al. 1998).

3.4 Conclusions and Future Research Directions

Tropical and subtropical coastal ecosystems are characterized by intense primary production and a high degree of carbon exchange on various spatial scales, which can be abiotically driven (flux of particulate and dissolved material) as well as biologically driven (animal movement and trophic relay). The past few decades have seen an increased awareness of the role of the tropical coastal zone in global carbon budgets. We are still far from being able to constrain this role in quantitative terms, however, because of: (1) the complexity of assessing material fluxes and combining this with information on the origin of the material considered, and (2) the diversity of ecosystems in the tropical coastal zone (estuaries, lagoons, mangroves, mudflats, seagrass beds, and coral reefs). These factors necessitate a range of approaches and analytical techniques to adequately address their biogeochemical functioning. Moreover, assessing the importance of biologically mediated carbon exchange is complex and has only rarely been attempted in quantitative terms.

Even carbon budgeting efforts for individual ecosystems are hampered by a striking scarcity of data on basic processes such as primary production (e.g., Bouillon et al. 2008b), water-atmosphere CO_2 fluxes (e.g., Borges et al. 2005), or carbon burial rates (Duarte et al. 2005). The collection of baseline data from a range of tropical coastal ecosystems thus remains important as a complement to state-of-the art analytical techniques to trace and quantify carbon exchange (see Section 12.4 in Chapter 12).

One area in which carbon pathways are yet to be used but should prove effective is as indicators of ecosystem health. The goal of conservation should be about more than species conservation, and should also conserve ecological processes. There is very little guidance in the aquatic conservation literature on what processes might really be important (or measurable). Carbon transfer and utilization is probably central; for example, the source of an animal's food is one of the central organizing themes in ecology (Polis et al. 1997), and a range of tracer tools are available to elucidate carbon pathways (see Chapter 12). As carbon pathways are better understood, it will be possible to detect changes in these pathways due to disturbances such as eutrophication, land-use change in catchments (C4 agriculture), clearing of coastal habitats such as mangroves, and accidental destruction of seagrass through dredging and land claims.

Carbon isotopes are already being used to study ecosystem health in tropical inland waters. In the headwaters of pristine rivers, food webs are supported predominantly by allochthonous input of riparian vegetation (the original river continuum concept by Vannote et al. 1980). In tropical streams, once riparian vegetation is removed, the fundamental pathways are altered, from the original reliance on allochthonous inputs of macrophytes to autochthonous in-stream production, usually of microalgae which rapidly increase production because of greater light availability (Douglas et al. 2005).

Degraded estuarine habitats are beginning to be restored in tropical areas, for example in the Florida Everglades restoration project. Such efforts usually incorporate monitoring of flora and fauna assemblages, but these can be poor indicators of ecological processes. The degree to which restored habitat mimics ecological processes in natural habitat is best measured directly. Again, carbon pathways are an obvious candidate, since they are relatively easily measured using chemical tracers and can be predicted from models based on data from other, less perturbed systems (Twilley et al. 1999).

References

Ahmad Adnan N, Loneragan NR, Connolly RM (2002) Variability of, and the influence of environmental factors on, the recruitment of postlarval and juvenile *Penaeus merguiensis* in the Matang mangroves of Malaysia. Mar Biol 141:241–251

Alongi DM (1987) The influence of mangrove-derived tannins on intertidal meiobenthos in tropical estuaries. Oecologia 71:537–540

Alongi DM (1990) Effect of mangrove detrital outwelling on nutrient regeneration and oxygen fluxes in coastal sediments of the central Great Barrier Reef lagoon. Estuar Coast Shelf Sci 31:581–598

Alongi DM, Christoffersen P (1992) Benthic infauna and organism–sediment relations in a shallow, tropical coastal area – influence of outwelled mangrove detritus and physical disturbance. Mar Ecol Prog Ser 81:229–245

Arthington AH, Bunn SE, Poff NL et al (2006) The challenge of providing environmental flow rules to sustain river ecosystems. Ecol Appl 16:1311–1318

Ayukai T, Miller D, Wolanski E et al (1998) Fluxes of nutrients and dissolved and particulate organic carbon in two mangrove creeks in northeastern Australia. Mangroves and Salt Marshes 2:223–230

Baker R, Sheaves M (2005) Redefining the piscivore assemblage of shallow estuarine nursery habitats. Mar Ecol Prog Ser 291:197–213

Baum A, Rixen T, Samiaji J (2007) Relevance of peat draining rivers in central Sumatra for the riverine input of dissolved organic carbon into the ocean. Estuar Coast Shelf Sci 73:563–570

Borges AV, Delille B, Frankignoulle M (2005) Budgeting sinks and sources of CO_2 in the coastal ocean: diversity of ecosystems counts. Geophys Res Lett 32, L14601, doi:10.1029/2005GL023053

Borges AV, Djenidi S, Lacroix G et al (2003) Atmospheric CO_2 flux from mangrove surrounding waters. Geophys Res Lett 30, 1558, doi: 10.1029/ 2003GL017143

Boto KG, Wellington JT (1988) Seasonal variations in concentrations and fluxes of dissolved organic and inorganic materials in a tropical, tidally dominated waterway. Mar Ecol Prog Ser 50:151–160

Bouillon S, Dahdouh-Guebas F, Rao AVVS et al (2003) Sources of organic carbon in mangrove sediments: variability and possible implications for ecosystem functioning. Hydrobiologia 495:33–39

Bouillon S, Moens T, Dehairs F (2004) Carbon sources sustaining benthic mineralization in mangrove and adjacent seagrass sediments (Gazi bay, Kenya). Biogeosciences 1:71–78

Bouillon S, Dehairs F, Velimirov B et al (2007a) Dynamics of organic and inorganic carbon across contiguous mangrove and seagrass systems (Gazi bay, Kenya). J Geophys Res 112, G02018, doi:10.1029/2006JG000325

Bouillon S, Dehairs F, Schiettecatte LS et al (2007b) Biogeochemistry of the Tana estuary and delta (northern Kenya). Limnol Oceanogr 52:46–59

Bouillon S, Middelburg JJ, Dehairs F et al (2007c) Importance of intertidal sediment processes and porewater exchange on the water column biogeochemistry in a pristine mangrove creek (Ras Dege, Tanzania). Biogeosciences 4:311–322

Bouillon S, Connolly R, Lee SY (2008a). Organic matter exchange and cycling in mangrove ecosystems: recent insights from stable isotope studies. J Sea Res 59:44–58

Bouillon S, Borges AV, Castañeda-Moya E et al (2008) Mangrove production and carbon sinks: a revision of global budget estimates. Glob Biogeochem Cycles 22, GB2013, doi: 10.1029/2007GB003052

Chong VC, Low CB, Ichikawa T (2001) Contribution of mangrove detritus to juvenile prawn nutrition: a dual stable isotope study in a Malaysian mangrove forest. Mar Biol 138:77–86

Chua TE (1992) Coastal aquaculture development and the environment: the role of coastal area management. Mar Pollut Bull 25:98–103

Cifuentes LA, Coffin RB, Solorzano L et al (1996) Isotopic and elemental variations of carbon and nitrogen in a mangrove estuary. Estuar Coast Shelf Sci 43:781–800

Cloern JE (2001) Our evolving conceptual model of the coastal eutrophication problem. Mar Ecol Prog Ser 210:223–253

Connolly RM, Gorman D, Guest MA (2005) Movement of carbon among estuarine habitats and its assimilation by invertebrates. Oecologia 144:684–691

Connolly RM, Lee SY (2007) Mangroves and saltmarsh. In: Connell SD, Gillanders BM (eds) Marine Ecology, pp. 485–512. Oxford University Press, Oxford

Coynel A, Seyler P, Etcheber H et al (2005) Spatial and seasonal dynamics of total suspended sediment and organic carbon species in the Congo River. Glob Biogeochem Cycles 19, GB4019, doi:10.1029/2004GB002335

Dai J, Sun M-Y (2007) Organic matter sources and their use by bacteria in the sediments of the Altamaha estuary during high and low discharge periods. Org Geochem 38:1–15

Daniel PA, Robertson AI (1990) Epibenthos of mangrove waterways and open embayments: community structure and the relationship between exported mangrove detritus and epifaunal standing stocks. Estuar Coast Shelf Sci 31:599–619

Darnaude AM, Salen-Picard C, Harmelin-Vivien ML (2004) Depth variation in terrestrial particulate organic matter exploitation by marine coastal benthic communities off the Rhone River delta (NW Mediterranean). Mar Ecol Prog Ser 275:47–57

Davis III SE, Childers DL, Day JW et al (2001) Wetland-water column exchanges of carbon, nitrogen, and phosphorus in a southern Everglades dwarf mangrove. Estuaries 24:610–622

de Boer WF (2000) Biomass dynamics of seagrasses and the role of mangrove and seagrass vegetation as different nutrient sources for an intertidal ecosystem. Aquat Bot 66:225–239

de Goeij JM, van Duyl FC (2008) Coral cavities are sinks of dissolved organic carbon (DOC). Limnol Oceanogr 52:2608–2617

Delesalle B, Buscail R, Carbonne J et al (1998) Direct measurements of carbon and carbonate export from a coral reef ecosystem (Moorea Island, French Polynesia). Coral Reefs 17: 121–132

Dittmar T, Lara RJ (2001a) Driving forces behind nutrient and organic matter dynamics in a mangrove tidal creek in North Brazil. Estuar Coast Shelf Sci 52:249–259

Dittmar T, Lara RJ (2001b) Do mangroves rather than rivers provide nutrients to coastal environments south of the Amazon River? Evidence from long-term flux measurements. Mar Ecol Prog Ser 213:67–77

Dittmar T, Lara RJ, Kattner G (2001) River or mangrove? Tracing major organic matter sources in tropical Brazilian coastal waters. Mar Chem 73:253–271

Dittmar T, Hertkorn N, Kattner G et al (2006) Mangroves, a major source of dissolved organic carbon to the oceans. Glob Biogeochem Cycles 20, GB1012, doi: 10.1029/2005GB002570

Douglas MM, Bunn SE, Davies MP (2005) River and wetland food webs in Australia's wet-dry tropics: general principles and implications for management. Mar Freshw Res 56:329–342

Duarte CM, Middelburg JJ, Caraco N (2005) Major role of marine vegetation on the oceanic carbon cycle. Biogeosciences 2:1–8

Eyre B (1998) Transport, retention and transformation of material in Australian estuaries. Estuaries 21:540–551

Ford P, Tillman P, Robson B et al (2005) Organic carbon deliveries and their flow-related dynamics in the Fitzroy estuary. Mar Pollut Bull 51:119–127

Fry B, Mumford PL, Robblee MB (1999) Stable isotope studies of pink shrimp (*Farfantepenaeus duorarum*) Burkenroad migrations on the southwestern Florida shelf. Bull Mar Sci 65:419–430

Galy V, France-Lanord C, Beysacc O et al (2007) Efficient organic carbon burial in the Bengal fan sustained by the Himalayan erosional system. Nature 450:407–410

Gaston TF, Schlacher TA, Connolly RM (2006) Flood discharges of a small river into open coastal waters: plume traits and material fate. Estuar Coast Shelf Sci 69:4–9

Gattuso JP, Frankignoulle M, Wollast R (1998) Carbon and carbonate metabolism in coastal aquatic ecosystems. Annu Rev Ecol Syst 29:405–434

Gillanders BM, Kingsford MJ (2002) Impact of changes in flow of freshwater on estuarine and open coastal habitats and the associated organisms. Oceanogr Mar Biol Annu Rev 40:233–309

Goñi MA, Monacci N, Gisewhite R et al (2006) Distribution and sources of particulate organic matter in the water column and sediments of the Fly River Delta, Gulf of Papua (Papua New Guinea). Estuar Coast Shelf Sci 69:225–245

Gonneea ME, Paytan A, Herrera-Silveira JA (2004) Tracing organic matter sources and carbon burial in mangrove sediments over the past 160 years. Estuar Coast Shelf Sci 61:211–227

Guest MA, Connolly RM (2004) Fine-scale movement and assimilation of carbon in saltmarsh and mangrove habitat. Aquat Ecol 38:599–609

Guest MA, Connolly RM, Lee SY et al (2006) Mechanism for the small-scale movement of carbon among estuarine habitats: organic matter transfer not crab movement. Oecologia 148:88–96

Guest MA, Connolly RM, Loneragan NR (2004) Carbon movement and assimilation by invertebrates in estuarine habitats occurring at a scale of metres. Mar Ecol Prog Ser 278:27–34

Harrigan P, Zieman JC, Macko SA (1989) The base of nutritional support for the grey snapper (*Lutjanus griseus*) – an evaluation based on a combined stomach content and stable isotope analysis. Bull Mar Sci 44:65–77

Hata H, Kudo S, Yamano H et al (2002) Organic carbon flux in Shiraho coral reef (Ishigaki Island, Japan). Mar Ecol Prog Ser 232:129–140

Hata H, Suzuki A, Maruyama T et al (1998) Carbon flux by suspended and sinking particles around the barrier reef of Palau, Western Pacific. Limnol Oceanogr 43:1883–1893

Hemminga MA, Mateo MA (1996) Stable carbon isotopes in seagrasses: variability in ratios and use in ecological studies. Mar Ecol Prog Ser 140:285–298

Hemminga MA, Nieuwenhuize J (1991) Transport, deposition and in situ decay of seagrasses in a tropical mudflat area (Banc d'Arguin, Mauretania). Neth J Sea Res 27:183–190

Hemminga MA, Slim FJ, Kazungu J et al (1994) Carbon outwelling from a mangrove forest with adjacent seagrass beds and coral reefs (Gazi Bay, Kenya). Mar Ecol Prog Ser 106:291–301

Hollingsworth A, Connolly RM (2006) Feeding by fish visiting inundated subtropical saltmarsh. J Exp Mar Biol Ecol 336:88–98

Hsieh HL, Chen CP, Chen YG et al (2002) Diversity of benthic organic matter flows through polychaetes and crabs in a mangrove estuary: delta C-13 and delta S-34 signals. Mar Ecol Prog Ser 227:145–155

Hung JJ, Huang MH (2005) Seasonal variations of organic-carbon and nutrient transport through a tropical estuary (Tsengwen) in southwestern Taiwan. Environ Geochem Health 27: 75–95

Hutchings PA, Saenger P (1987). Ecology of mangroves. University of Queensland Press, Brisbane.

Hyland J, Balthis L, Karakassis I et al (2005) Organic carbon content of sediments as an indicator of stress in the marine benthos. Mar Ecol Prog Ser 295:91–103

Kilar JA, Norris JN (1988) Composition, export, and import of drift vegetation on a tropical, plant-dominated, fringing-reef platform (Caribbean Panama). Coral Reefs 7:93–103

Kneib RT (1997) The role of tidal marshes in the ecology of estuarine nekton. Oceanogr Mar Biol Annu Rev 35:163–220

Kneib RT (2000) Saltmarsh ecoscapes and production transfers by estuarine nekton in the southeastern U. S. In: Weinstein MP, Kreeger DA (eds) Concepts and controversies in tidal marsh ecology, pp. 267–292. Kluwer Academic, Netherlands

Kristensen E, Bouillon S, Dittmar T, et al (2008) Organic carbon dynamics in mangrove ecosystems: a review. Aquat Bot 89:201–209

Kruitwagen G, Nagelkerken I, Lugendo BR et al (2007) Influence of morphology and amphibious life-style on the feeding ecology of the mudskipper *Periophthalmus argentilineatus*. J Fish Biol 71:39–52

Kuramoto T, Minagawa M (2001) Stable carbon and nitrogen isotopic characterization of organic matter in a mangrove ecosystem on the south-western coast of Thailand. J Oceanogr 57: 421–431

Lee SY (1995) Mangrove outwelling – a review. Hydrobiologia 295:203–212

Lee SY (1999) The effect of mangrove leaf litter enrichment on macrobenthic colonization of defaunated sandy substrates. Estuar Coast Shelf Sci 49:703–712

Lee SY (2000) Carbon dynamics of Deep Bay, eastern Pearl River estuary, China. II: Trophic relationship based on carbon- and nitrogen-stable isotopes. Mar Ecol Prog Ser 205:1–10

Loneragan NR, Bunn SE (1999) River flows and estuarine ecosystems: implications for coastal fisheries from a review and a case study of the Logan River, southeast Queensland. Aust J Ecol 24:431–440

Loneragan NR, Bunn SE, Kellaway DM (1997) Are mangroves and seagrasses sources of organic carbon for penaeid prawns in a tropical Australian estuary? A multiple stable isotope study. Mar Biol 130:289–300

Ludwig W, Probst JL, Kempe S (1996a) Predicting the oceanic input of organic carbon by conti-
    nental erosion. Glob Biogeochem Cycles 10:23–41
Ludwig W, Amiotte-Suchet P, Probst JL (1996b) River discharges of carbon to the world's oceans:
    determining local inputs of alkalinity and of dissolved and particulate organic carbon. C R Acad
    Sci Paris II 323:1007–1014
Lugendo BR, Nagelkerken I, Kruitwagen G et al (2007) Relative importance of mangroves as
    feeding habitats for fishes: a comparison between mangrove habitats with different settings.
    Bull Mar Sci 80:497–512
Machiwa JF (1999) Lateral fluxes of organic carbon in a mangrove forest partly contaminated with
    sewage wastes. Mangroves and Salt Marshes 3:95–104
Machiwa JF (2000) 13C signatures of flora, macrofauna and sediment of a mangrove forest partly
    affected by sewage wastes. Tanz J Sci 26:15–28
MacIntyre HL, Geider RJ, Miller DC (1996) Microphytobenthos: The ecological role of the "secret
    garden" of unvegetated, shallow-water marine habitats. 1. Distribution, abundance and primary
    production. Estuaries 19:186–201
Marguillier S, van der Velde G, Dehairs F et al (1997) Trophic relationships in an interlinked
    mangrove-seagrass ecosystem as traced by δ^{13}C and δ^{15}N. Mar Ecol Prog Ser 151:115–121
Martinelli LA, Ballester MV, Krusche AV et al (1999) Landcover changes and δ^{13}C composition
    of riverine particulate organic matter in the Piracicaba river basin (southeast region of Brazil).
    Limnol Oceanogr 44:1826–1833
Melville AJ, Connolly RM (2003) Spatial analysis of stable isotope data to determine primary
    sources of nutrition for fish. Oecologia 136:499–507
Melville AJ, Connolly RM (2005) Food webs supporting fish over subtropical mudflats are based
    on transported organic matter not in situ microalgae. Mar Biol 148:363–371
Meziane T, d'Agata F, Lee SY (2006) Fate of mangrove organic matter along a subtropical estuary:
    small-scale exportation and contribution to the food of crab communities. Mar Ecol Prog Ser
    312:15–27
Middelburg JJ, Nieuwenhuize J, Slim FJ et al (1996) Sediment biogeochemistry in an East African
    mangrove forest (Gazi Bay, Kenya). Biogeochemistry 34:133–155
Middelburg JJ, Barranguet C, Boschker HTS et al (2000) The fate of intertidal microphytobenthos
    carbon: an in situ ^{13}C-labeling study. Limnol Oceanogr 45:1224–1234
Milliman JD, Farnsworth KL, Albertin CS (1999) Flux and fate of fluvial sediments leaving large
    islands in the East Indies. J Sea Res 41:97–107
Nagelkerken I, van der Velde G (2004) Are Caribbean mangroves important feeding grounds for
    juvenile reef fish from adjacent seagrass beds? Mar Ecol Prog Ser 274:143–151
Odum EP (1968) Evaluating the productivity of coastal and estuarine water. Proceedings of the
    Second Sea Grant Conference, pp. 63–64. University of Rhode Island
Ovalle ARC, Rezende CE, Lacerda LD et al (1990) Factors affecting the hydrochemistry of a
    mangrove tidal creek, Sepetiba Bay, Brazil. Estuar Coast Shelf Sci 31:639–650
Pennings S, Bertness M (2001) Salt marsh communities. In: Bertness M, Gaines S, Hay M (eds)
    Marine community ecology, pp. 289–316. Sinauer, Mass
Pile AJ (2005) Overlap in diet between co-occurring active suspension feeders on tropical and
    temperate reefs. Bull Mar Sci 76:743–749
Polis GA, Anderson WB, Holt RD (1997) Toward and integration of landscape and food
    web ecology: the dynamics of spatially subsidized food webs. Annu Rev Ecol Syst 28:
    289–316
Poloczanska ES, Babcock RC, Butler A et al (2007) Climate change and Australian marine life.
    Oceanogr Mar Biol Annu Rev 45:407–478
Postel SL (2000) Entering an era of water scarcity: the challenges ahead. Ecol Appl 10:941–948
Ralison O, Dehairs F, Middelburg JJ, et al (2008) Carbon biogeochemistry in the Betsiboka estuary
    (northwestern Madagascar). Org Geochem 39:1649–1658
Rezende CE, Lacerda LD, Ovalle ARC et al (1990) Nature of POC transport in a mangrove ecosys-
    tem: a carbon stable isotopic study. Estuar Coast Shelf Sci 30:641–645

Rodelli MR, Gearing JN, Gearing PJ et al (1984) Stable isotope ratio as a tracer of mangrove carbon in Malaysian ecosystems. Oecologia 61:326–333

Romigh MA, Davis SE, Rivera-Monroy VH et al (2006) Flux of organic carbon in a riverine mangrove wetland in the Florida coastal Everglades. Hydrobiologia 569:505–516

Schefuß E, Versteegh GJM, Jansen JHF et al (2004) Lipid biomarkers as major source and preservation indicators in SE Atlantic surface sediments. Deep Sea Res I 51:1199–1228

Schlünz B, Schneider RR (2000) Transport of terrestrial organic carbon to the oceans by rivers: re-estimating flux- and burial rates. Int J Earth Sci 88:599–606

Schwendenmann L, Riecke R, Lara RL (2006) Solute dynamics in a North Brazilian mangrove: the influence of sediment permeability and freshwater input. Wetl Ecol Manage 14:463–475

Sheaves M, Molony B (2000) Short-circuit in the mangrove food chain. Mar Ecol Prog Ser 199: 97–109

Slim FJ, Hemminga MA, Cocheret de la Morinière E et al (1996) Tidal exchange of macrolitter between a mangrove forest and adjacent seagrass beds (Gazi Bay, Kenya). Neth J Aquat Ecol 30:119–128

Smith SV, Jokiel PL, Key GS (1978) Biogeochemical budgets in coral reef systems. Atoll Res Bull 220:1–11

Smith WO, Demaster DJ (1996) Phytoplankton biomass and productivity in the Amazon River plume – correlation with seasonal river discharge. Cont Shelf Res 16:291–319

Staunton-Smith J, Robins JB, Mayer DG et al (2004) Does the quantity and timing of fresh water flowing into a dry tropical estuary affect year-class strength of barramundi (*Lates calcarifer*)? Mar Freshw Res 55:787–797

Sutula MA, Perez BC, Reyes E et al (2003) Factors affecting spatial and temporal variability in material exchange between the Southern Everglades wetlands and Florida Bay. Estuar Coast Shelf Sci 57:757–781

Thimdee W, Deein G, Sangrungruang C et al (2003) Sources and fate of organic matter in Khung Krabaen Bay (Thailand) as traced by 13C and C/N atomic ratios. Wetlands 23:729–738

Twilley RR (1985) The exchange of organic carbon in basin mangrove forests in a southwest Florida estuary. Estuar Coast Shelf Sci 20:543–557

Twilley RR, Rivera-Monroy VH, Chen R et al (1999) Adapting an ecological mangrove model to simulate trajectories in restoration ecology. Mar Pollut Bull 37:404–419

Vance DJ, Haywood MDE, Heales DS et al (1998) Seasonal and annual variation in abundance of postlarval and juvenile banana prawns, *Penaeus merguiensis*, and environmental variation in two estuaries in tropical northeastern Australia: a six-year study. Mar Ecol Prog Ser 163:21–36

Vannote RL, Minhall GW, Cummins JW et al (1980) The river continuum concept. Can J Fish Aquat Sci 37:130–137

Wooller M, Smallwood B, Jacobson M, et al (2003) Carbon and nitrogen stable isotopic variation in *Laguncularia racemosa* (L.) (White mangrove) from Florida and Belize: implications for trophic level studies. Hydrobiologia 499:13–23

Xu Y, Jaffé R (2007) Lipid biomarkers in suspended particles from a subtropical estuary: assessment of seasonal changes in sources and transport of organic matter. Mar Environ Res 64: 666–678

Young M, Gonneea ME, Herrera-Silveira J et al (2005) Export of dissolved and particulate carbon and nitrogen from a mangrove-dominated lagoon, Yucatan Peninsula, Mexico. Int J Ecol Environ Sci 31:189–202

Part II
Ecological Linkages

.

Chapter 4
Dynamics of Reef Fish and Decapod Crustacean Spawning Aggregations: Underlying Mechanisms, Habitat Linkages, and Trophic Interactions

Richard S. Nemeth

Abstract Spawning migrations are an important life-history event for many species of commercially important tropical reef fishes and decapod crustaceans. Spawning aggregations are highly predictable events in which hundreds to thousands of individuals migrate across multiple habitats to converge on specific sites for reproduction. Species that undergo spawning migrations provide a potential mechanism to interlink and possibly influence local food webs along their migratory pathways and at aggregation sites. The rapidly declining condition of many aggregating species world-wide emphasizes the urgency with which we need to increase our understanding of how spawning aggregations function within complex coral reef and other tropical ecosystems. This chapter provides a comprehensive review of reef fish and decapod crustacean spawning aggregations, including mechanisms underlying their timing and periodicity, characteristics of spawning aggregation sites, and the spatial and temporal patterns of movement and migration. This overview provides the foundation for a discussion of the habitat linkages and potential trophic interactions that occur during migration and spawning, and highlights the existing gaps in our knowledge of how spawning aggregations function and their importance to ecological processes and fisheries sustainability.

Keywords Fish behavior · Spawning migration · Coral reefs · Predation · Spatial and temporal patterns

4.1 Introduction

The rhythms and movements of tropical reef fishes from home sites to spawning grounds have been known for centuries by native fisherman (Johannes 1978, 1981). Only during the last few decades have scientists become aware of the prevalence of

R.S. Nemeth (✉)
Center for Marine and Environmental Studies, University of the Virgin Islands, MacLean Marine Science Center, 2 John Brewer's Bay, St. Thomas, US Virgin Islands
e-mail: rnemeth@uvi.edu

I. Nagelkerken (ed.), *Ecological Connectivity among Tropical Coastal Ecosystems,*
DOI 10.1007/978-90-481-2406-0_4, © Springer Science+Business Media B.V. 2009

these migrations among coral reef fishes and decapod crustaceans. Most research on spawning aggregations has focused on cataloging where, when, and what species form aggregations (Johannes 1981, Domeier and Colin 1997), understanding the underlying mechanisms of timing and periodicity, site selection, and reproductive behavior of aggregating species (Johannes 1978, Robertson 1991), and conducting population assessments for management and conservation (Sadovy 1994, Levin and Grimes 2002, Nemeth 2005, Sadovy and Domeier 2005). Recently, with the aid of tagging and ultrasonic telemetry, studies have begun to reveal the complex behaviors associated with reef fish and decapod spawning migrations (Herrnkind 1980, Zeller 1998, Carr et al. 2004, Nemeth et al. 2007). The movement and migration patterns associated with spawning aggregations provide an important ecological component of connectivity across tropical habitats including nearshore and offshore coral reefs, mangroves, estuaries, and freshwater systems.

Spawning aggregations are characterized by the movement of hundreds to thousands of herbivorous or predatory reef fishes from large expanses of various habitats to specific spawning locations (Fig. 4.1). Species that undergo spawning migrations provide a potential mechanism to interlink and possibly influence local food webs along their migratory pathways. One can only imagine the effect of these migrations and ephemeral concentrations of fish and decapods on spatial and temporal fluctuations in biomass, transient changes in trophic ecology along migratory pathways and at spawning sites, and the transfer of energy from feeding grounds to spawning grounds through predation and release of gametes. The release of millions of fertilized eggs at spawning aggregation sites further enhances connectivity among complex coral reef ecosystems via larvae dispersal and settlement.

The complex biological processes and behavioral patterns that ensure reproductive success of aggregating species and contribute to the sustainability of local and regional populations are threatened by artisanal and commercial fishermen, who have long relied on spawning aggregations to supplement their annual incomes. Recent modernization of fishing vessels, gear, and technology, lack of regulations or enforcement, and the expansion of the live reef food-fish industry have accelerated the decline and disappearance of many reef fish spawning aggregations around the world (see references in Claydon 2004, Sadovy and Domeier 2005). Fishing pressure also threaten decapod breeding migrations and is exacerbated when migratory routes cross international boundaries (Ye et al. 2006, Hogan et al. 2007). The rapidly declining condition of many aggregating species world-wide (Sadovy de Mitcheson et al. 2008) emphasizes the urgency with which we need to increase our basic understanding of how spawning aggregations function within complex coral reef and other tropical ecosystems. The intention of this chapter is to provide an overview of existing information related to the classification of spawning aggregations, the mechanisms underlying their timing and periodicity, and the general characteristics of spawning aggregation sites. These sections will be followed by an analysis of spatial and temporal patterns of movement and migration associated with spawning aggregations, a discussion of the habitat linkages during adult migration, and the

Fig. 4.1 Pictures of spawning aggregations at Saba Bank, Netherland Antilles: (**a**) *Epinephelus guttatus*; and Grammanik Bank, St. Thomas, US Virgin Islands: (**b**) *Lutjanus jocu*, (**c**) *Mycteroperca tigris* male with white head courting a female, (**d**) *Mycteroperca venenosa* and a few *Epinephelus striatus* aggregating along shelf edge—note various color morphs, (**e**) *Lutjanus cyanopterus* migrating to spawning site. Photos by R Nemeth (a) and E Kadison (b–e)

potential impacts of aggregating species on local food webs. The last section will identify existing gaps in our understanding of how spawning aggregations function, highlight future research directions, and provided steps that can be taken to ensure their sustainability.

4.2 Resident and Transient Spawning Aggregations

The timing, duration, and location of spawning aggregations are important factors contributing to the potential ecological impact of connectivity between coral reefs by spawning migrations of fish and crustaceans. Spawning aggregations of tropical reef fishes occur on daily, lunar, and seasonal time periods, and peak spawning seasons can span from two to eight months (Sadovy 1996). Some species form aggregations frequently but only for a few hours, whereas others form infrequently over longer periods of time (i.e., days to weeks). Mode of reproduction can vary within and between species during spawning aggregations (Domeier and Colin 1997). Pair spawning consists of courtship and spawning by a single male and female within a group or harem (Fig. 4.1c). Group spawning consists of a spawning rush by a single female and two to 15 or more males which are often part of a larger spawning aggregation (Figs. 4.1b, d). Mass spawning occurs when the majority of groups or subgroups within an aggregation spawn simultaneously.

Domeier and Colin (1997) classified reef fish spawning aggregations into resident and transient aggregations based on several specific criteria (Table 4.1). The most significant differences include: (1) the frequency and duration of their spawning aggregations, (2) the proportion of reproductive effort that is allocated during a single aggregation event, (3) the distance and areas from which they migrate and the time required to reach the spawning site, (4) relative body size and trophic level of species, and (5) mating system characteristics. Species that form transient aggregations are typically larger, have greater relative fecundity, and lower instantaneous mortality rates (Thresher 1984, Sadovy 1996). Because larger species tend to have longer life spans and reproduce later in life than smaller species, the benefits of delaying reproduction to optimal times each year outweigh the costs of potentially dying before the next spawning period (Petersen and Warner 2002). Larger species may be more capable of migrating the distances required to reach distant aggregation sites while being less vulnerable to predation during migration (Thresher 1984). Two examples illustrate this point. Within the Caribbean epinephelid groupers, the smallest two species (<30 cm max. length: *Cephalopholis fulvus* and *C. cruentatus*) do not form aggregations, whereas the larger species (55–200 cm max. length: *Epinephelus guttatus, E. adscensionis, E. striatus*, and *E. itajara*) do form aggregations (Sadovy et al. 1994). Within the Indo-Pacific Acanthuridae, the smaller *Acanthurus* spp. form resident aggregations, whereas the larger *Naso* spp. form transient aggregations (Rhodes 2003 as cited in SCRFA 2004). However, the energetic costs associated with migration may prevent extensive migration of large herbivorous fishes (Thresher 1984).

Regardless of their resident or transient classification, the ability of spawning adults to synchronize timing of gamete release during narrow windows of opportunity is most strongly tied to the daily and monthly lunar orbit which predictably influences nighttime illumination and tidal and current strength at specific locations. Thus, while the differences between resident and transient aggregations can be quite pronounced, they also share a number of similarities: (1) both display strong fidelity to traditional spawning sites, (2) spawning occurs near steep drop-offs, over

Table 4.1 Characteristics of resident and transient spawning aggregations (modified from Domeier and Colin 1997). Functional migration area indicates the area in which species migrate to spawning aggregation sites and may interact with local food webs

Aggregation characteristics	Resident	Transient
Frequency of occurrence	Frequent and regular, often daily, occasionally monthly	Infrequent, annual peaks during specific times of year
Duration of spawning event	Hours (1–5 hrs)	Days (~2–10 d)
Portion of reproductive effort of single aggregation	Represents 0.25% (daily) to 8% (monthly) of annual reproductive effort	Represents 33% (spawns over three consecutive lunar cycles) to 100% (spawns during one lunar cycle) of annual reproductive effort
Migration distance	Within or nearby home range (<2 km)	Well outside home range (~2 to >100 km)
Functional migration area	Small (<10 km^2)	Large (<10–>500 km^2)
Time required to reach aggregation	Minutes to hours	Days to weeks
Size of aggregating species	Small to medium (~5–50 cm)	Medium to large (~30–>100 cm)
Trophic level of species	Herbivorous, omnivorous	Carnivorous, piscivorous
Potential impact on trophic ecology at spawning site	Low?	High?
Mating system within aggregation	Group and mass spawn	Pair, haremic, group, and mass spawn
Location of spawning	Known to spawn outside aggregation	Not known to spawn outside aggregation*
Representative families known to form spawning aggregations	Acanthuridae, Caesionidae, Carangidae, Labridae, Scaridae	Balistidae, Lethrinidae, Lutjanidae, Mugilidae, Mullidae, Serranidae, Siganidae, Sparidae

* although see Krajewski and Bonaldo (2005)

prominent reef structures or at the mouth of channels, (3) both occur at specific, predictable times (daily, monthly, or annually), and (4) size of aggregations can range from tens to thousands of individuals (Domeier and Colin 1997). Based on the criteria described in Table 4.1, decapod crustaceans can be classified as transient aggregations. They share many features listed above with reef fishes with two exceptions. Spawning aggregations of some decapod crustaceans are composed of only gravid females which undergo seasonal migrations to specific sites to release larvae (Herrnkind 1980, Tankersley et al. 1998, Carr et al. 2004) while others seem to undertake an ontogenetic migration which terminates at the spawning grounds (Ruello 1975, Bell et al. 1987).

At least twenty-one families (over 120 species) of tropical reef fishes are known to form resident or transient aggregations for reproduction. These families (# spp.) are: Acanthuridae (11), Balistidae (1), Caesionidae (1), Carangidae (7),

Carcharhinidae (1), Centropomidae (1), Gerreidae (2), Kyphosidae (3), Labridae (6), Lethrinidae (7), Lutjanidae (14), Mugilidae (6), Mullidae (3), Pangasiidae (1), Rhincodontidae (1), Scaridae (8), Scombridae (4), Serranidae (36), Siganidae (8), Sparidae (2), and Sphyraenidae (2) (Appendices 4.1, 4.2; see also SCRFA 2004). Another eight families (Albulidae, Belonidae, Chanidae, Clupeidae, Elopidae, Haemulidae, Holocentridae, Priacanthidae) have been reported to aggregate but spawning has not been confirmed. Families of coral reef fishes that do not form spawning aggregations or little is known about their reproductive behavior include, but are not limited to: Apogonidae, Antennariidae, Aulostomidae, Blennioidei, Brotulidae, Callionymidae, Carapidae, Chaetodontidae, Cheilodactylidae, Cirrhitidae, Diodontidae, Fistulatidae, Gobioidei, Grammistidae, Malacanthidae, Mugiloididae, Opistognathidae, Ostraciidae, Pempheridae, Plotosidae, Pomacentridae, Pteroidae, Psuedochromidae, Sciaenidae, Scorpaenidae, Synodontidae, Tetraodontidae, and Zanclidae (see Thresher 1984).

4.3 Underlying Mechanisms of Spawning Aggregations

Most coral reef fishes are relatively site-attached and exist as spatially divided subpopulations in a patchy environment (Mapstone and Fowler 1988, Sale 1991). Successful reproduction, therefore, requires an individual to either attract a suitable mate near its home range, search for a mate or mates within the larger habitat patch, or migrate considerable distances between habitats to spawn. For the latter two alternatives, the proximate and ultimate factors which lead to successful reproduction, will act most strongly on synchronizing the timing and location of spawning events. In evolutionary terms the optimum time and location for spawning must incorporate the relative costs and benefits of current versus future reproductive success (Helfman et al. 1997). The complex associations of the earth-moon orbit around the sun bring rhythmic environmental changes (i.e., photoperiod, temperature, tidal cycles, lunar light) that are used by fishes to synchronize reproductive activities (Takemura et al. 2004). The proximate cues for determining timing of reproduction can be separated into four factors including predictive, synchronizing, and terminating cues, and environmental modifying factors (Munro 1990). These factors, which are discussed in more detail below, are useful in understanding how fishes and other organisms synchronize gonad development, timing of migration, and spawning to ensure maximum reproductive success.

Predictive cues are periodic environmental events, such as changing day length, water temperature, or tidal strength, used to predict the approaching spawning period and initiate migration. For species that spawn daily during large portions of the year (i.e., resident aggregations), the predictive cues initiating spawning migrations are primarily time of sunrise or sunset, tidal cycle, or a combination of these cues (Robertson 1991, Mazeroll and Montgomery 1998). Small pelagic spawners like the bluehead wrasse *Thalassoma bifasciatum* follow the tidal cycle and spawn just after high tide facilitating dispersal of eggs off the reef with the outgoing tide (Warner 1988). Diel spawning is delayed about one hour each day to synchronize with daily tidal cycle, and typically occurs in the afternoon (Warner and Robertson 1978). The importance of keeping gametes off the reef during daytime hours was illustrated by Hamner et al. (1988), whose detailed study calculated that diel feeding by planktivorous fishes can remove most of the zooplankton from the water column near the reef face (see also Motro et al. 2005). Other species, such as *Sparisoma rubripinne*, begin to aggregate in the late afternoon then spawn during dusk every day (Randall and Randall 1963). Spawning at dusk is a common feature of spawning aggregations as well as many other non-aggregating species, and may reduce predation on eggs by planktonic feeders or predation on adults by piscivores (Thresher 1984, Sancho et al. 2000a).

The predictive cues initiating timing of migration for transient aggregations are most likely operating at large spatial scales via changes in seasonal patterns such as day length, water temperature, and current speed (Nemeth et al. 2006b). Moore and MacFarlane (1984) reported that the initiation of mass migration of rock lobster (*Panulirus ornatus*) was extremely regular over six years and corresponded to when water temperatures were at a minimum. Fish families as diverse as Sparidae from the Pacific to Caribbean Serranidae seem to aggregate and spawn during the months when annual seawater temperatures were lowest (Carter et al. 1994, Sheaves 2006, Nemeth et al. 2007). In the US Virgin Islands spawning of red hind (*Epinephelus guttatus*) occurs when both water temperature and current speed decline rapidly during specific lunar periods from December through February (Nemeth et al. 2007). Minimal currents during the spawning season may enhance fertilization success for group or mass spawning species (Kiflawi et al. 1998, Petersen et al. 2001) or reduce offshore dispersal so that larvae can complete their pelagic phase without being carried away from suitable juvenile habitats (Johannes 1978). Predictive cues which initiate migration for tropical organisms other than reef fish include seasonal changes in tidal amplitude or weather patterns. For example, spawning migrations of barramundi (*Lates calcarifer*) from freshwater rivers to coastal areas typically occur during spring high tides (Moore and Reynolds 1982). The tropical anadromous catfish *Pangasius krempfi* migrate in May and June each year at the beginning of the wet season when the Mekong River flow rate begins to increase (Hogan et al. 2007). Migrations by adult spiny lobster (*Panulirus argus*) to offshore reefs are triggered by increased currents caused by the first autumnal storms (Kanciruk and Herrnkind 1976, 1978).

Synchronizing cues act to ensure that adults are in the same state of reproductive readiness to optimize fertilization of eggs, and can operate at both long and short time scales. Synchronizing cues acting at large temporal scales (i.e., months) are important for transient aggregations, but also for resident aggregations which have seasonal spawning periods. Two examples illustrate the variability among species. Gonad development and spawning in Nassau grouper (*Epinephelus striatus*) in Belize correlated with seasonal changes in photoperiods and water temperatures (Carter et al. 1994). Alternatively, Fishelson et al. (1987) found that spawning of *Acanthurus nigrofuscus,* which forms resident aggregations in the Red Sea from May through September, was related to a seasonally available food source. Although factors such as temperature and day length were not considered,

Fishelson et al. (1987) reported that condition factor and subsequent gonad development of *A. nigrofuscus* were positively correlated with a dietary shift from red and brown turf algae in summer months to green fleshy algae, which becomes very abundant from November to April. Synchronizing cues operating at shorter time scales (i.e., hours, days, weeks) may include a specific lunar cycle, time of sunset (i.e., light levels), presence of mates displaying breeding coloration or courtship behavior (Figs. 4.1c, d), production of specific sounds or pheromones, or presence of appropriate spawning habitat or substrate. For decapod crustaceans, females can only be inseminated immediately after molting while the exoskeleton is soft (Quackenbush and Herrnkind 1981). The ideal spawning temperature for spiny lobster (*Panulirus argus*) is around 24 °C (Lyons et al. 1981), therefore spawning season and duration varies with latitude. Synchronizing cues for both resident and transient aggregations are most likely linked to current or tidal regimes, ambient light levels, courtship behaviors of males, and presence of females within the aggregation. The relative importance of tides versus ambient light levels for synchronizing spawning may vary between the Caribbean, which has low tidal amplitudes (<1 m), and the tropical Pacific, where tidal amplitude is much greater (P Colin, Coral Reef Research Foundation, Palau, pers. comm.). Most resident aggregations begin spawning each day in mid afternoon whereas transient aggregations, especially groupers, begin spawning activity at sunset during the week of spawning. The most common lunar period for spawning is around the full or new moon. Most genera of tropical Pacific and Caribbean Serranidae and Lutjanidae spawn around the full moon, although some exceptions due occur (reviewed by Johannes 1978, 1981, Domeier and Colin 1997). Based on interviews with local fishermen, Johannes (1978, 1981) reported that Lethrinidae and Siganidae typically spawn during the new moon. Time of sunset and changes in fish behavior, density, and coloration within the aggregation also act as brief, yet important, synchronizing cues (Johannes 1978, Sale 1980, Garcia-Cagide et al. 2001). Many grouper species, for example, display distinct gender-specific breeding coloration (Thresher 1984, Sadovy et al. 1994a, Domeier and Colin 1997) and some, such as the Nassau grouper, may also require a certain density or number of fish within the aggregation to stimulate courtship and subsequent group spawning (Colin 1992).

Because spawning conditions remain optimal for only a short time, terminating cues mark the end of the spawning period and may include changes in environmental conditions, depletion or viability of gametes, and/or departure or changes in behavior of conspecifics. Terminating cues for resident aggregations which spawn in the late afternoon are most likely linked to ambient light levels (Randall and Randall 1963) whereas important cues for day-spawning species may include current speed and direction (Sancho et al. 2000b). Sancho et al. (2000b) found that short-term current reversals temporarily interrupted courting and spawning behavior of *Chlorurus sordidus* (Scaridae), which spawned during the day, but not of *Zebrasoma flavescens* or *Ctenochaetus strigusus* which spawned in late afternoon or at dusk. Even though transient aggregations can occupy a spawning site for several weeks, termination of spawning and departure of fish can be quite abrupt (Fig. 4.2) and may be controlled by a number of factors including depletion of gravid females, or changes in current

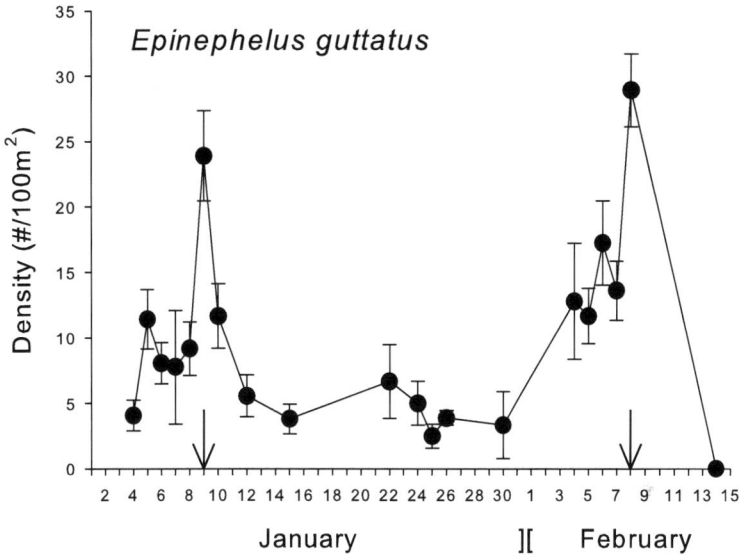

Fig. 4.2 Density (± S.E.) of red hind during 2001 spawning season in St. Thomas, US Virgin Islands. Arrows show timing of full moon in January and February (modified from Nemeth 2005, with permission from Inter-Research)

speed, temperature, and ambient lighting associated with the lunar cycle (Thresher 1984, Domeier and Colin 1997, Heyman et al. 2005, Nemeth et al. 2007, Starr et al. 2007).

Finally, environmental or biological modifying factors can cause intraspecific variation in spawning behavior or timing at different latitudes or in different habitats. For example, *Epinephelus guttatus* and *E. striatus* have a spawning temperature range between 25 and 26.5 °C. At lower latitudes in the Caribbean (e.g., Bahamas, Belize, Puerto Rico, Virgin Islands) these two species form aggregations in winter when water temperature cools to 26.5 °C (Carter 1987, Colin et al. 1987, Colin 1992, Carter et al. 1994, Nemeth et al. 2007), whereas at higher latitudes (e.g., Bermuda) aggregations form from May to July when water temperature warms to 25 °C (Burnett-Herkes 1975, Luckhurst 1998). This suggests that water temperature is a potential modifying factor which causes variation in timing of reproduction of *Epinephelus* species at different latitudes within the greater Caribbean. Likewise for decapods, seasonal temperatures and temperature changes regulate the time of spawning of spiny lobster (*Panulirus argus*) which spawns year-round near the equator, from spring through fall in the Bahamas, and is restricted to April–June in Florida (Kanciruk and Herrnkind 1976, Quackenbush and Herrnkind 1981, Marx and Herrnkind 1986). In the southern hemisphere *Plectropomus leopardus* spawns from August to December when water temperatures increase above 24 °C (Samoilys and Squire 1994, Samoilys 1997). *Acanthurus nigrofuscus* in the Red Sea also spawns during months when seawater temperatures range from 24 to 26 °C (Fishelson et al. 1987). The narrow temperature range of spawning may be an important cue for synchronizing reproduction in many aggregating species, but the physiological limitations of higher water temperatures may also inhibit vitellogenesis in certain species, preventing ovulation and halting egg cell development (Lam 1983). Finally, for some partially catadromous species (i.e., *Latus calcarifer*) high rainfall may stimulate and therefore regulate timing of spawning migrations from rivers to coastal spawning grounds (Milton and Chenery 2005).

The adaptive advantage of a breeding population migrating to a specific location, during specific time periods may result from multiple selective pressures affecting larval survival and/or adult survival and reproductive success. Most hypotheses have attempted to explain why reef fishes form spawning aggregations, and predict the timing of spawning and location of aggregation sites (Claydon 2004). Reef fishes and decapod crustaceans may form spawning aggregations to reduce predation on eggs and spawning adults through predator satiation, or spawning at times and locations when predation is minimal. Spawning aggregations may also facilitate mate selectivity in pair and haremic spawners, and allow assessment of breeding population sex ratios for hermaphroditic species (Shapiro et al. 1993). Finally, the location and timing of spawning aggregations may enhance larval survival and subsequent recruitment to suitable juvenile habitats. Claydon (2004) provided a thorough review of these hypotheses, but a few specific examples are provided below.

Thresher (1984) noted that most pelagic spawning fishes that have a lunar spawning cycle also migrate to specific spawning grounds. He suggested that certain selective pressures are acting equally strong on lunar periodicity and migratory behavior. These selective pressures function to situate eggs and larvae into presumably optimal oceanographic conditions which may either facilitate offshore dispersal, increase survival by minimizing predation of eggs, enhance growth by coupling larval hatching with availability of food supply, and/or increase retention and subsequent return to their natal reef by placing eggs and larvae into optimal currents (Cushing 1971, Johannes 1978, Robertson 1991). Moreover, the factors that affect spawning site selection of aggregating species are important for understanding the dispersal of eggs and larvae, subsequent recruitment patterns, and degree of connectivity among reef systems (Petersen and Warner 2002). For example, the mass spawning migration of the rock lobster (*Panulirus ornatus*) over 400–500 km from Torres Strait across the Gulf of Papua enables females to release larvae in oceanic currents that are favorable to recruitment back to the Torres Strait and the northeast coast of Queensland, Australia (Moore and MacFarlane 1984). However, *P. ornatus* seems to expend considerable energy on its migration, and most adults are in poor condition and probably die after spawning (Moore and MacFarlane 1984).

Potential selective mechanisms acting on the adult breeding population, instead of the larval phase of life, may also favor the formation of spawning aggregations at specific times and locations. These selective mechanisms may synchronize reproduction, optimize spawning conditions, facilitate mate choice, and minimize predation of adults (Robertson 1991, Claydon 2004). For example, the influence of the lunar cycle may help synchronize formation of spawning aggregations to increase reproductive success in species that occur in low densities or produce specific oceanographic conditions that optimize fertilization success (Robertson

1991, Petersen et al. 2001). Large aggregations may reduce adult mortality by temporarily overwhelming predators, whereas specific aggregation sites may provide greater protection from predation due to physical features of the reef (Shapiro et al. 1988). Of the few studies that have empirical data on predation rates, most have found differential mortality among males and females at spawning aggregation sites. Sancho et al. (2000a) calculated annual mortality rates of fishes at spawning aggregation sites could be as low as 1% for females, or as high as 18% for group-spawning males, which spawn multiple times per day. High mortality among males at spawning sites was also demonstrated for group-spawning parrotfishes in Panama (Clifton and Robertson 1993). In the case of estuarine blue crabs (*Callinectes sapidus*) and Australian mud crabs (*Scylla serrata*), larvae cannot survive in adult habitats (i.e., brackish estuaries), therefore adults must release larvae in higher salinity coastal areas outside the estuary (Tankersley et al. 1998). Based on these decapod studies and work by Herrnkind (1980) on the spiny lobster (*Panulirus argus*), the males seem to remain within feeding grounds while gravid females undergo spawning migrations to release larvae and, thus, may be exposed to greater risk of predation than males.

Ultimately, the benefits, in terms of enhanced reproductive success, of migrating 100's of meters to 100's of kilometers from feeding grounds to spawning aggregation sites must outweigh the associated costs (i.e., less foraging time, greater energy use, exposure to predators). Our knowledge of spawning aggregations is steadily increasing, but very little quantitative research has been conducted to systematically test these various hypotheses (Claydon 2004). Moreover, many features of spawning aggregations are interrelated and therefore their relative importance are not easily assessed (Claydon 2004). The suggestion that spawning aggregation sites may simply be ancient traditional locations first established when sea levels were significantly lower during the last ice age (Colin and Clavijo 1988), may further decrease our ability to construct predictive models of why, when, and where spawning aggregations occur. While no single hypothesis can explain the spawning patterns of all migrating species, those individuals that spawn at times and locations that enhance egg, larval, and adult survival will be more successful than those that spawn at nonoptimal times and locations.

4.4 Characteristics of Spawning Aggregation Sites

Tropical reef fishes that form resident and transient spawning aggregations typically migrate from inshore areas to offshore sites near steep drop-offs along the shelf break (Fig. 4.1d), at the mouth of tidal channels, over prominent reef structures, or sites with a combination of these reef features (Johannes 1978, Colin and Clavijo 1988, Claydon 2004). A handful of species are also known to migrate from rivers or estuaries to coastal marine habitats (i.e., Centropomidae, Mugilidae, and Sparidae), from offshore reefs to shallow lagoons or seagrass beds (i.e., Carcharhinidae, Rhincodontidae, Siganidae), or from coastal areas to inland rivers (i.e., Pangasiidae) for reproduction. Decapod spawning sites, for which there are few examples, seem to be located within 5 km of the mouth of estuaries or in deeper waters bordering oceanic currents (Kanciruk and Herrnkind 1978, Carr et al. 2004).

Resident spawning aggregations on nearshore reefs are usually located on the downcurrent end of a reef or patch reef often over individual coral heads or, as in the case of *Sparisoma rubripinne,* on the most seaward projection from a fringing reef (Randall and Randall 1963, Warner 1990b, Colin 1996). The best example of factors influencing spawning site selection in small pelagic spawning fishes comes from work on the bluehead wrasse *Thalassoma bifasciatum.* Warner (1988, 1990a, b) showed that female choice of spawning sites was based on specific physical features including location, structure, and oceanographic conditions. These sites were often located on the tallest promontories at the downcurrent end of a reef and may offer better protection from predation and dispersal of eggs off the reef, respectively (Robertson and Hoffman 1977, Warner 1988, Appeldoorn et al. 1994, Hensley et al. 1994). Long-term use of these spawning sites was maintained through cultural transmission by young females following older females. If a mating site was destroyed, by a storm for example, spawning activity shifted to the nearest potential site (Warner 1990b). The longevity of a spawning aggregation site is unpredictable. While some aggregations, such as Scaridae and Serranidae, have been known to use the same site for decades, others have been observed once or twice, then never seen again even after repeated searches (Colin 1996, Sadovy 1997, Eklund et al. 2000). However, when spawning is not verified, an observed aggregation may simply be a temporary staging area for a migrating group of fish *en route* to a spawning site.

Resident and transient aggregations which occur along the shelf break are often located at distinctive promontories (Carter et al. 1994, Garcia-Cagide et al. 2001). Spawning in transient aggregations often occurs on the upcurrent end of the reef or island (Colin et al. 1987, Luckhurst 1998, Nemeth et al. 2007), which is counterintuitive for offshore dispersal but may facilitate retention of larvae. These sites usually consist of well-developed deep coral reefs or contain many ledges, undercuts, or caves which are used for shelter by aggregating species (Carter et al. 1994, Sancho et al. 2000a, Nemeth 2005, Nemeth et al. 2006a, Kadison et al. 2009). Robertson and Hoffman (1977) and Sancho et al. (2000a) suggested pelagic spawners favor topographically complex sites (Fig. 4.1c—note reef structure) that offer a greater degree of protection from predation and allow them to seek shelter fairly rapidly. In Saba, Netherlands Antilles, *Epinephelus guttatus* aggregations are dispersed over several kilometers of old spur and groove habitat with areas of high fish density (Fig. 4.1a) associated with undercuts and higher coral cover (Kadison et al. 2009). Most spawning aggregation sites are located in water deeper than a species area of residence, but some exceptions do occur (Sadovy 1996). The reef-associated mullet spawn in shallow sandy areas nearshore (Helfrich and Allen 1975). The nurse shark *Ginglymostoma cirratum* and lemon shark *Negaprion brevirostris* both migrate to shallow seagrass lagoons for reproduction (Pratt and Carrier 2001, Feldheim et al. 2002). These shallow lagoons facilitate copulation and serve as juvenile nursery habitats (Feldheim et al. 2002). One of the few example of an anadromous tropical species is the Asian catfish *Pangasius krempfi* which migrates from coastal waters and estuaries to the upper reaches of the Mekong River in Laos (Hogan et al. 2007).

Within a location, there is often considerable overlap in use of the same areas for spawning (Johannes 1978, Moyer 1989, Colin and Bell 1991, Domeier and Colin 1997, Sancho et al. 2000a) although timing and mode of spawning may differ among species. Colin and Clavijo (1988) described a multi-species spawning aggregation site along the south coast of Puerto Rico which hosted a number of resident aggregating species, including *Scarus iserti, Clepticus parrai, Lachnolaimus maximus, Acanthurus coeruleus*, and *A. bahianus*. The scarid and labrid species spawned in different locations along the reef whereas the two acanthurids both spawned in the same location and during the same season. The only difference in these two species was that *A. coeruleus* showed a distinct lunar periodicity whereas *A. bahianus* did not (Colin and Clavijo 1978).

Species that form transient aggregations are also frequently observed using the same spawning site. The most common suite of aggregating groupers in the Caribbean include *Epinephelus striatus, Mycteroperca tigris*, and *M. venenosa* (Sadovy et al. 1994a, Nemeth et al. 2006b) and in the Pacific, *E. polyphekedion, E. fuscoguttatus*, and *Plectropomus areolatus* (Johannes et al. 1999, Rhodes and Sadovy 2002). Three species of snappers, *Lutjanus cyanopterus, L. apodus*, and *L. jocu* often use the same sites and may also co-occur with groupers and other species (Heyman et al. 2001, Heyman et al. 2005, Kadison et al. 2006). For instance, the Grammanik Bank, located on the shelf edge south of St. Thomas, US Virgin Islands, is a multi-species spawning aggregation site utilized by three species of groupers (*E. striatus, M. venenosa, M. tigris*) and three species of snappers (*L. cyanopterus, L. apodus, L. jocu*) over an eight month period (Kadison et al. 2006, Nemeth et al. 2006b). At this site *M. tigris* is a haremic spawner (Sadovy et al. 1994a) while *M. venenosa* and the three Lutjanidae are all group spawners (Fig. 4.1). *E. striatus*, whose spawning population is about 200 individuals here, has been observed showing courtship behavior within pairs and small groups but has not yet been observed spawning or attempting to form a large aggregation (Nemeth et al. 2006b). Spawning strategy in *E. striatus* may be a facultative response to population density (Colin 1992, Sadovy and Colin 1995, Sadovy 1996). Based on gonad histological evidence, Sadovy and Colin (1995) suggested that *E. striatus* may also pair spawn outside of aggregations as an alternative reproductive strategy. At the Red Hind Bank, another multi-species aggregation site 5 km to the west of the Grammanik Bank and about 300 m inshore of the shelf edge, *E. guttatus, M. tigris, L. analis*, and *L. apodus* all have been observed aggregating in very large numbers during winter and spring months (RS Nemeth pers. observ.). *Balistes vetula*, the queen triggerfish, has been observed to form spawning aggregations near *E. guttatus* spawning sites located in St. Croix and Saba (RS Nemeth pers. observ.). Spawning aggregations of *E. guttatus* and *B. vetula* occur during the same months (December–February), but *E. guttatus* spawns the week before the full moon while *B. vetula* spawns the week after the full moon. Moreover, male *E. guttatus*, a pelagic spawner, maintain territories on the top of the reef in areas of high coral cover and topographic complexity. *B. vetula*, a benthic spawner, forms aggregations just off the reef in sandy areas where male and female pairs dig shallow depressions for adhesive eggs, and together defend the nest. The triggerfish *Pseudobalistes flavimarginatus* also has a lek-like spawning behavior and displays biparental nest defense (Gladstone 1994). Nest defense is further facilitated through the combined efforts of all breeding pairs within the spawning aggregation site (Thresher 1984).

4.5 Spatial and Temporal Patterns of Connectivity

4.5.1 Habitat Linkages

The movements and migration patterns associated with spawning aggregations are an important aspect of connectivity among tropical marine habitats. Species representing both aggregation types shift from myriad feeding habitats to converge on specific spawning sites. In general most spawning migrations move from inshore reefs to offshore spawning grounds (Fig. 4.3, Appendices 4.1, 4.2). The majority (>70%) of resident and transient species migrate to spawn in reef pass channels or along shelf edge reefs with an additional 10% spawning on the seaward extension of fringing or mid-shelf reefs (Nemeth 2009). These movement patterns not only provide important connections between inshore and offshore reefs but also ensure connectivity between shallow and deep habitats. For example, *Epinephelus striatus* in Belize, which usually occupy shallow reefs from 10 to 25 m depth, were found to spend over half of their four month spawning season at depths ranging from 60 to 250 m (Starr et al. 2007). Studies within estuarine systems have found several species of Sparidae migrating from mangrove creeks out to channel mouths to spawn (Sheaves et al. 1999). In Papua New Guinea, barramundi (*Latus calcarifer*) adults live in rivers, estuaries, and marine environments and migrate to spawning sites along the coast (Moore and Reynolds 1982, Milton and Chenery 2005). The anadromous tropical catfish (*Pangasius krempfi*) is believed to migrate from coastal areas of the South China Sea up the Mekong River system (Hogan et al. 2007). Although undocumented, local fishermen in Palau and other Pacific islands claim that species living in lagoon or reef flat habitats tend to use predictable pathways during their seaward migration (references in Johannes 1981). Less research has been conducted on movements of tropical decapod crustaceans to spawning aggregation sites. However, there seem to be three general patterns that include species in which inseminated females migrate either from inland brackish estuaries or mangrove swamps to more saline areas near the mouth of an estuary (i.e., *Callinectes sapidus*), or from shallow coral reef and patch reef habitats to deeper reefs (i.e., *Panulirus argus*) to release larvae, and species in which both sexes undergo an ontogenetic migration during which time they continue to grow, mature, and eventually reproduce at specific spawning sites (i.e., *P. ornatus, Penaeus plebejus*) (Ruello 1975, Herrnkind 1980, Bell et al. 1987, Carr et al. 2004).

Comparing the distance and frequency of migration for aggregating species is important for documenting and understanding the extent and degree of connectivity among tropical reef environments. Resident aggregations move frequently across the reefscape often migrating up to two kilometers daily along the edge of a fringing reef or the shelf break to reach their spawning site (Robertson 1983, Colin and

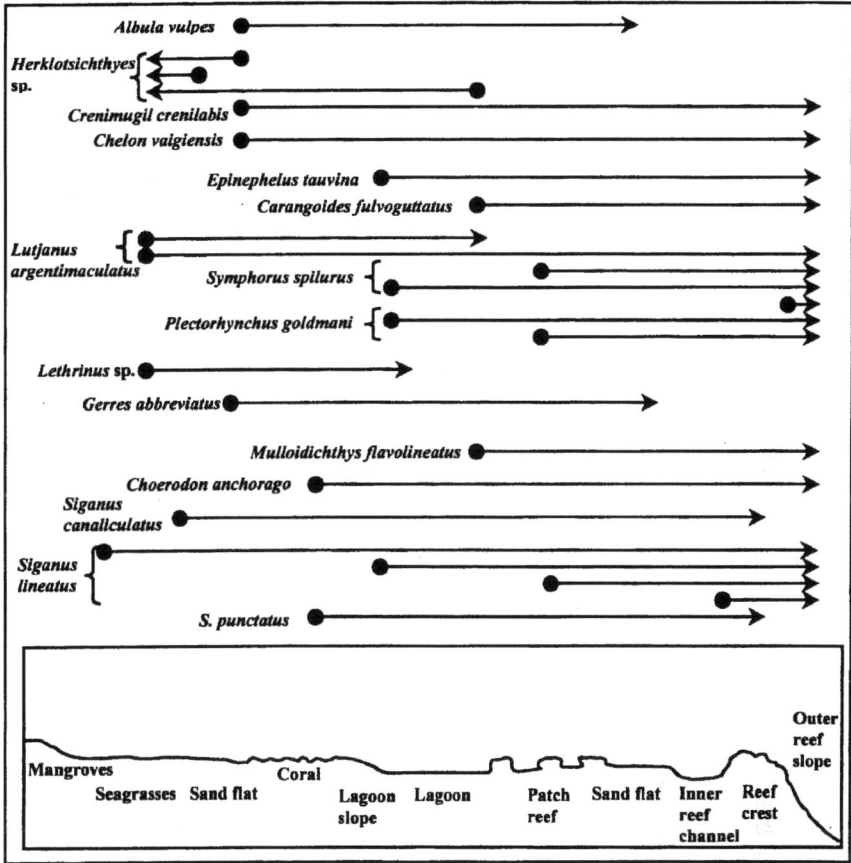

Fig. 4.3 Assumed spawning migrations of fish in Palau, Micronesia, from data collected through underwater observations and fishermen interviews (Johannes 1978). Arrows show the movement from usual habitat (•) to spawning sites (→). Actual distances traveled were not provided by Johannes (1978). Almost all species undertake a seaward spawning migration, with more than half of all species moving to the outer reef to spawn (reprinted from Pittman and McAlpine 2003, with permission from Elsevier)

Clavijo 1988, Colin 1996, Mazeroll and Montgomery 1998). Only a few species have detailed information related to spawning migrations. Fishelson et al. (1987) and Mazeroll and Montgomery (1998) reported *Acanthurus nigrofuscus* in the Red Sea migrated 500–1500 m from its shallow (2–3 m) feeding grounds along the shoreline to deeper spawning sites (10 m) at the edge of the fringing reef during the 5 month spawning season. Colin and Clavijo (1988) observed *A. bahianus* and *A. coeruleus* migrating 0.5–0.6 and 0.9–1.0 km, respectively, from inshore reef areas to deeper fore reef spawning sites. Colin (1996) reported on another *A. coeruleus* aggregation in the Bahamas, which migrated daily from May to October from inshore rocky reef to open rocky areas 20 m deep but that were >1 km from the shelf edge. *Acanthopagrus berda* and *A. australis* (Sparidae) which live in tropical estuaries, migrated 3 and 80 km, respectively, to spawning sites (Pollock 1984, Sheaves et al. 1999). Even the smallest migrating species, *Thalassoma bifasciatum*, has been known to travel as much as 1.5 km from its feeding grounds at the upcurrent end of a reef to its spawning site at the downcurrent end (Fitch and Shapiro 1990, Warner 1995).

Transient aggregations typically undergo annual migrations that can cover tens to hundreds of kilometers (Colin 1992, Carter et al. 1994). Milton and Chenery (2005) found that barramundi can migrate 15–300 km, but those freshwater populations, which must migrate the longest distance, may only spawn once or twice in their lifetime whereas coastal populations spawn annually. The longest migration distance recorded for a coral reef fish was for *Epinephelus striatus* which have been recaptured 100–240 km from where they were tagged (Colin 1992, Carter et al. 1994, Bolden 2000). These long-distance migrations by *E. striatus* (60–100 cm total length) occurred off the coast of Belize and in the Bahamas, both locations which have extensive areas of shelf. Red grouper (*E. morio*), which are smaller than *E. striatus*, have been tracked from 29 to 72 km (Moe 1969 in Sadovy 1994). Limitations to migration distances may simply be a function of insular shelf area and fish size. The depth of deep-water barriers may inhibit migration between islands and therefore restrict movements of migrating species to insular shelf areas. For example, *E. striatus* is not known to cross the >1,800 m deep channels separating the three Cayman Islands during spawning migrations and therefore may be restricted to maximum migration distances of 50 km on Grand Cayman, and 15 km on Little Cayman and Cayman Brac (Colin et al. 1987). However, in southern Puerto Rico a 27 cm *E. guttatus*, which was tagged at the spawning site, migrated over 18 km and crossed a 194 m deep channel between the main shelf and a seamount (Sadovy et al. 1994b). Nemeth et al. (2007) found that the migration distance of *E. guttatus* in St. Croix (mean: 9.4 km, range: 2–16 km) was significantly less than on St. Thomas (mean: 16.6 km, range: 6–33 km). The smaller St. Croix shelf (about 650 km^2) relative to the Puerto Rican shelf (about 18,000 km^2) on which St. Thomas is located, and the smaller size of *E. guttatus* in St. Croix vs. St. Thomas (mean: 32.5 vs. 38.5 cm total length), may have accounted for these differences in migration distance. However, within each island the only significant relationship between fish length and migration distance was opposite from what one would expect. On St. Croix the largest males remained on offshore reefs within 5 km of the spawning aggregation site, whereas the smallest males and females migrated to inshore reefs 10–15 km away (Nemeth et al. 2007). Little is known about the extent of spawning migrations in other commercial species. However, in April 2007, one *Lutjanus jocu* was acoustically tagged near its spawning aggregation site at the Grammanik Bank, St. Thomas, and migrated westward at least 18 km to beyond the limits of the last acoustic receiver (RS Nemeth unpubl. data). Identifying the natural boundaries and limitations to fish spawning migrations will improve our understanding of metapopulation dynamics and management of spawning aggregations.

Several species of decapod crustaceans can also migrate considerable distances from home sites to spawning sites. The record for the longest migration for an adult crustacean was 930 km (in 260 days) for the king prawn *Penaeus plebejus* of eastern Australia (Ruello 1975). Juvenile *P. plebejus* leave their estuary nursery grounds and migrate north to warmer waters to spawn (Ruello 1975, Glaister et al. 1987). Larvae are then carried from spawning grounds along the East Australian Current system back to estuaries in Southeast Australia. Elsewhere in Australia, female mud crabs (*Scylla serrata*) migrate up to 95 km from mangrove areas to offshore spawning sites just before the monsoon season (Hill 1994). Atlantic blue crab *Callinectes sapidus* migrate from low salinity estuarine habitats to high salinity coastal areas (up to 5 km offshore) to release larvae (Carr et al. 2005). Female blue crabs use ebb tide transport (Fig. 4.4) and can travel up to 10 km per day and cover distances from 10 to over 300 km (Millikin and Williams 1984, Tankersley et al. 1998, Carr et al. 2004). Although most research on blue crabs has been done along the Atlantic and Gulf coasts of the USA, similar migrations may also occur in the Caribbean. Gravid female spiny lobsters (*Panulirus argus*) migrate from shallow inshore areas to deep reefs bordering oceanic currents to release larvae (Fig. 4.5), which eventually settle onto mangrove roots or benthic algae as juveniles, and then begin to migrate into deeper water to complete the life cycle (Herrnkind 1980). In the Bahamas, *P. argus* males and females also undergo annual inshore to offshore migrations each autumn (Kanciruk and Herrnkind 1976). During these migrations, *P. argus* form single file lines and may walk over 10 km within a week (Herrnkind 1980), but these migrations are probably not related to spawning, but instead used to reach overwintering grounds in sheltered deep reef habitats. Adult rock lobsters *P. ornatus* transition across multiple habitats, beginning their mass migration in September from an area of extensive low-relief coral reef <20 m depth, south of

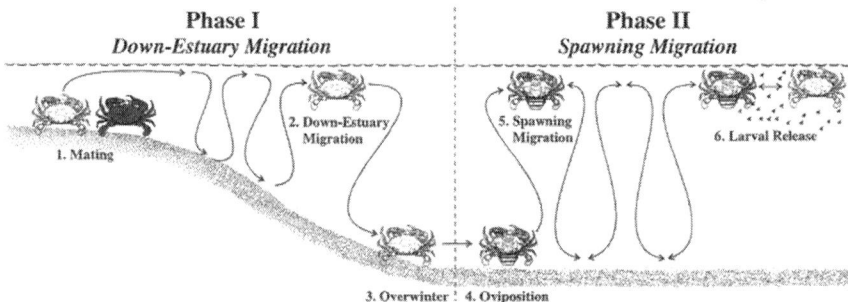

Fig. 4.4 Proposed migratory behavior of spawning female blue crabs (*Callinectes sapidus*). Females (light crabs) mate with males (dark crabs) in brackish estuaries from late spring to early fall, and then migrate during nocturnal ebb tides to high salinity areas where they may overwinter (Phase I). Female crabs with late-stage eggs continue to migrate on nocturnal ebb tides until they reach the mouth of the estuary, or enter coastal waters where larvae are released. Following larval release, female crabs re-enter the estuary during nocturnal flood tides (Figure 2 from Tankersley et al. 1998. Biol. Bull. 195:168-173. With permission from the Marine Biological Laboratory, Woods Hole, MA)

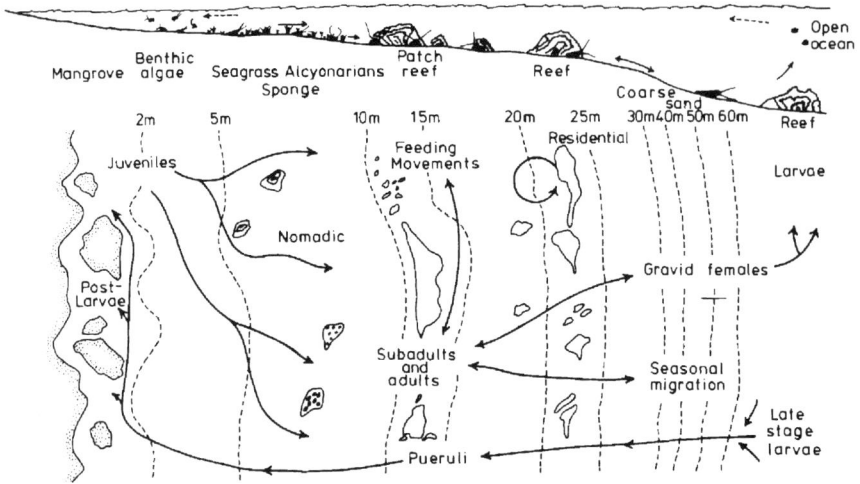

Fig. 4.5 Spatial aspects of the spiny lobster (*Panulirus argus*) life cycle. Gravid females migrate to reef edges near oceanic currents to release larvae. Postlarval pueruli settle on mangrove roots or benthic algae and then shift to seagrass beds and shallow lagoons as juveniles. Subadults gradually emigrate from these nursery habitats to shallows banks characteristic of their range. Adults also exhibit seasonal cycles of residency, small-scale movements, and inshore-offshore migrations. This figure was published in 'The biology and management of lobsters', Vol 1, Herrnkind WF, Spiny lobsters: patterns of movement, pp. 349-407, Copyright Elsevier (1980)

the Papua New Guinea mainland (Fig. 4.6). Tens of thousands of adults travel for 2–3 months across deep (<80 m) soft mud or hard coralline seabed of the Gulf of Papua to spawning grounds and eventually enter shallow (<15 m) reefs around Yule Island (Ruello 1975, Moore and MacFarlane 1984, MacFarlane and Moore 1986, Bell et al. 1987). Mating begins during migration (late October) and once migration has ended (November–December) females release up to three broods of larvae along the eastern boundary of the Gulf of Papua. Moore and MacFarlane (1984) suggested that *P. ornatus* experience high post-spawning mortality and that there is no return migration of adults to Torres Straight. Based on ocean circulation patterns, release of larvae in this location may be important for maintaining lobster stocks in the Torres Straight and Northern Queensland region via larval recruitment to juvenile and adult habitats (MacFarlane and Moore 1986).

4.5.2 Migration Behavior Associated with Spawning

Most aggregating species display strong site fidelity and visit their aggregation site multiple times during the spawning season. In transient aggregations, especially in serranids, males typically arrive earlier and stay longer than females (Johannes 1988, Zeller 1998, Rhodes and Sadovy 2002, Nemeth et al. 2007). Acoustically tagged *Plectropomus leopardus* showed strong spawning site fidelity

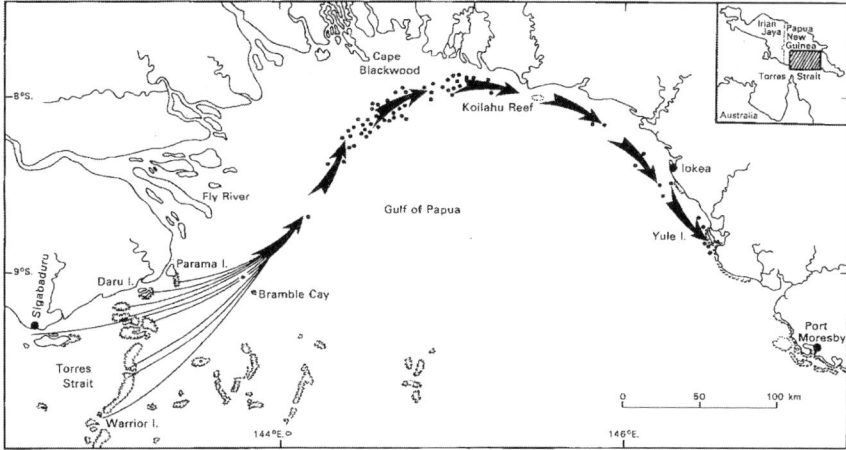

Fig. 4.6 Long-range movements and recapture locations (•) of *Panulirus ornatus* based on the recovery of 125 tagged rock lobsters that were released in Torres Straight (reprinted from Moore and MacFarlane 1984 with permission from CSIRO Publishing http://www.publish.csiro.au/nid/127/issue/2833.htm)

and did not necessarily migrate to the nearest spawning site, a pattern also found in *Epinephelus guttatus* (Sadovy et al. 1992, Sadovy et al. 1994b). *P. leopardus* males were shown to visit spawning sites repeatedly and returned to resident reefs between spawning periods (Zeller 1998). Residence times averaged 13 days for male *P. leopardus*, while females undertook only one trip over the two month spawning period and stayed an average of 1.5 days. Zeller (1998) recorded one male *P. leopardus* taking 10 separate trips (1,720 m round-trip) for a total of 17 km in 19 days. This pattern is similar to Caribbean groupers. Eighty percent of *E. striatus* males returned to spawning sites in Belize during two to three consecutive months, whereas 80% of females returned only one or two months (Starr et al. 2007). *E. guttatus* tagged during the first month of their spawning season (December) were mostly males and had significantly higher recapture rates (mean = 21.4%) than fish tagged in either January (4.2%) or February (1.5%), suggesting that fish arriving in December remain for the duration of the spawning season (Nemeth et al. 2007). This is also reflected in turnover rates of tagged *E. guttatus* where 50% of recaptured males returned for a second month but only 22% of tagged females were recaptured during the second month of the spawning season (Nemeth et al. 2007). Recent studies of *Plectropomus areolatus* in Pohnpei, Micronesia, and *Epinephelus polyphekadion* and *E. fuscoguttatus* in the Seychelles provide additional evidence that these sex-specific movement patterns are a common phenomenon of aggregating serranids worldwide (Rhodes and Tupper 2008, Robinson et al. 2008).

The ability of fish to return repeatedly to a spawning aggregation site is not only a function of migration distance, as described above, but also swimming speed. Several studies have reported migration speeds for resident and transient spawners and decapod crustaceans, and show that migration speeds to spawning aggregation sites can vary depending upon species, gender, and time of day. Average swim speeds for *Acanthurus nigrofuscus*, a resident spawner, was 1.5 km. h^{-1} (Mazeroll and Montgomery 1998). Average swim speeds for transient spawners can range from 0.9 to 1.9 km. h^{-1} for large groupers (e.g., *Epinephelus striatus, Mycteroperca venenosa*) to 0.13–0.6 km. h^{-1} for smaller groupers (e.g., *E. guttatus, Plectropomus leopardus*) (Colin 1992, Luckhurst 1998, Zeller 1998, Starr et al. 2007). Starr et al. (2007) found that *E. striatus* males swam significantly faster than females (2.0 vs 1.8 km. h^{-1}). They also found that *E. striatus* migrated mostly during the day, and had higher swim speeds during the day (1.96 km. h^{-1}) than at night (1.4 km. h^{-1}). King prawn (*Penaeus plebejus*) migration speed from tag-recapture data averaged 0.07 km. h^{-1} but could be as high as 0.23 km. h^{-1} (Ruello 1975). Migration speeds of lobster (e.g., *Panulirus argus, P. ornatus*) range from 0.06 to 0.25 km. h^{-1} (Herrnkind 1980, Moore and MacFarlane 1984). Female blue crabs (*Callinectes sapidus),* whose spawning migration is assisted by tidal transport, averaged 0.42 km. h^{-1} (Tankersley et al. 1998).

Although direct evidence and observations are very limited, adult reef fishes may travel in small or large clusters (Fig. 4.1e) sometimes composed of single-sex groups (e.g., Serranidae) which co-mingle at aggregation sites (Johannes 1978, Colin 1992, Johannes et al. 1999). Tag-recapture studies by Ruello (1975) also suggested that *Penaeus plebejus* maintain cohesive schools as they swim north along the western coast of Australia. Social interactions and gregarious behavior during migration may help reduce predation while simultaneously reinforcing use of traditional migratory pathways through behavioral transmission and recognition of unique reef structures (Colin 1992, Mazeroll and Montgomery 1998).

During its daily feeding and spawning migrations, *Acanthurus nigrofuscus* orient and mill about specific reef structures along their migration routes (Mazeroll and Montgomery 1998). When these landmarks were experimentally moved over short distances (<6 m) and in different directions relative to the migration route, groups of *A. nigrofuscus* continued to use these structures for orientation. Interestingly, when these landmarks were moved >7 m away from the original site, *A. nigrofuscus* continued to follow its normal migration route, but ceased to visit these structures even after they were returned to their original location (Mazeroll and Montgomery 1998). The use of redundant landmarks along migratory pathways may not always be necessary for successful orientation. In a study conducted in Bermuda, Luckhurst (1998) found that *Epinephelus guttatus*, which were captured on an aggregation site, then tagged and released at various distances (mean: 8.9 km), were able to relocate the spawning site within a few days. In the absence of learned behaviors or landmarks, innate orientation to currents, sun position, or the Earth's magnetic field may allow species to migrate in the correct direction to reach spawning aggregation sites. *Panulirus ornatus*, which may have terminal spawning migrations, is believed to orient with respect to the seasonal bottom currents in the Gulf of Papua (Moore and MacFarlane 1984).

Behavioral studies by Mazeroll and Montgomery (1995, 1998) provide some of the most compelling evidence of how social interactions influence migration behavior. *Acanthurus nigrofuscus*, a resident aggregating species in the Red Sea, migrate in groups of up to 200 fish from feeding grounds to spawning sites. Early in the spawning season *A. nigrofuscus* used at least 15 different routes to reach the spawning site, but by the end of the spawning season only three primary routes were followed. Three factors may have contributed to the increased efficiency and synchrony of migratory patterns. First, fish whose feeding grounds were further from the aggregation site initiated migration earlier in the afternoon and thus arrived simultaneously at the spawning aggregation site. Second, as these early groups passed through other feeding territories, several members of the migrating group split off to interact with individuals or small groups which then frequently joined the migration. This recruitment behavior resulted in an increase in the size of migrating groups as they reached the spawning site. This is similar to the migration behavior that was reported for rock lobster (*Panulirus ornatus*) from Papua New Guinea. Small groups of *P. ornatus* leaving their home reefs at the same time join together into large migrating aggregations which proceed as a migratory wave across the Gulf of Papua (Fig. 4.6). Finally, based on a tagging study in which individual fish could be recognized, Mazeroll and Montgomery (1995) suggested certain fish within groups may act as leaders and play an important roll in determining specific migration routes.

Similar examples of group migrating behavior have also been reported for several species of serranids which form large transient aggregations. *Epinephelus striatus* have been observed migrating in groups of 2–500 fish, swimming at 25 cm. s^{-1} at 15–40 m depth parallel to deep coral ridges or the shelf break (Colin et al. 1987, Whaylen et al. 2004, Aguilar-Perera 2006). In the Bahamas, Colin (1992) observed two large migrating groups (about 500 fish) swim past a known spawning site where other fish were present, and continued to swim eastward, presumably to a second spawning site several kilometers further along the shelf edge. In a recent acoustic study of *E. striatus* spawning aggregations off Glover's Reef, Belize, Starr et al. (2007) found that grouper movements along migration routes to spawning sites were rapid, consistent, and predictable. More unusual was that all tagged groupers descended from an average depth of 25 m to 72 m (maximum depth of 255 m) within one hour, seven days after the January full moon spawning period (Starr et al. 2007). During a three month period, most tagged fish migrated to home sites and back to spawning sites but all stayed at depths exceeding 50 m, then synchronously ascended within an hour about 10 d after the April full moon and remained at shallow depths for the remainder of the year. Starr et al. (2007) hypothesized that *E. striatus* occupy these deep water habitats to either spawn in currents more appropriate for larval survival, to feed and recover physiologically after spawning, or to escape shallow-water predators. Acoustically-tagged *E. striatus* and *Mycteroperca venenosa* in the US Virgin Islands also made consistent movements from the Grammanik Bank spawning aggregation site to nearby reefs along the shelf-edge and used these linear shelf-edge reefs as eastwest migration corridors when finally departing the aggregation site (RS Nemeth unpubl. data). However, unlike *E. striatus* in Belize, these fish did not make the synchronized depth change, most likely because they already occupied relatively deep reefs and aggregated to spawn in

35–50 m depth which is below the seasonal thermocline (RS Nemeth unpubl. data). Moreover, large sharks (*Negaprion brevirostris, Galeocerdo curvier, Carcharhinus leucas, C. perezii*, and *C. limbatus*) commonly occur along these deep offshore reefs (RS Nemeth pers. observ.), and therefore predator avoidance is probably not affecting the observed vertical movement patterns of *E. striatus* in Belize.

One final and exceptional example illustrating spawning migration patterns of reef fish involves two male red hind (*Epinephelus guttatus*), which were caught in the same fish trap on the spawning aggregation site, sequentially tagged, and released at the same time and location. Two months later, both fish were recaptured together on the same multiple-hook hand line about 3 km west of the spawning site (Kadison et al. 2009). These fish were obviously traveling together and may have remained within a few meters of each other for a considerable length of time. This example, as well as the studies by Mazeroll and Montgomery (1995, 1998), suggests that fish which occur in the same general feeding areas, migrate together as a group to their respective spawning aggregation site. For haremic spawning species such as the red hind, it also suggests that fish which migrate together may also occupy similar areas within the spawning aggregation site. This would certainly support the view that migration to specific spawning aggregation sites is a learned behavior maintained through tradition. It also highlights the potential for rapidly depleting localized populations and severing the traditional connections between local feeding grounds and spawning aggregation sites when aggregations are fished. These studies emphasize the complex and dynamic behavior of aggregating species. Considering the importance of synchronized spawning of aggregating species, gregarious behavior and the use of landmarks along traditional migratory pathways probably increases the efficiency of adults reaching aggregation sites. More detailed studies on migratory behaviors are needed to better understand the similarities and differences among sites and species.

4.5.3 Catchment and Functional Migration Areas

The terms catchment distance and functional migration area have been used to describe the area from which adult fish migrate to spawning aggregation sites (Zeller 1998, Nemeth et al. 2007). Here I expand upon the definition of functional migration area to include the complex biological processes that occur during migration and spawning within the catchment area. The functional migration area takes into account the mosaic of habitats through which fish migrate, the location of potential migration corridors, the daily movements and behavioral patterns of aggregating species during the spawning season, and the complex trophic interactions that may occur during migration and spawning. At smaller spatial scales within the functional migration area, a species will use a specific area for spawning. This spawning area may vary in size by several orders of magnitude depending upon species and reproductive strategy (FSA area in Appendices 4.1 and 4.2). For group-spawning species the area used for actual spawning each day will be considerably smaller than the area used for courtship during the lunar spawning period (Nemeth 2009). Around this courtship arena aggregating fishes may occupy an even larger staging area between spawning months for a variety of non-reproductive activities (Nemeth 2009). Documenting and mapping the various components within the functional migration area of a species adds to the complexity of spawning aggregation research, but are important for understanding connectivity and critical for ecosystem level management of multi-species tropical fisheries. Colin et al. (2003) and Heyman et al. (2005) outlined several useful and inexpensive techniques and methodologies for mapping migration pathways and spawning aggregation areas. Within the text and in Appendices 4.1 and 4.2, functional migration areas were reported when actual catchment areas were given for a species or were calculated when enough information was available within the primary reference on migration distance, direction, and/or width of reef or shelf. For example, when migration distances were reported and included north–south and east–west components, then an approximate catchment area could be calculated from the resulting polygon. When migratory pathways and habitat mosaics are described then functional migration areas can be constructed.

Transient aggregations of Serranidae and Lutjanidae, which form annually over several consecutive months, encompass functional migration areas of 500 km^2 for the smaller species (i.e., *Epinephelus guttatus*) to 2,500 km^2 or more for the largest (Colin 1992, Carter et al. 1994, Nemeth 2005). Detailed information on functional migration areas is limited to a few reef fish species that have been tracked with either traditional tag–recapture studies or acoustic tags (Appendices 4.1, 4.2). Based on a conventional tagging study in which over 4,000 *E. guttatus* were tagged and released in St. Thomas and St. Croix, US Virgin Islands, Nemeth et al. (2007) found *E. guttatus* populations have functional migration areas of at least 90 km^2 and 500 km^2, respectively (Fig. 4.7, Appendix 4.2). *E. guttatus* from both islands generally migrated against the prevailing currents from areas of high coral cover on mid-shelf reefs to spawning aggregation sites located on the top of well-developed coral reef ridges 300–500 m from the shelf edge. These patterns of migration, movement, and timing of the St. Thomas and St. Croix *E. guttatus* spawning aggregations were remarkably similar despite differences in population structure (Nemeth et al. 2006a). Examples of species from other locations that migrate upcurrent to spawning sites include *E. guttatus* in Bermuda (Luckhurst 1998), *E. striatus* in the Cayman Islands (Colin et al. 1987), *Lutjanus synagris* in Cuba (Garcia-Cagide et al. 2001), *Panulirus ornatus* in Papua New Guinea (MacFarlane and Moore 1986), and *Penaeus plebejus* in Australia (Ruello 1975). This behavior may be an adaptation to counter the downcurrent dispersal of eggs, pre-flexion fish larvae, or early-stage decapod zoeal larvae away from adult habitats, and may function to close these organisms' life cycles (Sinclair 1988, Claydon 2004).

Garcia-Cagide et al. (2001) provided information on potential functional migration areas and migratory pathways for two species of snapper in Cuba. Mutton snapper *Lutjanus analis* migrate from inner reef areas along the north coast of the Cuban shelf to certain reef promontories. *L. analis* are believed to migrate from Archipielago Los Canarreos west along the shelf to aggregate at Cabo Corrientes, a distance of about 120 km. A second spawning aggregation site, Corona de San Carlo, receives *L. analis* adults from eastern and western reef areas which

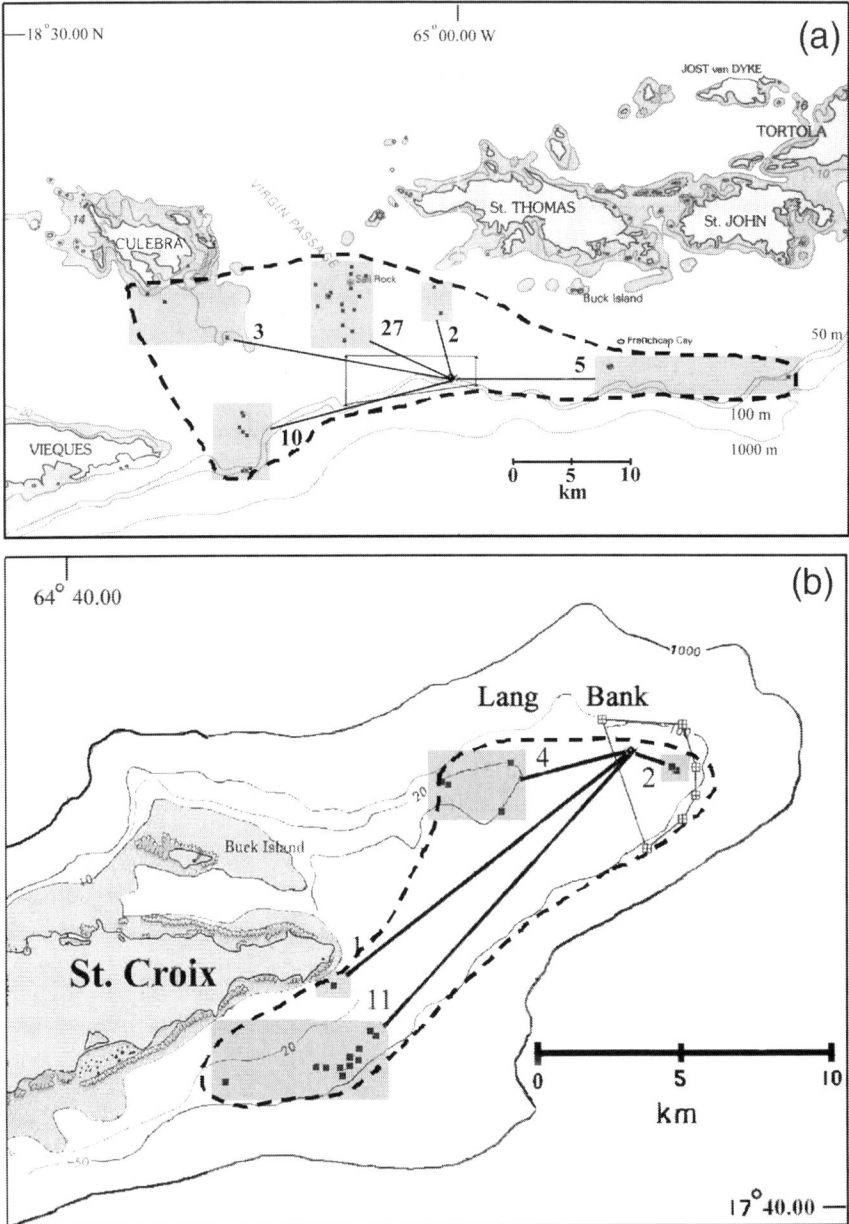

Fig. 4.7 Recapture locations (dots within shaded rectangles) of *Epinephelus guttatus* tagged at spawning aggregation sites (point where lines converge) on St. Thomas (**a**) and St. Croix (**b**), US Virgin Islands, which are located within the two protected areas shown as polygons. Numbers along radiating lines indicate total recaptured fish within each shaded area. Dashed lines represent the minimum functional migration area in which trophic interactions occur for migrating *E. guttatus* before, during, and after spawning (modified from Nemeth et al. 2007, with permission from Springer)

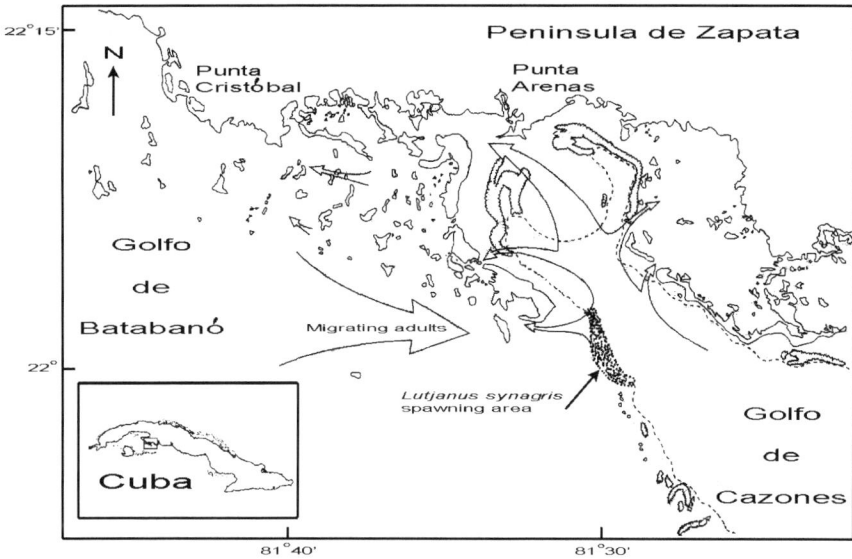

Fig. 4.8 Migratory pathways of adult *Lutjanus synagris* (large arrow) to primary spawning site (stippled area) and potential larval pathways (small arrows) into nearshore shallow-water areas of eastern Golfo de Batabanó, Cuba. Dashed line shows edge of Golfo de Cazones channel and white areas with scalloped edges indicate coral reefs (modified and reprinted from Garcia-Cagide et al. 2001 with permission)

encompass a functional migration area of at least 90 km^2 (Fig. 4.8). Lane snapper *L. synagris* migrate upcurrent from broad shallow areas of eastern Golfo de Batabanó to the western edge of Golfo de Cazones, where they aggregate to spawn (Fig. 4.9). Based on current patterns and ichthyoplankton surveys (references in Garcia-Cagide et al. 2001), these spawning sites may facilitate retention of larvae in oceanic waters near the shelf edge so that postlarvae can return to coastal waters close to their point of origin (Figs. 4.8, 4.9). Tarpon (*Megalops atlanticus*) which typically occupy inshore lagoons or coastlines where there is an abundance of bait fish, have been reported to migrate up to 25 km offshore to form spawning aggregations (Crabtree 1995, Garcia and Solano 1995). Leptocephali larvae of *M. atlanticus* in the Caribbean, and probably *M. cyprinoides* of the Indo-West Pacific (Coates 1987), then migrate to mangrove estuaries where they metamorphose into juveniles (Zerbi et al. 1999).

Resident aggregations migrate shorter distances between home ranges and spawning sites, and therefore have smaller functional migration areas. In a detailed acoustic study, Zeller (1998) found that *Plectropomus leopardus* typically migrated from 0.2 to 11 km to one of four known spawning sites around Lizard Island, Great Barrier Reef. Three other tagged fish, however, were later recaptured on outlying reef areas 3–11 km away and presumably made only a single trip to the spawning site. According to the results of this study, *P. leopardus*, which was classified as a resident rather than transient spawner (Domeier and Colin 1997), may actually use

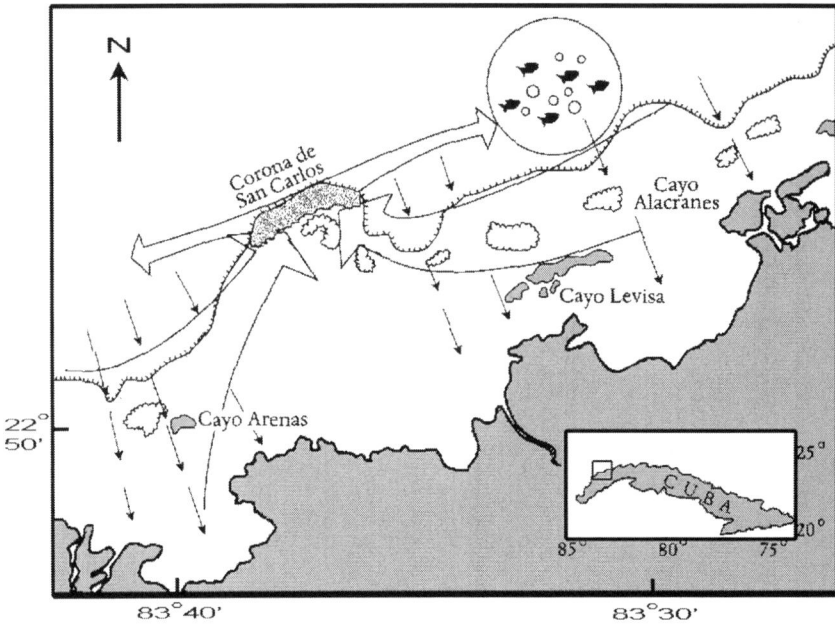

Fig. 4.9 Migratory pathways of adult *Lutjanus analis* (large arrows) to primary spawning site (stippled area), potential dispersal of eggs and larvae (medium arrows) along shelf, and recruitment of juveniles (thin arrows) into nearshore areas of northwestern Cuba at Corona de San Carlos. Spiked line indicates shelf edge and white areas with scalloped edges indicate coral reefs (modified and reprinted from Garcia-Cagide et al. 2001 with permission)

both strategies. Based on the reported migration distances and the high degree of spawning site fidelity, the functional migration area for *P. leopardus* was calculated to range from at least 5 km^2 for fish remaining on the Lizard Island shelf (resident spawners) to at least 80 km^2 for fish migrating from outlying reef areas (transient spawners?). Members of the Acanthuridae assemble aggregations of 6,000–20,000 fish each evening from a functional migration area of several square km (Colin and Clavijo 1988, Robertson et al. 1990, Kiflawi et al. 1998).

Appendices 4.1 and 4.2 list available data for most resident and transient aggregations for which some information is available on the spawning site, spawning aggregation size, migration distance, and functional migration area. There are many other species that form spawning aggregations for which spatial data are limited or do not exist. The limited information on migration distances to spawning aggregation sites for other fish families or decapods (Appendices 4.1, 4.2) highlights the need for more research in this important area. Existing information for most of these species can be accessed via the Society for the Conservation of Reef Fish Aggregations web site (http://www.scrfa.org/). All other species which pair spawn within their feeding territories and do not migrate, such as *Sparisoma viride* (van Rooij et al. 1996) were not included.

4.6 Trophic Interactions of Spawning Aggregations

The migration of fish to spawning aggregation sites provides an important component of connectivity across habitats. Besides the spatial and temporal fluctuations in fish biomass associated with resident and transient spawning migrations, the effect of these fishes on local food webs along their migratory pathways and at spawning sites has rarely been examined. The large-scale movements of spawning populations, especially predatory transient aggregating species, will have the effect of coupling local food webs within their functional migration areas (McCann et al. 2005). For example, the functional migration area of a spawning population of red hind (*Epinephelus guttatus*), a 30–50 cm long fish, is about 100–500 km^2, which is equivalent to the largest scale of food webs of top predators such as lions or sharks (Brose et al. 2005).

Several potential differences exist between the ecological effects of resident and transient aggregations at spawning aggregation sites. Resident aggregations are characterized by relatively rapid migrations of herbivorous or omnivorous species (i.e., Acanthuridae, Scaridae, Labridae) to spawning sites where they spend a few hours each day in courtship and spawning before returning to home sites. Many resident aggregating species such as the parrotfishes and wrasses show spawning activity year-round, while others (i.e., surgeonfishes) show seasonal patterns (Fishelson et al. 1987, Colin and Clavijo 1988). Resident aggregations are not known to feed during migration or spawning, but occasionally some individuals within an aggregation of the parrotfish *Sparisoma rubripinne* were observed grazing on benthic algae between spawning rushes (Randall and Randall 1963). Since resident aggregations do not require food resources during brief spawning episodes, they most likely contribute to local food webs through predation on spawning adults at spawning aggregation sites and possibly along migration routes, and predation on their newly released eggs (Fig. 4.10).

Successful predation attempts on spawning aggregations seem to be quite variable depending upon the species and location, but appear to be less common within the Caribbean (Colin and Clavijo 1988, Robertson et al. 1999) than in the Indo-Pacific (Robertson 1983, Moyer 1987, Sancho et al. 2000a). Robertson (1983) observed sharks (*Carcharhinus melanopterus*), groupers (*Cephalopholis argus*), snappers (*Lutjanus bohar*), and jacks (*Caranx melampygus*) attacking a spawning aggregation of *Acanthurus nigrofuscus* in Palau. Sharks have also been reported to attack spawning aggregations of mullet (*Crenimugil crenilabus*) in Enewetak and *Naso literatus* in Palau (Johannes et al. 1999). Spawning aggregations of *Sparisoma rubripinne* were occasionally attacked and captured by barracuda (*Sphyraena barracuda*) and kingfish (*Scomberomorous cavalla*) (Randall and Randall 1963). Predation events are probably more common than has been documented because the presence of divers will influence the behavior of both the aggregating species and potential predators (RS Nemeth pers. observ.). Detailed observations by Sancho et al. (2000a) at Johnston Atoll recorded 254 attacks by *Caranx melampygus* (Carangidae) and *Aphareus furca* (Lutjanidae) on five group-spawning species: *Chlorurus sordidus*, *Scarus psittacus* (Scaridae), *Acanthurus*

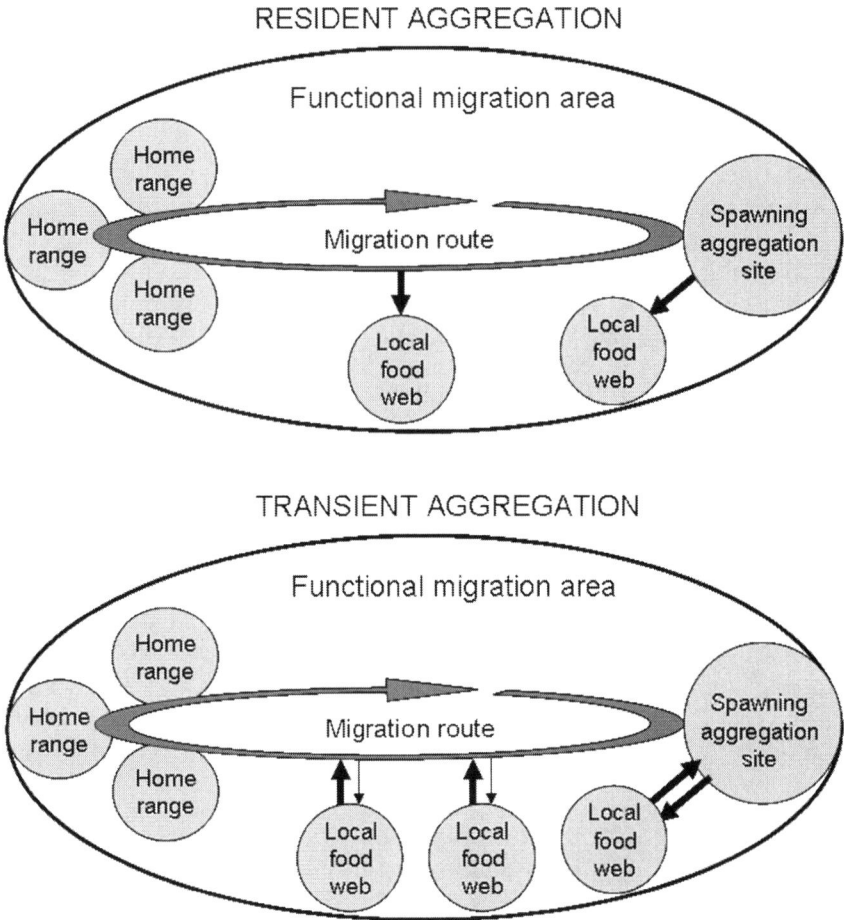

Fig. 4.10 Hypothetical interactions with local food webs within functional migration area, along migration route, and at spawning aggregation site. Resident aggregations, composed of herbivorous and omnivorous species, typically do not feed during migration or spawning so net flow of energy is from spawning adults and newly released eggs into local food webs via predation. Transient aggregations, composed of piscivores and other carnivores, may feed along migration route and at spawning aggregation site. Due to the large body size of transient aggregation species, predation along their migration routes may be minimal. However, concentrations of top predators (e.g., sharks) and planktivores at spawning sites may balance net energy flow between aggregating species and local food webs

nigroris, Zebrasoma flavescens, and *Ctenochaetus strigosus* (Acanthuridae). The majority of attacks (93%) were directed toward the two most abundant species (*Chlorurus sordidus* and *A. nigroris*) and occurred during spawning rushes (83%) or courtship (17%), but with no attacks observed on feeding or migrating fish (Sancho et al. 2000a). Both predators averaged 4% success rate in capturing spawning fish and preferred to attack group spawns containing more than four fish.

Egg predation rates on resident spawning aggregations can also be quite variable depend upon aggregating species and spawning location (Robertson 1983, Craig 1998). Sancho et al. (2000a) found that two triggerfish, *Melichthys niger* and *M. vidua*, were the most common egg predators on Johnston Atoll. These planktivores would quickly approach and bite within the center of a newly released gamete cloud. Attacks occurred more commonly during dusk (71%) and were directed at pair spawns (7.5%) more often than group spawns (0.3%). This is in contrast to Robertson (1983) who reported egg predators preferred attacking gamete clouds of group spawning (27–42%) rather then pair spawning (4–5%) Acanthuridae. The preference of the triggerfishes for attacking pair spawning species (*Aulostomus chinensis, Ostracion meleagris*, and *Bothus mancus*) may have been related to the large size of their eggs (>14 mm³) compared to the other three species (*Parupeneus bifasciatum, Chlorurus sordidus*, and *Acanthurus nigroris*) whose egg volumes were considerably smaller (<3 mm³) (Sancho et al. 2000a). Moreover the egg predators that Robertson (1983) observed consisted of a diverse group of planktivores including Pomacentridae (*Abudefduf saxatilis, Chromis caerulea, C. atripectoralis, Ambliglyphidodon curacao*), Labridae (*Thalassoma hardwicki, T. amlbycephalus*), Lutjanidae (*Caesio coerulaureus, C. erythrogaster, C. lunaris, Pterocaesio chrysozonus*), Scombridae (*Rastrelliger kanagurta*), Balistidae (*Melichthys vidua*), and two unidentified species within Exocoetidae and Dussumieridae.

Transient aggregations, on the other hand, are characterized by carnivorous and piscivorous species that can migrate considerable distances and spend several days to weeks at a spawning aggregation site. During this time, fish within transient aggregations may continue to feed along migration routes and in habitats surrounding the spawning aggregation site, and therefore may have a significant impact on local food webs (Fig. 4.10). Johannes et al. (1999) observed aggregating *Plectropomus areolatus* occasionally attacking schools of small caesionids in Palau. Between spawning peaks about 10–15% of the *Epinephelus guttatus* spawning population (both males and females) remain on the spawning site (Fig. 4.2), while the remainder move short distances (<1,000 m) from high density coral reef habitats to outlying patch reef areas presumably to feed (Nemeth et al. 2007). The gut contents of *E. guttatus* from spawning aggregations in the Virgin Islands commonly contain Brachyura (true crabs), Anomura (hermit crabs), and Palinura (lobster) crustaceans, and small reef fishes such as filefish (*Malacanthus tuckeri*) and even juvenile *E. guttatus* (RS Nemeth unpubl. data). Likewise, Samoilys (1997) suggested the daytime dispersal of a *Plectropomus leopardus* spawning aggregation was to feed in surrounding reef areas. This same daytime dispersal pattern has been observed by *Mycteroperca venenosa* and *E. striatus* spawning aggregations, albeit more extensive (Nemeth et al. 2006b). These two species can swim distances of 1–5 km during the day before returning to the spawning aggregation site by late afternoon and over 10 km per day between monthly spawning peaks (Fig. 4.11) (RS Nemeth unpubl. data). *M. venenosa* stomach contents contain a variety of reef fishes such as *Ocyurus chrysurus, Lutjanus buccanella, E. guttatus*, and *Clepticus parrae*, whereas *E. striatus* guts contain mainly crustaceans such as the red banded lobster (*Justitia longimanus*), but it is unclear when or where feeding takes place (RS

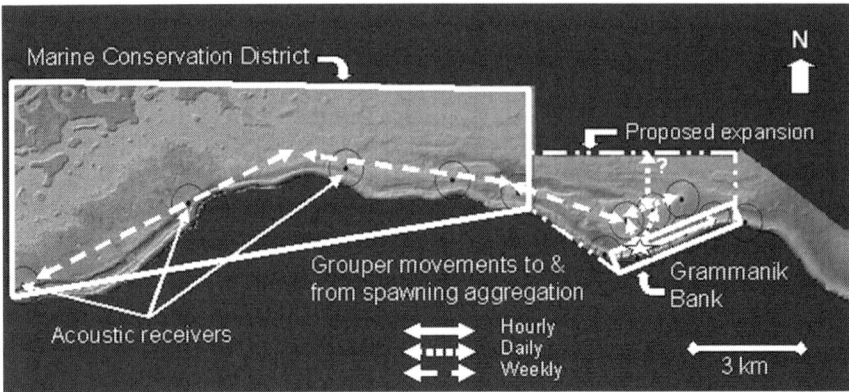

Fig. 4.11 The Marine Conservation District (MCD) and the Grammanik Bank (GB) are two marine protected areas that were established in the US Virgin Islands. A 2007 acoustic study was conducted to determine if the boundaries of the GB are appropriately placed to protect a multispecies spawning aggregation site (⋆). The first year of tracking Nassau (*Epinephelus striatus*) and yellowfin (*Mycteroperca venenosa*) grouper movements revealed that fish frequently swam outside the protected area during daily and weekly movements associated with the spawning aggregation, thus making them vulnerable to fishing mortality. These data suggest that the GB boundaries need to be expanded north to encompass daily movements associated with spawning and west to provide a safe migration corridor between the GB and MCD

Nemeth pers. observ.). However, when ovaries become hydrated, feeding ceases and *M. venenosa* and other aggregating species rarely take baited hooks or enter baited traps (Beets and Friedlander 1999, RS Nemeth pers. observ.). In other species such as *P. leopardus*, males may not feed during spawning since they are defending territories for incoming females and thus become very thin by the end of the aggregation period (Y Sadovy, University of Hong Kong, China, pers. comm.).

Predation on transient aggregations has been recorded in a number of studies. Johannes et al. (1999) observed a large (85–90 cm) *Epinephelus fuscoguttatus* eat a 50–55 cm long *E. polyphekadion*, both of which were aggregating in the same location in Palau. Sharks have also been reported attacking transient aggregations of *E. guttatus*, *E. striatus*, *Mycteroperca venenosa*, and *M. tigris* (Olsen and LaPlace 1978, Nemeth 2005). The number and variety of large predators (i.e., sharks: *Ginglymostoma cirratum*, *Negaprion brevirostrus*, *Caracharhinus perezii*, and *C. limbatus*; cubera snappers: *Lutjanus cyanopterus*; and moray eels: *Gymnothorax funebris*) are more common during spawning aggregations of *E. guttatus* in the Virgin Islands than at other times (Nemeth 2005). Colin (1992) observed two bull sharks (*Carcharhinus leucas*) following a group of 500 *E. striatus* which were presumably migrating to their spawning aggregation site. In a recent acoustic tagging study in the Virgin Islands, lemon (*Negaprion brevirostris*) and tiger (*Galeocerdo curvier*) sharks were detected more frequently at *E. guttatus* and *M. venenosa* spawning sites during the week of spawning than at other times (B Wetherbee, M Shivji, and RS Nemeth unpubl. data), suggesting that feeding migrations of predators may be synchronized with the seasonal spawning activity of transient aggregations. This is certainly the case in Belize, where eggs of *L. cyanopterus* and *L. jocu* are eaten by whale sharks (*Rhincodon typus*), yellowtail snapper (*Ocyurus chrysurus*), rainbow runner (*Elagatis bipinnulata*), and Atlantic spadefish (*Chaetodipterus faber*) (Heyman et al. 2001, Heyman et al. 2005). *R. typus* may undergo large-scale migrations and converge in large numbers to take advantage of this temporally patchy supply of fish eggs (Heyman et al. 2001, Hoffmayer et al. 2007). Samoilys (1997) observed planktivorous *Caesio* spp. frequently feeding on the eggs of *P. leopardus* with egg predation occurring on 27% of the spawning rushes. Aguilar-Perera and Aguilar-Davila (1996) reported small reef fish (i.e., *Thalassoma bifasciatum*, *Clepticus parrai*, *Sparisoma viride*, and *Melichthys niger*) becoming more numerous just prior to *E. striatus* aggregations in Mexico. On April 12, 2007, between 18:13 and 18:30 hrs, *L. jocu*, *L. analis* and *Elegatis bipinnulata* were observed feeding on newly released eggs of *M. venenosa* (RS Nemeth pers. observ.).

Based on the observations described above, transient aggregations not only extract resources but may also make a significant contribution to local food webs (Fig. 4.10). While they may incur lower mortality rates during migration due to their larger body size, the length of time they spend at spawning aggregation sites seems to attract a greater abundance of potential piscivorous and planktivorous predators that feed on high concentrations of spawning adults and eggs. The interactions between species that form spawning aggregations and the local food webs along migration routes and at spawning sites is poorly understood and requires further investigation. These types of studies will allow the construction of functional migration area maps for commercially important species and enhance our understanding of how spawning aggregations function within tropical ecosystems.

4.7 Summary and Future Directions

The majority of tropical reef fishes that form spawning aggregations migrate from inshore feeding areas to offshore spawning sites. Several reef fish families (i.e., Sparidae, Siganidae) and some decapod crustaceans which inhabit estuaries or mangroves as adults, migrate to coastal waters to spawn. A few species even migrate from coastal areas into shallow lagoons or freshwater systems for reproduction (i.e., Carcharhinidae, Pangasiidae, Rhincodontidae, Siganidae). Within the largest group of reef fishes undergoing spawning migrations, several general patterns emerge: (1) most species that migrate to traditional aggregation sites are pelagic spawners (exceptions include Balistidae and Siganidae), (2) spawning sites usually occur, but not always, on a submerged promontory at the outer edge of a reef, island, or shelf or at the mouth of a channel, near deep water, (3) migration and spawning of resident aggregations are often synchronized with tidal cycle and spawning tends to occur in the late afternoon or at high slack tide just before outgoing tides are strongest, (4) migration and spawning of transient aggregations are often synchronized around the new or full moon and spawning tends to occur at dusk, and (5) seasonal spawning peaks often correspond to when water temperature is near its annual minimum and oceanic currents are weakest.

While these patterns are synchronized with specific environmental cues and can be highly predictable in space and time, very little is known about the levels of connectivity in terms of the degree of larval dispersal or retention from spawning aggregation sites or the trophic interactions of resident and transient aggregations during migration and spawning. Using drifter vials, Domeier (2004) estimated that larvae from a *Lutjanus analis* spawning aggregation site in the Dry Tortugas may disperse over 500 km of coastline from the Florida Keys to Southeast Florida. Evidence for larval retention is also increasing (Jones et al. 1999, Swearer et al. 1999, Paris and Cowen 2004), but many questions still remain. For instance, to what extent do location, timing, and duration of spawning aggregations determine annual reproductive outputs and profiles of larval settlement events (Sadovy 1996)? More information is needed on synchrony or coupling of spawning with the physical oceanographic features at a spawning aggregation site (Nemeth et al. 2008) and the behaviors of late-stage larvae under different hydrographic conditions (Cowen 2002).

The timing and duration of a spawning aggregation and the degree of connectivity among habitats will influence the impact of aggregating species on trophic interactions at spawning sites and in areas through which they migrate. Reef fishes that migrate from feeding grounds to spawning aggregation sites will connect spatially distinct food webs across variable time scales. The level of interaction with these local food webs will be a function of the migrating species' trophic level and ecological requirements during migration and spawning, and the length of time they reside within the spawning aggregation area. Very little information exists on how other non-aggregating resident species (either predators or prey) respond to the spatial and temporal changes in trophic linkages caused by spawning aggregations. These gaps in knowledge lead to a number of testable hypotheses. For example, highly mobile predators may take advantage of spatial and temporal increases in prey abundance at spawning aggregation sites (Sancho et al. 2000a, Heyman et al. 2001). Does the feeding behavior of large predators, such as sharks, track the seasonal spawning patterns of aggregating species? Does the abundance of egg-feeding planktivores increase at spawning sites when spawning aggregations form? Alternatively, do prey species, which reside at the spawning aggregation sites of groupers and snappers, experience significantly higher mortality rates when aggregations are present than at other times of the year? If so can these smaller prey species temporarily emigrate away from spawning sites used by piscivorous species? The convergence of multiple species at spawning aggregation sites increases the complexity of these interactions and requires more detailed observations and new approaches. Is there a dominance hierarchy at multi-species spawning aggregations sites with the more aggressive species occupying higher-quality spawning habitat than subordinate species? At the Gammanik Bank, US Virgin Islands, the dominance hierarchy, based on observed interspecific aggression, seems to be as follows: *Lutjanus cyanopterus > Mycteroperca venenosa = Epinephelus striatus > M. tigris* (RS Nemeth pers. observ.). The dominance hierarchy at a multi-species grouper spawning aggregation site in Pohnpei, Micronesia, may be as follows: *Plectropomus areolatus > Epinephelus fuscoguttatus* > *E. polyphekadion* (Rhodes and Sadovy 2002), although inter-specific acts of aggression may differ among locations (Robinson et al. 2008). More extensive surveys and behavioral observations at spawning aggregation sites before, during, and after spawning may reveal complex predator–prey dynamics and inter-specific interactions that occur among aggregating species and resident and transient fish populations.

The ecological role of local food webs to the functioning of spawning migrations and aggregations is a completely unexplored area of research but is important for understanding and managing coral reefs at the ecosystem level. For example, do transient aggregations function more effectively and have greater reproductive out-put in areas which contain robust and intact food webs versus areas where food webs have been disrupted or prey populations depleted through fishing activity or habitat loss? More detailed comparisons of fish and invertebrate assemblage structure and food web linkages at aggregation and non-aggregation sites, and inside and outside permanently or seasonally protected areas, would be beneficial to understanding the role that trophic integrity has on population characteristics and reproductive success of aggregating species (Molloy et al. 2008).

The short duration and concentrated site-specific nature of spawning aggrega-tions makes them very vulnerable to even moderate fishing pressure (Sadovy and Domeier 2005). Worldwide greater than 25% of all known spawning aggregations have either disappeared or are in decline and the status of an equal number of aggregations remains uncertain (Sadovy de Mitcheson et al. 2008). Declines in catch per unit effort and fish size are some of the first signs of an aggregation in decline (Beets and Friedlander 1992, Graham et al. 2008) and many aggregations have been eliminated entirely (Olsen and LaPlace 1978, Sadovy 1997, Sadovy and Eklund 1999, Sala et al. 2001, Aguilar-Perera 2006). Aggregation fishing contin-ues throughout the world even though the benefits of protecting spawning aggre-gations sites for rebuilding breeding populations and/or improving fisheries have been documented for several reef fish species (Burton et al. 2005, Nemeth 2005). In the US Virgin Islands, a 40 km^2 Marine Conservation District was created to protect an *Epinephelus guttatus* spawning aggregation site that showed evidence of overfishing (Beets and Friedlander 1992). After 10 yrs of seasonal closure and 5 yrs of permanent protection, the *E. guttatus* spawning population showed signifi-cant increases in fish length, biomass, and density, and improved sex ratios (Nemeth 2005), and now is more abundant in the local fishery than before the closure (Pickert et al. 2006). Spawning sanctuaries and migration corridors have also been estab-lished for female blue crabs at a number of locations in the southeastern USA, and range in size from 18 to 2,400 km^2 (Lipcius et al. 2003). The benefits of these pro-tected areas for migrating decapod crustaceans have yet to be determined. Since spawning aggregations may represent the primary source of reproductive effort for aggregating species, eliminating an aggregation may severely deplete the local pop-ulation, hinder its ability to recover, and disrupt potentially important connections among coral reef habitats. Protecting spawning aggregations will require a detailed understanding of the movement and habitat-use patterns associated with aggregating species.

As Appendices 4.1 and 4.2 illustrate, spawning aggregation research is still in its infancy. The first step in filling some of these gaps is to establish well-designed tagging studies. Tagging and releasing fish at spawning aggregation sites, and then waiting for fishermen to recapture fish and return tags, is a useful way of determining migration distance and direction, finding resident habitats, and calculating catchment and functional migration areas so long as two assumptions are met: (1) fishermen provide accurate information on recapture location, and (2) enough time elapsed to ensure that a tagged fish had reached its home site before it was recaptured. Two factors that reduce the reliability of this method are: (1) fishermen do not participate in the study, and (2) most tag returns will be from locations where fishing effort is greatest leaving potential gaps in the spatial data. Acoustic studies provide more detailed information on frequency and timing of migration, small-scale movement patterns, habitat use, and length of time an individual fish spends at a particular site, and can contribute greatly to our understanding of a species functional migration area. However, determining directionality of movement can be difficult unless an appropriate acoustic receiver array is established (Domeier 2005), and the high cost of acoustic tags and receivers limits the number of fish that can be tagged and the area of reef that can be covered. Despite the weaknesses of these two methods, useful information can be obtained on the spatial and temporal aspects of migration which will be of great value for managing spawning aggregations (Rhodes and Tupper 2008).

Although the approximate location of many spawning aggregations is known, little information is available on the migration or movement patterns associated with aggregating species or the area occupied by fishes during spawning, courtship, or between spawning peaks (Nemeth 2009). This lack of information often results in the boundaries of proposed closures being drawn arbitrarily on a chart. These management initiatives are often rejected by local fishermen, who argue that the size of the proposed protected area is inappropriate. Another benefit of a well-designed acoustic tagging study is that the resulting spatial data can be used to identify migration corridors and determine biologically-relevant marine protected areas (Fig. 4.11), which maximize protection of aggregating species while minimizing economic impact of artisanal and commercial fishers. Several useful and inexpensive techniques and methodologies for mapping migration pathways and spawning aggregation areas on bathymetric and benthic habitat maps have been developed (Colin et al. 2003, Heyman et al. 2005). For deep reef or large geographic areas more sophisticated and expensive side-scan sonar and multi-beam technologies need to be employed (Rivera et al. 2005). This information will be critical for (1) identifying migration corridors and establishing boundaries of fishery protected areas, (2) estimating functional migration areas, (3) delineating potential subpopulations based on spawning aggregations, and (4) developing new ecosystem-level management approaches. Finally, understanding the degree of dispersal and retention of gametes from spawning aggregations and subsequent recruitment patterns (see Chapter 6) are essential for determining levels of connectivity among tropical coastal ecosystems. Developing novel techniques and methodologies which examine the full life cycle of aggregating species will result in informed management decisions that will increase compliance and effectiveness of fishery regulations while minimizing the economic impact on commercial fishers.

4.8 Acknowledgments

Support to RS Nemeth for preparation of this book chapter was provided by the Friday Harbor Laboratory's Helen Riaboff Whiteley Center, the University of the Virgin Islands, and the Virgin Islands-Experimental Program to Stimulate Competitive Research (VI-EPSCoR) under National Science Foundation (NSF) grant no. 0346483. Any opinions, findings, conclusions, or recommendations expressed in this paper are those of the author and do not necessarily reflect the official views of the NSF. Special thanks to P Colin, Y Sadovy, and M Domeier for providing useful comments that greatly improved this chapter. This is contribution # 53 from the University of the Virgin Islands' Center for Marine and Environmental Studies.

References

Aguilar-Perera A (1994) Preliminary observations of the spawning aggregation of Nassau grouper, *Epinephelus striatus*, at Mahahual, Quintana Roo, Mexico. Proc Gulf Caribb Fish Inst 43:112–122

Aguilar-Perera A (2006) Disappearance of a Nassau grouper spawning aggregation off southern Mexican Caribbean coast. Mar Ecol Prog Ser 327:289–296

Aguilar-Perera A, Aguilar-Davila W (1996) A spawning aggregation of Nassau grouper *Epinephelus striatus* (Pisces: Seranidae) in the Mexican Caribbean. Environ Biol Fishes 45:351–361

Appeldoorn RS, Hensley DA, Shapiro DY et al (1994) Egg dispersal in a Caribbean coral reef fish, *Thalassoma bifasciatum*. II. Dispersal off the reef platform. Bull Mar Sci 54:271–280

Beets J, Friedlander A (1992) Stock analysis and management strategies for red hind, *Epinephelus guttatus*, in the U.S. Virgin Islands Proc 42nd Gulf Caribb Fish Inst 42:66–79

Beets J, Friedlander A (1999) Evaluation of a conservation strategy: a spawning aggregation closure for red hind, *Epinephelus guttatus*, in the U.S. Virgin Islands. Environ Biol Fishes 55: 91–98

Bell LJ, Colin PL (1986) Mass spawning of *Caesio teres* (Pisces: Caesionidae) at Enewetak Atoll, Marshall Islands. Environ Biol Fishes 15:69–74

Bell RS, Channels PW, MacFarlane JW et al (1987) Movements and breeding of the ornate rock lobster, *Panulirus ornatus*, in Torres Strait and on the north-east coast of Queensland. Aust J Mar Freshw Res 38:197–210

Bolden SK (2000) Long-distance movement of a Nassau grouper (*Epinephelus striatus*) to a spawning aggregation in the central Bahamas. Fish Bull 98:642–645

Brose U, Pavao-Zuckerman M, Eklof A et al (2005) Spatial aspects of food webs. In: de Ruiter PC, Wolters V, Moore JC (eds) Dynamic food webs, pp. 463–470. Elsevier, Amsterdam

Brule T, Deniel C, Colas-Marrufo T et al (1999) Red grouper reproduction in the southern Gulf of Mexico. Trans Am Fish Soc 128:385–402

Burnett-Herkes J (1975) Contribution to the biology of the red hind, *Epinephelus guttatus*, a commercially important serranid fish from the tropical western Atlantic. Ph.D. thesis, University of Miami, Miami

Burton ML, Brennan KJ, Muñoz RC et al (2005) Preliminary evidence of increased spawning aggregations of mutton snapper (*Lutjanus analis*) at Riley's Hump two years after establishment of the Tortugas South Ecological Reserve. Fish Bull 103:404–410

Carr SD, Hench JL, Luettich RA, Forward RB et al. (2005) Spatial patterns in the ovigerous *Callinectes sapidus* spawning migration: results from a coupled behavioral-physical model. Mar Ecol Prog Ser 294:213–226

Carr SD, Tankersley RA, Hench JL et al (2004) Movement patterns and trajectories of ovigerous blue crabs *Callinectes sapidus* during the spawning migration. Estuar Coast Shelf Sci 60:567–579

Carter J (1987) Grouper sex in Belize. Nat Hist October:pp.60–69

Carter J, Perrine D (1994) A spawning aggregation of dog snapper, *Lutjanus jocu* (Pisces: Lutjanidae) in Belize, Central America. Bull Mar Sci 55:228–234

Carter JG, Marrow J, Pryor V (1994) Aspects of the ecology and reproduction of Nassau grouper, *Epinephelus striatus*, off the coast of Belize, Central America. Proc 43rd Gulf Caribb Fish Inst 43:65–111

Claro R, Lindeman KC (2003) Spawning aggregation sites of snapper and grouper species (Lutjanidae and Serranidae) on the insular shelf of Cuba. Gulf Caribb Res 14:91–106

Claydon J (2004) Spawning aggregations of coral reef fishes: characteristics, hypotheses, threats and management. Ocean Mar Biol Ann Rev 42:265–302

Claydon J (2006) Spawning aggregation of yellowfin mojarra. SCRFA newsletter 9:4–6

Clifton KE, Robertson DR (1993) Risks of alternative mating strategies. Nature 366:520

Coates D (1987) Observations on the biology of tarpon, *Megalops cyprinoides* (Broussonet) (Pisces: Megalopidae), in the Sepik River, northern Papua New Guinea. Aust J Mar Freshw Res 38:529–535

Coleman FC, Koenig CC, Collins LA (1996) Reproductive styles of shallow-water groupers (Pisces: Serranidae) in the eastern Gulf of Mexico and the consequences of fishing spawning aggregations. Environ Biol Fishes 47:129–141

Colin PL (1976) Filter-feeding and predation of the eggs of *Thalassoma* sp. by the scombrid fish *Rastrelliger kanagurta* Copeia 1976:596–597

Colin PL (1992) Reproduction of the Nassau grouper, *Epinephelus striatus* (Pisces: Serranidae) and its relationship to environmental conditions. Environ Biol Fishes 34:357–377

Colin PL (1994) Preliminary investigations of reproductive activity of the jewfish, *Epinephelus itajara* (Pisces: Serranidae). Proc Gulf Caribb Fish Inst 43:138–147

Colin PL (1996) Longevity of some coral reef fish spawning aggregations. Copeia 1996:189–192

Colin PL, Bell LJ (1991) Aspects of the spawning of labrid and scarid fishes (Pisces: Labroidei) at Eniwetak Atoll, Marshall Islands with notes on other families. Environ Biol Fishes 31:229–260

Colin PL, Clavijo IE (1978) Mass spawning by the spotted goatfish, *Pseudopeneus maculatus* (Bloch) (Pisces: Mullidae). Bull Mar Sci 28:780–782

Colin PL, Clavijo IE (1988) Spawning activity of fishes producing pelagic eggs on a shelf edge coral reef, southwestern Puerto Rico. Bull Mar Sci 43:249–279

Colin PL, Sadovy YJ, Domeier ML (2003) Manual for the study and conservation of reef fish spawning aggregations. Special Publication No. 1 (Version 1.0), pp.1–98. Society for the Conservation of Reef Fish Aggregations

Colin PL, Shapiro DY, Weiler D (1987) Aspects of the reproduction of two groupers, *Epinephelus guttatus* and *E. striatus*, in the West Indies. Bull Mar Sci 40:220–230

Cowen RK (2002) Larval dispersal and retention and consequences for population connectivity. In: Sale PF (ed) Coral reef fishes: dynamics and diversity in a complex ecosystem, pp. 149–170. Academic Press, London

Crabtree RE (1995) Relationship between lunar phase and spawning activity of tarpon, *Megalops atlanticus*, with notes on the distribution of larvae. Bull Mar Sci 56:895–898

Craig PC (1998) Temporal spawning patterns of several species of surgeonfishes and wrasses in American Samoa. Pac Sci 52:35–39

Cushing DH (1971) Upwelling and the production of fish. Adv Mar Biol 9:255–335

Daw T (2004) Reef fish aggregations in Sabah, East Malaysia. In: Western Pacific fisher survey series, Vol 5. Society for the Conservation of Reef Fish Aggregations

Domeier M (2005) Methods for the deployment and maintenance of an acoustic tag tracking array: an example from California's Channel Islands. Mar Tech Soc J 39:74–80

Domeier ML (2004) A potential larval recruitment pathway originating from a Florida marine protected area. Fish Oceanogr 13:287–294

Domeier ML, Colin PL (1997) Tropical reef fish spawning aggregations: defined and reviewed. Bull Mar Sci 60:698–726

Ebisawa A (1990) Reproductive biology of *Lethrinus nebulosus* (Pisces: Lethrinidae) around the Okinawa waters. Nippon Suisan Gakkai 56:1941–1954

Ebisawa A (1999) Reproductive and sexual characteristics in the Pacific yellowtail emperor, *Lethrinus atkinsoni*, in waters off the Ryukyu Islands. Ichthyol Res 46:341–358

Eklund AM, McClennal DB, Harper DE (2000) Black grouper aggregations in relation to protected areas within the Florida Keys National Marine Sanctuary. Bull Mar Sci 66:721–728

Feldheim KA, Gruber SH, Ashley MV (2002) The breeding biology of lemon sharks at a tropical nursery lagoon. Proc Royal Soc B 269:1655–1661

Fine JC (1990) Groupers in love. Sea Frontiers Jan–Feb, pp: 42–45

Fishelson L, Montgomery WL, Myberg AAJ (1987) Biology of surgeonfish *Acanthurus nigrofuscus* with emphasis on changeover in diet and annual gonadal cycles. Mar Ecol Prog Ser 39:37–47

Fitch WTS, Shapiro DY (1990) Spatial dispersion and nonmigratory spawning in the bluehead wrasse (*Thalassoma bifasciatum*). Ethology 85:199–211

Garcia-Cagide A, Claro R, Koshelev BV (2001) Reproductive patterns of fishes of the Cuban shelf. In: Claro R, Lindeman KC, Parenti LR (eds) Ecology of the marine fishes of Cuba, pp. 73–114. Smithsonian Institution Press, Washington, DC

Garcia CB, Solano OD (1995) *Tarpon atlanticus* in Columbia: big fish in trouble. Naga, ICLARM Quarterly: 18:47–49

Garratt PA (1993) Spawning of the riverbream, *Acanthopagrus berda*, in Kosi Estuary. S Afr J Zool 28:26–31

Gladstone W (1986) Spawning behavior of the bumphead parrotfish *Bolbometapon muricatum* at Yonge Reef, Great Barrier Reef. Jpn J Ichthyol 33:326–328

Gladstone W (1994) Lek-like spawning, parental care and mating periodicity of the triggerfish *Pseudobalistes flavimarginatus* (Balistidae). Environ Biol Fishes 39:249–257

Gladstone W (1996) Unique annual aggregation of longnose parrotfish (*Hipposcarus harid*) at Farasan Island (Saudi Arabia, Red Sea). Copeia 1996:483–485

Glaister JP, Lau T, McDonall VC (1987) Growth and migration of tagged Eastern Australian king prawns, *Penaeus plebejus* Hess. Aust J Mar Freshw Res 38:225–241

Graham RT, Carcamo R, Rhodes KL et al (2008) Historical and contemporary evidence of a mutton snapper (*Lutjanus analis* Cuvier, 1828) spawning aggregation fishery in decline. Coral Reefs 27:311–319

Graham RT, Castellanos DW (2005) Courtship and spawning of Carangid species in Belize. Fish Bull 103:426–432

Hamilton R (2003) A report on the current status of exploited reef fish aggregations in the Solomon Islands and Papua New Guinea – Choiseul, Ysabel, Bouganville and Manus Provinces. In: Western Pacific Fisher Survey Series, Vol 1. Society for the Conservation of Reef Fish Aggregations.

Hamilton R (2005) Indigenous ecological knowledge (IEK) of the aggregating and nocturnal spawning behavior of the longfin emperor, *Lethrinus erythropterus*. SPC Tradit Mar Resour Manage Knowl Inf Bull 18:9–17

Hamilton R, Matawai M, Potuku T et al (2005) Applying local knowledge and science to the management of grouper aggregation sites in Melanesia. SPC Tradit Mar Resour Manag Knowl Inf Bull 14:7–19

Hamner WM, Jones MS, Carlton JH et al (1988) Zooplankton, planktivorous fish, and water currents on a windward reef face: Great Barrier Reef, Australia. Bull Mar Sci 42:459–479

Hasse JJ, Madraisau BB, McVey JP (1977) Some aspects of the life history of *Siganus canalicula-tus* (Park) (Pisces: Siganidae) in Palau. Micronesica 13:297–312

Helfman GS, Collette BB, Facey DE (1997) Fishes as social animals: reproduction. In: Helfman GS, Collette BB, Facey DE (eds) The diversity of fishes, pp. 348–364. Blackwell Science Press, Malden, Massachusetts

Helfrich P, Allen PM (1975) Observations on the spawning of mullet, *Crenimugil crenilabis* (Forskol), at Enewetak, Marshall Island. Micronesica 11:219–225

Hensley DA, Appeldoorn RS, Shapiro DY et al (1994) Egg dispersal in a Caribbean coral reef fish, *Thalassoma bifasciatum*. I. Dispersal over the reef platform. Bull Mar Sci 54:256–270

Herrnkind WF (1980) Spiny lobsters: patterns of movement. In: Cobb JS, Phillips BF (eds) The biology and management of lobsters, Vol. 1, pp. 349–407. Academic Press, New York

Heyman W, Requena N (2002) Status of multi-species spawning aggregations in Belize. The Nature Conservancy, Punta Gorda, Belize

Heyman WD, Graham RT, Kjerfve B et al (2001) Whale sharks *Rhincodon typus* aggregate to feed on fish spawn in Belize. Mar Ecol Prog Ser 215:275–282

Heyman WD, Kjerfve B, Graham RT et al (2005) Spawning aggregations of *Lutjanus cyanopterus* (Cuvier) on the Belize Barrier Reef over a 6 year period. J Fish Biol 67:83–101

Hill BJ (1994) Offshore spawning by the portunid crab *Scylla serrata* (Crustacea: Decapoda). Mar Biol 120:379–384

Hoffmayer ER, Franks JS, Driggers WB et al (2007) Observations of a feeding aggregation of whale sharks, *Rhincodon typus*, in the north central Gulf of Mexico. Gulf Caribb Res 19: 69–73

Hogan Z, Baird IG, Radtke R et al (2007) Long distance migration and marine habitation in the tropical Asian catfish, *Pangasius krempfi*. J Fish Biol 71:818–832

Johannes RE (1978) Reproductive strategies of coastal marine fishes in the tropics. Environ Biol Fish 3:65–84

Johannes RE (1981) Words of the lagoon. Fishing marine lore in the Palau District of Micronesia. University of California Press, Los Angeles

Johannes RE (1988) Spawning aggregation of the grouper, *Plectropomus areolatus* (Ruppel) in the Solomon Islands. Proc 6th Int Coral Reef Symp 2:751–755

Johannes RE, Squire L, Graham T et al (1999) Spawning aggregations of groupers (Serranidae) in Palau Marine conservation research series publ # 1. The Nature Conservancy, 144 pp.

Johannes RE, Yeeting B (2001) I-Kiribati knowledge and management of Tarawa's lagoon resources. Atoll Res Bull 489:1–25

Jones GP, Milicich MI, Emslie MJ et al (1999) Self-recruitment in a coral reef fish population. Nature 402:802–804

Kadison E, Nemeth RS, Blondeau JE (2009) Assessment of an unprotected red hind (*Epinephelus guttatus*) spawning aggregation on Saba Bank in the Netherlands Antilles. Bull Mar Sci: in press

Kadison E, Nemeth RS, Herzlieb S et al (2006) Temporal and spatial dynamics of *Lutjanus cyanopterus* and *L. jocu* (Pisces: Lutjanidae) spawning aggregations on a multi-species spawn-ing site in the USVI. Rev Biol Trop 54:69–78

Kanciruk P, Herrnkind WF (1976) Autumnal reproduction of spiny lobster, *Panulirus argus*, at Bimini Bahamas. Bull Mar Sci 26:417–432

Kanciruk P, Herrnkind WF (1978) Mass migration of spiny lobster, *Panulirus argus* (Crustacea: Palinuridae): behavior and environmental correlates. Bull Mar Sci 28:601–623

Kiflawi M, Mazeroll AI, Goulet D (1998) Does mass spawning enhance fertilization suc-cess in coral reef fish? A case study of the brown surgeonfish. Mar Ecol Prog Ser 172:107–114

Koenig CC, Coleman FC, Collins LA et al (1996) Reproduction in Gag (*Mycteroperca microlepis*) (Pisces: Serranidae) in the Eastern Gulf of Mexico and the consequences of fishing spawning aggregations. ICLARM Conf Proc 48:307–323

Krajewski JP, Bonaldo RM (2005) Spawning out of aggregations: record of a single spawning dog snapper pair at Fernando de Noronha Archipelago, Equatorial Western Atlantic. Bull Mar Sci 77:165–167

Lam TJ (1983) Environmental influences on gonadal activity in fish. In: Hoar WS, Randall DJ, Donaldson EM (eds) Fish physiology, Vol 9 (A), pp. 65–116. Academic Press, New York

Levin PS, Grimes CB (2002) Reef fish ecology and grouper conservation and management. In: Sale PF (ed) Coral reef fishes. Dynamics and diversity in a complex ecosystem, pp. 377–389. Academic Press, London

Lindeman KC, Pugliese R, Waugh GT et al (2000) Developmental patterns within a multi-species reef fishery: management applications for essential fish habitats and protected areas. Bull Mar Sci 33:929–956

Lipcius RN, Stockhausen WT, Seitz RD et al (2003) Spatial dynamics and value of a marine protected area and corridor for the blue crab spawning stock in Chesapeake Bay. Bull Mar Sci 72:453–469

Luckhurst BE (1998) Site fidelity and return migration of tagged red hinds (*Epinephelus guttatus*) to a spawning aggregation site in Bermuda. Proc 50th Gulf Caribb Fish Inst 50:750–763

Lyons WG, Barber DG, Foster SM et al (1981) The spiny lobster, *Panulirus argus*, in the middle and upper Florida Keys: population structure, seasonal dynamics, and reproduction. FLA Mar Res 38:1–38

MacFarlane JW, Moore R (1986) Reproduction of the ornate rock lobster, *Panulirus ornatus* (Fabricius), in Papua New Guinea. Aust J Mar Freshw Res 37:55–65

Mackie M (2000) Reproductive biology of the half moon grouper, *Epinephelus rivulatus*, at Ningaloo reef, Western Australia. Environ Biol Fish 57:363–376

Mapstone BD, Fowler AJ (1988) Recruitment and the structure of assemblages of fish on coral reefs. Trends Ecol Evol 3:72–77

Marx JM, Herrnkind WF (1986) Species profiles: life histories and environmental requirements of coastal fishes and invertebrates (south Florida) – spiny lobster. U.S. Fish and Wildlife Service Biological Report # TR-EL-82-4. U.S. Army Corps of Engineers, 82(11–61), 21 pp.

Mazeroll AI, Montgomery WL (1995) Structure and organization of local migrations in brown surgeonfish (*Acanthurus nigrofuscus*). Ethology 99:89–106

Mazeroll AI, Montgomery WL (1998) Daily migrations of a coral reef fish in the Red Sea (Gulf of Aqaba, Israel): initiation and orientation. Copeia 1998:893–905

McCann K, Rasmussen J, Umbanhower J et al (2005) The role of space, time and variability in food web dynamics. In: de Ruiter PC, Wolters V, Moore JC (eds) Dynamic food webs, pp. 56–70. Elsevier, Amsterdam

Millikin MR, Williams AB (1984) Synopsis of biological data on the blue crab, *Callinectes sapidus* Rathbun. FAO Fisheries Synopsis, NOAA Technical Report, NMFS 1, 43 pp.

Milton DA, Chenery SR (2005) Movement patterns of barramundi *Lates calcarifer*, inferred from 87Sr/86Sr and Sr/Ca ratios in otoliths, indicate non-participation in spawning. Mar Ecol Prog Ser 270:279–291

Moe MA (1969) Biology of red grouper. Prof Pap Ser – Fla Dept Nat Resour, Mar Res Lab 10:1–95

Molloy PP, Reynolds JD, Gage MJG et al (2008) Links between sex change and fish densities in marine protected areas. Biol Conserv 141:187–197

Moore R, MacFarlane JW (1984) Migration of the ornate rock lobster, *Panulirus ornatus* (Fabricius) in Papua New Guinea. Aust J Mar Freshw Res 35:197–212

Moore R, Reynolds LF (1982) Migration patterns of barramundi *Lates calcarifer* in Papua New Guinea. Aust J Mar Freshw Res 33:671–682

Motro R, Ayalon I, Genin A (2005) Near-bottom depletion of zooplankton over coral reefs. III: vertical gradient of predation pressure. Coral Reefs. 24:95–98

Moyer JT (1987) Quantitative observations of predation during spawning rushes of the labrid fish, *Thalassoma cupido* at Miyake-Jima, Japan. Jpn J Ichthyol 34:76–81

Moyer JT (1989) Reef channels as spawning sites for fishes on the Shiraho coral reef, Ishigaki Island, Japan. Jpn J Ichthyol 36:371–375

Munro AD (1990) General introduction. In: Munro AD, Scott AP, Lam TJ (eds) Reproductive seasonality in teleosts: environmental influences. CRC press, pp. 1–12

Myers RF (1991) Micronesian reef fishes. Coral Graphics, Guam

Myrberg AA, Montgomery WL, Fishelson L (1988) The reproductive behavior of *Acanthurus nigrofuscus* (Forskal) and other surgeonfishes (Fam. Acanthuridae) off Eilat, Israel (Gulf of Aqaba, Red Sea). Ethology 79:31–61

Nemeth RS (2005) Population characteristics of a recovering US Virgin Islands red hind spawning aggregation following protection. Mar Ecol Prog Ser 286:81–97

Nemeth, RS (2009) Ecosystem aspects of species that aggregate to spawn. In: Sadovy Y, Colin P, (eds.) Reef fish spawning aggregations: biology, research and management. Springer, the Netherlands (in press)

Nemeth RS, Blondeau J, Herzlieb S et al (2007) Spatial and temporal patterns of movement and migration at spawning aggregations of red hind, *Epinephelus guttatus*, in the U.S. Virgin Islands. Environ Biol Fish 78:365–381

Nemeth RS, Herzlieb S, Blondeau J (2006a) Comparison of two seasonal closures for protecting red hind spawning aggregations in the US Virgin Islands. Proc 10th Int Coral Reef Symp, pp.1306–1313

Nemeth RS, Kadison E, Blondeau, JE et al (2008) Regional coupling of red hind spawning aggregations to oceanographic processes in the Eastern Caribbean. In: Grober-Dunsmore R, Keller BD (eds) Caribbean connectivity: implications for marine protected area management, pp.170–183. Proceedings of a Special Symposium, 9–11 November 2006, 59th Annual Meeting of the Gulf and Caribbean Fisheries Institute, Belize City, Belize. Marine Sanctuaries Conservation Series ONMS-08-07. U.S. Department of Commerce, National Oceanic and Atmospheric Administration, National Marine Sanctuary Program, Silver Spring, Maryland

Nemeth RS, Kadison E, Herzlieb S et al (2006b) Status of a yellowfin grouper (*Mycteroperca venenosa*) spawning aggregation in the US Virgin Islands with notes on other species. Proc 57th Gulf Caribb Fish Inst 57:543–558

Olsen DA, LaPlace JA (1978) A study of Virgin Islands grouper fishery based on a breeding aggregation. Proc 31st Gulf Caribb Fish Inst 31:130–144

Paris CB, Cowen RK (2004) Direct evidence of a biophysical retention mechanism for coral reef fish larvae. Limnol Oceanogr 49:1964–1979

Pelaprat C (2002) Observations on the spawning behaviour of the dusky grouper *Epinephelus marginatus* (Lowe, 1834) in the North of Corsica (France). Mar Life 91:59–65

Pet JS, Mous PJ, Muljadi AH et al (2005) Aggregations of *Plectropomis areolatus* and *Epinephelus fuscoguttatus* (groupers, Serranidae) in the Komodo National Park, Indonesia: monitoring and implications for management. Environ Biol Fish 74:209–218

Petersen CW, Warner RR (2002) The ecological context of reproductive behavior. In: Sale PF (ed) Coral reef fishes: dynamics and diversity in a complex ecosystem, pp. 103–118. Academic Press, London

Petersen CW, Warner RR, Shapiro DY et al (2001) Components of fertilization success in the bluehead wrasse, *Thalassoma bifasciatum*. Behav Ecol 12:237–245

Pickert P, Kelly T, Nemeth RS et al (2006) Seas of change: spawning aggregations of the Virgin Islands. In: Pickert P, Kelly T (eds) DVD documentary, Friday's Films, San Francisco

Pittman SJ, McAlpine CA (2003) Movement of marine fish and decapod crustaceans: process, theory and application. Adv Mar Biol 44:205–294

Pollock BR (1982) Movements and migrations of yellowfin bream *Acanthopagrus australis* (Gunther), in Moreton Bay, Queensland as determined by tag recoveries. J Fish Biol 20: 245–252

Pollock BR (1984) Relations between migration, reproduction and nutrition in yellowfin bream, *Acanthopagrus australis*. Mar Ecol Prog Ser 19:17–23

Pratt HL, Carrier JC (2001) A review of elasmobranch reproductive behavior with a case study on the nurse shark, *Ginglymostoma cirratum*. Environ Biol Fish 60:157–188

Quackenbush LS, Herrnkind WF (1981) Regulation of molt and gonadal development in the spiny lobster, *Panulirus argus* (Crustacea: Palinuridae): effect of eyestalk ablation. Comp Biochem Physiol 69A:523–527

Randall JE (1961a) A contribution to the biology of the convict surgeonfish of the Hawaiian Islands, *Acanthurus triostegus sandvicensis*. Pac Sci 15:215–272

Randall JE (1961b) Observations on the spawning of surgeonfishes (Acanthuridae) in the Society Islands. Copeia 1961:237–238

Randall JE, Randall HA (1963) The spawning and early development of the Atlantic parrot fish, *Sparisoma rubripinne*, with notes on other scarid and labrid fishes. Zoologica 48: 49–60

Rhodes KL (2003) Spawning aggregation survey: federated States of Micronesia. In: Western Pacific fisher survey series, Vol 2. Society for the conservation of reef fish aggregations, 32 pp.

Rhodes KL, Sadovy Y (2002) Temporal and spatial trends in spawning aggregations of camouflage grouper, *Epinephelus polyphekadion*, in Pohnpei, Micronesia. Environ Biol Fish 63:27–39

Rhodes K, Tupper MH (2008) The vulnerability of reproductively active squaretail coral grouper (*Plectropomus areolatus*) to fishing. Fish Bull 106:194–203

Rivera JA, Prada MC, Arsenault JL et al (2005) Detecting fish aggregations from reef habitats mapped with high resolution side scan sonar imagery. Natl Mar Fish Serv Prof Pap 5:88–104

Robertson DR (1983) On the spawning behavior and spawning cycles of eight surgeonfishes (Acanthuridae) from the Indo-Pacific. Environ Biol Fish 9:192–223

Robertson DR (1991) The role of adult biology in the timing of spawning of tropical reef fishes. In: Sale PF (ed) The ecology of fishes on coral reefs, pp. 356–370. Academic Press, Inc., San Diego

Robertson DR, Christopher WP, Brawn JD (1990) Lunar reproductive cycles of benthic-brooding reef fishes: reflections of larval biology or adult biology? Ecol Monogr 60:311–329

Robertson DR, Hoffman SG (1977) The roles of female mate choice and predation in the mating systems of some tropical Labroid fishes. Z Tierpsychol 45:298–320

Robertson DR, Swearer SE, Kaufmann K et al (1999) Settlement vs. environmental dynamics in a pelagic-spawning reef fish at Caribbean Panama. Ecol Monogr 69:195–218

Robinson J, Aumeeruddy R, Jörgensen TL et al (2008) Dynamics of camouflage (*Epinepuelus polyphekadion*) and brown marbled grouper (*Epinephelus fuscoguttatus*) spawning aggregations at a remote reef site, Seychelles. Bull Mar Sci 83:415–431.

Ruello NV (1975) Geographical distribution, growth and breeding migration of the Eastern Australian king prawn *Penaeus plebejus* Hess. Aust J Mar Freshw Res 26:343–354

Sadovy Y (1994) Grouper stocks of the western Atlantic: the need for management and management needs. Proc 43rd Gulf Caribb Fish Inst 43:43–65

Sadovy Y (1996) Reproduction of reef fishes. In: Polunin NVC, Roberts CM (eds) Reef fisheries, pp. 15–59. Chapman and Hall, London

Sadovy Y (1997) The case of the disappearing grouper: *Epinephelus striatus* (Pisces: Serranidae). J Fish Biol 46:961–976

Sadovy Y (2004) A report on the current status and history of exploited reef fish aggregations in Fiji. In: Western Pacific fisher survey series, Vol 4. Society for the conservation of reef fish aggregations.

Sadovy Y, Colin PL (1995) Sexual development and sexuality in the Nassau grouper. J Fish Biol 46:961–976

Sadovy Y, Colin PL, Domeier ML (1994a) Aggregation and spawning in the tiger grouper, *Mycteroperca tigris* (Pisces: Serranidae). Copeia 1994:511–516

Sadovy Y, Domeier M (2005) Are aggregation-fisheries sustainable? Reef fish fisheries as a case study. Coral Reefs 24:254–262

Sadovy Y, Eklund A-M (1999) Synopsis of biological data on the Nassau grouper, *Epinephelus striatus* (Bloch, 1792), and the Jewfish, *E. itajara* (Lichtenstein, 1822) NOAA Tech Rep NMFS 146

Sadovy Y, Figuerola M, Roman A (1992) Age and growth of red hind *Epinephelus guttatus* in Puerto Rico and St. Thomas. Fish Bull 90:516–528

Sadovy Y, Liu M (2004) Report on current status and exploitation history of reef fish spawning aggregations in Eastern Indonesia. In: Western Pacific fisher survey series, Vol 6. Society for the conservation of reef fish aggregations.

Sadovy Y, Colin PL, Domeier ML (1994a) Aggregation and spawning in the tiger grouper, *Mycteroperca tigris* (Pisces: Serranidae). Copeia 1994: 511–516

Sadovy Y, Rosario A, Roman A (1994b) Reproduction in an aggregating grouper, the red hind, *Epinephelus guttatus*. Environ Biol Fish 41:269–286

Sadovy de Mitcheson Y, Cornish A, Domeier M et al (2008) A global baseline for spawning aggregations of reef fishes. Conserv Biol 22:1233–1244

Sala E, Aburto-Oropeza O, Paredes G et al (2003) Spawning aggregations and reproductive behavior of reef fishes in the Gulf of California. Bull Mar Sci 72:103–121

Sala E, Ballesteros E, Starr RM (2001) Rapid decline of Nassau Grouper spawning aggregations in Belize: fishery management and conservation needs. Fisheries 26:23–30

Sale PF (1980) The ecology of fishes on coral reefs. Oceanogr Mar Biol Annu Rev 18:367–421

Sale PF (1991) Reef fish communities: open nonequilibrial systems. In: Sale PF (ed) The ecology of fishes on coral reefs, pp. 564–596. Academic Press, Inc., San Diego

Samoilys MA (1997) Periodicity of spawning aggregations of coral trout (*Plectropomous leopardus*) on the Great Barrier Reef. Mar Ecol Prog Ser 160:149–159

Samoilys MA, Squire LC (1994) Preliminary observations on the spawning behavior of coral trout, *Plectropomus leopardus* (Pisces: Serrandae), on the Great Barrier Reef. Bull Mar Sci 54:332–342

Sancho G, Petersen CW, Lobel PS (2000a) Predator-prey relations at a spawning aggregation site of coral reef fishes. Mar Ecol Prog Ser 203:275–288

Sancho G, Solow AR, Lobel PS (2000b) Environmental influences on the diel timing of spawning in coral reef fishes. Mar Ecol Prog Ser 206:193–212

SCRFA (2004) Spawning aggregation flobal database. Society for the conservation of reef fish aggregations

Shapiro DY, Hensley DA, Appeldoorn RS (1988) Pelagic spawning and egg transport in coral reef fishes: a skeptical overview. Environ Biol Fish 22:3–14

Shapiro DY, Sadovy Y, McGehee MA (1993) Size, composition, and spatial structure of the annual spawning aggregation of the red hind, *Epinephelus guttatus* (Pisces: Serranidae). Copeia 1993:399–406

Sheaves M (2006) Is the timing of spawning in sparid fishes a response to sea temperature regimes? Coral Reefs 25:655–669

Sheaves MJ, Molony BW, Tobin AJ (1999) Spawning migrations and local movements of a tropical sparid fish. Mar Biol 133:123–128

Sinclair M (1988) Marine populations: essay on population regulation and speciation. Washington Sea Grant Publication, Seattle

Sluka RD (2001) Grouper and Napoleon wrasse ecology in Laamu Atoll, Republic of Maldives: Part 2. Timing, location, and characteristics of spawning aggregations. Atoll Res Bull 492:1–17

Smith CL (1972) A spawning aggregation of Nassau grouper, *Epinephelus striatus* (Bloch). Trans Am Fish Soc 101:257–261

Squire L (2001) Live reef fish trade at M'burke Island, Manus Province: a survey of spawning aggregation sites, monitoring and management guidelines, The Nature Conservancy, 32 pp.

Starr RM, Sala E, Ballesteros E et al (2007) Spatial dynamics of the Nassau grouper *Epinephelus striatus* in a Caribbean atoll. Mar Ecol Prog Ser 343:239–249

Stone GS (2004) Phoenix islands. Natl Geogr February 2004:48–65

Swearer SE, Caselle JE, Lea DW et al (1999) Larval retention and recruitment in an island population of a coral-reef fish. Nature 402:799–802

Takemura A, Rahman MS, Nakamura S et al (2004) Lunar cycles and reproductive activity in reef fishes with particular attention to rabbitfishes. Fish Fish 5:317–328

Tankersley RA, Wieber MC, Sigala MA et al (1998) Migratory behavior of ovigerous blue crabs *Callinectes sapidus*: evidence for selective tidal-stream transport. Biol Bull 195:168–173

Thresher RE (1984) Reproduction in reef fishes. TFH Publication, Neptune City

Tucker JWJ, Bush PG, Slaybaugh ST (1993) Reproductive patterns of Cayman Islands Nassau grouper (*Epinephelus striatus*) populations. Bull Mar Sci 52:961–969

van Rooij J, Kroon F, Videler J (1996) The social and mating system of the herbivorous reef fish *Sparisoma viride*: one-male versus multi-male groups. Environ Biol Fish 47:353–378

Warner RR (1988) Traditionality of mating-site preferences in a coral reef fish. Nature 335:719–721

Warner RR (1990a) Male versus female influences on mating-site determination in a coral reef fish. Anim Behav 39:540–548

Warner RR (1990b) Resource assessment versus tradition in mating-site determination. Am Nat 135:205–217

Warner RR (1995) Large mating aggregations and daily long-distance spawning migrations in the bluehead wrasse, *Thalassoma bifasciatum*. Environ Biol Fish 44:337–345

Warner RR, Robertson DR (1978) Sexual patterns in the labroid fishes of the western Caribbean. I. The wrasses (Labridae). Smithson Contrib Zool 254:1–27

Whaylen L, Pattengill-Semmens CV, Semmens BX et al (2004) Observations of a Nassau grouper (*Epinephelus striatus*) spawning aggregation site in Little Cayman Island. Environ Biol Fish 70:305–313

White DB, Wyanski DM, Eleby BM (2002) Tiger grouper (*Mycteroperca tigris*): profile of a spawning aggregation. Bull Mar Sci 70:233–240

Wicklund R (1969) Observations on spawning of lane snapper. Underwater Nat 6:40

Ye Y, Prescott J, Dennis DM (2006) Sharing the catch of migratory rock lobster (*Panulirus ornatus*) between sequential fisheries of Australia and Papua New Guinea. http://www.cmar.csiro.au/e-print/open/yey_2006a.pdf. Sharing the fish – allocation issues in fisheries management, pp. 1–11. Department of Fisheries Western Australia, Perth, WA:1–11

Yogo Y, Nakazono A, Tsukahara H (1980) Ecological studies on the spawning of the parrotfish, *Scarus sordidus* Forsskal. Sci Bull Fac Agri Kyushu Univ 34:105–114

Zeller DC (1997) Home range and activity patterns of the coral trout *Plectropomus leopardus* (Serranidae). Mar Ecol Prog Ser 154:65–77

Zeller DC (1998) Spawning aggregations: patterns of movement of the coral trout *Plectropomus leopardus* (Serranidae) as determined by ultrasonic telemetry. Mar Ecol Prog Ser 162:253–263

Zerbi A, Aliaume C, Miller JM (1999) A comparison of two tagging techniques with notes on juvenile tarpon ecology in Puerto Rico. Bull Mar Sci 64:9–19

Appendix 4.1 Spatial and temporal aspects of species forming resident fish spawning aggregations (FSA) including maximum estimated number of fish in aggregation (FSA size), area occupied by aggregation (FSA area), migration distance (MD), and catchment or functional migration area (FMA). $\sim$ = value estimated from data in citation. * these species have been classified as also forming transient aggregations or are unspecified. § indicates that information is partly or entirely based on fishermen interviews or other undocumented sources and spawning has not been verified for all locations. Feeding and FSA habitats from citation or Myers 1991 (see footnote of Appendix 4.2 for references)

Species	Feeding habitat (adults)	FSA habitat	Geographic area (References)	FSA size (n)	FSA area in m² (max)	MD in km (range)	FMA in km² (max)
ACANTHURIDAE							
Acanthurus bahianus	5,6,12	12,14	Puerto Rico (1)	20,000	4,800	0.9–1	~2
A. coeruleus	5,6,12	12,14	Puerto Rico (1)	<7,000	1,500	0.6	~1.2
			Belize (64)	800	–	–	–
			Lee Stocking Island, Bahamas (29)	300	–	–	–
A. guttatus	11	9,10	American Samoa (72)	–	–	–	–
A. lineatus §	–	9,10,14	American Samoa (72)	–	–	–	–
			Papua New Guinea (78)	–	–	–	–
A. mata	14	4,11	Palau (40)	–	–	–	–
A. nigrofuscus	4,5,6	9,10,14	Eilat, Israel (42, 47, 60)	>2,500	–	0.5–1.5	~0.45
			Aldabra (43)	>3,000	2,500	0.3–2.0	–
			Palau (43)	300	500	0.4	–
			Lizard Island, Australia (43)	>200	–	–	–
A. nigroris	4,5,11	9	Johnston Atoll (54, 55)	200	175	–	–

Appendix 4.1 (continued)

Species	Feeding habitat (adults)	FSA habitat	Geographic area (References)	FSA size (n)	FSA area in m² (max)	MD in km (range)	FMA in km² (max)
A. triostegus	4,5,11	9,10 by 14	Hawaii (44)	—	—	—	—
			Society Island (45)	—	—	—	—
			Aldabra (43)	20,000	—	2	—
			American Samoa (72)	—	—	—	—
Ctenochaetus striatus §	4,5,11	9,10 by 14	Palau (23, 43, 78)	>2,000	—	—	—
			Lizard Island, Australia (43)	—	—	0.05	—
			Aldabra (43)	—	—	—	—
			French Polynesia (40)	—	—	—	—
			Eilat, Israel (40)	—	—	—	—
			Solomon Islands (78)	—	—	—	—
C. strigosus	4,5,11	9	Johnston Atoll (54, 55)	30	175	—	—
*Naso literatus** §	4,5,6,12	9,10 by 14	Palau (23, 79 in 78)	>1,000	—	—	—
			Federated States of Micronesia (23,78,79)	—	—	—	—
*N. lopezi** §	—	9,10	Federated States of Micronesia (79 in 78)	—	—	—	—
*N. unicornis** §	4,5,6,12	9,10 by 14	Palau (23)	>1,000	—	—	—
Paracanthurus hepatus	14	10,14	Escape Reef, Great Barrier Reef, Australia (43)	30	200	0.2	—

Appendix 4.1 (continued)

Species	Feeding habitat (adults)	FSA habitat	Geographic area (References)	FSA size (n)	FSA area in m^2 (max)	MD in km (range)	FMA in km^2 (max)
Zebrasoma flavescens	4,5,11	9	Johnston Atoll (54, 55)	100	175	–	–
Z. scopas	–	9,10	French Polynesia (40)	–	–	–	–
Z. veliferum	4,5,11	9,10 by 14	Palau (43)	20	500	0.3	–
			Aldabra (43)	–	–	–	–
CAESIONIDAE							
Caesio teres	12	10	Enewetak Atoll, Marshall Islands (69)	1,000	–	–	–
CARANGIDAE							
Caranx bartholomaei	12,13,14	14	Belize (110)	–	–	–	–
C. latus	7,12,13,14	14	Cayman Islands (80)	–	–	–	–
C. lugubris	12,13,14	14	Cayman Islands (80)	–	–	–	–
C. ruber	5,6,7,12,13,14	14	Cayman Islands (80)	–	–	–	–
C. sexfasciatus §	6,9,10,14	14	Gulf of California, Mexico (73)	1,500	10,000	–	–
C. tille §	–	9,10	Papua New Guinea (78)	–	–	–	–
			Papua New Guinea (81 in 78)	–	–	–	–
Decapterus macarellus	12,13,14	14	Cayman Islands (80)	–	–	–	–

Appendix 4.1 (continued)

Species	Feeding habitat (adults)	FSA habitat	Geographic area (References)	FSA size (n)	FSA area in m² (max)	MD in km (range)	FMA in km² (max)
Selaroides sp. §	–	7	Papua New Guinea (78)	–	–	–	–
Seriola lalandi	14	–	Gulf of California, Mexico (73)	80	10,000	–	–
Trachinotus falcatus	2,3,4,5	14	Belize (110)	500	–	–	–
GERREIDAE							
Gerres cinereus	1,2,3,5	9	Turks and Caicos (111)	100,000	–	–	–
Gerres sp.	3,6	6	Tarawa Atoll, Kiribati (82)	–	–	–	–
LABRIDAE							
Cheilinus undulatus §	6,9,10,14	9,10,14	Malaysia (78, 83)	–	–	–	–
			Palau (78, P Colin, Coral Reef Research Foundation, pers. comm.)	150	7,500	–	–
Choerodon anchorago §	4	14	Palau (2)	–	–	–	–
Clepticus parrai	13, 14	14	Puerto Rico (1)	~300	~600	–	–
Thalassoma amblycephalum §	6,14	14	Fiji (78. 84)	–	–	–	–
			Enewetak Atoll, Marshall Islands (85)				
			Papua New Guinea (78)				

Appendix 4.1 (continued)

Species	Feeding habitat (adults)	FSA habitat	Geographic area (References)	FSA size (n)	FSA area in m² (max)	MD in km (range)	FMA in km² (max)
T. bifasciatum	Upcurrent coral reef areas	5,7,13	San Blas, Panama (24)	200–400	<25	<1.5	~2.25
			US Virgin Islands (3)	80–100	–	–	–
			Bahamas (40)	–	–	–	–
T. hardwicki §	5	–	American Samoa (72)	–	–	–	–
			Palau (78)	–	–	–	–
			Papua New Guinea (78)	–	–	–	–
T. quinquevitta-tum	5,7,11	–	American Samoa (72)	–	–	–	–
SCARIDAE							
Bolbometopon muricatum §	4,5,14	9,10,14	Yonge Reef, Great Barrier Reef, Australia (90)	–	–	–	–
			Malaysia (83 in 78)	–	–	–	–
			Papua New Guinea (78)	–	–	–	–
Chlorurus frontalis §	5	5	Federated States of Micronesia (79 in 78)	–	–	–	–
*C. microrhinos** §	5	4,5,9	Marshall Islands (85)	–	–	–	–
			Federated States of Micronesia (79 in 78)	–	–	–	–
C. sordidus §	5,7	9,10,11,14	Johnston Atoll (54, 55)	300	175	–	–
			Japan (93 in 78)	–	–	–	–
			Marshall Islands (85)	–	–	–	–
			Palau (78)	–	–	–	–

Appendix 4.1 (continued)

Species	Feeding habitat (adults)	FSA habitat	Geographic area (References)	FSA size (n)	FSA area in m² (max)	MD in km (range)	FMA in km² (max)
Hipposcarus harid	4,5,11	5	Farasan Island, Saudi Arabia (117)	>200	~1,500	–	–
Hipposcarus longiceps §	3,6,7,8,14	4,5,9,10,12	Kiribati (94)	–	–	–	–
			Federated States of Micronesia (79 in 78)	–	–	–	–
			Palau (78)				
			Papua New Guinea (78)	–	–		
Scarus iserti	5,7,9,11,12	12,14	Puerto Rico (1)	100	~225	–	–
			Jamaica (29)	–	–	–	–
			Papua New Guinea (78)				
Scarus prasignathos	13	3	Papua New Guinea (78)	–	–	–	–
Sparisoma rubripinne §	4,5,11,12	3,12	St. John, US Virgin Islands (3, 29)	200	<1,000	–	–
			St. Thomas, US Virgin Islands (52)	100	500		
			Bermuda (78)	–	–	–	–
SPHYRAENIDAE							
Sphyraena barracuda * §	1,6,12,14	–	Papua New Guinea (81 in 78)	–	–	–	–
S. genie * §	10,14	–	Papua New Guinea (81 in 78)	–	–	–	–

Habitat types: 1. estuaries, mangrove creeks, or rivers; 2. seagrass bed; 3. sand flat; 4. coral reef flats; 5. shallow lagoonal reef slopes or back reef; 6. deep lagoon or embayments; 7. patch reefs or wrecks; 8. sand flats, rubble areas close to reef slopes, or fringing reef; 9. channel mouth or channel through fringing reef; 10. channel through barrier reef; 11. reef crest; 12. nearshore or mid-shelf fringing reef; 13. top of shelf edge fringing reef or exposed seaward reefs; 14. outer reef slope or shelf edge

Appendix 4.2 Spatial and temporal aspects of species forming transient spawning aggregations (FSA) including maximum estimated number of fish and decapods in aggregation (FSA size), area occupied by aggregation (FSA area), migration distance (MD), and catchment or functional migration area (FMA). ~ = value estimated from data in citation (table modified from Johannes 1978). * these species have been classified as also forming resident aggregations or are unspecified. § indicates that information is partly or entirely based on fishermen interviews or other undocumented sources and has not been verified for all locations (see footnote of Appendix 4.1 for habitat codes)

Species	Feeding habitat (adults)	FSA habitat	Geographic area (References)	FSA size (n)	FSA area in m² (max)	MD in km (range)	FMA in km² (max)
BALISTIDAE							
Balistes vetula	12,13	8	St. Croix, US Virgin Islands (52)	>100	1,000	–	–
			Saba Bank, Netherlands Antilles (52)	>500	5,000	–	–
Canthidermis sufflamen	14	14	Cayman Islands (80)	–	–	–	–
Pseudobalistes flavimarginatus	6	10	Yonge Reef, Great Barrier Reef, Australia (92)	–	–	–	–
			Palau (41)	10	600	–	–
CARCHARHINIDAE							
Negaprion brevirostris	12,14	2	Bimini, Bahamas (76)	–	–	–	–
CENTROPOMIDAE							
Lates calcarifer	1,12	12	Papua New Guinea (77)	–	–	15–300	–
ELOPIDAE							
Megalops atlanticus	1,5,6,9	14	Florida, USA (61, 62)	250	–	25	–

Appendix 4.2 (continued)

Species	Feeding habitat (adults)	FSA habitat	Geographic area (References)	FSA size (n)	FSA area in m² (max)	MD in km (range)	FMA in km² (max)
KYPHOSIDAE							
Kyphosus bigibbus §	14	14	Federated States of Micronesia (79 in 78)	–	–	–	–
K. cinerascens §	4,11,13	14	Federated States of Micronesia (79 in 78)	–	–	–	–
K. vaigensis §	4,11,13	14	Federated States of Micronesia (79 in 78)	–	–	–	–
LETHRINIDAE							
Lethrinus atkinsoni §	2,3,5,8	14	Ryukyu Island, Japan (86 in 78)	–	–	–	–
L. erythropterus §	14	9,10	Papua New Guinea (81 in 78)	–	–	–	–
			Solomon Islands (78,112)	10,000	20,000	–	–
L. harak §	2,3,5,7,10	14	Federated States of Micronesia (79 in 78)	–	–	–	–
L. nebulosus §	2,5,7	14	Egypt (78) Japan (87 in 78)	–	–	–	–
L. olivaceus §	5,14	9	Federated States of Micronesia (79 in 78)	–	–	–	–
L. xanthochilus §	2,3,5,7	9	Federated States of Micronesia (79 in 78)	–	–	–	–
Monotaxis grandoculis §	8,9,10	4	Federated States of Micronesia (79 in 78)	–	–	–	–

Appendix 4.2 (continued)

Species	Feeding habitat (adults)	FSA habitat	Geographic area (References)	FSA size (n)	FSA area in m² (max)	MD in km (range)	FMA in km² (max)
LUTJANIDAE							
Lutjanus analis	8	8,13,14	Belize (63, 64)	—	—	—	—
			St. Thomas, US Virgin Islands (52)	100	2,500	—	—
			St. Croix, US Virgin Islands (52)	—	—	—	—
			Cuba (28)	—	—	—	—
			Bahamas (78)	—	—	—	—
			Turks and Caicos (40)	—	—	—	—
			Dry Tortugas, USA (88, 91)	300	<40,000	—	—
			Florida Keys, USA (88)	—	—	—	—
L. apodus	12,13	13	St. Thomas, US Virgin Islands (52)	5,000	10,000	—	—
			Dry Tortugas, USA (88)	—	—	—	—
L. argentimaculatus §	1	6,14	Palau (2)	—	—	—	—
			Papua New Guinea (81 in 78)	—	—	—	—
L. argentiventris	1,5,6,12	12	Gulf of California, Mexico (73)	>30	500	—	—
L. bohar §	5,9,10,13	14	Papua New Guinea (81 in 78)	—	—	—	—
			Solomon Islands (81 in 78)	—	—	—	—
L. campechanus	14	14	Florida Keys, USA (88)	—	—	—	—
L. cyanopterus	13,14	13,14	Belize (31, 64)	10,000	45,000	—	—
			St. Thomas, US Virgin Islands (13)	1,000	6,000	—	—
			Cuba (28)	600	10,000	—	—
			Dry Tortugas, USA (88)	—	—	—	—
			Florida Keys, USA (88)	—	—	—	—
L. gibbus §	6,14	11,14	Palau (2)	—	—	—	—
			Papua New Guinea (81 in 78)	—	—	—	—
			Solomon Islands (81 in 78)	—	—	—	—

Appendix 4.2 (continued)

Species	Feeding habitat (adults)	FSA habitat	Geographic area (References)	FSA size (n)	FSA area in m² (max)	MD in km (range)	FMA in km² (max)
L. griseus	1,2,3,4,5,6,7	12,13,14	St. Thomas, US Virgin Islands (52)	250	2,500	–	–
			Cuba (28)	–	–	–	–
			Dry Tortugas, USA (88)	–	–	–	–
			Florida Keys, USA (88)	–	–	–	–
L. jocu	8	8,13,14	Belize (33, 34, 40, 64)	1,000	–	–	–
			St. Thomas, US Virgin Islands (13, 52)	1,000	2,000	>18	–
			Cuba (28)	400	10,000	–	–
			Cayman Islands (80)	–	–	–	–
			Dry Tortugas, USA (88)	–	–	–	–
L. novemfasciatus	1,5,6,12	12,14	Gulf of California, Mexico (73)	12	500	–	–
L. rivulatus §	4,12	9,10,14	Solomon Islands (81 in 78)	–	–	–	–
L. synagris	4,5,6,7	12,13,14	St. Thomas, US Virgin Islands (52)	300	2,500	–	–
			Cuba (28)	–	–	–	–
			Florida, USA (74)	–	–	–	–
			Dry Tortugas, USA (88)	–	–	–	–
L. vitta §	7,8	14	Solomon Islands (81 in 78)	–	–	–	–
Macolor niger	5,6,9,10	14	Palau (41)	300	–	–	–
Symphorichthys spilurus §	5,7	14	Palau (2)	–	–	–	–
			Papua New Guinea (81 in 78)	–	–	–	–
MUGILIDAE							
Chelon macrolepis	3	14	Kiribati (82)	–	–	–	–
Crenimugil crenilabis §	3	3,6,9,10,14	Palau (2)	–	–	–	–
			Marshall Islands (108 in 40)	1,500	–	–	–
			Papua New Guinea (81 in 78)	–	–	–	–

Appendix 4.2 (continued)

Species	Feeding habitat (adults)	FSA habitat	Geographic area (References)	FSA size (n)	FSA area in m² (max)	MD in km (range)	FMA in km² (max)
Liza vaigiensis §	3,6	14	Palau (2)	–	–	–	–
			Federated States of Micronesia (79 in 78)	–	–	–	–
Mugil cephalus §	3,6	1,12	Fiji (89 in 78)	–	–	–	–
Neomyxus leuciscus §	6	4	Federated States of Micronesia in 78)	–	–	–	–
Valamugil seheli	3,6	14	Kiribati (82)				–
MULLIDAE							
Mulloides flavolineatus §	3,6	3,14	Palau (2)	–	–	–	–
			Federated States of Micronesia (79 in 78)	–	–	–	–
M. vanicolensis §	4,6,12	3	Federated States of Micronesia (79 in 78)	–	–	–	–
Pseudopeneus maculatus	3,6,8,12	8	St. John, US Virgin Islands (1, 70)	<400	–	–	–
PANGASIIDAE							
Pangasius krempfi	1	1	South China Sea/Mekong River (109)	–	–	>720	–
RHINCODONTIDAE							
Ginglymostoma cirratum	12,14	2	Dry Tortugas, USA (74)	–	–	–	–
SCOMBRIDAE							
Acanthocybium solandri §	13,14	13,14	Palau (2)	–	–	–	–
Grammatorcynus bicarinatus §	13,14	13,14	Palau (2)	–	–	–	–

Appendix 4.2 (continued)

Species	Feeding habitat (adults)	FSA habitat	Geographic area (References)	FSA size (n)	FSA area in m² (max)	MD in km (range)	FMA in km² (max)
Rastrelliger kanagurta §	13,14	13,14	Papua New Guinea (78)	–	–	–	–
Scomberomorus commersoni §	12,13,14	13,14	Palau (2)	–	–	–	–
SERRANIDAE							
Cephalopholis argus §	5,12,13	4,12	Federated States of Micronesia (79 in 78)	–	–	–	–
C. boenak §	5	7,8	Papua New Guinea (81 in 78)	–	–	–	–
C. miniata §	9,10,14	7,8,9,10	Papua New Guinea (81 in 78)	–	–	–	–
C. sexmaculata §	14	7,8,9,10	Papua New Guinea (81 in 78)	–	–	–	–
C. sonnerati §	6,14	7,8	Papua New Guinea (81 in 78)	–	–	–	–
C. urodeta §	5,12	7,8,9,10	Papua New Guinea (81 in 78)	–	–	–	–
Epinephelus adscensionis	5,12	14	Puerto Rico (15)	–	–	–	–
			Peter Island, British Virgin Islands (52)	100	500	–	–
E. coioides §	1,5	3,8	Papua New Guinea (81 in 78)	–	–	–	–
E. corallicola §	1,5,12	14	Indonesia (96 in 78)	–	–	–	–
			Papua New Guinea (81 in 78)	–	–	–	–
E. cyanopodus §	5,7	9,10	New Caledonia (78)	–	–	–	–

Appendix 4.2 (continued)

Species	Feeding habitat (adults)	FSA habitat	Geographic area (References)	FSA size (n)	FSA area in m² (max)	MD in km (range)	FMA in km² (max)
E. fuscoguttatus §	12	9,10 by 14	Palau (23)				
			Ngerumekaol	350	~54,000	—	—
			Ebiil	350	~40,000	—	—
			West Entrance	185	12,000	—	—
			Pohnpei (67)	—		—	—
			Seychelles (68)	1,050	6,900	—	—
			Indonesia (71, 96 in 78)	82		—	—
			Fiji (89 in 78)	—		—	—
			Malaysia (83 in 78)	—		—	—
			Federated States of Micronesia (79 in 78)	—		—	—
			New Caledonia (78)	—		—	—
			Papua New Guinea (81 in 78)	—		—	—
			Solomon Islands (81 in 78)	—		—	—
E. guttatus	6,7,12,13	9,10,12,13	Puerto Rico (8, 15, 18, 27)	3,000	40,000	18	~250
			Bermuda (9, 37)	—		5–20	<1,000
			St. Thomas, US Virgin Islands (10, 11)	80,000	360,000	6–32	500
			St. Croix, US Virgin Islands (11, 12)	3,000	15,000	2–16	90
			Saba Bank, Netherlands Antilles (30)	10,000	52,000	—	—
			Belize (32, 40)	—		—	—
E. itajara	1,5,6	7,8	Florida (66, 98)	12		—	—
E. lanceolatus §	5,7,12	14	Indonesia (96 in 78)	—		—	—

Appendix 4.2 (continued)

Species	Feeding habitat (adults)	FSA habitat	Geographic area (References)	FSA size (n)	FSA area in m² (max)	MD in km (range)	FMA in km² (max)
E. maculatus §	4,5,7	–	Federated States of Micronesia (79 in 78)	–	–	–	–
E. malabaricus §	8,9,10	9,10	New Caledonia (78)	–	–	–	–
E. marginatus	5,12	12	Corsica, France (97)	–	–	–	–
E. merra §	5	4,9,10	Malaysia (83 in 78)	–	–	–	–
			Federated States of Micronesia (79 in 78)				–
			Solomon Islands (81 in 78)				–
E. multinotatus §	12	9,10	Papua New Guinea (81 in 78)	–			–
E. morio §	12	14	Campeche Bank, Mexico (25)	–			–
			Florida, USA (26, 38)			29–72	–
			Cuba (28)	–	–	–	–
E. ongus §	1,5,12	9,10,14	Malaysia (83 in 78)	–			–
			Papua New Guinea (81 in 78)	–	–	–	–
E. polyphekadion §	4,5,7	9,10 by 14	Pohnpei (7)	<20,000	5,000		–
			Palau (23)				–
			Ngerumekaol	2,300	~54,000	>10	–
			Ebiil	1,000	~40,000		–
			West Entrance	500	12,000		–
			Seychelles (68)	2,000	6,900		–
			Fiji (89)	–	–	–	–
			Indonesia (96 in 78)	–			–
			Malaysia (83 in 78)	–			–
			Federated States of Micronesia (79 in 78)	–	–	–	–
			New Caledonia (78)	–			–
			Papua New Guinea (81 and 114 in 113)	–			–

Appendix 4.2 (continued)

Species	Feeding habitat (adults)	FSA habitat	Geographic area (References)	FSA size (n)	FSA area in m² (max)	MD in km (range)	FMA in km² (max)
E. striatus	5,7,12,13	14	St. Thomas, US Virgin Islands historic site (14, 99)	2,000	10,000	–	–
			St. Thomas, US Virgin Islands Grammanik Bank (11)	200	10,000	20	~800
			Bahamas (21, 35, 78, 98)	3,000	25,000	110	–
			Cayman Islands (15, 16, 80)	5,200	–	–	–
			Mexico (34, 39, 115)	1,000	70,000	–	–
			Belize (20, 32, 53, 64, 78)	3,000	15,000	100–250	~7,500
			Cuba (28, 98)	150,000	–	–	–
			Bermuda (19, 98)	–	–	–	–
			Honduras (36, 78)	–	–	–	–
			Dominican Republic (99)	–	–	–	–
			Puerto Rico (98)	–	–	–	–
			Turks and Caicos (78)	–	–	–	–
E. tavina	5,14	9,10 by 14	Palau (2)	–	–	–	–
E. trimaculatus §	12	4	Solomon Islands (81 in 78)	–	–	–	–
Mycteroperca bonaci §	12	9,10,14	Belize (20, 32, 33, 64)	140	15,000	–	–
			Bahamas (19)	–	–	–	–
			Cayman Islands (80)	–	–	–	–
			Florida Keys, USA (100)	–	–	–	–
			Honduras (36, 78)	–	–	–	–
M. jordani	12	12	Gulf of California, Mexico (73)	–	–	–	–
M. microlepis	1,2,12	–	Gulf of Mexico, USA (101 in 78)	–	–	–	–
M. phenax	1,12,14	–	Gulf of Mexico, USA (101 in 78)	–	–	–	–
M. prionura	12	8,12,14	Gulf of California, Mexico (73)	100	600	–	–
M. rosacea	12	12,13,14	Gulf of California, Mexico (73)	400	10,000	–	–

Appendix 4.2 (continued)

Species	Feeding habitat (adults)	FSA habitat	Geographic area (References)	FSA size (n)	FSA area in m² (max)	MD in km (range)	FMA in km² (max)
M. tigris	13	9,10,13,14	Vieques, Puerto Rico (17, 102)	>5,000	250,000	—	—
			St. Thomas, US Virgin Islands (11)	<100	5,000	—	—
			St. Thomas, US Virgin Islands (52)	200	5,000	—	—
			Belize (32, 64)	500	15,000	—	—
			Cayman Islands (80)	250	—	—	—
			Honduras (36)	—	—	—	—
			Turks and Caicos (78)	—	—	—	—
M. venenosa	13	14	St. Thomas, US Virgin Islands (11)	900	10,000	>12	~500
			Bahamas (21)	100	10,000	—	—
			Belize (20, 32, 64)	900	17,000	—	—
			Cayman Islands (80)	200	15,000	—	—
			Cuba (28)	—	—	—	—
			Honduras (36)	—	—	—	—
			Turks and Caicos (78)	—	—	—	—
Paranthias colonus	12,13,14	12,13,14	Gulf of California, Mexico (73)	>1,000	10,000	—	—
Plectropomus areolatus §	5,12	9,10 by 14	Palau (23)	—	—	—	—
			Ngerumekaol	400	~54,000	—	—
			Ebiil	500	~40,000	—	—
			West Entrance	1,200	12,000	—	—
			Pohnpei (67)	—	—	—	—
			Indonesia (71, 96 in 78)	77	—	—	—
			Federated States of Micronesia (79 in 78)	—	—	—	—
			Fiji (89 and 96 in 78)	—	—	—	—
			Malaysia (83 in 78)	—	—	—	—
			Maldives (103)	—	—	—	—
			Papua New Guinea (81 in 78)	—	—	—	—
			Philippines (78)	—	—	—	—
			Solomon Islands (81 in 78, 116)	350	~45,000	—	—

Appendix 4.2 (continued)

Species	Feeding habitat (adults)	FSA habitat	Geographic area (References)	FSA size (n)	FSA area in m² (max)	MD in km (range)	FMA in km² (max)
P. laevis §	5,12	12	Papua New Guinea (81 in 78)	–	–	–	–
P. leopardus* §	5,7,12	12,14 by 9 and 10	Palau (2)	–	–	–	80
			Lizard Island, Great Barrier Reef, Australia (5, 6)	60	1,000	<1–5.2	
			Scott Reef, Great Barrier Reef, Australia (4, 22)	128	1,700	–	–
			Elford Reef, Great Barrier Reef, Australia (4, 22)	59	3,200	–	–
			Indonesia (96 in 78)	–			–
			Malaysia (83 in 78)	–			–
			Papua New Guinea (81 in 78)	–			–
			Solomon Islands (81 in 78)	–			–
P. maculatus §	5,7,12	9,10	Malaysia (83 in 78)	–			–
			Papua New Guinea (81 in 78)	–			–
P. oligacanthus §	9,10,14	9,10,12	Malaysia (83 in 78)	–			–
			Papua New Guinea (81 in 78)	–			–
SIGANIDAE							
Siganus argenteus	5,8	–	Marshall Islands (40)	–			–
			Federated States of Micronesia (40)	–			–
S. canaliculatus §	1,2	9,11	Palau (2, 46)	–			–
			Papua New Guinea (81 in 78)	–			–
S. guttatus* §	1	3	Malaysia (83 in 78)	–			–
S. lineatus §	1,5,7,9,10	On sand at 14	Palau (2)	–			–
S. puellus §	5,12	4	Federated States of Micronesia (79 in 78)	–			–

Appendix 4.2 (continued)

Species	Feeding habitat (adults)	FSA habitat	Geographic area (References)	FSA size (n)	FSA area in m² (max)	MD in km (range)	FMA in km² (max)
S. punctatus §	4	11	Palau (2)	–	–	–	–
S. randalli §	1,5,8	4	Federated States of Micronesia (79 in 78)	–	–	–	–
S. spinus §	2,4,8	1,4	Federated States of Micronesia (79 in 78)	–	–	–	–
			Fiji (89 in 78)	–	–	–	–
S. vermiculatus §	1,5,12	1,4	Federated States of Micronesia (79 in 78)	–	–	–	–
			Fiji (89 in 78)	–	–	–	–
SPARIDAE							
Acanthopagrus australis	1	9	Australia (50, 51)	–	–	80	–
A. berda	1	9	Australia (48)	–	<1,000	0.5–3.1	<3
			South Africa (49)	–	–	–	–
Pagrus auratus*	12	3	Australia (78)	–	–	–	–
DECAPODA							
Callinectes sapidus	1	9	Western Atlantic (58, 59)	–	–	10–300	>900
Panulirus argus	7,12	13,14	Bahamas (56)	–	–	–	–

Appendix 4.2 (continued)

Species	Feeding habitat (adults)	FSA habitat	Geographic area (References)	FSA size (n)	FSA area in m^2 (max)	MD in km (range)	FMA in km^2 (max)
P. ornatus	4	5,12	Papua New Guinea (95, 104, 105)	>60,000	–	500	25,000
Penaeus plebejus	–	–	Australia (106, 107)	–	–	930	10,000
Scylla serrata	1	9,12	Australia (57)	–	–	–	–

References: 1. Colin and Clavijo 1988; 2. Johannes 1978 and 1981; 3. Randall and Randall 1963; 4. Samoilys 1997; 5. Zeller 1997; 6. Zeller 1998; 7. Rhodes and Sadovy 2002; 8. Sadovy et al. 1994b; 9. Luckhurst 1998; 10. Nemeth 2005; 11. Nemeth et al. 2006b; 12. Nemeth et al. 2007; 13. Kadison et al. 2006; 14. Olsen and LaPlace 1978; 15. Colin et al. 1987; 16. Tucker et al. 1993; 17. Sadovy et al. 1994a; 18. Shapiro et al. 1993; 19. Smith 1972; 20. Carter et al. 1994; 21. Colin 1992; 22. Samoilys and Squire 1994; 23. Johannes et al. 1999; 24. Warner 1995; 25. Brule et al. 1999; 26. Coleman et al. 1996; 27. Sadovy et al. 1992; 28. Claro and Lindeman 2003; 29. Colin 1996; 30. Kadison et al. 2009; 31. Heyman et al. 2005; 32. Sala et al. 2001; 33. Carter and Perrine 1994; 34. Anguilar-Perera and Aguilar-Davila 1996; 35. Bolden 2000; 36. Fine 1990; 37. Burnett-Herkes 1975; 38. Moe 1969; 39. Anguilar-Perera 1994; 40. Domeier and Colin 1997; 41. Myers 1991; 42. Myrberg et al. 1988; 43. Robertson 1983; 44. Randall 1961a; 45. Randall 1961b; 46. Hasse et al. 1977; 47. Fishelson et al. 1987; 48. Sheaves et al. 1999; 49. Garratt 1993; 50. Pollock 1982; 51. Pollock 1984; 52. RS Nemeth unpubl. data; 53. Starr et al. 2007; 54. Sancho et al. 2000a; 55. Sancho et al. 2000b; 56. Herrnkind 1980; 57. Hill 1994; 58. Millikin and Williams 1984; 59. Carr et al. 2005; 60. Mazeroll and Montgomery 1998; 61. Garcia and Solano 1995; 62. Crabtree 1995; 63. Heyman et al. 2001; 64. Heyman and Requena 2002; 65. Mackie 2000; 66. Colin 1994; 67. Rhodes and Tupper 2008 68. Robinson et al. 2008; 69. Bell and Colin 1986; 70. Colin and Clavijo 1978; 71. Pet et al. 2005; 72. Craig 1998; 73. Sala et al. 2003; 74. Wicklund 1969; 75. Pratt and Carrier 2001; 76. Feldheim et al. 2002; 77. Moore and Raynolds 1982; 78. SCRFA 2004; 79. Rhodes 2003; 80. Whaylen et al. 2004; 81. Hamilton 2003; 82. Johannes and Yeeting 2001; 83. Daw 2004; 84. Colin 1976; 85. Colin and Bell 1991; 86. Ebisawa 1999; 87. Ebisawa 1990; 88. Lindeman et al. 2000; 89. Sadovy 2004; 90. Gladstone 1986; 91. Burton et al. 2005; 92. Gladstone 1994; 93. Yogo et al. 1980; 94. Stone 2004; 95. Bell et al. 1987; 96. Sadovy and Liu 2004; 97. Pelaprat 2002; 98. Sadovy and Eklund 1999; 99. Sadovy 1997; 100. Eklund et al. 2000; 101. Koenig et al. 1996; 102. White et al. 2002; 103. Sluka 2001; 104. Moore and MacFarlane 1984; 105. MacFarlane and Moore 1986; 106. Ruello 1975; 107. Glaister et al. 1987; 108. Helfrich and Allen 1975; 109. Hogan et al. 2007; 110. Graham and Castellanos 2005; 111. Claydon 2006; 112. Hamilton 2005; 113. Hamilton et al. 2005; 114. Squire 2001; 115. Anguilar-Perera 2004; 116. Johannes 1988; 117. Gladstone 1996

Chapter 5
The Senses and Environmental Cues Used by Marine Larvae of Fish and Decapod Crustaceans to Find Tropical Coastal Ecosystems

Michael Arvedlund and Kathryn Kavanagh

Abstract Almost all demersal tropical teleost fishes have pelagic larvae that may disperse, in common with most tropical marine decapod larvae. The degree to which behavior and sensory abilities of the larvae influence or control dispersal, and thus the spatial scale of connectivity, is largely unknown, but emerging evidence indicates that this influence is large. Until recently, the established opinion was that the sensory abilities of tropical larval fishes and decapods were mainly irrelevant for the location of the first benthic settlement habitat. However, an increasing number of studies show that pre-settlement coral reef fishes are not only capable swimmers but also show directed swimming in relation to the location of nearby relevant habitat. Many species of tropical decapod larvae and postlarvae also seem capable of detecting environmental habitat cues and may use this ability to move toward a suitable habitat. In this chapter, we review studies on the topics of senses and environmental cues used by marine fish and decapod crustacean larvae to find tropical coastal ecosystems.

Keywords Behavioral ecology · Settlement mechanisms · Navigation · Orientation · Sensory ecology

5.1 Introduction

Tropical marine demersal teleost fishes and decapod crustaceans typically have complex life histories, beginning with a benthic or pelagic embryonic phase, followed by a pelagic larval phase, and then a return to the benthos for a juvenile-adult phase (Montgomery et al. 2001, Kingsford et al. 2002, Leis and McCormick 2002, Jeffs et al. 2005, Anger 2006). Because the population-level connectivity between tropical benthic habitats, e.g., coral reefs, estuaries, mangroves, and seagrass beds,

M. Arvedlund (✉)
Reef Consultants, Rådmand Steins Allé 16A, 2-208, 2000 Frederiksberg, Denmark
e-mail: arvedlund@speedpost.net

I. Nagelkerken (ed.), *Ecological Connectivity among Tropical Coastal Ecosystems*,
DOI 10.1007/978-90-481-2406-0_5, © Springer Science+Business Media B.V. 2009

depends to a large extent on pelagic larval movements between locations, research on larval capabilities and environmental cues mediating the transition between pelagic and benthic phases has received increasing attention in the past decade. Many late-stage larvae are competent swimmers (Jeffs et al. 2005, Leis 2006, 2007). However, competent swimming is only effective for finding suitable habitat if larvae also possess navigational capabilities. Evidence that fish and decapod larvae actively orientate is rapidly emerging. Here we review some of the latest research on orientation, early sensory development, and the role of specific environmental cues in the recruitment process of fish and decapods in tropical marine ecosystems.

Demographic connectivity is a key parameter in models of marine population dynamics and, therefore, in the management of fisheries and marine parks (Cowen et al. 2000, Kingsford et al. 2002, Leis and McCormick 2002, Jeffs et al. 2005, Leis 2006, 2007 and references therein). Making broad generalizations about larval capabilities in these models would be a mistake, as there is a large range of sizes, ages, and competence levels among settling fishes and decapods. For example, in decapod larvae the pelagic phase may last from a few days (Bradbury and Snelgrove 2001) up to as much as 18 months (Phillips and Sastry 1980). Likewise in fishes, pelagic phases last from a week in anemonefishes (family Pomacentridae) to more than 120 days in some labrids (family Labridae) (Brothers et al. 1983, Wellington and Victor 1989, Leis and McCormick 2002) and at the extreme end, some porcupine fish (family Diodontidae) have a juvenile pelagic phase of more than 64 weeks (Ogden and Quinn 1984). Such differences have consequences on individual capabilities as well as the geographic range of connectivity among sites, and as such, they call for using taxon-specific rather than overarching generalizations about the abilities of pelagic larvae when modeling connectivity.

The aim of this chapter is to provide a summary of studies on this topic since the last reviews from 2001 to 2002 (Montgomery et al. 2001, Leis and McCormick 2002, Kingsford et al. 2002, Myrberg and Fuiman 2002). Sound as an orientation cue for the pelagic larvae of reef fishes and decapod crustaceans was recently reviewed comprehensively by Montgomery et al. (2006), and therefore we review this subject only briefly here. Likewise, because of Jeffs et al.'s (2005) review on how spiny lobster (family Palinuridae) larvae (phyllosomes) and post-larvae (pueruli) find the coast, we primarily discuss studies published after 2004 on the use of environmental cues to find tropical ecosystems by crustacean decapod larvae.

The main message of the last series of reviews from 2001 to 2002 was that species from many taxa have senses that let them distinguish variation in water chemistry, sound and vibration, white light gradients, polarized light, current direction, magnetism, and water pressure. It was reported that some aquatic organisms can detect multiple cues, and sensory responses are probably widespread; however, only the visual, olfactory, and auditory senses are known to be functional in the few (mainly Indo-Pacific) reef fish larvae and decapod crustaceans ever examined. Generally, the development of the senses of tropical coral reef fishes and decapod crustaceans is poorly understood. Many potential settlers are efficient swimmers, and there is evidence for orientation to stimuli and navigation over short (centimeters to meters) and broad (tens of meters to kilometers) spatial scales in these two groups (Montgomery et al. 2001, Leis and McCormick 2002, Kingsford et al. 2002, Myrberg and Fuiman 2002).

Of other relevant previous reviews on the issues covered in this chapter, we would like to mention: Atema et al. (1988) on the sensory biology of aquatic animals, and Lenz et al. (1997) on the sensory ecology and physiology of zooplankton. Hadfield and Paul (2001) published a comprehensive review on the issue of natural chemical cues for settlement and metamorphosis of marine invertebrate larvae. Collin and Marshall (2003) focused on sensory processing in aquatic environments. A theme review in the journal Marine Ecology Progress Series (2005) focused on marine sensory biology and how to link the internal and external ecologies of marine organisms (coordinated and initiated by M Weissburg and H Browman). We also recommend Levin's (2006) review on larval dispersal based on insights gained from physical modeling, chemical tracking, and genetic approaches.

5.2 The Senses

The ability of organisms to use sensory cues at different spatial scales depends on the presence of relevant sense organs, the sensitivity of those organs, the aptitude to decide direction, the behavioral responses to cues, and the mobility of larva (Kingsford et al. 2002 and references therein). In this section, we review recent progress of the morphology and neuroanatomy of the senses of tropical fish larvae, and decapod crustacean zoea and megalopa (i.e., decapod crustacean larvae; see Hadfield and Paul (2001, p. 435) and Anger (2006) for details about definitions of decapod crustacean zoea and megalopa).

5.2.1 Olfaction

Olfaction, not taste, is normally involved in remote chemoreception tasks such as navigation and orientation towards a suitable habitat (Basil et al. 2000). For detailed information about the olfactory system in adult fish compared to crustaceans, see Caprio and Derby (2008). For recent reviews on fish chemosenses see, e.g., Hara (1992, 1994a,b) and Reutter and Kapoor (2005), and for decapod crustaceans see Derby et al. (2001). In addition, we recommend Farbman's (1992) comprehensive review on cell biology of olfaction.

There are four components required for a functional olfactory system in fishes: (1) olfactory receptor neurons lining the olfactory epithelium, (2) axons from the olfactory receptor cells forming the olfactory nerve, (3) synaptic connections between the olfactory nerve fascicles and mitral cells in the olfactory bulb, and finally (4) connections between mitral cells and the telencephalon (Hara and Zielinski 1989, Farbman 1992). However, the presence of these four components does not say which kinds of odors can be detected and at what concentration. To address these questions requires electrophysiological (e.g., Wright et al. 2005, 2008) or behavioral bioassays (e.g., Murata et al. 1986, Kasumyan 2002).

The olfactory epithelium shapes a multi-lamellar olfactory rosette in many adult teleosts. In Acanthopterygii, there are examples of olfactory epithelia that are flat, single, double, or triple folded. The olfactory chamber is ventilated with a single accessory nasal sac in most teleost taxa, whereas the presence of two sacs is confined to species within the Acanthopterygii (Hansen and Zielinski 2005). Three different types of olfactory receptor neurons are found in the olfactory epithelium: ciliated, microvillous, and crypt. Each type is set with specialized receptors. G-proteins can be found spread in an apparently overlapping construction in the olfactory epithelium (Hamdani and Døving 2007). Each type of olfactory receptor neuron expresses a particular class of odorant receptors. The ciliated olfactory receptor neuron responds to bile salts (hypothetically of significance in fish migration) and to alarm substances secreted from fish skin, the crypt olfactory receptor neuron respond to sex pheromones, and the microvillous olfactory receptor neuron to food odors (Hamdani and Døving 2007). This list of biochemical compounds to which these receptors respond is based on studies of salmons (Salmonidae), pacific mackerels (*Scomber scombrus*, Scombridae), zebrafish (*Brachydanio rerio*, Cyprinidae), and goldfish (*Carassius auratus*, Cyprinidae) (Hamdani and Døving 2007 and references therein).

The axons of the olfactory receptor neurons leave in three bundles via the olfactory bulb to the telencephalon (Hamdani and Døving 2007). Studies in salmonids (Salmonidae) and cyprinids (Cyprinidae) have shown that both ciliated olfactory receptor neurons and microvillous olfactory receptor neurons respond to amino acid odorants (Hansen and Zielinski 2005). Bile acids stimulate ciliated olfactory receptor neurons, and nucleotides activate microvillous olfactory receptor neurons. G-protein-coupled odorant receptor molecules (OR-, V1R-, and V2R-types) have been identified in several teleost species (Hansen and Zielinski 2005). Ciliated olfactory receptor neurons express the G-protein subunit $G_{\alpha olf/s}$, which activates cyclic AMP during transduction. Localization of G-protein subunits $G_{\alpha 0}$ and $G_{\alpha q/11}$ to microvillous or crypt olfactory receptor neurons varies among different species (Hansen and Zielinski 2005). All teleost species appear to have microvillous and ciliated olfactory receptor neurons (examples of such receptors in Figs. 5.1e, 5.2b, and 5.6b). The recently discovered crypt olfactory receptor neuron (see Fig. 5.6c) is likewise found broadly (Hansen and Zielinski 2005), including at least one species of coral reef fish (*Paragobiodon xanthosomus*, Gobiidae) (Arvedlund et al. 2007). There is a surprising diversity of olfactory cell types during ontogeny (i.e., some hours after fertilization and in the subsequent stages). In some species, olfactory receptor neurons and supporting cells derive from placodal cells; in others, supporting cells develop from epithelial (skin) cells. In some species, epithelial cells covering the developing olfactory epithelium degenerate; in others, these retract. Similarly, there are different mechanisms for nostril formation (Hansen and Zielinski 2005). Olfactory receptor neurons in fish and crustaceans generally have low spontaneous spiking activity, which is more usually

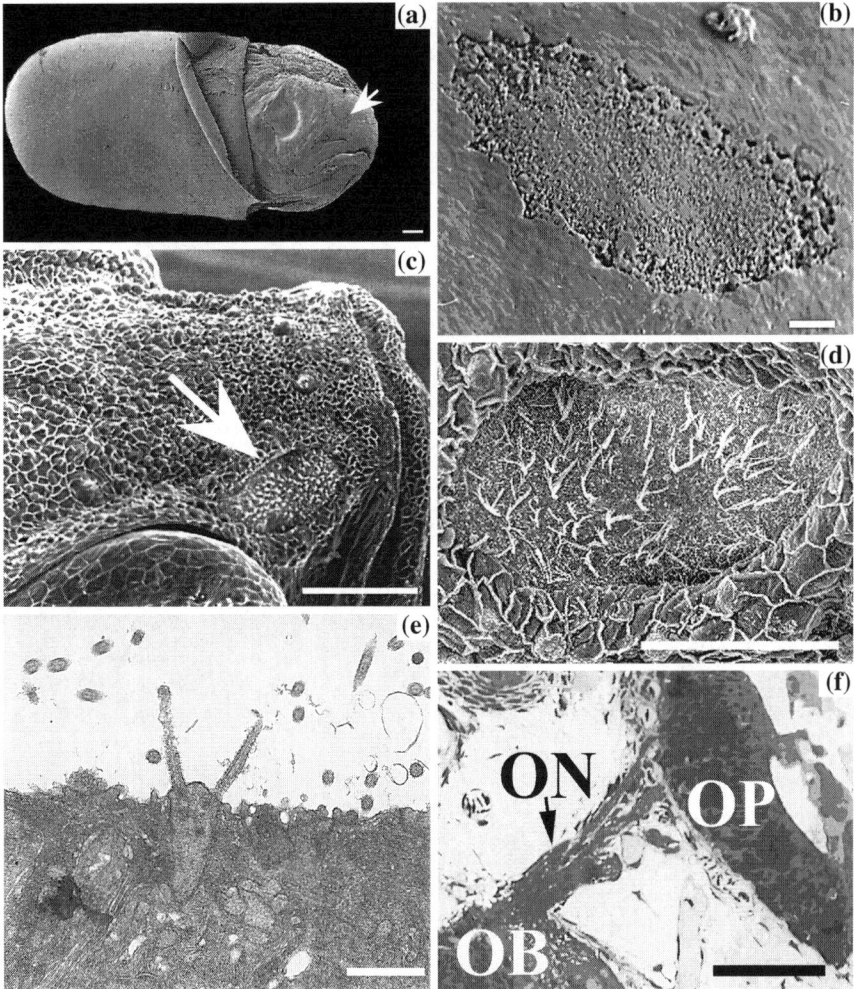

Fig. 5.1 (a) SEM of the embryonic development of the anemonefish *Amphiprion melanopus*: view of whole embryo partly in chorion day seven post-fertilization. The arrow points to the olfactory placode. Scale bar = 100 μm. (b) Embryo day six post-fertilization of the anemonefish *A. melanopus*. The depression dominating in the image is a preliminary olfactory placode. Scale bar = 10 μm. (c) SEM of the snout of the anemonefish *A. melanopus* day nine post-fertilization, 5 min post-hatching. The arrow points to the olfactory placode. Scale bar = 100 μm. (d) SEM close-up of the olfactory placode in (c). Scale bar = 50 μm. (e) Transmission electron microscopy (TEM) of the same type olfactory receptor neuron as shown in Fig. 5.2b. Scale bar = 4 μm. (f) TEM overview of neuron bundle from the olfactory placode to the olfactoric bulb next to the forebrain. Scale bar = 250 μm. OB = olfactoric bulb, ON = olfactoric neuron bundle, OP = olfactory placode. Images from Arvedlund et al. (2000a), with kind permission from the Journal of the Marine Biological Association of the United Kingdom

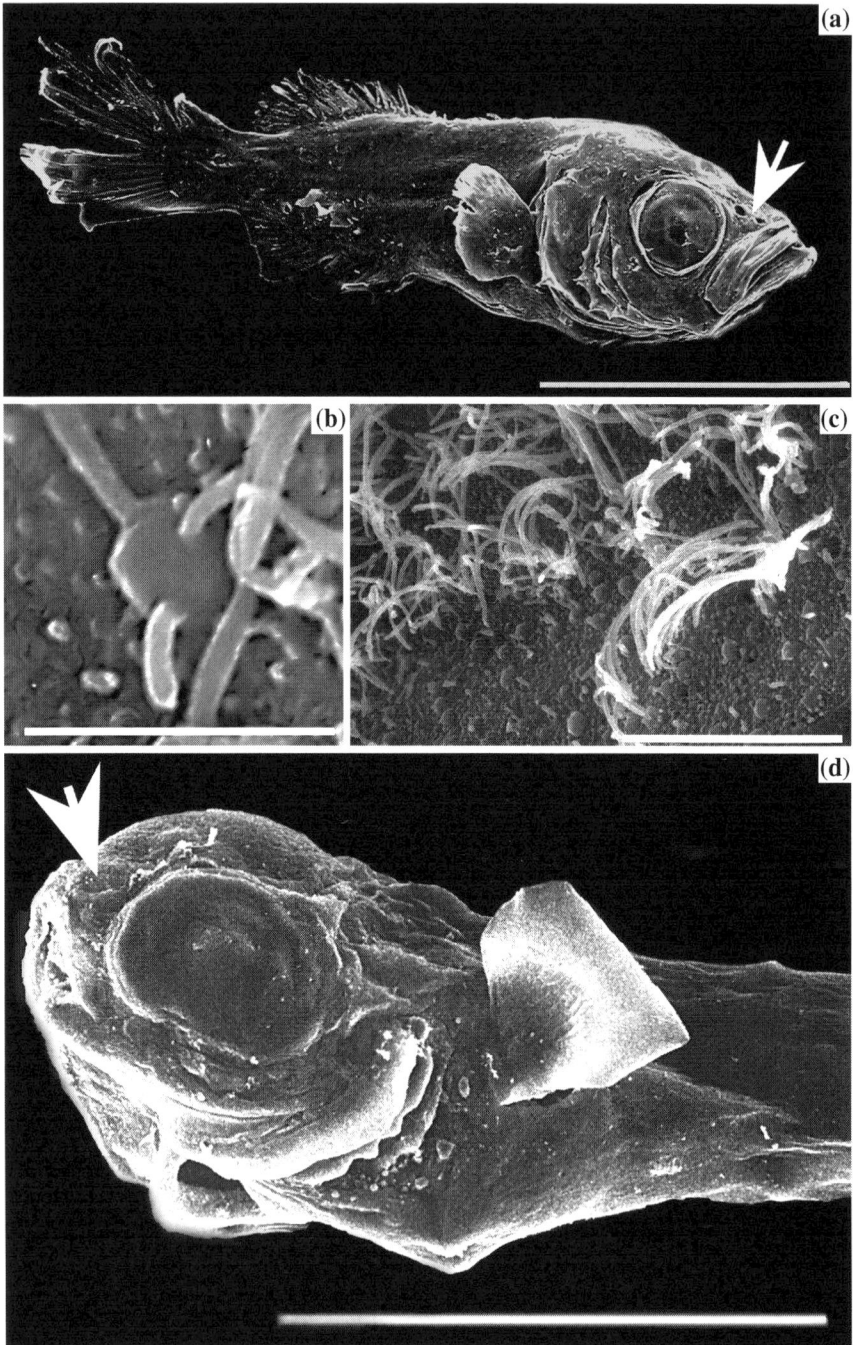

Fig. 5.2 (a) SEM of larval anemonefish *Amphiprion clarkii* day six post-hatching. The arrow points to the olfactory inlet (at the front of the snout) and to the outlet (a little further back).

increased but sometimes decreased by chemical stimulation (Caprio and Derby 2008). The population of olfactory receptor neurons on an olfactory organ responds to many chemicals, with small nitrogenous compounds, such as amino acids, amines, and nucleotides being the most effective. However, individual olfactory receptor neurons differ in their response specificities, which can be quite narrow (Caprio and Derby 2008). The response of olfactory receptor neurons increases in a concentration-dependent manner. Crustacean olfactory receptor neurons can follow pulses of chemicals up to at least four to five per second (Caprio and Derby 2008).

Although the nasal olfactory organs of some fish taxa have been studied for more than 100 years (reviews in, e.g., Døving et al. 1977, Yamamoto 1982, Hara and Zielinski 1989, Zeiske et al. 1992, Hara 1994a, b, Hansen and Reutter 2004, Hansen and Zielinski 2005), in only few taxa have there been studies of their embryogenesis and of their developmental state in larvae and juvenile stages. Apart from a very brief description of the gross morphology of the olfactory organs in larvae of the tropical marine milkfish *Chanos chanos* (Kawamura 1984) there were no published morphological studies, to our knowledge, of the olfactory senses of any stages of tropical shallow-water coral reef fishes until 2000 (Leis and McCormick 2002, Myrberg and Fuiman 2002).

At least some coral reef fish develop their olfactory organs soon after fertilization (Arvedlund et al. 2000b, Kavanagh and Alford 2003, Lara 2008) almost as fast as (captive-reared) zebrafish *Brachydanio rerio* (Cyprinidae) (Whitlock and Westerfield 2000, Hansen and Zeiske 1993). Embryos of captive-reared anemonefish *Amphiprion melanopus* (Pomacentridae) at day six post-fertilization possess a preliminary olfactory placode, although without any olfactory receptor neurons visible by the aid of scanning electron microscopy (SEM) (Arvedlund et al. 2000b, Figs. 5.1a, b). Right after hatching *A. melanopus* larvae possess two olfactory placodes, one on each side of the snout (Figs. 5.1c, d) with ciliated olfactory receptor neurons lining the epithelium among non-sensory cilia (Fig. 5.1e). Neuronal axons project from the placodes into the olfactory bulb in newly hatched *A. melanopus* larvae (Fig. 5.1f) and neuron bundles can be found in the deeper part of the olfactory placode near the basal lamina (Arvedlund et al. 2000b). From another study with rainbow trout (Salmonidae: *Oncorhynchus mykiss*), we know that an olfactory organ at this stage of development is likely to be functional (Zielinski and

---

Fig. 5.2 (continued) Scale bar = 2 mm. (**b**) SEM close-up of an olfactory receptor neuron with stiff cilia of larval anemonefish *A. clarkii* day eight post-hatching. Scale bar = 2 μm. (**c**) SEM of a part of the olfactory placode of the anemonefish *A. melanopus* day ten post-fertilization, i.e., the day of settling and metamorphosis. Notice several immature olfactory receptor neurons with stiff cilia in the lower part of the image. At the upper half of the image, shading over some of the immature olfactory receptor neurons, there are several long slender non-sensory cilia. Scale bar = 10 μm. (**d**) SEM overview of the damselfish *Chrysiptera cyanea*, 5 min post-hatching, with a visible olfactory placode, indicated by the arrow. Images by Michael Arvedlund, with kind permission from Reef Consultants

Hara 1988). The functional olfactory placode enables newly-hatched *A. melano-pus* larvae to use chemical cues for orientation, away from the coral reef, as they also are able to swim (speed: 12.4 body lengths[-1], Bellwood and Fisher 2001), and to imprint to chemical cues secreted from the host sea anemone of their parents (Murata et al. 1986, Arvedlund and Nielsen 1996, Arvedlund et al. 1999). In captive-reared larval anemonefish *A. clarkii,* day six post-hatching, there is an olfactory inlet leading into the olfactory placode with an outlet nearby further posterior (Arvedlund unpubl. data; Fig. 5.2a). Olfactory receptor neurons are present in *A. clarkii* from hatching (Figs. 5.2b, c). Some damselfishes (Pomacentridae) other than anemonefishes also possess well-developed olfactory placodes at hatching (Fig. 5.2d).

It is currently poorly understood whether species that have a slow larval development also develop their olfactory organs slowly, compared to, for example, the fast developing damselfishes. However, a preliminary study of the slow developing nursery species, the spangled emperor *Lethrinus nebulosus* (Lethrinidae) was recently conducted (Arvedlund unpubl. data). In *L. nebulosus*, the mouth does not open until several days after hatching, contrary to damselfishes, and the larval phase is longer (six weeks) than for damselfishes (from a few days to 1–2 weeks). However, at hatching the *L. nebulosus* larvae have olfactory placodes with ciliated olfactory receptor neurons in the olfactory epithelium (Arvedlund unpubl. data), suggesting an early functioning olfactory system despite delayed development in other systems. Since the olfactory system is functional even prior to the larvae needing to find prey (i.e., the mouth is not open), one might speculate that early olfaction may be needed for another purpose, such as orientation towards a suitable habitat.

Kavanagh and Alford (2003) conducted a comparative study of rates of growth and development of captive-reared larvae of four coral reef damselfishes (Fig. 5.3): *Chromis atripectoralis, Pomacentrus amboinensis, Premnas biaculeatus,* and *Acanthochromis polyacanthus* (all Pomacentridae). They found that the rate of olfactory development (Fig. 5.4) was remarkably consistent among species that settled at very different ages and sizes, suggesting olfaction capability is not tightly correlated with settlement or any particular habitat. The clear exception was *Premnas biaculeatus*, an anemonefish whose olfactory development started earlier and developed at an intrinsically faster rate the other three damselfishes (Figs. 5.4, 5.8). The authors suggested that this exceptionally early olfactory development may be an adaptation to facilitate the host-imprinting phenomenon of newly hatched anemonefishes (Arvedlund and Nielsen 1996, Arvedlund et al. 1999, Arvedlund et al. 2000a, b).

Recently, Lara (2008) examined the peripheral olfactory system of wild-caught late-stage larvae, early juveniles, and some adults of 14 species of Caribbean wrasse (Labridae), parrotfishes (Scaridae), and damselfish (Pomacentridae) by the aid of SEM. The separation of the anterior and posterior nares occurred before settlement in the labrids but, in some specimens of scarids, this separation was not complete by the time of settlement. It is generally believed that the olfactory receptor sites in fishes are located on the membranes of olfactory cilia (Hara 1994b). Therefore, an increase in the density of receptors, and/or of the total area covered by

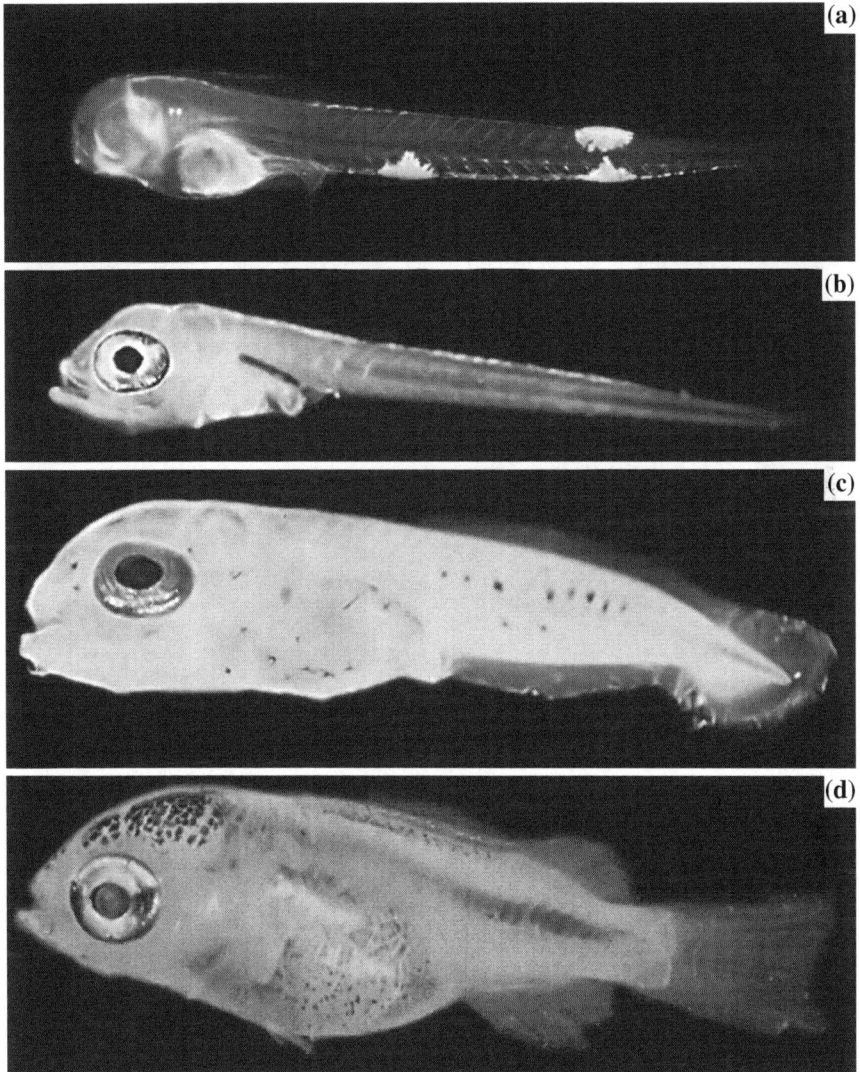

Fig. 5.3 Hatchlings of four Pomacentridae species displaying the range of variability in developmental stage at hatching within the family: (**a**) *Chromis atripectoralis*, (**b**) *Pomacentrus amboinensis*, (**c**) *Premnas biaculeatus*, and (**d**) *Acanthochromis polyacanthus*. The egg stage durations of these species are (in order) 2, 4, 7, and 16 days at 28 °C. From Kavanagh and Alford (2003), with kind permission from Wiley-Blackwell Publishing

receptors increases the total receptive area of the olfactory epithelium. Densities of ciliated receptor cells in several reef fish (Labridae) larvae studied by Lara (2008) ranged from 0.389 μm^{-2} in juvenile *Thalassoma bifasciatum* to 0.0057 μm^{-2} in juvenile *Bodianus rufus*, and of microvillous receptor cells from 0.038 μm^{-2} in

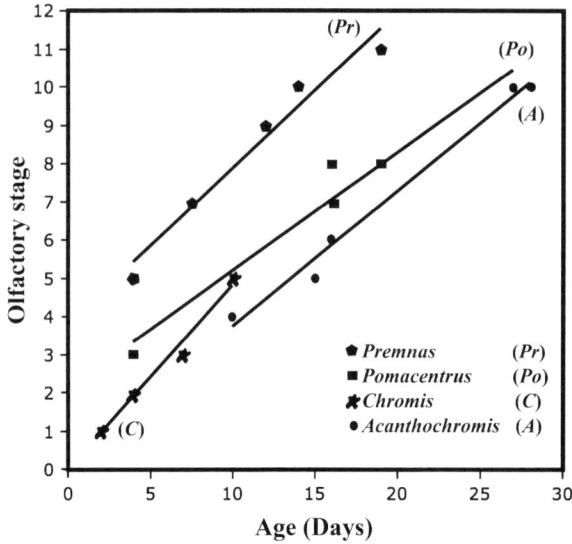

Fig. 5.4 Developmental rates of the olfactory systems of four pomacentrid species from fertilization until settlement. Linear regression lines are shown for each species. Age = days after fertilization. Embryos and larvae were reared in the laboratory at 28 °C. Stages for each developmental measure are based on the following external morphological traits: 1. Thin row of cilia, 2. Broad stripe of cilia, 3. Round patch of cilia, 4. Pit formation starts, 5. Shallow single pit, 6. Deep single pit, 7. Elongation of pit, 8. Pit division starts 'pinching in', 9. Nares divided, suture visible, 10. Nares cleanly divided, 11. Asymmetry, one opening enlarges. Figure modified from Kavanagh and Alford (2003), with kind permission from Wiley-Blackwell Publishing

a *Clepticus parrae* juvenile to 0.266 μm^{-2} in a juvenile *Doratonotus megalepis*. In comparison, fishes known to be highly sensitive to olfactory cues, such as the European eel *Anguilla anguilla* (Anguillidae) have an olfactory receptor density of 0.075 μm^{-2}. The cherry salmon *Oncorhyncus masou* (Salmonidae), known for their migratory long distance navigational abilities involving olfactory senses, have an olfactory receptor density of 0.110 μm^{-2} (Yamamoto 1982). The labrid larvae have even higher densities of olfactory receptors than these highly sensitive taxa, and Lara (2008) concluded that the olfactory organ in labrids is well developed prior to settlement and is comparable to that of adults.

Late stage pelagic larvae and newly-settled reef fishes have well-developed olfactory organs (Atema et al. 2002, Arvedlund and Takemura 2006, Arvedlund et al. 2007, Figs. 5.2, 5.5, 5.6). Atema et al. (2002) conducted a comprehensive study that, among other results, showed that wild-caught, late-stage, pelagic larvae of the cardinalfishes *Apogon guamensis* (10 and 11 mm standard length (SL)) and *A. doederleini* (10.5 and 12 mm SL) (Apogonidae) contain well-developed noses with in- and outflow nares and accessory sacs typical for efficient nose ventilation. The olfactory rosette consists of 2–3 lamellae covered with sensory epithelium, innervated by olfactory nerves connecting them to prominent olfactory bulbs, rostral-ventral to the

Fig. 5.5 (a) SEM of the intact snout of a newly metamorphosed juvenile *Paragobiodon xanthosomus*. OC = olfactory chambers, E = eye, M = mouth, CLLS = cephalic lateral line system. Scale bar = 500 μm. (**b**) SEM of the inlet (I) and outlet (O) to the nasal cavity. Scale bar = 200 μm. Classification of the inlet and the outlet is based on morphology of other fish species (Zeiske et al. 1992). Images from Arvedlund et al. (2007), with kind permission from Elsevier

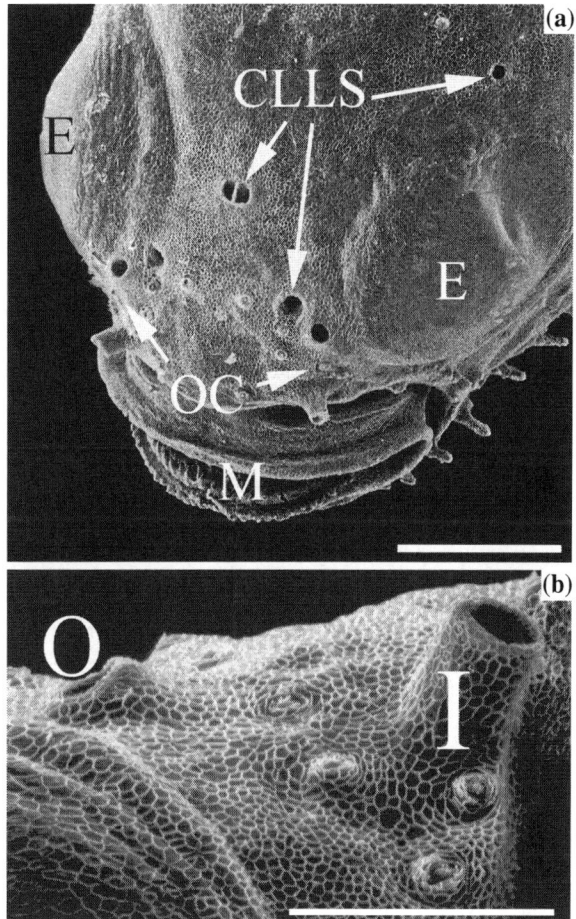

telencephalon. By micropipetting small dye clouds near an inflow naris of unanaesthetized apogonids *(Apogon* sp. 12–15 mm SL), Atema et al. (2002) showed that the cardinalfish larvae inhale continuously: puffs of dye emerged from the outflow naris synchronous with 3–4 Hz gill ventilation movements.

Arvedlund and Takemura (2006) examined the morphology and neuroanatomy of the olfactory organs of captive-reared spangled emperors *Lethrinus nebulosus* (Lethrinidae) at day 53 post-hatching, i.e., one week after metamorphosis. *L. nebulosus* exhibited a well-developed pair of nasal olfactory organs, positioned in nares on the dorsal side of the head. These organs were elliptical radial rosettes, one in each of the olfactory chambers, each comprising 12 lamellae, six on each side of a midline raphe, which were totally covered with sensory and non-sensory cilia, except for the margins. This type of cilia distribution is thought to indicate an acute sense of olfaction (Yamamoto 1982).

Fig. 5.6 (a) Light microscopy of the snout of a newly metamorphosed juvenile *Paragobiodon xanthosomus*. OP = olfactory placode, S = snout, FB = forebrain. Magnification = ×40. (**b**) TEM of parts of the olfactory placode. MV = olfactory receptor neuron with microvilli attached to the dendritic knob, CI = olfactory receptor neuron with cilia attached to the dendritic knob. Scale bar = 2 μm. (**c**) TEM of a crypt olfactory chemoreceptor (CORN). Scale bar = 2 μm. Images from Arvedlund et al. (2007), with kind permission from Elsevier

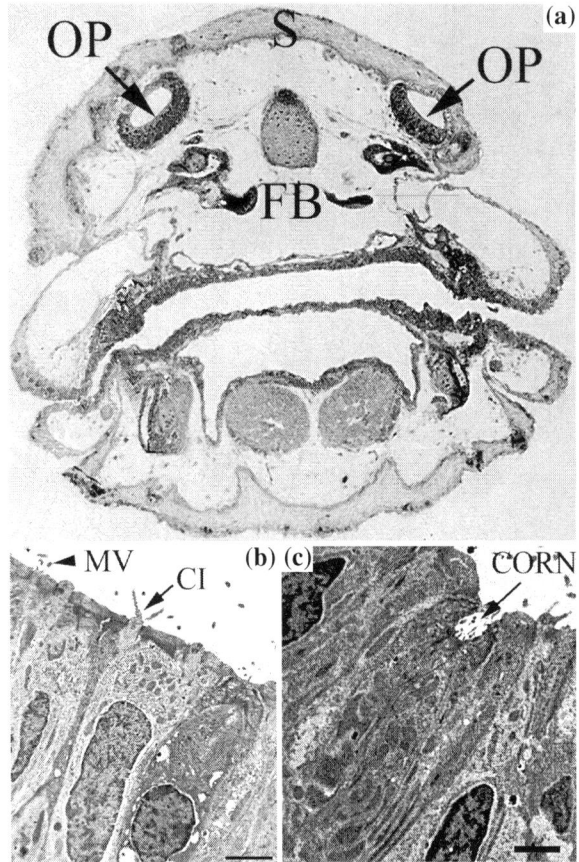

Arvedlund et al. (2007) examined the peripheral olfactory organ in newly metamorphosed coral-dwelling gobies *Paragobiodon xanthosomus* (Gobiidae) (SL ± standard deviation = 5.8 ± 0.8 mm, N = 15) by the aid of electron microcopy (scanning and transmission) and light microscopy (Figs. 5.5, 5.6). Two bilateral olfactory placodes were present in each fish. They were oval-shaped and located medio-ventrally, one in each of the olfactory chambers. Each placode had a continuous cover of cilia. The placode epithelium contained three different types of olfactory receptor neurons: ciliated, microvillous, and crypt cells. The latter type was rare. After a pelagic larval phase, *P. xanthosomus* settle to the reef and form an obligate association with one species of coral, *Seriatopora hystrix*. Based on the field studies by Sweatman (1985) and Elliott et al. (1995), which showed that some damselfish recruits can detect and orientate towards their microhabitat at short distances (less than ten meters: Elliott et al. 1995) by the aid of conspecific or microhabitat olfactory cues, and a recent study by Arvedlund (unpubl. data) of the morphology of the olfactory organs of anemonefish larvae (Fig. 5.2a–c), the well-developed olfactory organs of *P. xanthosomus* likely enable larvae to detect chemical cues on short distances (<10 m) and may assist in navigating towards and selecting appropriate coral habitat at settlement.

In summary, recent studies show that coral reef fish larvae develop their olfactory organs rapidly (including olfactory receptor neurons). This developmental pattern includes both species with a short larval phase (i.e., from 1 to 2 weeks) such as anemonefishes (Pomacentridae) and coral gobies (Gobiidae) as well as some species with a prolonged larval phase (i.e., >2 weeks) such as emperors (Lethrinidae) and wrasses (Labridae). However, this conclusion is based on few studies and includes studies of captive-reared rather than wild-caught fishes. Captive-reared larval fishes may hypothetically develop deformed olfactory organs: Mana and Kawamura (2002) found that captive-reared red sea bream *Pagrus major* and black sea bream *Acanthopagrus schlegelii* (Sparidae), developed an abnormal nasal opening and irregularities in the distribution patterns of the olfactory receptor neurons.

Methodological bias in studying such delicate organ development must be considered in future studies. For example, SEM does not always reveal every olfactory receptor neuron, with some remaining hidden underneath the numerous long non-sensory cilia (Arvedlund unpubl. data, see also Fig. 5.2c). Studies using transmission electron microscopy and immunocytochemistry against olfactory-receptor-coupled G-proteins should be included (Belanger et al. 2003) in order to reveal precisely which types of receptors are present in what densities, and subsequently what type of chemical compounds a fish may be able to detect. To determine whether one specific receptor density makes one fish species more sensitive to olfactory cues than another fish species with a lower receptor density, the ultrastructural methods used by Belanger et al. (2003) or by Hansen and Zeiske (1998) should be coupled with electrophysiological (e.g., Wright 2005) or behavioral bioassays (e.g., Kasumyan 2002).

Last, but not least, a challenge in morphological studies of wild-caught reef fish larvae, compared to captive-reared specimens, is the time of transportation from a collection site to a laboratory with appropriate facilities for analysis. Samples for morphological and ultrastructural studies must be examined within few weeks (best within days) from collection, to avoid possible artifacts (Arvedlund unpubl. data, Hayat 2000). Artifacts can appear after just one month of sample storage (Arvedlund unpubl. data). One possible solution is to carry out morphological studies on field stations equipped with histological and electron microscopy facilities.

Decapod crustaceans have chemosensory receptors over most of their body. These receptors play a key role in settlement to the benthic habitat (see recent reviews in Derby et al. 2001, Jeffs et al. 2005, Caprio and Derby 2008 and references therein). Decapod crustaceans have an exoskeleton (see Fig. 5.9), and consequently, their chemosensory neurons are packaged into thin extensions of the cuticle, called setae or sensilla (Caprio and Derby 2007). The first antennae, or antennules, of decapod crustaceans are major chemosensory organs and considered a functional unit that acts as a 'compound nose' (Derby et al. 2001). Aesthetascs are the most thoroughly studied antennular chemoreceptor sensilla (Derby et al. 2001).

Aesthetascs are located on the paired first antennae (antennules) but only on the distal end of the lateral flagellum of the antennule. Chemical stimuli pass through the porous cuticle of the aesthetascs and bind to receptor sites on the olfactory receptor neuron dendrites (Caprio and Derby 2007). The olfactory receptor neurons of the aesthetascs project ipsilaterally to the paired olfactory lobes (OL). The OLs are thought to receive input almost exclusively from aesthetasc olfactory receptor neurons. The OLs have a glomerular organization generally similar to first-order olfactory neuropils of other animals (Caprio and Derby 2007). For more details see Derby et al. (2001) and Caprio and Derby (2007). Integumentary sensory structures are common and numerous on the entire body surface of decapod larvae (Anger 2001). Many decapod larvae have specific chemosensors, but most sensors are bimodal, i.e., also mechanochemoreceptors as well as chemosensors (Anger 2001). There is an almost continuous array of pinnate setae along the flagella of the antennae of both the larval phase called pueruli in spiny lobsters and in early juveniles, but these are absent from late stage larvae. Similar arrays of chemosensory setae are seen in other decapods without a shoreward migrating lifecycle phase (Jeffs et al. 2005). The dorsal surface of megalopa of decapod crustaceans (but not in adults) bears a unique organ, arranged in the shape of a central aperture with four pits and nipples, called the 'dorsal organ' (Laverack 1988 and references therein). The function is unknown but believed to be involved with chemoreception (Laverack 1988). Keller et al. (2003) found that adult blue crabs *Callinectes sapidus* (Portunidae; presumably adult, no size class was mentioned) use both cephalic and thoracic appendages for olfactory-mediated orientation. Because blue crab larvae use olfactory cues for orientation when settling to the first benthic habitat (e.g., Forward et al. 2003) it may be possible that they use both cephalic and thoracic appendages for this behavior.

In summary, the morphology and neuroanatomy of the olfactory senses in the early phases of tropical decapod crustaceans are poorly understood (C Derby, Georgia State University, pers. comm.), but indirect behavioral evidence indicates that olfactory senses are clearly present in several species (e.g., Forward et al. 2001, 2003). Until further progress on larval stages is available, we recommend readers to see Grünert and Ache (1988) on the ultrastructure of the aesthetasc (olfactory) sensilla of the (adult) spiny lobster *Panulirus argus* (Palinuridae), Hallberg et al. (1992) on the aesthetasc concept, that is, structural variations of putative olfactory receptor cell complexes in Crustacea, and Hallberg et al. (1997) on olfactory sensilla in crustaceans, covering morphology, sexual dimorphism, and distribution patterns. Other useful references include Derby et al. (2001) on the functional and morphological development of the compound nose in spiny lobsters, Derby et al. (2003) which conducted a comparative study of turnover in the olfactory organ of the early juvenile and adult stages of Caribbean spiny lobsters, Laverack (1988) on the diversity of chemoreceptors, including crustacean receptors, Anger (2001) on the biology of crustacean larvae, Caprio and Derby (2007) on the olfactory system in adult fish compared to crustaceans, Kennedy and Cronin (2007) on the biology of the blue crab, Phillips (2006) on lobster biology, and Lavalli and Spanier (2007) on the biology and fisheries of the slipper lobster. The last three references contain, among other topics, reviews on the general morphology of the early stages. Santos et al. (2004) described the complete larval development of the captive-reared subtropical partner shrimp *Periclimenes sagittifer* (Palaemonidae).

5.2.2 Auditory Senses

Our current understanding of the auditory senses in teleosts was reviewed in detail by Popper and Fay (1999) and Popper et al. (2003). The auditory senses in crustaceans were reviewed by Popper et al. (2001) and by Montgomery et al. (2006). The latter reference is a comprehensive review comparing fish and decapod crustacean larval ability to use sound for detecting tropical ecosystems. Montgomery et al. (2006) concluded that sound cues are available for orientation, and that fishes and crustaceans hear sound and orient to it in a manner that is consistent with their use of sound to guide settlement onto reefs.

Teleost fishes have a pair of inner ears that lie inside the cranium on either side of the head at approximately the level of the hindbrain (Popper et al. 2003, Popper and Fay 1999). The fish inner ear consists of three semicircular canals and their associated sensory epithelium or cristae, and three otolith organs (for examples of otolith organs in coral reef fishes, see Fig. 5.7). The sensory epithelium in all these organs is composed of mechanosensory hair cells and support cells (Popper et al. 2003, Popper and Fay 1999). The presence of a swim bladder, or other gas filled compartments, may provide such fishes with an ability to detect sound pressure in water (Popper et al. 2003, Popper and Fay 1999).

Embryonic anemonefishes (Pomacentridae: *Amphiprion ephippium* and *A. rubrocinctus*) can detect sound (Simpson et al. 2005b). The frequency range of detected sounds and the sensitivity of the response both increase through the embryonic period. This means that at least some reef fishes develop hearing abilities early, i.e., they must be present already during embryogenesis (Simpson et al. 2005b). Damselfishes and anemonefishes use sound extensively in courtship and agonistic and territorial behaviors, with some species even developing regional dialects (Parmentier et al. 2005) and thus these reef fishes have evolved auditory receptive ability for these functions in addition to the possibility of sound use for locating suitable settlement habitat.

Recently Gagliano et al. (2008) showed that tropical coral reef fish larvae with asymmetrical ears not only encountered greater difficulties in detecting suitable settlement habitats, but suffered significantly higher rates of mortality. Further, they demonstrated that ear asymmetries arising early in the embryonic stage were not corrected by any compensatory growth mechanism in the larval stage.

There are considerably more studies on the sensory morphology and physiology of fishes than crustaceans, and therefore our understanding of hearing in crustaceans is presently rather poor (Montgomery et al. 2006). Receptors have been identified that may be able to respond to underwater sound such as hydrodynamic flows, particle motion, and pressure changes, however, their operation, sensory thresholds, and

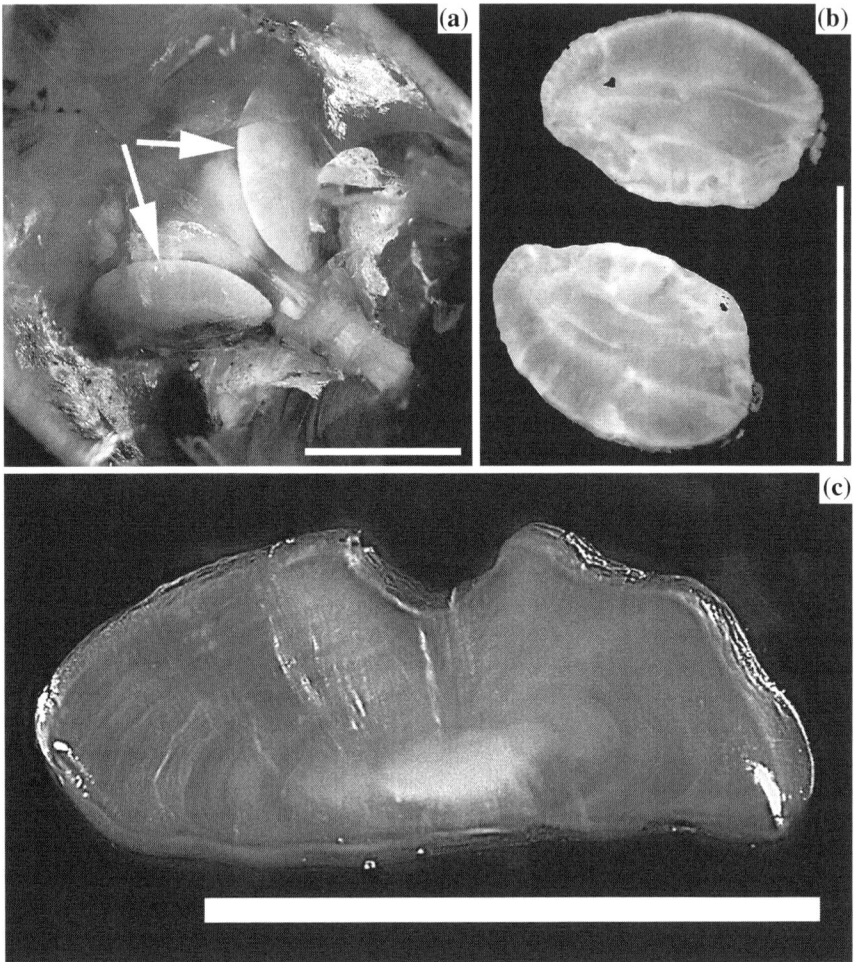

Fig. 5.7 Auditory senses in coral reef fishes. (**a**) Otoliths (arrows) partly dissected from a juvenile cardinalfish *Apogon cyanosoma* 25 mm SL (Apogonidae). Scale bar = 5 mm. (**b**) Otoliths separated from the cardinalfish *A. cyanosoma* in (a). Scale bar = 5 mm. Images by Michael Arvedlund with kind permission from Reef Consultants. (**c**) Otolith, transverse, thin-sectioned and polished, from an adult damselfish *Pomacentrus moluccensis* (Pomacentridae). Annual growth rings are visible. Scale bar = 3 mm. Image by Thea Marie Brolund with kind permission

range of sensitivity are not well defined (Popper et al. 2001). In at least some tropical fish larvae auditory sensory structures are present, because several species of reef fishes in the pelagic phase are more attracted to light traps that are enhanced with reef biosound (Montgomery et al. 2006). In addition, Wright et al. (2005, 2008) showed that pre- and post-settlement stages of the damselfish *Pomacentrus nagasakiensis* (Pomacentridae) and settlement-stage larvae of the coral trout *Plectropomus leopardus* (Serranidae) could detect sounds at several frequencies.

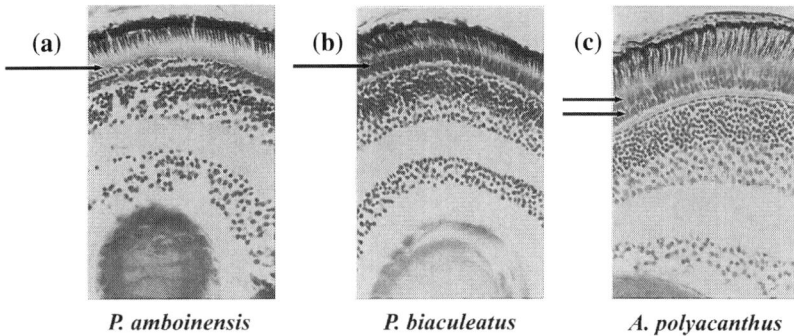

P. amboinensis   *P. biaculeatus*   *A. polyacanthus*

Fig. 5.8 Vertical histological sections, by the aid of light microscopy, through eyes of seven-day-old larvae or embryos of three species of pomacentrids demonstrating variation in developmental rate of retinae. The layers of photoreceptor cell nuclei increase (arrows), as do the external and internal nuclear layers (located below arrows). *Pomacanthus amboinensis* (**a**) has a thin layer of cone cell nuclei, while *Premnas biaculeatus* (**b**) has a thicker layer of cone cell nuclei, and *Amphiprion polyacanthus* (**c**) has a double layer of cone cell nuclei developing, even at this young age. Figure from Kavanagh and Alford (2003), with kind permission from Wiley-Blackwell Publishing

5.2.3 Vision

In fish larvae, while vision is clearly an important sense for prey detection and predator avoidance, it is the least likely to be important in larval orientation towards a reef from any large distance. That is, the use of vision directly in orientation toward reefs from a distance of kilometers, or even 100 meters, seems unreasonable, with the possible exception of celestial or sun orientation, as observed in sea turtles but as yet undiscovered in tropical fish larvae. However, the highly successful use of light traps, in which a light inside a transparent box is hung offshore at night (Doherty 1987), to specifically attract pre-settlement reef fish larvae indicates that these settling larva are at least phototactic at this stage.

As larvae approach the reef environment, however, vision becomes increasingly useful for detecting appropriate habitat for settlement and for eliciting species-specific behaviors such as schooling. In adults, because of the clarity of the waters around reefs, visual cues in this environment have evolved greater importance than in other habitats (Myrberg and Fuiman 2002). Furthermore, while reef fish species vary substantially in the age or size at which they settle to the reef, in nearly every case, the retina undergoes significant morphological and functional changes correlated quite precisely with settlement, indicating evolutionary adaptation of vision to this shift from pelagic to benthic environment. Here we review recent advances in our understanding of visual development of reef fish and decapod larvae.

The basic developmental trajectory of retinal development is fairly well stereotyped in all reef fishes thus far examined. Eyes of reef fish larvae develop early, become functional within a few days of fertilization, and initially have retinae consisting solely of cones. As they approach settlement or metamorphosis, they begin to

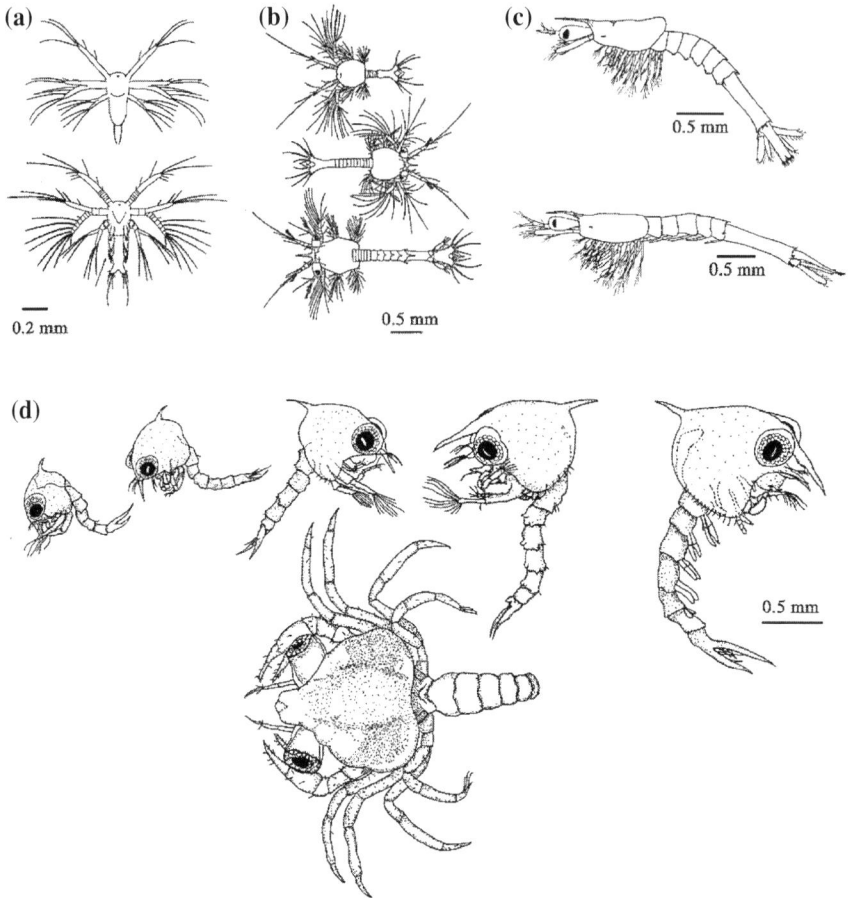

Fig. 5.9 Some typical crustacean larvae. (**a–c**) Larvae of dendrobranchiate shrimps: (**a**) *Farfantepenaeus brasiliensis*, nauplius stages I and V, (**b**) *Penaeus esculentus*, protozoeal stages I-III, and (**c**) *F. brasiliensis*, mysis stages I and III. (**d**) Larvae of a brachyuran crab, *Uca thayeri*, zoeal stages I-V, megalopa; all drawings after Anger (2006) with kind permission from Balaban Science Publishers

develop rods, cone density and diversity increases, the cone mosaic becomes more organized, and larvae obtain the ability to light-adapt their eye by moving the pigment layer (Kavanagh and Alford 2003). The rate of retinal development is known to vary among related coral reef fish species, correlating with life history differences (Kavanagh and Alford 2003). Below we review several recent studies which measure in late-stage reef fish larvae (1) visual acuity or resolving power, (2) spectral sensitivity, including color and UV wavelengths, and (3) ability to see polarized light.

Recent studies have measured the visual acuity, or resolving power, of reef fish larval retinae, and its changes with development, by both anatomical and behavioral means. Anatomical studies of the retina by Lara (2001) calculated minimum separable angles (MSA) based on cone density and focal length as a measure of acuity in 12 coral reef labrid and scarid species over the pre- to post-settlement period. Her study indicated that labroids and especially scarids have comparatively low acuity compared with other species at settlement. However, she estimates that a settling labrid larva has sufficient acuity to distinguish a 30 cm coral head from a distance of 12–30 m in daylight (Lara 2001). Calculated MSA for the anemonefish *Premnas biaculeatus* indicate an unusually high visual acuity compared with other larvae (Job and Bellwood 1996, Lara 2001). In addition to these anatomical measures, several different types of behavioral tests of acuity have been reported. Job and Bellwood (1996) used video-recorded measurements of reactive distances to prey, and determined that *Premnas* only react to prey at 1–2 body lengths of distance. They argue from their laboratory study that the acuity estimate based on eye anatomy significantly overestimates the realized visual acuity in larvae (e.g., Lara's estimate above). However, contrasting results from a behavioral study in the field (Lecchini et al. 2005c) indicate that damselfish (*Chromis viridis*) larvae can detect suitable habitat, even when olfactory capabilities are impaired, at a distance of up to 375 cm. The presence of conspecifics may be significant for visual as well as olfactory sensory detection. These highly contrasting conclusions about the realized visual capabilities of reef fish larvae call for additional carefully designed studies to determine the relationship between anatomical and behavioral acuity, and what level of acuity is needed to detect important habitat characteristics.

Ontogenetic changes in spectral sensitivity, or the ability to see different colors or wavelengths of light, have been investigated by several researchers recently. In young larvae the first photoreceptors to develop are single cones, yet Shand et al. (2002) found in a study of an estuarine-dwelling species that different opsin proteins in single cones allowed sensitivity in two wavelengths peaks, 425 and 534 nm in retinae of black bream *Acanthopagrus butcheri*. The larvae then developed double cones and sensitivity increased with age, shifting upwards in range. Ultraviolet (UV) light is more likely to be important in clear (rather than turbid) tropical waters where the short wavelengths can penetrate into the water column. Although UV-sensitive pigments have been found in young larvae of a wide range of northern and temperate marine fishes (Britt et al. 2001, Loew et al. 1997), UV-sensitive photoreceptors have not been found in young larvae of tropical reef fishes, but have been found in juveniles and adults (Hawryshyn et al. 2003, McFarland and Loew 1994). Concordant with the morphological data on reef fishes, behavioral evidence supports the idea that younger larvae cannot use UV light while older pre-settlement larvae can. Job and Shand (2001) showed in laboratory behavioral tests that general spectral sensitivity increases with age and size, and in a comparison of three coral reef fish species (*Apogon compressus, Pomacentrus amboinensis, Premnas biaculeatus*), concluded that late-stage larvae, but not young larvae, can feed in UV light (365 nm). They also concluded that larvae from all three species have average spectral sensitivity in a similar range (493–507 nm), despite some species living at greater depths than others. In another study on the same three species, Job and Bellwood (2007) found again that older larvae could feed successfully in solely

UV-A wavelength light (365 nm), but younger larvae fed only in wavelengths of 400 nm or greater. They conducted a field experiment using UV versus white light in light traps, which demonstrated the wide taxonomic range of pre-settlement reef fishes that detect and move toward UV light in the field. The traps with UV light collected 16 families of reef fishes (as well as unidentified crustaceans), while the white light trap collected 21 families. Although UV light is clearly useful on the reef (Losey et al. 1999), there is a hypothesized tradeoff where, on the one hand, retinal sensitivity to UV light can give enhanced ability to distinguish prey against background, but on the other hand, allowing UV light to penetrate the eye can cause tissue damage (Siebeck and Marshall 2007). Siebeck and Marshall (2007) examined the transmission properties of ocular media to see if UV light could pass through the media to the retina in larvae and adults of a large range of reef fish species or, alternatively, if UV light was blocked during transmission. They found that UV light could pass through ocular media in about half the species examined (the cutoff transmission wavelength varied), and that there was a significant difference among families in the pattern of ontogenetic change in UV transmission through ocular media. For example, in some fish taxa, UV was allowed in only the larvae and not the adults, and in other taxa, UV was allowed only in adults and not in larvae. Further comparative analysis of taxonomic variability in UV sensitivity may be insightful in understanding its functional significance. In summary, the morphological and behavioral evidence to date suggests that UV light is not used by young pelagic larvae, but as larvae approach settlement, many species develop the ability to see UV light. It would be interesting as an 'evolutionary test' to assess the ability of the hatchlings of the coral reef pomacentrid *Acanthochromis polyacanthus* to detect UV light, as they lack pelagic stages, hatch directly on the reef, and are known to have accelerated eye development (Kavanagh and Alford 2003, Pankhurst and Pankhurst 2002, see retinal sections of coral reef fishes in Fig. 5.8). One would expect them to have accelerated UV sensitivity to coincide with hatching if habitat is driving the difference between larvae and adults.

The ability to use polarized light is also potentially useful in clear tropical waters, as it eliminates the cloudiness that can occur by scattered light in shallow water. Juvenile damselfishes have been found to have 3 and 4-channel polarization, the most complex polarization sensitivity recorded for any vertebrate (Hawryshyn et al. 2003). Decapods have 2-channel polarization ability (see below). More research is needed in this area to see if this capability may be useful for orientation or navigation, and whether it is found in fish larvae in addition to adults.

Decapod crustaceans are similar to fishes in their bipartite life history, but undergo a more complete metamorphosis through several larval and juvenile stages (Fig. 5.9). Larval and adult decapods both have compound eyes, but they change the relative position of the eyes on the head as they undergo metamorphosis. Several studies have analyzed the ontogenetic change from appositional eyes in larvae to superpositional eyes in adult decapods, with suggestions regarding functional changes (Douglas and Forward 1989, Mishra et al. 2006). Decapod larval vision has been presumed to be important for vertical orientation in the water column and avoidance of predators (Cronin and Jinks 2001, Huang et al. 2005). A comparison of several decapod crustaceans provided evidence of conservation among taxa in early larval development of eyes (Harzsch et al. 1999) presumably because they all live in open pelagic waters and are subject to similar environments. However, as metamorphosis approaches, differences in eye development among species become more obvious and reflect their divergent habitats (Cronin and Jinks 2001). Compound eyes of crustaceans have small compact retinae, and, in the larvae of any given species, only a single class of photoreceptor has been found (maximum absorption between 450–500 nm); however, behavioral studies have suggested that UV receptors may be active as well as those in the green/blue spectrum (Cronin and Jinks 2001).

The ability to use polarized light is found widely in decapod adults and larvae. Polarized light was found to affect the dorsal light reflex of the crayfish in a manner somewhat similar to that of the 'sun compass' used by grass shrimp for determining the direction of deeper water (Glantz and Schroeter 2007) although it has not been tested specifically in orientation. Mishra et al. (2006) suggests that the eye of the phyllosoma larva of the spiny lobster *Jasus edwardsii* is capable of e-vector discrimination and thus can be useful in orientation. These studies suggest that decapod crustacean larvae may have navigational abilities requiring vision.

Finally, local migrations of benthic adults may account for some connectivity among adjacent tropical habitats and may involve vision for guidance. Recent evidence from behavioral experiments on adult benthic decapods suggests that visual cues are important for adult host- or shelter-seeking behavior in symbiotic shrimp and crabs (Huang et al. 2005, Baeza and Stotz 2003).

In summary, vision is likely to be most useful in local habitat choice or navigation rather than navigation across large distances. Measures of visual acuity are highly variable depending on the method of measurement. Settlement-age larvae of both tropical fish and decapods use at least colors in the green/blue spectrum and often use short-wavelength UV light too, with increasing evidence coming from morphological and behavioral analyses. The ability to detect polarized light has been observed in adults of some decapods and damselfish, but needs to be tested in younger stages.

5.2.4 Lateral Line and Electroreception

The lateral line organ is a superficial sensory system found in amphibians and fishes, similar to the mechanoreceptor system in decapod crustaceans, which detects near-field water movement relative to the skin's surface (Maruska 2001). Because the lateral line can detect movement only in the near field, it is unlikely to be of any use over scales of >10 m. However, the mechanoreceptor system plays a major role in rheotaxis possibly facilitating detection of the first benthic habitat during settlement of reef fish and decapod larvae (Baker and Montgomery 1999, Coombs et al. 2001).

In fishes this system consists of solitary sensory units, neuromasts, which are scattered over most areas of the body of the fish. Some neuromasts are freestanding, while others are embedded in lateral line canals. A neuromast consists of a group of sensory hair cells that respond to the deflection of their cilia. Neuromasts therefore provide information about the local water flow (Ghysen and Dambly-Chaudière 2007). The ciliary bundles of lateral line hair-cells project into a gelatinous cupula, which has a flat, ribbon- or rod-like shape (Mogdans and Bleckmann 2001). The differing morphological designs of the peripheral lateral line are hypothesized to be adaptations to the hydrodynamic conditions in the habitat of a given species. The general physiology, however, is remarkably similar in different species of fish (recent reviews in Coombs et al. 1988, 1989, Mogdans and Bleckmann 2001, Myrberg and Fuiman 2002 (focusing entirely on coral reef fish senses), Mogdans et al. 2004, Ghysen and Dambly-Chaudière 2007). For a review on the function of the free neuromasts of marine teleost larvae see Blaxter and Fuiman (1989). These organs are divided classically into two main different types, ordinary and specialized, whose functions are mechanoreceptive and electroreceptive, respectively (see review in Cernuda-Cernuda and Garcí-Fernández 1996 for details on the structural diversity of these two receptor types).

Lateral line receptors develop from six pairs of placodes (Fuiman 2004 and references therein). Although the lateral line system has been studied in many species of fish from several taxa (e.g., Coombs et al. 1989), at the time of writing there are no published studies on the morphology of the lateral line system of tropical teleost reef fish larvae (JF Webb, University of Rhode Island, pers. comm.). Recent reviews on the ontogeny of the lateral line system in fishes are Webb (1999), Northcutt et al. (2000), and Fuiman (2004).

Diaz et al. (2003a) examined the development of the lateral line system in the temperate/subtropical European sea bass *Dicentrarchus labrax* (family Moronidae) from embryo to adult, and in the absence of detailed studies of tropical species its development may be a useful guide to further studies. Using light and electron microscopy, Diaz et al. (2003b) found that the first free neuromasts appeared on the head shortly before hatching and multiplied during the larval stage. Free neuromasts were aligned on the head and trunk in a pattern that corresponded to the location of future canals. The transition to the juvenile stage marked the start of important anatomical changes during which head and trunk canals were formed successively.

Lateral line canals are not present in young fish larvae; e.g., they were never observed in larvae of the anemonefishes *Amphiprion melanopus* and *A. clarkii*, when examined by SEM (Arvedlund unpubl. data, Fig. 5.10b). However, the cephalic lateral line system was clearly present in newly-metamorphosed coral gobies of the species *Paragobiodon xanthosomus* (Gobiidae; Fig. 5.5a) (Arvedlund et al. 2007) and in spangled emperors *Lethrinus nebulosus* (Lethrinidae) about six days before metamorphosis, i.e., day 38 post-hatching (Fig. 5.10a). Lara (1999) examined, by the aid of SEM, the cephalic lateral line neuromasts and canals of settlement-stage larvae and early juveniles of several species of Caribbean labrids (Labridae) from the genera *Halichoeres*, *Clepticus, Doratonotus, Xyrichtys*, and *Thalassoma*, and some species of scarids (Scaridae). She found that canals develop around a line of free neuromasts, which lie in the position of the future canal. Canal pores start to appear at one or both ends of the neuromast line. After the canal encloses these free neuromasts, pores continue to be added along the canal at least through the

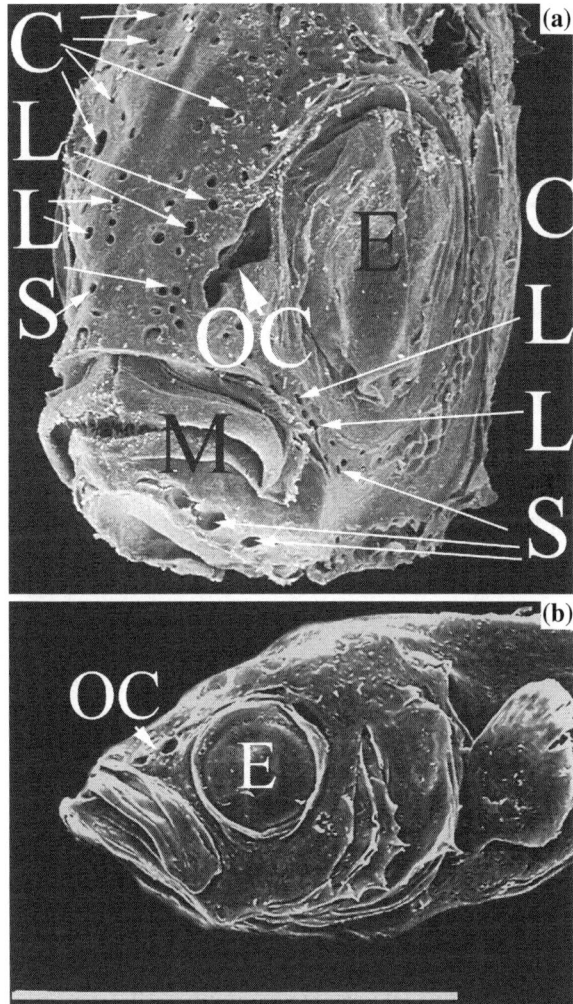

Fig. 5.10 Differences in the development of the cephalic lateral line system in reef fish larvae. (**a**) The snout of a nursery reef fish, *Lethrinus nebulosus*, 38 days after hatching, about six days before metamorphosis (day 43–44). At this stage *L. nebulosus* has a well-developed cephalic lateral line system. CLLS = cephalic lateral line system, OC = olfactory chamber, E = eye, M = mouth. (**b**) The snout of an anemonefish larva, *Amphiprion clarkii*, six days after hatching, and about four days before metamorphosis. *A. clarkii* has no cephalic lateral line system at this stage. Images by Michael Arvedlund with kind permission from Reef Consultants

juvenile stage. Newly formed pores often appear wider than pores in older juveniles and adults. Earlier stages have fewer, wider pores in their canals and more canals or portions of canals still composed of exposed neuromasts. Later stages have a higher proportion of their canals completely enclosed, with more and smaller pores than seen in the earlier stages. The labrid *Halichoeres maculipinna* appeared to be the most developed of the labrid species studied at the time of settlement: settlement-stage *Halichoeres maculipinna* possessed the most enclosed canals and the largest number of pores in those canals at this stage. The remaining *Halichoeres*, *Clepticus*, and *Doratonotus* reached a similar state of development at settlement and all were slightly less developed than *H. maculipinna*. *Xyrichtys* and *Thalassoma* appeared slightly paedomorphic at settlement, resembling earlier stages of the other labrid species. *Thalassoma* settles before the preoperculum is fully formed, the orbit of the eye is indistinct in this species at settlement, and the epithelium of settlement-stage larvae of *Xyrichtys* lack free neuromasts, though post-settlement juveniles possess them. The scarid larvae appeared to be much less developed than any of the labrid species. Scarids were smaller at settlement than any species of settlement stage labrid collected. Settlement-stage scarids have indistinct orbits, few lateral line pores and an incompletely formed olfactory organ in the case of *Scarus* sp. These studies suggests that many tropical marine teleost larvae have no lateral line system at hatching and through the early larval stage, but only free neuromasts for a tactile sense. Some species develop lateral lines during the pelagic phase. However, more comparative studies are needed before any general conclusions can be drawn about lateral line development in tropical teleosts.

Mechanoreceptors in decapod crustaceans can be divided into at least two types: vibration receptors and touch receptors. The mechanosensory cells are believed to be very similar, and both types are generally recognized by five distinct ultra-structural characteristics, which separate them from chemosensory cells (Garm and Høegh 2006 and references therein).

Most of the work on mechanoreception in decapods has been done with species from temperate waters (AG Jeffs, University of Auckland, pers. comm.). We are not aware of published results on the morphology of mechanoreceptors in tropical decapod crustaceans since 2001. A study on the external morphology and distribution of the integumental organs of the final-stage phyllosoma of the rock lobster *Jasus edwardsii* was conducted by Nishida and Kittaka (1992). A comparison of the morphology of these organs with the sense organs of other decapods with known function suggests that the dorsal surface of the body trunk is one of the major sites of reception of near-field water movement in *Jasus phyllosoma*. Seven types of organs were recognized on the integument of the body trunk, antennules, and antennae: (1) plumose setae, (2) simple setae, (3) porous setae, (4) aesthetasc setae, (5) simple pores, (6) dorsal cuticular organs, and (7) dome-shaped structures. The plumose setae and simple pores were abundant on the dorsal surface of the cephalosome, abdomen, and telson. The dorsal cuticular organs were present only on the dorsal surface of the cephalosome, and the aesthetasc setae were restricted to the antennule.

Electroreception occurs in most non-teleost fishes, especially Agnatha, Elasmobranchii, Holocephali, Chondrostei, Polypteri, and Dipnoi, and in four orders of teleosts: the siluriforms (catfishes), the gymnotiforms (knifefishes), the mormyriforms (elephant nose fishes), and in one subfamily (Xenomystinae) of the osteoglossiforms (Zupanc and Bullock 2005 and references therein). For a comprehensive recent review covering all aspects of electroreception see Bullock et al. (2005). In addition, see also Myrberg and Fuiman (2002, pp. 140–143) who summarize published results on coral reef fish electroreception until 2001. There is no evidence for the presence of electroreception in marine invertebrates (Jeffs et al. 2005).

The perception of electric signals is mediated by two distinct classes of specific electroreceptor neurons: ampullary and tuberous. The first type is found in most non-teleost fishes (except in Myxiniformes and Holostei), and in four orders of Teleostei. Tuberous receptors have only been identified in two teleostean order, the Gymnotiformes and the Mormyriformes (Zupanc and Bullock 2005 and references therein). At the time of writing we are not aware of any published results since 2001 on the issue of electroreceptors in larvae of tropical fishes. For a recent review on the early development of electroreceptors see Northcutt (2005).

5.2.5 *Magnetic, Thermal, and Other Senses*

From conditioning and orientation experiments *ex situ*, evidence exists that some teleost fishes, as well as spiny lobsters, have a magnetic sense. In marine animals three types of detection of magnetic fields are currently known to exist: 1) magnetic field detection based on magnetite particles, 2) based on photopigments, and 3) based on electrical induction (see reviews by Walker and Dennis 2005, Cain et al. 2005 and references therein). For a summary on the physical principles of magnetic orientation see Kalmijn (2003). At the time of writing we are not aware of published results on the use of magnetic senses in larvae of tropical fishes and decapod crustaceans.

While it has long been established that fish and decapods can detect and respond to local temperature differences (e.g., Doudoroff 1938), the use of a thermal sense for orientation or navigation toward suitable habitat in tropical environments has not been studied. However, the large-scale ocean gradients in temperature relative to coastal regions (Fig. 5.11) suggest that this is a potential navigational cue and should be investigated.

Salinity gradients are another possibility for navigational cues for oceanic teleost and decapod larvae over a wide range of distances. It was recently shown that polyvalent cation receptor proteins (CaRs) act as salinity sensors in fish (Nearing et al. 2002). CaRs allow fish to sense and respond to alterations in water salinity based on changes in Ca^{2+}, Mg^{2+}, and Na^{2+} concentrations found in freshwater, brackish water, and seawater. Likewise, changes in plasma Ca^{2+}, Mg^{2+}, and Na^{2+} occur when fish move from freshwater to seawater, and probably serve as salinity sensing cues for CaRs positioned within fish internal organs (Nearing et al. 2002).

Dufort et al. (2001) found that the primary receptors responsible for detecting reductions in salinity in adult lobsters *Homarus americanus* (family Nephropidae) are located within or near the branchial chambers and are primarily sensitive to chloride ions.

5.3 The Cues

The use of the multiple sensory systems described above for navigation requires the presence of environmental cues to guide larvae to benthic habitats. Below we discuss recent progress in understanding what cues are available for teleost and decapod larvae in this context.

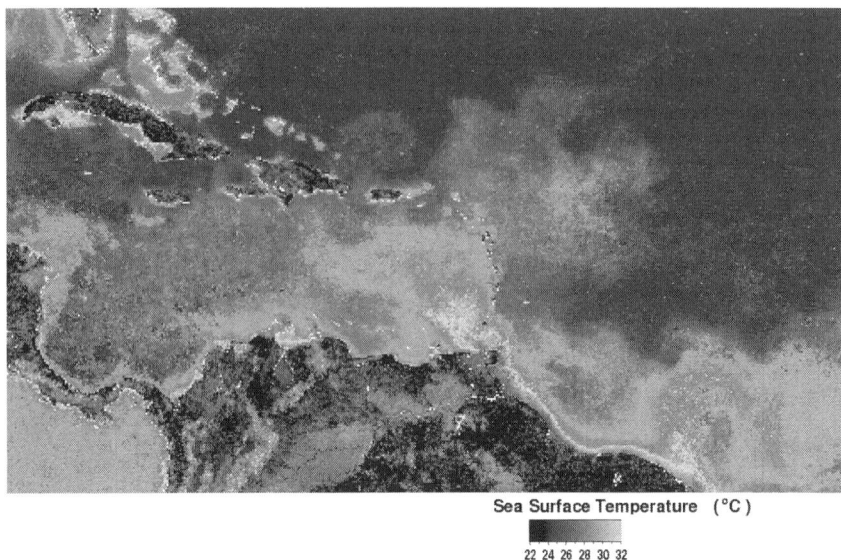

Sea Surface Temperature (°C)

22 24 26 28 30 32

Fig. 5.11 Sea surface temperatures in the southern Caribbean, showing large scale gradients as potential cues. Image provided with kind permission by the SeaWiFS Project, NASA/Goddard Space Flight Center and GeoEye

5.3.1 Olfactory Cues

Coral reef fish and decapod larvae appear to use gradients of olfactory environmental cues when settling to the first benthic habitat. For tropical fish larvae, see the recent reviews by Leis and McCormick (2002) and Montgomery et al. (2001). For decapod larvae, see Forward et al. (2001) and Gebauer et al. (2003a). For a review that covers both tropical fish and invertebrates see Kingsford et al. (2002).

Atema et al. (2002) provided the first evidence that wild-caught larval reef fish (primarily cardinalfishes, Apogonidae) near the time of settlement to the first benthic habitat prefer lagoon water over ocean water. They described ebb tide plumes of lagoon water that extend several kilometers from reefs, providing olfactory cues for dispersal and settlement of larvae of tropical fish and decapod crustaceans. Atema et al. (2002) argue that their result provides support for the reef fish chemical habitat imprinting hypothesis (Arvedlund and Nielsen 1996, Arvedlund et al. 1999, 2000a, b), which proposes that fishes as embryos or early larvae may imprint to reef odors (secreted from, e.g., cnidarians) and that this could facilitate both retention near the natal reef and navigation toward reefs from greater distances. Hypothetically, imprinting to conspecifics may also be a possibility for some reef fishes (Atema et al. 2002, Gerlach et al. 2007): some apogonids are mouth brooders (Job and Bellwood 2000) with the male carrying the fertilized eggs and later hatched embryos. This close contact during embryogenesis may enable such species to imprint to conspecific cues.

Isolation and identification of a chemical conspecific cue used by reef fish juveniles of the damselfish species *Chromis viridis* (Pomacentridae) was conducted by Lecchini et al. (2005a). By applying high performance liquid chromatography (HPLC) analyses of seawater containing *C. viridis* juveniles and isolating high concentrations of several organic compounds to be used in subsequent laboratory trials, they demonstrated that *C. viridis* larvae responded positively to only one of several organic compounds. This compound was characterized by a weak polarity and was detected at 230 nm with a 31-min retention time in HPLC. The same year, Wright et al. (2005) succeeded in pinpointing the olfactory abilities of pre- and post-settlement stages of the damselfish *Pomacentrus nagasakiensis*, by aid of the electro-olfactogram (EOG). No difference in olfactory ability was found between the two developmental stages: both showed olfactory responses to conspecific chemical cues as well as L-alanine. Therefore, the olfactory sense has similar capabilities in both ontogenetic stages. These results show that larvae of *P. nagasakiensis* that are ready to settle to the first benthic habitat can smell biochemical coral reef cues, but it is unclear as to what extent these fish larvae use such cues when locating settlement sites. Lecchini et al. (2005c) also conducted a series of *ex* and *in situ* tests using larvae of the coral reef fish *Chromis viridis* to determine ecological determinants of settlement choice (conspecifics versus heterospecifics versus coral substrates), sensory mechanisms (visual, acoustic/vibratory, olfactory) underlying settlement choice, and sensory abilities (effective detection distances of habitat) under field conditions. *C. viridis* larvae responded positively to visual, acoustic/vibratory, and olfactory cues expressed by conspecifics. Overall, larvae chose compartments of experimental arenas containing conspecifics in 75% of trials, and failed to show any significant directional responses to heterospecifics or coral substrates. In field trials, *C. viridis* larvae detected reefs containing conspecifics using visual and/or acoustic/vibratory cues at distances <75 cm; detection distances increased to <375 cm when olfactory capacity was present (particularly for reefs located up-current).

The first short-range evidence showing that a tropical seagrass-settling fish can use chemical environmental cues in selecting its first benthic habitat was provided by Arvedlund and Takemura (2006) in their study of the spangled emperor *Lethrinus nebulosus* (Lethrinidae). Huijbers et al. (2008) found a similar olfactory ability with juvenile french grunts (*Haemulon flavolineatum*, Haemulidae). *H. flavolineatum* is also strongly associated with mangroves and seagrass beds during the juvenile life stage.

It is unknown what type of chemical cues emitted from seagrass bed habitats are used for habitat detection by settling *L. nebulosus* and *H. flavolineatum*. In their experimental scenario Arvedlund and Takemura (2006) used pieces of complete seagrass bed habitat that included seagrass, sand and, possibly, algae, bacteria, and silent organisms (i.e., with no biosound). Chemical cues could stem from any one of these components of the seagrass bed habitat in both of these experiments, in addition to chemicals from plant tannins and related phenolic substances, which are produced by submerged vascular plants, emergent saltmarsh vegetation and mangroves, and brown algae (Arnold and Targett 2002). These molecules have many secondary roles, such as antimicrobial agents, herbivore deterrents/attractants, digestion reducers, and defense-related messengers (Arnold and Targett 2002), but in addition we now consider they may provide chemical cues that enable settling fishes to find a seagrass bed. Murata et al. (1986) showed that some chemical cues attracting settling anemonefish may be secreted by the sea anemone host (amphikuemin and analogs), but that others, also significant although weaker, can be secreted by dinoflagellates found in the sea anemone epithelium. As such, dinoflagellates in seagrass beds may also provide chemical cues to settling fishes and decapods (Arvedlund and Takemura 2006).

In addition to plants and microorganisms, chemical cues may also stem from prey organisms, which secrete amino acids. It is known that dissolved amino acids commonly provide fish with chemical cues to food (Ishida and Kobayashi 1992) and thus the same cues could be used to find appropriate habitat. Wright et al. (2005) showed that damselfish *Pomacentrus nagasakiensis* individuals were able to detect the odors of amino acids, both before and after they settled. Other chemical cue sources possibly used by fish and decapod larvae to find a suitable habitat are biogenic trace gases such as dimethyl sulphide (DMS), organohalogens, and non-methane hydrocarbons that may function as oceanic chemical signals for some plankton organisms. Several functions of DMS have been described, including a role as a chemosensory attractant and deterrent (Steinke et al. 2002 and references therein). The olfactory response of settlement stage larvae of the coral trout *Plectropomus leopardus* (Serranidae) to amino acids was tested electrophysiologically (Wright et al. 2008). The response was similar for the two amino acids tested and for the water conditioned by conspecifics. The authors concluded that the olfactory abilities of *P. leopardus* are well developed at settlement stage and apparently sufficient to detect olfactory cues from reefs.

Recently, several studies have shown that some coral reef fishes 'home' to their natal reef when settling to the first benthic habitat (Jones et al. 1999, Swearer et al. 1999, Robertson 2001, Jones et al. 2005, Almany et al. 2007, Gerlach et al. 2007) perhaps similar to salmon homing (e.g., Stabell 1984, 1992, Dittman and Quinn 1996). Although entirely a speculation, chemical imprinting to habitat cues (i.e., ecological imprinting; for a definition of ecological and other types of imprinting see Immelmann 1975a, 1975bb), or conspecific cues, may play a role for species that have homing patterns when settling to the first benthic habitat.

There are several examples of decapod crustaceans whose settlement and/or metamorphosis (see further below in this section) is aided by chemical cues. Such a mechanism may also exist in coral reef fish (McCormick 1999). Presettlement larvae of the manini *Acanthurus triostegus* (Acanthuridae) are capable of delaying metamorphosis in the absence of proximity to a benthic environment (McCormick 1999). Because of the ability of reef fish to distinguish between lagoon and oceanic water (e.g., Atema et al. 2002), it should be investigated whether the metamorphosis of pre-settlement manini larvae is influenced specifically by the presence of chemical environmental or conspecific cues or the presence of visual cues.

Burgess et al. (2007) recently identified the presence of eddies at One Tree Island, Southern Great Barrier Reef, East Australia. They sampled pre-settlement fishes in surface waters based on the presence or absence of eddies as predicted from a calibrated hydrodynamic model of the Capricorn-Bunker region. Higher concentrations of pre-settlement fishes, mostly mullids (goatfishes) were found close to the reef in locations where eddies were known to form rather than in locations without eddies, and this was consistent among days and tidal cycles and in varying wind conditions. Locations where eddies were not predicted to form consistently had low concentrations of pre-settlement fishes. There was evidence for an effect of the windward side of the reef, but areas with eddies maintained high concentrations even when on the leeward side. Higher concentrations were not necessarily found in the eddy itself; rather, they occurred at locations where eddies were predicted to form on the flood or ebb tide. Eddies increase the probability that pre-settlement fishes will stay near reefs through retention, in some cases their natal reef. Burgess et al. (2007) finally concluded that eddies may also increase behavioral interactions among marine animals and assist in the detection of reefs that may elicit settlement behavior.

In summary, contemporary studies show that larvae of several coral reef fish species use chemical cues for settling to the first benthic habitat, including species that settle into nursery habitats such as mangroves and seagrass beds. In addition, some reef fishes may use chemical (or visual) cues for inducement of the metamorphosis from pelagic larvae to benthic juvenile. However, these conclusions are based on very few studies.

The interesting findings of possible 'homing' in some species of coral reef fish should be investigated further to determine whether the larvae demonstrate homing in the traditional sense, where individuals are transported away and then find their way back to the starting point, or whether larvae simply remain close to the reef. In fact, there is no evidence that the self-recruited larvae ever moved away from the immediate vicinity of the natal reef. New technology could help address this question by the aid of remote controlled underwater miniature submarines equipped with a camera and an eddy current sensor probe (the MIDAS submarine, Sjo et al. 1988) or the submarine designed by Bokser et al. (2004). Such vehicles may be able to follow fish or decapod larvae *in situ* for many hours or even days.

When settling to the benthic habitat, many species of decapod crustaceans, particularly crabs, are known to use chemical cues that are usually linked to the adult habitat or to the presence of conspecific adults (Forward et al. 2001, Gebauer et al. 2003b, Keller et al. 2003, Jeffs et al. 2005 and references therein). Gebauer et al. (2002) studied the impact of intra- and interspecific chemical settlement cues from adults in captivity-reared larvae of the semi-terrestrial tropical saltmarsh/mangrove crab *Sesarma curacaoense* (Sesarmidae). They showed that the presence of substrate did not significantly influence the time to metamorphosis, but did reduce the mortality rate. Development was consistently fastest when larvae were put in water conditioned with conspecific adult odor, and also responded significantly to water conditioned with adults of congeneric crab species. These response patterns suggest that chemically similar factors (presumably pheromones) are produced by closely related species (Gebauer et al. 2002). Forward et al. (2003) found that chemical cues from seagrass beds could provide cues that allow orientation to nursery habitat for premoult megalopae of the blue crab *Callinectes sapidus*. van Montfrans et al.

(2003) found that initial non-random distribution of blue crabs in Chesapeake Bay may be deterministic and due to active habitat-selection behavior by megalopae. Moksnes and Heck (2006) found additional supporting evidence for these ideas, showing that the habitat-specific distribution of juvenile blue crab is dictated by active habitat selection in blue crab megalopae and early juveniles.

When environmental cues are absent, 'competent' invertebrate larvae (i.e., those being physiologically and morphologically ready for settlement and metamorphosis) may delay the initiation of these developmental processes, remaining as plankton for an extended period. In the tropical semi-terrestrial crab, *Sesarma curacaoense,* this mechanism is limited to exposure approximately during days 4–6 of the molting cycle. This may be crucial as a temporal window of receptivity for the cue as this period coincides with the transition between intermoult and premoult (Gebauer et al. 2005). In relation to induced settlement and metamorphosis, Hadfield and Paul (2001) discuss whether settlement and metamorphosis are induced by the same compound or by two separate compounds. Settlement and metamorphosis are usually considered separately (Hadfield and Paul 2001). Settlement is defined as the behavioral performance of pelagic larvae leaving the plankton, approaching the benthos, and moving upon a substratum with or without attachment to the latter (Hadfield and Paul 2001). Metamorphosis includes loss of larva-specific organs and emergence of juvenile/adult-specific structures. Hadfield and Paul (2001) concluded that there are examples of both one and two separate cues inducing these processes within marine invertebrates, but this topic remains largely unexplored.

Remote chemicals can act as habitat cues for pelagic larvae of benthic invertebrates by stimulating settlement and metamorphosis. To test whether chemical cue effectiveness declines with increasing distance from the source, O'Connor and Judge (2004) examined whether the ability of seawater to stimulate metamorphosis (molting) of the tropical fiddler crab *Uca minax* megalopae is restricted to water overlying saltmarshes. The results showed that chemical cues for molting of fiddler crab megalopae originate in marshes and decline in effectiveness within a short (<15 m) distance from the marsh habitat.

O'Connor and Van (2006) showed that adult-associated chemical cues can stimulate settlement and metamorphosis of invertebrate larvae into habitats with an enhanced likelihood of juvenile and adult survival. For example, sediments from adult fiddler crab habitat stimulate fiddler crab megalopae to metamorphose (molt) sooner than sediments without adult cues. A similar stimulation of molting occurs after exposure to waterborne chemical cues from adult habitats and to exudates and extracts of adult crabs. They tested whether sediments from habitats without adult *Uca pugnax*, which do not stimulate molting of their megalopae, could become stimulatory through brief exposure to adult crabs. Results suggest that the chemical cues that adult crabs release are retained by sediments and consequently stimulate molting of megalopae, regardless of the nature of the sediments themselves. O'Connor and Van (2006) concluded that the absence of chemical cues may delay colonization of newly created or heavily disturbed habitats that are otherwise suitable settlement and adult habitat. Support for O'Connor and Van's results (2006) came from a study conducted by Diele and Simith (2007) which showed that the megalopa of the semi-terrestrial mangrove crab *Ucides cordatus*, mainly settle in areas populated by conspecific crabs and/or muddy habitats. Chemical cues from conspecifics for settling may also override salinity stress. An experimental laboratory study with the megalopa stage of *Armases roberti*, a freshwater inhabiting species of crab from the Caribbean region, combined the effects of odors from conspecific adults and of stepwise salinity reductions. The duration of development to metamorphosis was significantly (by about 25%) shortened, when odors from conspecific adult crabs were present, regardless of the salinity conditions. Such results indicate that the metamorphosis-stimulating effect of chemical cues from an adult population of *A. roberti* is far stronger than the potentially retarding effect of increasing hypo-osmotic stress (Anger et al. 2006).

The overall conclusion from these studies is that chemical cues from conspecifics seem to play a major role for decapod crustaceans settling to the first benthic habitat. Although based on few studies, evidence is accumulating for a similar effect in reef fishes.

5.3.2 Auditory Cues

Sound is available as an orientation cue in the ocean. The physical properties of sound in water differ from sound in air, with sound waves traveling faster and propagating farther in water, creating a noisy background against which to detect a directional signal. Despite this potential problem, evidence is accumulating that larvae of fishes and crustaceans do hear sound and orient to sound in a way that is consistent with their use of sound to guide settlement onto reefs. Recent field experiments, including improvement of light trap catches by replayed reef sound, *in situ* observations of behavior, and sound-enhanced settlement rate on patch reefs, together show that sound is used as a navigation and settlement cue for at least some late larval stages (Montgomery et al. 2006; see this reference and Popper et al. (2001) for a recent comprehensive review on sound as a settlement cue for fish and decapod larvae).

The question of what sound frequencies might be used for navigation has been addressed by several recent studies. By using the electrophysiological method of auditory brainstem response, a technique originally used in mammalian audition studies and later adapted for audition studies on fishes (Corwin et al. 1982, Kenyon et al. 1998), Wright et al. (2005) tested pre- and post-settlement larvae of the damselfish *Pomacentrus nagasakiensis* (Pomacentridae) for their hearing abilities of reef sound cues. For both pre- and post-settlement fish, the sensitivity was greatest at 100 Hz, followed by 200 and 600 Hz. The auditory threshold for both pre- and post-settlement fish increased from 100 Hz to 400 and 500 Hz, and then dropped at 600 Hz. Thereafter, thresholds increased with increasing frequency from 700 Hz to 2,000 Hz. The hearing of post-settlement fish was significantly more sensitive than their pre-settlement counterparts at only two frequencies: 100 Hz and 600 Hz. Thresholds of post-settlement fish were 8 dB lower than thresholds of pre-settlement fish at these two frequencies.

Wright et al. (2008) examined the auditory abilities of settlement-stage larvae of the coral trout *Plectropomus leopardus* (Serranidae) electrophysiologically to determine if these senses are sufficiently developed to aid larvae in detection of settlement habitats on coral reefs. *P. leopardus* larvae detected sounds from 100 to 2,000 Hz with hearing most sensitive at the frequencies of 100, 200, and 600 Hz. Wright et al. (2008) concluded that the auditory abilities of *P. leopardus* are well developed at settlement stage and apparently sufficient to detect auditory cues from reefs.

Damselfishes (Pomacentridae) and cardinalfishes (Apogonidae) in particular may use sound for orientation to the first benthic habitat, as shown by Simpson et al. (2005a) in a field experiment using patch reef enhanced with biosound at Lizard Island, Great Barrier Reef. This result is important because Simpson et al. (2005a) did not use light traps enhanced with biosound as some similar past studies did (e.g., Tolimieri et al. 2000). Light traps may create biased results on a taxonomic basis because they are highly selective in which species that are caught (Leis and McCormick 2002 and references therein). Later, Simpson et al. (2008) used light traps to measure the response of a diverse range of settlement-stage fishes to the filtered 'high' (570–2,000 Hz) and 'low' (<570 Hz) frequency components of reef noise, and compared these catches with those from control 'silent' traps. Of the seven families represented by >10 individuals, four (Pomacentridae, Apogonidae, Lethrinidae, and Gobiidae) were caught in significantly greater numbers in the high frequency traps than either the low frequency or the silent traps. The Syngnathidae preferred high to low frequency traps, while the Blenniidae preferred high frequency to silent traps. The remaining family (Siganidae) showed no preference between any of the sound treatments. The results of this study suggest that many settlement-stage fishes may select the higher frequency audible component of reef noise, which is produced mainly by marine invertebrates, as a means of selectively orienting toward suitable settlement habitats. Combined with the auditory brainstem response studies by Wright et al. (2005, 2008), it suggests that the 'best frequencies' are those between 570 and 1,000 Hz, because sensitivity is very poor at >1,000 Hz.

Egner and Mann (2005) found that the damselfish *Abudefduf saxatilis* has poor hearing sensitivity in comparison to other hearing generalists including other species of pomacentrids. They used the auditory brainstem response technique to measure hearing of the sergeant major damselfish *A. saxatilis* of a size range between 11 and 121 mm. Significant effects of standard length on hearing thresholds at 100 and 200 Hz were detected. At these lower frequencies, thresholds increased with an increase in size. All fish were most sensitive to the lower frequencies (100–400 Hz). The frequency range over which fish could detect sounds was dependent upon the size of the fish; the larger fish (>50 mm) were more likely to respond to higher frequencies (1,000–1,600 Hz). Egner and Mann (2005) concluded that because of the high hearing thresholds found in their study in comparison to recorded ambient reef noise, it is unlikely that sound plays a significant role in the navigation of the pelagic larvae of sergeant majors returning to the reef from long distances (>1 km), but it may play a role in short-range orientation (<1 km). This conclusion was supported by the result by Mann et al. (2007), which showed that larval fishes in acoustically unbounded habitats most probably cannot detect the ambient noise of particle motion at distances >1 km.

Many recent studies point in the direction that underwater sound can provide important environmental cues for the pelagic stages of decapod crustaceans when settling to the first benthic habitat (Jeffs et al. 2005, Montgomery et al. 2006). However, experimental evidence for the orientation of larvae to underwater sound has been hard to secure due to the difficulties of conducting field experiments, and of controlling sound in experimental aquaria. Radford et al. (2007) developed an effective method for using a binary choice chamber coupled with an artificial source of underwater sound to conduct *in situ* behavioral experiments on crab postlarvae, at night in coastal waters of Omaha Bay, New Zealand. Postlarvae of five common coastal crabs from around New Zealand were used: *Notomithrax ursus* (Majidae), *Plagusia chabrus, Cyclograpsus lavauxi, Hemigrapsus edwardsii* (Grapsidae), and *Pagurus* sp. (Paguridae). The postlarvae of all five crab species showed an orientation response towards the sound source confirming that the binary choice chamber can be used as a reliable experimental tool for determining directional swimming behavior of postlarvae in response to sound cues. Radford et al. (2007) concluded that orientation to a sound cue is widespread among crab species and that this behavior could be of considerable ecological importance in influencing the settlement success of coastal crustaceans.

5.3.3 Visual Cues

It is unknown whether marine larvae use visual cues to orientate from the open ocean to appropriate benthic habitats. Light attracts pre-settlement reef fish larvae to light traps (Doherty 1987), but the adaptive reason for such a phototactic response is not known. In decapod crustaceans, phototaxis is also associated with vertical migration, which may be an adaptation to take advantage of stratified tidal currents bringing larvae towards estuarine habitats (Forward et al. 2007, but see Webley and Connolly 2007). A more likely use for visual cues in environmental selection involves microhabitat choice when settling larvae are already close to the substrate. For this situation, several studies have concluded that using visual cues to identify specific habitat type or conspecifics can facilitate finding appropriate habitat (Lecchini et al. 2005a, b, 2007b). Certainly the close correlation between retinal developmental changes and settlement to the substrate, where the spectral environment is very different than that of the open ocean, suggests that vision is a key adaptive sense for life on the bottom. However, visual acuity is broadly useful in feeding, avoiding predators, and many other functions, and as such the specific adaptive link between visual characteristics and choosing habitat is difficult to isolate. Marshall et al. (2006) analyzed the visual signals from colorful reef fishes as they would be received by other reef fish retinae, and concluded that the dichromatism of the retinal pigments of most reef fishes aids in discrimination and contrast of fish against background. This communication between fishes may be useful for settling larvae to detect appropriate habitat, but that is still an untested hypothesis. Lecchini conducted a series of studies examining the use of particular senses in the specific periods when returning larvae are selecting microhabitats on the reef. Lecchini et al. (2007b) found in laboratory experiments that visual cues (as well as other cues; see other sections in this chapter) were useful in detecting and moving toward conspecifics, but only in post-settlement wrasses, not in presettlers. In field experiments, settling larvae chose habitats with conspecifics from a moderate distance of 75–375 cm, even when the olfactory system was ablated (Lecchini et al. 2005c). Thus, the color patterns or behavior of conspecifics seem to act as a visual cue aiding the choice of habitat once a reef fish is settled. A detailed analysis of the visual environment of the coral reef and the visual systems of adult reef fishes is found in a series of papers (Losey et al. 2003, Marshall et al. 2003a, b, 2006). Such details are not reported for other tropical habitats or for settling decapods.

5.3.4 Rheotactic and Electric Cues

New results suggest that rheotactic cues are used in conjunction with other cues. Consequently, this is discussed in Section 5.3.6 (the use of two or more cues). At the time of writing we are not aware of published studies on the use of electric cues in larvae of tropical fishes and decapod crustaceans.

5.3.5 Solar, Magnetic, Wave, Thermal, Salinity, and Other Cues

The only evidence so far that tropical marine fish larvae use a solar compass comes from a field study of larval behavior during different weather conditions. Leis and Carson-Ewart (2003) studied *in situ* orientation, in daylight hours, of settlement-stage reef-fish larvae and younger reef-fish larvae. Larvae were taken 100–1,000 m offshore from coral reefs in water 10–40 m deep, at Lizard Island, Great Barrier Reef, then released and observed by divers. Depending on area, time, and species, 80–100% of larvae swam directionally. For example, three species *(Chromis atripectoralis, Neopomacentrus cyanomos*, and *Pomacentrus lepidogenys)* were observed in morning and late afternoon at the leeward area, and all swam in a more westerly direction in the late afternoon. Also, in the afternoon, *C. atripectoralis* larvae were highly directional in sunny conditions, but nondirectional and individually more variable in cloudy conditions. These correlations suggest that damselfish larvae utilized a solar compass, but controlled field and laboratory experiments to further evaluate these results are needed.

   Leis and Carson-Ewart (2003) also demonstrated generally that many species of reef fish larvae can detect the direction of the island's reefs by an unspecified cue, and that currents did not influence their orientation. For example, two species of butterflyfishes *Chaetodon plebeius* and *Chaetodon aureofasciatus* (Chaetodontidae) consistently swam away from the island, indicating that they could detect the island's reefs. *Caesio cuning* (Caesionidae) and *Pomacentrus lepidogenys* was non-directional overall, but their swimming direction differed with distance from the reef, implying the reef was detected by these species. Net movement by larvae of six of the seven species differed from that of currents in either direction or speed, demonstrating that larval behavior can result in non-passive dispersal, at least near the end of the pelagic phase. A similar experimental method using young (11–15 days), small (8–10 mm) larvae of *Amblyglyphidodon curacao* found that even larvae of this early stage appear to orient directionally in relation to the reef (Leis et al. 2007).

The use of magnetic fields as a navigational cue for tropical marine fish or decapod larvae is completely unknown, but has been explored in adult tropical decapods. Recent experiments (Boles and Lohmann 2003) have confirmed a study by Creaser and Travis (1950), that (adult) spiny lobsters (*Panulirus argus*) orient consistently toward capture areas when displaced to unfamiliar sites, even when deprived of all known orientation cues en route (Boles and Lohmann 2003). To test the hypothesis that lobsters exploit positional information in the Earth's magnetic field, they were exposed to fields replicating those that exist at specific locations in their environment. Lobsters tested in a field that exists north of the capture site oriented southward, whereas those tested in a field like one that exists south of the capture site oriented northward (Boles and Lohmann 2003). These results parallel those with turtles and provide evidence that spiny lobsters possess a magnetic map which facilitates navigation toward specific habitats. Whether larvae of spiny lobsters possess the same ability is a question that remains to be addressed.

Many other cues are possibilities, but have not been explored in any specific study. The direction of ocean swell presents one potentially useful orientation cue (Lewis 1979, cited in Montgomery et al. 2006). The vertebrate inner ear is capable of sensing wave direction in the ocean through the detection of the orbital motion of the wave (Montgomery et al. 2006). Salinity or temperature gradients are also potentially useful cues for orientation toward a suitable habitat (Fig. 5.11). In fact these large-scale gradients should be explored as a long-distance cue for navigation toward coastal regions in general. For example, it would be a useful research program to explore larval ability to detect salinity gradients in coastal water near riverine output, or lagoon water which may be hypersaline due to evaporation, as well as temperature variation coming from eddies or lagoon water on outgoing tides, or temperature changes correlated with depth and major currents. However, to our knowledge, there are no published results on those cues at the time of writing.

5.3.6 The Use of Two or More Cues

Given the availability of so many potential cues, it seems likely that multiple cues are used by settling larvae. Recent studies have indeed demonstrated that some reef fish larvae can use a range of sensory mechanisms effective over different (small, i.e., a few meters) spatial scales to detect and choose settlement sites (Lecchini et al. 2005b). This finding supports previous hypotheses of multiple sensory use by settling reef fishes, suggested by Montgomery et al. (2001), Leis and McCormick (2002), Kingsford et al. (2002), and Lara (2001). Specifically, Lecchini et al. (2005b) conducted experiments *ex situ*. Larvae were captured with crest nets and were then introduced into experimental tanks that allowed testing of each type of cue separately (visual, chemical, or mechanical cues). Among the 18 species studied, 13 chose their settlement habitat due to the presence of conspecifics and not based on the characteristics of coral habitat, and five species did not move toward their settlement habitat (e.g., *Scorpaenodes parvipinnis* (Scorpaenidae) and *Apogon novemfasciatus* (Apogonidae)). Among the different sensory cues tested, two species used three types of cues (i.e., *Parupeneus barberinus* (Mullidae) and *Ctenochaetus striatus* (Acanthuridae): visual, chemical, and mechanical cues), six species used two types of cues (e.g., *Myripristis pralinia* (Holocentridae): visual and chemical cues; *Naso unicornis* (Acanthuridae): visual and mechanical cues), and five species used one type of cue (e.g., *Chrysiptera leucopoma* (Pomacentridae): visual cues; *Pomacentrus pavo* (Pomacentridae): chemical cues). Thus, the results regarding multiple cue use, even in a controlled, short-range environment, are complex and underscore the difficulties in predicting larval navigation even among closely-related species.

Gardiner and Atema (2007) in a study of multiple cues for orientation (in the shark *Mustelus canis* from temperate waters) pointed out that there are several different types of rheotaxis: (1) orientation to the large-scale flow field (olfaction, vision, and superficial lateral line), (2) eddy chemotaxis: tracking the trail of small-scale and (3) odor-flavored turbulence (olfaction and lateral line canals), and (4) pinpointing the source of the plume (lateral line canals and olfaction). 'Odor-gated rheotaxis', combining olfaction and mechanoreception, has been useful in studies of other reef invertebrates (Pasternak et al. 2004). Experimental designs such as these, employed by researchers exploring similar questions in other taxa, may help in designing studies for tropical marine larval fishes and decapods.

5.4 Future Directions

When reviewing unresolved problems and suggestions for future research directions in the last series of reviews (Montgomery et al. 2001, Myrberg and Fuiman 2002, Kingsford et al. 2002, Leis and McCormick 2002, Collin and Marshall 2003, Jeffs et al. 2005, Montgomery et al. 2006), it is clear that many questions from those reviews remain to be addressed. In addition, new questions have appeared due to an improved understanding of the mechanisms of settlement in tropical fish and decapod larvae. Indeed, at present there appear to be more papers speculating about the functional development of tropical reef fish and decapod senses and what cues are used to find a suitable habitat, than there are actual experimental studies. Below, we list a number of selected unresolved problems to help guide future research.

(1) Clearly, in order to progress in this field, experimental and descriptive studies of the morphology, neuroanatomy, and functional development of all senses, from fertilization to the adult stage, of tropical fish and Decapoda are highly warranted.

(2) There has been recent research progress on a few of the environmental cues, such as olfactory and auditory cues, but studies need to expand to other potentially important senses and cues. In particular, we believe potential long-range cues such as magnetism, and temperature and salinity gradients are worth exploring.

(3) Recent results showed, somewhat surprisingly, that various reef fishes may 'home' to their natal habitat. If substantiated, the impact of such a finding could be highly influential, and, as such, we believe the phenomenon of imprinting to habitat or conspecific cues in reef fish or decapod species is worth exploring. Using salmonid homing as a model may be a good place to start. It is important to address the question of whether reef fish that are said to be 'homing' are actually leaving the reef and returning, or whether they have been retained in the immediate vicinity of the natal reef.

(4) Studies *ex situ* of the use of environmental chemical cues in tropical decapods have been executed under both apparent darkness (red light) and in daylight (e.g., Díaz et al. 2003). Nocturnal studies of tropical reef fishes of the same sort are uncommon but highly desirable because in nature, settlement often happens at night. Leis and McCormick (2002) also indicate the importance of finding 'innovative means' to study the behavior of larvae at night.

(5) The scales, from meters to kilometers, at which individual cues may be detected (see Fig. 5.13) are poorly understood. Knowing these limits will help with eventual modeling of larval movements.

(6) The settlement mechanisms, particularly the use of environmental cues when settling to the first benthic habitat, are poorly understood for nursery fishes, i.e., fishes that settle into seagrass or mangrove after the pelagic larvae phase before moving later to the coral reef.

(7) Similarly, the senses and cues used by fishes and decapods that live in, or on, corals, sponges, or other animals are not known. As specialists, these species may employ different mechanisms.

(8) In many marine invertebrate larvae, environmental cues associated with the adult habitat can induce metamorphosis and settlement, however, such a mechanism is not known in coral reef fishes. Previous work demonstrated that pre-settlement larvae of the manini *Acanthurus triostegus* (Acanthuridae) are capable of delaying metamorphosis in the absence of proximity to a benthic environment (McCormick 1999). However, not all individuals in this study delayed metamorphosis, and so there may be a minimum threshold level of cue necessary to trigger metamorphosis. McCormick's experiment must have been close to this level. This developmental response, combined with the ability of some reef fishes to distinguish between lagoon and oceanic water by the aid of chemical cues (Atema et al. 2002), suggests an experiment to test (for

example) whether pre-settlement manini larvae metamorphose because of the presence of particular chemical cues as opposed to visual or other cues.

(9) Mana and Kawamura (2002) found that captive-reared subtropical sea bream *Acanthopagrus schlegeli* (Sparidae) developed an abnormal nasal opening and irregularities in the distribution patterns of the olfactory receptor neurons. Thus, the morphology and ultrastructure of the olfactory organs of captive-reared tropical coral reef fishes and crustacean decapods may differ from those of their wild counterparts. However, this hypothesis has not been specifically tested. Another hypothesis that must be examined is whether captive-rearing of fish and decapods has any impact on the functionality of the senses. Whether senses in general develop abnormally in captive-reared tropical fish and decapod crustaceans compared to wild species is unknown. Because several captive-reared species are released in large numbers into the wild, e.g., the spangled emperor *Lethrinus nebulosus* (Arvedlund and Takemura 2006, p. 120), their survival may be compromised by poor sensory function.

(10) The capacity to distinguish chemical cues from preferred habitat is anticipated to be particularly well-developed among fish species that associate closely with specific habitat types ('habitat specialists'). For example, larvae of the anemonefishes *Amphiprion ocellaris* and *A. melanopus* remember the odor secreted by their species of host sea anemone during the embryonic stage. The larvae may use these chemical cues for orientation when selecting to the first benthic microhabitat (Arvedlund and Nielsen 1996, Arvedlund et al. 1999, 2000a, b). This host imprinting hypothesis should be tested with additional species (Atema et al. 2002, Jones et al. 2005, Gerlach et al. 2007) including mouth brooders such as apogonids (Job and Bellwood 2000). Here, it may be of help to apply similar methods used by Hasler, Scholtz, Wisby, and co-workers (Scholtz et al. 1976, but see also review in Dittman and Quinn 1996), which showed that salmon imprint to chemical cues. Scholtz et al. (1976) succeeded in artificial imprinting of salmon to morpholine and p-alcohol. Can anemonefishes (and other relevant candidate species) imprint to those compounds? Electrophysiological studies similar to Wright et al. (2005), or behavioral bioassays similar to Kasumyan (2002) are also needed. Last but not least, it will be strong support for the imprinting hypothesis if labeled chemical cues from the host can be shown to attach to olfactory receptors in anemonefish embryos.

(11) Murata et al. (1986) identified and described, and Konno et al. (1990) synthesized, the molecular structure of the pyridinium compound named amphikuemin (and analogs), that initiates symbiosis between the anemonefish *Amphiprion perideraion* and the sea anemone *Heteractis crispa*. Many more similar compounds secreted from tropical cnidarians may exist. The isolation and (partial) identification of a chemical conspecific cue used by reef fish juveniles of the damselfish *Chromis viridis* (Lecchini et al. 2005c) also suggests more similar conspecific compounds exists. Identifying these specific compounds would be a great advance.

(12) Each morphological type of sensory neuron expresses a particular class of odorant receptors (Hamdani and Døving 2007). Ciliated olfactory receptor neurons respond to bile salts and alarm substances found in the skin, the crypt cell olfactory receptor neurons respond to sex pheromones, and the microvillous olfactory receptor neurons to food odors (Hamdani and Døving 2007). To which olfactory receptor neuron type does a chemical compound like amphikuemin, the cnidarian symbiosis chemical (Murata et al. 1986, Konno et al. 1990), affiliate to?

(13) Environmental chemical contamination is now affecting every ecosystem. Ward et al. (2008) recently showed that acute exposure to low, environmentally relevant dosages of the ubiquitous contaminant 4-nonylphenol (found in rivers and estuaries throughout the world), can seriously affect social recognition and ultimately social organization in fishes. A one-hour $0.5 \ \mu g.l^{-1}$ dose was sufficient to alter the response of members of a shoaling fish species (juvenile banded killifish, *Fundulus diaphanus*) to conspecific chemical cues. Dosages of $1-2 \ \mu g.l^{-1}$ caused killifish to orient away from dosed conspecifics, in both a flow channel and an arena. What impact does 4-nonylphenol and similar compounds have on tropical reef fish and decapod crustacean's ability to use environmental cues to detect tropical ecosystems?

(14) To better understand the distances over which larval fishes can detect sounds from reefs, more studies on larval fish hearing and reef noise are needed. Larval fish hearing measurements need to independently distinguish sensitivities to particle motion and acoustic pressure. Likewise, independent measurements of particle motion around reefs are required (Mann et al. 2007).

(15) Simpson et al. (2005b) demonstrated that in embryonic anemonefishes (*Amphiprion ephippium* and *A. rubrocinctus*) the heart rate is influenced by sound. However, this detection of sound waves was not necessarily via the ears, and therefore, the sensory system used in the response may not have been hearing in the conventional sense: the state of development of the ears at this stage is unknown. Additional tests are needed to validate this hypothesis of early embryonic imprinting of sounds, and to determine the developmental and functional capabilities of auditory systems at these early stages.

(16) The vast difference in conclusions stemming from anatomical versus behavioral measures of visual acuity indicates a need to resolve how to measure visual acuity accurately and appropriately for a given question. Furthermore, future studies must also consider the use to which vision is put—'acuity' measured by feeding behavior may be very different from that measured for avoiding predators, or finding a settlement site.

(17) Studies are needed to understand the importance of ultraviolet light discrimination in larval environments, in both fish and decapod larvae.

(18) The ability to use polarized light is a potentially useful adaptation in clear tropical waters, and known to occur in adults. Studies are needed to ascertain whether larvae of fish or decapods have the ability to detect polarized light.

5.5 Final Thoughts

Since the recognition that pelagic larvae may possess the capability to orientate toward and very specifically choose benthic settlement habitat, significant progress has been made in understanding sensory abilities and the potential environmental cues that might guide larvae. Despite this, the entire field can be considered to be still in its infancy; we are quite far from having any practical level of predictability about the use of a given cue by a marine larva at a given time and place. To move toward this goal, we need to expand sensory studies to include *a broader range of taxa, more early life stages, more wild-caught specimens, and additional senses, especially those sensing potential long-range cues.* Functional considerations should include *establishing cue thresholds, cue sensitivity differences among taxa (e.g., specialists versus generalists), and use of combined cues.* Environmental factors in need of further analysis include *time of day (especially night-time), the scale at which a given cue is useful, less-studied habitats (e.g., seagrass, mangroves), and other cues that have not been studied in this context (e.g., temperature, salinity gradients).* The use of new technologies is needed to overcome major barriers in studying navigation in such tiny oceanic organisms. For example, remotely operated vehicles able to observe such small larvae *in situ* could provide behavioral information otherwise unavailable, and DNA fingerprinting to identify wild-caught larvae could help with understanding species-level differences.

The marine environment is rife with potential cues and gradients for larvae to detect and follow, and those cues differ substantially from the primary cues used by animals (including humans) in the terrestrial environment. Gerlach et al. (2007) state the idea in this way: *'Only olfaction can provide information on the identity of the water mass encountered. Compare navigating the New York subway system with no external frame of reference: it works only when one knows what train to take. Humans' trains are labeled with visual signs whereas the water currents are labeled with olfactory signals.'* Our human perspective may have biased the initial level of concentration on some types of cues (e.g., vision and hearing) versus others that may be more important from a marine larva's perspective. For example, while humans would rarely use temperature as a navigational cue, small ectotherms would be highly sensitive to temperature gradients. Thus, broadening the range of possible cues in our investigations will be important. It is also worth restating that larvae may use multiple cues at any given time and/or a series of different cues as they move among different areas of the ocean environment, or from offshore to onshore (Fig. 5.12). The limit of usefulness of each of these cues is also very much still open for debate.

Finally, we now know adaptive evolutionary change can happen quickly even in large marine populations (Conover and Present 1990) and so we should be cognizant of the possibility of local adaptation of sensory developmental rates and/or cue usage in different regions, even among populations. Indeed, one might reasonably argue that adaptive lability of such a selectively advantageous capability is likely. Taking this idea of local adaptation and system lability further, we might predict some level of coordination between sensory system developmental patterns

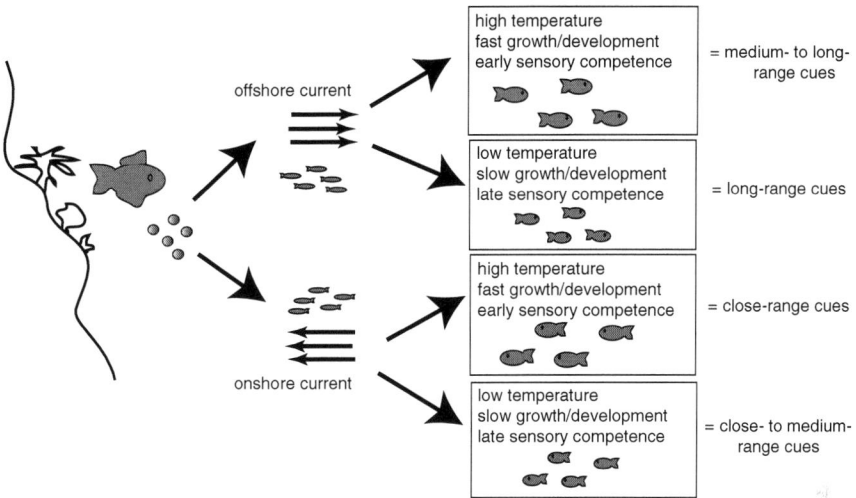

Fig. 5.12 Why larvae might need competence in multiple senses in early life. Simple changes in current direction and temperature can create different scenarios of larval developmental environments that require use of different types of cues. For example, young larvae encountering a cold offshore current would develop their sensory systems more slowly while being dispersed farther offshore from the natal habitat. Under these conditions, the larvae may need to detect long-range cues in order to return to or find suitable habitat. On the other hand, in a scenario with a warm onshore current, larvae would develop sensory competence quickly and would be retained close to shore. In this case, they may only need close-range cues to find suitable habitat

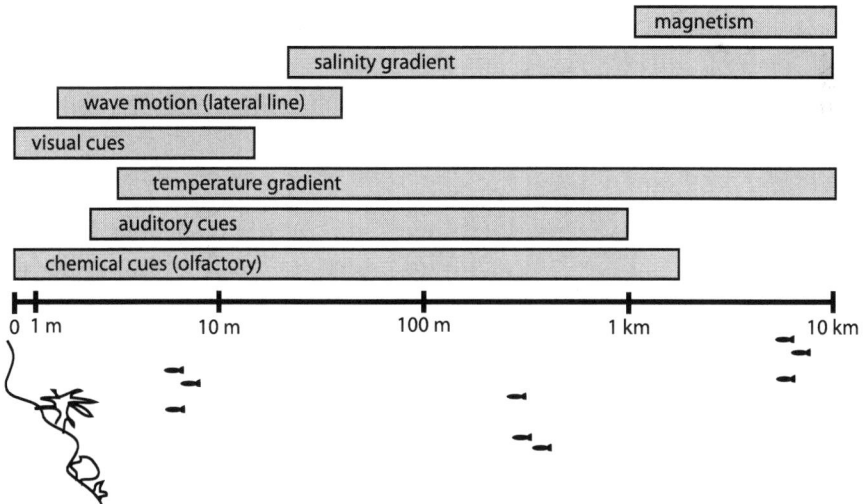

Fig. 5.13 Associations between sensory stimuli and the spatial scales over which they are likely to be detected

and the importance of a given cue at a given stage/time. On the other hand, the stochasticity inherent in the system—that is, in the possible locations where a given larva might end up—is large, and might mean that a better strategy is to generally develop sensory capabilities for all possible cues (Figs. 5.12, 5.13). Phenotypic plasticity in developmental pattern and rates of sensory development in the presence of a given environmental cue is another possibility for dealing with stochastic variation, but this phenomenon has not been documented in tropical marine fish or decapod larvae. Perhaps the 'rearing artifacts' mentioned above may give a clue to sensory system developmental variation.

In summary, like any phenotypic trait, larval sensory systems are subject to adaptive ecological and evolutionary pressures from their environment. The ever-present limitation in studies of marine larvae is the difficulty in obtaining good quality, identified larvae of known age and developmental environment (either wild-caught or captive-reared) for analysis. Many exciting discoveries are yet to be made in this field. We have been surprised over and over again when it comes to tropical marine animals and their ecology. Most likely we are in for more surprises.

Acknowledgments   Many thanks to K Anger, C Derby, L Fishelson, LA Fuiman, M Gagliano, A Garm, A Jeffs, MR Lara, JM Leis, J Mogdans, JC Montgomery, CA Radford, P Steinberg, SD Simpson, JF Webb and K Wright for help with relevant literature and papers in press, to K Larsen and JM Leis for taxonomic advice, to MR Lara for constructive comments to an early version of the manuscript, to two referees for excellent comments that further improved the manuscript and to the Journal of the Marine Biological Association of the United Kingdom, Wiley-Blackwell Science, Balaban Publishers, the SeaWiFS Project, NASA/Goddard Space Flight Center and GeoEye, Reef Consultants and TM Brolund for kind permission to use copyrighted figures.

References

Almany GR, Berumen ML, Thorrold SR et al (2007) Local replenishment of coral reef fish populations in a marine reserve. Science 316:742–744

Anger K (2001) The biology of decapod crustacean larvae. Crustacean issues, 14 Balkema Publishers, the Netherlands

Anger K (2006) Contributions of larval biology to crustacean research: a review. Invertebr Reprod Dev 49:175–205

Anger K, Torres G, Giménes L (2006) Metamorphosis of a sesarmid river crab, *Armases roberti*: stimulation by adult odours versus inhibition by salinity stress. Mar Freshw Behav Physiol 39:269–278

Arnold TM, Targett NM (2002) Marine tannins: the importance of a mechanistic framework for predicting ecological roles. J Chem Ecol 28:1919–1934

Arvedlund M, Bundgaard I, Nielsen LE (2000a) Host imprinting in anemonefishes – does it dictate spawning site preferences? Environ Biol Fish 58:201–211

Arvedlund M, Larsen K, Windsor H (2000b) The embryonic development of the olfactory system in *Amphiprion melanopus* in relation to the anemonefish host imprinting hypothesis. J Mar Biol Assoc UK 80:1103–1109

Arvedlund M, McCormick MI, Fautin DG et al (1999) Host recognition and possible imprinting in the anemonefish *Amphiprion melanopus* (Pisces: Pomacentridae). Mar Ecol Prog Ser 188:207–218

Arvedlund M, Munday P, Takemura A (2007) The morphology and ultrastructure of the peripheral olfactory organ in newly metamorphosed coral-dwelling gobies, *Paragobiodon xanthosomus* Bleeker (Gobiidae, Teleostei). Tissue Cell 39:335–342

Arvedlund M, Nielsen LE (1996) Do the anemonefish *Amphiprion ocellaris* (Pisces: Pomacentridae) imprint themselves to their host sea anemone *Heteractis magnifica* (Anthozoa: Actinidae)? Ethology 102:197–211

Arvedlund M, Takemura A (2006) The importance of chemical environmental cues for juvenile *Lethrinus nebulosus* Forsskål (Lethrinidae, Teleostei) when settling into their first benthic habitat. J Exp Mar Biol Ecol 338:112–122

Atema J, Kingsford MJ, Gerlach G (2002) Larval reef fish could use odour for detection, retention and orientation to reefs. Mar Ecol Prog Ser 241:151–160

Atema J, Richard RF, Popper AN et al (1988) Sensory biology of aquatic animals. Springer-Verlag, New York

Baeza JA, Stotz W (2003) Host-use and selection of differently colored sea anemones by the symbiotic crab *Allopetrolisthes spinifrons*. J Exp Mar Biol Ecol 284:25–39

Baker CF, Montgomery JC (1999) Lateral line mediated rheotaxis in the antarctic fish, *Pagothenia borchgrevinki*. Polar Biol 21:305–309

Basil JA, Hanlon RT, Sheikh SI et al (2000) Three-dimensional odor tracking by *Nautilus pompilius*. J Exp Biol 203:1409–1414

Belanger RM, Smith CM, Corkum LD et al (2003) Morphology and histochemistry of the peripheral olfactory organ in the round goby, *Neogobius melanostomus* (Teleostei: Gobiidae). J Morph 257:62-71

Bellwood D, Fisher R (2001) Relative swimming speeds in reef fish larvae. Mar Ecol Prog Ser 211:299–303

Blaxter JHS, Fuiman LE (1989) Function of the free neuromasts of marine teleost larvae. In: Coombs P, Goerner HM (eds) The mechanosensory lateral line: neurobiology and evolution, pp. 481–499. Springer-Verlag, New York

Boles LC, Lohmann KJ (2003) True navigation and magnetic maps in spiny lobsters. Nature 421:60–63

Bokser V, Oberg C, Sukhatme G et al (2004) A small submarine robot for experiments in underwater sensor networks. Center for Embedded Network Sensing. Paper 519. http://repositories.cdlib.org/cens/wps/519

Bradbury IR, Snelgrove PVR (2001) Contrasting larval transport in demersal fish and benthic invertebrates: the roles of behaviour and advective processes in determining spatial pattern. Can J Fish Aquat Sci 58:811–823

Britt LL, Loew ER, McFarland WN (2001) Visual pigments in the early life history stages of Pacific northwest marine fishes. J Exp Biol 204:2581–2587

Brothers EB, Williams DM, Sale PF (1983) Length of larval life in twelve families of fishes at "One Tree Lagoon", Great Barrier Reef, Australia. Mar Biol 76:319–324

Bullock TH, Hopkins CD, Popper AN et al (2005) Electroreception. Springer Science+ Business Media, New York

Burgess SC, Kingsford MJ, Black KP (2007) Influence of tidal eddies and wind on the distribution of presettlement fishes around One Tree Island, Great Barrier Reef. Mar Ecol Prog Ser 341:233–242

Cain SD, Boles LC, Wang JH et al (2005) Magnetic orientation and navigation in marine turtles, lobsters, and mollusks: concepts and conundrums. Integr Comp Biol 45:539–546

Caprio J, Derby CD (2008) Aquatic animal models in the study of chemoreception. In: Shepherd GM, Smith DV, Firestein S (eds) The senses: a comprehensive reference. Vol 4. Olfaction and taste. Elsevier, New York

Cernuda-Cernuda R, Garcí-Fernández JM (1996) Structural diversity of the ordinary and specialized lateral line organs. Mic Res Tech 34:302–312

Collin SP, Marshall NJ (2003) Sensory processing in aquatic environments. Springer-Verlag, New York

Conover DO, Present TMC (1990) Countergradient variation in growth rate: compensation for length of the growing season among Atlantic silversides from different latitudes. Oecologia 83:316–324

Coombs S, Braun CB, Donovan B (2001) The orienting response of Lake Michigan mottled sculpin is mediated by canal neuromasts. J Exp Biol 204:337–348

Coombs S, Görner P, Münz H (1989) The mechanosensory lateral line. Neurobiology and evolution. Springer-Verlag, New York

Coombs S, Janssen J, Webb JF (1988) Diversity of the lateral line system In: Atema J, Fay RR, Popper AN, Tavolga WN (eds) Sensory biology of aquatic animals, pp. 553–593. Springer-Verlag, New York

Corwin JT, Bullock TH, Schweitzer J (1982) The auditory brainstem response in five vertebrate classes. Electroencephalogr Clin Neurophysiol 54:629–641

Cowen RK, Lwiza KMM, Sponaugle S et al (2000) Connectivity of marine populations: open or closed? Science 287:857–859

Creaser EP, Travis D (1950) Evidence of a homing instinct in the Bermuda spiny lobster. Science 112:169–170

Cronin, TW, Jinks R (2001) Ontogeny of vision in marine crustaceans. Am Zool 41:1098–1107

Derby CD, Cate HS, Steullet P et al (2003) Comparison of turnover in the olfactory organ of early juvenile stage and adult Caribbean spiny lobsters. Arthropod Struct Dev 31:297–311

Derby CD, Steullet P, Cate HS et al (2001) A compound nose: functional organization and development of aesthetasc sensilla. In: Wiese K (ed) The crustacean nervous system. Springer-Verlag, Berlin

Díaz H, Orihuela B, Forward Jr RB et al (2003a) Orientation of juvenile blue crabs, *Callinectes sapidus* Rathbun, to currents, chemicals, and visual cues. J Crust Biol 23:15–22

Diaz JP, Prie-Granie M, Kentouri M et al (2003b) Development of the lateral line system in the sea bass. J Fish Biol 62:24–40

Diele K, Simith DJB (2007) Effects of substrata and conspecific odour on the metamorphosis of mangrove crab megalopae, *Ucides cordatus* (Ocypodidae) J Exp Mar Biol Ecol 348:174–182

Dittman AH, Quinn TP (1996) Homing in Pacific salmon: mechanisms and ecological basis. J Exp Biol 199:83–91

Dufort CG, Jury SH, Newcomb JM et al (2001) Detection of salinity by the lobster, *Homarus americanus*. Biol Bull 201:424–434

Doherty PJ (1987) Light traps: selective but useful devices for quantifying the distributions and abundances of larval fishes. Bull Mar Sci 41:423–431

Doudoroff P (1938) Reactions of marine fishes to temperature gradients. Biol Bull 75:494–509

Douglass JK, Forward RB Jr (1989) The ontogeny of facultative superposition optics in a shrimp eye: hatching through metamorphosis. Cell Tissue Res 258:289–300

Døving KB, Dubois-Dauphin M, Holley A et al (1977) Functional anatomy of the olfactory organ of fish and the ciliary mechanism of water transport. Acta Zool 58:245–255

Egner SA, Mann DA (2005) Auditory sensitivity of sergeant major damselfish *Abudefduf saxatilis* from post-settlement juvenile to adult. Mar Ecol Prog Ser 285:213–222

Elliott JK, Elliott JM, Mariscal RN (1995) Host selection, location, and association behaviours of anemonefishes in field settlement experiments. Mar Biol 122:377–389

Farbman AI (1992) Cell biology of olfaction. Cambridge University Press, New York

Forward RB Jr, Diaz H, Ogburn MB (2007) The ontogeny of the endogenous rhythm in vertical migration of the blue crab *Callinectes sapidus* at metamorphosis. J Exp Mar Biol Ecol 348:154–161

Forward RB, Tankersley RA, Rittschof D (2001) Cues for metamorphosis of brachyuran crabs: an overview. Am Zool 41:1108–1122

Forward RB, Tankersley RA, Smith KA et al (2003) Effects of chemical cues on orientation of blue crab, *Callinectes sapidus*, megalopae in flow: implications for location of nursery areas. Mar Biol 142:747–756

Fuiman LA (2004) Changing structure and function of the ear and lateral line system of fishes during development. Am Fish Soc Symp 40:117–144

Gagliano M, Depczynski M, Simpson SD et al (2008) Dispersal without errors: symmetrical ears tune into the right frequency for survival. Proc R Soc B 275:527–534

Gardiner JM, Atema J (2007) Sharks need the lateral line to locate odor sources: rheotaxis and eddy chemotaxis. J Exp Biol 210:1925–1934

Garm A, Høegh JT (2006) Ultrastructure and functional organization of mouthpart sensory setae of the spiny lobster *Panulirus argus*: new features of putative mechanoreceptors. J Morph 267:464–476

Gebauer P, Paschke K, Anger K (2002) Metamorphosis in a semiterrestrial crab, *Sesarma curacaoense:* intra- and interspecific settlement cues from adult odors. J Exp Mar Biol Ecol 268:1–12

Gebauer P, Paschke K, Anger K (2003a) Delayed metamorphosis in decapod crustaceans: evidence and consequences. Rev Chil Hist Nat 76:169–175

Gebauer P, Paschke K, Anger K (2003b) Metamorphosis in a semiterrestrial crab, *Sesarma curacaoense*: intra- and interspecific settlement cues from adult odors. J Exp Mar Biol Ecol 268:1–12

Gebauer P, Paschke KA, Anger KA (2005) Temporal window of receptivity and intraspecific variability in the responsiveness to metamorphosis-stimulating cues in the megalopa of a semiterrestrial crab, *Sesarma curacaoense*. Invertebr Reprod Dev 47:39–50

Gerlach G, Atema J, Kingsford MJ et al (2007) Smelling home can prevent dispersal of reef fish larvae. Proc Natl Acad Sci USA 104:858–863

Ghysen A, Dambly-Chaudiere C (2007) The lateral line microcosmos. Gene Dev 21:2118–2130

Glantz RM, Schroeter JP (2007) Orientation by polarized light in the crayfish dorsal light reflex: behavioral and neurophysiological studies. J Comp Physiol A 193:371–384

Grünert U, Ache BW (1988) Ultrastructure of the aesthetasc (olfactory) sensilla of the spiny lobster, *Panulirus argus*. Cell Tissue Res 251:95–103

Hadfield MG, Paul VJ (2001) Natural chemical cues for settlement and metamorphosis of marine-invertebrate larvae. In: McClintock JB (ed) Marine chemical ecology. CRC Press, Boca Raton

Hallberg E, Johansson KUI, Elofsson R (1992) The aesthetasc concept: structural variations of putative olfactory receptor cell complexes in Crustacea. Microsc Res Tech 22:325–335

Hallberg E, Johansson KUI, Wallén R (1997) Olfactory sensilla in crustaceans: morphology, sexual dimorphism, and distribution patterns. Int J Insect Morphol Embryol 26:173–180

Hamdani EH, Døving KB (2007) The functional organization of the fish olfactory system. Prog Neurobiol 82:80–86

Hansen A, Reutter K (2004) Chemosensory systems in fish: structural, functional and ecological aspects. In: von der Emde G, Mogdans J, Kapoor BG (eds) The senses of fish. Kluwer Academic Publishers, Dordrecht Holland and Narosa Publishing House, New Delhi

Hansen A, Zeiske E (1998) The peripheral olfactory organ of the zebrafish, *Danio rerio*: an ultrastructural study. Chem Senses 23:39–48

Hansen A, Zeiske E (1993) Development of the olfactory organ in the zebrafish, *Brachydanio rerio*. J Comp Neurol 333:289–300

Hansen A, Zielinski B (2005) Diversity in the olfactory epithelium of bony fishes: development, lamellar arrangement, sensory neuron. Cell types and transduction components. J Neurocytol 34:183–208

Hara TJ (ed) (1992) Fish chemoreception. Springer-Verlag, New York

Hara TJ (1994a) Olfaction and gustation in fish – an overview. Acta Physiol Scand 152:207–217

Hara TJ (1994b) The diversity of chemical stimulation in fish olfaction and gustation in fish. Rev Fish Biol Fish 4:1–35

Hara TJ, Zielinski BS (1989) Structural and functional developments of the olfactory organ in teleost. Trans Am Fish Soc 118:183–194

Harzsch S, Benton J, Dawirs RR et al (1999) A new look at embryonic development of the visual system in decapod crustaceans: neuropil formation, neurogenesis, and apoptotic cell death. J Neurobiol 39:294–306

Hawryshyn CW, Moyer HD, Allison WT et al (2003) Multidimensional polarization sensitivity in damselfishes. J Comp Physiol A 189:213–220

Hayat MA (2000) Principles and techniques of electron microscopy. Biological applications. 4th edn. Cambridge University Press, Cambridge

Huang HD, Rittschof D, Jeng MS (2005) Visual orientation of the symbiotic snapping shrimp *Synalpheus demani*. J Exp Mar Biol Ecol 326:56–66

Huijbers CM, Mollee EM, Nagelkerken I (2008) Post-larval French grunts (*Haemulon flavolineatum*) distinguish between seagrass, mangrove and coral reef water: implications for recognition of potential nursery habitats. J Exp Mar Biol Ecol 357:134–139

Immelman K (1975a) Ecological significance of imprinting and early learning. Annu Rev Ecol Syst 6:15–37

Immelman K (1975b) The evolutionary significance of early experience. In: Baerends G, Beer C, Manning A (eds) Function and evolution in behaviour. Clarendon Press, Oxford

Ishida Y, Kobayashi H (1992) Stimulatory effectiveness of amino acids on the olfactory response in an algivorous marine teleost, the rabbitfish *Siganus fuscescens* Houttuyn. J Fish Biol 41:737–748

Jeffs AG, Montgomery JC, Tindle CT (2005) How do spiny lobster post-larvae find the coast? N Z J Mar Freshw Res 39:605–617

Job SD, Bellwood DR (1996) Visual acuity and feeding in larval *Premnas biaculeatus*. J Fish Biol 48:952–963

Job SD, Bellwood DR (2000) Light sensitivity in larval fishes: implications for vertical zonation in the pelagic zone. Limnol Oceanogr 45:362–371

Job SD, Bellwood DR (2007) Ultraviolet photosensitivity and feeding in larval and juvenile coral reef fishes. Mar Biol 151:495–503

Job SD, Shand J (2001) Spectral sensitivity of larval and juvenile coral reef fishes: implications for feeding in a variable light environment. Mar Ecol Prog Ser 214:267–277

Jones GP, Milicich MJ, Emslie MJ et al (1999) Self-recruitment in a coral-reef fish population. Nature 402:802–804

Jones GP, Planes S, Thorrold SR (2005) Coral reef fish larvae settle close to home. Curr Biol 15:1314–1318

Kalmijn AJ (2003) Physical principles of electric, magnetic, and near-field acoustic orientation. In: Collin SP, Marshall NJ (eds) Sensory processing in aquatic environments. Springer-Verlag, New York

Kasumyan, AO (2002) Sturgeon food searching behaviour evoked by chemical stimuli: a reliable sensory mechanism. J Appl Ichthyol 18:685–690

Kavanagh DK, Alford R (2003) Sensory and skeletal development and growth in relation to the duration of the embryonic and larval stages in damselfishes (Pomacentridae). Biol J Linn Soc 80:187–206

Kawamura G (1984) The sense organs and behavior of milkfish fry in relation to collection techniques. In: Juario JV, Ferraris RP, Benitez LV (eds) Advances in milkfish biology and culture. Island Publ, Manila

Keller TA, Powell I, Weissburg MJ (2003) Role of olfactory appendages in chemically mediated orientation of blue crabs. Mar Ecol Prog Ser 261:217–231

Kennedy VS, Cronin LE (eds) (2007) The blue crab: *Callinectes sapidus*. Maryland Sea Grant, Maryland Sea Grant College, University System of Maryland, Maryland

Kenyon TN, Ladich F, Yan HY (1998) A comparative study of hearing ability in fishes: the auditory brainstem response approach. J Comp Physiol A 182:307–318

Kingsford MJ, Leis JM, Shanks A et al (2002) Sensory environments, larval abilities and local self-recruitment. Bull Mar Sci 70:309–340

Konno K, Qin G, Nakanishi K (1990) Synthesis of amphikuemin and analogs: a synomone that mediates partner recognition between anemonefish and sea anemone. Heterocycles 30:247–251

Lara MR (1999) Sensory development in settlement-stage larvae of Caribbean labrids and scarids – a comparative study with implications for ecomorphology and life history strategies. PhD Dissertation, College of William and Mary, Virginia

Lara MR (2001) Morphology of the eye and visual acuities in the settlement intervals of some coral reef fishes (Labridae, Scaridae). Environ Biol Fish 62:365–378

Lara MR (2008) Development of the nasal olfactory organs in the larvae, settlement-stages and some adults of 14 species of Caribbean reef fishes (Labridae, Scaridae, Pomacentridae). Mar Biol 154:54–64

Lavalli K, Spanier E (eds) (2007) The biology and fisheries of the slipper lobster. CRC Press, Taylor and Francis Group, Boca Raton

Laverack MS (1988) The diversity of chemoreceptors. In: Atema J, Fay RR, Popper AN, Tavolga W (eds) Sensory biology of aquatic animals. Springer-Verlag, New York

Lecchini D (2005a) Spatial and behavioural patterns of reef habitat settlement by fish larvae. Mar Ecol Prog Ser 301:247–252

Lecchini D, Osenberg CW, Shima JS et al (2007b) Ontogenetic changes in habitat selection during settlement in a coral reef fish: ecological determinants and sensory mechanisms. Coral Reefs 26:423–432

Lecchini D, Planes S, Galzin R (2005b) Experimental assessment of sensory modalities of coral-reef fish larvae in the recognition of their settlement habitat. Behav Ecol Sociobiol 58:18–26

Lecchini D, Shima J, Banaigs B et al (2005c) Larval sensory abilities and mechanisms of habitat selection of a coral reef fish during settlement. Oecologia 143:326–334

Leis JM (2006) Are larvae of demersal fishes plankton or nekton? Adv Mar Biol 51:57–141

Leis JM (2007) Behaviour of fish larvae as an essential input for modelling larval dispersal: behaviour, biogeography, hydrodynamics, ontogeny, physiology and phylogeny meet hydrography. Mar Ecol Prog Ser 347:185–193

Leis JF, McCormick MI (2002) The biology, behavior and ecology of the pelagic, larval stage of coral reef fishes. In: Sale PF (ed) Coral reef fishes. Dynamics and diversity in a complex ecosystem, pp. 171–199. Academic Press, Elsevier Science, San Diego

Leis JM, Carson-Ewart BM (2003) Orientation of pelagic larvae of coral-reef fishes in the ocean. Mar Biol Prog Ser 252:239–253

Leis JM, Wright KJ, Johnston RN (2007) Behaviour that influences dispersal and connectivity in the small, young larvae of a reef fish. Mar Biol 153:103–117

Lenz P, Hartline DK, Purcell J et al (eds) (1997) Zooplankton: sensory ecology and physiology. CRC Press, Taylor and Francis Group, London

Levin LA (2006) Recent progress in understanding larval dispersal: new directions and digressions. Integr Comp Biol 46:282–297

Lewis D (1979) We, the navigators. University Press of Hawaii, Honolulu

Loew ER, McAlary FA, McFarland WN (1997) Ultraviolet visual sensitivity in the larvae of two species of marine atherinid fishes. In: Lenz PH, Hartwell DK, Purcell JE, Macmillan DL (eds) Zooplankton: sensory ecology and physiology. CRC Press, London

Losey GS, Cronin TW, Goldsmith TH et al (1999) The UV visual world of fishes: a review. J Fish Biol 54:921–943

Losey GS, McFarland WN, Loew ER et al (2003) Visual biology of Hawaiian coral reef fishes. I. Ocular transmission and visual pigments. Copeia 3:433–454

Mana RR, Kawamura G (2002) A comparative study on morphological differences in the olfactory system of red sea bream (*Pagrus major*) and black sea bream (*Acanthopagrus schlegeli*) from wild and cultured stocks. Aquaculture 209:285–306

Mann DA, Casper BM, Boyle KS et al (2007) On the attraction of larval fishes to reef sounds. Mar Ecol Prog Ser 338:307–310

Marshall NJ, Jennings K, McFarland WN et al (2003a) Visual biology of Hawaiian coral reef fishes. II. Colors of Hawaiian coral reef fish. Copeia 3:455–466

Marshall NJ, Jennings K, McFarland WN et al (2003b) Visual biology of Hawaiian coral reef fishes. III. Environmental light and an integrated approach to the ecology of reef fish vision. Copeia 3:467–480

Marshall J, Vorobyev M, Siebeck UE (2006) What does a reef fish see when it sees a reef fish? Eating "Nemo" © In: Ladlich F (ed) Communication in fishes. Science Publishers, Enfield, NH

Maruska KP (2001) Morphology of the mechanosensory lateral line system in elasmobranch fishes: ecological and behavioral considerations. Environ Biol Fish 60:47–75

McCormick MI (1999) Delayed metamorphosis of a tropical reef fish (*Acanthurus triostegus*): a field experiment. Mar Ecol Prog Ser 176:25–38

McFarland WN, Loew ER (1994) Ultraviolet visual pigments in marine fishes of the family Pomacentridae. Vision Res 34:1393–1396

Mishra M, Jeffs A, Meyer-Rochow VB (2006) Eye structure of the phyllosoma larva of the rock lobster *Jasus edwardsii* (Hutton, 1875): how does it differ from that of the adult? Invertebr Reprod Dev 49:213–222

Mogdans J, Bleckmann H (2001) The mechanosensory lateral line of jawed fishes. In: Kapoor BG, Hara TJ (eds) Sensory biology of jawed fishes. New Insights Science Publishers Inc, Enfield (NH)

Mogdans J, Kröther S, Engelman J (2004) Neurobiology of the fish lateral line: adaptations for the detection hydrodynamic stimuli in running water. In: von der Emde G, Mogdans J, Kapoor BG (eds) The senses of fish. Adaptations for the reception of natural stimuli. Kluwer Academic Publishers, Dordrecht

Moksnes and Heck (2006) Relative importance of habitat selection and predation for the distribution of blue crab megalopae and young juveniles. Mar Ecol Prog Ser 308:166–181

Montgomery JC, Jeffs A, Simpson SD et al (2006) Sound as an orientation cue for the pelagic larvae of reef fishes and decapod crustaceans. Adv Mar Biol 51:143–196

Montgomery JC, Tolimieri N, Haine OS (2001) Active habitat selection by pre-settlement reef fishes. Fish Fish 2:261–277

Murata M, Miyagawa-Koshima K, Nakanishi K et al (1986) Characterisation of compounds that induce symbiosis between sea anemone and anemonefish. Science 234:585–587

Myrberg AA Jr, Fuiman LA (2002) The sensory world of coral reef fishes. In: Sale PF (ed) Coral reef fishes. Dynamics and diversity in a complex ecosystem. Academic Press, Elsevier Science, San Diego

Nearing J, Betka M, Quinn S et al (2002) Polyvalent cation receptor proteins (CaRs) are salinity sensors in fish. Proc Natl Acad Sci USA 99:9231–9236

Nishida S, Kittaka J (1992) Integumental organs of the phyllosoma larva of the rock lobster *Jasus edwardsii* (Hutton). J Plankton Res 14:563–573

Northcutt RG (2005) Ontogeny of electroreceptors and their neural circuitry In: Bullock TH, Hopkins CD, Popper AN, Fay RR (eds) Electroreception. Springer Science+Business Media, New York

Northcutt RG, Holmes PH, Albert JS (2000) Distribution and innervation of lateral line organs in the channel catfish. J Comp Neurol 421:570–592

O'Connor NJ, Judge ML (2004) Molting of fiddler crab *Uca minax* megalopae: stimulatory cues are specific to salt marshes. Mar Ecol Prog Ser 282:229–236

O'Connor NJ, Van BT (2006) Adult fiddler crabs *Uca pugnax* (Smith) enhance sediment-associated cues for molting of conspecific megalopae. J Exp Mar Biol Ecol 335:123–130

Ogden JC, Quinn TP (1984) Migration in coral reef fishes: ecological significance and orientation mechanisms. In: McCleave JD, Arnold GP, Dodson JJ, Neill WH (eds) Mechanisms of migration in fishes. Plenum Press, New York

Pankhurst PM, Pankhurst NW (2002) Direct development of the visual system of the coral reef teleost, the spiny damsel, *Acanthochromis polyacanthus*. Environ Biol Fish 65:431–440

Parmentier E, Lagardere JP, Vandewalle P et al (2005) Geographical variation in sound production in the anemonefish *Amphiprion akallopisos*. Proc Royal Soc B 272:1697–1703

Pasternak Z, Blasius B, Achituv Y et al (2004) Host location in flow by larvae of the symbiotic barnacle *Trevathana dentata* using odor-gated rheotaxis. Proc R Soc Lond B 271:1745–1750

Phillips BF (2006) Lobsters: Biology, management, aquaculture and fisheries. Wiley-Blackwell Publishing, Oxford

Phillips BF, Sastry AN (1980) Larval ecology. In: Cobb JS, Phillips BF (eds) The biology and management of lobsters. Vol II. Academic Press, New York

Popper AN, Fay RR (1999) The auditory periphery in fishes. In: Fay RR, Popper AN (eds) Comparative hearing: fish and amphibians. Springer-Verlag, New York

Popper AN, Fay RR, Platt C et al (2003) Sound detection mechanisms and capabilities of teleost fishes. In: Collin SP, Marshall NJ (eds) Sensory processing in aquatic environments. Springer-Verlag, New York

Popper AN, Salmon M, Horch KW (2001) Acoustic detection and communication by decapod crustaceans. J Comp Physiol 187:83–89

Radford CA, Jeffs AG, Montgomery JC (2007) Directional swimming behavior by five species of crab postlarvae in response to reef sound. Bull Mar Sci 80:369–378

Reutter K, Kapoor BG (2005) Fish chemosenses. Science Publishers Inc, Enfield, NH

Robertson DR (2001) Population maintenance among tropical reef fishes: inferences from small-island endemics. Proc Nat Acad Sci USA 98:5667–5670

Scholtz AT, Horall RM, Cooper JC et al (1976) Imprinting to olfactory cues: the basis for home stream selection in salmon. Science 192:1247–1249

Santos A, Ricardo C, Bartilotti C et al (2004) The larval development of the partner shrimp. Helgol Mar Res 58:129–139

Siebeck UE, Marshall NJ (2007) Potential ultraviolet vision in pre-settlement larvae and settled reef fish – a comparison across 23 families. Vision Res 47:2337–2352

Simpson SD, Meekan MG, Jeffs A et al (2008) Settlement-stage coral reef fishes prefer the higher frequency invertebrate-generated audible component of reef noise. Anim Behav, in press

Simpson SD, Meekan MG, Montgomery J et al (2005a) Homeward sound. Science 308:221

Simpson SD, Yan HY, Wittenrich ML et al (2005b) Response of embryonic coral reef fishes (Pomacentridae: *Amphiprion* spp.) to noise. Mar Ecol Prog Ser 287:201–208

Simpson SD, Meekan MG, Jeffs A et al (2008) Settlement-stage coral reef fishes prefer the higher frequency invertebrate-generated audible component of reef noise. Anim Behav 75:1861–1868

Sjo T, Andersson E, Kornfeldt H (1988). MIDAS: a miniature submarine for underwater inspections. 9th Int Conf Nondestr Eva Nuc Ind, Tokyo

Stabell OB (1984) Homing and olfaction in salmonids: a critical review with special reference to the Atlantic salmon. Biol Rev 59:33–388

Stabell OB (1992). Olfactory control of homing behaviour in salmonids. In: Papi F (ed) Animal homing, pp. 249–271. Chapman & Hall, London

Steinke M, Malin G, Liss PS (2002) Trophic interactions in the sea: an ecological role for climate relevant volatiles? J Phycol 38:630–640

Swearer SE, Caselle JE, Lea DW et al (1999) Larval retention and recruitment in an island population of a coral-reef fish. Nature 402:799–802

Sweatman HPA (1985) The influence of adults of some coral-reef fishes on larval recruitment. Ecol Mon 55:469–485

Tolimieri N, Jeffs A, Montgomery J (2000) Ambient sound as a cue for navigation by the pelagic larvae of reef fishes. Mar Ecol Prog Ser 207:219–224

van Montfrans J, Ryer CH, Orth RJ (2003) Substrate selection by blue crab *Callinectes sapidus* megalopae and first juvenile instars. Mar Ecol Prog Ser 260:209–217

Ward AJW, Duff AJ, Horsfall JS et al (2008) Scents and scents-ability: pollution disrupts chemical social recognition and shoaling in fish. Proc R Soc Lond B 275:101–105

Walker MM, Dennis TE (2005) Role of the magnetic sense in the distribution and abundance of marine animals. Mar Ecol Prog Ser 287:295–300

Webley JAC, Connolly RM (2007) Vertical movement of mud crab megalopae (*Scylla serrata*) in response to light: doing it differently down under. J Exp Mar Biol Ecol 341:196–203

Webb JF (1999) Diversity of fish larvae in development and evolution. In: Hall BK, Wake MH (eds) Origin and evolution of larval forms, pp. 109–158. San Diego, Academic Press

Wellington GM, Victor B (1989) Planktonic larval duration of one hundred species of Pacific and Atlantic damselfishes (Pomacentridae). Mar Biol 101:557–567

Weissburg MJ, Browman HI (2005) Sensory biology: linking the internal and external ecologies of marine organisms. Mar Ecol Prog Ser 287:263–307

Whitlock KE, Westerfield M (2000) The olfactory placodes of the zebrafish form by convergence of cellular fields at the edge of the neural plate. Development 127:3645–3653

Wright KJ, Higgs DM, Belanger AJ et al (2005) Auditory and olfactory abilities of pre-settlement larvae and post-settlement juveniles of a coral reef damselfish (Pisces: Pomacentridae). Mar Biol 147:1425–1434

Wright KJ, Higgs DM, Belanger AJ et al (2008) Auditory and olfactory abilities of larvae of the Indo-Pacific coral trout *Plectropomus leopardus* at settlement. J Fish Biol, 72:2543–2556

Yamamoto M (1982) Comparative morphology of the peripheral olfactory organ in teleosts. In: Hara TJ (ed) Chemoreception in fishes. Elsevier, Amsterdam

Zeiske E, Theisen B, Breucker H (1992) Structure, development, and evolutionary aspects of the peripheral olfactory system. In: Hara TJ (ed) Fish chemoreception. Chapman & Hall, London

Zielinski B, Hara TJ (1988) Morphological and physiological development of olfactory receptor cells in rainbow trout (*Salmo gairdneri*) embryos. J Comp Neurol 271:300–311

Zupanc GKH, Bullock TH (2005) From electrogenesis to electroreception: an overview. In: Bullock TH, Hopkins CD, Popper AN, Fay RR (eds) Electroreception. Springer Science + Business Media, New York

Chapter 6
Mechanisms Affecting Recruitment Patterns of Fish and Decapods in Tropical Coastal Ecosystems

Aaron J. Adams and John P. Ebersole

Abstract The early benthic life history of fishes and decapods in tropical coastal ecosystem can be partitioned into three main stages—settlement, post-settlement transition, post-settlement stage—which culminate in recruitment. Although most species go through these early life history stages, not all species follow the same strategy. Life history strategies occur in three general categories: habitat specialists, habitat generalists, and ontogenetic shifters. Despite this variation in life history strategy, common processes affect the early life history stages of tropical marine fishes and decapods. The life history transition from planktonic larva to benthic post-larva connects oceanic and coastal habitats. However, benthic features and benthic processes affect early life history stages so that settlement and post-settlement distributions are not perfect reflections of larval supply patterns. The abundances and distributions of settlement and post-settlement life history stages result from complex interactions of larval supply, larval behavior, and the interactions of early settlers with the benthic environment. Since much of the very high mortality that occurs during settlement and early post-settlement appears to be due to predation, the direct effects of predators may be the most important factors acting on these early life history stages. Habitat selection, priority effects, predator avoidance, inter- and intraspecific competition, and aggression during and after settlement are also important influences on abundances and distributions of settlement and post-settlement fishes and decapods. The connection between nursery habitat availability and adult population abundances has been demonstrated, so it is likely that these other interactions of early life history stages with the benthic environment have demographic implications that are not yet understood.

Keywords Habitat selection · Mortality · Post-settlement · Priority effects · Settlement

A.J. Adams (✉)
Center for Fisheries Enhancement, Habitat Ecology Program, Mote Marine Laboratory, Charlotte Harbor Field Station, P.O. Box 2197, Pineland, FL 33945, USA
e-mail: aadams@mote.org

I. Nagelkerken (ed.), *Ecological Connectivity among Tropical Coastal Ecosystems*, DOI 10.1007/978-90-481-2406-0_6, © Springer Science+Business Media B.V. 2009

6.1 Introduction

Some three decades ago, tropical marine ecologists were struck with the fact that profound ignorance of recruitment mechanisms hampered understanding of population dynamics and community ecology in tropical ecosystems. Knowledge of the processes affecting larval-through-juvenile life stages still lags well behind knowledge of adults, but much has been learned. Through a review of this knowledge, we can begin to discern some of the patterns of recruitment, surmise some underlying processes, and think of how the focus of future research might be sharpened on critical recruitment issues.

The transition from larval to benthic life history phase connects oceanic and coastal habitats. The distribution of settlers in benthic habitats depends on the distribution of larvae in oceanic habitats, but benthic distributions are not a simple reflection of oceanic distributions. Distributions of settlers reflect a complex interaction of larval supply, larval behavior, and the interactions of early settlers with benthic features. In addition, the condition of settling larvae during the planktonic phase influences post-settlement growth and survival, further connecting oceanic and benthic processes. Oceanic and early benthic processes have demographic consequences because unsuccessful individuals will not join adult populations. Finally, different benthic habitats are often connected as early life history stages undergo ontogenetic habitat shifts. Thus, recruitment-associated ontogenetic processes connect the habitat mosaic of nearshore tropical systems, with habitat-associated features of coastal tropical systems having a powerful influence on recruitment.

6.2 Defining Recruitment

Since definitions of early life history stages of fish and decapods have been ambiguous in the peer-reviewed literature, some of the conflicting results among studies may be simply semantic. Here we provide a distinct definition of the early life history stages that culminate in recruitment. Recruitment occurs at the end of the post-settlement stage, and incorporates effects of larval, settlement, and post-settlement processes. Recruitment is characterized by entrance into a period of lower mortality, and marks the first record of an individual in the juvenile stage. Thus, it is the early juvenile stage when many recruitment surveys occur (references for and definitions of pre-recruitment stages are in Table 6.1).

Most tropical marine fishes and invertebrates have a two-phase life cycle that decouples local reproduction from recruitment into the local population. For these species, larvae are planktonic, and juveniles and adults are demersal. As a general rule, fertilization for these species is external and eggs are buoyant. Duration of the planktonic period varies among species, and also depends on environmental conditions. At the end of the planktonic phase, larvae search for appropriate settlement habitats and enter the demersal portion of their life history, many undergoing metamorphosis as they settle out of the water column. The processes affecting the transi-

Table 6.1 Summary of early life history stage definitions, adapted from Adams et al. (2006). Source citations are the references from which these definitions were derived

Term	Definition	Source citations
Settlement	The initial establishment of larvae onto a benthic substrate. Includes only larval processes. Important factors include larval condition and size.	Calinski and Lyons (1983), Kaufman et al. (1992), Guttierez (1998)
Post-settlement transition	Occurs during and immediately following settlement. Late stage larvae explore and evaluate benthic habitats (and may re-enter the pelagic environment several times), undergo metamorphosis, and join the benthic population. Priority effects* are especially important during this stage.	Kaufman et al. (1992), McCormick and Makey (1997), Sancho et al. (1997)
Post-settlement stage	Time period directly after metamorphosis. A period of high benthic mortality. Duration of this stage varies among species, in part due to different susceptibility to predation. Important factors include density-dependent mortality and competition, modified by habitat complexity.	Doherty and Sale (1985), Sogard (1997), Almany (2004), Almany and Webster (2004)
Recruitment	Occurs at the end of the post-settlement stage, and incorporates effects of larval and post-settlement processes. Characterized by entrance into a period of lower mortality. First record of an individual in the juvenile stage. Stage when many recruitment surveys occur.	Doherty and Sale (1985), Kaufman et al. (1992), Booth and Brosnan (1995), Guttierez (1998), McCormick and Hoey (2004)

* Priority effect: the process by which the presence of one species in a habitat decreases the probability of invasion by another. One species can lower the recruitment of another via competition (adult and subadult residents or settling juveniles can interfere with larval settlement; i.e., interference competition), or preempting resources, or predation (predatory adult and subadult residents can decrease settlement directly by preying on settlers or indirectly by inducing settlers to choose other sites, or predatory juveniles can prevent settlement by prey species; Shulman et al. 1983)

tion from larva to juvenile, and the early period of the juvenile stage have important demographic implications that could be critical to population regulation.

Knowledge of the early life history of a species is critical to understanding the mechanisms affecting recruitment. To some extent, the comparative importance of these mechanisms remains unclear because the literature often does not sufficiently partition the early life history stages of fishes and invertebrates. In many cases, for example, the term 'juvenile' is used to refer to all life stages after larval settlement and before maturity (e.g., St. John 1999), even though recent research has shown that the relative importance of different mechanisms influencing early life histories changes as individuals grow (Jones 1991). Predation is typically most important within 48 hrs of settlement (Almany 2004b, Almany and Webster 2004), whereas competition may be more important in later stages (Risk 1998).

Research would benefit from a clearly delineated nomenclature describing early life history of fishes and decapods. Common use of terms will clarify discussion, promote testing of findings and formulation of predictions, and provide a framework for applying these results on a larger scale. Moreover, a common delineation of early life history stages will help bring about a common structure to future discussion and research. It is the overall structure proposed here, rather than the early life history delineations themselves, that is new. Heretofore, each definition of an early life history stage has, for the most part, stood alone, independent of the context of other early stages. For our discussion on recruitment, we follow a structure assembled from four definitions proposed by Adams et al. (2006), since it provides a clear and convenient breakdown of the life history stages that contribute to fish and decapod recruitment in tropical ecosystems (Table 6.1).

6.3 Defining Early Life History Strategies

Within the general and ubiquitous life history strategy of planktonic larvae and demersal juveniles and adults are three categories based on the patterns of habitat use by the demersal juvenile and adult life stages (Table 6.2). In Strategy I, *habitat specialists*, planktonic larvae settle into the same location they will remain throughout their demersal life stages. In Strategy II, *habitat generalists*, larvae can stay, or move among, numerous habitats, and are not site attached. In Strategy III, *ontogenetic shifters*, larvae tend to settle in habitats and locations different from those used by adults, and undergo ontogenetic transitions to the adult life stage habitat.

This categorization of early life history strategies underscores the limitations of applying species-specific research findings at the community level. For example, much early research on the implications of early life history processes focused on the lottery hypothesis (Sale 1977, 1978), which emphasized the chanciness of larval settlement, and postulated that natural selection must produce habitat generalists to maximize the probability of finding appropriate settlement sites. To a great extent, the lottery hypothesis was based upon the prevalence of research on site-attached species, such as territorial pomacentrids (e.g., Doherty 1983). Since that time, research has shown extreme variation among species in habitat use during early life stages (reviewed in Adams et al. 2006), plasticity in early life stage habitat use within species (reviewed in Adams et al. 2006), and previously undetected changes in habitat use by pre-recruitment fishes (e.g., Kaufman et al. 1992). In addition, habitat variability greatly modifies species interactions such as competition and predation (e.g., Anderson 2001, Almany 2004a).

Given the categorization of life history strategies, and the need to find general patterns that apply within these categories, it is useful to first determine which species fall into which categories. Unfortunately, the strategy used does not appear to be phylogenetically constrained, in that life history category can vary within the family, and even the genus level, so species-specific data are needed. Among eight species of labrids studied by Green (1996), two exhibited ontogenetic shifter patterns, whereas six species were habitat generalists. Similarly, McGehee (1995) found that three species (*Stegastes planifrons, S. variabilis,* and *S. partitus*) of poma-

Table 6.2 Summary of life history strategies. Strategy definitions adapted from Adams et al. (2006). Example citations are studies that focus on species that fit this strategy category, and are listed by family

Term	Definition	Example citations
Habitat specialists	Larval settlement and the juvenile and adult benthic stages occur in the same location. These species tend to be site-attached (e.g., Pomacentridae). Ontogenetic shifts that may occur are relatively minor, likely microhabitat changes (e.g., juveniles use microhabitats within adult habitats). Microhabitat shifts may occur to occupy areas of different complexity to reduce predation, but they occur within the same site.	Alpheidae: Knowlton and Keller (1986) Palaemonidae: Preston and Doherty (1990) Pomacentridae: Doherty (1983), Bergman et al. (2000), Lirman (1994), Nemeth (1998), Schmitt and Holbrook (1999b)
Habitat generalists	Larval settlement and the juvenile and adult benthic stages of an individual may occur in the same location, but the species is able to settle and stay, or move among, numerous habitat types (e.g., *Halichoeres bivittatus*) Species are generally not site-attached (but see McGehee 1995 – species site-attached but can use and move among many habitat types). To the extent that ontogenetic shifts occur, they do not follow a well-defined pattern and/or are minor compared to ontogenetic shifters (e.g., Labridae).	Xanthidae: Beck (1995, 1997) Labridae: Green (1996)
Ontogenetic shifters	These species exhibit complex habitat, behavioral, and diet shifts during transitions from settlement through late juvenile stages, and again into adult. Larvae tend to settle into habitats distinct from adults and undergo notable ontogenetic shifts Larval settlement areas may differ from juvenile habitats	Panuliridae: Herrnkind et al. (1994), Childress and Herrnkind (2001); Labridae: Green (1996); Serranidae: Eggleston (1995), Dahlgren and Eggleston (2000), St. John (1999); Acanthuridae: Robertson (1988), Risk (1997, 1998), Adams and Ebersole (2002, 2004), Parrish (1989)

centrids exhibited high site fidelity (habitat specialists), while the fourth species (*S. leucostictus*) showed poor site fidelity (habitat generalist). One method that may be useful in predicting life history strategy is to use patterns of larval metamorphosis as predictors (McCormick and Makey 1997), which can be done via larval collection and laboratory observations.

At the community level, Gratwicke et al. (2006) determined that 47% of species surveyed in a study of non-estuarine lagoons and adjacent reefs in the Caribbean exhibited habitat use patterns indicative of ontogenetic habitat shifts. Similarly,

Nagelkerken and van der Velde (2002) found evidence for ontogenetic shifter strategy for 21 of the 50 (42%) most common reef species. In a survey of juvenile and adult densities of 17 species of nocturnal reef fishes at Moorea Island, Lecchini (2006) found that 47% showed ontogenetic habitat shifts. In contrast, when Adams and Ebersole (2002), working in the Caribbean, examined all fishes within lagoon and backreef habitats, they found clear ontogenetic lagoon (juvenile)—reef (adult) division for only 22 of 96 (23%) species. Differences in these estimates were due largely to the suite of species surveyed, but also to assignment of different size classes to ontogenetic stages.

Characterizing the life history stages of a species can be difficult. For many species the early life history stages are not clearly defined (but see Shulman and Ogden 1987 for a clear depiction of ontogeny by size class for *Haemulon flavolineatum*). When ontogeny is inadequately represented by defining life history stages only by size or characterizing all immature fishes as juveniles, the inference that a given habitat is a nursery for some species may be incorrect. More research elucidating early life history is needed to better evaluate factors affecting recruitment.

The following sections review the growing knowledge base of fish and decapod recruitment, set in a framework that follows the definitions set forth above. We hope that this review will contribute to a synthesis of research already completed and help focus the design of future research.

6.4 Larval Settlement (Departure from the Pelagic Environment and Entrance into Benthic Habitats)

Oceanographic processes influence settlement of coral reef fishes and decapods by transporting and influencing the survival of larvae (Choat et al. 1988, Acosta and Butler 1999). That larval settlement patterns can be temporally consistent across space at multiple scales (e.g., Fowler et al. 1992, Caselle and Warner 1996, Acosta and Butler 1997, Tolimieri et al. 1998, Vigliola et al. 1998, Schmitt and Holbrook 1999b) demonstrates the important impact of oceanographic processes on larval supply. In Barbados, the occurrence of late stage larvae in light traps corresponded with the first appearance of juveniles of these species on reefs, suggesting that larval supply was a good indicator of settlement (Sponaugle and Cowen 1996). Moreover, larvae of some fishes (e.g., *Stegastes partitus* and *Acanthurus bahianus*) were consistently associated spatially and temporally, suggesting these species were influenced similarly by oceanographic processes such as prevailing currents, tidal currents, wind-induced water flow, and large-scale externally forced events. Similarly, appearance of post-larval spiny lobsters in the Florida Keys, USA, is strongly related to tides, and less strongly related to favorable winds (Acosta and Butler 1997, Eggleston et al. 1998). Post-larvae of brachyuran crabs in Barbados also responded to tidal influences, generally producing the greatest supply of post-larvae at third-quarter moons with minimal tidal amplitude (Reyns and Sponaugle 1999).

6.4.1 Modification of Larval Supply

Patterns of larval supply established by oceanographic processes can not be expected to persist through settlement. Butler and Herrnkind (1992) found that spatial patterns of benthic settlement for spiny lobster (*Panulirus argus*), for example, differed from abundance patterns of planktonic larvae at local scales in Florida Bay and in the Florida Keys, USA (though the patterns sometimes agreed at regional scales). Settling larvae begin to interact with the bottom well before they actually reach it (Choat et al. 1988). Attacks of benthic-dwelling predators provide but one example of how benthic-associated processes may directly influence the abundance of settling larvae (Choat et al. 1988, Fowler et al. 1992, Booth and Beretta 1994, Gibson 1994, Booth and Brosnan 1995, Tolimieri 1998a, Tolimieri et al. 1998). Behavioral responses to such features by settlement-phase fishes and decapods also serve to modify the pattern of larval supply (for fishes, see Sweatman 1988, Booth and Beretta 1994, Fernandez et al. 1994, Elliot et al. 1995, Shanks 1995, Leis and Carson-Ewart 1999, Almany 2003, Garpe and Ohman 2007; for decapods, see Forward 1974, 1976, Knowlton 1974, Forward and Hettler 1992, Welch et al. 1997, Gimenez et al. 2004, Gimenez 2006). Overall, complex interactions of larval supply, larval behavior, and benthic features ultimately determine patterns of settlement. Furthermore, consistency in spatial patterns of larval settlement is usually not the case for fishes or decapods (e.g., Fowler et al. 1992, Green 1998, Tolimieri et al. 1998, Vigliola et al. 1998, Montgomery and Craig 2005) because the relative importance of oceanographic vs. benthic processes varies among species, among locations, and over time.

Habitat dispersion interacts with larval behavior to influence the distribution of settlers. Larvae are not passive particles. Fish (e.g., Stobutzki and Bellwood 1997, Stobutzki 1998, Leis and Carson-Ewart 1999) and decapod (e.g., Fernandez et al. 1994, Shanks 1995) larvae are capable of active swimming for considerable distances. Swimming ability, however, differs among species (e.g., Stobutzki and Bellwood 1997, Stobutzki 1998), and their behavior influences timing and location of settlement. Furthermore, competent larvae are capable of active searching for appropriate settlement habitats based on a variety of criteria (for fishes, see Sweatman 1988, Booth and Beretta 1994, Elliot et al. 1995, Leis et al. 2002, Almany 2003, Leis and Lockett 2005, Garpe and Ohman 2007; for decapods, see Knowlton 1974, Welch et al. 1997, Gimenez et al. 2004, Gimenez 2006). The combination of larval supply, larval behavior, and availability of settlement habitats determines patterns of settlement.

Recent research has shown that settlement-stage larvae of many species use a variety of cues to find reefs and appropriate settlement habitats. Sound is important for finding reefs (e.g., Tolimieri et al. 2000, reviewed in Montgomery et al. 2001, Leis and Lockett 2005), whereas olfactory senses are important for settlement site selection (e.g., Sweatman 1988, Butler and Herrnkind 1991, Elliot et al. 1995, Harvey 1996, reviewed in Montgomery et al. 2001, Atema et al. 2002, Horner et al. 2006). Settlement cues are discussed in detail in Chapter 5.

6.4.2 Behavior of Settlement-stage Larvae

Regardless of the mechanisms used by larvae to find suitable settlement habitats, larval behavior during the settlement process modifies patterns of settlement so that they differ from patterns of offshore larval distribution (e.g., Sponaugle and Cowen 1996, Cruz et al. 2007). Settlement patterns can be used to make strong inferences about settlement behavior. For example, working in French Polynesia, Schmitt and Holbrook (1999b) found consistent patterns of settlement by three species of *Dascyllus* (all habitat specialists): at the island scale, one species settled primarily on the north end of the island, whereas the other two species tended to settle toward the south end of the island. At the lagoon scale, they found that *D. trimaculatus* settled throughout the lagoon, whereas *D. aruanus* settled on habitats in the nearshore, and *D. flavicaudus* in the offshore, portions of the lagoon. Since they used standardized, initially empty, settlement habitats specific to each species, habitat availability was not a factor. The implications are clear: for *D. trimaculatus*, finding suitable settlement habitats was highly likely because of the widespread pattern of settlement. However, for the other two species, larval behavior and habitat availability likely limit the extent to which settlement would be successful—if suitable habitats are not present in nearshore (*D. aruanus*) or offshore (*D. flavicaudus*) areas, these species will be absent regardless of larval supply. Working in Hawaii, Kobayashi (1989) also found species-specific differences in larval behavior. Larvae of two gobiid species used visual cues to remain near reef settlement habitats, whereas larvae of *Foa brachygramma* (Apogonidae—cardinalfishes) and *Encrasicholina purpurea* (Engraulidae—anchovies) were most abundant at off-reef sample stations. Similar effects of larval behavior are evident at the community level, where, overall, more species settle on offshore than nearshore areas (Planes et al. 1993, Hamilton et al. 2006).

Although cues used at settlement by ontogenetic shifters and habitat generalists have not been examined as closely as in habitat-specialist damselfishes, consistent settlement patterns suggest active settlement site selection. In the Caribbean, for example, larvae of many reef fishes (*Acanthurus* spp., Adams and Ebersole 2002, 2004; *Epinephelus striatus*, Eggleston 1995, Dahlgren and Eggleston 2000; *Haemulon flavolineatum*, McFarland 1980, Shulman 1985a) and decapod crustaceans (e.g., Panuliridae—spiny lobsters, Acosta and Butler 1999) pass over reef habitats to reach lagoon habitats, which suggests that these larvae are using some cue to find these habitats, and olfactory differentiation of lagoonal versus oceanic water has been shown for some fishes (Atema et al. 2002, Huijbers et al. 2008).

Active habitat selection also implies that non-reef habitats provide advantages toward successful recruitment that compensate for the fitness costs of additional energy expenditure and predation risk experienced by the incoming larvae, and by the juveniles that must later move again to adult habitats on the reef. Moreover, it is likely that many species use the rather large target of non-reef benthic habitats as settlement habitat, and then move to more suitable post-settlement micro-habitats within the lagoon (e.g., Herrnkind 1980, McFarland 1980, Marx and Herrnkind 1985a, 1985b, Herrnkind and Butler 1986, Robertson 1988, reviewed in Parrish

1989, Adams and Ebersole 2004). Gratwicke et al. (2006) found that 67% of the fishes of reefs and non-estuarine lagoons in the British Virgin Islands exhibited patterns of ontogenetic habitat partitioning between lagoon and reef habitats, so such a strategy may be widespread.

6.4.3 Mortality and Larval Condition

Mortality is extremely high during larval settlement (and post-settlement transition—see Section 6.5). For example, Doherty et al. (2004) estimated that 61% of the nocturnally settling fish larvae entering their study area in Moorea Island (French Polynesia) were lost by morning. Similarly high rates of predation on settling and recently-settled spiny lobsters have been found by Acosta and Butler (1999) in the Florida Keys, USA. Although the high mortality rate continues for days (or longer, depending on species), mortality during settlement makes a significant contribution to overall mortality of the recruitment phase.

Since mortality is extremely high during the settlement and post-settlement process (Almany 2004a, Almany and Webster 2004, Doherty et al. 2004), any advantage provided by good condition has survival implications. For example, in a laboratory experiment, McCormick and Molony (1992) found that reef fish larvae at their study site near Lizard Island, Great Barrier Reef, Australia, receiving more food were larger and in better condition, so were able to settle faster than were low condition fish. They conjectured that these advantages would provide greater flexibility in the timing of settlement and hence increased opportunity to select habitat at settlement. Lower condition larvae were able to recover rapidly, feeding at rates similar to those in the high feed treatment and settling soon after, but it is unclear whether this compensatory ability would offset advantages of reduced exposure to predation in the water column and larger size at settlement that the well-fed larvae experienced.

Does high lipid content translate directly to high larval fitness (Sponaugle and Grorud-Colvert 2006)? Settlement-stage larvae with greater lipid stores are able to swim greater distances (Stobutzki 1998), enabling them to search more widely for suitable settlement habitat. Positive connections between nutritional condition, larval growth, and juvenile survival have also been established for decapod crustaceans (Knowlton 1974, Gimenez et al. 2004, Gimenez 2006). Larvae entering the recruitment process in poor condition are more likely to be preyed upon (Hoey and McCormick 2004, McCormick and Hoey 2004), though this is not always the case: in some cases increased aggression (Jones 1987) or size-selective predation (Sogard 1997) is focused on the largest individuals. Even small-scale variation in larval growth and condition can be important (McCormick 1994).

The degree to which larval size and condition are environmentally versus genetically controlled is unclear. In the Caribbean/western Atlantic, Sponaugle and Grorud-Colvert (2006) and Sponaugle et al. (2006) used growth as a proxy for condition (condition traditionally measured by physiological metrics such as the amount of lipid stores), and found that environmental variability affected growth (and thus implied condition) of bluehead wrasse (*Thalassoma bifasciatum*) larvae, and that these larval characteristics influenced survival. In the Florida Keys, USA, Sponaugle et al. (2006) examined otoliths of post-settlement bluehead wrasse to examine the effect of water temperature on larval growth, pelagic larval duration, and other factors, and the effect of these factors on size-at-settlement. Size-at-settlement, a function of larval growth and pelagic larval duration, was greatest at intermediate temperatures. They concluded that larvae grow fastest at metabolically optimal temperatures. Perhaps most important, larval growth was positively correlated with early juvenile growth (Fig. 6.1 also in Vigliola and Meekan 2002, Nemeth 2005), and mortality was lower for fish with good larval condition and high early juvenile growth (Sponaugle and Grorud-Colvert 2006), which likely increased survival.

Although they worked on post-settlement stages, Vigliola et al.'s (2007) results from western Australia are also applicable here. They suggest that some of the variation in traits exhibited by settlement-stage larvae may be inherited. They found size-selective post-settlement mortality, with smaller, slower-growing individuals suffering highest mortalities in all cohorts they examined. This size-selective mortality was so severe, in fact, that it affected the genetic composition of juvenile populations as measured by mtDNA haplotypes. To the extent that these traits were linked to the condition and size of settling larvae, their results have both demographic and population-level genetic effects because of the apparent links between larval condition at settlement and subsequent juvenile survival. To the extent that

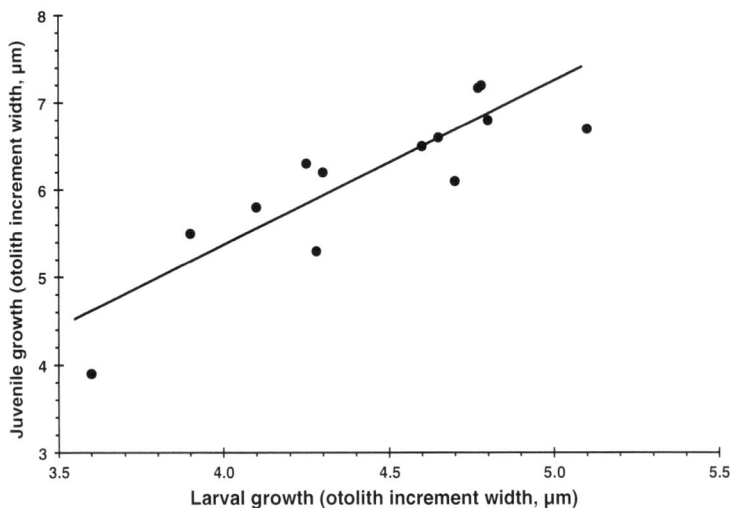

Fig. 6.1 Relationship between mean juvenile growth (mean otolith increment width) during the first four days on the reef and larval growth (mean otolith increment width) during the entire larval period for 13 cohorts of *Thalassoma bifasciatum* that recruited to the upper Florida Keys, USA. Figure reproduced from Sponaugle et al. (2006), with kind permission from Inter-Research Science Centre

larval growth rate is a heritable trait, these factors are especially important to examining effects of settlement-stage larval condition on juvenile survival because coral reef fishes invest in rapid growth as an opportunistic bet hedging strategy to achieve a selective advantage in the highly competitive and structurally complex coral reef environments (Fonseca and Cabral 2007).

Larval size at settlement is not always a good indicator of condition. McCormick and Molony (1993) found a poor correlation between condition and fish length for a tropical goatfish (*Upeneus tragula*) in Australia, and age at settlement was not correlated with standard measures of condition (carbohydrate content, lipid content, burst swimming speed). Their results indicate that a multi-faceted analysis of settlement-stage larval condition is necessary to make predictions on post-settlement survival and recruitment. In addition, in some cases increased aggression (Jones 1987) or size-selective predation (Sogard 1997) can be focused on the largest individual. So bigger is not always better, which underscores the need to examine species individually.

6.5 Post-settlement Transition (A Distinct Behavioral Phase During Which Individuals are Associated with the Benthos but are not yet Functioning as Juveniles)

6.5.1 Delayed Metamorphosis

During the post-settlement transition, larvae of many species are able to re-enter the pelagic phase to search for better habitat (Kaufman et al. 1992). Reviewing observations of larval settlement behaviors by others, Kaufman et al. (1992) estimated that the post-settlement transition phase applied to 68 species of coral reef fishes in the US Virgin Islands, showing the phenomenon to be widespread. Transitional individuals often have specialized behavioral and morphological characteristics particular to this phase. The post-settlement transitional phase may last hours to weeks (McCormick and Makey 1997) depending on the species, and metamorphosis may occur before, during, or after the post-settlement transition (Kaufman et al. 1992). In general, transition individuals differ in appearance from larval and juvenile conspecifics.

The ability to delay or accelerate metamorphosis associated with settlement appears to be common in fishes and decapods, and this may influence post-settlement processes. For example, Butler and Herrnkind (1991) have shown that pueruli larvae of spiny lobster (*Panulirus argus*) of the Caribbean and western Atlantic accelerate metamorphosis slightly when exposed to the red alga that is the preferred settlement habitat. In a laboratory study of three Florida hermit crabs (*Pagurus maclaughlinae*, *Paguristes tortugae*, and *Clibanarius guttatus*), Harvey (1996) found that exposure to water previously inhabited by conspecific adults inhabiting shells accelerated metamorphosis in two species, and all three species delayed metamorphosis in the absence of empty shells. Kaufman et al. (1992)

observed a four-fold difference in size of post-settlement *Acanthurus* larvae in the Caribbean, which they attributed to the species' ability to delay metamorphosis. The largest individual initially settled, but then re-entered the water column, presumably to continue searching for a settlement site. Using cages to place settlement-stage *Acanthurus triostegus* larvae on the benthos or suspended in the upper water column in French Polynesia, McCormick (1999) showed that many of the pelagic-caged larvae were able to delay metamorphosis, whereas all of the benthic-caged fish completed metamorphosis within five days. However, the individuals that delayed metamorphosis still deposited a settlement mark on their otoliths, indicating competency to settle. It is unclear whether this settlement mark is deposited for other species that delay settlement, and the extent to which this might influence estimates of post-settlement growth rates. Leis and Carson-Ewart (1999) captured settlement-stage larvae of the coral trout (*Plectropomus leopardus*) at night, and observed their swimming behavior during daylight hours. Many of the released larvae (26–32%) exhibited their ability to delay metamorphosis and swam toward open water away from the reef, presumably to attempt settlement the next night. Other larvae searched for settlement locations on the reef.

6.5.2 Habitat Selection

A post-settlement transitional phase implies that selective settlement is occurring, but does it occur in many species? Though spiny lobster pueruli tend to stick with the *Laurencia* algae clump where they first settled if that clump is isolated, they rapidly emigrate from the clump when other clumps form a more continuous mat, and are especially likely to leave when food is scarce on the first clump (Marx and Herrnkind 1985b; Chapter 7). Sancho et al. (1997) observed transitional surgeon-fish (acanthurid) larvae (*Ctenochaetus strigosus*) swimming upcurrent in search of suitable settlement habitat. When the habitats explored were not suitable or already occupied, the transitional larvae resumed swimming upcurrent. During their observations of coral trout larvae (*Plectropomus leopardus*, Serranidae—sea basses) at Lizard Island, Great Barrier Reef, Leis and Carson-Ewart (1999) noted active swimming to search for settlement sites, with avoidance of areas with predators, but no selection of specific settlement habitats.

Given that finding appropriate settlement habitat is challenging, species with less restrictive settlement habitat requirements may have an initial advantage. For example, Robertson (1988) and Parrish (1989) suggested that lagoon seagrass, algal plain, and other common non-reef habitats provide large target areas for larval settlement, with subsequent movement to nearby suitable recruit and juvenile habitats such as rubble, patch reef, mangroves, or back-reef. This settle-and-move strategy would allow post-settlement fishes to respond to benthic processes such as priority effects, competition, and predation. Priority effects refer to the process by which the presence of one species in a habitat decreases the probability of colonization by another. One species can reduce recruitment of another via interference competition (adult and subadult residents or settling juveniles can interfere with larval settlement), or preemption of resources, or predation (predatory adult and subadult residents can decrease local settlement directly by preying on settlers, or indirectly by inducing settlers to choose other sites (Shulman et al. 1983)). Adams and Ebersole (2004) conjectured that observed juvenile abundance patterns in St. Croix, US Virgin Islands for the surgeonfish *Acanthurus chirurgus* and grunts (*Haemulon* spp.) may have resulted from this settlement strategy. Use of this 'settle and move' strategy is also suggested by the patterns of habitat use by the early juveniles of spiny lobsters (*Panulirus argus*; reviewed in Lipcius and Eggleston 2000) and Nassau groupers (*Epinephelus striatus*; Eggleston 1995).

A likely reason for use of non-reef habitats by juveniles of species with reef-associated adults is reduced inter-specific interactions, especially reduced predation. In tethering experiments, Acosta and Butler (1999) found much higher predation on recently-settled transparent larvae and pigmented post-larvae of Caribbean spiny lobsters on coral reefs than on inshore vegetated habitats (Fig. 6.2). The plant stems inhabited by post-larval brown shrimp (*Penaeus aztecus*) protect them from predators (Minello and Zimmerman 1983a, 1983b, Zimmerman and Minello 1984, Zimmerman et al. 1984, Minello and Zimmerman 1985).

Predation on juvenile French grunts (*Haemulon flavolineatum*) in the US Virgin Islands in seagrass beds decreased with distance from the backreef (Shulman 1985a). Similarly, predator encounter rates for juvenile surgeonfish (*Acanthurus chirurgus*) in seagrass beds in the Caribbean decreased with distance from patch reefs, as did aggression from territorial herbivores (Sweatman and Robertson 1994).

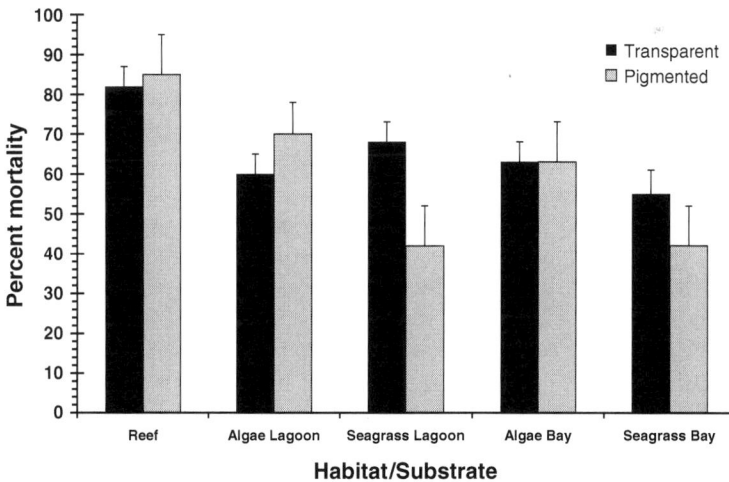

Fig. 6.2 Predation on transparent and pigmented post-larval lobsters in benthic habitats during the day along an offshore to nearshore transport path. Post-larvae were tethered in coral crevices on reefs, and in seagrass and macroalgae in the coastal lagoon and bay. Bars are standard errors of mean percent mortality. Figure reproduced from Acosta and Butler (1999), with kind permission from the American Society of Limnology and Oceanography, Inc. (copyright 2008)

Moreover, effects of predators, and the suite of predators that impact recruits, change with degree of patch reef isolation (Overholtzer-McCleod 2006). Marx and Herrnkind (1985b) concluded that recently-settled spiny lobsters in Florida (USA) choose to live in clumps of *Laurencia* algae because this habitat provides both food and protection from predators, and the predation experienced by newly settled Caribbean lobsters in the mangrove prop roots they choose is less than they would encounter with coral shelter (Acosta and Butler 1997); microhabitat (shelter size) and habitat location continue to be important to survival of juvenile spiny lobsters (Eggleston and Lipcius 1992, Mintz et al. 1994).

Searches for suitable habitats by post-settlement transition individuals may be influenced by saturation of habitats (Shulman et al. 1983, Forrester 1995, 1999, Schmitt and Holbrook 1999b). In comparisons of juvenile abundance on lagoon and back-reef habitats of the US Virgin Islands, Adams and Ebersole (2002) suggested that suitable juvenile habitats on the back-reef became saturated early during the summer (high settlement season), so that later-arriving fishes settle on lagoon habitats that are not yet saturated. In this scenario, lagoon habitats attracted more settlers in summer because resources (food, shelter, and space) were more available than on the back-reef, which is crowded with fish of all ages competing for these resources. In winter, when the density of fishes is lowest, incoming larvae may settle on the first appropriate habitat they encounter, which is the back-reef. These findings were similar to those of Munro et al. (1973) and Shulman (1985a). In this scenario, post-settlement transition individuals seek out alternative habitats where fish densities are lower, evening out the per capita use of resources among habitats.

Much of the habitat selection by settlement-stage larvae likely takes place during the post-settlement transition. During this period, larvae are associated with the benthos, but have not yet taken on full occupancy of benthic habitats or juvenile behaviors, so individuals may be able to make additional assessments of potential settlement sites. The post-settlement transition is also when most priority effects (e.g., Shulman et al. 1983, Almany 2003, 2004b) take place. Post-settlement often involves competitive and aggressive interactions (Booth and Brosnan 1995), but how these interactions act to modify patterns established at settlement varies among species (Almany 2003, 2004a). For example, Caribbean post-larval spiny lobsters in the 'algal' phase (so-named because clumps of red *Laurencia* are a preferred habitat at this stage) are solitary and fiercely agonistic toward conspecifics (Andree 1981, Marx 1983, Marx and Herrnkind 1986). Sancho et al. (1997) observed that transitional individuals of the Pacific surgeonfish *Ctenochaetus strigosus* were rebuffed by conspecifics as they explored potential settlement sites at Johnston Atoll, Central Pacific. Moreover, this competitive/aggressive priority effect likely had an indirect effect on survival, since predation on schools of transitional *C. strigosus* searching for settlement sites was also observed. Territorial damselfishes (pomacentrids) are particularly prone to using aggression to inhibit settlement of hetero- and conspecifics (e.g., Shulman et al. 1983, Sweatman 1985, Risk 1998). For example, in an experiment that manipulated the presence of adults on experimental reefs in the Caribbean, Almany (2003) found that adult beaugregory (*Stegastes leucostictus*) reduced conspecific recruitment. Settlement and post-settlement persistence of surgeonfishes (*Acanthurus* spp.) was reduced by the presence of the beaugregory dam-selfish (*S. leucostictus*) (Shulman et al. 1983, Risk 1998). Priority effects such as these may be strictly hierarchical, as in the consistent exclusion of post-settlement *Acanthurus* surgeonfishes by beaugregory damselfishes (Shulman et al. 1983) or the consistent effects imposed by interspecific competition among *Dascyllus* dam-selfishes found by Schmitt and Holbrook (1999b). In contrast, Munday (2004a) found no competitive hierarchy between two coral-dwelling gobies in the Pacific.

6.5.3 Predation

Predation is often a strong influence on recruitment, especially during the first 48 hrs of settlement (Webster 2002, Almany 2004b, Almany and Webster 2004, see Chapter 7 for details on Decapoda), which is well within the post-settlement tran-sition window for many species (e.g., *Acanthurus triostegus* post-settlement meta-morphosis takes up to five days; McCormick 1999). Almany (2003) manipulated the presence of resident piscivores, and found that piscivores reduced settlement of the beaugregory damselfish (*Stegastes leucostictus*). However, on reefs where adults of other damselfish species were also present, the piscivores had no effect on recruit-ment, suggesting that interspecific aggression was indirectly influencing settlement. The effects of resident piscivores on recruitment of another pomacentrid (*S. parti-tus*) were similar, but were not significant because of overall low larval supply. In the same study, Almany (2003) found that resident piscivores also reduced recruitment of the surgeonfish *A. coeruleus*.

Almany's (2003) study is especially pertinent because he surveyed his exper-imental reefs on a daily basis, so he was able to observe post-settlement transi-tional individuals. Although Almany concluded that his results could be explained by post-settlement mortality, with much of this mortality occurring within hours of settlement (often before his daily visual censuses), his visual censuses included new settlers and post-settlement individuals, and so mixed settlement (site selection by larvae) and post-settlement effects. Shulman et al. (1983) and Tupper and Juanes (1999) found that settlement of Caribbean grunts (*Haemulon* spp.) was lower where juveniles of predatory fishes such as snappers (lutjanids) had already settled, sug-gesting that some species select settlement sites to avoid potential future predation. Webster (2002) and Almany (2004b) found similar results on the Great Barrier Reef, where presence of resident piscivores reduced recruitment of most fishes, with losses occurring mostly during the first 48 hrs after settlement. However, as with Almany's (2003) findings in the Caribbean, the relative effects of resident piscivores on recruitment varied among fish species. Some species (e.g., the damselfish *Pomacen-trus amboinensis*) experienced density-dependent mortality, whereas others (e.g., the damselfish *Neopomacentrus cyanomos*) experienced density-independent mor-tality. Density-independent mortality resulted in recruitment that reflected larval supply, but density-dependent mortality modified patterns of larval supply.

The examples listed above on differences in settlement site selection suggest that caution must be used when applying findings across species and families of fishes and decapods. The community-level effects of differential predation on post-settlement fishes are underscored in a study by Almany and Webster (2004). They censused post-settlement fishes of 20 species on the Great Barrier Reef, and 15 species in the Bahamas, on reefs with and without predators. The predators were serranids (sea basses) and pseudochromids (dottybacks) in Australia, and serranids and muraenids (morays) in the Bahamas. The species they surveyed for recruitment represented Acanthuridae (surgeonfishes), Chaetodontidae (butterflyfishes), Labridae (wrasses), Pomacentridae (damselfishes), Pomacanthidae (angelfishes), and Siganidae (rabbitfishes). Using rarefaction analysis to examine whether predator effects on recruitment directionally changed fish community composition, they found that recruitment species richness was higher on reefs without piscivores, predators had a greater effect on relatively rare species, and some species were present only on reefs without predators. Although they acknowledged that their study did not discriminate between settler avoidance of reefs with piscivores versus predation of post-settlers, they cited previous research (Almany 2003) which used caged piscivores to demonstrate that piscivore presence did not effect settlement (i.e., effects were due to post-settlement predation). Although they measured effects over a period of 44–50 days, most settlement modification via predation occurred within the 48 hrs of settlement that generally includes the post-settlement transition.

6.6 Post-settlement Stage (Time Period Directly After Metamorphosis, and One of Total Benthic Association—a Period of High Benthic Mortality)

The post-settlement transition merges into the post-settlement stage, with the rate of progress varying among species. Post-settlement stage fishes and decapods are entirely benthic-oriented, yet remain within the high mortality period extending from settlement. As stated in previous sections, the first days of association with benthic habitats are a period of extreme mortality (e.g., Acosta and Butler 1999, Minello et al. 1989, Webster 2002, Almany 2004a, Almany and Webster 2004, McCormick and Hoey 2004, Doherty et al. 2004; see Chapter 7 for details on Decapoda).

6.6.1 Mortality

Predation is a primary cause of mortality for post-settlement fishes and decapods, but effects vary among species. Following cohorts of post-larval and juvenile brown shrimp (*Penaeus aztecus*) inside and outside predator-exclusion cages allowed Minello et al. (1989) to determine that high mortality in these stages in coastal Texas is due almost entirely to predation, with declines in mortality as the shrimp grow older and larger.

Doherty et al. (2004) followed cohorts of settling unicornfish (*Naso unicornis*) to determine mortality rates over time on coral reefs of Moorea. Initial mortality rates of approximately 61% during the first night of settlement were density-independent, and the density-dependent loss of post-settlement fishes on the first day after settlement ranged from only 9–20%, depending on post-settler abundance. Since fish were censused on all available habitats throughout the entire lagoon, all the mortality could be attributed to predation, rather than emigration or re-settlement away from the study area.

Webster (2002) also manipulated resident predators to examine effects of predation on post-settlement mortality of seven species and three family groups at Lizard Island, Great Barrier Reef. In predator-absent treatments, mortality was density-independent. Resident predators negatively affected survival of all species, primarily within two days of settlement, and in contrast with predator-absent treatments, mortality was density-dependent for most species (Fig. 6.3). Although the intensity of predator impacts varied among species, mortality rates ranged from 1.1–3.7 times higher than in treatments without predators, with predators causing complete recruitment failure for some rarer species (e.g., Chaetodontidae—butterflyfishes).

The influence of post-settlement density-dependent mortality (presumably due to predation) may even vary among cohorts of a single species in a single reef system. Schmitt and Holbrook (1999c) found that the majority of mortality occurred very soon after settlement, but density-dependent mortality was not evident in all cohorts in French Polynesia. Rather, earlier-arriving cohorts experienced density-independent mortality, but their presence induced density-dependent mortality in later-arriving cohorts. These results demonstrate the need to incorporate time and space into studies of post-settlement processes to include inherent variability both within (e.g., Schmitt and Holbrook 1999c) and among (Webster 2002) species.

Density-dependent mortality probably has its greatest impact during the post-settlement stage (Hixon and Webster 2002), although density-dependence may not be apparent (Osenberg et al. 2002). Although their research occurred over a longer time period than recruitment, Hixon and Jones (2005), building upon previous experimentation, showed that competition and predation interacted to cause density-dependent mortality of fishes at Lizard Island, Great Barrier Reef. Although competition generally did not appear to cause mortality directly, the eventual result of competitive exploitation and aggression was predation (Fig. 6.4).

6.6.2 Competition

Competition leading to slower growth rates may be a particular problem for late arriving settlers, since small size is likely to place them lower in competitive hierarchies than earlier-arriving conspecifics. High densities of post-settlement individuals also create competition for resources, and may decrease growth rates (Jones 1991). Such competition may result in predation since slower growth rates often result in higher mortality (Jones 1991), but this is not always the case. For example,

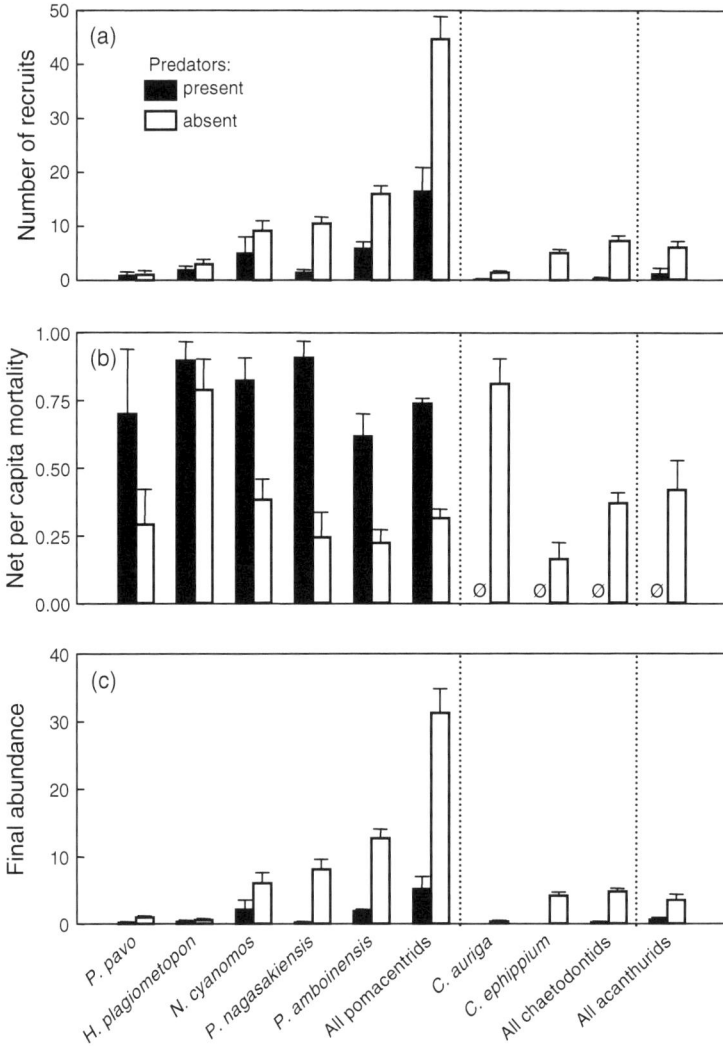

Fig. 6.3 The effect of predators on (**a**) total recruitment, (**b**) mortality, and (**c**) final abundance (mean + standard error) for all species and families over 50 days. Dotted vertical lines separate taxonomic families. ∅ indicates that net per capita mortality values could not be calculated due to insufficient data. Figure reproduced from Webster (2002), with kind permission from Springer Science+Business Media

Forrester (1990) found that although growth rates were lower at high densities because food was less available, this did not influence survival. Moreover, since some predation is selective toward a particular size (Sogard 1997), rapid growth rates resulting in larger individuals may not be a universally positive trait.

The importance of competition probably increases greatly in the post-settlement stage. During settlement, selecting structurally suitable microhabitat is generally

Physical prey refuges present?

yes | no

Competition for prey refuges?

Density-dependent predation faster than competition?

yes | no | no | yes

3: Predation & Competition (simultaneous)

Competition for non-refuge resources?

4: Predation only

yes | no

Predation density-dependent?

Competition if predators removed?

yes | no | yes | no

5: Predation & Competition (separate)

6: Competition only

2: potential Competition (normally precluded)

1: None (extreme recruitment limitation)

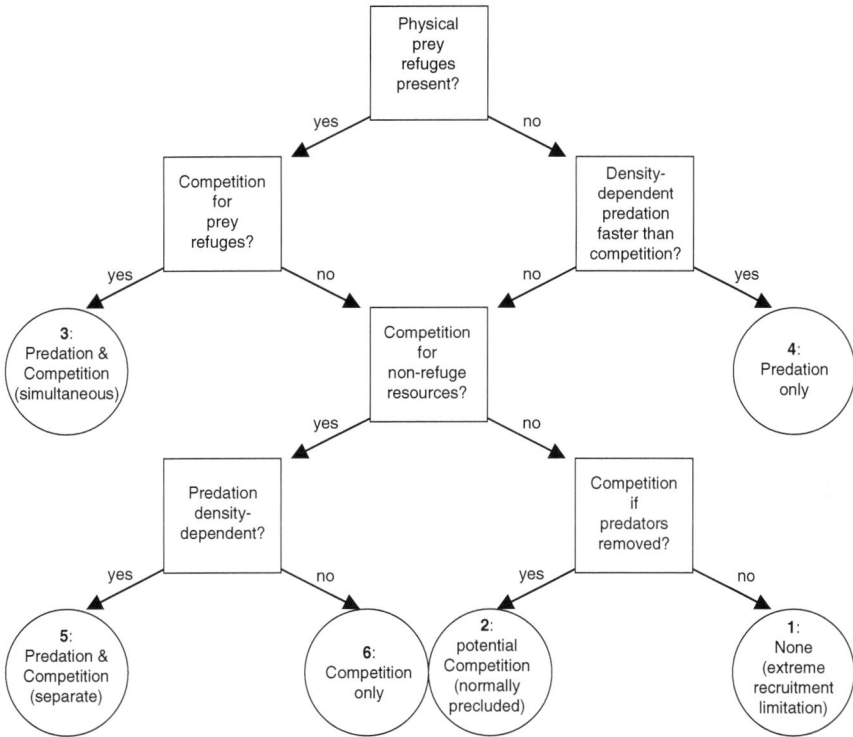

Fig. 6.4 Synthetic flowchart explaining how the interplay of competition, predation, and prey refuges determines the source of density-dependent mortality in demersal marine fishes. Figure reproduced from Hixon and Jones (2005), with kind permission from the Ecological Society of America

more important than avoiding competition (Jones 1991). During the post-settlement transition, and continuing into the post-settlement phase, competition for space—either among settlers or between settlers and occupants—becomes important. The change in relative importance of competition with progressing life phases is exemplified by the tropical Atlantic surgeonfish *Acanthurus bahianus*. Initial settlement of *A. bahianus* is higher in the presence of conspecifics, but post-settlement individuals later compete for limited nocturnal shelter (Risk 1998).

Fishes and decapods fare better when they use shelter appropriate to their body size. Beets (1997) and Hixon and Beets (1993), for example, found that survival of post-settlement Caribbean fishes was lower on artificial reefs with shelter holes large enough for predators, and that survival increased on reefs with shelter holes sized appropriately for post-settlement fishes. Similarly, Nemeth (1998) found that survival of juvenile bi-colored damselfish (*Stegastes partitus*) was higher in habitats with more and smaller crevices they could use as shelter. Since the main source of mortality was encounters with predators, appropriate-size shelters were probably a limiting resource.

Competition for appropriate-sized shelter likely leads to increased mortality, mainly due to predation (Hixon and Menge 1991, Eggleston and Lipcius 1992, Friedlander and Parrish 1998). Competition for suitable shelter is important in shaping post-settlement abundances. Settlement of two *Dascyllus* damselfishes in French Polynesia, for example, was suppressed 80–90% by insufficient supply of suitable microhabitats (Schmitt and Holbrook 2000), and post-settlement intra- and interspecific competition for shelter strongly influenced juvenile abundance (Schmitt and Holbrook 1999b). Aggressive interactions between adult and juvenile damselfishes can make the juveniles more susceptible to predation (Holbrook and Schmitt 2002, Almany 2003). In numerous site-attached species, ontogenetic partitioning of habitat reduces competition between life stages, and enables settlement. For example, juvenile three-spot damselfish (*Stegastes planifrons*) escaped competition for space from territorial adults by settling in dead coral heads (primarily *Agaricia tenuifolia*), and avoiding the adults that were mostly in live *A. tenuifolia* (Lirman 1994). Similarly, *Pomacentrus sulfureus* appeared to undergo ontogenetic microhabitat shifts (Bergman et al. 2000). Juveniles were associated with branching corals, and were mostly associated with the benthos. In contrast, adult abundance was negatively related to substrate diversity, indicating more general habitat requirements. In addition, adults spent most of their time in the water column.

The appropriate shelter size changes with body size. In the Bahamas, habitat complexity was not a factor in mortality of the damselfish *Stegastes leucostictus* during and immediately after settlement: resident predators (i.e., Serranidae—sea basses, Muraenidae—morays) and competitors (i.e., territorial adult damselfishes *S. leucostictus* and *S. partitus*) negatively affected survival on both high and low habitat complexity reefs (Almany 2004a). This was attributed to the ability of competitors and small predators to access the shelter holes available to post-settlers. Over time, as post-settlers grew, survival became higher on reefs with higher complexity.

6.6.3 Movement Among Habitats

Type of predator and habitat dispersion can interact to influence the effects of predation. Overholtzer-McLeod (2006) experimentally examined the effect of habitat dispersion on interactions between predators and juveniles on the damselfish *Stegastes leucostictus* and the wrasse *Halichoeres garnoti* in the Bahamas. Both species experienced density-dependent mortality on spatially-dispersed patch reefs (separated by 50 m), and density-independent mortality on aggregated patch reefs (separated by 5 m). She attributed high (nearly 100%) mortality on aggregated reefs to visits by transient predators that occurred independent of prey densities. In contrast, most of the predation on spatially-dispersed reefs was caused by resident predators (primarily the small grouper *Cephalopholis fulva*). Although the resident predators were also present on the aggregated reefs, their impact on mortality was swamped by the transient predators that generally ate all juveniles present.

Unfortunately, Overholtzer-McCleod (2006) discounted movement among reefs by *Stegastes leucostictus* as an effect on mortality estimates—this damselfish shows relatively low site fidelity (McGehee 1995) and the ability to colonize apparently isolated habitats during all benthic life phases (Adams and Ebersole 2002)—but her results reflect an emerging view that the interaction between species vagility and habitat contiguity has community-level effects (Ault and Johnson 1998). The importance of contiguous habitats to community-level processes was emphasized by Ault and Johnson (1998). They found that vagile fishes are able to move among isolated habitats in response to resource availability, whereas more sedentary species seem to require contiguous habitats for larger scale movement. Habitat contiguity, in conjunction with predator behavior, may also facilitate an aggregative response by predators (e.g., Anderson 2001), with subsequent increases in post-settlement mortality.

6.7 Recruitment (Occurs at the End of the Post-settlement Stage, and Incorporates Effects of Larval and Post-settlement Processes)

The recruitment phase of early life history of fishes and decapods, when individuals can properly be called juveniles, is generally when most surveys occur or begin, so much of the information we have on recruitment has been gathered during this stage. Depending on the species and locations, it has been argued that post-recruitment surveys may provide accurate assessments of larval supply (reviewed in Jones 1991). As seen in previous sections, however, more recent evidence points to very strong modification of abundances from initial larval supply due to settlement through post-settlement processes, which would negate the use of post-recruitment surveys as assessments of larval supply. Although recruitment is characterized by entry into a period of lower mortality compared to that of the settlement through post-settlement stages, some of the same factors influencing those earlier stages—predation, competition for shelter and food resources, and aggression—are also influential for juveniles in the recruitment phase.

Habitat quality and individual condition may interact to influence recruit condition and growth rates of recruits, and subsequently affect density-dependent mortality. In other words, processes observed in recruits are connected backward to the larval settlement stage. A series of studies on recruitment of *Thalassoma hardwicke* in French Polynesia suggested that individual traits (as measured by growth and lipid content) and habitat quality have synergistic effects on survival. Condition (as measured by lipid content) and growth rates of individuals were high at high quality habitats, and these habitats also had more strongly density-dependent mortality (Shima et al. 2006). The combination of higher settlement and condition at higher quality sites and density-dependent synergy of habitat quality produce extra high recruitment, which may be augmented further by the reduced predation associated with good condition (Booth and Hixon 1999). The overall recruitment patterns that result are complex, however, and the extent to which habitat quality and individual intrinsic traits contribute to differences in density-dependent mortality and survival remains unclear.

6.7.1 Growth and Shelter Size

Survival also depends on growth, especially during the recruitment phase. In a study of the planktivorous damselfish *Neopomacentrus filamentosus* in western Australia from settlement to three months after settlement, faster growing recruits had higher survival, with size-selective mortality (presumably due to predation) causing the loss of the smallest and slowest growing recruits (Vigliola et al. 2007). Moreover, the intensity of this size-selective mortality was higher in the more numerous of two sequential cohorts, indicating density-dependent mortality. This intense natural selection for large size is indicated by the genetic differences between settlers and recruits that resulted from the intense size-selective mortality. Genetic differences were less evident for a second, less numerous cohort (20% the size of the first cohort), suggesting that natural selection was reduced at the lower density. Alternatively, size-selective mortality in the second, less numerous cohort may have been hard to detect because the mortality rates were so low (Sogard 1997). In any event, the ultimate effect of such size-selective mortality on evolutionary trajectories is unclear.

Changes in habitat requirements with recruit size underscore the importance of appropriately-sized shelter. Small crevices that were suitable for post-settlement individuals become too small to provide shelter as the individuals grow. The need for size-specific shelter means that several different life stages may be forced through population bottlenecks. By providing more shelters, Shervette et al. (2004) identified such a bottleneck for juvenile stone crabs (*Menippe adina*) of coastal Mississippi. In bays of coastal Florida, USA, Beck (1995) found that larger stone crabs (*M. mercenaria*) grew slowly, molted infrequently, and were slow to produce eggs when appropriately sized shelter holes were not available.

Post-settlement Caribbean spiny lobster (*Panulirus argus*) use structurally complex microhabitats, such as macro-algae, and move as juveniles to larger, less complex crevice habitats in sponges, soft corals, coralline algae, seagrass, and rock-rubble habitats as they grow too large for the shelter provided by the macro-algae (reviewed in Lipcius et al. 1998 and Chapter 7). Aggregation becomes part of the anti-predator defense of crevice-dwelling juveniles, since detection of predators and repulsion by antenna-lashing are more effectively accomplished by groups (Eggleston and Lipcius 1992), and juveniles use olfactory cues to find crevices with conspecifics (Nevitt et al. 2000). This association of body size with crevice size continues in later stages, as older juveniles and adults utilize ever-larger crevice-type habitats for shelter. Within the context of this early life history ontogeny, Lipcius et al. (1998) tethered juvenile lobsters of two size classes in experimental plots of varying algal biomass (their proxy for habitat structure) in seagrass beds, and derived a habitat-survival function (HSF) to describe the effect of habitat structure and size on juvenile lobster survival. There was a large increase in survival of both large and small juvenile lobster associated with moderate increases in algal biomass, until an asymptote was attained as algal biomass increased further. Notably, survival of small juveniles was significantly higher than survival of large juveniles. The authors conjectured that this inverted size effect resulted from habitat-body size scaling, such that the algal habitat did not provide appropriate refuge for large juveniles, and the functional habitat area diminishes as individuals grow larger.

Changes in habitat requirements with body size were also evident in an examination of tradeoffs between growth and predation for juvenile Nassau grouper (*Epinephelus striatus*) in the Caribbean. Eggleston and colleagues (Eggleston 1995, Eggleston et al. 1998, Grover et al. 1998, Dahlgren and Eggleston 2001) documented ontogenetic habitat and diet shifts for Nassau grouper from post-settlement to late juvenile stages. In a study that compared habitats in shallow, protected areas that received grouper larvae, Eggleston (1995) identified previously undocumented juvenile habitats: grouper settled exclusively in clumps of macroalgae and not in seagrass or on sand, post-settlement fish (25–35 mm Total Length) resided within the algae clumps, early juveniles (60–150 mm TL) resided adjacent to the algae, and juveniles >150 mm TL colonized natural and artificial patch reefs in areas apparently removed from the post-settlement and early juvenile habitat. Adult Nassau groupers are associated with deeper, high-relief reefs (Sluka et al. 1998).

With ontogenetic habitat and diet shifts as a foundation, Dahlgren and Eggleston (2000) used caging and tethering to examine the tradeoffs between growth and predation that might underlie the observed habitat use patterns. They found a dynamic trade-off between predation risk and growth, where the relative costs and benefits changed over time; overall, minimal predation risk with maximal growth was achieved through ontogenetic shifting of habitat. Although their research focused on ontogenetic shifts of later-stage juveniles, their findings should be generally applicable to examining habitat-growth interactions for earlier life stages and other species (Fig. 6.5).

6.7.2 Competition

Interspecific competition may become more important as individuals reach the recruit phase. Shervette et al. (2004) found that stone crabs in the Mississippi Sound faced competition from mud crabs (*Eurypanopeus depressus* and *Panopeus simpsoni*) for available shelter. The intensity of competitive relationships may change with competitor size. For example, the aggression elicited by intruders from the territorial damselfish *Stegastes leucostictus* is directly related to their potential to consume algal food resources, so that larger intruders tend to elicit a stronger aggressive reaction (Ebersole 1977). In a more focused study, Risk (1998) found that aggression from *S. leucostictus* toward intruding juvenile *Acanthurus bahianus* was greatest for the largest intruders, and this aggression was sufficient to decrease the

Fig. 6.5 Habitat-specific
values of the mortality-risk/
growth-rate ratio calculated
for each size class of Nassau
grouper (*Epinephelus
striatus*). Figure reproduced
from Dahlgren and Eggleston
(2000), with kind permission
from the Ecological Society
of America

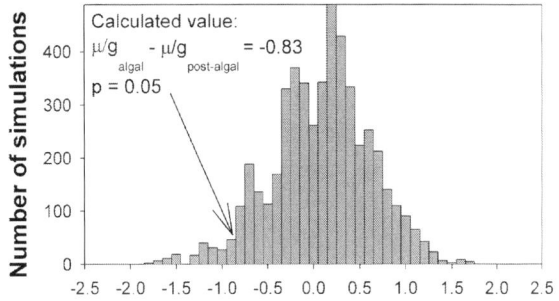

A. Small fish

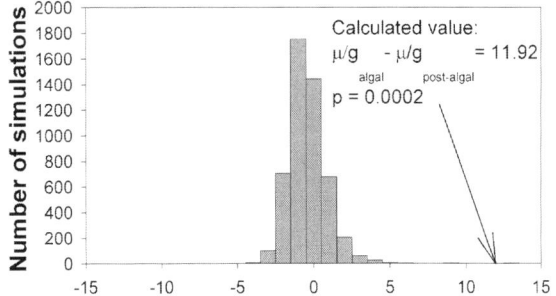

B. Medium fish

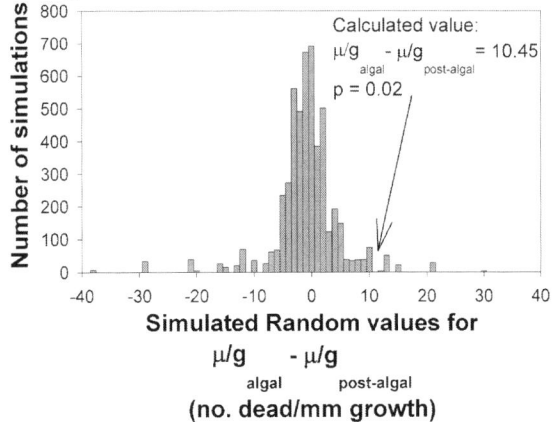

C. Large fish

persistence of juvenile *A. bahianus* in *S. leucostictus* territories. Territorial pomacentrids have similar negative effects on recruits of many herbivorous species (Almany
2003).

Overholtzer and Motta (1999) observed inter- and intraspecific aggressive
behavior in mixed species aggregations of juvenile parrotfishes (Scaridae) in the
Caribbean, focusing on *Scarus iserti*, *Sparisoma aurofrenatum*, and *Sparisoma
viride*. Interspecific aggression occurred among the focal species, and between the
focal species and damselfishes (Pomacentridae), grunts (Haemulidae), and wrasses

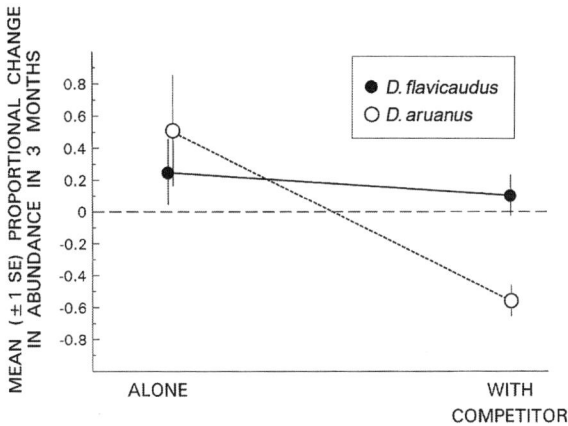

Fig. 6.6 The effects of interspecific competition on the population growth of *Dascyllus flavicaudus* and *D. aruanus*. The figure shows the results of a field experiment where the presence and absence of the congener were manipulated among different coral heads. The data are mean (± standard error) change in population size over 72 days for *D. flavicaudus* (solid circle) and *D. aruanus* (open circle). The horizontal dashed line denotes no net change in population size. Figure reproduced from Schmitt and Holbrook (1999b), with kind permission from Springer Science+Business Media

(Labridae). They concluded that the aggressive interactions would likely have implications for later life stages because these interactions would act as precursors to territoriality as adults. This is an example of the potential demographic implications of mechanisms occurring during recruitment (i.e., connectivity between recruitment and adult populations).

The effects of interspecific interactions on the abundance and distribution of juveniles may not be reciprocal. Manipulating presence/absence of congeneric planktivorous damselfishes to examine effects of competition on abundance of recruits, Schmitt and Holbrook (1999b) found that groups of *Dascyllus aruanus* that would show a 50% increase in numbers over three months in the absence of *D. flavicaudus* experienced a 55% decline when the congeneric competitor was present. However, this strong negative impact of *D. flavicauda* on *D. aruanus* recruits was entirely one-sided; the presence of *D. aruanus* had no discernible effect on the population growth rates of *D. flavicaudus* groups (Fig. 6.6). Accordingly, the modification of larval settlement pattern to recruitment pattern in abundance and distribution were greater for *D. aruanus* than *D. flavicaudus*.

6.8 Use of Reef and Non-reef Areas as Recruitment Habitats

As summarized in preceding sections of this chapter, individual fish and decapod larvae, post-settlers, and recruits are capable of selecting habitat and moving among habitats. Ontogenetic shifters, especially, undergo directional habitat selections and movements, using multiple habitats to maximize survival to adulthood. Until recently, few researchers had examined use of multiple lagoon and reef habitats, despite the observed presence of juvenile fishes and decapods in multiple habitats. Such studies are needed to determine the importance of reef and non-reef habitats as essential nursery habitats (Beck et al. 2001) or effective juvenile habitats (Dahlgren et al. 2006).

6.8.1 Habitat Mosaics

Dispersion of habitat types within coastal tropical ecosystems is an important habitat component, affecting habitat selection as well as survival and growth of early life stages. Habitat dispersion is the spatial distribution of habitat types within a defined area, and includes the entire habitat mosaic to which recruits may be exposed (see Chapter 14). An important feature of this definition is the differentiation of habitat types from habitats: 'habitat type' describes a distinct feature in some general way (e.g., mangrove, seagrass, algal plain, reef), whereas 'habitat' elaborates on the description of a habitat type by including its location in the overall habitat mosaic, thereby taking into account the contiguity or isolation of habitat types (e.g., continuous reef versus small patch reefs in a seagrass matrix).

Although the typical definition of essential nursery habitat (e.g., Beck et al. 2001) implies that some early life stage of a species depends upon a single habitat type, recent research has shown that recruits of many species depend upon a mosaic of contiguous habitat types. For example, although conventional sampling indicates that recruits of Caribbean/western Atlantic snappers (Lutjanidae) and grunts (Haemulidae) use mangroves as shelter during the day, stable isotope and gut content analyses show that these fishes feed in adjacent seagrass beds at night (Harrigan et al. 1989, Serafy et al. 2003, Kieckbush et al. 2004, Nagelkerken and van der Velde 2004). Risk (1998) saw low persistence for post-settlement surgeonfish (*Acanthurus bahianus*) that suffered aggression from territorial damselfish (*Stegastes leucostictus*), but attributed their eventual absence to movement to other locations rather than (immediate) mortality. Whether post-settlement fishes risk such early life history movement to escape site-specific problems (e.g., aggression or competition for resources) is an open question.

Habitat mosaics are important to the early life history stages of non-reef-associated individuals as well. Laegdsgaard and Johnson (2001) conducted experiments in Australia to examine factors affecting use of mangrove habitats by juvenile fishes, and found that use of complex mangrove prop-root habitats and less complex adjacent habitats changed with growth. Artificial mangrove prop-root structure plus fouling algae attracted juvenile fishes of more species and more total individuals than bare structure. In laboratory experiments, the use of artificial mangrove prop-root shelters by small juveniles increased in the presence of predators, but this effect was not evident for larger individuals. Small juveniles also fed most effectively within mangrove habitat, whereas their larger counterparts fed at higher rates on adjacent mud flats. Thus, habitat use can change as juveniles grow and develop even when habitat associations are unchanged. Community-level sampling of mangrove and seagrass habitats in tropical Australia corroborates these experiments, with significantly more juvenile fish and crustaceans captured in mangroves than seagrass habitats (Robertson and Duke 1987), indicating the importance of mangrove roots as shelter for small fishes.

The 'settle-and-move' strategy of ontogenetic shifting species (Robertson 1988, reviewed in Parrish 1989, Sweatman and Robertson 1994, Adams and Ebersole 2004), whereby larvae settle into an extensive habitat (e.g., mangrove prop roots, seagrass, algal beds) and move later to microhabitats (e.g., patch reef, rubble) within the settlement habitat, exemplifies the importance of habitat as a mix of habitat types in close proximity (i.e., a habitat mosaic). These species may briefly settle in habitats different from post-settlement habitats, and then move quickly to habitats that provide better resources. For example, Caribbean *Haemulon* spp. settle to algal plain and sparse seagrass, but later, as they grow, move to structure located near seagrass such as rubble patches, patch reefs, or back-reefs located near seagrass, where many researchers first note their presence (McFarland 1980, Shulman and Ogden 1987, Adams and Ebersole 2004). The common Caribbean French grunt *Haemulon flavolineatum* provides a detailed example of the importance of habitat mosaics to early life history, since it has a well-described ontogenetic pattern of habitat use. Larvae settle in seagrass, algal plain, or soft-bottom habitats and then move (via numerous ontogenetic shifts) to adult habitats on reefs (Shulman and Ogden 1987, Adams and Ebersole 2002). Through most of their benthic existence (excluding the early post-settlement stage) French grunts feed on benthic invertebrates living in soft-bottom. They feed primarily at night, and spend the day in close association with hard-bottom structure (rubble, patch reef, larger coral reef, mangrove). Juvenile French grunts appear to use structure within their settlement habitats (e.g., tiny patch reefs or queen conch shells in seagrass beds) before moving to adult habitats on larger reefs. Using benthic maps developed from GIS and high-resolution aerial photos and *in situ*, day-time censuses of fish to examine influence of habitat distribution on juvenile French grunt abundance, Kendall et al. (2003) found that the abundance of juveniles (post-recruits) on hard-bottom was inversely related to the distance from soft-bottom feeding areas, and that when refuge and feeding habitats were in close proximity, refuges near larger feeding areas had more juveniles (Fig. 6.7). The extent to which habitat mosaic also influenced post-settlers and recruits is unclear. Similarly, a mixture of seagrass or algae (settlement habitats) and patch reefs (juvenile habitat) increases movement and enhances survival of juvenile spiny lobsters (*Panulirus argus*) (Acosta and Butler 1999).

Pollux et al. (2007) and Adams and Ebersole (2002) each used a comparative approach in the Caribbean to demonstrate that the importance of non-reef habitat dispersion varies among species. Pollux et al. (2007) conducted post-settlement censuses of *Acanthurus bahianus*, *Ocyurus chrysurus*, and *Lutjanus apodus* along transects on coral reef, seagrass, and mangrove habitats. They found that each species exhibited habitat-specific patterns that they attributed to settlement preferences: *A. bahianus* to coral reef (and to a lesser extent seagrass), *O. chrysurus* to seagrass and mangroves, and *L. apodus* exclusively to mangroves. Since the habitat-specific patterns they observed were similar to those reported for later juvenile stages for

Fig. 6.7 (**a**) Logistic regression of juvenile *Haemulon flavolineatum* presence by distance (m) from hard bottom census site to soft bottom. $X^2 = 5.11$, p = 0.024. (**b**) Logistic regression of juvenile *H. flavolineatum* presence/absence by area of soft bottom within 100 m. $X^2 = 4.75$, p = 0.029. Figure reproduced from Kendall et al. (2003), with kind permission from Springer Science+Business Media

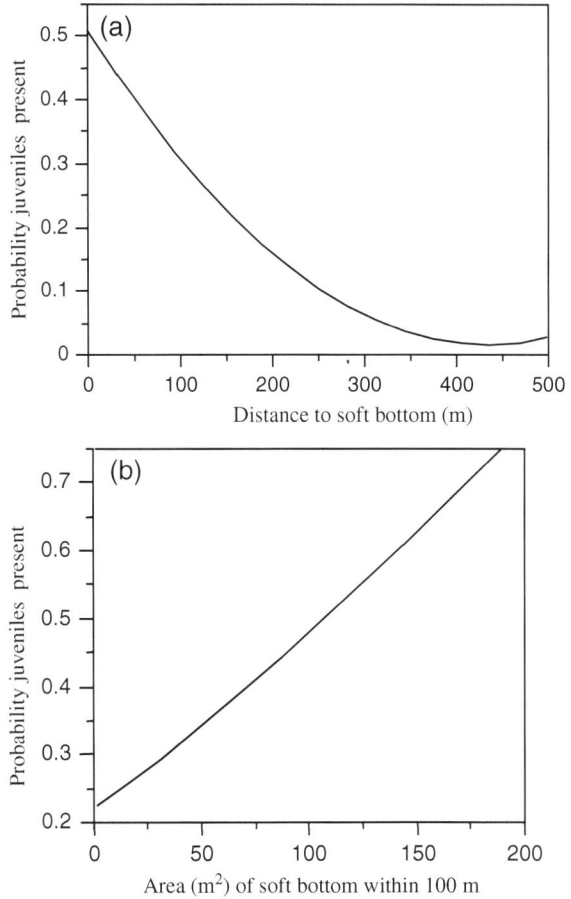

these species, they concluded that juvenile habitat use patterns resulted primarily from settlement patterns. Adams and Ebersole (2002) censused post-settlement fishes and recruits in backreef habitat and five lagoonal habitats (seagrass, rubble, patch reef, sand, algal plain), and found two general patterns of juvenile habitat use among fishes with reef-associated adults: one group (e.g., *Acanthurus* spp., *Haemulon* spp.) used lagoon patch reef and rubble as juvenile habitats, whereas a second group (e.g., *Scarus iserti*, *Sparisoma aurofrenatum*) used backreef as juvenile habitat.

6.8.2 Recruitment Habitat Quality

For some organisms that shift habitats ontogenetically, abundances of adult populations may depend on input from lagoon nurseries, and thus show a strong connectivity to post-settlement and recruitment habitats. Finding such relationships is likely to be difficult, since variable mortality rates associated with the movement between habitats may obscure them. Robertson (1988) posited that post-settlement relocation played a large role in the difficulties of establishing a correlation between abundance of recruits and abundances of adults for three acanthurids in the Caribbean. However, Adams and Ebersole (2004) developed a Lagoon Quality Index (LQI) to quantify recruit habitat availability and examine the relationship between recruitment habitat availability in seagrass lagoons (inshore of bank barrier reefs) and adult abundance on the nearby reefs for two genera of ontogenetic shifters in the Caribbean. An LQI that combined availability and use of recruit habitats was calculated for each of six sites for small (<3 cm) and medium (3–5 cm) juvenile *Haemulon* spp. and for juvenile *Acanthurus* spp. (small and medium size classes were combined since settlement occurs at approximately 2.5–3 cm):

$$\text{LQI}_{ij} = \sum a_{ix} \cdot \text{P}_{jx}$$

where x = a given habitat type (e.g., patch reef, rubble, algal plain, seagrass, sand), a_{ix} = mean density of species i on habitat x_j (computed from values pooled from all six sites), and P_{jx} = relative cover of habitat x in lagoon j. Least squares regression showed that the Lagoon Quality Index was a good predictor of adult densities on nearby reefs for *Acanthurus* spp. and for small *Haemulon* spp. (Figs. 6.8, 6.9). The LQI was calculated separately for each lagoon by pooling two years of census

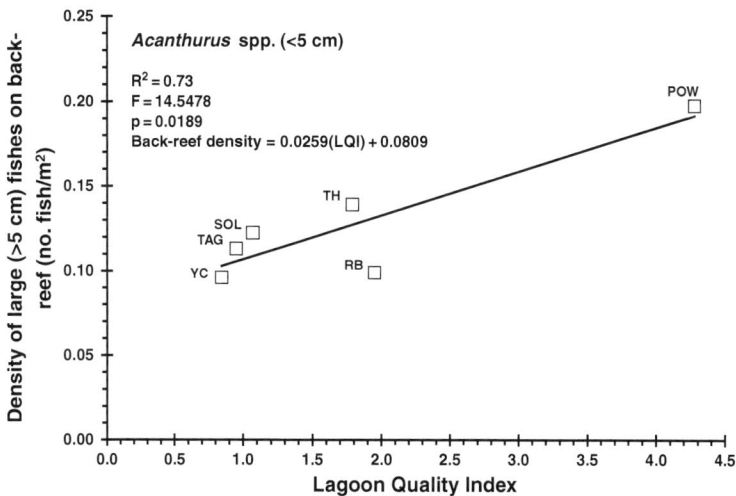

Fig. 6.8 Relationship between density of large (>5 cm) *Acanthurus* spp. on back reefs and the Lagoon Quality Index (LQI) for small (<3 cm) and medium (3–5 cm) *Acanthurus* spp. combined. Site abbreviations: YC = Yellowcliff Bay, TAG = Teague Bay, SOL = Solitude Bay, TH = Turner Hole, RB = Rod Bay, POW = Pow Point. Figure reproduced from Adams and Ebersole (2004), with kind permission from Bulletin of Marine Science

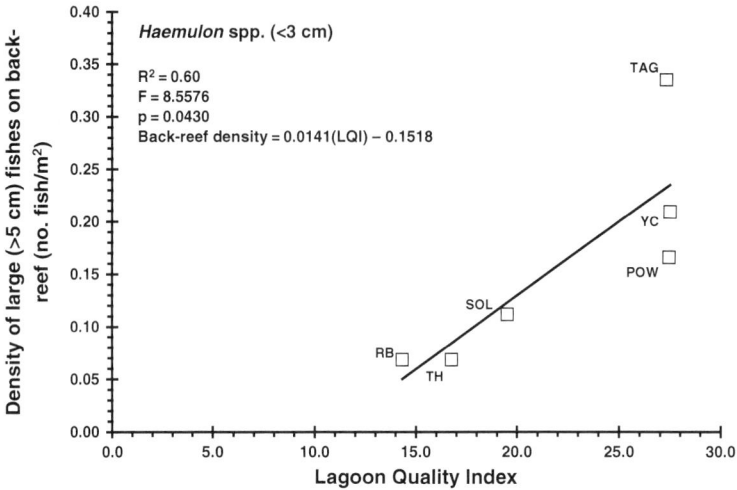

Fig. 6.9 Relationship between density of large (>5 cm) *Haemulon* spp. on back reefs and the Lagoon Quality Index (LQI) for small (<3 cm) *Haemulon* spp. Site abbreviations: YC = Yellowcliff Bay, TAG = Teague Bay, SOL = Solitude Bay, TH = Turner Hole, RB = Rod Bay, POW = Pow Point. Figure reproduced from Adams and Ebersole (2004), with kind permission from Bulletin of Marine Science

data to reduce the impact of the storage effect (Warner and Chesson 1985) that tends to obscure the relationship between abundances of recruits and adults in fishes with ontogenetic shifts (Tolimieri 1998b). The LQI for the medium size class of *Haemulon* spp. was not a good predictor because *Haemulon* spp. undergo ontogenetic shifts beginning at a small size, and by the time they reach the medium size class, they are already transitioning to reef habitats (McFarland et al. 1985, Shulman and Ogden 1987, Adams and Ebersole 2004). Similar relationships between availability of recruitment habitat and adult population density have been found elsewhere for these and other fishes (Nagelkerken et al. 2001, 2002, Mumby 2006), spiny lobsters (Butler and Herrnkind 1997), and stone crabs (*Menippe mercenaria*, Beck 1995, 1997).

6.8.3 Recruitment Habitat Proximity

There may be a proximity threshold for the influence of non-reef nurseries on reef populations. In a study of mangrove shorelines in a lagoon adjacent to a fringing reef in the US Virgin Islands, for example, Adams and Tobias (1999) found that abundance of juveniles of species with reef-associated adults (e.g., surgeonfish *Acanthurus chirurgus*) decreased from the mangrove shorelines closest to the reef to the interior mangrove shorelines, whereas juveniles of species with a higher affinity for mangrove habitats (e.g., mangrove snapper *Lutjanus apodus*) were equally abundant

Fig. 6.10 Mean fish density of several nursery species (**a, b**) and of *Stegastes dorsopunicans* (**a**) on the coral reef as a function of distance to the mouth of the bay. All reef sites are located down-current of the bay, and in this area other lagoons are absent. Figure from Nagelkerken et al. (2000b), with kind permission from Inter-Research Science Centre

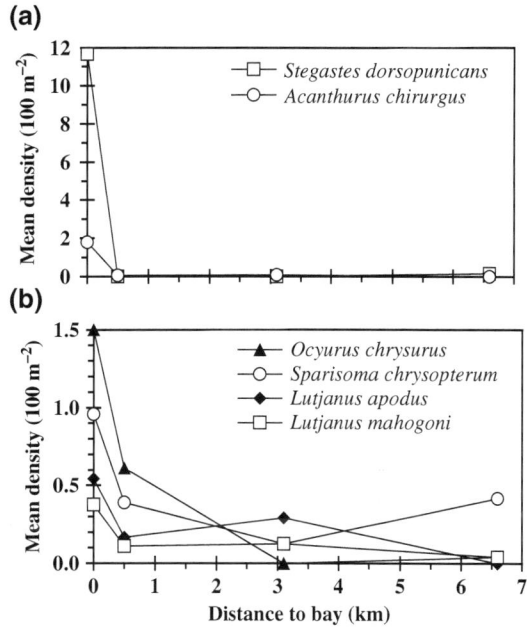

throughout the mangrove lagoon. Similarly, in a lagoon studied by Nagelkerken et al. (2000b), juvenile densities of reef fish species with reef-associated adults rapidly decreased with distance from the coral reef. Densities of many species that used lagoon habitats as putative nurseries, for example, were higher on habitats nearer the reef (Fig. 6.10), whereas species with affinities for lagoon habitats were most abundant in habitats farther from the reef. Ley et al. (1999) also found differences in fish community composition across an estuarine gradient. They examined fish communities in a mangrove-fringed subtropical estuary in Florida, USA, to determine fish community composition. Juveniles and sub-adults of reef-associated species were present only in downstream, higher-salinity locations. In contrast, euryhaline residents (e.g., Poeciliidae—livebearers, Cyprinodontidae (Fundulidae)—killifishes) dominated community assemblages throughout the study area, regardless of salinity.

6.8.4 Recruit–Adult Connectivity

On a regional scale, the correlations between nursery availability and adult populations on the reef are reflected in associations between nurseries and community structure on reefs. Mumby et al. (2004) examined the distribution of mangroves and reefs in the Caribbean, and found an association between mangroves and fish species composition on reefs. Reefs with (more) nearby mangrove habitats had higher adult abundances of species that used mangroves as juveniles. Moreover, for many of these species, mangroves served as intermediary nursery habitat that may have increased survivorship of juvenile fishes. The effects of mangroves as juvenile habitat was very pronounced for some fishes of commercial value, with biomass greater on reefs with nearby mangroves than reefs with fewer or no mangroves nearby. Effects were most dramatic for species with the greatest dependence on mangrove habitats: *Scarus guacamaia*, functionally dependent on mangroves, is now locally extinct in areas with high mangrove loss. The relationship between adult populations on reefs and contiguity of non-reef juvenile habitats has also been shown for seagrass (e.g., Dorenbosch et al. 2004, 2006) in the Caribbean.

Examination of fish abundances and sizes in Australia suggests a similar connection between species composition on inshore (or shallow) and offshore (or deeper) habitats. For example, at least 11 fish species in northern Australia use estuaries exclusively as juveniles and then use offshore habitats as adults (Blaber et al. 1989), suggesting selection of settlement habitats followed by ontogenetic shifts to adult habitats. Similarly, juveniles of at least 14 species of eastern Australia occur in estuarine and adjacent, shallow, turbid mangrove and seagrass-bed habitats, while the adults are found in deeper water or ocean habitats (Blaber and Blaber 1980). Sheaves (1995) also found that at least 14 species of fish with adult stages on reefs or in deeper offshore waters had juvenile stages that used estuarine mangrove habitats.

6.9 Effects of Disturbance on Recruitment

6.9.1 Tropical Cyclones/Hurricanes

The most prominent natural disturbances in tropical marine ecosystems are tropical cyclones/hurricanes. The relatively few studies of hurricane effects on coral reef fishes generally report temporary, or no, measurable effects (Kaufman 1983, Lassig 1983, Letourneur et al. 1993, Aronson et al. 1994, Bouchon et al. 1994, Adams 2001, Adams and Ebersole 2004), with the most notable effects on juvenile fishes. However, working in the Pacific, Lassig (1983) found high juvenile mortality and redistribution of subadults immediately after a tropical cyclone that occurred during the settlement season. Given the frequency of cyclones during the settlement season and the influence of recruitment on adult abundance, he postulated that cyclones might be important factors in population structure. In contrast, comparing fish abundances before and after a cyclone in the Caribbean that occurred at the end of the settlement season, Adams and Ebersole (2004) found no short-term effects on abundances or size distributions of fishes. They postulated that larger juveniles present at the end of the settlement season may be of sufficient size to be resistant to hurricane disturbances. This, however, is counter to Lassig's (1983) findings for subadult fishes. Differences in findings between these studies may have resulted from the combination of the high storm frequency and general reef degradation in the Caribbean, which may have combined to favor a community that is relatively resistant to disturbances (i.e., caused a phase shift; Done 1992, Jones and Syms 1998). Alternatively, Caribbean fish assemblages may be more resistant to hurricanes, since little or no measurable impact are common observations (e.g., Kaufman 1983, Aronson et al. 1994).

Tropical cyclones may also serve to increase larval supply to tropical estuaries and coral reefs for some species. During a multi-year study of larval tarpon (*Megalops atlanticus*), Shenker et al. (2002) experienced a tropical cyclone pass directly over their study location in Florida, USA. The abundance of *M. atlanticus* larvae was higher in association with that tropical cyclone event than during all other sampling periods. Abnormally high settlement of juvenile honeycomb grouper *Epinephelus merra* occurred following a tropical cyclone on reefs of Réunion Island (Letourneur et al. 1998). However, when observations were continued after the cyclone, density-dependent processes were found to decrease densities to more typical levels (Letourneur et al. 1998).

6.9.2 Habitat Disturbance and Degradation

Other types of disturbance also impact recruitment. Butler et al. (1995), for example, found that the distribution of juvenile Caribbean spiny lobsters *Panulirus argus* changed dramatically when one of their preferred habitats, sponges, suffered a massive die-off in Florida Bay, USA. Lobster density decreased in areas where few alternative shelters existed, but sites where artificial shelters were added experienced an increase in lobster density. Whether density-dependent factors acting at another stage evened out these density differences eventually, as in the honeycomb grouper, is an open question.

Studies of larger scale disturbances that impact recruitment are becoming increasingly important. These disturbances tend to cause wholesale changes in habitats, and whether it is habitat loss or degradation, loss of habitat integrity affects recruitment. Much of this habitat loss and degradation is anthropogenic. For example, the worldwide loss of mangroves between 1980 and 2000 exceeds 34% (Valiela et al. 2001). Since many species of tropical fishes and decapods use estuarine and marine mangroves as recruitment habitats, loss of these habitats likely has population-level implications that require study.

Increasingly, general degradation of coral reefs is hindering recruitment. Examination of coral reef degradation on a gradient (healthy–stressed–dead coral–algae dominated–habitat structure changed) reveals a clear picture of the impacts of habitat integrity loss on fish and decapods. Coral stress (the partial degradation of coral colonies) does not appear to effect larval settlement and subsequent recruitment of fishes (Feary et al. 2007). With increasing degradation, as coral dies, larval fish recruitment decreases (Jones et al. 2004, Munday 2004b, Feary et al. 2007), with specialists more impacted than generalists (Munday 2004b). Reef-wide coral bleaching, coral death, and change in benthic cover (e.g., live coral converts to algae) reduce fish recruitment (Garpe and Ohman 2003, 2007, Garpe et al. 2006), moderately in the short term and severely in the long term (Garpe and Ohman 2003, 2007, Garpe et al. 2006). Initially, loss of coral habitats causes decreases in habitat specialists such as corallivores and coral-associated territorial species (Garpe and Ohman 2003). In the long term, a much wider array of species are impacted, including invertivores and planktivores, with Pomacentridae (damselfishes), Chaetodontidae (butterflyfishes), and Pomacanthidae (angelfishes) particularly impacted (Garpe and Ohman 2003).

6.10 Greatest Knowledge Gaps

6.10.1 Connecting Larval and Juvenile Traits

Although recent research has shown the importance of settlement-stage larval condition on growth, survival, and mortality of post-settlers and juveniles (e.g., Searcy and Sponaugle 2001, Vigliola and Meekan 2002, McCormick and Hoey 2004, Nemeth 2005), the interactions between habitat quality and individual condition are less well studied. As mentioned previously, for example, Shima et al. (2006) found that juvenile *Thalassoma hardwicke* with higher lipid levels and higher growth rates were also associated with better quality habitats. These habitats had higher levels of *Pocillopora* spp. coral that provided cover for juveniles and shelter from predators. In addition, the types of predators (resident vs. transient) and dispersion of habitats (Overholtzer-McCleod 2006) also influence the survival of juveniles, and are independent of larval condition at settlement. Moreover, post-settlement competition, food and habitat limitation, cohort size, timing of settlement, and priority effects all interact and contribute to recruitment success or failure. Combined, these results emphasize the need for future studies to incorporate the interactions of individual traits, habitat quality, and numerous post-settlement mechanisms influencing early life history of fishes and decapods.

6.10.2 Partitioning Mortality and Emigration

Implicit in the measures of post-settlement mortality is that the decline in post-settler abundance results from mortality (mostly from predation). For fishes and decapods with ontogenetic shifter or generalist life history strategies, future research should focus on partitioning mortality and emigration. The ability of juveniles to locate specific habitats (e.g., Blackmon and Eggleston 2001), and the 'settle-and-move' strategy (Robertson 1988, reviewed in Parrish 1989, Sweatman and Robertson 1994, Adams and Ebersole 2004) of many species suggests that the mobility of some species may influence effects of density-dependent mortality. Habitat shifts early in life history may be driven by site-specific densities (e.g., competition for resources) such that recruits and juveniles risk moving in search of other (better) locations. For example, Overholtzer-McCleod (2004) found that loss of *Halichoeres garnoti* from patch reefs resulted from both mortality and emigration to neighboring reefs. In subsequent work (Overholtzer-McCleod 2005), she found that the degree of isolation of patch reefs was a strong factor determining emigration rates; more isolated reefs had less emigration. These results suggest caution when interpreting causes of abundance declines for recruits and juveniles of species with inherent mobility. Similarly, although not dealing with recruits, Lewis (1997a, b) also found that abundances of many fish species on isolated patch reefs were strongly influenced by post-recruitment migrations among habitats.

6.10.3 Conclusion

Clearly, much has been learned in recent decades about the processes influencing fish and decapod recruitment. For example, the planktonic life stage of marine fishes was once thought of as a black box, and recruitment was merely a reflection of larval supply. This view was modified, with Choat et al. (1988), for example, suggesting that habitat-associated variables filter larval supply, slightly modifying the settlement patterns determined by oceanographic processes. It is becoming clear with more recent research, however, that the filter effect of benthic-associated processes active during the recruitment phase can be extremely selective and severe, and requires additional study.

References

Acosta CA, Butler MJ IV (1997) The role of mangrove habitat as nursery for juvenile spiny lobster, *Panulirus argus*, in Belize. Mar Freshw Res 48:721–728

Acosta CA, Butler MJ IV (1999) Adaptive strategies that reduce predation on Caribbean spiny lobster postlarvae during onshore transport. Limnol Oceanog 44:494–501

Adams AJ (2001) Effects of a hurricane on two assemblages of coral reef fishes: multiple-year analysis reverses a false "snapshot" interpretation. Bull Mar Sci 69:341–356

Adams AJ, Dahlgren CP, Kellison GT et al (2006) Nursery function of tropical back-reef systems. Mar Ecol Prog Ser 318:287–301

Adams AJ, Ebersole JP (2002) Use of back-reef and lagoon habitats by coral reef fishes. Mar Ecol Prog Ser 228:213–226

Adams AJ, Ebersole JP (2004) Processes influencing recruitment inferred from distributions of coral reef fishes. Bull Mar Sci 75:153–174

Adams AJ, Tobias WJ (1999) Red mangrove prop-root habitat as a finfish nursery area: a case study of Salt River Bay, St. Croix, USVI. Proc Gulf Carib Fish Inst 46: 22–46

Almany GR (2003) Priority effects in coral reef fish communities. Ecology 84:1920–1935

Almany GR (2004a) Does increased habitat complexity reduce predation and competition in coral reef fish assemblages? Oikos 106:275–284

Almany GR (2004b) Priority effects in coral reef fish communities of the Great Barrier Reef Great Barrier Reef. Ecology 85:2872–2880

Almany GR, Webster MS (2004) Odd species out as predators reduce diversity of coral-reef fishes. Ecology 85:2933–2937

Anderson TW (2001) Predator responses, prey refuges, and density-dependent mortality of a marine fish. Ecology 82:245–257

Andree SW (1981) Locomotory activity patterns and food items of benthic post-larval spiny lobsters, *Panulirus argus*. M.Sc. thesis, Florida Florida State University, Tallahassee

Aronson RB, Sebens KP, Ebersole JP (1994) Hurricane Hugo's impact on Salt River Submarine Canyon, St. Croix, U.S. Virgin Islands. In: Ginsburg RN (compiler) Proceedings of the colloquium on global aspects of coral reefs: health, hazards, and history, 1993

Atema J, Kingsford MJ, Gerlach G (2002) Larval reef fish could use odour for detection, retention and orientation to reefs. Mar Ecol Prog Ser 241:151–160

Ault TR, Johnson CR (1998) Spatial variation in fish species richness on coral reefs: habitat fragmentation and stochastic structuring processes. Oikos 82:354–364

Beck MW (1995) Size-specific shelter limitation in stone crabs: a test of the demographic bottleneck hypothesis. Ecology 76:968–980

Beck MW (1997) A test of the generality of the effects of shelter bottlenecks in four stone crab populations. Ecology 78:2487–2503

Beck MW, Heck KL Jr, Able KW et al (2001) The identification, conservation, and management of estuarine and marine nurseries for fish and invertebrates. BioScience 51:633–641

Beets J (1997) Effects of a predatory fish on the recruitment and abundance of Caribbean coral reef fishes. Mar Ecol Prog Ser 148:11–21

Bergman KC, Ohman MC, Svensson S (2000) Influence of habitat structure on *Pomacentrus sulfureus*, a western Indian Ocean reef fish. Environ Biol Fish 59:243–252

Blaber SJM, Blaber TG (1980) Factors affecting the distribution of juvenile estuarine and inshore fish. J Fish Biol 17:143–162

Blaber SJM, Brewer DT, Salini JP (1989) Species composition and biomasses of fishes in different habitats of a tropical northern Australian estuary: their occurrence in the adjoining sea and estuarine dependence. Estuar Coast Shelf Sci 29:509–531

Blackmon DC, Eggleston DB (2001) Factors influencing planktonic, post-settlement dispersal of early juvenile blue crabs (*Callinectes sapidus* Rathbun). J Exp Mar Biol Ecol 257:183–203

Booth D, Beretta GA (1994) Seasonal recruitment, habitat associations and survival of pomacentrid reef fish in the U.S. Virgin Islands. Coral Reefs 13:81–89

Booth DJ, Brosnan DM (1995) The role of recruitment dynamics in rocky shore and coral reef fish communities. Adv Ecol Res 26:309–385

Booth DJ, Hixon MA (1999) Food ration and condition condition affect early survival of the coral reef damselfish, *Stegastes partitus*. Oecologia 121:364–368

Booth DJ, Wellington G (1998) Settlement preferences in coral-reef fishes: effects on patterns of adult and juvenile distributions, individual fitness, and population structure. Aust J Ecol 23:274–279

Bouchon C, Bouchon-Navarro Y, Louis M (1994) Changes in the coastal fish communities following Hurricane Hugo in Guadeloupe Island (French West Indies). Atoll Res Bull 422:1–13

Butler MJ IV, Hunt JH, Herrnkind WF et al (1995) Cascading disturbances in Florida Bay, USA: cyanobacteria blooms, sponge mortality, and implications for juvenile spiny lobster *Panulirus argus*. Mar Ecol Prog Ser 129:119–125

Butler MJ IV, Herrnkind WF (1991) Effect of benthic microhabitat cues on the metamorphosis of pueruli of the spiny lobster *Panulirus argus*. J Crust Biol 11:23–28

Butler MJ IV, Herrnkind WF (1992) Spiny lobster recruitment in south Florida: field experiments and management implications. Proc Gulf Caribb Fish Inst 41:508–515

Butler MJ IV, Herrnkind WF (1997) A test of recruitment limitation and the potential for artificial enhancement of spiny lobster (*Panulirus argus*) populations in Florida. Can J Fish Aquat Sci 54:452–463

Calinski MD, Lyons WG (1983) Swimming behavior of the puerulus of the spiny lobster *Panulirus argus* (Latreille, 1804) (Crustacea, Palinuridae). J Crust Biol 3:329–335

Caselle JE, Warner RR (1996) Variability in recruitment of coral reef fishes: the importance of habitat at two spatial scales. Ecology 77:2488–2504

Childress MJ, Herrnkind WF (2001) Influence of conspecifics on the ontogenetic habitat shift of juvenile Caribbean spiny lobsters. Mar Freshw Res 52:1077–1084

Choat, JH, Ayling AM, Schiel DR (1988) Temporal and spatial variation in an island fish fauna. J Exp Mar Biol Ecol 121:91–111

Connell JH (1980) Diversity and coevolution of competitors, or the ghost of competition competition past. Oikos 35:131–138

Cruz R, Lalana R, Báez-Hidalgo M et al (2007) Gregarious behaviour of juveniles of the spiny lobster, *Panulirus argus* (Latreille, 1804) in artificial shelters. Crustaceana 80:577–595

Dahlgren C, Eggleston DB (2000) Ecological processes underlying ontogenetic habitat shifts in a coral reef fish. Ecology 81:2227–2240

Dahlgren CP, Eggleston DB (2001) Spatiotemporal variability in abundance, distribution and habitat associations of early juvenile Nassau grouper. Mar Ecol Prog Ser 217:145–156

Dahlgren CP, Kellison GT, Adams AJ et al (2006) Marine nurseries and effective juvenile habitats: concepts and applications. Mar Ecol Prog Ser 312:291–295

Dahlgren CP, Marr J (2004) Back reef systems: important but overlooked components of tropical marine ecosystems. Bull Mar Sci 75:145–152

Doherty PJ (1983) Tropical territorial damselfishes: is density limited by aggression or recruitment? Ecology 64:176–190

Doherty PJ, Dufour V, Galzin R et al (2004) High mortality mortality during settlement is a population bottleneck for a tropical surgeonfish. Ecology 85:2422–2428

Doherty PJ, Sale PF (1985) Predation on juvenile coral reef fishes: an exclusion experiment. Coral Reefs 4:225–234

Done T (1992) Constancy and change in some Great Barrier Reef coral communities. Am Zool 32:655–662

Dorenbosch M, van Riel MC, Nagelkerken I et al (2004) The relationship of reef fish densities to the proximity of mangrove and seagrass nurseries. Estuar Coast Shelf Sci 60:37–48

Dorenbosch M, Grol MGG, Nagelkerken I et al (2006) Seagrass beds and mangroves as potential nurseries for the threatened Indo-Pacific humphead wrasse, *Cheilinus undulatus* and Caribbean rainbow parrotfish, *Scarus guacamaia*. Biol Conserv 129:277–282

Ebersole JP (1977) The adaptive significance of interspecific territoriality in the reef fish *Eupomacentrus leucostictus*. Ecology 58:914–920

Eggleston DB (1995) Recruitment in Nassau grouper *Epinephelus striatus*: post-settlement post-settlement abundance, microhabitat features, and ontogenetic habitat shifts. Mar Ecol Prog Ser 124:9–22

Eggleston DB, Lipcius RN (1992) Shelter selection by spiny lobster under variable predation risk, social conditions and shelter size. Ecology 73:992–1011

Eggleston DB, Lipcius RN, Marshall LS et al (1998) Spatiotemporal variation in postlarval recruitment of the Caribbean spiny lobster in the central Bahamas: lunar and seasonal periodicity, spatial coherence, and wind forcing. Mar Ecol Prog Ser 74:33–49

Elliot JK, Elliot JM, Mariscal RN (1995) Host selection, location and association behaviors of anemonefishes in field settlement experiments. Mar Biol 122:377–389

Feary DA, Almany GR, Jones GP et al (2007) Habitat choice, recruitment and the response of coral reef fishes to coral degradation. Oecologia 153:727–737

Fernandez M, Iribarne OO, Armstrong DA (1994) Swimming behavior of Dungeness crab, *Cancer magister* Dana, megalopae in still and moving water. Estuaries 17:271–275

Fonseca VF, Cabral HN (2007) Are fish early growth and condition patterns related to life-history strategies? Rev Fish Biol Fish 17:545–564

Forrester GE (1990) Factors influencing the juvenile demography of a coral reef fish. Ecology 71:1666–1681

Forrester GE (1995) Strong density-dependent survival and recruitment regulate the abundance of a coral reef fish. Oecologia 103:275–282

Forrester GE (1999) The influence of adult density on larval settlement in a coral reef fish, *Coryphopterus glaucofraenum*. Coral Reefs 18:85–89

Forward RB Jr (1974) Negative phototaxis in crustacean larvae larvae: possible functional significance. J Exp Mar Biol Ecol 16:11–17

Forward RB Jr (1976) A shadow response in a larval crustacean. Biol Bull 151:126–140

Forward RB Jr, Hettler WF Jr (1992) Effects of feeding and predator exposure on photoresponses during diel vertical migration of brine shrimp larvae larvae. Limnol Oceanogr 37:1261–1270

Fouqurean JW, Rutten LM (2004) The impact of Hurricane Georges on soft-bottom, back reef communities: site- and species-specific effects in South Florida Florida seagrass beds. Bull Mar Sci 75:239–257

Fowler AJ, Doherty PJ, Williams DMcB (1992) Multi-scale analysis of recruitment of a coral reef fish on the Great Barrier Reef Great Barrier Reef. Mar Ecol Prog Ser 82:131–141

Friedlander AM, Parrish JD (1998) Temporal dynamics of fish communities on an exposed shoreline in Hawaii. Environ Biol Fish 53:1–18

Garpe KC, Ohman MC (2003) Coral and fish distribution patterns in Mafia Island Marine Park, Tanzania: fish-habitat interactions. Hydrobiologia 498:191–211

Garpe KC, Ohman MC (2007) Non-random habitat use by coral reef fish recruits in Mafia Island Marine Park, Tanzania. Afr J Mar Sci 29:187–199

Garpe KC, Yahya SAS, Lindahl U et al (2006) Long-term effects of the 1998 coral bleaching event on reef fish assemblages. Mar Ecol Prog Ser 315:237–247

Gibson RN (1994) Impact of habitat quality and quantity on the recruitment of juvenile flatfishes. Neth J Sea Res 32:191–206

Gillanders BM, Able KW, Brown JA et al (2003) Evidence of connectivity between juvenile and adult habitats for mobile marine fauna: an important component of nurseries. Mar Ecol Prog Ser 247:281–295

Gimenez L, Anger K, Torres G (2004) Linking life history traits in successive phases of a complex life cycle: effects of larval biomass on early juvenile development in an estuarine crab *Chasmagnathus granulata*. Oikos 104:570–80

Gimenez L (2006) Phenotypic links in complex life cycles: conclusions from studies with decapod crustaceans. Integr Comp Biol 46:615–622

Gratwicke B, Petrovic C, Speight MR (2006) Fish distribution and ontogenetic habitat preferences in non-estuarine lagoons and adjacent reefs. Environ Biol Fish 76:191–210

Green AL (1996) Spatial, temporal and ontogenetic patterns of habitat use by coral reef fishes (Family Labridae). Mar Ecol Prog Ser 133:1–11

Green AL (1998) Spatio-temporal patterns of recruitment of labroid fishes (Pisces: Labridae and Scaridae) to damselfish territories. Environ Biol Fish 51:235–244

Grover JJ, Eggleston DB, Shenker JM (1998) Transition from pelagic to demersal phase in early-juvenile Nassau grouper, *Epinephelus striatus*: pigmentation, squamatation, and ontogeny of diet. Bull Mar Sci 62:97–113

Guttierez L (1998) Habitat selection by recruits establishes local patterns of adult distribution in two species of damselfishes: *Stegastes dorsopunicans* and *S. planifrons*. Oecologia 115:268–277

Hamilton SL, White JW, Caselle JE et al (2006) Consistent long-term spatial gradients in replenishment for an island population of a coral reef fish. Mar Ecol Prog Ser 306:247–256

Harrigan P, Zeiman JC, Macko SA (1989) The base of nutritional support for the gray snapper (*Lutjanus griseus*): an evaluation based on a combined stomach content and stable isotope analysis. Bull Mar Sci 44:65–77

Harvey AW (1996) Delayed metamorphosis in Florida hermit crabs: multiple cues and constraints (Crustacea: Decapoda: Paguridae and Diogenidae). Mar Ecol Prog Ser 141:27–36

Heck KL Jr, Hays G, Orth RJ (2003) Critical evaluation of the nursery role hypothesis for seagrass meadows. Mar Ecol Prog Ser 253:123–136

Helfman GS, Meyer JL, McFarland WN (1982) The ontogeny of twilight migration patterns in grunts (Pisces: Haemulidae). Anim Behav 30:317–326

Herrnkind WF (1980) Movement patterns of palinurid lobsters. In: Cobb JS, Phillips BF (eds) The biology and management of lobsters. Vol. I. Physiology and behavior. Academic Press, New York

Herrnkind WF, Butler MJ IV (1986) Factors regulating postlarval settlement and juvenile microhabitat use by spiny lobsters, *Panulirus argus*. Mar Ecol Prog Ser 34:23–30

Herrnkind WF, Jernakoff P, Butler MJ IV (1994) Puerulus and post-puerulus ecology. In: Phillips BF, Cobb JS, Kittaka J (eds) Spiny lobster management. Blackwell Scientific, Oxford

Hixon MA, Beets JP (1993) Predation, prey refuges, and the structure of coral-reef fish assemblages. Ecol Monogr 63:77–101

Hixon MA, Jones GP (2005) Competition, predation, and density-dependent mortality mortality in demersal marine fishes. Ecology 86:2847–2859

Hixon MA, Menge BA (1991) Species diversity: prey refuges modify the interactive effects of predation and competition. Theor Popul Biol 39:178–200

Hixon MA, Webster MS (2002) Density dependence in reef fishes: coral-reef populations as model systems. In: Sale PF (ed) Coral reef fishes: dynamics and diversity in a complex ecosystem. Academic Press, San Diego

Hoey AS, McCormick MI (2004) Selective predation for low body condition at the larval-juvenile transition of a coral reef fish. Oecologia 139:23–29

Holbrook SJ, Schmitt RJ (2002) Competition for shelter space causes density-dependent predation mortality in damselfishes. Ecology 831:2855–2868

Horner AJ, Nickles SP, Weisburg MJ et al (2006) Source and specificity of chemical cues mediating shelter preference of Caribbean spiny lobsters (*Panulirus argus*). Biol Bull 211:128–139

Huijbers CM, Mollee EM, Nagelkerken I (2008). Post-larval French grunts (*Haemulon flavolineatum*) distinguish between seagrass, mangrove and coral reef water: implications for recognition of potential nursery habitats. J Exp Mar Biol Ecol 357:134–139

Jones GP (1987) Some interactions between residents and recruits in two coral reef fishes. J Exp Mar Biol Ecol 114:169–182

Jones GP (1991) Post recruitment processes in the ecology of coral reef fish populations: a multifactorial perspective. In: Sale PF (ed) The ecology of fishes on coral reefs. Academic Press, New York

Jones GP, McCormick MI, Srinivasan M et al (2004) Coral decline threatens fish biodiversity in marine reserves. Publ Nebraska Acad Sci 101:8251–8253

Jones GP, Syms C (1998) Disturbance, habitat structure and the ecology of fishes on coral reefs. Austral J Ecol 23:287–297

Kaufman LS (1983) Effects of hurricane hurricane Allen on reef fish assemblages near Discovery Bay, Jamaica. Coral Reefs 2:42–47

Kaufman LS, Ebersole JP, Beets J et al (1992) A key phase in the recruitment dynamics of coral reef fishes: post-settlement transition. Environ Biol Fish 34:109–118

Kendall MS, Christensen J, Hillis-Starr Z (2003) Multi-scale data used to analyze the spatial distribution of French grunts, *Haemulon flavolineatum*, relative to hard and soft bottom in a benthic landscape. Environ Biol Fish 66:19–26

Kieckbush DK, Koch MS, Serafy JE et al ((2004) Trophic linkages of primary producers and consumers in fringing mangroves of subtropical lagoons. Bull Mar Sci 74:271–285

Knowlton N, Keller BD (1986) Larvae which fall far short of their potential: highly localized recruitment in an alpheid shrimp with extended larval development. Bull Mar Sci 39:213–223

Knowlton R (1974) Larval developmental processes and controlling factors in decapod Crustacea, with emphasis on Caridea. Thalassia Jugosl 10:139–58

Kobayashi DR (1989). Fine-scale distribution of larval fishes: patterns and processes adjacent to coral reefs in Kaneohe Bay, Hawaii. J Mar Biol 100:285–293

Laegdsgaard P, Johnson CR (2001) Why do juvenile fish preferentially utilise mangrove habitats? J Exp Mar Biol Ecol 257:229–253

Lassig BR (1983) The effects of a cyclonic storm on coral reef fish assemblages. Environ Biol Fish 9:55–63

Layman CA, Arrington DA, Blackwell M (2005) Community-based collaboration restores tidal flow to an island estuary (Bahamas). Ecol Restor 23:58–59

Lecchini D (2006) Highlighting ontogenetic shifts in habitat use by nocturnal coral reef fish. CR Biol 329:265–270

Leis JM, Carson-Ewart BM (1999) In situ swimming and settlement behaviour of larvae of an Indo-Pacific coral-reef fish, the coral trout *Plectropomus leopardus* (Pisces: Serranidae) Mar Biol 134:51–64

Leis JM, Carson-Ewart BM, Webley J (2002) Settlement behaviour of coral-reef fish larvae at subsurface artificial-reef moorings. Mar Freshw Res 53:319–327

Leis JM, Lockett MM (2005) Localization of reef sounds by settlement-stage larvae of coral-reef fishes (Pomacentridae). Bull Mar Sci 76:715–724

Letourneur Y, Chabanet P, Vigliola L et al (1998) Mass settlement and post-settlement mortality of *Epinephelus merra* (Pisces: Serranidae) on Reunion coral reefs. J Mar Biol Assoc UK 78: 307–319

LetourneurY, Harmelin-Vivien M, Galzin R et al (1993) Impact of Hurricane Firinga on fish community structure on fringing reefs of Reunion Island, SW Indian Ocean. Environ Biol Fish 37: 109–120

Lewis AR (1997a) Recruitment and post-recruitment immigration affect the local population size of coral reef fishes. Coral Reefs 16:139–149

Lewis AR (1997b) Effects of experimental coral disturbance on the structure of fish communities on large patch reefs. Mar Ecol Prog Ser 161:37–50

Ley JA, McIvor CC, Montague CL (1999) Fishes in mangrove prop-root habitats of northeastern Florida Bay: disinct assemblages across an estuarine gradient. Estuar Coast Shelf Sci 48: 701–723

Lipcius RN, Eggleston DB (2000) Ecology and fishery biology of spiny lobster. Spiny lobsters: fisheries and culture, pp.1–41. In: Phillips BF, Kittaka J (eds)., Fishing News Books, Blackwell Science.

Lipcius RN, Eggleston DB, Miler DL et al (1998) The habitat-survival function for Caribbean spiny lobster: an inverted size effect and non-linearity in mixed algal and seagrass habitats. Mar Freshw Res 49:807–816

Lirman D (1994) Ontogenetic shifts in habitat preferences in the three-spot damselfish, *Stegastes planifrons* (Cuvier), in Roatan Island, Honduras. J Exp Mar Biol Ecol 180:71–81

Marx JM (1983) Macroalgal communities as habitat for early benthic spiny lobsters, *Panulirus argus*. M.Sc. thesis, Florida State University, Tallahassee

Marx JM, Herrnkind WF (1985a) Macroalgae (Rhodophyta: *Laurencia* spp.) as habitat for young juvenile spiny lobsters, *Panulirus argus*. Bull Mar Sci 86:423–431

Marx JM, Herrnkind WF (1985b). Factors regulating microhabitat use by young juvenile spiny lobsters. J Crust Biol 5:650–657

Marx JM, Herrnkind WF (1986) Species profiles: life histories and environmental requirements of coastal fishes and invertebrates (South Florida) – spiny lobster. U.S. Fish and Wildlife Service Biol Rep 82(11.61), 21pp.

McCormick MI (1994) Variability in age and size at settlement of the tropical goatfish *Upeneus tragula* (Mullidae) in the northern Great Barrier Reef lagoon. Mar Ecol Prog Ser 103:1–15

McCormick MI (1999) Delayed metamorphosis of a tropical reef fish (*Acanthurus triostegus*): a field experiment. Mar Ecol Prog Ser 176:25–38

McCormick MI, Hoey AS (2004) Larval growth history determines juvenile growth and survival in a tropical marine fish. Oikos 106:225–242

McCormick MI, Makey LJ (1997) Post-settlement transition in coral reef fishes: overlooked complexity in niche shifts. Mar Ecol Prog Ser 153:247–257

McCormick MI, Molony BW (1992) Effects of feeding history on the growth characteristics of a reef fish at settlement. Mar Biol 114:165–173

McCormick MI, Molony BW (1993) Quality of the reef fish *Upeneus tragula* (Mullidae) at settlement: is size a good indicator of condition condition? Mar Ecol Prog Ser 98:45–54

McFarland WN (1980) Observations on recruitment in haemulid fishes. Proc Gulf Caribb Fish Inst 3:132–138

McFarland WN, Brothers EB (1985) Recruitment patterns in young French grunts, *Haemulon flavolineatum* (family Haemulidae), at St. Croix, Virgin Islands. Fish Bull 83:151–161

McGehee MA (1995) Juvenile settlement, survivorship and in situ growth rates of four species of Caribbean damselfishes in the genus *Stegastes*. Environ Biol Fish 44: 393–401

Minello TJ, Zimmerman R J (1983b) Selection for brown shrimp, *Penaeus aztecus*, as prey by the spotted sea trout *Cynoscion nebulosus*. Contrib Mar Sci 27:159–167

Minello TJ, Zimmerman RJ (1983) Differential selection for vegetative structure between juvenile brown shrimp (*Penaeus aztecus*) and white shrimp (*P. setiferus*) and implications in predator-prey relationships. Estuar Coast Shelf Sci 20:707–716

Minello TJ, Zimmerman RJ (1983a) Fish predation predation on juvenile brown shrimp, *Penaeus aztecus* Ives: the effect of simulated *Spartina* structure on predation rates. J Exp Mar Biol Ecol 72:211–231

Minello TJ, Zimmerman RJ (1985) Differential selection for vegetative structure between juvenile brown shrimp (*Penaeus aztecus*) and white shrimp (*P. setiferus*), and implications in predator-prey relationships. Estuar Coast Shelf Sci 20:707–716

Minello TJ, Zimmerman RJ, Martinez EX (1989) Mortality of young brown shrimp *Penaeus aztecus* in estuarine nurseries. Trans Am Fish Soc 118:693–708

Mintz JD, Lipcius RN, Eggleston DB et al (1994) Survival of juvenile Caribbean spiny lobster: effects of shelter size, geographic location and conspecific abundance. Mar Ecol Prog Ser 12:255–266

Montgomery JC, Tolimieri N, Haine OS (2001) Active habitat selection by pre-settlement reef fishes. Fish Fish. 2:261–277

Montgomery SS, Craig JR (2005) Distribution and abundance of recruits of the eastern rock lobster (*Jasus verreauxi*) along the coast of New South Wales, Australia. N Z J Mar Freshw Res 39:619–628

Mumby PJ, Edwards AJ, Arias-Gonzalez JE et al (2004) Mangroves enhance the biomass of coral reef fish communities in the Caribbean Caribbean. Nature 427:533–536

Mumby PJ (2006) Connectivity of reef fish between mangroves and coral reefs: algorithms for the design of marine reserves at seascape scales. Biol Conserv 128:215–222

Munday PL (2004a) Competitive coexistence of coral-dwelling fishes: the lottery hypothesis revisited. Ecology 85:623–628

Munday PL (2004b). Habitat loss, resource specalization, and extinction on coral reefs. Glob Chang Biol 10:1642–1647

Munro JL, Gaut VC, Thompson R et al (1973) The spawning seasons of Caribbean reef fishes. J Fish Biol 5:69–84

Nagelkerken I, van der Velde G (2002) Do non-estuarine mangroves harbour higher densities of juvenile fish than adjacent shallow-water and coral reef habitats in Curaçao (Netherlands Antilles)? Mar Ecol Prog Ser 245:191–204

Nagelkerken I, van der Velde G (2004) Relative importance of interlinked mangroves and seagrass beds as feeding habitats for juvenile reef fish on a Caribbean island. Mar Ecol Prog Ser 274:153–159

Nagelkerken I, Dorenbosch M, Verberk WCEP et al (2000a) Day–night shifts of fishes between shallow-water biotopes of a Caribbean bay, with emphasis on the nocturnal feeding of Haemulidae and Lutjanidae. Mar Ecol Prog Ser 194:55–64

Nagelkerken I, Dorenbosch M, Verberk WCEP et al (2000b) Importance of shallow-water biotopes of a Caribbean bay for juvenile coral reef fishes: patterns in biotope association, community structure and spatial distribution. Mar Ecol Prog Ser 202:175–192

Nagelkerken I, Kleijnen S, Klop T et al (2001) Dependence of Caribbean reef fishes on mangroves and seagrass beds as nursery habitats: a comparison of fish faunas between bays with and without mangroves/seagrass beds. Mar Ecol Prog Ser 214:225–235

Nagelkerken I, Roberts CM, van der Velde G et al (2002) How important are mangroves and seagrass beds for coral reef fish? The nursery hypothesis tested on an island scale. Mar Ecol Prog Ser 244:299–305

Nemeth RS (1998) The effect of natural variation in substrate architecture on the survival of juvenile bicolor damselfish. Environ Biol Fish 53:129–141

Nemeth RS (2005) Linking larval history to juvenile demography in the bicolor damselfish *Stegastes partitus* (Perciformes: Pomacentridae). Rev Biol Trop 53:155–163

Nevitt G, Pentcheff ND, Lohmann KJ et al (2000) Den selection by the spiny lobster *Panulirus argus*: testing attraction to conspecific odors in the field. Mar Ecol Prog Ser 203:225–231

Ogden JC, Ehrlich PR (1977) The behavior of heterotypic resting schools of juvenile grunts (Pomadasyidae). Mar Biol 42:273–280

Osenberg CW, St. Mary CM, Schmitt RJ (2002) Rethinking ecological inference: density dependence in reef fishes. Ecol Lett 5:715–721

Overholtzer KL, Motta P (1999) Comparative resource use by juvenile parrotfishes in the Florida Keys. Mar Ecol Prog Ser 69:177–187

Overholtzer-McCleod KL (2004) Variance in reef spatial structure masks density dependence in coral reef fish populations on natural vs artificial reefs. Mar Ecol Prog Ser 276:269–280

Overholtzer-McCleod KL (2005) Post-settlement emigration affects mortality estimates for two Bahamian wrasses. Coral Reefs 24:283–291

Overholtzer-McCleod KL (2006) Consequences of patch reef spacing for density-dependent mortality of coral-reef fishes. Ecology 87:1017–1026

Parrish JD (1989) Fish communities of interacting shallow habitats in tropical ocean regions. Mar Ecol Prog Ser 58:143–160

Planes S, Lecaillon G (2001) Caging experiment to examine mortality during metamorphosis of coral reef fish larvae. Coral Reefs 20:211–218

Planes S, Levefre A, Legendre P et al (1993) Spatio-temporal variability in fish recruitment to a coral reef (Moorea, French Polynesia). Coral Reefs 12:105–113

Pollux BJA, Verberk WCEP, Dorenbosch M et al (2007) Habitat selection during settlement of three Caribbean coral reef fishes: indications for direct settlement to seagrass beds and mangroves. Limnol Oceanogr 52:903–907

Preston NP, Doherty PJ (1990) Cross-shelf patterns in the community structure of coral-dwelling Crustacea in the central region of the Great Barrier Reef Great Barrier Ref. I. Agile shrimps. Mar Ecol Prog Ser 66:47–61

Reyns N, Sponaugle S (1999) Patterns and processes of brachyuran crab settlement to Caribbean coral reefs. Mar Ecol Prog Ser 185:155–170

Risk A (1997) Effects of habitat on the settlement and post-settlement success of the ocean surgeonfish *Acanthurus bahianus*. Mar Ecol Prog Ser 161:51–59

Risk A (1998) The effects of interactions with reef residents on the settlement and subsequent persistence of ocean surgeonfish, *Acanthurus bahianus*. Environ Biol Fish 51:377–389

Robertson AI, Duke NC (1987) Mangroves as nursery sites: comparisons of the abundance and species composition of fish and crustaceans in mangroves and other nearshore habitats in tropical Australia. Mar Biol 96:193–205

Robertson DR (1988) Abundances of surgeonfishes on patch-reefs in Caribbean Panama: due to settlement, or post-settlement events? Mar Biol 97:495–501

Robertson DR (1991) Increases in surgeonfish populations after mass mortality of the sea urchin *Diadema antillarum* in Panama indicate food limitation. Mar Biol 111:437–444

Rodriguez RW, Webb RMT, Bush DM (1994) Another look at the impact of Hurricane Hugo on the shelf and coastal resources of Puerto Rico, USA. J Coast Res 10:278–296

Rooker JR, Dennis GD (1991) Diel, lunar and seasonal changes in a mangrove fish assemblage off southwestern Puerto Rico. Bull Mar Sci 49:684–698

Sale PF (1977) Maintenance of high diversity in coral reef fish communities. Am Nat 111:337–359

Sale PF (1978) Coexistence of coral reef fishes: the lottery for living space. Environ Biol Fish 3:85–102

Sancho G, Ma D, Lobel PS (1997) Behavioral observations of an upcurrent reef colonization event by larval surgeonfish *Ctenochaetus strigosus* (Acanthuridae). Mar Ecol Prog Ser 153:311–315

Schmitt RJ, Holbrook SJ (1999a) Temporal patterns of settlement of three species of damselfish of the genus *Dascyllus* (Pomacentridae) in the coral reefs of French Polynesia. Proc 5th Indo-Pac Fish Confific, pp. 537–551

Schmitt RJ, Holbrook SJ (1999b) Settlement and recruitment of three damselfish species: larval delivery and competition for shelter space. Oecologia 118:76–86

Schmitt RJ, Holbrook SJ (1999c). Mortality of juvenile damselfish: implications for assessing processes that determine abundance. Ecology 80:35–50

Schmitt RJ, Holbrook SJ (2000) Habitat-limited recruitment of coral reef damselfish. Ecology 81:3479–3494

Searcy S, Sponaugle S (2001) Selective mortality during the larval-juvenile transition in two coral reef fishes. Ecology 82:2452–2470

Serafy JE, Faunce CH, Lorenz JJ (2003) Mangrove shoreline fishes of Biscayne Bay, Florida. Bull Mar Sci 72:161–180

Shanks AL (1995) Mechanisms of cross-shelf dispersal of larval invertebrates and fish. In: McEdward L (ed) Ecology of marine invertebrate Larvae. CRC-Press, Boca Raton, Florida. Mar Sci Ser 6

Sheaves M (1995) Large lutjanid and serranid fishes in tropical estuaries: are they adults or juveniles? Mar Ecol Prog Ser 129:31–40

Shenker JM, Cowie-Mojica E, Crabtree RE et al (2002) Recruitment of tarpon (*Megalops atlanticus*) leptocephali into the Indian River Lagoon, Florida. Contrib Mar Sci 35:55–69

Shervette VR, Perry HM, Rakocinski CF et al (2004) Factors influencing refuge occupation by stone crab *Menippe adina* juveniles in Mississippi Sound. J Crust Biol 24:652–665

Shima JS, Osenberg CW, St. Mary CM et al (2006) Implication of changing coral communities: do larval traits or habitat features drive variation in density-dependent mortality and recruitment of juvenile reef fish? Proc 10th Int Coral Reef Symp, pp. 226–231

Shulman MJ (1985a) Recruitment of coral reef fishes: distribution of predators and shelter. Ecology 66:1056–1066

Shulman MJ (1985b) Variability in recruitment of coral reef fishes. J Exp Mar Biol Ecol 89: 205–219

Shulman MJ, Ogden JC (1987) What controls tropical reef fish populations: recruitment or mortality mortality? An example in the Caribbean reef fish *Haemulon flavolineatum*. Mar Ecol Prog Ser 39:233–242

Shulman MJ, Ogden JC, Ebersole JP et al (1983) Priority effects in the recruitment of coral reef fishes. Ecology 64:1508–1513

Simberloff D (1983) Competition theory, hypothesis-testing and other community ecological buzzwords. Am Nat 122:626–635

Sluka R, Chiappone M (1998) Density, species and size distribution of groupers (Serranidae) in three habitats at Elbow Reef, Florida Keys. Bull Mar Sci 62:219–228

Sogard SM (1997) Size-selective mortality in the juvenile stage of teleost fishes: a review. Bull Mar Sci 60:1129–157

Sponaugle S, Cowen RK (1996) Nearshore patterns of coral reef fish larval supply to Barbados, West Indies. Mar Ecol Prog Ser 133:13–28

Sponaugle S, Grorud-Colvert K (2006) Environmental variability, early life-history traits, and survival of new coral reef fish recruits. Integr Comp Biol 46:623–633

Sponaugle S, Grorud-Colvert K, Pinkard D (2006) Temperature-mediated variation in early life history traits and recruitment success of the coral reef fish *Thalassoma bifasciatum* in the Florida Keys. Mar Ecol Prog Ser 308:1–15

St. John J (1999) Ontogenetic changes in the diet of the coral reef grouper *Plectropomus leopardus* (Serranidae): patterns in taxa, size and habitat of prey. Mar Ecol Prog Ser 180: 233–246

Stobutzki IC (1998) Interspecific variation in sustained swimming ability of late pelagic stage reef fish from two families (Pomacentridae and Chaetodontidae). Coral Reefs 17:111–119

Stobutzki IC Bellwood DR (1997) Sustained swimming abilities of the late pelagic stages of coral reef fishes. Mar Ecol Prog Ser 149:35–41

Stoner AW (2003) What constitutes essential nursery habitat for a marine species? A case study of habitat form and function for queen conch. Mar Ecol Prog Ser 257:275–289

Sweatman HP (1985) The influence of adults of some coral reef fishes on larval recruitment. Ecol Monogr 55:469–485

Sweatman HP (1988) Field evidence that settling coral reef fish larvae larvae detect resident fishes using dissolved chemical cues. J Exp Mar Biol Ecol 124:163–174

Sweatman HP, Robertson DR (1994) Grazing halos and predation on juvenile Caribbean surgeon-fishes. Mar Ecol Prog Ser 111:1–6

Tolimieri N (1998a) The relationship among microhabitat characteristics, recruitment and adult abundance in the stoplight parrotfish, *Sparisoma viride*, at three spatial scales. Bull Mar Sci 62:253–268

Tolimieri N (1998b) Contrasting effects of microhabitat use on large-scale adult abundance in two families of Caribben reef fishes. Mar Ecol Prog Ser 167:227–239

Tolimieri N, Jeffs A, Montgomery JC (2000) Ambient sound as a cue for navigation by the pelagic larvae of reef fishes. Mar Ecol Prog Ser 207:219–224

Tolimieri N, Sale PF, Nemeth RS et al (1998) Replenishment of populations of Caribbean reef fishes: are spatial patterns of recruitment consistent through time? J Exp Mar Biol Ecol 230:55–71

Tupper M, Juanes F (1999) Effects of a marine reserve on recruitment of grunts (Pisces: Haemuli-dae) at Barbados, West Indies. Environ Biol Fish 55:53–63

Valiela I, Bowen JL, York JK (2001) Mangrove forests: one of the world's most threatened major tropical environments. BioScience 51:807–815

Vigliola L, Doherty PJ, Meekan MG et al (2007) Genetic identity determines risk of post-settlement mortality of a marine fish. Ecology 88:1263–1277

Vigliola L, Harmelin-Vivien ML, Biagi F et al (1998) Spatial and temporal patterns of settlement among sparid fishes of the genus *Diplodus* in the northwestern Mediterranean. Mar Ecol Prog Ser 168:45–56

Vigliola L, Meekan MG (2002) Size at hatching and planktonic growth determine post-settlement survivorship of a coral reef fish. Oecologia 131:89–93

Warner RR, Chesson PL (1985) Coexistence mediated by recruitment fluctuations: a field guide to the storage effect. Am Nat 125:769–787

Webster MS (2002) Role of predators in the early postsettlement demography of coral-reef fishes. Oecologia 131:52–60

Welch JM, Rittschof D, Bullock TM et al (1997) Effects of chemical cues on settlement behavior of blue crab *Callinectes sapidus* postlarvae. Mar Ecol Prog Ser 154:143–153

Zimmerman RJ, Minello TJ, Zamora G Jr (1984) Selection of vegetated habitat by brown shrimp, *Penaeus aztecus*, in a Galveston Bay salt marsh. Fish Bull 82:325–336

Chapter 7
Habitat Shifts by Decapods—an Example of Connectivity Across Tropical Coastal Ecosystems

Michael D.E. Haywood and Robert A. Kenyon

Abstract Decapod life cycles are complex and many utilize a range of habitats throughout their development. Many species tend to settle on shallow (often vegetated) inshore habitats and commonly move offshore into deeper water as they grow. The species that exhibit an inshore/offshore life history are often large individuals and may support commercial fisheries. In this chapter the habitats of a range of tropical decapods are described and likely mechanisms underlying habitat shifts are discussed. It is generally accepted that in most animals these mechanisms are aligned with maximizing the animal's fitness. Possible mechanisms include minimizing mortality risk (μ), maximizing absolute growth rates (g), or a trade-off in which the animal chooses the habitat that minimizes the ratio of mortality risk to growth rate (minimize μ/g). There do not appear to be any studies that address these issues for tropical decapods and this is identified as an important topic for future research. Similarly, studies that have explicitly demonstrated habitat shifts in tropical decapods are rare; most shifts have been implied by comparing length frequencies in different habitats and it is recommended that future studies consider the use of natural and artificial tags to assist in more accurate characterization of connectivity between coastal habitats.

Keywords Decapoda · Ontogenetic shift · Habitat · Life history

7.1 Introduction

An organism's ability to utilize resources and reduce predation risk are related to its body size and habitat, and as a consequence, many organisms (including tropical

M.D.E Haywood (✉)
CSIRO Division of Marine and Atmospheric Research, P.O. Box 120, Cleveland, 4163, Queensland, Australia
e-mail: mick.haywood@csiro.au

We would like to dedicate this chapter to the memory of Burke Hill whose passion for decapod research was inspiring to many of those working in the field.

decapods) shift their distribution during their development as a function of varying predation risk and potential growth (Werner and Gilliam 1984, Pardieck et al. 1999, Dahlgren and Eggleston 2000). The transition of organisms through different habitats during their life cycle results in a substantial transfer of organic matter, nutrients, and energy across a variety of ecosystems ranging from estuarine to oceanic waters (Deegan 1993, Fairweather and Quinn 1993). In this chapter we describe a range of ontogenetic habitat shifts in tropical decapods, discuss reasons for making habitat shifts, and highlight some areas for future research. A large portion of the published information on tropical decapod life history characteristics is concerned with commercially important lobsters, prawns, and crabs and so most of the examples described here come from this rather narrow range of species. Hermit and land crabs are also discussed because they represent what is probably the most extreme example of an ontogenetic habitat shift—moving from the marine to the terrestrial environment.

Decapods have rather complex life cycles, usually involving several pelagic larval stages, which settle as post-larvae, develop into juveniles and ultimately, adults (Figs 7.1, 7.2; see Chapter 6). Many species undergo migrations to different habitats as they develop, and some undergo extensive migrations as adults to spawn. In most cases, the eggs usually share their mother's habitat because on extrusion they are typically attached to her pleopods with a cementing material where they remain until hatching. Penaeids provide the exception to this rule as their eggs are shed

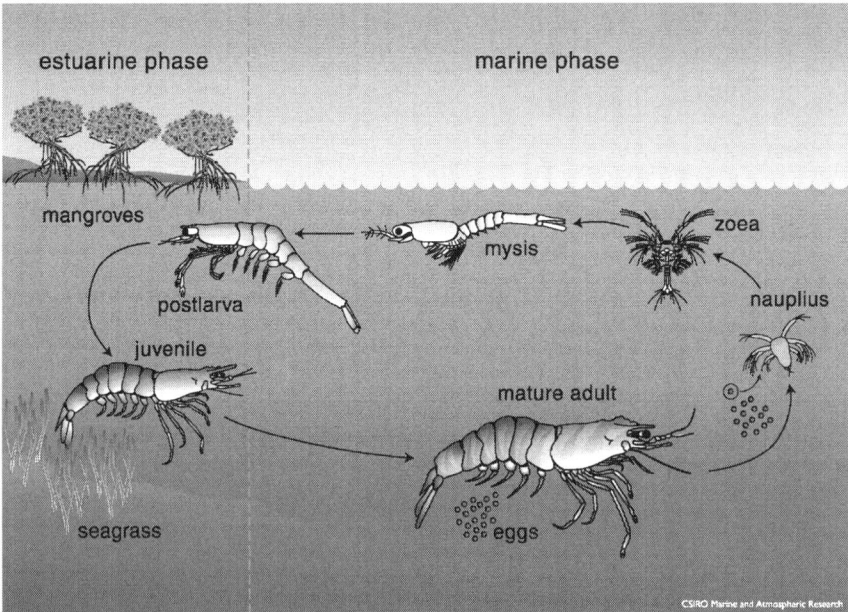

Fig. 7.1 Life cycle of a typical penaeid, with kind permission of CSIRO Marine and Atmospheric Research

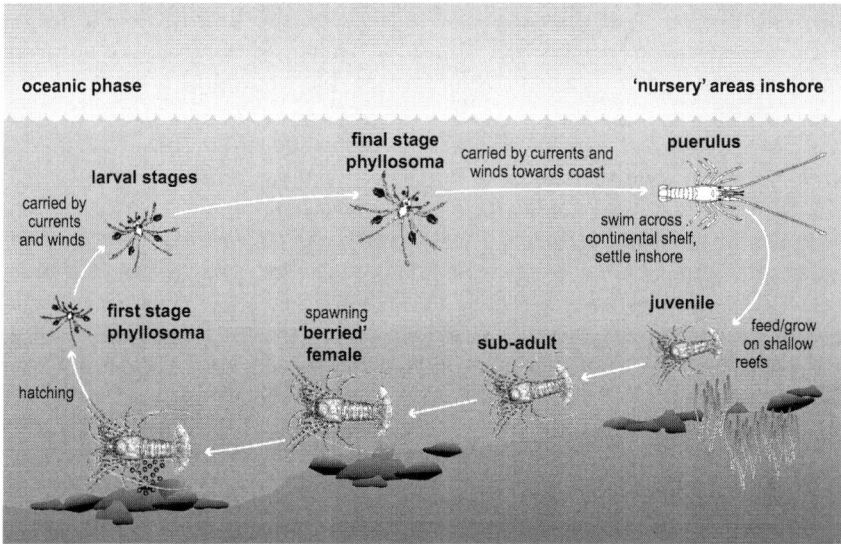

Fig. 7.2 Life cycle of the ornate rock lobster (*Panulirus ornatus*)

directly into the water column where they may sink to the seabed or in some species the eggs are pelagic (Dall et al. 1990).

The hatching stage of decapods varies greatly. In some penaeids and sergestids, the eggs hatch as nauplia, metanauplia, or protozoea larvae, whereas in virtually all other marine decapods the first stage is a protozoea or zoea as the nauplial stage takes place within the egg (Barnes 1974; Table 7.1).

Table 7.1 Types of larval stages of Decapoda (after Waterman and Chase 1960)

Group	Larval forms
Suborder Natantia	
Family Penaeidae	Nauplius → protozoea → mysis (zoea) → mastigopus (post-larva)
Family Sergestidae	Nauplius → elaphacaris (protozoea) → acanthosoma (zoea) → post-larva
Section Caridea	Protozoea → zoea → parva (post-larva)
Section Stenopodidea	Protozoea → zoea → post-larva
Suborder Reptantia	
Section Macrura	
Superfamily Scyllaridae	Phyllosoma (zoea) → puerulus (post-larva)
Superfamily Nephropsidae	Mysis (zoea) → post-larvae
Section Anomura	Zoea → glaucothöe in pagurids, grimnothea in others
Section Brachyura	Zoea → megalopa (post-larva)

Despite the variety of different larval stages in decapods, all seem to be plank-tonic and spend between days (e.g., Penaeidae; Dall et al. 1990) and months (e.g., some species of Scyllaroidea and Palinuridae; Chittleborough and Thomas 1969) in the pelagic environment. Advection of these stages to nursery habitat may be facilitated by selective tidal stream transport (Forward and Tankersley 2001, Jeffs et al. 2005). Whether the late stages are able to actively choose suitable benthic habitats based on environmental triggers has been investigated, yet remains mostly unknown. Experimental techniques, that demonstrate a distinct response to a trig-ger, have proved difficult to develop, and settlement cues remain a topic for future study (see Chapter 5). Settlement onto the seabed occurs after metamorphosis into the post-larval stage and in some species the settlement habitat is where the animal remains. More commonly, however, tropical decapods undergo a number of habitat shifts as they develop. Pittman and McAlpine (2003) characterized coastal fish and decapods as having a tri-phasic life cycle involving:

(1) Movement of planktonic eggs and larvae to nursery areas;
(2) A range of shelter and foraging movements that maintain a home range;
(3) Spawning migrations away from the home range to close the life cycle.

In many tropical decapods the second stage can be divided into two: post-larvae set-tle in a particular habitat and may remain there as early-stage juveniles, but subse-quently move to an alternative habitat as they grow into late stage juveniles (Staples and Vance 1986, Butler and Herrnkind 2000).

While a larval stage is seen as a strategy to facilitate dispersal, each phase in the life history has evolved to exploit suitable habitats (see Chapter 6). Pittman and McAlpine (2003) discuss a 'multi-phase ontogeny, in which each life history is char-acterized by changes in morphology, physiology, and behavior'. For example, Dall et al. (1990) suggest penaeid life history strategies can be characterized into four types according to the ecosystem occupied by the various life history stages: Type 1: wholly estuarine, Type 2: estuarine/offshore, moving through inshore habitats, Type 3: inshore/offshore, Type 4: wholly offshore.

Ontogenetic changes in habitat occur as part of the multi-phase life history of decapods. For animals that live in both estuarine and shallow continental shelf waters during different stages of their life cycle, the shift in habitat represents a shift from one regime of ecosystem processes to another. Such a major habitat change is typical of the penaeid prawn with a Type 2 or 3 life history strategy (estuar-ine/inshore/offshore; Dall et al. 1990).

For example, the pelagic larval/post-larval phase of many prawn species under-goes a significant ontogenetic habitat shift from an offshore pelagic habitat, to a shallow inshore pelagic habitat, to a shallow inshore epibenthic habitat. The shift from a pelagic to an epibenthic habitat is associated with the metamorphosis to the juvenile stage. The juvenile phase then undergoes an ontogenetic habitat shift usu-ally associated with a subtle change in the use of inshore habitats and an increase in water depth that the individual occupies as it grows. Many species undergo another habitat shift associated with the transition from juvenile to adult. This often involves movement from inshore vegetated areas to unvegetated areas in deeper water. The adults then undergo further habitat shift as they increase in size, usually associated with an offshore migration from nearshore to deeper waters.

From an evolutionary perspective, the difference in strategy between species that do not shift ecosystem during their life cycle to those that do is significant. For species that move from offshore to inshore during their larval phase (e.g., Type 2 and 3 penaeids; Dall et al. 1990), the risk of predation while exposed in the water column during inshore advection is high. Subsequently, juveniles moving from the estuary/inshore to the offshore environment are also exposed to predators. In species that spend their entire life cycle either wholly offshore or inshore (e.g., Type 1 and 4 penaeids; Dall et al. 1990), the larval and juvenile phases do not move far, probably remaining demersal for the majority of time and consequently are less exposed to predation.

Yet, species with a Type 2 and 3 life history strategy have evolved to migrate inshore/offshore. These species include the large commercially-harvested penaeid prawns. These animals have a relatively large body size and fast growth rates. They access the refugia that occur in vegetated inshore habitats as post-larvae/juveniles, at a developmental stage when they are less able to evade predators (Kenyon et al. 1995). As larger individuals, they move offshore to access an ecosystem that offers greater opportunity for large animals to forage and grow. From an evolutionary perspective, this strategy must provide an advantage for the species compared to, for example, an exclusive estuarine life history. The advantage provided must be greater than the risk associated with the migrations between the estuarine and off-shore ecosystems.

Few ontogenetic habitat shifts are more exceptional than those demonstrated by terrestrial crabs, and the shifts best demonstrate the evolutionary component of an ontogenetic change in habitat. The hermit and land crabs have evolved the strategy of a terrestrial existence to their benefit. They avoid marine predators and access food and shelter on land. Many of the tropical terrestrial crabs are found on small islands that were created as geographically isolated entities over geologic time (millions of years). The islands remained separated from major land masses, surrounded by ocean which acted as a barrier to the development of a diverse terrestrial fauna. Both colonization from adjacent land masses and a radiation of species (including marine crabs) through evolutionary processes to colonize niches in unoccupied terrestrial habitat contribute current species to island fauna (Hedges 1989, Schubart et al. 1998).

7.2 Early Juvenile Habitats

Decapods undergo a dramatic habitat shift when the pelagic larval stage metamorphoses and adopts a benthic existence; in some species this may also involve a substantial physical transformation. For example, after spending between 9 and 12 months in the plankton the palinurid lobster *Panulirus cygnus* changes from a transparent leaf-like phyllosoma larva to a puerulus which resembles the adult in general body form (Chittleborough and Thomas 1969). The pueruli become pigmented and settle on inshore reefs where they molt into benthic-phase juveniles. In contrast, the mysis larva of penaeids superficially resembles the post-larval and juvenile form. Post-larval penaeids settle on inshore and estuarine areas, often amongst macrophytes or mangroves (Staples et al. 1985). Many decapods also undergo a habitat shift during the juvenile phase. The first habitat is usually similar to the settlement habitat, whereas the second may involve a relatively subtle movement into deeper water, e.g., tiger prawns (*Penaeus*[1] *semisulcatus* and *P. esculentus*) (Loneragan et al. 1998), or a distinct shift such as that exhibited by spiny lobsters which move from a macrophyte environment (Marx and Herrnkind 1985b) to crevices in the reef (Herrnkind et al. 1975).

The habitats of recently-settled juvenile tropical decapods are often dominated by vegetation of some sort—mangroves, seagrasses, or algae. These will now be discussed in more detail.

7.2.1 Pelagic to Algae, Seagrass, or Saltmarsh

In southern Florida, USA, Caribbean spiny lobsters (*Panulirus argus*) settle onto algal (generally the red algae *Laurencia*) or seagrass-dominated reefs. The puerulus metamorphoses within days to the first benthic instar which has a carapace length (CL) of between 6 and 7 mm (Herrnkind and Butler 1986)[2]. Their small size and disruptive coloration makes them cryptic amongst the algae or seagrass and unlike their larger conspecifics, they maintain a solitary existence (Marx and Herrnkind 1985b). In some regions such as Belize, juvenile *P. argus* do not settle in algal beds, instead they use seagrass, corals, or the fouled prop roots of mangroves (Acosta and Butler 1997). Algal-stage juveniles of the ornate rock lobster (*Panulirus ornatus*) also tend to favor macrophyte-dominated reefs and are found sheltering in solution holes in limestone reef, surrounded by either macroalgae (mainly *Sargassum* sp. or *Padina* sp.) or seagrass (*Cymodocea rotundata*, *Syringodium isoetifolium*, *Halophila ovalis*, and *H. spinulosa*) (Dennis et al. 1997; Fig. 7.3). Similarly, recently-settled juveniles of the western rock lobster (*Panulirus cygnus*) are also found in small holes or crevices on limestone reef surrounded by macroalgae or seagrass (Fitzpatrick et al. 1989, Jernakoff 1990).

The camouflaged nature of their coloration and the fact that they are normally solitary and widely dispersed following settlement makes algal-stage spiny lobsters

---

[1] The subgenera of *Penaeus* were elevated to genera by Pérez-Farfante and Kensley (1997). However, as there is some controversy over this revision we have chosen to use the old names in this paper (Lavery et al. 2004, W Dall, CSIRO Marine and Atmospheric Research, pers. comm.).

[2] Various researchers around the world have adopted different terminologies for the juvenile lobster stages. In this article we have chosen to adopt the North American terms: algal (generally <15 mm CL) = post puerulus, post-algal (15–45 mm CL) = early stage juvenile, subadult (45–80 mm CL), adult (>80 mm CL).

Fig. 7.3 An algal-stage ornate rock lobster (*Panulirus ornatus*) sheltering in a solution hole. Note that the hole diameter is scaled to the lobster's body size

very difficult to study in the wild and so information on their specific habitat requirements, behavior, and population dynamics is limited (Marx and Herrnkind 1985b, Butler and Herrnkind 2000).

The early juvenile stages of many species of penaeid also utilize algal or seagrass habitats. In the Gulf of Mexico, juvenile brown shrimp (*Penaeus aztecus*) are abundant on seagrasses (*Thalassia testudinum, Halodule wrightii, Syringodium filiforme, Halophila engelmanni,* or *Ruppia maritime*) and saltmarsh (*Spartina alterniflora*) edge (Minello 1999). White shrimp (*P. setiferus*) were found in highest densities on the saltmarsh (edge and inner), mixed marsh edge vegetation, and bare substrate, while pink shrimp (*P. duorarum*) were found on saltmarsh edge and seagrasses (Minello 1999). In the Embley River, in tropical northern Australia, grooved (*P. semisulcatus*) and brown (*P. esculentus*) tiger prawns were found on seagrass, whereas endeavor prawns (*Metapenaeus endeavouri*) were most common on seagrass, but also occurred on algal beds and mangrove-lined mud banks (Staples et al. 1985; Fig. 7.4). The red-spot king prawn (*Penaeus longistylus*) also settles onto seagrass beds, although generally only on seagrass beds that overlie coral reef platforms (Coles et al. 1987).

On occasion tiger prawn post-larvae may not be obligate selectors for seagrass at settlement, although they do show a preference for shallow inshore areas. Post-larval tiger prawns in waters around Groote Eylandt (northern Australia) tended to concentrate in intertidal and shallow (<2.0 m) waters (Loneragan et al. 1994). All tiger prawn post-larvae were found within 200 m of the high water mark, many on seagrass, but some on bare substrate, despite that fact that high biomass seagrass beds were nearby, in only slightly deeper water (2.5 m; Loneragan et al. 1994). The post-larvae found on bare substrate were small (≤1.9 mm CL) recently-settled

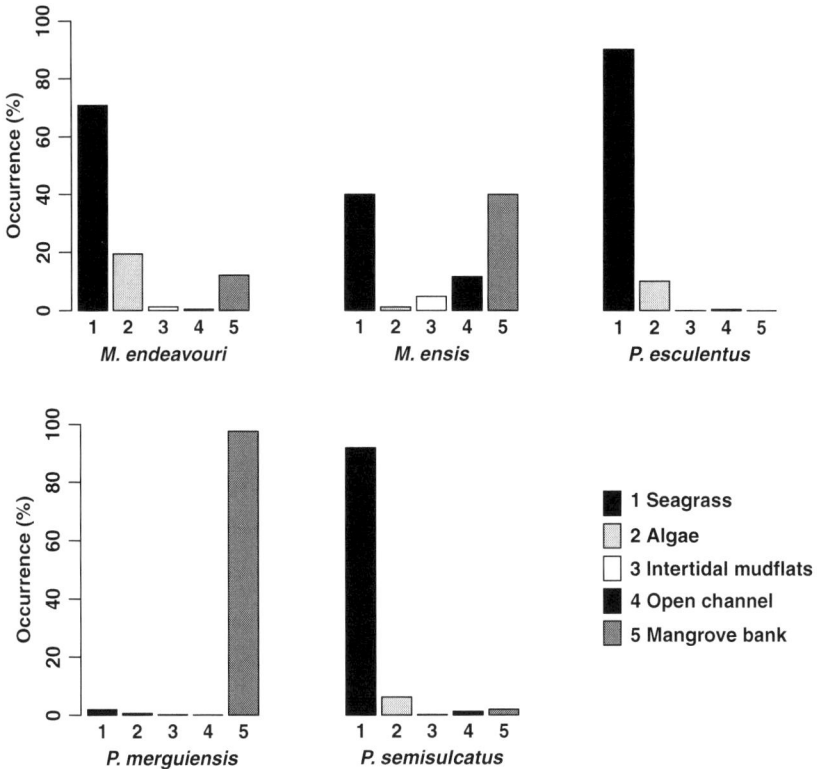

Fig. 7.4 Percentage distribution of the catch of the five major commercial prawn species in each of five habitats in the Embley River, northern Australia 1981–1982. Redrawn from Staples et al. (1985), with kind permission of CSIRO Marine and Atmospheric Research

post-larvae; there were few larger (2–2.9 mm CL) post-larvae and no juveniles were found on the bare substrate. These results suggest the recently-settled prawns either moved or perished on the bare substrate (Loneragan et al. 1998). The field studies support laboratory studies that show that small post-larvae do not develop an affinity for seagrass until they attain a CL >1.7 mm (Liu and Loneragan 1997). Though very small tiger prawn post-larvae may settle on bare substrate, trawls on seagrass or algae habitat and bare substrate over many locations and many years show high abundances of post-larvae and juvenile tiger prawns on vegetated habitats compared to bare substrates (Staples et al. 1985, Poiner et al. 1993, Haywood et al. 1995).

Recently-settled xanthid and portunid crabs also commonly live on seagrass or algal beds. Early juvenile stone crabs (*Menippe mercenaria*) in the Gulf of Mexico are found on seagrass beds although they tend to inhabit sponges and gorgonians rather than residing directly amongst the seagrass itself (Bert et al. 1986). Juvenile (<70 mm carapace width (CW)) sand crabs (*Portunus pelagicus*) are found in intertidal pools amongst macroalgae and seagrass (Williams 1982) and recently-settled mud crabs (*Scylla serrata*) have also been found amongst algae and seagrass in India (Chandrasekaran and Natarajan 1994), although this seems to be unusual.

7.2.2 Pelagic to Mangroves

Unlike India, in northern Australia, recently-settled mud crabs (*Scylla serrata*) were found in the intertidal zone, amongst mangroves (Hill et al. 1982). These young juveniles appear to be restricted to the mangrove zone and do not venture out onto intertidal banks (Hill et al. 1982). The first benthic stages (5 mm CW) of the closely related species *Scylla paramamosian* were found foraging amongst pneumatophores along the seaward edge of a mangrove (*Sonneratia*) forest in Vietnam, in the Mekong Delta (Walton et al. 2006). Like *S. serrata*, at this size, they do not move far from the mangrove fringe (Hill et al. 1982, Le Vay et al. 2007).

In the Embley River, in tropical northern Australia, the recently-settled post-larval (1–2 mm CL) banana prawns (*Penaeus merguiensis*) settled in the upper reaches of mangrove-lined creeks, where their abundance was nearly five times that in the nearby mangrove-lined main river channel (Vance et al. 1990; Fig. 7.5). A closely related species, the red-legged banana prawn (*P. indicus*) also favors a similar environment in the Joseph Bonaparte Gulf (Loneragan et al. 2002, Kenyon et al. 2004).

Juvenile decapods that rely on shelter in intertidal areas are presented with a choice when the tide recedes: remain in the intertidal and deal with the possibility of desiccation or migrate into the subtidal where they may be at greater risk of pre-

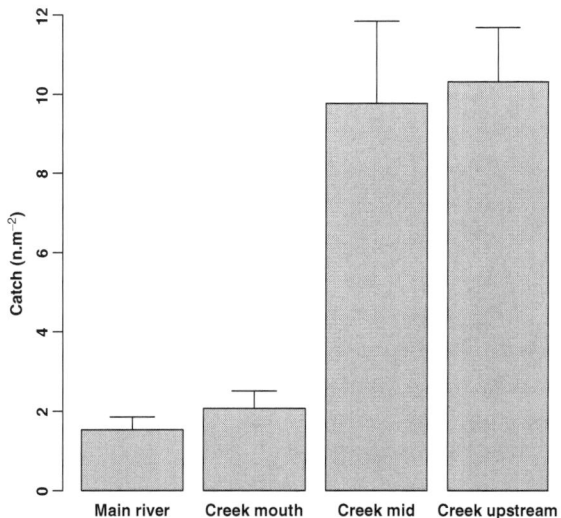

Fig. 7.5 Three-year mean catch (+ SE) of *Penaeus merguiensis* for river and creek sites in the Embley River, northern Australia. Redrawn from Vance et al. (1990), with permission from Elsevier

dation. In Australia, juvenile mud crabs (*Scylla serrata*: 20–99 mm CW) remain amongst mangrove roots and in burrows during low tide (Hill et al. 1982) and juvenile brown and grooved tiger prawns (*Penaeus esculentus* and *P. semisulcatus*) bury into soft mud at the bottom of shallow intertidal pools on seagrass beds in tropical Queensland, Australia (D Heales, CSIRO Division of Marine and Atmospheric Research, pers. comm.). The majority of decapods, however, do appear to move into the subtidal as the water recedes (Vance et al. 1996, Rönnbäck et al. 1999, Johnston and Sheaves 2007)

7.2.3 Pelagic (or Freshwater) to Terrestrial

Hermit crabs of the family Cenobitidae, and other land crabs have evolved a strategy that incorporates the most extreme decapod ontogenetic habitat shift: a shift from an aquatic to a terrestrial existence. Most species hatch from eggs spawned in the shallows, close to the shore, spend their larval phase at sea, and then move from the marine environment to the terrestrial as small juveniles.

Adult coconut crabs (*Birgus latro*) are found on Indo-Pacific islands and are the world's largest terrestrial arthropod. The females release their eggs into the sea at high tide where they hatch immediately on contact with the water. The larvae spend about a month as pelagic phytotrophs, undergoing four zoeal stages, before they metamorphose into benthic glaucothoe post-larvae that use empty mollusk shells to protect their soft abdomen (Schiller et al. 1992). The amphibious post-larvae metamorphose into juveniles and continue to carry the mollusk shells for about nine months. The crabs become terrestrial as early juveniles and eventually the exoskeleton becomes strong enough for the crabs to dispense with the mollusk shells (Brown and Fielder 1992, Schiller et al. 1992). Many hermit crabs employ the same ontogenetic shifts in habitat, from a marine existence as larvae and post-larvae, to a terrestrial existence as small juveniles (Brodie 2002, Barnes 2003).

The red crab of Christmas Island (*Gecarcoidea natalis*—Gecarcinidae) also migrate to the coast to cast their eggs into the ocean. The larvae spend 3–4 weeks at sea before they gather in nearshore pools as megalopae (2–3 days duration). They then metamorphose as juveniles and return to the land as 5 mm CW crabs. The small crabs move inland to the forested island plateau over about nine days and become cryptic for the first three years of life. The juvenile red crabs probably live in burrows and crevices on the forest floor to shelter from desiccation (Adamczewska and Morris 2001a).

Some land crabs (Grapsidae) endemic to Jamaica have adapted their water-dependent larval phase to freshwater habitat in the terrestrial ecosystem, e.g., bromeliad crabs brood their larvae in the water-filled bromeliad leaf axils (Schubart et al. 1998). These Jamaican endemics are examples of crabs that have evolved to dispense with a reliance on the marine ecosystem altogether (Schubart et al. 1998). The bromeliad crabs are the only known crabs to exhibit brood behavior to support their larvae.

7.3 Late Juvenile/Subadult Habitat

In many tropical decapods there is an ontogenetic habitat shift from the early to late juvenile stage and the common theme across many species is a general movement from the shallow early juvenile stage habitat to deeper water (usually offshore).

7.3.1 Algal to Crevice

In Florida (USA), the Bahamas, and Caribbean, juvenile *Panulirus argus* maintain a solitary existence, spending a variable period (usually a few months) in their initial settlement habitat (vegetation or small solution holes). After reaching between 15 and 20 mm CL they move offshore into deeper water where they occupy crevice shelters: under ledges, rock outcrops, solution holes, gorgonians, or large sponges (Kanciruk 1980, Herrnkind and Butler 1986, Forcucci et al. 1994) where they become socially gregarious (Berrill 1975, Lozano-Alvarez et al. 2003). Gregariousness is driven, in part, by the interacting effects of body size, the temporal shift in the release of a chemical attractant by the lobsters, and the sheer volume of the attractant (Ratchford and Eggleston 1998, Ratchford and Eggleston 2000). In the Caribbean, however, macroalgal habitat is rare and it is not clear whether juvenile spiny lobsters undergo an ontogenetic habitat shift here (Acosta and Butler 1997). Caribbean islands are commonly fringed by mangroves with seaward beds of seagrass and algae. These habitats, along with crevice shelters, are utilized by both small and large juvenile spiny lobsters. Large juveniles are also found on coral patch reefs, where small lobsters suffer a high mortality rate (Acosta and Butler 1997).

In Torres Strait (Australia), ornate rock lobsters (*Panulirus ornatus*) undergo a habitat shift at approximately the same size as *P. argus* in southern Florida. Late stage juveniles (40–70 mm CL) move from their small algal-covered holes in reef pavement to shelter in bare holes and crevices in limestone pavement and consolidated rubble (Dennis et al. 1997) in deeper water in areas between the reefs, often amongst epibenthic gardens (Trendall and Bell 1989, Pitcher et al. 1992a). They shelter in their dens during the day and forage at night. For this size class of *P. ornatus* in Torres Strait, Pitcher et al. (1992b) found that lobster abundance was significantly related to the nature of the seabed. The amount of rock and rubble accounted for about 58% of the variation in lobster abundance. Lobster abundance was positively correlated with rock and rubble and negatively correlated with the amount of sand. Lobster abundance was also positively correlated with the density of epibenthic macrofauna (Skewes et al. 1997).

Similarly, in western Australian littoral habitats approximately a year after settlement, *P. cygnus* move from their solitary existence in small holes and fissures to shelter in caves and under ledges with conspecifics (Jernakoff et al. 1993). This generally occurs during summer (December–January) each year as a mass migration involving recently molted juveniles known as 'whites' because of their pale color (Chittleborough 1970). After dark, the juveniles leave their dens to forage on sparse seagrass (*Heterozostera* and *Halophila*) beds away from the reefs (Edgar 1990a). This is in contrast to the small, recently-settled juveniles that tend to feed on *Amphibolis* and macroalgal beds on the reefs (Jernakoff et al. 1993).

7.3.2 Intertidal to Subtidal

Many species that settle onto intertidal habitats such as mangroves and seagrass beds also tend to move into deeper water as they develop. In Vietnam, as mud crabs (*Scylla paramamosain*) grow, they move deeper into the mangrove forest and away from the outer mangrove fringe where they settled as first benthic instars. When they reach a CW of 45 mm, they either begin digging burrows or they move into subtidal waters and then back into the mangroves with each tide to feed (Walton et al. 2006). Tanzanian *S. serrata* do not begin digging burrows until they reach a much larger size (70 mm CW; Barnes et al. 2002). Large juvenile mud crabs (*S. serrata* 80–130 mm CW) leave the mangroves occupied by their younger conspecifics and move to subtidal waters, however, they make excursions onto intertidal flats at high tide to feed (Hill et al. 1982). *S. paramamosain* also move into deeper subtidal waters, away from the mangroves by the time they reach a size of 125 mm CW (Walton et al. 2006).

Numerous species of tropical penaeids also move seaward as they grow (*Penaeus setiferus*, *P. aztecus*, *P. duorarum*, Williams 1955; *P. merguiensis*, Vance et al. 1990; *P. semisulcatus*, *P. esculentus*, Loneragan et al. 1994). Banana prawn (*P. merguiensis*) post-larvae settle in shallow intertidal mangrove-lined creeks (Vance et al. 1990) or the upper reaches of rivers and may be found large distances upstream (up to 72 km from the river mouth, Staples 1980b). As the prawns grow, they gradually move downstream from the creeks into the main river (Vance et al. 1990). Usually these prawns will emigrate from the estuary to coastal waters during the wet season, although during years of very low rainfall they may overwinter within the estuary (Staples 1980a, Staples and Vance 1986).

As tiger prawns (*Penaeus semisulcatus* and *P. esculentus*) develop, their distribution within the nursery habitat (seagrass and algal beds) also changes. While small post-larval tiger prawns were widely distributed over a range of seagrass types (and sometimes on bare substrate), larger tiger prawns were found in higher densities among high-biomass, tall, long-leaved seagrasses (*Enhalus acoroides*) than among low-biomass, short, thin-leaved seagrasses (Loneragan et al. 1998). The authors were not able to establish whether this was because of migration from other habitats or differential mortality. Interestingly, laboratory studies have shown that the behavior of small *P. semisulcatus* changes as they grow: small (<1.7 mm CL) post-larvae showed no preference between seagrass and bare substrate, whereas larger post-larvae spent more time perched on seagrass leaves (Fig. 7.6) compared to bare substrate (Liu and Loneragan 1997).

Tiger prawns also tend to move seawards as they grow. In a study around Groote Eylandt, northern Australia, Loneragan et al. (1994) sampled seagrass

Fig. 7.6 *Penaeus esculentus.*
Small juvenile prawn perched
on a blade of seagrass
(*Zostera marina*)

(*Cymodocea serrulata, Syringodium isoetifoleum, Thalassia hemprichii, Halophila ovalis, H. spinulosa,* and *Halodule uninervis*) and juvenile tiger prawns at a range of depths from 1 m (100 m from shore) to 7 m ($\sim$ 1 km from shore). They found that the mean CL of both species of tiger prawn increased with water depth at all sites (Fig. 7.7).

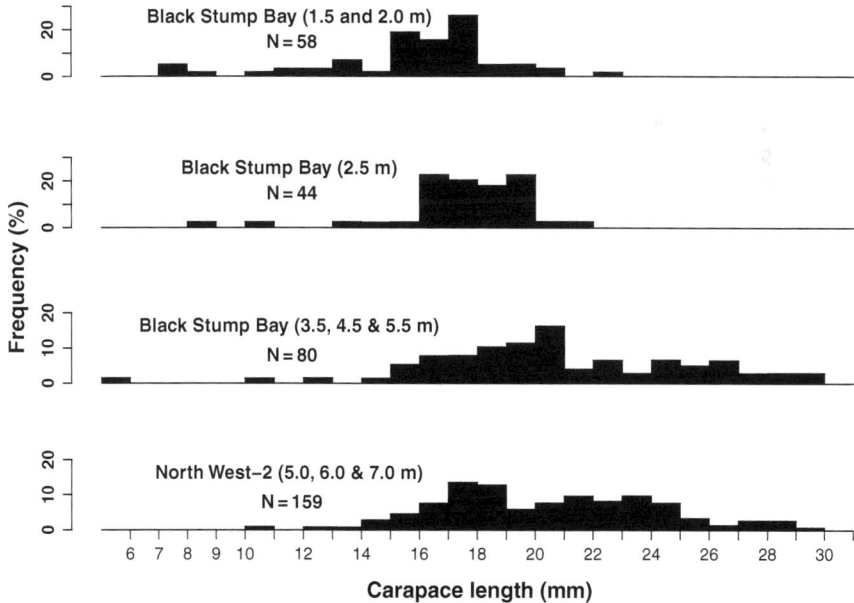

Fig. 7.7 *Penaeus esculentus.* Carapace-length frequency distributions of prawns caught in large beam-trawls at two sites around northwestern Groote Eylandt, Gulf of Carpentaria, Australia, between August 1983 and August 1984. Mean depths of stations in parentheses. Redrawn from Loneragan et al. (1994), with kind permission of Springer Science+Business Media

7.4 Adult Habitats

7.4.1 Estuary to Offshore

The movement of late-stage juvenile banana prawns (*Penaeus merguiensis*) to their adult habitat can be quite sudden. Northern Australia has a distinct wet season during which >90% of the annual rainfall occurs between the months of November and March often resulting in an abrupt decrease in salinity (Australian Bureau of Meteorology 2008). Banana prawns are stimulated to emigrate from the estuary out to sea by rainfall (Staples 1980a, Staples and Vance 1986) and the amount of rain affects the size and number of emigrating prawns (Fig. 7.8). Early in the wet season, low amounts of rain result in a small number of large prawns leaving the estuary. During periods of high rainfall a mass emigration of prawns of all sizes may occur on ebbing tides (Staples 1980a, Haywood and Staples 1993). Rainfall alone explained 70% of the observed variation in numbers of *P. merguiensis* emigrating from the Norman River in northern Queensland, Australia (Staples and Vance 1986). While resident in the creeks and rivers, juvenile banana prawns are demersal, but when emigrating they swim near the surface and may be spread laterally across the full width of the river (Staples 1980a). It is not known how long or how far from

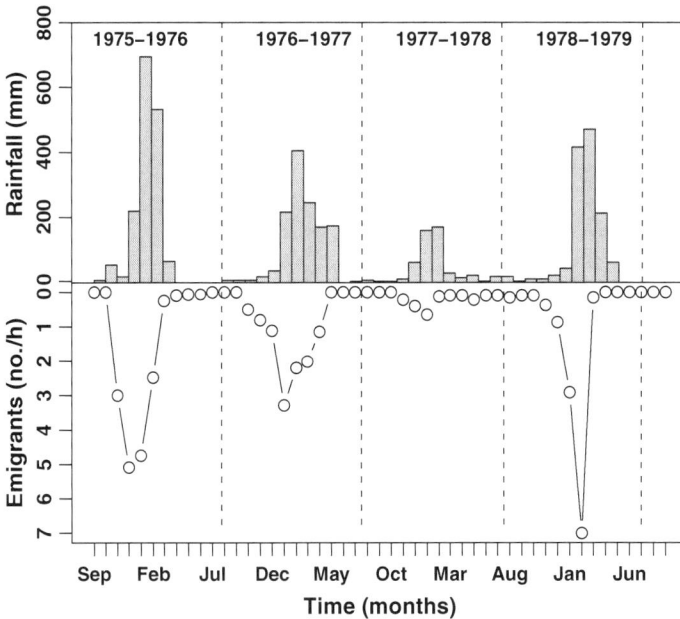

Fig. 7.8 *Penaeus merguiensis.* Changes in monthly rainfall (solid histogram) and mean monthly emigration rate of juvenile prawns from the Norman River between September 1975 and September 1979. Redrawn from Staples and Vance (1986), with kind permission of Inter-Research Science Centre

the estuary banana prawns remain swimming close to the surface, but presumably they resume their demersal existence once they reach coastal waters. Adult banana prawns tend to remain relatively close to shore in waters <20 m deep and are found on muddy sediments (Somers 1987, Somers 1994).

Unlike banana prawns, tiger prawns do not seem to undertake dramatic emigrations from their nursery grounds to coastal waters in response to environmental stimuli. Instead their movement towards subadult/adult habitats seems to be based on size (Loneragan et al. 1994). Very few of the juvenile *Penaeus semisulcatus* and *P. esculentus* sampled from seagrass beds around Groote Eylandt and Weipa in northern Australia were >10.5 mm CL (Loneragan et al. 1994, Vance et al. 1994), suggesting that they moved off the seagrass and into coastal waters at about this size. Further south in the subtropical waters of Moreton Bay, *P. esculentus* emigrated at a larger size (16 mm CL, O'Brien 1994b). After leaving the seagrass beds, tiger prawns disperse widely; for instance, the highest catch rate of juvenile *P. esculentus* on seagrass around Groote Eylandt was 18.2 juveniles.100 m^{-2} compared to 6 juveniles.100 m^{-2} in water deeper than 2.5 m off the seagrass (Loneragan et al. 1994) and <0.1 adults.100 m^{-2} on the fishing grounds offshore (Somers et al. 1987a). Despite the fact that both species of tiger prawns share the same nursery habitat (except in the case of algal beds; Haywood et al. 1995), once they move offshore, they prefer different habitats that can be discriminated in terms of depth and sediment type. *Penaeus semisulcatus* prefer mud or sandy mud in deep (>35 m) water and *P. esculentus* move to relatively shallow inshore waters comprising sand or sandy-mud sediments (Somers 1987, Somers 1994). It is not known why different species prefer different sediment types. For animals like prawns that bury into the substrate it is possible that it is related to the prawn's ability to maintain water circulation across the gills whilst buried (Gray 1974; Fig. 7.9). However, in laboratory experiments monitoring the burying behavior of the mud-preferring *Penaeus semisulcatus* (Hill 1985) and *P. merguiensis* (BJ Hill, CSIRO Division of Marine and Atmospheric Research, pers. comm.) a sand substrate was used effectively, so the reasons must be more complex.

Fig. 7.9 *Penaeus semisulcatus*. Juvenile buried in the substrate

There is little information regarding the movements of juvenile red-spot king prawns (*Penaeus longistylus*) except that they disappear from their juvenile habitat (reefal seagrass beds) at between 15–20 mm CL, and subadults appear in deeper inter-reefal waters (Dredge 1990) where they are commonly associated with coralline sand sediment (Somers et al. 1987b, Gribble et al. 2007).

7.4.2 Shallow to Deep

In mud crabs and many spiny lobsters there is some overlap in the distribution of the subadults and adults although there is a general trend for older individuals to move into deeper water. Most species of spiny lobsters remain in the reef crevice dwelling habitat until they mature and undertake a spawning migration. Lobsters tend to shelter in their dens during the day, and make foraging excursions at night. They tend to move around on the reef and may not return to the same den each morning, but movement between reefs is uncommon (Herrnkind et al. 1975, Smale 1978, Moore and MacFarlane 1984, Trendall and Bell 1989).

The distribution and habitat of adult mud crabs (*Scylla serrata*) overlaps that of the subadults; they forage on intertidal banks at high tide and retreat to subtidal waters as the tide recedes (Hill et al. 1982). Large *S. serrata* were found sheltering in elliptical-shaped burrows exposed by spring tides at Inhaca, Mozambique (MacNae 1968). *Scylla* burrows are also common on mangrove banks in Australia (Hyland et al. 1984), although in the Kowie River estuary in South Africa, large mud crabs tend to bury themselves in the mud rather than construct burrows (Hill 1978). Hill (1978) postulated that living in a burrow may offer protection from predation and desiccation, however, burrows restrict the area over which a crab can forage and so they may only choose to use burrows in areas with adequate food supply.

Ornate rock lobsters (*Panulirus ornatus*) are the exception to this trend—instead the older lobsters move into shallower water. Large *P. ornatus* in their third and fourth year following settlement (2+ and 3+) move from the deeper reefal waters to occupy dens on the shallow parts of the reef (Skewes et al. 1997; Fig. 7.10). The 3+ lobsters are all males as the females leave the reefs the previous year on their spawning migration. *P. ornatus* tend to favor the lee side of the reef so that they are found in higher abundance on the north-western reef edge during winter when the prevailing wind is from the south-east and they move to the south-eastern edge during the north-westerly monsoon season (Skewes et al. 1997).

7.4.3 Lobster Dens

Palinurid lobsters do not build their dens, they occupy existing crevices and holes in the substrate (Kanciruk 1980). In contrast to their younger conspecifics, den occupying lobsters can be quite gregarious. During the Tektite II program conducted in the US Virgin Islands, divers found that in some areas up to 80% of the *Panulirus*

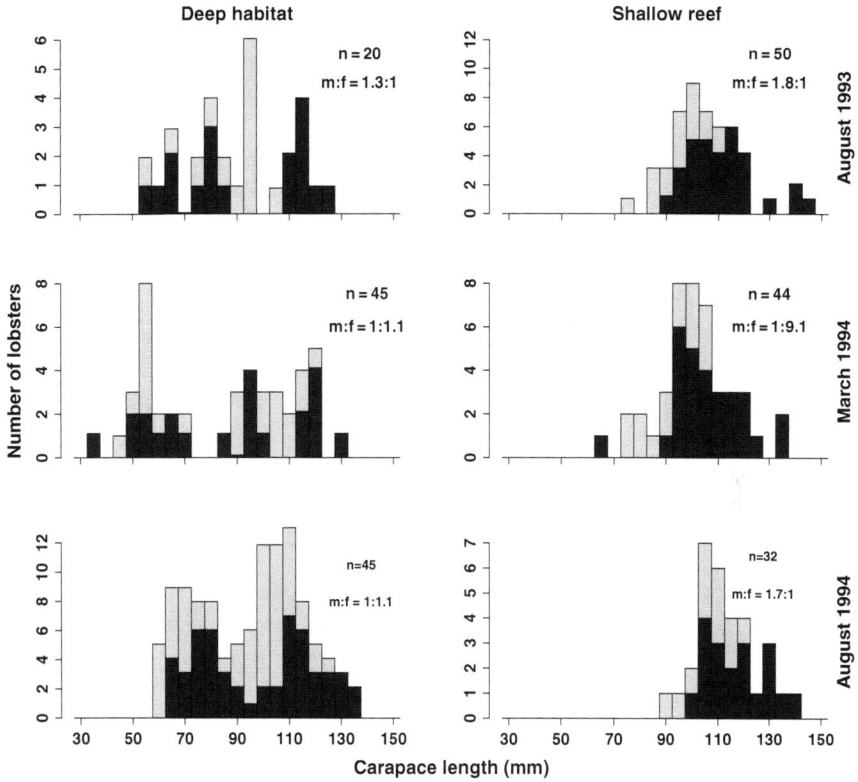

Fig. 7.10 Size distribution of ornate rock lobsters (*Panulirus ornatus*) sampled in deep and shallow habitats around Warraber Island, Torres Strait, Australia. Black columns: males; gray columns: females. Sample size and sex ratio (male:female) also shown. Redrawn from Skewes et al. (1997) with kind permission of CSIRO Publishing (http://www.publish.csiro.au/nid/127/issue/128.htm)

argus observed were cohabiting dens with other lobsters (Herrnkind et al. 1975). In a study on den occupation of the ornate rock lobster, Trendall and Bell (1989) noted that dens composed of loose rubble were only occupied by solitary lobsters, whereas rock (caves or crevices in lumps of coral or rock) dens commonly housed multiple lobsters. Moreover, rubble dens were only used temporarily; there was little evidence of continuity of occupancy. Although more than 70% of observed lobsters were found in groups of two or more, dens having more than 10 lobsters were very rare (Trendall and Bell 1989).

Dens that have multiple occupants tend to be occupied more frequently than others and the higher the mean number of occupants, the higher the frequency of occupancy (Herrnkind et al. 1975). While the favored dens of *Panulirus argus* identified during the Tektite II study all tended to offer good protection from predators, other similar structures existed elsewhere in the area, but were often not used by lobsters. Herrnkind et al. (1975) felt that it was the den's proximity to foraging grounds and possibly areas where reproductive activities take place that determined den attractiveness. In addition to food, habitat use by *P. argus* in back reef areas is dictated by the interacting effects of shelter scaling, predation risk, and potential for gregariousness (Eggleston et al. 1990, Eggleston and Lipcius 1992, Eggleston et al. 1992). In Torres Strait, *P. ornatus* also tend to favor particular dens, with some dens consistently containing more lobsters than others, but the individuals changed from day to day (Trendall and Bell 1989). Childress and Herrnkind (2001) used a combination of laboratory and field experiments to determine how lobsters might benefit from den sharing. Their results showed that lobsters sharing dens did not have increased survival compared to solitary lobsters in dens, suggesting that co-operative group defense was not important. Lobsters in areas of high conspecific density did not have improved survival over those residing in areas of low conspecific density, suggesting the dilution effect was probably not important. Conversely, Mintz et al. (1994) found that survival of *P. argus* increased with density of conspecifics, suggesting a survival benefit to gregarious behavior. The presence of a lobster in a den appears to act as an attractant to conspecifics. In laboratory experiments, nearly twice as many lobsters searching for a shelter were able to locate it when it already contained another lobster compared to when it was empty. Moreover, on average, the occupied shelter was located in a third of the time required to locate the unoccupied shelter (Childress and Herrnkind 2001). Childress and Herrnkind (2001) termed this the 'guide effect' and postulated that it benefited the shelter-seeking individuals by reducing their time exposed to risk of predation.

7.4.4 Land Crabs

Land crabs (Gecarcinidae and Grapsidae) and some hermit crabs (Cenobitidae) occupy terrestrial forest and upland habitats. Often the crabs are cryptic during their terrestrial juvenile phase and are extremely difficult to locate. As early juveniles (<10 mm thoracic length), they do not respond to baited-trap attempts to locate them, possibly because of competition with hermit crabs which are attracted to the baits (Fletcher et al. 1992b). However, as they grow they become more robust and are easy to locate in the habitats that they occupy. The islands that they inhabit often have few large terrestrial predators to avoid.

In undisturbed populations, the terrestrial coconut crab *Birgus latro*, is active diurnally, while in some areas the crabs are nocturnal and adults are readily found outside their day refuges around sunset (Brown and Fielder 1992). They are omnivorous and forage in the forest and along the coast for a variety of food. They burrow in order to molt and may remain in the burrow for up to 16 weeks (Fletcher et al. 1992). Hermit crabs live a terrestrial existence as adults, often in tropical humid environments where they have adapted to life on land by developing highly specific behavior and morphology with each life stage to avoid predators and water loss (Brodie 1999, Brodie 2002, Barnes 2003). However, some hermit crabs occupy the upper extremities of exposed rock formations in the wet-dry tropics of northern

Australia ($\sim 12°$ S latitude) and during the dry season their environment is extremely hot and dry (R Kenyon pers. observ.).

Christmas Island red crabs (*Gecarcoidea natalis*) inhabit the central plateau of Christmas Island and are cryptic for the first three years of life. The largest male crabs are found furthest inland in their rainforest habitat (Adamczewska and Morris 2001a). The red crabs live in burrows where they shelter from desiccation and only emerge when humidity is >85% (Adamczewska and Morris 2001b). They can remain inactive in their burrows for months. Their spawning migration to the coast is cued by the monsoon season; rain and high humidity allows the crabs to travel $\sim$ 1 km.d^{-1} without loosing excessive moisture to reach the coast (Adamczewska and Morris 2001a).

7.5 Why Change Habitats?

A decapod's level of predation risk (Eggleston et al. 1990, Lipcius et al. 1998), foraging needs (Marx and Herrnkind 1985a, Heales 2000), and reproductive condition change during ontogeny and it has been hypothesized that animals shift habitats to meet their changing requirements (Werner and Mittlebach 1981, Werner and Gilliam 1984). In the following section we discuss these mechanisms underlying ontogenetic habitats shifts in relation to tropical decapods.

7.5.1 Protection from Predation

Habitat structural complexity provided by marine vegetation (Crowder and Cooper 1982) or substrate (Lipcius and Hines 1986, Eggleston et al. 1990) is thought to offer protection from predation by limiting the ability of predators to locate their prey, either by acting as a visual or physical barrier (Rooker et al. 1998). Numerous studies have demonstrated that predation rates on tropical decapods sheltering amongst submerged macrophytes, mangrove roots, or pneumatophores are reduced in comparison with those on bare substrate (Heck and Thoman 1981, Heck and Wilson 1987, Haywood and Pendrey 1996, Primavera 1997). For many tropical juvenile penaeids, mangroves and seagrasses offer shallow water, structural complexity, and often have high turbidity and fine sediment which facilitates burrowing. These attributes, in combination with the prawn's behavior should convey some protection against predation by reducing their visibility and lowering their encounter rates with potential predators (Minello and Zimmerman 1983, Laprise and Blaber 1992, Kenyon et al. 1995).

For all of the examples provided hereafter, the key issue is that the juvenile stages gain greater protection from predators by residing amongst complex vegetated and reef habitats than if they remained in the pelagic environment of their larvae or the demersal environment of the adults of the species.

Algal-stage Caribbean spiny lobsters are particularly vulnerable to predation because of their small size and they remain almost exclusively on clumps of red algae (*Laurencia*) for 3–5 months (6–17 mm CL) (Marx and Herrnkind 1985b, Herrnkind and Butler 1986, Smith and Herrnkind 1992). They remain on the clump on which they first settled if the clump is isolated, but rapidly move to other clumps if they are adjacent, particularly when food is scarce on the initial clump (Marx and Herrnkind 1985a). These early-stage lobsters are at their most vulnerable when exposed in the open, moving between algal clumps. Childress and Herrnkind (1994) speculated that during the cooler months, food on a clump should last longer because the lobster's metabolic rate will be lower, and this might offset the increased predation risk associated with moving to a new clump at a time of year when the abundance of clumps is also reduced. In contrast to the sedentary nature of algal-stage *Panulirus argus*, *P. cygnus* in western Australia appear to be far more mobile. Jernakoff (1990) found that as many as 50% of algal-stage *P. cygnus* moved from their holes in limestone reefs within 24 h. However, it appears that this mobility comes at a cost: lobsters of this size (8–15 mm CL) suffered disproportionately high predation levels from fish (Howard 1988). Perhaps ironically, the increased mobility of this and other species is likely due to the need to find shelter of suitable dimensions, to reduce the risk of predation. Studies on *P. cygnus* (Fitzpatrick et al. 1989, Jernakoff 1990), *P. japonicus* (Yoshimura and Yamakawa 1988), *P. versicolor* (George 1968), *P. argus* (Eggleston et al. 1990, Eggleston et al. 1992), and *P. ornatus* (Dennis et al. 1997) found that dimensions of holes selected by juvenile lobsters closely matched those of the lobsters; presumably this makes it difficult for predators to extract them.

The degree to which habitat complexity confers protection appears to be species-specific and determined to some degree by the hunting ability of the predator. Juvenile banana prawns (*Penaeus merguiensis*) in northern Australia moved well into the mangrove forests at high tide, gaining refuge because large predatory fish were found to be restricted to the forest margins (Vance et al. 1996), although see Sheaves (2001) for exceptions to this. In laboratory studies designed to test the effect of simulated smooth cordgrass (*Spartina alterniflora*) on the predation rates of four estuarine fish on juvenile brown shrimp (*Penaeus aztecus*), Minello and Zimmerman (1983) found that vegetation reduced the predation rates of two species (pinfish *Lagodon rhomboides* and Atlantic croaker *Micropogonias undulatus*), but had no effect on the remaining two species of fish (red drum *Sciaenops ocellatus* and speckled trout *Cynoscion nebulosus*). Similarly, Primavera (1997) found that while the predation efficiency of a snapper (*Lutjanus argentimaculatus*) on juvenile prawns was reduced amongst pneumatophores and leaf bracts, the efficiency of the relatively aggressive sea bass (*Lates calcarifer*) was not affected by the presence of vegetation.

One popular hypothesis has been that the higher the density or biomass of vegetation, the greater the level of protection from predation. However, this hypothesis appears to be an oversimplification. Although this hypothesis has been supported by some studies, e.g., long, wide seagrass leaves offered more protection to juvenile brown tiger prawns (*Penaeus esculentus*) than thin, short leaves (Kenyon et al. 1995) and recently-settled juveniles actively select long-leaved over short-leaved seagrass (Kenyon et al. 1997), it is not by others. For example, Heck and Wilson

(1987) found that the number of blue crabs (*Callinectes sapidus*) taken by predators was not affected by seagrass biomass. Other studies have found that there may be a threshold in macrophyte density with respect to the degree of protection from predation. For example, in laboratory experiments, predation of juvenile prawns was greater on bare substrate compared to medium density, but not high density pneumatophores (Primavera 1997).

The structural complexity offered in subtidal areas, particularly in the channels of tropical estuaries, is likely to be less than that offered by an intertidal seagrass bed or mangrove forest. In addition to this, the draining of water from the intertidal has the effect of concentrating predators and prey into a smaller volume (Krumme et al. 2004, Johnston and Sheaves 2007), potentially increasing encounter rates between predator and prey. A recent study in tropical Australia demonstrated that crustaceans that had moved out from among the mangroves at low tide tended to concentrate in the small channels that drained the mangroves rather than remaining in areas with little habitat complexity (Johnston and Sheaves 2007). Concentrating in the drainage channels may provide some protection from predation, however, this behavior may also be a mechanism to enable access back into the mangroves once the tide had turned or provide access to greater food (Johnston and Sheaves 2007).

Increasingly, studies have demonstrated that the protective value of vegetation (or other habitat structure) is a function of the relative scaling between the animals and the habitat structure (Ryer 1988, Eggleston et al. 1990, Palmer 1990, Beck 1995). Small juvenile Caribbean spiny lobsters had significantly higher survival rates when sheltering amongst algal beds than their larger conspecifics, apparently because of the lack of suitable-sized refugia for larger juveniles (Lipcius et al. 1998). Tethering studies on post-algal (Smith and Herrnkind 1992) and subadult juveniles (Eggleston et al. 1992, Mintz et al. 1994) showed that the relative rate of predation is significantly reduced when these larger lobsters are able to utilize crevice shelters (Fig. 7.11). These results suggest that as they outgrow the protective value of macroalgae and are exposed to an elevated predation risk, larger juveniles are stimulated to undergo an ontogenetic habitat shift from algal beds to adjacent coral reefs (Lipcius et al. 1998, Childress and Herrnkind 2001).

It is difficult to invoke increased risk of predation as a stimulus for shifting habitat in large juvenile banana prawns (*Penaeus merguiensis*), which shelter and forage amongst mangroves at high tide (Vance et al. 1996). Apart from the small, narrow pneumatophores of mangroves of the genus *Avicennia*, the buttress roots and prop roots of most other mangroves are large enough that they should be able to provide refuge for subadult and adult penaeids. In fact if wet season rains fail, juvenile banana prawns do not emigrate offshore from their juvenile estuarine mangrove habitat; instead they overwinter in the estuary moving offshore the following spring (Staples and Vance 1986). In addition to this, a movement from the relative protection of inshore mangroves generally does not provide relief from predation pressure as predation of subadult and adult penaeids offshore is very high (Euzen 1987, Brewer et al. 1991, Salini et al. 1994).

In contrast to banana prawns, is possible that juvenile tiger prawns (*Penaeus semisulcatus* and *P. esculentus*) outgrow the protective value conferred by

Fig. 7.11 Total predation on different size classes of juvenile lobsters over a 24-hr period by habitat type. Color break indicates relative amount of predation which occurred over a 10-hr nighttime period. Blank space above color break shows relative amount of daytime predation. N = 134 for open habitat, N = 113 for shelter habitat. Algal = algal stage (5–15 mm CL), Trans = transitional stage (16–25 mm CL), and P. algal = post-algal stage (26–35 mm CL) lobsters. Redrawn from Smith and Herrnkind (1992), with permission from Elsevier

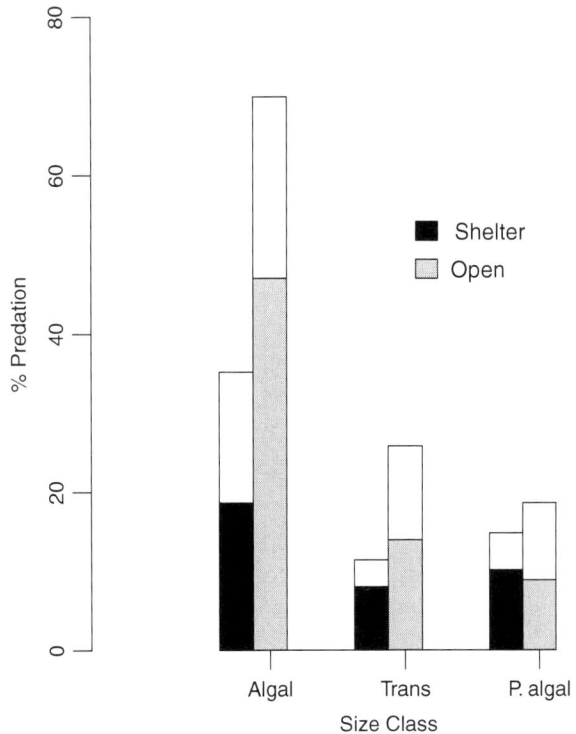

seagrasses. Tiger prawns cling to seagrass leaves to camouflage themselves from predators. The degree of protection is related to the scaling of leaves and prawns, i.e., longer, wider leaves confer more protection to larger prawns (Kenyon et al. 1995). However, at some point the prawns grow to a point where even the largest leaves no longer offer protection and they move into deeper water offshore from the seagrass (Loneragan et al. 1994, O'Brien 1994b).

In northern Australia, juvenile (20–99 mm CW) mud crabs (*Scylla serrata*) are resident amongst the mangroves and remain there at low tide. Subadults migrate out onto intertidal flats to feed at high tide and retreat to subtidal areas at low tide; adults tend to spend more time subtidally and only occasionally venture into intertidal habitats (Hill et al. 1982). Whilst inundated, the intertidal flats are foraged by flathead (Platycephalidae; Baker and Sheaves 2006) and herons (Mukherjee 1971), both of which are potential predators of juvenile *S. serrata*. The larger subadult mud crabs would not be as vulnerable to predation and are thus able to utilize the intertidal foraging grounds with less risk.

Terrestrial crabs also provide an indicator that predation may be a driver for ontogenetic habitat shifts over evolutionary time. Tropical land crabs occupy terrestrial habitats on isolated islands that initially were 'empty' ecosystems. Because of the isolation, few terrestrial predators have been able to colonize these islands providing the land crabs with an environment that is free of predation pressure, in contrast to the marine environment they inhabited as larvae.

Cannibalism is a special case of predation and it is possible that in some cases ontogenetic shifting has evolved as a mechanism to reduce cannibalism. In juvenile crab populations under high densities where conspecific encounters are frequent, cannibalism can be significant (Perkins-Visser et al. 1996, Ut et al. 2007). Cannibalism is common among penaeids (Rosales Juarez 1976, Otazu-Abrill and Ceccaldi 1981, M Haywood pers. observ.) and has also been reported in the Caribbean spiny lobster *Panulirus argus* (Lipcius and Herrnkind 1982) and stone crabs (genus *Menippe*; Bert 1986).

There is a significant risk associated with the traverse from a pelagic habitat ~ 50 km offshore in ~ 30 m depth of water to a demersal existence in a structured habitat in 1 m depth of water 200 m from a shoreline in a littoral seagrass bed. The benefit to the individual accrues once they reach the refuge provided by the inshore habitat in the juvenile phase. From an evolutionary perspective, a key indicator of the benefit of inshore migration is a comparison of mortality in inshore and offshore environments. Natural mortality of penaeids has been measured in nursery habitats and offshore habitats (together with fishing mortality). Mortality for small juveniles can be as high as 0.89 wk^{-1}, declining markedly for large juveniles (0.02 wk^{-1}) (O'Brien 1994b, Wang and Haywood 1999). These rates generally are higher than mortality estimates for adult prawns offshore (~ 0.05–0.07 wk^{-1}) (Lucas et al. 1979). Yet the mortality of recently-settled post-larvae and juveniles may be much higher for individuals that do not access structured habitats. Given the high fecundity of penaeids (300,000–500,000 eggs per individual), it is clear that the mortality of the larval and post-larvae phases must be very high. Over evolutionary time, the mortality of post-larvae offshore may be much higher than those of post-larvae reaching structured habitats inshore. High mortality may favor the evolution of an inshore/offshore strategy. However, due to their small size and dispersion in coastal seas, estimates of mortality of early life stages are a significant gap in our knowledge.

7.5.2 Food

It is possible that the food and nutritional requirements of decapods change as they grow, and developing animals may undertake a habitat shift to satisfy these needs, but for most tropical decapods there is little evidence to support this hypothesis. There are three characteristics of prey items that are important in relation to a decapod's access to different food resources as a result of an ontogenetic habitat shift: the taxa of prey consumed, the size of the prey items, and the density of individual prey. Many studies indicate that there is a high degree of dietary overlap in prey taxa consumed by juvenile and adult decapods (e.g., Briones-Fourzán et al. 2003, Sara et al. 2007). However, it is not only the type of prey available in different habitats that is important, it is the ability of an individual decapod to locate and capture food in the two habitats. This is a function of the density of the prey and the density of the decapods in each habitat.

Sara et al. (2007) quantified and identified the gut contents of juvenile, sub-adult, and adult mud crabs (*Scylla serrata*). All size classes were found to be feeding on crustaceans, mollusks, fish (probably scavenged), and algae in similar proportions, although the diet in all sizes varied seasonally, reflecting the availability of the various prey. Hill (1976) examined the gut contents of South African and Australian *S. serrata* ranging from small juveniles to adults, but unfortunately the results were pooled over all sizes. Mollusks and crustaceans dominated the gut contents of crabs from both areas. Different size classes of the sand crab (*Portunus pelagicus*) also had similar diets (Williams 1982). Interestingly, ontogenetic differences in crab diets have been documented in other crabs, but only for temperate species, e.g., *Cancer magister* (Gotshall 1977) and *Callinectes sapidus* (Tagatz 1968).

Algal-stage Caribbean spiny lobsters (*Panulirus argus*) forage amongst the red algae *Laurencia* sp. and feed on a wide range of epifauna including gastropods, bivalves, amphipods, ostracods, isopods, and echinoderms (Herrnkind and Butler 1986, Herrnkind et al. 1988). Briones-Fourzán et al. (2003) studied the diets of three stages of juveniles (algal, post-algal, and sub-adult); all prey items tended to be slow-moving or sessile invertebrates or plants. Dominant categories were crustaceans, gastropods with small amounts of sponge, echinoderm, polychaetes, bryozoans, tunicates, seagrass, macroalgae, and coralline algae. There was a high degree of overlap between all stages—algal, post-algal, and sub-adult—and the diet was also similar to that of the adults although the adults had a higher proportion of mollusks in their diet (Herrnkind et al. 1975, Cox et al. 1997).

The post-pueruli of the western rock lobster (*Panulirus cygnus*) forage amongst seagrass (*Amphibolis*) and macroalgae on limestone reefs, although they do not move onto the seagrass beds adjacent to the reefs (Jernakoff et al. 1993). Their diet consists primarily of coralline algae, mollusks, and crustaceans, and is similar to that of the juveniles (>45 mm CL) which forage in sparse *Heterozostera* and *Halophila* meadows away from the reefs (Joll and Phillips 1984, Jernakoff 1987, Edgar 1990a) where small benthic prey is more abundant than on the reefs (Edgar 1990a, Edgar 1990b). The difference between diets of post-pueruli and older juveniles were relatively subtle and consisted of a variation in the relative proportions of the dietary components. For example, small juveniles (25–30 mm CL) ate less coralline algae and more polychaetes, and larger juveniles (40–60 mm CL) ate fewer mollusks, a similar portion of coralline algae, and had a higher proportion of 'other plants' and polychaetes (Edgar 1990a, Jernakoff et al. 1993). Given the opportunistic nature of the foraging habits of palinurids and the fact that their diet often reflects the abundance of the available prey (Edgar 1990a), it is unlikely that these differences in dietary requirements would be significant enough to initiate an ontogenetic habitat shift. Perhaps the high degree of dietary overlap between different size classes of conspecifics results in intraspecific competition for food that initiates a habitat shift. Although this also seems unlikely given that in general, prey size is correlated with the size of the predator (Edgar 1990a, Brewer et al. 1991, de Lestang et al. 2000, Mantelatto and Christofoletti 2001).

There is some evidence for an ontogenetic shift in the diet of banana prawns (*Penaeus merguiensis*). Juvenile banana prawns forage amongst the mangrove forests when the mangroves are inundated. Prawns collected as they left the mangrove forest on a receding tide had full guts, whereas 2–3 hrs after the water had receded from the mangroves, their guts were almost empty, indicating they had not been feeding after leaving the mangrove forest (Wassenberg and Hill 1993). Juvenile banana prawns eat a wide variety of food items including crustaceans, bivalves, gastropods, polychaetes, mangrove detritus, fish, insects, foraminiferans, and diatoms (Chong and Sasekumar 1981, Robertson 1988, Wassenberg and Hill 1993). Foraminiferans, copepods, algae, and nematodes featured less in the diet of adult banana prawns, whereas they ate more bivalves, gastropods, and polychaetes (Wassenberg and Hill 1993). Wassenberg and Hill (1993) postulated that the absence of the small prey (e.g., foraminiferans, copepods, and nematodes) from the adult diet may have been due to the inability of the adults to manipulate such small items.

Despite undergoing a major habitat change from shallow inshore seagrass bed to relatively deep water muddy offshore substrates, the diet of grooved tiger prawns (*Penaeus semisulcatus*) does not change markedly. Small (2–5 mm CL) juvenile grooved tiger prawns foraged amongst seagrass and algae and ate predominately copepods, diatoms, and filamentous algae (Heales et al. 1996, Heales 2000). Larger juveniles (also foraging on seagrass) ate a high proportion of bivalves, gastropods, and crustaceans with smaller amounts of ophiuroids and copepods. The adults, feeding offshore away from the seagrass beds had a similar diet in composition and relative proportions, although they did not feed on copepods (Wassenberg and Hill 1987).

Unlike grooved tiger prawns, there is some evidence for an ontogenetic shift in diet in brown tiger prawns (*Penaeus esculentus*). Small (2–5 mm CL) juvenile brown tiger prawns sampled from seagrass beds ate the same items as *P. semisulcatus*, as well as gastropods, ostracods, decapods, amphipods, and small amounts of miscellaneous items such as bryozoans and foraminiferans (O'Brien 1994a). Larger juveniles ate mainly gastropods and copepods, whereas subadult and adult brown tiger prawns added bivalves and ophiuroids to this diet (Wassenberg and Hill 1987).

Given the positive correlation between prey and predator size it is possible that food plays a more important role initiating ontogenetic habitat shifts than is suggested by the literature. Optimal foraging theory predicts that predators will select prey of a size that will maximize the net energetic gain, i.e., larger prey usually means maximizing net energetic intake (Werner and Mittlebach 1981). It is also likely that as predators grow they are less able to manipulate small prey (Sousa 1993) and may need to shift habitats to hunt for larger prey. Studies on gut contents rarely give data on prey size and generally are only able to supply relatively coarse taxonomic detail on prey and consequently only coarse differences in diet between ontogenetic stages are distinguishable.

As the consumption and availability of prey can be difficult to measure, it may be useful to consider growth as an indicator of prey availability in different habitats. Juvenile prawns, for example, grow at about 0.63–2.10 mm CL wk^{-1} in their inshore vegetated habitats (Haywood and Staples 1993, O'Brien 1994b). Adult prawns grow at 0.82–1.53 mm CL wk^{-1} in their offshore habitats (Somers and Kirkwood 1991, Loneragan et al. 2002). An adult tiger or banana prawn that grows from 35 to 40 mm CL adds approximately 15 g in weight and, conservatively, the process takes five weeks. In five weeks, a juvenile tiger prawn in an inshore seagrass bed might also grow 5 mm CL, from say 4 to 9 mm CL. Yet the weight gain for this individual is <1 g. It is possible that food limitation initiates the ontogenetic shift in this case. In an estuarine environment, would large adult prawns have access to the food resources to support a weight gain of 15 g. individual^{-1} in the hypothetical situation where they did not move to offshore habitats? Although we are not aware of any data on this subject, there are anecdotal observations that support this hypothesis. In years when the amount of rainfall is insufficient to cause large juvenile *Penaeus merguiensis* to emigrate from the estuaries, they overwinter in their nursery habitat and seem to grow much more slowly than similarly-aged prawns that have moved offshore during years of sufficient rainfall (M Haywood pers. observ.).This may provide an indication of why penaeid prawns adopt a life cycle with estuarine/offshore phases. Offshore, the density of prawns is much lower than in juvenile habitats; yet they have access to abundant prey even though they might be found at low densities.

The dearth of information on prey size and prey abundance/availability for decapods in different habitats is a significant information gap in our knowledge of these species. It is worth noting that the penaeid species that close their life cycle wholly within estuaries are smaller species than the offshore commercial prawns (Grey et al. 1983). They mature at a much smaller size and presumably grow at a rate that incurs lower overall weight increment in the estuarine habitat. Perhaps a wholly estuarine life cycle restricts the size of penaeids. If it does, then food availability provides a strong driver for ontogenetic habitats shifts in the larger species.

Terrestrial crabs provide a strong indicator that food has cued ontogenetic habitat shifts over evolutionary time. Tropical land crabs occupy terrestrial habitats on small islands that were created by geologic processes as isolated land masses that initially were 'empty' ecosystems (Schubart et al. 1998). Islands were created by uplifting of the earth's crust, volcanic eruption, or hydrologic and sedimentation processes forming islands in reef ecosystems. They were devoid of terrestrial biota and surrounded by ocean, and offered a unique opportunity to exploit new food resources that became available as these land masses were colonized by biota. Molecular evidence shows that in some cases decapods colonized the new terrestrial habitats by a radiation of species from a common marine ancestor (Schubart et al. 1998). Food resources on the land would have been different and abundant relative to the food items that were available in the marine habitat. Additionally, the terrestrial fauna on these island ecosystems is often depauperate relative to large, more ancient land masses in the same region; few reptiles and mammals inhabit the ecosystem to compete for food with the decapods that evolved to exploit the abundant habitats.

7.5.3 Reproduction

A number of tropical palinurids shift habitat either just prior to, or during their reproductive phase. In spring (September–November) in the Torres Strait, Australia, many male and almost all female *Panulirus ornatus* in their third year after hatching

(2+) leave the reefs *en mass* and migrate eastwards, walking across an open sand-mud seabed for three months (Moore and MacFarlane 1984). Some move as far as 500 km to the eastern Gulf of Papua where they breed from November to March (Moore and MacFarlane 1984, MacFarlane and Moore 1986; Fig. 7.12). Female *P. ornatus* move temporarily into deepwater to hatch their eggs during full moons, returning to the reef post-hatching (Dennis et al. 1992). There is no return migration to Torres Strait following the spawning season; instead the lobsters suffer a severe loss of condition and subsequently a high mortality (Moore and MacFarlane 1984). Consequently, the movement between habitats may best be described as a spawning migration, rater than an ontogenetic habitats shift (Pittman and McAlpine 2003).

Reproduction in *Panulirus argus* begins during spring and extends through summer (Davis 1977). The sexes move differentially, with the females migrating into areas on the reef occupied by adult males (Cooper et al. 1975, Herrnkind 1980, Davis and Dodrill 1989, Puga et al. 1996). After mating, the females may move offshore to the reef edge or the continental shelf where they incubate their eggs and release the larvae (Buesa Mas 1965 in Herrnkind 1980). Like *P. ornatus*, after their larvae have been released, the females move back inshore (Olsen et al. 1975). There seem to be regional differences in movements associated with reproduction

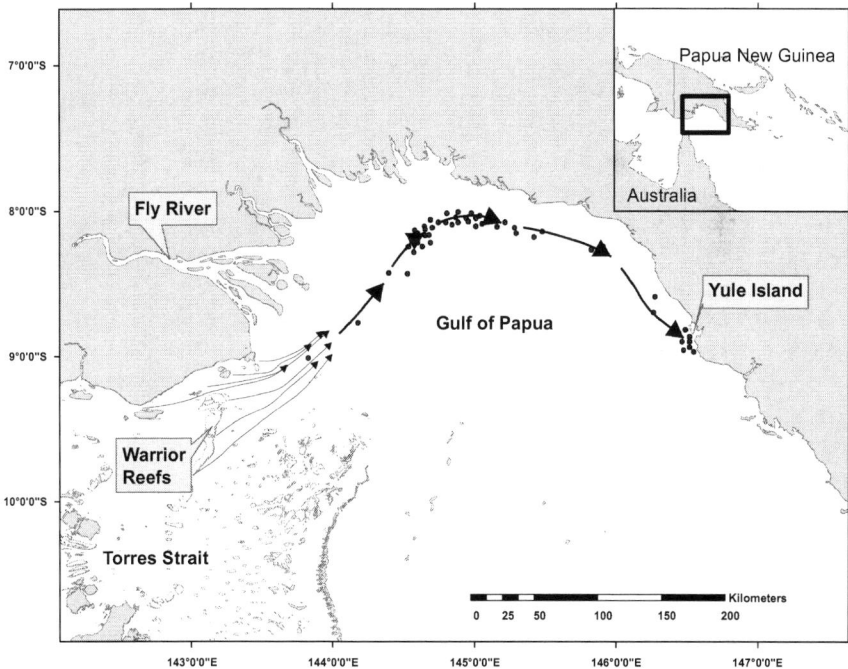

Fig. 7.12 Long-range movements of *Panulirus ornatus* based on the recovery of 125 tagged lobsters that were released in Torres Strait. ● Recapture site for lobsters released in Torres Strait. Redrawn from MacFarlane and Moore (1986) with kind permission of CSIRO Publishing (http://www.publish.csiro.au/nid/127/issue/2844.htm)

of *P. argus*, however. In a study on the distribution of *P. argus* on Glover's Reef in Belize, Acosta and Robertson (2003) found females with egg masses on the fore reef, but they also observed them on shallow reefs in the lagoon. Earlier tagging studies also showed that the movement of adult Caribbean spiny lobsters between shallow lagoon and deep reef habitats was random through the year (Acosta 1999, Acosta 2002). Evans and Lockwood (1994) speculated that the female spotted lobster (*Panulirus guttatus*) also migrate from shallow to deep reefs for reproduction, but during a tagging study conducted over two years at Looe Key marine sanctuary, Florida, USA, Sharp et al. (1997) found no evidence to support this hypothesis.

It has been postulated that the main reason for reproductive migrations in palinurids is to enable the females to release their larvae into an area exposed to oceanic currents to enhance larval dispersal (Moore and MacFarlane 1984). This hypothesis would explain the regional differences in the degree of migration associated with reproduction in palinurids. By moving across the shallow seabed of Torres Strait, to the eastern edge of the Gulf of Papua, *Panulirus ornatus* have access to an area of favorable currents in the Coral Sea. Importantly, these currents provide a transport mechanism enabling larvae to return to the adult habitat. By comparison the population of *P. argus* on Glover's Reef, Belize, have access to deep water (400–2000 m) of the Caribbean Sea within 2 km from the reef crest (Acosta and Robertson 2003).

The dependence of land crabs on a marine larval phase provides a classic example of a spawning migration from their upland forest habitats to the coastal zone to release their larvae in the ocean. The crabs retain their fertilized eggs on their abdomen where they mature, and the crabs release the larvae on contact with water. Male Christmas Island red crabs (*Gecarcoidea natalis*) dig burrows close to the land/sea interface where they shelter from desiccation and lure a female crab to mate. The male crabs move back inland while the female remains in the burrow until the eggs are mature and the environmental conditions facilitate successful release into the ocean (Adamczewska and Morris 2001a).

Several species of portunid crab spend most of their lives in estuarine waters, but apparently all need access to the sea to spawn (Norse 1977, Le Vay 2001) because the larvae do not survive low salinities (Hill 1974) and to facilitate larval dispersal (Hill 1994). Hill (1975) studied reproduction in *Scylla serrata* in two South African estuaries, one of which was usually closed from the sea by a sand bar. Females migrated out of the open estuary about one month after mating. Large numbers of females gathered in the shallows behind the sand bar of the other estuary, and at high tide when occasional waves broke over the sand bar several crabs left the water and walked over the sandbar and into the sea (Hill 1975). In northern Australia, the ovigerous females begin migrating in September–October and can move a considerable distance offshore (up to 95 km), and are found between 10 and 60 m depth (Hill 1994). After spawning at sea, the females return to estuaries and inshore waters (Heasman et al. 1985, Hill 1994).

Migrations to spawning grounds are not as dramatic in penaeids as they are in some other decapods. There is a general movement into deeper water with increased age in many species and some authors have characterized this as a spawning migration, but it is more likely this represents the ontogenetic shift from juvenile to adult habitats (Dall et al. 1990). In most cases the spawning grounds have been identified as being a subset of the adult distribution. Adult *Penaeus merguiensis* are distributed relatively close to shore (5–20 m depth) while the spawning females tend to concentrate in the deeper extent of their range (15–20 m depth; Crocos and Kerr 1983), although on the east coast of Australia some females move inshore to spawn (Dredge 1985). Similarly, the spawning grounds of the tiger prawns *P. esculentus* and *P. semisulcatus* in the northwestern Gulf of Carpentaria, Australia, were inshore and offshore, respectively, and within the general distribution identified for the adults of each species (Crocos 1987b). The peak spawning in *P. semisulcatus* in the Gulf of Carpentaria occurs during late winter to early spring (August–September; Crocos 1987a), but prior to this males and females undertake an extensive migration offshore into deep water (>40 m) before returning to spawn (CSIRO unpubl. observ., Crocos and van der Velde 1995). The reason for this migration is unknown.

In general, most of the movement identified to facilitate reproduction is best described as a spawning migration, rather than an ontogenetic habitats shift. The migration of lobsters over 100s of kilometers to spawn near oceanic deeps is the classic case. The offshore movement of penaeids occurs from the late-juvenile to early-adult phase and is an ontogenetic habitat shift. They may move within the offshore environment as adults and some species possibly move inshore to spawn (van der Velde et al., CSIRO Division of Marine and Atmospheric Research, unpubl. data).

7.6 Future Research

In many of the studies discussed in this review ontogenetic habitat shifts have been inferred by (sometimes separate) studies of juvenile and adult temporal and spatial distributions. Increasingly, researchers are recognizing the need to move beyond correlative studies between life history stages and to use markers to measure shifts and linkages between different habitats directly (Fairweather and Quinn 1993, Gillanders et al. 2003, Pittman and McAlpine 2003). The use of traditional artificial tags on decapods can be challenging because they may be shed when the animal molts and until recently tags small enough to be used effectively on juvenile decapods were not available. However, tags have been used successfully in some cases to directly demonstrate links between juvenile and adult habitats (see Chapters 10 and 13). For example, Frusher (1985) tagged juvenile *Penaeus merguiensis* in shallow waters in the Gulf of Papua; adults were subsequently captured offshore by the commercial fishery. In a more novel approach, Owens (1983) used the presence of the bopyrid parasite *Epipenaeon ingens* on adult *P. merguiensis* as a biological tag. *E. ingens* was only able to infect juvenile *P. merguiensis* whilst they were in their nursery grounds within estuaries and *E. ingens* was only present in certain estuaries. The parasite remains with the host for life and so Owens was able to determine that infected adults must have originated from the particular estuaries that contained the bopyrid.

More recently, new tagging technologies have become available that are small enough to be used on very small juvenile decapods, and because they are applied internally they are retained through successive molts. Davis et al. (2004) compared the growth, survival, and tag retention of coded wire and fluorescent elastomer tags in very small (<25 mm CL) juvenile blue crabs (*Callinectes sapidus*). Both techniques had different advantages: initial mortality from elastomer tags was less than that from coded wire, but tag retention was higher for coded wire tags. The fact that these techniques are able to be applied to very small animals and that they have excellent retention rates in a frequently molting decapod makes them ideal for application in studies concerned with demonstrating ontogenetic habitat shifts.

Of more general application are the relatively new techniques that use stable isotopes or genetic markers to demonstrate ontogenetic links between habitats (see Chapter 13 for a detailed review of methods). Fry (1981) successfully used ratios of $^{13}C:^{12}C$ to determine that brown shrimp (*Penaeus aztecus*) had moved from inshore seagrass beds into offshore waters. However, results from other studies have not been as clear (e.g., Loneragan et al. 1997) and further research into this area is warranted. In particular, there appears to be little information on the spatial variability of isotopic signatures within a habitat (see Chapter 3).

As part of an investigation into the feasibility of enhancing the natural population of *Penaeus esculentus* in Exmouth Gulf, western Australia, Preston et al. (1996) developed a stable isotope tag to provide a short-term chemical tag to monitor the fate of juvenile prawns over the first few weeks following release into the seagrass nursery areas. Unfortunately, isotope tags are only temporary: the longer an organism remains in a new isotopically distinct habitat, the more its original isotopic signature is diluted as the organism takes on the signature of the new habitat. DNA markers have also been developed to monitor the fate on released prawns sharing habitat with their wild-spawned conspecifics (Rothlisberg et al. 1999, Loneragan et al. 2003). DNA markers have the advantage of being permanent, compared to the stable isotopes and are ideal for tracking the movements of decapods from juvenile through to adult habitats. DNA markers also would allow the determination of the mortality of hard-to-study early life phases of decapods; the mortality of larvae, post-larvae, and early juveniles is a significant gap in our knowledge. For example, post-larvae or early juvenile penaeid prawns from parents with particular identifiable genetic tags could be released *en masse* and with subsequent sampling of the population, their abundance among the wild population and the mortality of the released individuals could be measured (Loneragan et al. 2003). Estimates of mortality of post-larval prawn would enhance our understanding of the drivers underlying ontogenetic habitat shifts.

Tracking organisms between habitats directly rather than merely inferring the links will assist fisheries and conservation managers. Ontogenetic connectivity is important when considering fisheries management and marine protected area planning; adult populations may be dependent upon subadult or juvenile populations elsewhere and so local management actions may prove ineffective (Roberts 1997, Cowen et al. 2005). In some cases it may not be only the various life history stages that require conservation, protection of corridors used by organisms transferring between habitats is also important. Apart from the larval stage most decapods make their ontogenetic shifts by moving across the seabed and may not cross certain seabed types. For example, Acosta (1999) found that fields of rubble acted as barriers to the dispersal of all life history stages (except adults) of the Caribbean spiny lobster *Panulirus argus*.

7.7 Summary

All decapods have complex life cycles involving several different stages and most undergo at least one habitat shift during their ontogeny. Several planktonic larval stages culminate in post-larvae that undergo their first habitat shift, settling from the pelagic environment to adopt a benthic existence. For many tropical decapods the initial habitat of the settling post-larvae and small juveniles comprises aquatic vegetation, e.g.; Caribbean spiny lobsters (*Panulirus argus*) settle onto red algae (*Laurencia* spp., Marx and Herrnkind 1985b), and grooved tiger prawns (*Penaeus semisulcatus*) settle onto seagrass (Staples et al. 1985). After spending a variable period in the initial settlement habitat most decapods then undergo a habitat shift as late-stage juveniles. For most species this entails a move into deeper water or from the intertidal into the subtidal. This trend of movement to deeper water continues with the development of the juveniles into adults. For example banana prawns (*Penaeus merguiensis*) move from their estuarine juvenile habitat where they live amongst mangroves, to the nearshore coastal zone (Staples 1980a) where they complete their life cycle. The exception to this general trend of a move from inshore to offshore is provided by the terrestrial crabs which spend the majority of their life cycle on land, only returning to the sea to spawn; the young return to land as small juveniles (Schiller et al. 1992, Brodie 2002, Barnes 2003).

It has been hypothesized that the mechanisms underlying habitat shifts in most animals are concerned with maximizing the fitness of the animal and that these may include minimizing mortality risk (μ), maximizing absolute growth rates (g), or a trade-off in which the animal chooses the habitat that minimizes the ratio of mortality risk to growth rate (minimize μ/g) (Werner and Gilliam 1984). Proof of which of these mechanisms underpins ontogenetic shifts in tropical decapods requires careful experimentation. We were unable to find examples of these sorts of experiments being done on tropical decapods, however, there are several cases that provide evidence to support these hypotheses. Examples of decapod ontogenetic habitat shifts that appear to function by minimizing mortality risk are:

(1) Late-stage juvenile Caribbean spiny lobsters (*Panulirus argus*) when they shift from algal to crevice habitat (Eggleston et al. 1992, Smith and Herrnkind 1992, Mintz et al. 1994);

(2) Late-stage juvenile grooved and brown tiger prawns (*Penaeus semisulcatus* and *P. esculentus*) moving offshore from inshore seagrass beds (Loneragan et al. 1994, O'Brien 1994b, Kenyon et al. 1995).

Decapods that appear to be making habitat shifts as a means of maximizing growth are:

(1) Late-stage juvenile banana prawns (*Penaeus merguiensis*) moving offshore from mangrove-lined rivers and creeks (M Haywood pers. observ.);
(2) Early-stage juvenile terrestrial crabs moving from the sea to the land (Adamczewska and Morris 2001a, Brodie 2002).

Without the appropriate experiments it is difficult to speculate on examples of decapods that might be shifting in order to minimize mortality risk to growth rates. One approach that could be used to help determine the mechanisms underlying decapod habitat shifts would be to follow the model of Dahlgren and Eggleston (2000). They conducted field experiments designed to quantify habitat-specific growth and mortality rates of range of size classes of Nassau grouper (*Epinephelus striatus*). The size range investigated spanned that over which the ontogenetic shift takes place. They then used habitat-specific mortality and growth rates in a cost-benefit analysis to test which of the three mechanisms (maximizing growth rates, minimizing mortality risk, or minimizing the ratio of mortality risk to growth rate) applied. It is only through the use of carefully designed experiments such as these that we will be able to determine the mechanisms underlying ontogenetic habitat shifts in tropical decapods.

Acknowledgments Thanks to Darren Dennis and to two anonymous referees who provided helpful suggestions and constructive criticisms that helped to improve this manuscript and to Lea Crosswell who did the artwork for the life cycle figures.

References

Acosta CA (1999) Benthic dispersal of Caribbean spiny lobsters among insular habitats: implications for the conservation of exploited marine species. Conserv Biol 13:603–612

Acosta CA (2002) Spatially explicit dispersal dynamics and equilibrium population sizes in marine harvest refuges. ICES J Mar Sci 59:458–468

Acosta CA, Butler MJ, IV (1997) Role of mangrove habitat as a nursery for juvenile spiny lobster, *Panulirus argus*, in Belize. Mar Freshw Res 48:721–728

Acosta CA, Robertson DN (2003) Comparative spatial ecology of fished spiny lobsters *Panulirus argus* and an unfished congener *P. guttatus* in an isolated marine reserve at Glover's Reef atoll, Belize. Coral Reefs 22:1–9

Adamczewska AM, Morris S (2001a) Ecology and Behavior of *Gecarcoidea natalis*, the Christmas Island red crab, during the annual breeding migration. Biol Bull 200:305–320

Adamczewska AM, Morris S (2001b) Metabolic status and respiratory physiology of *Gecarcoidea natalis*, the Christmas Island red crab, during the annual breeding migration. Biol Bull 200:321–335

Australian Bureau of Meteorology (2008) Climate statistics for Australian sites – Queensland. http://www.bom.gov.au/climate/averages/tables/ca_qld_names.shtml. Accessed 30 July 2008.

Baker R, Sheaves M (2006) Visual surveys reveal high densities of large piscivores in shallow estuarine nurseries. Mar Ecol Prog Ser 323:75–82

Barnes DKA (2003) Ecology of subtropical hermit crabs in SW Madagascar: short-range migrations. Mar Biol 142:549–557

Barnes DKA, Dulvy NK, Priestley SH et al (2002) Fishery characteristics and abundance estimates of the mangrove crab *Scylla serrata* in southern Tanzania and northern Mocambique. Afr J Mar Sci 24:19–25

Barnes RD (1974) Invertebrate Zoology. W.B. Saunders Co.

Beck MW (1995) Size-specific shelter limitation in stone crabs: a test of the demographic bottleneck hypothesis. Ecology 76:968–980

Berrill M (1975) Gregarious behavior of juveniles of the spiny lobster, *Panulirus argus* (Crustacea: Decapoda). Bull Mar Sci 25:515–522

Bert TM (1986) Speciation in western Atlantic stone crabs (genus: *Menippe*): the role of geological processes and climatic events in the formation of species. Mar Biol 93:157–170

Bert TM, Tilmant J, Dodrill J et al (1986) Aspects of the population dynamics and biology of the stone crab (*Menippe mercenaria*) in Everglades and Biscayne National Parks as determined by trapping. National Park Service, South Florida Research Center, Everglades National Park Homestead, Florida, 77 pp.

Brewer DT, Blaber SJM, Salini JP (1991) Predation on penaeid prawns by fishes in Albatross Bay, Gulf of Carpentaria. Mar Biol 109:231–240

Briones-Fourzán P, Castañeda-Fernández de Lara V, Lozano-Álvarez E et al (2003) Feeding ecology of the three juvenile phases of the spiny lobster *Panulirus argus* in a tropical reef lagoon. Mar Biol 142:855–865

Brodie RJ (1999) Ontogeny of shell-related behaviors and transition to land in the terrestrial hermit crab, *Coenobita compressus* (Crustacea: Coenobitidae). J Exp Mar Biol Ecol 241:67–80

Brodie RJ (2002) Timing of the water-to-land transition and metamorphosis in the land hermit crab *Coenobita compressus* H. Milne Edwards: evidence that settlement and metamorphosis are de-coupled. J Exp Mar Biol Ecol 272:1–11

Brown IR, Fielder DR (1992) Project overview and literature survey. In: Brown IR, Fielder DR (eds) The Coconut Crab: aspects of the biology and ecology of *Birgus latro* in the Republic of Vanuatu. ACIAR Monograph, pp. 1–11

Butler MJ, IV, Herrnkind WF (2000) Puerulus and juvenile ecology. In: Phillips BF, Kittaka J (eds) Spiny lobsters: fisheries and culture, pp. 276–301. Fishing News Books, Oxford

Chandrasekaran VS, Natarajan R (1994) Seasonal abundance and distribution of seeds of mud crab *Scylla serrata* in Pichavaram Mangrove, Southeast India. J Aquacult Trop 9:343–350

Childress MJ, Herrnkind WF (1994) The behavior of juvenile Caribbean spiny lobster in Florida Bay: seasonality, ontogeny and sociality. Bull Mar Sci 54:819–827

Childress MJ, Herrnkind WF (2001) The guide effect influence on the gregariousness of juvenile Caribbean spiny lobsters. Anim Behav 62:465–472

Chittleborough RG (1970) Studies on recruitment in the Western Australian rock lobster *Panulirus longipes cygnus* Geroge: density and natural mortality of juveniles. Aust J Mar Freshw Res 21:131–148

Chittleborough RG, Thomas LR (1969) Larval ecology of the Western Australian marine crayfish, with notes upon other panulirid larvae from the eastern Indian Ocean. Aust J Mar Freshw Res 20:199–224

Chong VC, Sasekumar A (1981) Food and feeding habits of the white prawn *Penaeus merguiensis*. Mar Ecol Prog Ser 5:185–191

Coles RG, Lee Long WJ, Squire BA et al (1987) Distribution of seagrasses and associated juvenile commercial penaeid prawns in north-eastern Queensland waters. Aust J Mar Freshw Res 38:103–119

Cooper RA, Ellis R, Serfling S (1975) Population dynamics, ecology and behavior of spiny lobsters, *Panulirus argus*, of St. John, U.S.V.I. (III) Population estimation and turnover. Bull Nat Hist Mus Los Angeles Cty 20:23–30

Cowen RK, Paris CB, Srinivasan A (2005) Scaling of connectivity in marine populations. Science 310:522–527

Cox C, Hunt JH, Lyons WG et al (1997) Nocturnal foraging of the Caribbean spiny lobster (*Panulirus argus*) on offshore reefs of Florida, USA. Mar Freshw Res 48:671–680

Crocos PJ (1987a) Reproductive dynamics of the grooved tiger prawn, *Penaeus semisulcatus*, in the north-western Gulf of Carpentaria, Australia. Mar Freshw Res 38:79–90

Crocos PJ (1987b) Reproductive dynamics of the tiger prawn, *Penaeus esculentus*, and a comparison with *P. semisulcatus*, in the north-western Gulf of Carpentaria, Australia. Aust J Mar Freshw Res 38:91–102

Crocos PJ, Kerr JD (1983) Maturation and spawning of the banana prawn *Penaeus merguiensis* de man (Crustacea:Penaeidae) in the Gulf of Carpentaria, Australia. J Exp Mar Biol Ecol 69: 37–59

Crocos PJ, van der Velde TD (1995) Seasonal, spatial and interannual variability in the reproductive dynamics of the grooved tiger prawn *Penaeus semisulcatus* in Albatross Bay, Gulf of Carpentaria, Australia: the concept of effective spawning. Mar Biol 122:557–570

Crowder LB, Cooper WE (1982) Habitat structural complexity and the interaction between bluegills and their prey. Ecology 63:1802–1813

Dahlgren CP, Eggleston DB (2000) Ecological processes underlying ontogenetic habitat shifts in a coral reef fish. Ecology 81:2227–2240

Dall W, Hill BJ, Rothlisberg PC et al (1990) The biology of the Penaeidae. Adv Mar Biol 27: 1–461

Davis GE (1977) Effects of recreational harvest on a spiny lobster, *Panulirus argus*, population. Bull Mar Sci 27:223–276

Davis GE, Dodrill JW (1989) Recreational fishery and population dynamics of spiny lobsters, *Panulirus argus*, in Florida Bay, Everglades National Park, 1977–1980. Bull Mar Sci 44: 78–88

Davis JLD, Young-Williams AC, Hines AH et al (2004) Comparing two types of internal tags in juvenile blue crabs. Fish Res 67:265–274

de Lestang S, Platell ME, Potter IC (2000) Dietary composition of the blue swimmer crab *Portunus pelagicus* L. Does it vary with body size and shell state and between estuaries? J Exp Mar Biol Ecol 246:241–257

Deegan LA (1993) Nutrient and energy transport between estuaries and coastal marine ecosystems by fish migration. Can J Fish Aquat Sci 50:74–79

Dennis DM, Pitcher CR, Prescott JH et al (1992) Severe mortality in a breeding population of ornate rock lobster *Panulirus ornatus* (Fabricius) at Yule Island, Papua New Guinea. J Exp Mar Biol Ecol 162:143–158

Dennis DM, Skewes TD, Pitcher CR (1997) Habitat use and growth of juvenile ornate rock lobsters, *Panulirus ornatus* (Fabricius, 1798), in Torres Strait, Australia. Mar Freshw Res 48: 663–670

Dredge MCL (1985) Importance of estuarine overwintering in the life cycle of the banana prawn, *Penaeus merguiensis*. In: Rothlisberg PC, Hill BJ, Staples DJ (eds) Second Australian national prawn seminar pp. 115–123. NPS2, Kooralbyn, Queensland

Dredge MCL (1990) Movement, growth and natural mortality rate of the red spot king prawn, *Penaeus longistylus* Kubo, from the Great Barrier Reef Lagoon. Aust J Mar Freshw Res 41:399–410

Edgar GJ (1990a) Predator-prey interactions in seagrass beds. I. The influence of macrofaunal abundance and size-structure on the diet and growth of the western rock lobster *Panulirus cygnus* George. J Exp Mar Biol Ecol 139:1–22

Edgar GJ (1990b) Predator-prey interactions in seagrass beds. II. Distribution and diet of the blue manna crab *Portunus pelagicus* Linneaus at Cliff Head, Western Australia. J Exp Mar Biol Ecol 139:23–32

Eggleston DB, Lipcius RN (1992) Shelter selection by spiny lobster under variable predation risk, social conditions, and shelter size. Ecology 73:992–1011

Eggleston DB, Lipcius RN, Miller DL (1992) Artificial shelters and survival of juvenile Caribbean spiny lobster *Panulirus argus*: spatial, habitat, and lobster size effects. Fish Bull 90:691–702

Eggleston DB, Lipcius RN, Miller DL et al (1990) Shelter scaling regulates survival of juvenile Caribbean spiny lobster *Panulirus argus*. Mar Ecol Prog Ser 62:79–88

Euzen O (1987) Food habits and diet composition of some fish of Kuwait. Kuwait Bull Mar Sci 9:65–85

Evans CR, Lockwood APM (1994) Population field studies of the guinea chick lobster (*Panulirus guttatus* Latreille) at Bermuda: abundance, catchability and behavior. J Shellfish Res 13:393–415

Fairweather PG, Quinn GP (1993) Seascape ecology: the importance of linkages. In: Battershill CN, Schiel DR, Jones GP, Creese RG, McDiarmid AB (eds) Proceedings of the 2nd international temperate reef symposium. NIWA Marine, Auckland, New Zealand

Fitzpatrick JJ, Jernakoff P, Phillips BF (1989) An investigation of the habitat requirements of the post-puerulus stocks of the western rock lobster. Final report to the Fishing Industry Research and Development Council, 80, Canberra

Fletcher WJ, Brown IR, Fielder DR et al (1992) Moulting and growth characteristics. In: Brown IR, Fletcher WJ (eds) The Coconut Crab: aspects of the biology and ecology of *Birgus latro* in the Republic of Vanuatu. ACIAR Monograph 8: 136 pp.

Forcucci D, Butler MJ, IV, Hunt JH (1994) Population dynamics of juvenile Caribbean spiny lobster, *Panulirus argus*, in Florida bay, Florida. Bull Mar Sci 54:805–818

Forward RBJ, Tankersley RA (2001) Selective tidal-stream transport of marine animals. Oceanogr Mar Biol: Annu Rev 39:305–353

Frusher SD (1985) Tagging of *Penaeus merguiensis* in the Gulf of Papua, Papua New Guinea. In: Rothlisberg PC, Hill BJ, Staples DJ (eds) Second Australian national prawn seminar, pp. 65–70. NPS2, Kooralbyn, Queensland

Fry B (1981) Natural stable carbon isotope tag traces texas shrimp migrations. Fish Bull 79:337–345

George RW (1968) Tropical spiny lobsters, *Panulirus* spp., of Western Australia (and the Indo-west Pacific). J R Soc West Aust 51:33–38

Gillanders BM, Able KW, Brown JA et al (2003) Evidence of connectivity between juvenile and adult habitats for mobile marine fauna: an important component of nurseries. Mar Ecol Prog Ser 247:281–295

Gotshall DW (1977) Stomach contents of northern California dungeness crabs, *Cancer magister*. Calif. Fish. Game 63:43–51

Gray JS (1974) Animal-sediment relationships. Oceanogr Mar Biol: Annu Rev 12:223–261

Grey DL, Dall W, Baker A (1983) A guide to the Australian penaeid prawns. Department of Primary Production of the Northern Territory, Darwin.

Gribble NA, Wassenberg TJ, Burridge C (2007) Factors affecting the distribution of commercially exploited penaeid prawns (shrimp) (Decapod: Penaeidae) across the northern Great Barrier Reef, Australia. Fish Res 85:174–185

Haywood MDE, Pendrey RC (1996) A new design for a submersible chronographic tethering device to record predation in different habitats. Mar Ecol Prog Ser 143:307–312

Haywood MDE, Staples DJ (1993) Field estimates of growth and mortality of juvenile banana prawns (*Penaeus merguiensis*). Mar Biol 116:407–416

Haywood MDE, Vance DJ, Loneragan NR (1995) Seagrass and algal beds as nursery habitats for tiger prawns (*Penaeus semisulcatus* and *P. esculentus*) in a tropical Australian estuary. Mar Biol 122:213–223

Heales DS (2000) The feeding of juvenile grooved tiger prawns *Penaeus semisulcatus* in a tropical Australian estuary: a comparison of diets in intertidal seagrass and subtidal algal beds. Asian Fish Soc 13:97–104

Heales DS, Vance DJ, Loneragan NR (1996) Field observations of moult cycle, feeding behaviour, and diet of small juvenile tiger prawns *Penaeus semisulcatus* in the Embley River, Australia. Mar Ecol Prog Ser 145:45–51

Heasman MP, Fielder DR, Shepherd RK (1985) Mating and spawning in the mudcrab, *Scylla serrata* (Forskal) (Decapoda: Portunidae), in Moreton Bay, Queensland. Aust J Mar Freshw Res 36:773–783

Heck KL, Jr., Thoman TA (1981) Experiments on predator-prey interactions in vegetated aquatic habitats. J Exp Mar Biol Ecol 53:125–134

Heck KL, Jr., Wilson KA (1987) Predation rates on decapod crustaceans in latitudinally separated seagrass communities: a study of spatial and temporal variation using tethering techniques. J Exp Mar Biol Ecol 107:87–100

Hedges SB (1989) An island radiation: Allozyme evolution in jamaican frogs of the genus *Eleutherodactylus* (Leptodactylidae). Caribb J Sci 25:123–147

Herrnkind WF (1980) Spiny lobsters: patterns of movement. In: Cobb JS, Phillips BJ (eds) Biology and management of lobsters, pp. 349–407. Academic Press, New York

Herrnkind WF, Butler MJ, IV (1986) Factors regulating postlarval settlement and juvenile micro-habitat use by spiny lobsters *Panulirus argus*. Mar Ecol Prog Ser 34:23–30

Herrnkind WF, Butler MJ, IV, Tankersley RA (1988) The effects of siltation on recruitment of spiny lobsters, *Panulirus argus*. Fish Bull 86:331–338

Herrnkind WF, Vanderwalker J, Barr L (1975) Population dynamics, ecology and behavior of the spiny lobster, *Panulirus argus*, of St. John, US Virgin Islands: habitation and pattern of movements and general behavior. Sci Bull, Nat Hist Mus, Los Angeles Cty 20:31–45

Hill BJ (1974) Salinity and temperature tolerance of zoeae of the portunid crab *Scylla serrata*. Mar Biol 25:21–24

Hill BJ (1975) Abundance, breeding and growth of the crab *Scylla serrata* in two South African estuaries. Mar Biol 32:119–126

Hill BJ (1976) Natural food, foregut clearance-rate and activity of the crab *Scylla serrata*. Mar Biol 34:109–116

Hill BJ (1978) Activity, track and speed of movement of the crab *Scylla serrata* in an estuary. Mar Biol 47:135–141

Hill BJ (1985) Effect of temperature on duration of emergence, speed of movement, and catchability of the prawn *Penaeus esculentus*. In: Rothlisberg PC, Hill BJ, Staples DJ (eds) Second Australian national prawn seminar, pp. 77–84. NPS2, Kooralbyn, Queensland

Hill BJ (1994) Offshore spawning by the portunid crab *Scylla serrata* (Crustacea: Decapoda). Mar Biol 120:379–384

Hill BJ, Williams MJ, Dutton P (1982) Distribution of juvenile, subadult and adult *Scylla serrata* (Crustacea: Portunidae) on tidal flats in Australia. Mar Biol 69:117–120

Howard RK (1988) Fish predators of the western rock lobster (*Panulirus cygnus* George) in a nearshore nursery habitat. Mar Freshw Res 39:307–316

Hyland SJ, Hill BJ, Lee CP (1984) Movement within and between different habitats by the portunid crab *Scylla serrata*. Mar Biol 80:57–61

Jeffs AG, Montgomery JC, Tindle CT (2005) How do spiny lobster post-larvae find the coast? N Z J Mar Freshw Res 39:605–617

Jernakoff P (1987) Foraging patterns of juvenile western rock lobsters, *Panulirus cygnus*. J Exp Mar Biol Ecol 113:125–144

Jernakoff P (1990) Distribution of newly settled western rock lobsters *Panulirus cygnus* Mar Ecol Prog Ser 66:63–74

Jernakoff P, Phillips BF, Fitzpatrick JJ (1993) The diet of post-puerulus western rock lobster, *Panulirus cygnus* George, at Seven Mile Beach, Western Australia. Mar Freshw Res 44: 649–655

Johnston R, Sheaves M (2007) Small fish and crustaceans demonstrate a preference for particular small-scale habitats when mangrove forests are not accessible. J Exp Mar Biol Ecol 353: 164–179

Joll LM, Phillips BF (1984) Natural diet and growth of juvenile western Rock Lobsters *Panulirus cygnus* George. J Exp Mar Biol Ecol 75:145–169

Kanciruk P (1980) Ecology of juvenile and adult Palinuridae (Spiny Lobsters). In: Cobb JS, Phillips BF (eds) The biology and management of lobsters, pp. 59–95. Academic Press, New York

Kenyon RA, Loneragan NR, Hughes JM (1995) Habitat type and light affect sheltering behaviour of juvenile tiger prawns (*Penaeus esculentus* Haswell) and success rates of fish predators. J Exp Mar Biol Ecol 192:87–105

Kenyon RA, Loneragan NR, Hughes JM et al (1997) Habitat type influences the microhabitat preference of juvenile tiger prawns (*Penaeus esculentus* Haswell and *Penaeus semisulcatus* De Haan). Estuar Coast Shelf Sci 45:393–403

Kenyon RA, Loneragan NR, Manson FJ et al (2004) Allopatric distribution of juvenile red-legged banana prawns (*Penaeus indicus* H. Milne Edwards, 1837) and juvenile white banana prawns (*Penaeus merguiensis* De Man, 1888), and inferred extensive migration, in the Joseph Bonaparte Gulf, northwest Australia. J Exp Mar Biol Ecol 309:79–108

Krumme U, Saint-Paul U, Rosenthal H (2004) Tidal and diel changes in the structure of a nekton assemblage in small intertidal mangrove creeks in northern Brazil. Aquat Living Resour 17:215–229

Laprise R, Blaber SJM (1992) Predation by moses perch, *Lutjanus russelli*, and blue spotted trevally, *Caranx bucculentus*, on juvenile brown tiger prawn, *Penaeus esculentus*: effects of habitat structure and time of day. J Fish Biol 40:627–635

Le Vay L (2001) Ecology and management of mud crab *Scylla* spp. Asian Fish Sci 14:101–111

Le Vay L, Ut VN, Walton M (2007) Population ecology of the mud crab *Scylla paramamosain* (Estampador) in an estuarine mangrove system: a mark-recapture study. Mar Biol 151: 1127–1135

Lipcius RN, Eggleston DB, Miller DL et al (1998) The habitat-survival function for Caribbean spiny lobster: an inverted size effect and non-linearity in mixed algal and seagrass habitats. Mar Freshw Res 49:807–816

Lipcius RN, Herrnkind WF (1982) Molt cycle alterations in behavior, feeding and diel rhythms of a decapod crustacean, the spiny lobster *Panulirus argus*. Mar Biol 68: 241–252

Lipcius RN, Hines H (1986) Variable functional responses of a marine predator in dissimilar homogeneous microhabitats. Ecology 67:1361–1371

Liu H, Loneragan NR (1997) Size and time of day affect the response of postlarvae and early juvenile grooved tiger prawns *Penaeus semisulcatus* De Haan (Decapoda: Penaeidae) to natural and artificial seagrass in the laboratory. J Exp Mar Biol Ecol 211:263–277

Loneragan N, Die D, Kenyon R et al (2002) The growth, mortality, movements and nursery habitats of red-legged banana prawns (*Penaeus indicus*) in the Joseph Bonaparte Gulf. CSIRO Marine Research, FRDC 97/105 Cleveland, Australia

Loneragan NR, Bunn SE, Kellaway DM (1997) Are mangroves and seagrasses sources of organic carbon for penaeid prawns in a tropical Australian estuary? A multiple stable isotope study. Mar Biol 130:289–300

Loneragan NR, Kenyon RA, Crocos PJ et al (2003) Developing techniques for enhancing prawn fisheries, with a focus on brown tiger prawns (*Penaeus esculentus*) in Exmouth Gulf. CSIRO, Cleveland, 287 pp.

Loneragan NR, Kenyon RA, Haywood MDE et al (1994) Population dynamics of juvenile tiger prawns (*Penaeus esculentus* and *P. semisulcatus*) in seagrass habitats of the western Gulf of Carpentaria, Australia. Mar Biol 119:133–143

Loneragan NR, Kenyon RA, Staples DJ et al (1998) The influence of seagrass type on the distribution and abundance of postlarval and juvenile tiger prawns (*Penaeus esculentus* and *P. semisulcatus*) in the western Gulf of Carpentaria, Australia. J Exp Mar Biol Ecol 228:175–195

Lozano-Alvarez E, Briones-Fourzan P, Ramos-Aguilar ME (2003) Distribution, shelter fidelity, and movements of subadult spiny lobsters (*Panulirus argus*) in areas with artificial shelters (casitas). J Shellfish Res 22:533–540

Lucas C, Kirkwood G, Somers I (1979) An assessment of the stocks of the banana prawn *Penaeus merguiensis* in the Gulf of Carpentaria. Aust J Mar Freshw Res 30:639–652

MacFarlane JW, Moore R (1986) Reproduction of the ornate rock lobster, *Panulirus ornatus* (Fabricius), in Papua New Guinea. Aust J Mar Freshwat Res 37:55–65

MacNae W (1968) A general account of the fauna and flora of mangrove swamps and forests in the Indo-West-Pacific region. Adv Mar Biol 6:73–270

Mantelatto FLM, Christofoletti RA (2001) Natural feeding activity of the crab *Callinectes ornatus* (Portunidae) in Ubatuba Bay (São Paulo, Brazil): influence of season, sex, size and molt stage. Mar Biol 138:585–594

Marx J, Herrnkind W (1985a) Factors regulating microhabitat use by young juvenile spiny lobsters, *Panulirus argus*: food and shelter. J Crustacean Biol 5:650–657

Marx JM, Herrnkind WF (1985b) Macroalgae(Rhodophyta: *Laurencia* spp.) as habitat for young juvenile spiny lobsters, *Panulirus argus*. Bull Mar Sci 36:423–431

Minello TJ (1999) Nekton densities in shallow estuarine habitats of Texas and Louisiana and the identification of essential fish habitat. Am Fish Soc Symp 22:43–75

Minello TJ, Zimmerman RJ (1983) Fish predation on juvenile brown shrimp, *Penaeus aztecus* Ives: the effect of simulated *Spartina* structure on predation rates. J Exp Mar Biol Ecol 72: 211–231

Mintz JD, Lipcius RN, Eggleston DB et al (1994) Survival of juvenile Caribbean spiny lobster: effects of shelter size, geographic location and conspecific abundance. Mar Ecol Prog Ser 112:255–266

Moore R, MacFarlane JW (1984) Migration of the ornate rock lobster, *Panulirus ornatus* (Eabricius), in Papua New Guinea. Aust J Mar Freshwat Res 35:197–212

Mukherjee AK (1971) Food habits of water birds of the Sundarban, West Bengal. II Herons and bitterns. J Bombay Nat Hist Soc 68:37–64

Norse EA (1977) Aspects of the zoogeographic distribution of *Callinectes* (Brachyura: Portunidae). Bull Mar Sci 27:440–447

O'Brien CJ (1994a) Ontogenetic changes in the diet of juvenile brown tiger prawns *Penaeus esculentus*. Mar Ecol Prog Ser 112:195–200

O'Brien CJ (1994b) Population dynamics of juvenile tiger prawns *Penaeus esculentus* in south Queensland, Australia. Mar Ecol Prog Ser 104:247–256

Olsen D, Herrnkind WF, Cooper R (1975) Population dynamics, ecology and behavior of the spiny lobster, *Panulirus argus*, of St. John, US Virgin Islands: introduction. Results of the Tektite Program. Nat Hist Mus, Los Angeles Cty Sci Bull 20:11–16

Otazu-Abrill M, Ceccaldi HJ (1981) Contribution to the study of the behavior of the reared *Penaeus japonicus* (Crustacea: Decapoda) opposite to light and sediment. Tethys 10:149–156

Owens L (1983) Bopyrid parasite *Epipenaeon ingens nobili* as a biological marker for the banana prawn *Penaeus merguiensis* de Mann. Aust J Mar Freshw Res 34:477–481

Palmer AR (1990) Predator size, prey size, and the scaling of vulnerability: hatchling gastropods vs. barnacles. Ecology 71:759–775

Pardieck RA, Orth RJ, Diaz RJ et al (1999) Ontogenetic changes in habitat use by postlarvae and young juveniles of the blue crab. Mar Ecol Prog Ser 186:227–238

Perkins-Visser E, Wolcott TG, Wolcott DL (1996) Nursery role of seagrass beds: enhanced growth of juvenile blue crabs (*Callinectes sapidus* Rathbun). J Exp Mar Biol Ecol 198:155–173

Pitcher CR, Skewes TD, Dennis DM et al (1992a) Distribution of seagrasses, substratum types and epibenthic macrobiota in Torres Strait, with notes on pearl oyster abundance. Aust J Mar Freshw Res 43:409–419

Pitcher CR, Skewes TD, Dennis DM et al (1992b) Estimation of the abundance of the tropical lobster *Panulirus ornatus* in Torres Strait, using visual transect-survey methods. Mar Biol 113:57–64

Pittman SJ, McAlpine CA (2003) Movements of marine fish and decapod crustaceans: process, theory and application. Adv Mar Biol 44:205–294

Poiner I, Conacher C, Loneragan N et al (1993) Effects of cyclones on seagrass communities and penaeid prawn stocks of the Gulf of Carpentaria. CSIRO, FRDC Projects 87/16 and 91/45 Cleveland

Preston NP, Smith DM, Kellaway DM et al (1996) The use of enriched 15 N as an indicator of the assimilation of individual protein sources from compound diets for juvenile *Penaeus monodon*. Aquaculture 147:249–259

Primavera JH (1997) Fish predation on mangrove-associated penaeids. The role of structures and substrates. J Exp Mar Biol Ecol 215:205–216

Puga R, de Leon ME, Cruz R (1996) Fishery of the spiny lobster *Panulirus argus* (Lattrille, 1804), and implications for management (Decapoda, Palinuridea). Crustaceana 69:703–718

Ratchford SG, Eggleston DB (1998) Size- and scale-dependent chemical attraction contribute to an ontogenetic shift in sociality. Anim Behav 56:1027–1034

Ratchford SG, Eggleston DB (2000) Temporal shift in the presence of a chemical cue contributes to a diel shift in sociality. Anim Behav 59:793–799

Roberts CM (1997) Connectivity and management of Caribbean coral reefs. Science 278: 1454–1457

Robertson AI (1988) Abundance, diet and predators of juvenile banana prawns, *Penaeus merguiensis*, in a tropical mangrove estuary. Aust J Mar Freshw Res 39:467–478

Rönnbäck P, Troll M, Kautsky N et al (1999) Distribution pattern of shrimps and fish among *Avicennia* and *Rhizophora* microhabitats in the Pagbilao mangroves, Philippines. Estuar Coast Shelf Sci 48:223–234

Rooker JR, Holt GJ, Holt SA (1998) Vulnerability of newly settled red drum (*Sciaenops ocellatus*) to predatory fish: is early-life survival enhanced by seagrass meadows? Mar Biol 131: 145–151

Rosales Juarez FJ (1976) Feeds and feeding of some species of the *Penaeus* genus. Proceedings symposium on the biology and dynamics of prawn populations Memorias. symposio sobre biologia y dinamica poblacional de camarones Instituto Nacional de Pesca, Guaymas, Son., Mexico

Rothlisberg PC, Preston NP, Loneragan NR et al (1999) Approaches to reseeding penaeid prawns. In: Howell BR, Moksness E, Svasand T (eds) Stock enhancement and sea ranching. Fishing News Books, Blackwell Science, Oxford, UK

Ryer CH (1988) Pipefish foraging: effects of fish size, prey size and altered habitat complexity. Mar Ecol Prog Ser 48:37–45

Salini JP, Blaber SJM, Brewer DT (1994) Diets of trawled predatory fish of the Gulf of Carpentaria, Australia, with particular reference to predation on prawns. Aust J Mar Freshw Res 45: 397–411

Sara L, Aguilar RO, Laureta LV et al (2007) The natural diet of the mud crab (*Scylla serrata*) in Lawele Bay, southeast Sulawesi, Indonesia. Philippine Agric Sci 90:6–14

Schiller C, Fielder DR, Brown IR et al (1992) Reproduction, early life history and recruitment. In: Brown IW, Fielder DR (eds) The Coconut Crab: aspects of the biology and ecology of *Birgus latro* in the Republic of Vanuatu. ACIAR Monograph 8:13–33

Schubart CD, Diesel R, Blair Hedges S (1998) Rapid evolution to terrestrial life in Jamaican crabs. Nature 393:363–365

Sharp WC, Hunt JH, Lyons WG (1997) Life history of the spotted spiny lobster, *Panulirus guttatus*, an obligate reef-dweller. Mar Freshw Res 48:687–698

Sheaves M (2001) Are there really few piscivorous fishes in shallow estuarine habitats? Mar Ecol Prog Ser 222:279–290

Skewes TD, Dennis DM, Pitcher CR et al (1997) Age structure of *Panulirus ornatus* in two habitats in Torres Strait, Australia. Mar Freshw Res 48:745–750

Smale M (1978) Migration, growth and feeding in the natal rock lobster *Panulirus homarus* (Linnaeus). South African Association for Marine Biological Research, 56 pp.

Smith KN, Herrnkind WF (1992) Predation on early spiny lobsters *Panulirus argus* (Latrille): influence of size and shelter. J Exp Mar Biol Ecol 157:3–18

Somers IF (1987) Sediment type as a factor in the distribution of commercial prawn species in the Western Gulf of Carpentaria, Australia. Aust J Mar Freshw Res 38:133–149

Somers IF (1994) Species composition and distribution of commercial peneaid prawn catches in the Gulf of Carpentaria, Australia, in relation to depth and sediment type. Aust J Mar Freshw Res 45:317–335

Somers IF, Crocos PJ, Hill BJ (1987a) Distribution and abundance of the tiger prawns *Penaeus esculentus* and *P. semisulcatus* in the north-western Gulf of Carpentaria, Australia. Aust J Mar Freshw Res 38:63–78

Somers IF, Kirkwood GP (1991) Population ecology of the Grooved Tiger Prawn, *Penaeus semisulcatus*, in the North-western Gulf of Carpentaria, australia: Growth, movement, age structure and infestation by the bopyrid parasite *Epipenaeon ingens*. Aust J Mar Freshw Res 42:349–367

Somers IF, Poiner IR, Harris AN (1987b) A study of the species composition and distribution of commercial penaeid prawns of Torres Strait. Aust J Mar Freshw Res 38:47–61

Sousa WP (1993) Size-dependent predation on the salt-marsh snail *Cerithidea californica* Haldeman. J Exp Mar Biol Ecol 166:19–37

Staples DJ (1980a) Ecology of juvenile and adolescent banana prawns, *Penaeus merguiensis*, in mangrove estuary and adjacent off-shore area Gulf Carpentaria. II. Emigration, population structure and growth of juveniles. Aust J Mar Freshw Res 31:653–665

Staples DJ (1980b) Ecology of juvenile and adolescent banana prawns, *Penaeus merguiensis*, in mangrove estuary and adjacent off-shore area Gulf of Carpentaria. I. Immigration and settlement of postlarvae. Aust J Mar Freshw Res 31:635–652

Staples DJ, Vance DJ (1986) Emigration of juvenile banana prawns *Penaeus merguiensis* from a mangrove estuary and recruitment to offshore areas in the wet-dry tropics of the Gulf of Carpentaria, Australia. Mar Ecol Prog Ser 27:239–252

Staples DJ, Vance DJ, Heales DS (1985) Habitat requirements of juvenile penaeid prawns and their relationship to offshore fisheries. In: Rothlisberg PC, Hill BJ, Staples DJ (eds) Second Australian national prawn seminar, pp. 47–54. NPS2, Kooralbyn, Queensland

Tagatz ME (1968) Biology of the blue crab, *Callinectes sapidus* Rathbun, in St. Johns River, Florida. Fish Bull 67:17–26

Trendall J, Bell S (1989) Variable patterns of den habitation by the ornate rock lobster, *Panulirus orantus*, in the Torres Strait. Bull Mar Sci 45:564–573

Ut VN, Le Vay L, Nghia TT et al (2007) Development of nursery culture techniques for the mud crab *Scylla paramamosain* (Estampador). Aquac Res 38:1563–1568

Vance DJ, Haywood MDE, Heales DS et al (1996) How far do prawns and fish move into mangroves? Distribution of juvenile banana prawns *Penaeus merguiensis* and fish in a tropical mangrove forest in nothern Australia. Mar Ecol Prog Ser 131:115–124

Vance DJ, Haywood MDE, Staples DJ (1990) Use of a mangrove estuary as a nursery area for postlarval and juvenile banana prawns, *Penaeus merguiensis* de Man, in northern Australia. Estuar Coast Shelf Sci 31:689–701

Vance DJ, Heales DS, Loneragan NR (1994) Seasonal, diel and tidal variation in beam-trawl catches of juvenile grooved tiger prawns, *Penaeus semisulcatus* (Decapoda: Penaeidae), in the Embley River, north-eastern Gulf of Carpentaria, Australia. Aust J Mar Freshw Res 45:35–42

Walton ME, Le Vay L, Truong LM et al (2006) Significance of mangrove mudflat boundaries as nursery grounds for the mud crab, *Scylla paramamosain*. Mar Biol 149:1199–1207

Wang YG, Haywood MDE (1999) Size-dependent natural mortality of juvenile banana prawns *Penaeus merguiensis* in the Gulf of Carpentaria, Australia. Mar Freshw Res 50:313–317

Wassenberg TJ, Hill BJ (1987) Natural diet of the tiger prawns, *Penaeus esculentus* and *P. semisulcatus*. Aust J Mar Freshw Res 38:169–182

Wassenberg TJ, Hill BJ (1993) Diet and feeding behaviour of juvenile and adult banana prawns *Penaeus merguiensis* in the Gulf of Carpentaria, Australia. Mar Ecol Prog Ser 94:287–295

Waterman TH, Chase FA Jr. (1960) General crustacean biology. In: Waterman TH (ed.) The physiology of the Crustacea. Vol. 1. Metabolism and growth. Academic Press, New York and London, 670 pp.

Werner EE, Gilliam JF (1984) The ontogenetic niche and species interactions in size-structured populations. Annu Rev Ecol Syst 15:393–425

Werner EE, Mittlebach GG (1981) Optimal foraging: field tests of diet choice and habitat switching. Am Zool 21:813–829

Williams AB (1955) A contribution to the life histories of commercial shrimps (Penaeidae) in North Carolina. Bull of Mar Sci Gulf Caribb 5:116–146

Williams MJ (1982) Natural food and feeding in the commercial sand crab *Portunus pelagicus* Linnaeus, 1766 (Crustacea : Decapoda : Portunidae) in Moreton Bay, Queensland. J Exp Mar Biol Ecol 59:165–176

Yoshimura T, Yamakawa H (1988) Microhabitat and behaviour of settled pueruli and juveniles of the Japanese spiny lobster *Panulirus japonicus* at Kominato, Japan. J Crustacean Biol 8:524–531

Chapter 8
Diel and Tidal Movements by Fish and Decapods Linking Tropical Coastal Ecosystems

Uwe Krumme

Abstract Short-term movements of fishes and decapods can lead to regular changes in biomass, diversity, mortality, predation, and flux of energy between adjacent ecosystems. At low latitudes the day-night cycle is relatively stable and uniformly affects activity rhythms of marine organism at all longitudes. In contrast, tidal ranges and tidal types differ significantly between coasts and regions. On coasts with weak tides, twilight migrations connect adjacent habitats. On tidal coasts, migrations are tightly coupled to the interactive effect of the diel and tidal cycles which results in complex but predictable patterns of change within and between ecosystems. Diel and tidal migrations share several similarities (connection of resting and feeding sites, sequence of species and size groups, site fidelity, homing, constant pathways). The spring-neap tide cycle and its interaction with the diel cycle is a key factor influencing regular short-term variations on tidal coasts. The home range of a species on a macrotidal coast may be an order of magnitude greater than that of conspecifics from a microtidal coast, suggesting a need for larger marine parks on macrotidal coasts. Regional comparisons, e.g., between the Caribbean and the Indo-West Pacific, often disregard the significant tidal differences inherent to the ecosystems. It is suggested here that broad-scale comparisons must be redefined; regional comparisons should focus on geographical regions with similar tidal regimes, or on systems with different tidal regimes but with similar species communities.

Keywords Shallow-water fishes · Twilight migration · Lobsters · Shrimps · Portunid crabs

8.1 Introduction

The shallow waters of tropical coasts are home to unique ecosystems such as coral reefs and mangrove forests. Where coral reefs, mangrove forests, and seagrass beds co-occur, the ecosystems are usually connected with one another through the

U. Krumme (✉)
Leibniz-Center for Tropical Marine Ecology (ZMT), Fahrenheitstrasse 6, 28359
Bremen, Germany
e-mail: uwe.krumme@zmt-bremen.de

I. Nagelkerken (ed.), *Ecological Connectivity among Tropical Coastal Ecosystems*,
DOI 10.1007/978-90-481-2406-0_8, © Springer Science+Business Media B.V. 2009

movements of organisms, nutrients, and other materials. In many tropical regions, coral reefs and seagrass beds form a mosaic of patches within a matrix of sandy sediments. Extensive mangrove forests grow within a complex network of sublittoral channels and intertidal creeks, mudflats, and sand banks. For some species, one habitat within a complex seascape setting is sufficient to complete its life cycle. For most other species, however, one habitat cannot satisfy the changing needs. Mobility is the solution to local deficiencies and the species move between different available habitats at different temporal and spatial scales. These movements can occur on a longer time scale such as on a seasonal basis or once during ontogeny (see Chapters 6, 7, 10), or on a shorter time scale according to the lunar, diel, or tidal cycle. When adjacent habitats are used on the short-term scale, the movements greatly affect the everyday life of an organism and likely influence growth and survival. In this chapter, the focus is on the short-term movements of fish and decapods in tropical shallow-water environments in relation to the diel and tidal cycles. It should be noted that all nektonic organisms (organisms living in the water column that can swim strongly enough to move counter to modest water currents) display movement, but not all migrate. For the purpose of this review, I refer to the general definition of migration as discussed by Dingle (1996): 'Migratory behavior is persistent and straightened-out movement effected by the animal's own locomotory exertions or by its active embarkation on a vehicle. It depends on some temporary inhibition of station-keeping responses, but promotes their eventual disinhibition and recurrence'. In the special case of this review, migratory movements connect adjacent habitat types or ecosystems, and involve a regular directional and temporal component.

A migration persists as an evolutionary stable strategy when the benefits exceed the costs. The ability to move enables species to optimize the use of resources in more than one ecosystem. A particular species exploits an ecosystem when benefits are high and avoids it when the benefits are low relative to other available ecosystems. The pay-off of a migration depends on the benefit provided by the present ecosystem, the cost of movement to another ecosystem, and by the expected conditions in an alternative ecosystem. For example, movement to a potentially profitable adjacent ecosystem would not pay off when the ecosystem is too far away, or when the risk of predation precludes its use at a given time of day.

Short-term migrations between adjacent ecosystems usually serve at least one of five functions (Gibson 1992, 1996, 1999, Gibson et al. 1998, Rountree and Able 1993): (1) feeding, (2) shelter or reducing the risk of predation, (3) avoidance of inter- and/or intraspecific competition, (4) reproduction, and (5) searching for a physiologically optimum environment.

Mobile organisms that shuttle between ecosystems influence each of the ecosystems used. Short-term migrations (i) change the species diversity and abundance counts in a given ecosystem (Thompson and Mapstone 2002), (ii) are a vector for the export of organic matter and nutrients from feeding to resting sites (Meyer et al. 1983, Meyer and Schultz 1985), (iii) regularly change the biomass of organisms in an ecosystem (e.g., Nagelkerken et al. 2000), (iv) shape patterns in herbivory and mortality in the feeding grounds (e.g., Ogden and Zieman 1977), and (v) shape the ecological value to the organisms of a given ecosystem. If a given ecosystem is out of reach for short-term migrants due to an unfavorable seascape configuration, the ecosystem cannot perform its potential values (Baelde 1990, Dorenbosch et al. 2007). The connectivity between shallow-water ecosystems caused by short-term migrations of fish and decapods has received increasing attention (e.g., Sheaves 2005), but with the major drawback that studies have been mainly conducted in the Caribbean where the influence of the tide is virtually absent. Surprisingly few studies on the topic come from Australia and the Indo-Pacific region and even less from African coasts.

Understanding the spatial and temporal dynamics caused by mobile organisms that use different habitat types at different times of day and at different tidal stages is crucial to sampling design, interpretation of ecological studies, and ecosystem management (Pittman and McAlpine 2003, Beck et al. 2001, Adams et al. 2006). Migrations can lead to faulty or incomplete population censuses or confounding effects in catch rates from a single ecosystem (e.g., Wolff et al. 1999). Optimized sampling strategies designed to measure long-term changes have to account for the short-term effects of migrations, e.g., considering the variation caused by the spring-neap tide cycle. The spatial and temporal patterns in short-term migrations determine the routine, everyday movements within the home range of mobile species and the connectivity between adjacent ecosystems. This information is required for population dynamics, spatial population models, and resource management (Cowen et al. 2006).

Fishes and most decapods are nektonic organisms. Many are commercially important, and as they are often caught together and interact ecologically (e.g., in predator-prey relationships), these species are addressed jointly in this chapter. For more information on the movements of their larvae see Sale (2006) for coral reef fishes, and Dall et al. (1990) for Penaeidae.

In this chapter an applied overview of the effects by which diel and tidal cycles influence the activity and use patterns of fish and decapods in the tropics will be given. The description of the distribution of tidal ranges and tidal types on tropical coasts highlights the diversity and regional differences in tidal regimes between coasts and regions. The review on the diel movements of fish is focused on examples from the Caribbean, the best studied and largest tropical area with negligible tides. The review of diel movements of decapods covers diel changes in activity and foraging ranges of lobsters, shrimps, and swimming crabs. The tidal movements of fish provides an overview on the variety of responses displayed by individuals, size groups, sexes, species, populations, and assemblages in response to the tide-time of day interactions. The section provides a comparison of the similarities and differences between diel and tidal migrations. The tidal movements of shrimps and swimming crabs highlight the interactive influence of the diel and tidal cycles on activity patterns. A regional comparison contributes to the debate on differences in connectivity between the Caribbean and the Indo-West Pacific. The final sections cover two aspects which have been largely overlooked in previous studies: the possibly greater home ranges of species in areas of greater tidal range, and the need to consider the influence of the tides when studying and comparing ecosystem functions and biodiversity patterns.

8.2 The Diel Cycle

The length of the diel cycle is 24 hrs, or the time it takes for the Earth to make one complete rotation around its axis. Due to the inclination of the Earth's axis by 23.5°, day length and solar irradiation differ with latitude and season. The diel cycle in the tropics differs from that in higher latitudes in two important aspects: (1) in lower latitudes, there is a relatively fixed cycle of approximately 12 hrs light and 12 hrs darkness year-round, whereas in temperate regions day length can vary between 16 hrs in summer and 8 hrs in winter, and (2) in the tropics the transition or twilight period at sunrise and sunset is relatively short (approx. 1 hr) whereas these periods can last for hours at higher latitudes. Consequently, in the tropics changes in illumination levels are highly consistent throughout the year and consequently, changes in the activity of most tropical organisms are well synchronized with the diel cycle. Four distinct diel periods can be distinguished that structure the activity patterns of most organisms exposed to light: sunrise, daytime, sunset, and night. Helfman (1993) classified the diel activity of fish families into diurnal, nocturnal, crepuscular (active at dusk and dawn), and two groups without clear activity periods. One notable example of the influence of the twilight period on fish behavior and distribution is the regular species changeover before and after the 'quiet period' in the clear water environment of coral reefs (Hobson 1972). However, even in the very turbid waters of mangrove estuaries, upsurges in the activity of fishes occur during twilight (Krumme and Saint-Paul 2003).

Sunlight is also reflected by the moon. The effect of moonlight on the activity of aquatic organisms depends on the lunar phase, cloud cover, water clarity, and water depth of an organism. Artificial light originating from fisheries and coastal construction, which brightly illuminates coastal areas for hours each night, may also be considered. With the exception of sea turtles, our understanding of the effects of light pollution on changes in the activity patterns of marine species is still in its infancy.

8.3 The Tides and Tidal Currents

This section does not present an exhaustive account of tides. The focus is on tidal characteristics which influence, or are in some way relevant to, the movement of nektonic organisms. General information on tides can be found in oceanographic textbooks (e.g., Dietrich 1980).

Tides are a complex natural phenomenon (e.g., Kvale 2006), which may be described as the periodic rise and fall of the sea surface level. A tidal current refers to the horizontal flow. Tides are mainly generated by gravitational forces of the moon, and to a smaller extent by the sun. Several factors such as the shape of the sea floor, local bathymetry, coastal morphology, the Coriolis effect, and changes in freshwater discharge, wind or air pressure act together to form a local tidal regime (Dietrich 1980). For nektonic organisms, tidal currents provide a mode of free transport. Tidal currents regularly reverse and can be used to selectively travel in a particular direction (selective tidal stream transport) or to commute between low and high tides, enabling an organism to move back and forth.

8.3.1 Short-term Patterns

A flood tide is defined as a rising of the water level (incoming tide); the ebb tide is the fall of the water level (outgoing tide). The point at which current speed and current direction are at zero and the tide turns from flood to ebb tide is termed slack high water. A tidal cycle lasts from one phase of the tide to the recurrence of the same phase. The tidal range is the difference in water level between low tide and high tide. Tides are of a semidiurnal, diurnal, or mixed type. (1) The semidiurnal tide is the most common tide (Fig. 8.1a) and is characterized by the biggest ranges and fastest current speeds. Two tidal cycles, each of 12 hrs 25 min duration, are observed on the coast each day, with small differences between successive high and low water levels. (2) Diurnal tides have only one tidal cycle per day (24 hrs 50 min) (Fig. 8.1b). (3) Mixed tides are predominantly diurnal (Fig. 8.1c) or semidiurnal (Fig. 8.1d). Mixed tides often display large differences in the heights of high or low water, or in both.

The fact that the lunar day period lasts 24 hrs and 50 min means that each tidal cycle is completed with a time delay in relation to the diel cycle (Fig. 8.1). Thus, in a sense each tidal cycle is a unique event and cannot be replicated. For example, if today slack high tide was at midday, then after six, twelve and fourteen days slack high tide will occur at 17:00, 22:00, and 23:40 hrs, respectively. The delay from tide to tide causes significant interactions between the tidal and diel cycles, with far-reaching consequences for the activity patterns of coastal organisms. The investigation of these complex interactions requires sophisticated sampling designs (see, e.g., Kleypas and Dean 1983, Krumme et al. 2004). However, on a few coasts the tidal cycle is in phase with the diel cycle and low and high waters occur at approximately the same time each day (e.g., Indonesia, South Pacific, and Adelaide/Australia; American Practical Navigator 2002).

8.3.2 Spring–Neap Tide Cycle

The tidal range changes following the lunar cycle. Usually there is a delay of one or two days between the lunar phase and the effect of the tide. At spring tides (at approximately full and new moon when the sun, moon, and earth are aligned), tidal

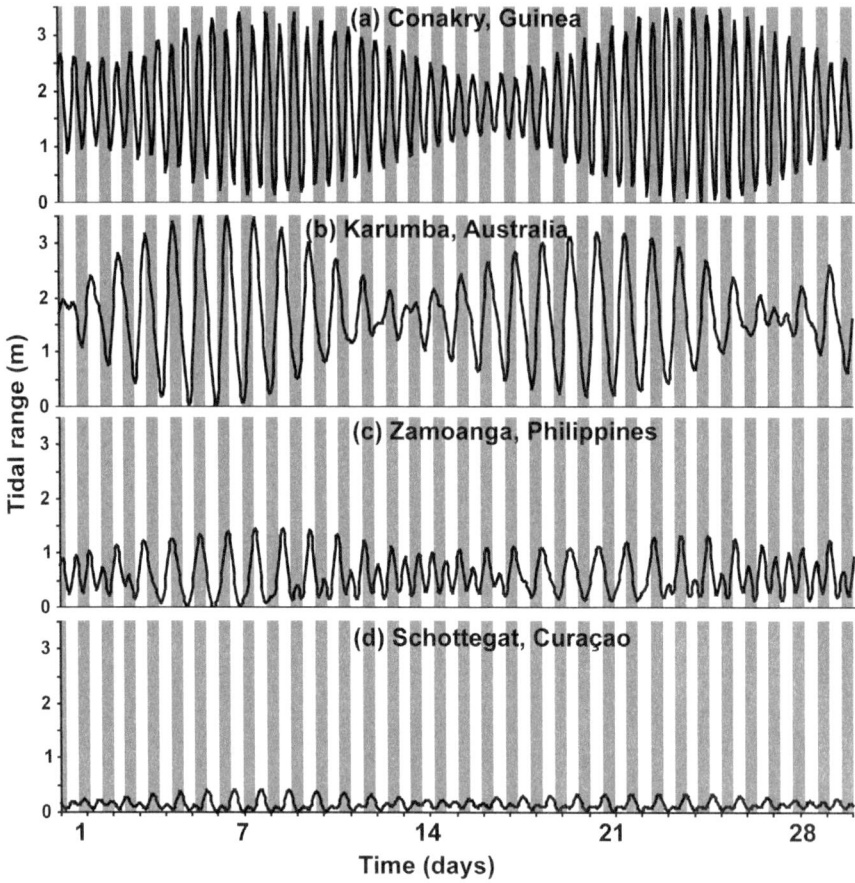

Fig. 8.1 Interaction of the diel and tidal cycles during one lunar cycle for a (**a**) semidiurnal tide (Conakry, Guinea; 13°43'W, 9°30'N), (**b**) diurnal tide (Karumba, Gulf of Carpentaria, Australia; 140°50'E, 17°30'S), (**c**) mixed-diurnal tide (Zamboanga, Mindanao, Philippines; 122°4'E, 6°54'N), and (**d**) mixed-semidiurnal tide (Schottegat, Curaçao; 68°56'W, 12°7'N). Shaded columns indicate night. Tide data from www.wxtide32.com

ranges and the current speeds reach a maximum. The slack high tide is extremely high and the slack low tide is extremely low. During neap tide (at the waxing and waning of the moon, when the earth and moon are perpendicular to each other) tidal ranges are significantly reduced and current speeds are much weaker (Kvale 2006).

There are, however, exceptions to these norms, for example on the south eastern Gulf of Carpentaria, Australia, the lunar phases are unrelated to the spring–neap tide cycle (Munro 1975). Comparative investigations of coasts with more unusual tidal characteristics would contribute to our overall understanding of tidal movement patterns of nektonic organisms.

8.3.3 Extreme Tides

Extreme spring tides cause large-scale changes on our coasts. Mobile coastal organisms may synchronize large-scale movements such as home range relocation or ontogenetic shifts to other ecosystems. They occur on a regular basis: (1) At about equinox (each March 21 and September 21) spring and neap tides are extremely strong and weak, respectively. (2) Every 7.5 orbits (or every 221 days) the moon comes closest to the earth (perigree), either at full or new moon. Then, particularly strong tidal forces result in strong perigrean tides. (3) The nodal tide caused by variations in the moon's declination results in extremely high tides approximately every 18.6 yrs. Thus, very strong tides occur on a fortnightly (regular spring tides), seasonal, annual, and decadal basis, providing means of transport for organisms throughout their development stages, for both short- and long-lived coastal species.

On coasts with weak tides, meteorological effects may sometimes exceed the tidal range. For instance, in shallow water annual tides are often wind-driven. This can lead either to exceptionally high or low tides. In the Red Sea, catastrophic seasonal low tides can expose the reef flats to the air for hours in the summer (Loya 1972, Sheppard et al. 1992; Fig. 8.2). In the Gulf of Mexico extreme high tide periods due to meteorological and climatic events lead to saltmarsh accessibility for nektonic organisms (e.g., Rozas 1995).

Fig. 8.2 Reef flat exposed due to extremely low seasonal tide in the coral reef of Eilat, Red Sea. Picture from Loya (1972), with kind permission from Springer Science + Business Media and Y Loya

8.3.4 Predicted and Observed tides

Tide tables are only predictions. The observed tide can deviate considerably from the predicted tide. Site-to-site differences can be significant, especially in mangrove or coral reef areas where the tidal currents are channelized (Wolanski et al. 1992,

Wolanski 1994). Three examples highlight the need to understand the tide at a study site:

(1) The duration of flood and ebb tide may be unequal, e.g., due to local topography or intertidal vegetation. Not only the time but also the current speeds differ between flood and ebb because a fixed volume of water flows faster during a shorter period, and slower during a longer period. This results in either flood or ebb-dominated systems. Tidal asymmetries most likely are reflected in the timing of tidal movements and usage patterns of nektonic organisms. Intertidal land reclamation and mangrove loss can modify the flood-ebb tide asymmetry and hence, the use patterns of nektonic organisms.

(2) In theory, flood and ebb current speeds and the rate of rise and fall reach a maximum halfway between slack high and slack low water. Nektonic organisms may respond specifically to this time window of maximum currents because it provides the greatest potential for transport. It may further coincide with the period of greatest turbidity, i.e., lowest visibility, and hence decreased risk of predation when moving with the tide. However, current speeds and water level change may differ during flood and ebb tide. In addition, a distinct current peak is often difficult to identify, especially with weaker tides. A distinct maximum may be undetectable, or a peak may be earlier or later than halfway between low and high tide. In interconnected channel systems, momentary stops or current reversals may occur during weak flood tides, leading to two or possibly even three distinct current speed maxima.

(3) In estuaries and channel systems tides can be extremely complex. The tidal currents in channels are characterized by significant vertical and horizontal gradients that may vary with estuarine location and tidal stage. Ebb tide currents are usually greatest close to the surface in the centre of a channel whereas flood tide currents may be stronger at subsurface depths. Furthermore, the currents tend to turn earlier close to the shore than in the midchannel. In layered estuaries flood tides may begin between a few minutes to >1 hr earlier at the bottom.

8.3.5 Distribution of Tidal Types and Tidal Ranges on Tropical Coasts

Unlike the diel cycle which has a clear latitudinal gradient and appears the same at any given longitude, tides vary strongly with coastal region, both in type and range. Davies (1972) and Hayes (1975) classified the hydrographic regime of marine coasts into micro-, meso-, macrotidal using spring tide ranges of <2 m, 2–4 m, and >4 m, respectively. Others use slightly different subdivisions and the relevance of this classification scheme for the movements of nektonic organisms is still unclear.

Figures 8.3 and 8.4 illustrate the complex modern pattern of tidal types and ranges on tropical coasts. Semidiurnal tides are characteristic of the Atlantic coasts and many coastlines are meso- or macrotidal. The enclosed Caribbean is

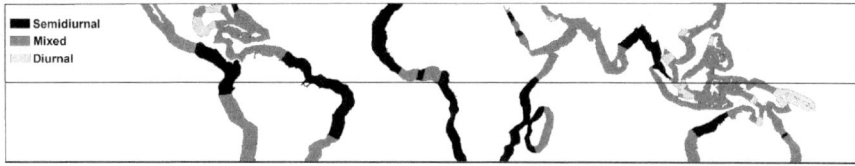

Fig. 8.3 Distribution of tidal types along the tropical coasts of the world. Mixed tides are not further separated into mixed diurnal and mixed semidiurnal tides. Adapted from Fig. 1.2 in 'Distribution of tidal types along the world's coast' in Davies (1972), Geographical Variation in Coastal Development, Oliver Boyd. Additional information used: Dietrich (1980), Eisma (1998), Admiralty Co-Tidal Atlas (2001), and www.wxtide32.com

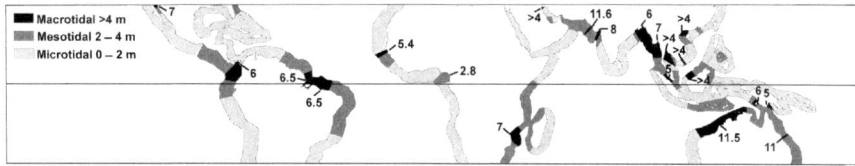

Fig. 8.4 Distribution of tidal ranges (in meters) along tropical coasts of the world. Adapted from Kelletat (1995). Additional information used: Dietrich (1980), Eisma (1998), Admiralty Co-Tidal Atlas (2001), and www.wxtide32.com

an exception in having mixed and diurnal tides with a tidal range of <1 m (e.g., in Curaçao, Fig. 8.1d).

In the Indian Ocean, tides are semidiurnal (e.g., East Africa, Bay of Bengal, Andaman Sea, Strait of Malacca, Northwest Australia) or mixed semidiurnal (e.g., Arabian Sea, Coromandel Coast, West Sumatra, South Java). The Red Sea is almost totally enclosed and tides are negligible, although in the Gulfs of Aqaba and Suez, spring tide ranges may reach 1–1.5 m (Sheppard et al. 1992).

The Pacific Ocean is dominated by mixed semidiurnal tides, for example the Great Barrier Reef with tidal ranges >3 m (Wolanski 1994). Semidiurnal and mixed semidiurnal tides characterize the East Pacific coast. Small tidal ranges are characteristic of islands in the open ocean. In the Indo-Malayan Archipelago the patterns are complex. Two regions can be differentiated: (1) The South China Sea and adjacent basins are dominated by mixed diurnal tides with tidal ranges from <1 m to >2 m (e.g., Zamboanga, Fig. 8.1c). However, within this area there are cells with distinct diurnal tides (e.g., western Gulf of Thailand, Gulf of Tonkin, north coast of Java), and cells with mixed semidiurnal tides (e.g., Singapore, Mekong Delta, Sarawak). (2) East of this area mixed semidiurnal tides dominate (e.g., east coast of the Philippines, Celebes Sea, northern New Guinea). Tidal ranges can be <1 m (Sulawesi), or exceed 2 m (east Indonesian islands, East Kalimantan). Again, within this area there is a diurnal cell at the southeast coast of West Papua.

In conclusion, tides are a common and regular natural disturbance of many tropical coastlines, except for the Caribbean, the Red Sea, a few cells in the West Pacific (e.g., the Java Sea), and many oceanic islands. The triangle of marine biodiversity between Indonesia, the Philippines, and Papua New Guinea coincides with the greatest diversity of tidal types and tidal ranges. The overall tidal ranges of the Indo-West Pacific are not exceptionally high, but undoubtedly are higher than the microtidal coasts of the Caribbean.

If we imagine the global distribution of mangroves, coral reefs, and seagrass beds (not covered here, but refer to Spalding et al. 1997, Spalding et al. 2001, Larkum et al. 2006) and superimpose it in the minds' eye on the global distribution of tidal ranges (Fig. 8.4), it is apparent that the coexistence of the troika mangrove-seagrass-coral reef is restricted to coasts with weak or intermediate tidal ranges and minor freshwater input. In contrast, the world's largest contiguous mangrove stands (e.g., the Sundarbans, the coasts north and south of the Amazon mouth, South Papua, and West Sumatra) grow on coastal plains where large rivers enter the sea and where tidal ranges are large. Due to coastal estuarization and high sediment loads in these areas, coral reefs and seagrass beds are excluded. Each ecosystem can thrive in the absence of the others, but where environmental conditions facilitate their spatial overlap, biodiversity and productivity can be significantly enhanced (Nagelkerken et al. 2002, Mumby et al. 2004, Dorenbosch et al. 2005).

8.4 Diel Movements of Fish

Most fish will alter their activity and patterns of movements following the day-night cycle. Diel movements of tropical coastal fish are most adequately studied where the influence of the tides is negligible. In the absence of tidal currents, distances between sites have to be covered by active movements. Hence, the benefits of moving (e.g., finding a rich food patch) have to outweigh not only the potential costs that incur also on tidal coasts (e.g., increased risk of predation) but also the cost of increased energy expenditure.

Daily twilight migrations are common among fishes of heterogeneous tropical marine seascapes (Table 8.1). Often these migrations connect one micro-habitat or habitat type that provides shelter with another that provides food. Diurnal reef-associated families such as Acanthuridae, Chaetodontidae, Labridae, Pomacentridae, and Scaridae move from their daytime foraging sites to nighttime shelter in crevices and cavities of rocks and corals, in seagrass or sediment. Nocturnal families such as Apogonidae, Haemulidae, Lutjanidae, Holocentridae, Pempheridae, and Sciaenidae shelter in or near complex structured resting sites through the day and migrate to adjacent feeding grounds at night (e.g., Hobson 1965, 1968, 1972, 1974, Starck and Davis 1966, Randall 1967, Collette and Talbot 1972).

8.4.1 Haemulidae

Certainly the best documented diel movement of fish is the twilight migration of grunts (Haemulidae/Pomadasyidae) between daytime resting sites on patch reefs and nightly feeding sites in adjacent seagrass beds (Figs. 8.5, 8.6). Grunts are

Table 8.1 Fish families and species that display regular diel movements between tropical coastal shallow-water habitats. Underlined habitats indicate the time of day when feeding occurs. Foraging range: distance between resting and feeding sites. Superscript numbers refer to the respective literature reference

| Fish family | Species | Residence at | | Foraging range | Region |
		Day	Night		
Carcharhinidae	Carcharhinus leucas[1]	Offshore	Reef	–	Caribbean (Florida)
Acanthuridae	Prionurus punctatus[2]	Offshore rocks	Reef	–	Gulf of California
Acanthuridae	Naso hexacanthus[2]	Offshore	Reef	–	Hawaii
Anomalopidae	Photoblepharon palpebratum[3]	Reef	Away from reef	–	Red Sea
Apogonidae[9, 10, 11, 12, 13, 14]	Apogon cyanosoma[4] Apogon doerderleini[4] Apogon exostigma[4]	Reef	Above sand	–	Great Barrier Reef
Apogonidae	Apogon fuscus[5]	Reef	Seagrass	–	Gulf of Thailand
Apogonidae	Apogon aureus[6] Apogon cyanosoma[6]	Reef	Seagrass	–	Negros Oriental, Philippines
Apogonidae	Apogon hartzfeldti[7] Apogon hoevenii[7] Sphaeramia orbicularis[7]	Reef	Seagrass	–	Sulawesi, Indonesia
Apogonidae	Apogon affinis[8] Apogon quadrisquamatus[8]	Reef	Sand	–	Caribbean (St. Croix)
Apogonidae	Apogon aurolineatus[1, 15] Apogon binotatus[1, 15] Phaeoptyx conklini[1, 15]	Reef	Seagrass, midwater	–	Caribbean (Florida, St. Croix)
Atherinidae[14]	Hypoatherina harringtonensis[1, 15]	Reef	Seagrass, beyond the reef	<800 m[1]	Caribbean (Florida, St. Croix)

Table 8.1 (continued)

Fish family	Species	Residence at		Foraging range	Region
		Day	Night		
Atherinidae	Pranesus insularum[16]	Reef, schooling in shallow water or along the edge of channels	Off-reef, water surface	—	Hawaii
Atherinidae	Pranesus pinguis[17]	Reef	Offshore	1.2 km[18]	Marshall Islands
Aulostomidae	Aulostomus maculatus[13]	Reef	Seagrass	—	Caribbean (Panama)
Carangidae	Caranx hippos[19] Caranx spp.[12] Scomberomorus spp.[12]	Seagrass, reef, mangrove	Seagrass, reef	—	Caribbean (St. Croix)
Carangidae	Selar crumenophthalmus[20]	Close to shore	Sand	—	Gulf of California
Clupeidae[14]	Harengula thrissina[20]	Over rocks	Offshore sand	>500 m[18]	Gulf of California
Clupeidae	Harengula humeralis[1,15] Jenkinsia majua[1,15] Jenkinsia lamprotaenia[1,15] Sardinella anchovia[1]	Reef, back-reef area	Seagrass, offshore	J. majua >800 m[1] S. anchovia >2 km[1]	Caribbean (Florida, St. Croix)
Clupeidae	Jenkinsia lamprotaenia[19,21]	Mangrove	Windward side of the bay	—	Caribbean (Puerto Rico)
Clupeidae	Several species[1,22]	Reef	Away from reef	—	Caribbean, Gulf of California
Congridae	Conger cinereus[6]	Reef	Seagrass	—	Negros Oriental, Philippines
Diodontidae	Chilomycterus schoepfi[1]	Reef	Seagrass	—	Caribbean (Florida)
Diodontidae	Diodon holocanthus[6]	Reef	Seagrass	—	Negros Oriental, Philippines
Engraulidae[14]	Anchoa cayorum[15] Anchoa lamprotaenia[15]	Reef	Seagrass	—	Caribbean (St. Croix)

Table 8.1 (continued)

Fish family	Species	Residence at		Foraging range	Region
		Day	Night		
Fistulariidae	*Fistularia tabacaria*[1, 13]	Reef	Seagrass, sand	–	Caribbean (Panama, Florida)
Gerreidae	*Eucinostomus* spp.[19] *Gerres cinereus*[19]	Mangrove	Sandflat, seagrass	–	Caribbean (Puerto Rico)
Haemulidae[1, 10, 13, 14, 20, 22, 25]	*Haemulon aurolineatum*[1, 15, 19] *Haemulon chrysargyreum*[1] *Haemulon flavolineatum*[19] *Haemulon parra*[19] *Haemulon plumierii*[8, 12, 19, 26] *Haemulon sciurus*[8, 15, 19]	Patch reef, mangrove, gorgonians[27]	Seagrass, sand	*H. aurolineatum* 800 m[1] *H. flavolineatum* 1.6 km[1] *H. sciurus* 400 m[1], 200–300 m[28], >1 km[12]	Caribbean (St. Croix)
Haemulidae	*Haemulon flavolineatum*[23]	Mangrove/Seagrass	Seagrass	–	Caribbean (Curaçao)
Haemulidae	No species names given[24]	Reef	Gorgonian site	–	Caribbean (St. Croix)
Haemulidae	*Anisotremus interruptus*[20]	Reef	Shallow water	–	Gulf of California
Haemulidae	*Haemulon sexfasciatum*[20] *Microlepidotus inornatus*[20] *Haemulon flavigutattum*[20] *Haemulon maculicauda*[20]	Reef, rock, inshore sandy bottom	Offshore sand	–	Gulf of California
Holocentridae[9, 10, 11, 12, 13]	*Holocentrus adscensionis*[15] *Sargocentron coruscum*[1, 8, 15] *Myripristis jacobus*[15] *Holocentrus marianus*[8] *Holocentrus rufus*[15] *Holocentrus vexillarius*[1, 15]	Reef	Seagrass, sand, rubble, rock	–	Caribbean (Florida, Panama, St. Croix)

Table 8.1 (continued)

Fish family	Species	Residence at		Foraging range	Region
		Day	Night		
Holocentridae	Sargocentron rubrum[5]	Reef	Seagrass	–	Gulf of Thailand
Holocentridae	Myripristis argyromus[2] Myripristis berndti[2]	Reef	Away from reef	–	Hawaii
Holocentridae	Species not identified[6]	Reef	Seagrass	–	Negros Oriental, Philippines
Holocentridae	Holocentrus sp.[29]	Reef	Dispersed at dawn	–	Great Barrier Reef
Labridae	Halichoeres chloropterus[5]	Seagrass	Reef	–	Gulf of Thailand
Lutjanidae[1,2,10,12,13,22,30]	Lutjanus apodus[1,8,19] Lutjanus griseus[1,8] Lutjanus synagris[1,8]	Reef, mangrove	Seagrass, sand, rubble, algal flats	L. griseus 1.6 km[1]	Caribbean
Lutjanidae	Lutjanus argentiventris[20]	Reef	Rocky areas near shore	–	Gulf of California
Mullidae[10,12,13,18,31]	Mulloides flavolineatus[32,33]	Reef	Sand flat	600 m[33]	Hawaii
Mullidae	Parupeneus porphyreus[34]	Reef	Sand, coral rubble	–	Hawaii
Muraenidae[14]	Species not identified[6]	Reef	Seagrass	–	Negros Oriental, Philippines
Muraenidae[13,35]	Unspecified	Reef	Seagrass/sand	–	Pacific Ocean, Caribbean (Panama)
Ophichtidae[14]	Leiuranus semicinctus[6] Muraenichthys macropterus[6]	Reef	Seagrass	–	Negros Oriental, Philippines
Pempheridae[14]	Pempheris oualensis[36]	Reef	To the sea	–	Red Sea
Pempheridae	Pempheris schomburgkii[1,37]	Back-reef	Fore-reef	1 km[37]	Caribbean (Florida, St. Croix)
Plotosidae	Plotosus lineatus[6]	Reef	Seagrass	–	Negros Oriental, Philippines
Pomacentridae[2,38,39]	Pomacentrus tripunctatus[5]	Seagrass	Reef	–	Gulf of Thailand

Table 8.1 (continued)

Fish family	Species	Residence at		Foraging range	Region
		Day	Night		
Priacanthidae	*Priacanthus cruentatus*[2]	Reef	Away from reef	—	Hawaii
Scaridae[40]	*Sparisoma radians* (<15 cm)[12]	Seagrass	Reef	—	Caribbean (St. Croix)
	Scarus guacamaia (>40 cm)[12]				
Sciaenidae[20]	*Pareques viola*	Reef	Sand	—	Gulf of California
Sciaenidae[35]	*Sciaena* spp.[18]	Reef	Seagrass/sand/rubble	—	Pacific Ocean
Sciaenidae	*Equetus acuminatus*[27]	Reef	Sand	—	Caribbean (Florida)
Scorpaenidae[41]	*Scorpaena grandicornis*[1]	Reef	Seagrass	—	Caribbean (Florida)
Serranidae[9, 13, 41]	*Epinephelus merra*[41]	Reef	Seagrass	—	Caribbean, Madagascar
Siganidae[42]	Unspecified	In or over coral heads	Away from coral heads	—	Unspecified
Sparidae	*Archosargus rhomboidalis*[19]	Mangrove	Sandflat, seagrass	—	Caribbean (Puerto Rico)
Sphyraenidae	*Sphyraena barracuda*[12]	Seagrass, reef	Seagrass, reef	—	Caribbean
Tetraodontidae	*Arothron stellatus*[6]	Reef	Seagrass	—	Negros Oriental, Philippines

References: [1]Starck and Davis 1966, [2]Hobson 1972, [3]Morin et al. 1975, [4]Marmane and Bellwood 2002, [5]Sudara et al. 1992, [6]Kochzius 1999, [7]Unsworth et al. 2007a, [8]Collette and Talbot 1972, [9]Randall 1963, [10]Randall 1967, [11]Vivien and Peyrot-Clausade 1974, [12]Ogden and Zieman 1977, [13]Weinstein and Heck 1979, [14]Helfman 1993, [15]Robblee and Zieman 1984, [16]Major 1977, [17]Hobson and Chess 1973, [18]Hobson 1973, [19]Rooker and Dennis 1991, [20]Hobson 1965, [21]Radakov and Silva 1974, [22]Hobson 1968, [23]Verweij et al. 2006, [24]Wolff et al. 1999, [25]Starck 1971, [26]Ogden and Ehrlich 1977, [27]Longley and Hildebrand 1941, [28]Quinn and Ogden 1984, [29]Domm and Domm 1973, [30]Hiatt and Strasburg 1960, [31]Jones and Chase 1975, [32]Hobson 1974, [33]Holland et al. 1993, [34]Meyer et al. 2000, [35]Hobson 1975, [36]Fishelson et al. 1971, [37]Gladfelter 1979, [38]Doherty 1983, [39]Foster 1987, [40]Ogden and Buckman 1973, [41]Harmelin-Vivien and Bouchon 1976, [42]Meyer et al. 1983

abundant Caribbean reef fish and their diel migrations mark a major link among shallow-water ecosystems. The most detailed studies originate from St. Croix, US Virgin Islands (Ogden and Ehrlich 1977, Ogden and Zieman 1977, McFarland et al. 1979, Quinn and Ogden 1984, Robblee and Zieman 1984, Beets et al. 2003), and

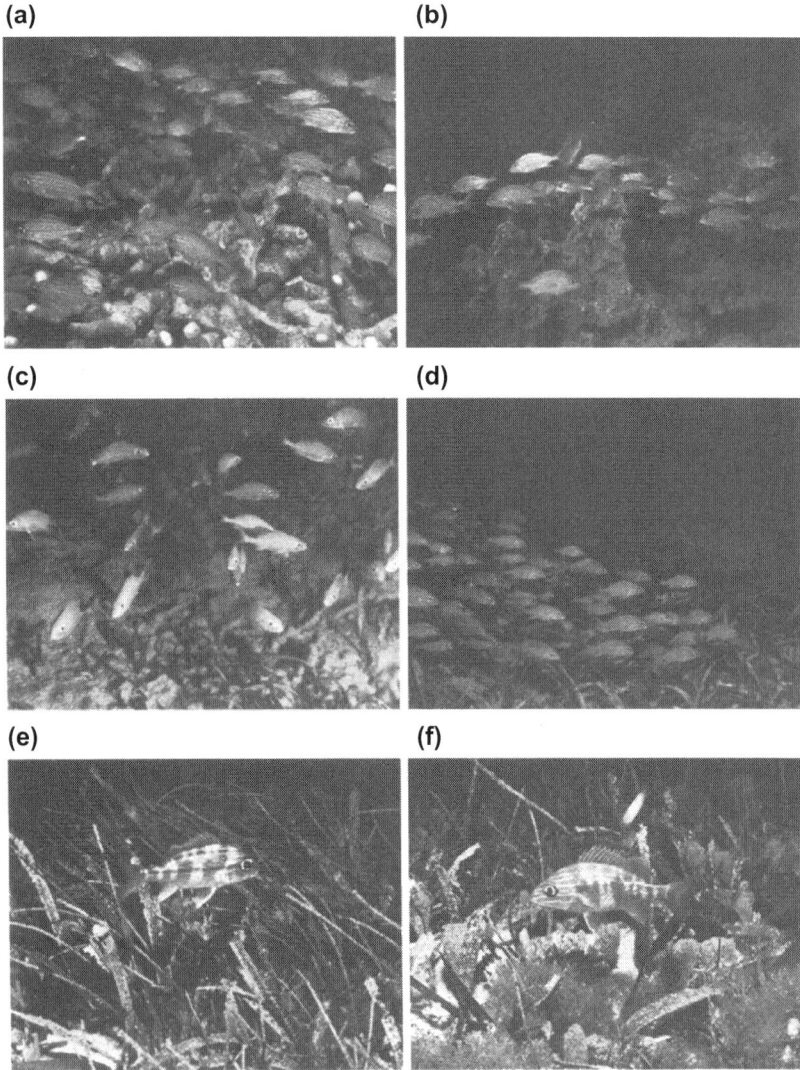

Fig. 8.5 Juvenile grunts in various phases of daily behavior in St. Croix, US Virgin Islands: (a) schooling on a patch reef during the day, (b) an 'assembly' at the staging area, (c) 'ambivalence' at the staging area, (d) evening migration of *H. flavolineatum* and *H. plumierii*, (e) solitary *H. flavolineatum* on grass bed at night, (f) solitary *H. plumierii* on grass bed at night. Pictures from McFarland et al. (1979), with kind permission from Springer Science + Business Media and JC Ogden

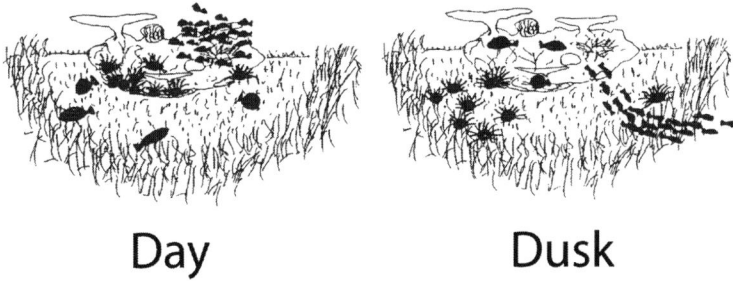

Day   Dusk

Fig. 8.6 Illustration of a Caribbean patch reef at daytime and at dusk. Picture from Ogden (1997), with kind permission from Springer Science + Business Media and JC Ogden

numerous studies have shown that the twilight migrations of juvenile grunts in St. Croix occur in heterogeneous seascapes throughout the Caribbean (e.g., Ogden and Ehrlich 1977, Panama: Weinstein and Heck 1979, Puerto Rico: Rooker and Dennis 1991, Tulevech and Recksiek 1994, Guadeloupe: Kopp et al. 2007, Curaçao: Nagelkerken et al. 2000, Belize: Burke 1995, Florida: Tulevech and Recksiek 1994).

These studies found that during the day mixed schools of juvenile grunts are inactive, resting on patch reefs surrounded by seagrass meadow. The dominant species were French grunts *Haemulon flavolineatum* and white grunts *H. plumierii*. Starck and Davis (1966) mention several other haemulid species involved in regular diel migrations to the reef adjacencies (Table 8.1). The start of the feeding migration off the reef at dusk is highly ritualized, comprising four behaviors (Fig. 8.5): (1) separate schools begin to stream along the reef surface (restlessness), (2) merge with other schools (assembly), and (3) finally concentrate on the reef edge (ambivalence) from where they (4) start the evening migration (departure) along fixed and constant corridors into the adjacent seagrass meadow (Ogden and Ehrlich 1977, Helfman et al. 1982). After up to 50 m linear movement away from the reef, small groups begin to disassociate from the main group and disperse in a dendritic pattern in the seagrass bed. Typically, the active migration takes the grunts to places 100–200 m away from the reef (Quinn and Ogden 1984), but sometimes >1 km (Ogden and Zieman 1977). The grunts forage solitarily for benthic invertebrates through the night. French grunts and bluestriped grunts (*H. sciurus*) seem to forage preferably in sandy areas, whereas white grunts forage on both sandy and grassy areas (Starck and Davis 1966, McFarland et al. 1979, Burke 1995), suggesting space and diet partitioning by the different species. Grunts display considerable flexibility in foraging ground use. They were observed foraging in seagrass, mangrove, sand, rubble, gorgonian habitat, and algal beds (Starck and Davis 1966, Ogden and Ehrlich 1977, Wolff et al. 1999, Nagelkerken et al. 2000). However, once established, nocturnal feeding territories are maintained over time (McFarland and Hillis 1982). Using acoustic telemetry, Beets et al. (2003) found high site fidelity for *H. sciurus* to nocturnal foraging sites in seagrass beds up to 767 m from diurnal resting sites.

It is not known if the grunts display a stereotyped behavior during their morning migration back to the reef. Usually the fish schools stream a few centimeters above the seagrass bed on the same pathway used during the evening migration to quickly arrive at their typical daytime resting position (McFarland et al. 1979). The proportion of the population of grunts that carry out twilight migrations is usually around 100%. Meyer et al. (1983) and Meyer and Schultz (1985) quantified the fertilizing effect on the coral heads where grunts aggregate at daytime. Daytime resting sites are not restricted to corals. In fact, grunts seem to shelter at or near any available structurally complex habitat (i.e., boulders, channel, crevices, mangrove, long-leaved seagrass; Rooker and Dennis 1991, Nagelkerken et al. 2000, Verweij and Nagelkerken 2007). However, similar to nocturnal feeding territories, once established daytime resting sites are used for extended periods. Verweij and Nagelkerken (2007) found that juvenile grunts displayed high site fidelity to daytime resting site core areas of <200 m for >1 yr.

The diel migrations are precisely timed and strongly correlated with changes in light intensity. The migrations to and from the seagrass beds occur at a specific light intensity (McFarland et al. 1979), although the timing is adaptive and can respond to cloud cover-related changes in light intensity. The 'quiet period' is an important time window in the twilight migrations of grunts. The 'quiet period' is the twilight period between the shelter-seeking of diurnal fishes and the emergence of nocturnal fishes in the evening. In the morning the sequence is reversed. Most fishes are close to the substrate during this period and the activity of crepuscular piscivores peaks because they are visually superior to their prey during twilight periods (Hobson 1972, McFarland and Munz 1976). The juvenile grunts leave the reef before the evening 'quiet period' starts and return just after it ends in the morning at identical light intensities.

Many fish exhibit color changes between day and night (Figs. 8.5e, f). In French grunts, color changes are associated with the twilight migration. Unlike white grunts, French grunts migrate in their daytime color dress to and from the reef but forage with a colorless appearance at night which likely provides crypsis in the seagrass (Starck and Davis 1966, McFarland et al. 1979). Thus, the grunts combine foraging and reduced risk of predation while feeding alone in seagrass beds through the night. There is an ontogenetic switch from diurnal to nocturnal activity in juvenile grunts (Helfman et al. 1982), and there are size and age differences in the precision of the diel migration in grunts. Younger specimens (15–30 mm total length) set out later in the evening and return earlier in the morning to the patch reefs (i.e., stay longer in areas of shelter) than larger-sized specimens (40–120 mm). In other words, larger juveniles depart and arrive on the patch reefs in stronger light than the smaller juveniles (McFarland et al. 1979). Similar size-related differences in the timing of twilight activities are known from other marine and freshwater fishes (see Hobson 1972, Helfman 1979, 1981). Factors such as the development of the visual apparatus and predation pressure may lead to less variable diel migration activities with increasing fish size (Helfman et al. 1982). However, Tulevech and Recksiek (1994) found that the twilight migrations of adult *H. plumierii* were less regular than those of juveniles, suggesting that there is a relaxation in the timing of the migratory behavior after maturity.

The start of the nocturnal feeding migration in grunts seems to differ between locations and between grunt species (Rooker and Dennis 1991). In shallow embayments with reduced abundance of piscivores, juvenile grunts still forage during the morning or start foraging in the afternoon in adjacent mangroves or seagrass beds (Rooker and Dennis 1991, Verweij et al. 2006, Verweij and Nagelkerken 2007). McFarland et al. (1979) and Helfman et al. (1982) suggested that piscivores such as lizardfish (Synodontidae) play an important role in shaping the twilight migration of juvenile grunts. This assumption still needs to be tested, e.g., by comparing the migratory patterns of grunts under conditions of higher risk of predation vs. lower risk of predation. Potential experiments might include comparisons of the migratory patterns in marine parks (i.e., higher risk due to more predators) vs. fishing zones (lower risk due to less predators) (see, e.g., Tupper and Juanes 1999), or be based on the older comparison studies of grunts. However, Randall (1963) already mentioned overfishing in the US Virgin Islands. In addition to predation, biological factors such as changes in prey availability, parasite activity, and/or intra- and interspecific competition may influence diel migratory behavior (Helfman 1993).

Active diel migrations follow well-defined pathways between patch reef resting and seagrass feeding sites. The migration routes of smaller specimens persist over several months; those of larger specimens can be re-visited even after years (Ogden and Ehrlich 1977, McFarland et al. 1979). The size-specific stability of migratory routes is likely related to differences in age composition of smaller and larger specimens. The smaller sizes come from a single cohort and cannot rely on experience of others when deciding which migration route to take. The larger grunts are composed of overlapping cohorts, and knowledge of resting sites and migration routes is likely transmitted as a process of social transmission and learning (Helfman et al. 1982, Helfman and Schultz 1984). Quinn and Ogden (1984) provided evidence that juvenile grunts use compass orientation for their regular diel migrations. They concluded that landmarks were unimportant for orientation although they did not explicitly test this. Fish are readily able to generate spatial maps of their environment which are used to guide their movements (Braithwaite and Burt de Perera 2006).

Grunts are primarily obligate diel migrants and dependent on a heterogeneous seascape around reefs. Their diversity and abundance is reduced where either diurnal shelter sites or back-reef forage habitat are lacking (Starck and Davis 1966 p. 352, Gladfelter et al. 1980, Baelde 1990). Although grunts are mobile species, they display high site fidelity to feeding and shelter core areas. This makes them considerably vulnerable to selective small-scale habitat change, e.g., the loss of particular patch reefs or single mangrove stands.

8.4.2 Other Families and Species

The diel migrations of a few other species have been studied in some detail (Table 8.1). *Pempheris schomburgkii* (Pempheridae) in St. Croix (US Virgin Islands) migrate around sunset from daytime back-reef refuges to nocturnal fore-reef feeding sites (Gladfelter 1979). At sunset, a complex sequence of behaviors begins (e.g., appearance at crevice entrance, 'flashing', ambivalence, emergence, aggregation, swimming along pathways) which takes the schools along complex pathways to the fore-reef where groups split off and fishes forage for meroplanktonic crustaceans singly or in small groups through the night. In the morning, *P. schomburgkii* return along the same pathways but at lower light levels. *Acropora* landmarks are important, and the timing is age-specific and triggered by light intensity, similar to grunts. The twilight migration covers distances of almost 1 km but is limited to the reef structure.

Likewise, Apogonidae maintain daytime resting sites close to reef structures from where they migrate to nocturnal feeding grounds in different near-reef habitats such as open water, sand, seagrass habitats, and the reef proper, partitioning niches spatially (Collette and Talbot 1972, Vivien 1975, Marnane and Bellwood 2002). They may return to the same resting sites each morning for months or years (e.g., Kuwamura 1985, Okuda and Yanagisawa 1996, Marnane 2000, Ménard et al. 2008) and thus regularly transport nutrients and energy to and from their reefal resting sites. However, their foraging range is relatively short (30 m in Marnane and Bellwood 2002).

Hobson (1968) was able to chart the evening route of *Harengula thrissina* (Clupeidae) in the Gulf of California which took the fish more than 500 m offshore. In the Marshall Islands, *Pranesus pinguis* (Atherinidae) rested in schools nearshore during the day and followed 'the same route each evening' to disperse offshore and feed on plankton up to 1.2 km away from the diurnal schooling sites (Hobson and Chess 1973). In Hawaii, the mullid *Mulloides flavolineatus* formed daytime resting schools and moved distances between 75 and 600 m to nearby sand flats to forage (Holland et al. 1993). Site fidelity was extremely high. Individuals were recaptured after up to 531 days. The foraging range was restricted to $\sim$ 13–14 ha (own estimate from a map in Holland et al. 1993). Acoustically-tagged *Parupeneus porphyreus* (Mullidae) showed consistent diel patterns of behavior, taking refuge in holes in the reef by day and moving over extensive areas of sand and coral rubble habitat at night (Meyer et al. 2000).

In the Caribbean, juvenile snappers such as *Lutjanus apodus* also carry out twilight migrations from their daytime resting sites in mangroves or protected rocky shorelines (Verweij et al. 2007) to nighttime soft bottom foraging grounds, such as seagrass beds (Starck and Davis 1966, Rooker and Dennis 1991, Nagelkerken et al. 2000). Nocturnal Lutjanidae might be important predators of foraging Haemulidae at night (Starck and Davis 1966). Acoustic tracking of lane snapper *L. synagris* on St. John (US Virgin Islands) showed sun-synchronous nocturnal migrations with a regular departure from the eastern side of the Lameshur Bay after sunset and a return before sunrise, and strong daytime site fidelity for a period of 268 days (Fig. 8.7).

Similar twilight migrations can also be observed on the Pacific coast of Colombia (G Castellanos-Galindo, Universidad del Valle, Colombia, pers. comm.). Evidence for short-term migrations is often inferred from day-night shifts in species compositions at given sites or accidental observations, but real-time tracking of movements and behavior remains insufficient. Portable GPS tracking of *L. decussatus*

Fig. 8.7 Plot of receiver detections for an individual lane snapper (*Lutjanus synagris*, 29 cm total length) at Lameshur Bay, St. John (US Virgin Islands) between 12 July 2006 and 5 April 2007 (bottom graph). Upper figure: locations of nine receivers in the bay showing the 300-m-radius detection buffer (circles) and station detection overlap. Receiver 6: inner bay site with patchy seagrass at 17 m water depth; receiver 2 and 3: outer bay sites at ~ 22 m water depth. Adapted from Friedlander and Monaco (2007), with kind permission of S Pittman (NOAA). Fish artwork commissioned from D Peebles by Florida Fish and Wildlife Conservation Commission

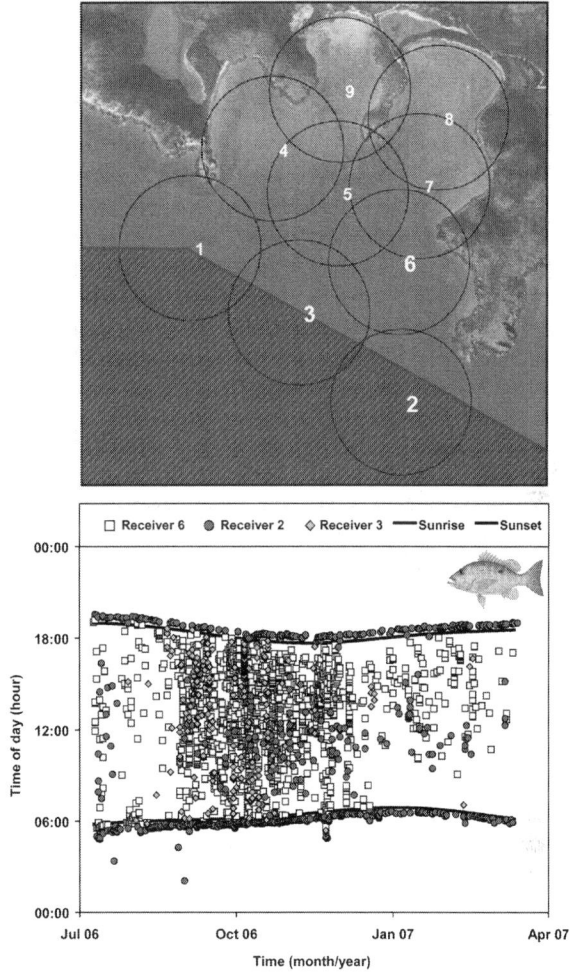

during daylight hours in an Okinawan coral reef revealed high daytime site fidelity (Nanami and Yamada 2008). The size of daytime home ranges ranged between 93 and 3638 m^2.

8.4.3 Feeding Guilds

Diel activity seems to have a strong phylogenetic background and may be a familial characteristic (Helfman 1993). Since the species of many fish families feed on similar food organisms, a generalization of the diel migrations in terms of feeding guilds is possible. Note, however, that fish are opportunistic, and activity patterns can vary in response to numerous biological and abiotic factors.

Nocturnal invertebrate feeders usually migrate from daytime shelter at dusk, feed through the night, and return to seek cover at dawn (e.g., Haemulidae, Mullidae). Seagrass beds are often preferred as nighttime habitats because during darkness invertebrate abundance is high (e.g., Robertson and Howard 1978), and higher than in adjacent habitats (e.g., Nagelkerken et al. 2000). However, French grunts, for example, are facultative nocturnal foragers and were observed feeding also at daytime in mangrove and seagrass beds (Verweij et al. 2006).

Herbivores are clearly diurnal (Pomacentridae, Scaridae) (Helfman 1993). The latter feed on seagrass, and also on coral and filamentous algae growing on dead coral rock during daytime, and hide in the reef at night (Ogden and Zieman 1977). If moonlight is sufficient, herbivores may also be active at night (Hobson 1965).

Piscivores may adopt one of two strategies. They can hide in the reef at daytime, emerge after the 'quiet period', roam from reefs into seagrass meadows and forage at night (e.g., moray eels, snake eels, Lutjanidae). They can also show opportunistic behavior with activity at day and/or night in response to prey availability (e.g., Sphyraenidae or Carangidae in the Caribbean, Ogden and Zieman 1977; Lutjanidae, Belonidae, and Carangidae in Pacific Colombian rocky shores, G Castellanos-Galindo, Universidad del Valle, Colombia, pers. observ.).

Zoo- and phytoplanktivores usually aggregate in dense, relatively inactive reef-associated schools to fan out from their reefal resting sites in the horizontal and vertical scale to search for food in the adjacencies. At dusk, schools of nocturnal planktivores such as Apogonidae or Pempheridae disperse to feed at night, often after moving a considerable distance offshore (Hobson 1965, Fishelson et al. 1971). At dawn, diurnal planktivores such as Pomacentridae disperse and forage on near-reef plankton. It is debatable whether the reefal migrations and those of many other families and species mentioned in Table 8.1 really connect different ecosystems, or whether they are movements in the reef–sand interface and restricted to the sphere of influence of a reef. Nevertheless, these movements lead to the regular transfer of non-reef production to the reefal resting sites and may thus in fact be considered diel migrations between adjacent ecosystems.

8.5 Diel Movements of Decapods

In the marine realm the order Decapoda is represented by three groups: shrimps, lobsters, and crabs (Ruppert and Barnes 1994). Given the differences in their biology, the groups are treated in order following their ability to swim, which may affect the potential to make extensive diel movements.

8.5.1 Lobsters

After settlement, lobsters are benthic animals and avoid swimming. Our knowledge on short-term movements of lobsters mainly stems from clawed (Nephropidae and Homaridae) and spiny lobsters (Palinuridae) that are remarkably similar in morphology, ecology, and behavior. Lobsters are mostly nocturnal. They spend most of their daytime sheltering in subtidal crevices and caverns of reefs and erosional edges of seagrass patches ('blowouts'), and leave their dens at night to forage in surrounding areas (e.g., Herrnkind 1980, Cobb 1981, Joll and Phillips 1984, Phillips et al. 1984, Jernakoff and Phillips 1988, Jernakoff et al. 1993, Acosta 1999). Where different ecosystems form a heterogeneous seascape, nightly walks clearly connect adjacent ecosystems.

Detailed studies come mostly from subtropical or temperate coasts. For instance in Western Australia, *Palinurus cygnus* leave their dens at night and forage in the seagrass beds around the reefs (Cobb 1981). Home range sizes are usually <500 m and lobsters display high site fidelity. Juvenile *P. cygnus* forage over small areas, usually within a 20 m radius, but sometimes as far as 50 m from their den (Chittleborough 1974). Juvenile *P. cygnus* usually move at a rate of ca. 1 m.min⁻¹ at night. When walking over bare sand, they travel at speeds of up to 18 m.min⁻¹ (Jernakoff 1987). Similarly, in the tropics subadult *P. argus* walked between 25 and 416 m overnight, and distances walked were similar after one week (Lozano-Alvarez et al. 2003). The emergence patterns, however, change ontogenetically. Early benthic-phase *P. argus* (<15 mm carapace length CL) rarely leave their shelters, juveniles (30–62 mm CL) leave their shelter 2–30 times per night, usually with excursions of <10 min, and adults (>80 mm CL) walked 'for extended periods of time during the night' (Weiss et al. 2008). The home range of *P. guttatus* ranges within a radius of 100 m (Lozano-Alvarez et al. 2002). All *P. versicolor* were recaptured within 500 m of their original den (Frisch 2007).

Foraging activity varies between sexes and size groups (Weiss et al. 2008). The activity is constant throughout the night (Jernakoff 1987, Frisch 2007), displays peaks, often in the hours after sunset (e.g., Fiedler 1965), or ceases several hours before dawn when the lobsters return to their dens (Herrnkind 1980). While walking through their home range, the lobsters use geomagnetic fields, water movements (Creaser and Travis 1950, Herrnkind and McLean 1971, Lohmann 1985), and structural cues such as blowouts for orientation (Cox et al. 1997).

Light levels and turbidity at the onset of light or darkness seem to control the movements (Herrnkind 1980). *Jasus lalandii* feed at maximum rate a few hours after sunset, closely matching the locomotory patterns (Fiedler 1965). In juvenile *P. cygnus* most foraging activity begins in response to changes in light levels associated with dusk and not by diurnal changes in water temperature or currents. Similarly, juvenile lobsters return from foraging at about dawn when light levels begin to increase (Jernakoff 1987).

Juvenile lobsters are likely important predators; their feeding ecology probably affects the structure of the benthic community in their home range (Joll and Phillips 1984), but due to their mixed and diverse diet (e.g., Briones-Fourzan et al. 2003) it is difficult to quantify their contribution to overall trophic flows.

Nightly forays of lobsters can be restricted to a single reef (e.g., Chittleborough 1974) or can include traveling to reefs several kilometers away where they stay for some time before returning to the original home range (Herrnkind 1980). These movements, however, seem to be infrequent and usually below 20 km (Trendall and Bell 1989). Vegetated substrates may function as movement corridors between insular habitats and facilitate dispersal, besides being important settlement areas. Consequently, protection of areas with a heterogeneous and vegetated seascape is important to fulfill the changing requirements of species with a complex life cycle, such as lobsters (Acosta 1999).

8.5.2 Penaeid Shrimps

Shrimps are usually bottom dwellers and intermittently use their pleopods to swim. Burial and activity of penaeid shrimps is influenced by (1) the diel cycle, (2) moonlight, (3) turbidity, and (4) the tidal cycle (Dall et al. 1990). The typical life cycle of penaeid shrimps connects the offshore areas (adult stock, reproduction) with the innermost areas of estuaries and embayments (nursery ground) (see Chapter 7). Short-term movements that connect different shallow-water ecosystems are restricted to the postlarvae and juvenile stage in the nursery grounds.

If regular diel movements of shrimp connect adjacent ecosystems, their activity patterns most likely follow one of three light-dependent activity types that Penn (1984) classified according to the shrimps' burrowing behavior. (1) Shrimps in clear water may burrow at day or in bright moonlight and emerge only at night. (2) Shrimps in slightly turbid water may be nocturnal but occasionally emerge during the day. (3) Shrimps in more turbid water seldom if ever burrow. The home range size of individual shrimps after settlement in a nursery ground has rarely been identified, most likely due to obvious problems in mark-recapture experiments (see Schaffmeister et al. 2006). Evidence for short-term movements between adjacent ecosystems by shrimps mainly comes from tidal coasts (see Section 8.8.2).

8.5.3 Crabs

Most crabs cannot swim and their benthic movements are unlikely to regularly connect adjacent ecosystems. Portunid crabs, however, are agile swimmers. The last pair of legs terminates in paddle-like swimming legs. The legs resemble figure eights in their movement, similar to a propeller. The forth pair of legs counter-beat and act as stabilizers. Nevertheless, they are mostly benthic and swim only intermittently (Ruppert and Barnes 1994). Information is available primarily from commercially important species. In a South African estuary with a maximum tidal range of 1.4 m, nocturnal foraging movements of the subtidal mud crab *Scylla serrata* ranged from 219–910 m (Hill 1978). Interactions of the movements with the tidal cycle were not mentioned. They tended to stay in the same general area although they were capable of moving at least 800 m along the length of the estuary at night (Hill 1978). By continuously shifting their general area, mud crabs can cover distances of several kilometers within a few weeks (Hyland et al. 1984). Thus, mud crabs are readily able to connect adjacent ecosystems but no attention has been paid as to whether this occurs on a regular diel or tidal basis.

The blue swimmer crabs *Portunus pelagicus* are opportunistic, bottom-feeding carnivores and scavengers (Kangas 2000). They are most active in foraging and feeding at sunset (Grove-Jones 1987, Smith and Sumpton 1987, Wassenberg and Hill 1987). Although *P. pelagicus* generally forage in the habitat in which they rest diurnally, they readily move to other habitats and have a wide-ranging foraging strategy (Edgar 1990). Due to their strong swimming ability, *P. pelagicus* are capable of moving substantial distances, with one recorded as traveling 20 km in one day in Moreton Bay, Queensland (Sumpton and Smith 1991). However, tagging studies in Moreton Bay showed that small-scale movement of crab populations are more common. Of the recaptures, 79% were caught <2 km from their release points, and only 4% were recaptured >10 km from their release point (Potter et al. 1991). Similarly, recaptures within 4 km of release sites have occurred for *Scylla serrata* (Hyland et al. 1984) and *Callinectes sapidus* (Mayo and Dudley 1970). Feeding of *C. arcuatus* in Pacific Mexico occurs mostly at dusk (Paul 1981). *Thalamita crenata* in Kenya forages both during the day and night, however, in significant interaction with the tidal cycle (Vezzosi et al. 1995, Cannicci et al. 1996). Unfortunately, detailed biological information for many other portunid species of the genera *Charybdis*, *Portunus*, *Scylla*, and *Thalamita*, particularly from Southeast Asia, is not available.

In conclusion, there are no accounts of synchronized short-term mass movements of decapods to particular habitats from any tropical coast with weak tides that are comparable to the diel migrations reported for numerous reef fish families and species. However, the diel movements of lobsters unambiguously connect adjacent ecosystems. The putative connectivity between adjacent ecosystems by diel migrations of shrimps and portunid crabs still awaits more robust evidence. The timing of diel movements in shallow-water decapods is also related to the twilight period but compared to fish, the movements seem to be less strictly structured in time. Overall diel foraging ranges of lobsters, and particularly of portunid crabs, can be similar to those of fish.

8.6 Tidal Movements of Fish

On marine coasts, diel activity is usually tightly coupled to the additional stimulus of the tides. On coasts with low tidal range, e.g., the Red Sea or the Caribbean, the activity patterns of fishes are primarily synchronized with the day-night cycle (see Section 8.4). On most other coasts, rhythmic behavior of the coastal organisms is synchronized more strongly with the tide, but still in close interaction with the diel cycle (Gibson 1993). The few studies available on fish species that live on both tidal and non-tidal coasts have shown that the rhythmic behavior reflects the relative importance of the tide in their respective environment (the gobiid *Pomatoschistus minutus*, the pleuronectid *Platichthys flesus*: Gibson 1982; the ariid *Arius felis*: Steele 1985, Sogard et al. 1989).

Previous reviews on tidal movements of fish have focused on rocky shores, sandy beaches, saltmarshes, or migrations in the open sea (Gibson 1969, 1982, 1988, 1992, 1993, 1999, 2003, Kneib 1997, Harden Jones 1968, Metcalfe et al. 2006). These reviews refer mostly to the higher latitudes of the northern hemisphere but their findings provide a robust background to investigate the tidal movements of tropical shallow-water fishes. References from outside the tropics are cited whenever examples from the tropics are unavailable.

8.6.1 Transients and Residents

Intertidally migrating fish transport energy from the intertidal into the subtidal area, both short-term and long-term (e.g., seasonal emigration or ontogenetic movements, see Chapter 6). Therefore, the sublittoral is treated here as an ecosystem different from the intertidal, and virtually all intertidal migrants are considered to connect different ecosystems.

Gibson (1969, 1988) classified intertidal fishes into residents (fish that live in the intertidal) and transients (temporary visitors during times of intertidal inundation that return to the subtidal). He further classified the transients according to the regularity of intertidal use into tidal, lunar, seasonal, and accidental visitors. The residents are not considered in this section because their movements do not connect adjacent ecosystems. In contrast to the intertidal migrants that move up and down the shore with the rise and fall of the tide (i.e., strong upshore component), there are the subtidal migrants which principally use tidal currents to move between subtidal habitats and avoid entering the intertidal zone (i.e., strong alongshore component). These classifications, however, have smooth transitions. Migratory patterns can differ between individuals, size groups, sexes, species, and populations (Gibson 1999, 2003), resulting in complex patterns of seascape use by mobile organisms.

8.6.2 Tidal Migrations in Tropical Habitats

During high tide, transients make temporary use of a wide range of accessible tropical habitats. When immersed, numerous fish colonize mangroves and mangrove-lined creeks (e.g., Robertson and Duke 1987, Little et al. 1988, Chong et al. 1990, Robertson and Duke 1990, Sasekumar et al. 1992, Laroche et al. 1997, Kuo et al. 1999, Rönnbäck et al. 1999, Tongnunui et al. 2002, Krumme et al. 2004, Vidy et al. 2004) and mudflats (e.g., Abou-Seedo et al. 1990, Chong et al. 1990). Transient fish move onto sandy beaches (e.g., Brown and McLachlan 1990, Abou-Seedo et al. 1990, Yamahira et al. 1996) and forage on rocky shores (e.g., Castellanos-Galindo et al. 2005, Gibson 1999). In Australian seagrass beds, midwater feeders move from adjacent habitats to the water column above the seagrass at high tide (Robertson 1980). The rock flathead *Platycephalus laevigatus* use seagrass beds together with their main prey *Nectocarcinus integrifons* during nightly or evening high tides (Klumpp and Nichols 1983). Sogard et al. (1989) inferred from continuous gill net catches that fish moved in seagrass banks primarily around high tide in Florida Bay, USA. On the Marshall Islands, Central Pacific, Bakus (1967) observed the mass migration of algal grazing surgeonfish (*Acanthurus triostegus, A. guttatus*) on and off reef flats with the tide. The herbivorous surgeonfish *A. lineatus* re-established intertidal territories each morning on a reef flat in Samoa (Craig 1996). At some sites in the Gulf of Aqaba, schools of the surgeonfish *A. nigrofuscus* migrated daily on a routine pathway 500–600 m from nocturnal reefal resting sites to intertidal daytime feeding sites (Fishelson et al. 1987). In South Sulawesi, Unsworth et al. (2007b) assumed that *Caranx melampygus, Hemiramphus far*, and *Lutjanus* spp. moved from reef to seagrass habitat at high tide. Bray (1981) observed large *Chromis punctipinnis* consistently foraging at the incurrent end of a reef in southern California and moving to the opposite end of the reef when the current turned.

8.6.3 Functions of Intertidal Migrations

Feeding (function 1) and shelter (function 2, i.e., avoidance of predation) are likely the two most prominent functions of tidal movements (Gibson 1999). Tidal migrations often connect low-water resting sites with high-water feeding sites. Most transient fish seem to enter the intertidal with the flood tide, feed around high tide, and return at ebb tide to subtidal resting sites, thereby avoiding stranding (Robertson and Duke 1990, Krumme et al. 2004). During low tide the fish rest and digest. Differences in the local environmental settings, such as intertidal habitat accessibility, may influence the importance of intertidal feeding grounds. Results from Lugendo et al. (2007) suggest that mangroves close to subtidal resting areas (mangrove-lined creeks) are more important feeding sites for fishes than mangroves that drain completely (fringing mangrove).

Figure 8.8 illustrates the intertidal migration of the four-eyed fish *Anableps anableps*. The fish ride the early flood tide towards the upper reaches of mangrove-lined creeks where they feed around high tide. The fish return with the late ebb tide to the subtidal channel where they rest near the channel banks through the low-water period (Brenner and Krumme 2007). Amphibious mudskippers (Gobiidae) show a reverse tidal migration pattern; they rest at high tide, move downshore at ebb tide, and feed during the low-tide period to again retreat at flood tide (e.g., Colombini et al. 1996).

Examples from temperate coasts show that the rhythmic pattern of ingestion in transients is reflected in quantitative changes, with fuller stomachs at ebb tide vs. flood tide (e.g., Weisberg et al. 1981, Kleypas and Dean 1983, Hampel and Cattrijsse 2004), and in qualitative changes during tidal cycles (Ansell and Gibson 1990). During high tide the intertidal accessibility and visibility peak, and slower current speeds likely facilitate maneuvering, particularly for benthic invertebrate feeders and herbivores (Brenner and Krumme 2007). In addition to high-tide feeding, phyto- and zooplanktivores may take advantage of plankton naturally concentrated at low tide, e.g., in dead-ending channels (Krumme and Liang 2004). Likewise, piscivores may feed primarily during ebb tide at the mouths of creeks and channels, preying upon returning fish. Hoeinghaus et al. (2003) inferred such a strategy for piscivores from samples taken in the Venezuelan floodplains.

Fig. 8.8 Intertidal migration of the surface-swimming four-eyed fish *Anableps anableps* in a mangrove creek in north Brazil (neap tide, 29 June 2005). Simultaneous visual censuses on four transects of increasing distance upstream from the creek mouth (bottom to top). Falling water levels (solid lines) and decreased duration of inundation reflect topographic height of transect. Grey: upstream fish movements, black: downstream fish movements. Note different Y-axes (U Krumme, unpubl. data)

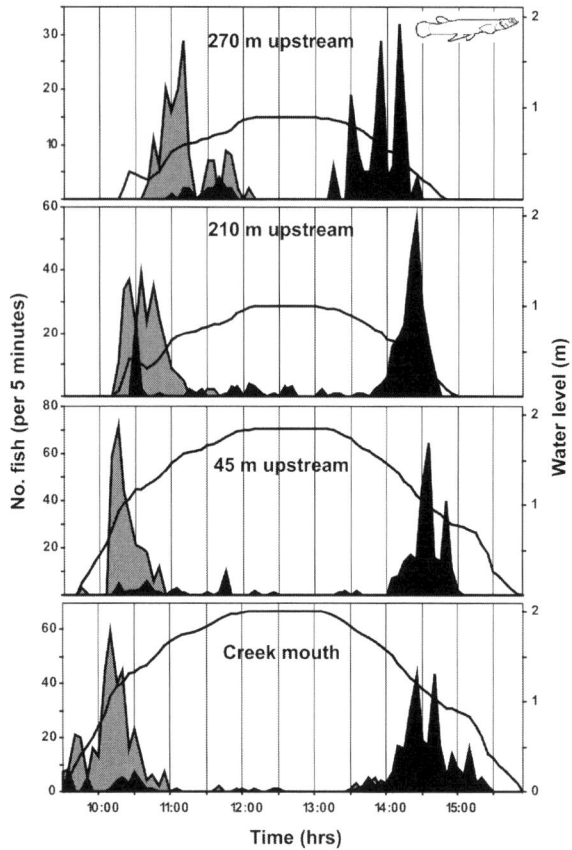

Fish swimming in shallow water reduce the risk of predation by piscivores which also enter at flood tide from deeper waters (c.g., Ruiz et al. 1993). Transients may move to high-tide sites to avoid predation in the area occupied during low tide (e.g., Dorenbosch et al. 2004). Reduction of the risk of predation is suggested when mouth-breeding and fasting male catfish *Cathorops* sp. only enter the intertidal at evening spring tides (Krumme et al. 2004) or – an example from a temperate coast – when fish move to the intertidal although the food supply is richer in the subtidal (Ansell and Gibson 1990).

Convincing evidence is lacking for tidal migrations (function 3) of fish as the result of inter- or intraspecific competition (see, e.g., Hill et al. 1982 for an example of the swimming crab *Scylla serrata*). In field experiments, it will be difficult to exclude the potential effect of piscine predators on differences between species or age groups.

Besides using the intertidal for feeding and shelter, several species carry out tidal migrations (function 4) to spawn (see Gibson 1992, 1999, De Martini 1999) or undertake regular small-scale or seasonal larger-scale movements to spawning sites, often following the lunar or spring tide cycle (e.g., Shapiro 1987, Zeller 1998; Chapter 4). A fifth function of tidal movements, the search by fish for physiologically appropriate environmental conditions (Gibson 1999), remains unclear because changes in water temperature, salinity, or oxygen content are highly correlated with tidal changes in water level. In the tropics, water temperature fluctuations are relatively small and estuarine fish are mostly euryhaline. Extreme salinities or hypoxia usually build up gradually and fish try to move out from affected areas, however, not necessarily with the tide (e.g., Shimps et al. 2005).

8.6.4 Sequence of Species and Size Groups

The little evidence available suggests that the intertidal movements of dominant nektonic organisms during flood and ebb tide are structured, both on the species level and among size groups. Inter- and intraspecific differences in factors such as minimum water level requirements, the relative location of the low-water resting sites, presence of predators, or foraging efficiency in the intertidal may lead to an ordered sequence of species and size groups entering and leaving the eulittoral. The most robust results come from eastern US saltmarsh creeks. Kneib and Wagner (1994) found that the number of species and individuals generally peak at high tide in flume weir samples compared to shallow flood and ebb tide samples. Their data suggested that smaller fish and shrimp travel shorter distances onto the marsh than larger conspecifics. Bretsch and Allen (2006a) used a sweep flume to quantify the migrations of nektonic species into and out of saltmarsh intertidal creeks. The migrations were nonrandom and structured; residents entered early at flood tide while transient species entered later at higher water levels. A species' water depth at peak migration increased as the species grew during summer. For north Brazilian mangrove creeks, Krumme et al. (2004) suggested that the emigration routes of the mangrove creek transients split inter- and intraspecifically at ebb tide, and took different species and life stages to specific resting sites. Data of Giarrizzo and Krumme (unpubl.) show that smaller *Colomesus psittacus* (Tetraodontidae) enter earlier at flood and leave later at ebb tide than larger conspecifics. Similarly, the youngest four-eyed fish *Anableps anableps* immigrate a few minutes earlier and return at few minutes later at lower water levels than the larger specimens (U Krumme unpubl. data). Thus, the more vulnerable and smaller fish maximize the time spent in the intertidal. It remains to be tested whether the smaller sizes stay longer to avoid predation or because they are still less efficient foragers.

8.6.5 Cues of Tidal Migrations

Tidal migrations are precisely timed, to achieve movement in appropriate conditions at flood tide and to avoid stranding when the tide recedes. The outcome of studies can differ considerably depending, e.g., on when and where samples are taken and what type of sampling gear is used (Gibson 1999). Knowledge about the cues fish use to migrate with the tide is crucial to define replicable sampling intervals in accordance with parameters relevant to the fish. Experiments have shown that fish respond to changes in underwater pressure (Gibson 1971, 1982, Gibson et al. 1978), fluctuations in water level (Ishibashi 1973), and hunger state (Nishikawa and Ishibashi 1975). Temperate flatfish may migrate with the tide by simply maintaining a constant depth (Gibson 1973). Bretsch and Allen (2006a) found that species use the same water depth to enter and leave the creeks, thus supporting Gibson's hypothesis. Likewise, the tidal migration of *Anableps anableps* is controlled by water level and not by a particular time interval before or after high or low tide (Brenner and Krumme 2007) (Fig. 8.8). Bretsch and Allen (2006b) further showed experimentally that mummichog (*Fundulus heteroclitus*) and grass shrimp (*Palaemonetes* spp.) selected shallower water depths to migrate in the presence of other predatory and non-predatory fish species, i.e., the timing of tidal migrations was in response to abiotic factors and to multiple-taxa effects, e.g., in attempts to reduce the risk of predation (Gibson and Robb 1996). Most likely tidal transients use several cues, both abiotic and biotic, simultaneously. The influence of other possible tide-related stimuli such as current speed, temperature, the sound generated by the current, or other biotic interactions awaits experimental testing.

To ensure comparability between samples, it is necessary to sample at slack low water and/or high water when assemblage compositions are most stable. Flood and ebb tides are fairly dynamic periods when the nektonic community is reshuffled. Sampling at flood or ebb tide likely increases variation and can lead to unwanted bias.

Studies that comparatively evaluate habitat types should take samples both at high and low tide in all habitat types to avoid results biased by tidal movements. For instance, high-tide samples from adjacent mangrove and seagrass habitats may suggest that mangroves support higher fish biomass than seagrass beds, but in actuality the fish may use the seagrass bed at low tide – which may result in higher low-tide fish biomass in the seagrass habitat – and move to the mangroves at high tide.

8.6.6 Movements and Foraging Ranges

The intertidal movement patterns and foraging ranges are of particular interest in attempts to identify appropriate marine park limits. Knowledge of the tidal movements of tropical fish is scarce. In Hawaiian atolls, the top predator *Aprion virescens* (Lutjanidae) was seasonally site-attached to core activity areas of up to 12 km in length, and ranged up to 19 km across atolls. Within their core areas, tagged *A. virescens* exhibited diel and tidal habitat shifts, with the latter resulting in round trips of up to 24 km in 24 hrs despite a tidal range <1 m (Meyer et al. 2007a). Fish moved along the barrier reef at flood tide and returned at ebb tide. A similar home range size of up to 29 km per day was determined for giant trevally *Caranx ignobilis* (Meyer et al. 2007b). The Hawaiian atolls lack vegetated ecosystems so that the fish do not migrate between mangroves and seagrass habitats, but the studies highlight the relatively broad scale of movements that subtidal top predators can cover during short-term migrations, even in microtidal areas.

In examples from temperate estuarine species, ultrasonically-tagged flatfish covered distances from 0.1–1.5 km (Wirjoatmodjo and Pitcher 1984, Szedlmayer and Able 1993) while *Liza ramada* (Mugilidae) even covered a median distance of 6,245 m during a complete tidal cycle (Almeida 1996). When tidally migrating, the fish may use selective tidal stream transport at flood or ebb tide (Forward and Tankersley 2001). Kleypas and Dean (1983) and Krumme (2004) showed that intertidal fish ride the flood tidal currents to arrive at their foraging grounds.

Strong tidal currents can also limit the activity of fishes. *Labroides dimidiatus* (Labridae) adapted the position of its cleaning station in response to tidal currents (Potts 1973). Flatfish such as *Pleuronectes platessa* bury in the sediment when currents are too strong (Arnold 1969).

8.6.7 Site Fidelity and Homing

Given the dynamics of the tides, one could suggest that fish on tidal coasts are organized in 'mobile stocks without attachment to particular locations' (Sogard et al. 1989). Evidence is mounting, however, that shallow-water fish center their short-term activities in core areas, display site fidelity, and home to familiar sites. This makes the fish particularly vulnerable to local exploitation on the one hand, but also likely to receive adequate protection with the establishment of no-fishing zones on the other. Knowledge of the surrounding seascape structure and topography is certainly beneficial for the fish to optimize the use of resting and feeding sites in complex environments. Fish may use physical features such as landmarks to navigate in a complex dynamic 3d-environment (Gibson 1999, Braithwaite and Burt De Perera 2006, Brown et al. 2006). Dorenbosch et al. (2004) suggested homing and site fidelity in tagged juvenile *Lutjanus fulviflamma* and *L. ehrenbergii* in Zanzibar. During daytime the fish apparently moved with the tide from a low-tide to a high-tide resting habitat (channel to notches), probably to avoid predation. Fishelson et al. (1987) and Craig (1996) provide evidence for site fidelity to high-tide feeding sites in reef systems (see also McFarland and Hillis 1982, Kuwamura 1985, Okuda and Yanagisawa 1996, Marnane 2000, Beets et al. 2003).

8.7 Comparison Between Diel and Tidal Migrations

8.7.1 Analogies and Differences Between Diel and Tidal Migrations

From the aforementioned patterns in diel and tidal migration it has become apparent that there are remarkable analogies between these short-term migrations, but

Table 8.2 Analogies and differences between diel and tidal migrations in fish[a]

	Feature	Diel migration[b]	Tidal migration[c]
Analogies			
1	Resting site	Structurally complex habitats	Subtidal, low water
2	Feeding site	Reef adjacencies (seagrass, sand, etc.)	Intertidal, high water
3	Migration period	Dusk and dawn	Flood and ebb tide
4	Sequence in species	Yes, according to light intensity	Yes, according to water depth
5	Sequence in size groups	Yes	Yes
6	Smaller stay longer in shelter	i.e., resting site	i.e., intertidal zone
7	Site fidelity	Resting and feeding sites	Resting and feeding sites
8	Homing	Yes	Yes
9	Migratory pathways	Constant over time	Constant over time
Differences			
1	Timing	Light intensity	Tidal cues and light intensity
2	Movements	Active	Partly gratis; riding the tide, selective tidal stream transport, active
3	Duration of migration	Short (min); twilight periods	Longer (hrs); flood and ebb tide
4	Predictability	Temporal variation lower	Temporal variation higher
5	Max. no. of migrations day^{-1}	One	Two (semidiurnal tide)
6	Foraging range	A few 100 m, rarely >1 km	Several 100 m to a few km

[a] Transients *sensu* Gibson (1969, 1988)
[b] Pure diel migrations are restricted to coasts with negligible tidal range (Caribbean, Red Sea, oceanic islands, and several areas in South and Southeast Asia; see Section 8.3)
[c] On all other coasts the tidal and diel cycles are tightly coupled

that there are also differences (Table 8.2). The analogies (numbered 1–5 below) certainly do not apply to all species, size groups, and locations, but the comparison may emphasize underlying constituents of short-term migrations subject to different ambient cycles.

(1) Both diel and tidal migrations usually connect resting areas with feeding areas. In diel migrations which only occur on coasts with weak tides, the fish rest at daytime and forage at night or vice versa, whereas in tidal migrations fish usually rest at low tide and forage at high tide.

(2) The structurally complex resting sites of diel migrants (e.g., caves, crevices, or mangroves) may correspond to low-water resting sites in the subtidal (burial in soft sediment, shelter in subtidal structure). Subtidal structure provided by dead plant material (Daniel and Robertson 1990) or vegetation such as seagrass can provide significant benefits to tidally migrating species (Irlandi and Crawford 1997).

(3) The diel migrations at sunset and sunrise may correspond to the immigration at flood tide which takes the fish to their foraging grounds, and the emigration at ebb tide which takes them back to their resting sites.

(4) Diel and tidal migrations are characterized by a sequence of species and size groups in the departure from and return to resting sites. Different species and size groups have different requirements and respond differently to changes in light intensity or water depth which results in temporally and spatially more or less structured short-term migrations. Smaller diurnal individuals and species returned earlier to nocturnal resting sites at dusk and emerged later at dawn than larger individuals and species (Hobson 1972, Helfman 1981). In nocturnal species, smaller size-groups migrated at lower light levels (McFarland et al. 1979). In tidal migrations, smaller individuals or species often travel at shallower depths than larger species and size groups. In general, smaller individuals seem to stay longer in the shelter site than larger individuals, either in the resting site or in the inundated littoral. Ontogenetic changes in the risk of predation seem to be reflected in the timing of both diel and tidal migrations.

(5) Site fidelity, homing, and constant migratory pathways, i.e., the use of core areas, has been evidenced in both diel and tidal migrations. Knowledge of the surrounding seascape structure is certainly useful to optimize shelter use and food search.

Differences between diel and tidal migrations (numbered 1–6 below) are apparent and mostly related to the different durations and physical features of the underlying cycles (Table 8.2).

(1) Changes in light intensity trigger diel migrations, whereas tidal cues control tidal migrations. On tidal coasts, primarily diurnal species center their foraging activities on daytime high tides, and nocturnal species on nightly high tides.

(2) The twilight migration is a relatively short event, often completed in <0.5 hr, whereas the movements to and from the intertidal foraging grounds may last more than 1 hr at flood and ebb tide, respectively.

(3) In diel migrations the distance between resting and feeding sites has to be covered by active swimming. Tidal migrations also involve active movements but are significantly facilitated by the tidal conveyor belt. Fish may ride the tide and use selective tidal stream transport to move to their destinations. Locomotory expenditure is reduced and saved energy can directly be transferred into increased growth and survival.

(4) Due to the relatively constant diel cycle in the tropics, diel movements are highly predictable but restricted to the relatively short twilight period, and therefore movements are precisely timed and show low variation. In contrast, the tidal cycle is subject to considerable variations due to astronomical, meteorological, and topographical reasons. Therefore, patterns in tidal migrations are likely more variable and more difficult to untangle than those of diel migrations.

(5) Diel migrations are restricted to dusk and dawn so that only one round trip per day is possible. Tidal migration on semidiurnal coasts can be carried out up to twice daily.

(6) Given our current knowledge, the foraging range of diel migrations may extend a few 100 m from the resting sites, sometimes exceeding 1 km, but are rarely larger than 2 km. Much less is known about foraging ranges of tidally migrating species but evidence suggests that distances are larger, maybe by one order of magnitude. Fish seem to travel distances of several 100 m to several kilometers each tide. It is remarkable that Hawaiian reef top predators move >20 km per day with the tide despite a tidal range of only 1 m.

8.7.2 Spring–Neap Tide Alternation

One feature characterizing tidal migrations but lacking in diel migrations is the spring-neap tide alternation. The quasi-weekly pulse of greater and smaller tidal ranges and current speeds is likely to have profound consequences for intertidal organisms. Tidal coasts are characterized by a vertical zonation of benthic organisms. The higher the tide, the more vertical zones are accessible and the more profitable is a visit by intertidal transients, and vice versa. Consequently, at spring tides usually more fish use the intertidal area than at neap tides (e.g., Davis 1988, Laegdsgaard and Johnson 1995, Wilson and Sheaves 2001, Krumme et al. 2004). The possibility of more extensive tidal migrations during spring tides was referred to above (see Section 8.3). The spring-neap tide cycle is reflected in cycles of food intake (e.g., Colombini et al. 1996, Brenner and Krumme 2007, Krumme et al. 2008), growth in intertidal fish (Rahman and Cowx 2006), and likely in cycles of mortality in the prey organisms. Thus, many tidal coasts are systems of two states, characterized by their different levels of tidal disturbance (Brenner and Krumme 2007). At neap tides the interaction between system compartments is relatively low (low inundation and low current speeds) compared to the highly dynamic spring tide periods (high inundation high and current speeds).

Mangrove coasts feature an additional transport mechanism for nearshore organisms. Floating mangrove litter is exported particularly during spring ebb tides (Schories et al. 2003), and provides structure, shade, and transport for larval and juvenile fishes and decapods (Daniel and Robertson 1990, Wehrtmann and Dittel 1990, Schwamborn and Bonecker 1996).

Furthermore, the spring-neap alternation is correlated with the lunar phases and changes in moonlight intensity. Moonlight intensity can change the activity patterns of fish, but the effect is apparently negligible in turbid estuaries (e.g., Quinn and Koijs 1981, Krumme et al. 2004, Krumme et al. 2008) and more relevant on clear water coasts (Hobson 1965). Untangling the effect of moonlight and spring tide, however, is a formidable task due to statistical considerations. Lunar cycles only recur monthly. The need to sample several lunar cycles automatically adds the effects of month and/or season. In addition, tides can cause unexpected co-variation, e.g., consistently higher or lower tidal ranges at a certain lunar phase, so that the effects can be inextricably correlated.

8.7.3 Interaction Between Tide and Time of Day

Another particularity, absent from coasts with weak tides, is the fact that the diel and the tidal cycle act in concert and neither of the two factors can be studied without considering the other. Let us assume the most common case of a semidiurnal tide, and a neap high-tide occurring around 12:00 and 00:00 hrs. Due to the retardation from tide to tide a week later, at spring tide, high tides occur around 18:00 and 06:00 hrs. In-between, mid high-tides would occur around 15:00 and 03:00 hrs. Each of the six groups is characterized by a particular combination of light intensities, tidal heights, and current speeds. These unique combinations recur, however, on a weekly or fortnightly basis. Intertidal fish assemblages and penaeid shrimp (see Section 8.8.2) respond strongly to these interacting factors.

Laroche et al. (1997) and Krumme et al. (2004) found recurring fish assemblages following particular combinations between spring-neap tide and day-night. For a given site, the nektonic community is predictably reshuffled each tide. Given certain environmental conditions, mainly determined by the interplay of the diel (light intensity) and tidal cycles (water depth, current speeds), a specific assemblage temporarily colonizes the intertidal. The assemblages alter in a characteristic pattern that not only involves species presence or absence, but also proportional differences in the intertidal occurrence among dominant species. Consequently, results from one of these short-term combinations are not fully representative for the other combinations, and care should be taken against making premature conclusions when the full set of short-term assemblage combinations is not known. The variation caused by the interaction of the diel and tidal cycles can be equal to seasonal variations in tropical estuarine fish assemblages (Krumme et al. 2004). Therefore, long-term monitoring programs on (meso- and macro-) tidal coasts should seriously consider the short-term variation caused by the interactive effects of the diel and tidal cycles.

It is apparent that fishes do not use each tide to migrate and that considerable variation can occur between individuals (e.g., Szedlmayer and Able 1993), size groups (Bretsch and Allen 2006a), sexes (Krumme et al. 2004), species, and regions (Gibson 1973, van der Veer and Bergman 1987). Results from temperate coasts have shown that species such as the plaice *Pleuronectes platessa* change their migratory behavior during ontogeny (Gibson 1997). In juvenile plaice there are examples for each of the three high-tide distributional patterns for a population: (1) complete population shift to the intertidal (Kuipers 1973), (2) only partial spread (Edwards and Steele 1968, Ansell and Gibson 1990), or (3) separation in intertidal and subtidal fish populations at high tide (Berghahn 1987). The study of such variations can provide insight into the mechanisms controlling migrations.

8.8 Tidal Movements of Decapods

The responses of decapods to tidal currents vary from avoidance of displacement, to intermittently walking and swimming, and selective tidal stream transport (Forward and Tankersley 2001). Connectivity by short-term movements between adjacent tropical ecosystems is most evident in the intertidal migrations of penaeid shrimps and swimming crabs.

8.8.1 Lobsters

Settled lobsters are not adapted to swim in tidal currents. The tides rather confine than foster the activity and movements of lobsters. Their natural behavior enables them to live on coasts with high tidal velocities. They shelter in areas of reduced flow on the sea bed or bury themselves in soft sediment. Instead of using the tide to move, lobsters have to reduce their mobility during stronger current periods to avoid displacement (Howard and Nunny 1983). In fact, lobsters of British coastal water approached baits only during the period of slack water (Howard 1980).

8.8.2 Penaeid Shrimps

Shallow, tidally influenced, and often turbid waters commonly provide essential nursery grounds for many commercially important penaeid shrimps. Juvenile penaeid shrimps are frequent visitors in intertidal mangroves at high tide (e.g., Staples and Vance 1979, Robertson and Duke 1987, Chong et al. 1990, Vance et al. 1990, Mohan et al. 1995, Primavera 1998, Rönnbäck et al. 1999, Krumme et al. 2004), as well as mudflats (e.g., Bishop and Khan 1999) and seagrass beds (Schaffmeister et al. 2006). On short-term migrations within the nursery ground the shrimps regularly transfer energy from the littoral to the sublittoral. Ontogenetic movements as part of their life cycle export the accumulated energy to the coastal ocean via reproductive offshore migrations (see Chapters 4 and 7). Kneib (1997) described this successive and stepwise export of energy via tidal and ontogenetic movements of fish and decapods in saltmarshes and aptly named it a 'trophic relay'.

Catchability of juvenile shrimps by trawls is often highly variable in space and time. The availability of shrimps depends on the species, behavior (i.e., buried or not, active or not), and response to the sampling gear (Vance and Staples 1992). Shrimp species such as *Penaeus merguiensis* are most active during nightly high tides (Dall et al. 1990), but are most catchable by trawls at low tides (e.g., Vance and Staples 1992). Other species such as *P. semisulcatus* and *P. esculentus* are more catchable during nightly high tides.

On a short-term scale the migratory behavior of shrimps closely responds to the interactive effects of the diel and tidal cycles as shown both in the field (Staples and Vance 1979, Vance and Staples 1992) and in laboratory experiments (e.g., Hindley 1975, Natajaran 1989a, b, Vance 1992). The relative strength of the response to the tide and light cycle is species-specific (Vance and Staples 1992). The interaction of these factors can lead to confounding effects that make the establishment of standard sampling programs difficult (Staples and Vance 1979, Bishop and Khan 1999). Laboratory studies further suggest that the activity of shrimps may change with turbidity, moonlight, salinity, and temperature. Figure 8.9 illustrates the complex

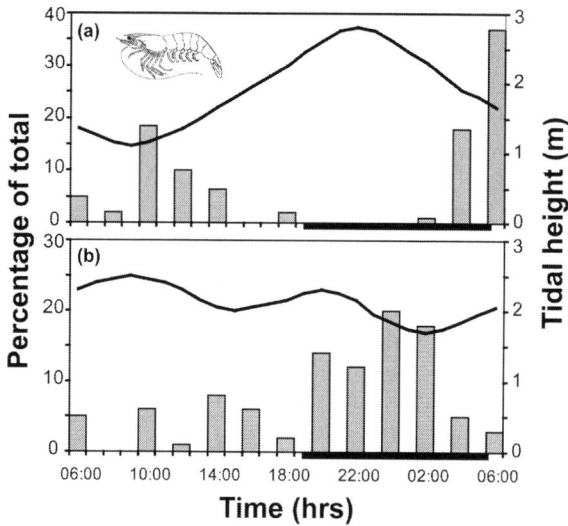

Fig. 8.9 Effect of the tidal (solid lines) and diel cycles during (**a**) a diurnal spring tide ($N = 529$), and (**b**) a semidiurnal neap tide ($N = 638$), on catchability (vertical bars) of juvenile *Penaeus merguiensis* in the Embly River, Gulf of Carpentaria, tropical Australia. Black horizontal bars on X-axes indicate hours of darkness. Modified after Staples and Vance (1979), with kind permission of DJ Vance. Reproduced with permission from the Australian J Mar Freshw Res 30(4):511–519. Copyright CSIRO (1979). Published by CSIRO PUBLISHING, Melbourne, Australia

interaction of the tidal and the diel cycles and, in particular, the influence of different tidal types on the catchability of *Penaeus merguiensis* near Weipa, eastern Gulf of Carpentaria, Australia. The tidal stage was more important than the time of day as suggested by a unimodal distribution during a diurnal tide and a bimodal distribution in catches at a semidiurnal tide.

Similar to many tidally migrating fish, there is information on the end points of the migration, i.e., the resting and feeding sites, but information on the movements connecting the end points is scarce. Some shrimp species such as *Penaeus merguiensis* are known to congregate in shallow water during the low-tide period, often close to the water edge (e.g., Hindley 1975, Hill 1985, Vance et al. 1990). Others such as the tiger prawn *P. monodon* bury and do not congregate near the water's edge. The smallest juvenile shrimps often inhabit more shallow water and the larger individuals live at greater depths (e.g., Staples and Vance 1979). With the flood tide the shrimps move upstream and enter intertidal mangrove-lined creeks.

Inundated mangroves provide a number of microhabitats for shrimps, but the high-tide distribution of shrimps in mangroves is highly variable (e.g., Rönnbäck et al. 1999, Vance et al. 2002, Meager et al. 2003). Factors such as local currents, topography, habitat type, and site-specific water clarity as determined by water depth and turbidity may play a role in influencing the distribution of shrimps in these intertidal microhabitats. Quinn and Koijs (1987) and Vance et al. (2002) have suggested that the movements of the shrimps are strongly influenced by the local currents. Krumme et al. (2004) found a significant positive relationship between the high-tide level and the abundance and catch weight of *Penaeus subtilis* from intertidal mangrove creeks. Depending on the local topography, tidal movements can take shrimps as far as 200 m into the mangrove forests (Vance et al. 1996, Rönnbäck et al. 1999, Vance et al. 2002). Feeding seems to occur mainly during high tide (e.g., Robertson 1988). Vance et al. (1990) suggested that at ebb tide the shrimps move downstream by both active and passive movements. They may, however, control downstream displacement at ebb tide by near-bottom activity, and return to the subtidal at very low intertidal water levels and concentrate along the turbid water edge during low tide, from where they may or may not enter with the next flood tide.

Little is known about differences or changes in the proportions of tidally migrating shrimps in a population. Bishop and Khan (1999) distinguished between subtidal and intertidal mudflat shrimps. Subtidal shrimps are unlikely to regularly connect adjacent ecosystems. Schaffmeister et al. (2006) caught juvenile and subadult *Palaemon elegans* from seagrass ponds at low tide in Mauritania and marked them with bright nail polish. Sample size and the recapture rates were low but the results suggested that juveniles left the ponds at flood tide to forage in the surrounding seagrass, and that some returned to the previously occupied pond while others were found in adjacent ponds at low tide. Subadults remained in their home pond at high and low tide. A better understanding of the fine-scale intertidal movements of decapods requires localized studies following the movements of individual shrimps while migrating with the tides.

8.8.3 *Portunid Crabs*

Tidal movements are likely common in tropical swimming crabs, but surprisingly little information has been published. Important parameters that determine migratory activity are age, sex, and molting stage. Hill et al. (1982) found that juvenile mud crab *Scylla serrata* were resident in the intertidal mangrove zone, similar to juvenile *Portunus pelagicus* that remained in intertidal pools at low tide (Williams 1982). The majority of subadult and on occasion adult mud crabs moved in the intertidal zone only during high tides and retreated to the subtidal zone at low tide (Hill et al. 1982). Sublittoral estuarine adult *S. serrata* 'live a free-ranging non-territorial existence' (Hill 1978). They may stay in the same area (<1 km) for longer periods or move larger distances downstream (>10 km) within weeks. Hill et al. (1982) suggested that reduction of intraspecific competition and feeding are the main reasons for intertidal migrations of *S. serrata* and the intertidal residence of juveniles, thus indicating that tidal migrations possibly serve as a means of avoiding intraspecific competition (see Section 8.6.3).

Regular movements to and from the intertidal with the rising and falling tide, respectively, have been reported for *Thalamita crenata* (Cannicci et al. 1996), *Callinectes sapidus* (Nishimoto and Herrnkind 1978), *Cancer magister* (Williams 1979), and *Carcinus maenas* (Dare and Edwards 1981). *T. crenata* showed greatest activity when the intertidal water level was between 10 and 40 cm high (Vezzosi et al. 1995), and used landmarks to locate its refuges and was able to home (Vannini and Cannicci 1995, Cannicci et al. 2000).

On temperate tidal coasts, juvenile blue crabs *Callinectes sapidus* enter the intertidal zone with the flood tide but usually do not venture far into the saltmarsh (<100 m; Fitz and Wiegert 1991, Kneib 1995). At low tide they may bury in shallow water (van Montfrans et al. 1991). *C. sapidus* stomachs were fullest at high tide, indicating that the immigration during flood tide is used as an active feeding period (Ryer 1987). Thus, the tidal cycle may result in cycles of food intake in the blue crab (Weissburg and Zimmer-Faust 1993, 1994, Zimmer-Faust et al. 1995, 1996, Weissburg et al. 2003). Cannicci et al. (1996) reported greater feeding of *Thalamita crenata* at spring than at neap tides.

8.9 Comparison of the Degree of Habitat Connectivity among Geographic Regions

The available literature suggests that heterogeneous seascapes are often tightly connected by short-term, i.e., diel and tidal movements of nektonic organisms (for population connectivity due to ontogenetic migrations refer to Chapters 6, 7, 10). The degree of connectivity among habitats may, however, differ between regions.

One important factor for regional differences in short-term habitat connectivity is hydrology. The tidal ranges in the Indo-West Pacific are generally greater than in the Caribbean, which may facilitate connections between adjacent ecosystems. Short-term movements need not be restricted to the diel cycle, i.e., there is only one round trip in 24 hrs. On coasts with semidiurnal tides, two round trips in 24 hrs are possible; as a consequence, subtidal habitats can house both diel and tidal visitors. Evidence for diel and tidal habitat connectivity in the Indo-West Pacific either originates from just one habitat (seagrass) and only infers connectivity to adjacent habitats (e.g., Kochzius 1999, Unsworth et al. 2007a) or is in fact based on results from several adjacent habitats (Nakamura and Sano 2004, Dorenbosch et al. 2005, Unsworth et al. 2007b, Unsworth et al. 2008).

Due to the negligible tidal pulse, twilight movements are the major driver of short-term habitat linkages in the Caribbean. The link between Caribbean mangroves and seagrass and reef fish fauna may be relatively strong because the latter two habitats can occur sufficiently close to mangroves to allow diel fish connectivity. However, on a global scale, the Caribbean mangroves are an exception rather than the rule. Unlike most other tropical mangrove coasts of the world, the Caribbean patch mangroves thrive in this clear-water environment with relatively little terrestrial runoff, are continuously inundated, and thus are always accessible to nektonic organisms. Commonly, mangroves and muddy mangrove-lined channels and creeks are intertidal, and access is restricted to periods of inundation.

Another factor is species richness and the composition of functional groups. In the Indo-West Pacific, more species are potentially involved in short-term inter-habitat migrations. All functional fish groups have more species here than in the

Caribbean (Bellwood et al. 2004). Particularly, invertebrate feeders, and diurnal and nocturnal planktivores are more diverse in the West Pacific. Haemulidae are the dominant diel migrants of Caribbean reefs and are more diverse here than in the Indo-West Pacific. However, there are several other families whose inter-habitat connectivity has been demonstrated (Table 8.1) that have greater species richness in the Great Barrier Reef (Bellwood and Wainwright 2006), e.g., Pomacentridae, Apogonidae, Holocentridae, Lutjanidae, Mullidae, and Siganidae. Apogonidae are the dominant nocturnal planktivores of the Indo-Pacific reefs, both in terms of abundance and species diversity. Parrish (1989) suggested the connecting function of Caribbean Haemulidae may be substituted by Lethrinidae, though this family is less dominant in the Indo-Pacific than Haemulidae in the Caribbean. More qualitative and quantitative field evidence of short-term inter-habitat linkages is needed for the majority of the families listed in Table 8.1. The Indo-West Pacific in particular is lacking in this kind of data.

8.10 Tidal Range and Home Range Size

Pittman and MacAlpine (2003) suggested that there is unlikely to be a strong linear relationship between fish body size and home range size due to geographical and high intra- and interspecific variability in fish behavior. Large reefal top predators can be both highly mobile (Meyer et al. 2007a, b) or extremely sedentary (Zeller 1997, Kaunda-Arara and Rose 2004, Popple and Hunte 2005). Tiny fish can occupy territories extremely small in size, yet migrate vast distances to find plankton patches, such as anchovies. Thus, habitat connectivity due to migration is not a simple function of fish size.

Habitat connectivity is likely greatest where multiple habitat types are coexisting in close proximity. Yet, when profitable habitats are more distant, increased tidal ranges, i.e., higher current speeds, may facilitate traveling to otherwise remote resources and shelter sites. It is postulated here that the home range size of the same or cognate species is greater when the tidal range and currents are greater and that therefore, habitat connectivity by short-term movements is likely greater than on coasts with negligible tides.

Two examples illustrate the potential increase in home range size due to increased tidal range. In Trinidad, where the tidal range is ~0.5 m (Wothke and Greven 1998), the four-eyed fish *Anableps anableps* occupies a home range with a maximum distance of <100 m (H Greven, University of Duesseldorf, Germany, pers. comm.). In north Brazilian mangrove creeks, where the spring tide range is between 3 and >4 m, the same species may travel >1.5 km between low-water resting and high-water feeding sites each tide, i.e., >3 km per day (U Krumme, unpubl. data). Likewise, the ariid catfish *Sciades herzbergii* may occupy home ranges <1 km in the southern Caribbean (A Acero Pizarro, INVEMAR, Colombia, pers. comm.) where tidal ranges are ~1 m, whereas the tidal movements of the same species in north Brazil may also cover distances >1.5 km per tide (U Krumme, pers. observ.). It should be noted, however, that the increase in home range size in the two examples by more than one order of magnitude between a micro- and macrotidal area is also due to differences in topography between the two sites. The larger the tidal range and the flatter the intertidal, the greater the possible distance of intertidal upshore excursions and the greater the home range occupied by a population of transients. In the case of subtidal migrations, increased foraging ranges would largely be a result of greater current speeds that allow for greater distances covered each tide.

Where different ecosystems co-occur, a much greater proportion of the heterogeneous ecosystem is accessible to nektonic organisms when current speeds are increased. When a greater tidal range increases foraging range, use of more distant sites becomes profitable and habitat connectivity increases. Alternatively, nektonic species at macrotidal coasts may simply undertake longer migrations to a similar number of sites that are, however, more profitable than the restricted number of sites accessible in a microtidal setting. Accessibility of more profitable sites should result in faster growth, reduced mortality, and greater recruitment of juveniles to the adult stock. If 'the greater the tidal range, the greater a species home range' holds true, it is evident that marine parks on macrotidal coasts need to be much larger than those on microtidal coasts.

8.11 Tides—an Overlooked Component of Variation Between Coasts

The diversity of tidal pulses – as briefly outlined in Section 8.3 – is a component of variation between coasts that seems to be fairly overlooked in large scale comparisons of biodiversity or productivity among coastal regions. Tides are the principal pulse for exchange processes linking adjacent coastal ecosystems on the short- to medium-term, and are the key engineers of coastal processes that determine ecosystem productivity and functioning in the long term. Tides regularly expose the intertidal which is particularly rich in epifauna and flora and provides the nursery grounds for various marine species. Tides create currents that mix the sediment and resuspend nutrients that enhance plankton production, fostering the production of higher trophic levels. Tides transport plankton to sessile filter feeders that provide food and shelter to other organisms.

Clearly, tides add a significant level of natural disturbance to a coastal system. Systems under different regimes likely have different natural levels of habitat connectivity, vulnerability, and resilience against disturbance. According to the intermediate disturbance hypothesis (Connell 1978), which proposes that the highest diversity is maintained at intermediate levels of disturbance, meso- or macrotides may favor a greater habitat connectivity and resilience, and a lower level of vulnerability of coastal ecosystems. (i) In systems with weak tides such as the Caribbean, short-term exchange processes are restricted to ocean currents and active animal movements related to light intensity. A given set of species lives under these conditions and exhibits a certain level of habitat connectivity between the coastal systems. Local disturbances are barely buffered by adjacencies. (ii) In intermediate tide systems, life is more dynamic. Exchange processes are facilitated by the tide (e.g., for filter feeders and higher trophic levels). Habitat connectivity is increased by tidal movements. Disturbances can be buffered from adjacent areas. Species benefit from the increased movement of the water. (iii) In systems with strong tides, life is very dynamic and habitat connectivity may be high. However, certain species may be excluded (e.g., frequent sediment rearrangement excludes long-lived sessile organisms). Disturbances have to be high to add to the naturally high level of disturbance of a system with macrotides.

It is reasonable to assume that tidal range and tidal type influence species diversity and ecosystem functioning. Besides biogeographical differences in species richness and composition of functional groups in the Indo-Pacific coasts (see, e.g., Bellwood et al. 2004), intermediate tides may favor greater habitat connectivity and resilience, and a lower level of vulnerability in Indo-Pacific coastal ecosystems compared to the Caribbean. Unsworth et al. (2007b) noticed considerable variation between the seagrass fish fauna in Indonesia and other Indo-West Pacific regions which might be due to different tidal regimes.

Given that the tidal ranges on the Caribbean coasts are very small, biological studies comparing the Caribbean and the Indo-West Pacific are only appropriate when areas with weak tides of similar seascape configuration are compared. Consequently, due to a tidal range >3 m in the Great Barrier Reef, comparisons with the Caribbean are inherently faulty because they compare two systems with different levels of natural disturbance. Consequently, to reduce the likely variation between data sets and thereby increase our understanding of the variation caused by different tidal regimes, future studies should (1) compare systems in different geographical regions but of similar tidal regimes (e.g., coasts with weak tides in the Caribbean vs. coasts with weak tides in the Indo-West Pacific), or (2) compare systems from similar geographical regions, i.e., with similar species communities, but of different tidal regimes (e.g., Caribbean vs. Brazilian coast, numerous study comparisons would be possible in the Indo-West Pacific region). In Recife, East Brazil, fishermen report that different age groups of different species move between specific sites in a mangrove/seagrass/coral reef seascape according to the interactive combination of tide and time of day (S Schwamborn, Universidade do Estado da Bahia, Brazil, pers. comm.). This results in more complex patterns of habitat connectivity in heterogeneous seascapes exposed to meso- and macrotides than in microtidal areas such as the Caribbean.

Several hypotheses remain untested. Are there overall differences in life history patterns (migrations, growth performance, or natural morality) within a species from similar micro-, meso-, and macrotidal coasts? Do different tidal regimes lead to detectable differences in the functioning of ecosystems?

If tidal range and the functioning of ecosystems in fact significantly interact, the scope of broad-scale comparisons must be redefined with a new focus on variation caused by differences in tidal regimes. There is certainly a need for enhanced international cooperation which should include multi-national projects, with standardized methods and sample designs to allow comparisons between results, in the search for global patterns and improved conservation of tropical marine resources.

Acknowledgments I am grateful to I Nagelkerken for the invitation to contribute to this book and I thank two anonymous reviewers, the editor, and G Castellanos-Galindo for their helpful comments.

References

Abou-Seedo F, Clayton DA, Wright JM (1990) Tidal and turbidity effects on the shallow-water fish assemblage of Kuwait Bay. Mar Ecol Prog Ser 65:213–223

Acosta CA (1999) Benthic dispersal of Caribbean spiny lobsters among insular habitats: implications for the conservation of exploited marine species. Conserv Biol 13:603–612

Adams AA, Dahlgren CP, Todd Kellison G et al (2006) Nursery function of tropical back-reef systems. Mar Ecol Prog Ser 318:287–301

Admiralty Co-Tidal Atlas (2001) South-East Asia. NP 215, ed. 1–1979. Admiralty Charts and Publications, UK Hydrographic Office

Almeida PR (1996) Estuarine movement patterns of adult thin-lipped grey mullet, *Liza ramada* (Risso) (Pisces, Mugilidae), observed by ultrasonic tracking. J Exp Mar Biol Ecol 202:137–150

American Practical Navigator (2002) Digital navigation publication No. 9. National Geospatial-Intelligence Agency. www.nga.mil

Ansell AD, Gibson RN (1990) Patterns of feeding and movement of juvenile flatfish on an open sandy beach. In: Barnes M, Gibson RN (eds) Trophic relationships in the marine environment, pp. 191–207. Aberdeen University Press

Arnold GP (1969) The reaction of the plaice (*Pleuronectes platessa* L.) to water currents. J Exp Biol 51:681–697

Baelde P (1990) Differences in the structures of fish assemblages in *Thalassia testudinum* beds in Guadeloupe, French West Indies, and their ecological significance. Mar Biol 105:163–173

Bakus GJ (1967) The feeding habits of fishes and primary production at Eniwetok, Marshall Islands. Micronesia 3:135–149

Beck MW, Heck KL, Able KW et al (2001) The identification, conservation, and management of estuarine and marine nurseries for fish and invertebrates. BioScience 51:633–641

Beets J, Muehlstein L, Haught K et al (2003) Habitat connectivity in coastal environments: patterns and movement of Caribbean coral reef fishes with emphasis on bluestriped grunt, *Haemulon sciurus*. Gulf Caribb Res 14:29–42

Bellwood DR, Hughes TP, Folke C et al (2004) Confronting the coral reef crisis. Nature 429:827–833

Bellwood DR, Wainwright PC (2006) The history and biogeography of fishes on coral reefs. In: Sale PF (ed) Coral reef fishes: dynamics and diversity in a complex ecosystem, pp. 5–32. Academic Press, San Diego

Berghahn R (1987) Effects of tidal migration on growth of 0–group plaice (*Pleuronectes platessa* L.) in the North Friesian Wadden Sea. Meeresforschung 31:209–226

Bishop JM, Khan MH (1999) Use of intertidal and adjacent mudflats by juvenile penaeid shrimps during 24-h tidal cycles. J Exp Mar Biol Ecol 232:39–60

Braithwaite VA, Burt de Perera T (2006) Short-range orientation in fish: how fish map space. Mar Freshw Behav Physiol 39:37–47

Bray RN (1981) Influence of water currents and zooplankton densities on daily foraging movements of blacksmith, *Chromis punctipinnis*, a planktivorous reef fish. Fish Bull 78:829–841

Brenner M, Krumme U (2007) Tidal migration and patterns in feeding of the four-eyed fish *Anableps anableps* L. in a north Brazilian mangrove. J Fish Biol 70:406–427

Bretsch K, Allen DM (2006a) Tidal migrations of nekton in salt marsh creeks. Estuar Coasts 29:479–491

Bretsch K, Allen DM (2006b) Effects of biotic factors on depth selection by salt marsh nekton. J Exp Mar Biol Ecol 334:130–138

Briones-Fourzan P, Castaneda-Fernandez de Lara V, Lozano-Alvarez E et al (2003) Feeding ecology of the three juvenile phases of the spiny lobster *Panulirus argus* in a tropical reef lagoon. Mar Biol 142:855–865

Brown AC, McLachlan A (1990) Ecology of sandy shores. Elsevier, Amsterdam

Brown C, Laland KN, Krause J (2006) Fish cognition and behavior. Fish and Aquatic Resources 11, Wiley-Blackwell, Oxford

Burke NC (1995) Nocturnal foraging habitats of French and bluestriped grunts, *Haemulon flavolineatum* and *H. sciurus*, at Tobacco Caye, Belize. Environ Biol Fish 42:365–374

Cannicci S, Barellia C, Vannini M (2000) Homing in the swimming crab *Thalamita crenata*: a mechanism based on underwater landmark memory. Anim Behav 60:203–210

Cannicci S, Dahdouh-Guebas F, Anyona D et al (1996) Natural diet and feeding habits of *Thalamita crenata* (Decapoda: Portunidae). J Crust Biol 16:678–683

Castellanos-Galindo GA, Giraldo A, Rubio EA (2005) Community structure of an assemblage of tidepool fishes in a Tropical eastern Pacific rocky shore, Colombia. J Fish Biol 67:392–408

Chittleborough RG (1974) Home range, homing and dominance in juvenile western rock lobsters. Aust J Mar Freshw Res 25:227–234

Chong VC, Sasekumar MUC, Cruz RD (1990) The fish and prawn communities of Malaysian coastal mangrove systems, with comparisons to adjacent mudflats and inshore waters. Estuar Coast Shelf Sci 31:703–722

Cobb JS (1981) Behaviour of the Western Australian spiny lobster, *Panulirus cygnus* George in the field and in the laboratory. Aust J Mar Freshw Res 32:399–409

Collette BB, Talbot FH (1972) Activity patterns of coral reef fishes with emphasis on nocturnal-diurnal changeover. In: Results of the Tektite program: ecology of coral reef fishes. Nat Hist Mus Los Angeles Count Sci Bull 14:98–124

Colombini I, Berti R, Nocita A et al (1996) Foraging strategy of the mudskipper *Periophthalmus sobrinus* Eggert in a Kenyan mangrove. J Exp Mar Biol Ecol 197:219–235

Connell JH (1978) Diversity in tropical rain forests and coral reefs. Science 199:1302–1310

Cowen RK, Paris CB, Srinivasan A (2006) Scaling of connectivity in marine populations. Science 311:522–527

Cox C, Hunt JH, Lyons WG et al (1997) Nocturnal foraging of the Caribbean spiny lobster (*Panulirus argus*) on offshore reefs of Florida, USA. Mar Freshw Res 48:671–680

Craig P (1996) Intertidal territoriality and time-budget of the surgeonfish, *Acanthurus lineatus*, in American Samoa. Environ Biol Fish 46:27–36

Creaser EP, Travis D (1950) Evidence of a homing instinct in the Bermuda spiny lobster. Science 112:169–170

Dall W, Hill BJ, Rothlisberg PC et al (1990) The biology of the Penaeidae. Adv Mar Biol 27:1–489

Daniel PA, Robertson AI (1990) Epibenthos of mangrove waterways and open embayments: community structure and the relationship between exported mangrove detritus and epifaunal standing stocks. Estuar Coast Shelf Sci 31:599–619

Dare PJ, Edwards DB (1981) Underwater television observations on the intertidal movements of shore crabs, *Carcinus maenas*, across a mudflat. J Mar Biol Ass UK 61:107–116

Davies JL (1972) Geographical variation in coastal development. Oliver Boyd, Edinburgh

Davis TLO (1988) Temporal changes in the fish fauna entering a tidal swamp system in tropical Australia. Environ Biol Fish 21:161–172

De Martini EE (1999) Intertidal spawning. In: Horn MH, Martin KLM, Chotkowski MA (eds) Intertidal fishes: life in two worlds, pp. 143–164. Academic Press, San Diego

Dietrich G (1980) General oceanography: an introduction. 2nd ed., translated by Roll S, Roll HU. Wiley-Interscience, New York

Dingle H (1996) Migration: the biology of life on the move. Oxford University Press, New York

Doherty PJ (1983) Tropical territorial damselfishes: is density limited by aggression or recruitment? Ecology 64:176–190

Domm SB, Domm AJ (1973) The sequence of appearance at dawn and disappearance at dusk of some coral reef fishes. Pacific Sci 27:128–135

Dorenbosch M, Grol MGG, Christianen MJA et al (2005) Indo-Pacific seagrass beds and mangroves contribute to fish density and diversity on adjacent coral reefs. Mar Ecol Prog Ser 302:63–76

Dorenbosch M, Verberk WCEP, Nagelkerken I et al (2007) Influence of habitat configuration on connectivity between fish assemblages of Caribbean seagrass beds, mangroves and coral reefs. Mar Ecol Prog Ser 334:103–116

Dorenbosch M, Verweij MC, Nagelkerken I et al (2004) Homing and daytime tidal movements of juvenile snappers (Lutjanidae) between shallow-water nursery habitats in Zanzibar, western Indian Ocean. Environ Biol Fish 70:203–209

Edgar GJ (1990) Predator-prey interactions in seagrass beds. II. Distribution and diet of the blue manna crab *P. pelagicus* Linnaeus at Cliff Head, Western Australia. J Exp Mar Biol Ecol 139:23–32

Edwards RRC, Steele JH (1968) The ecology of 0-group plaice and common dabs at Lochewe. I. Population and food. J Exp Mar Biol Ecol 2:215–238

Eisma D (1998) Intertidal deposits: river mouths, tidal flats, and coastal lagoons. CRC Press, Boca Raton

Fiedler DR (1965) The spiny lobster *Jasus lalandii* in South Australia. III. Food, feeding and locomotor activity. Aust J Mar Freshw Res 16:351–367

Fishelson L, Montgomery WL, Myrberg AH Jr (1987) Biology of surgeonfish *Acanthurus nigrofuscus* with emphasis on changeover in diet and annual gonadal cycles. Mar Ecol Prog Ser 39:37–47

Fishelson L, Popper D, Gunderman N (1971) Diurnal cyclic behaviour of *Pempheris oualensis* Cuv. and Val. (Pempheridae, Teleostei). J Nat Hist 5:503–506

Fitz HC, Wiegert RG (1991) Utilization of the intertidal zone of a salt marsh by the blue crab, *Callinectes sapidus*: density, return frequency, and feeding habits. Mar Ecol Prog Ser 76:249–260

Forward RB Jr, Tankersley RA (2001) Selective tidal stream transport of marine animals. Oceanogr Mar Biol An Rev 39:305–353

Foster SA (1987) Diel and lunar patterns of reproduction in the Caribbean and Pacific sergeant major damselfishes *Abudefduf saxatilis* and *A. troschelii*. Mar Biol 95:333–343

Friedlander AM, Monaco ME (2007) Acoustic tracking of reef fishes to elucidate habitat utilization patterns and residence times inside and outside marine protected areas around the island of St. John, USVI. NOAA Technical Memorandum NOS NCCOS 63

Frisch AJ (2007) Short- and long term movements of painted lobster (*Panulirus versicolor*) on a coral reef at Northwest Island, Australia. Coral Reefs 26:311–317

Gibson RN (1969) The biology and behavior of littoral fish. Oceanogr Mar Biol Ann Rev 7:367–410

Gibson RN (1971) Factors affecting the rhythmic activity of *Blennis pholis* L. (Teleostei). Anim Behav 19:336–343

Gibson RN (1973) The intertidal movements and distribution of young fish on a sandy beach with special reference to the plaice (*Pleuronectes platessa* L.). J Exp Mar Biol Ecol 12:79–102

Gibson RN (1982) Recent studies of the biology of intertidal fishes. Oceanogr Mar Biol Ann Rev 20:363–414

Gibson RN (1988) Patterns of movement in intertidal fishes. In: Chelazzi G, Vanini M (eds) Behavioural adaptions to intertidal life, pp. 55–63. NATO ASI Series Life Sciences Vol. 151. Plenum Press, London

Gibson RN (1992) Tidally-synchronised behaviour in marine fishes. In: Ali MA (ed) Rhythms in fishes, pp. 63–81. NATO ASI Series Life Sciences Vol. 236. Plenum Press, New York

Gibson RN (1993) Intertidal teleosts: life in a fluctuating environment. In: Pitcher TJ (ed) Behaviour of teleost fishes, pp. 513–536, 2nd ed. Fish and Fisheries Series 7. Chapman and Hall, London

Gibson RN (1996) Tidal, diel and longer term changes in the distribution of fishes on a Scottish sandy beach. Mar Ecol Prog Ser 130:1–17

Gibson RN (1997) Behaviour and the distribution of flatfishes. J Sea Res 37:241–256

Gibson RN (1999) Methods for studying intertidal fishes. In: Horn MH, Martin KLM, Chotkowski MA (eds) Intertidal fishes: life in two worlds, pp. 7–25. Academic Press, San Diego

Gibson RN (2003) Go with the flow: tidal migration in marine animals. Hydrobiologia 503:153–161

Gibson RN, Blaxter JHS, De Groot SJ (1978) Developmental changes in the activity rhythms of the plaice (*Pleuronectes platessa* L.). In: Thorpe JE (ed) Rhythmic activity of fishes. pp. 169–186. Academic Press, London

Gibson RN, Pihl L, Burrows MT et al (1998) Diel movements of juvenile plaice *Pleuronectes platessa* in relation to predators, competitors, food availability and abiotic factors on a microtidal nursery ground. Mar Ecol Prog Ser 165:145–159

Gibson RN, Robb L (1996) Piscine predation on juvenile fishes on a Scottish sandy beach. J Fish Biol 49:120–138

Gladfelter WB (1979) Twilight migrations and foraging activities of the copper sweeper, *Pempheris schomburgki* (Teleostei, Pempheridae). Mar Biol 50:109–119

Gladfelter WB, Ogden JC, Gladfelter EH (1980) Similarity and diversity among coral reef fish communities: a comparison between tropical western Atlantic (Virgin Islands) and tropical central Pacific (Marshall Islands) patch reefs. Ecology 61:1156–1168

Grove-Jones R (1987) Catch and effort in the South Australian blue crab (*Portunus pelagicus*) fishery. South Australian Department of Fisheries, discussion paper, September 1987, 45 pp.

Hampel H, Cattrijsse A (2004) Temporal variation in feeding rhythms in a tidal marsh population of the common goby *Pomatoschistus microps* (Kroyer, 1838). Aquat Sci 66:315–326

Harden Jones FR (1968) Fish migration. Arnold, London

Harmelin-Vivien ML, Bouchon C (1976) Feeding behavior of some carnivorous fishes (Serranidae and Scorpaenidae) from Tulear (Madagascar). Mar Biol 37:329–340

Hayes MO (1975) Morphology of sand accumulation in estuaries: an introduction to the symposium. In: Cronin LE (ed) Estuarine research, Vol. 2, Geology and Engineering, pp. 3–22. Academic Press, New York

Helfman GS (1979) Twilight activities of yellow perch, *Perca flavescens*. J Fish Res Board Can 36:173–179

Helfman GS (1981) Twilight activities and temporal structure in a freshwater fish community. Can J Fish Aquat Sci 38:1405–1420

Helfman GS (1993) Fish behaviour by day, night and twilight. In: Pitcher TJ (ed) Behaviour of teleost fishes, pp. 479–512, 2nd ed. Fish and Fisheries Series 7. Chapman and Hall, London

Helfman GS, Meyer JL, McFarland WN (1982) The ontogeny of twilight migration patterns in grunts (Pisces: Haemulidae). Anim Behav 30:317–326

Helfman GS, Schultz ET (1984) Social transmission of behavioural traditions in coral reef fish. Anim Behav 32:379–384

Herrnkind WF (1980) Spiny lobsters: patterns of movement. In: Cobb JS, Phillips BF (eds) The biology and management of lobsters, pp. 349–407. Academic Press, New York

Herrnkind WF, McLean RB (1971) Field studies of homing, mass emigration, and orientation in the spiny lobster, *Panulirus argus*. Ann N Y Acad Sci 188:359–377

Hiatt RW, Strasburg DW (1960) Ecological relationships of the fish fauna on coral reefs on the Marshall Islands. Ecol Monogr 30:65–127

Hill BJ (1978) Activity, track and speed of movement of the crab *Scylla serrata* in an estuary. Mar Biol 47:135–141

Hill BJ (1985) Effects of temperature on duration of emergence, speed and movement and catchability of the prawn, *Penaeus esculentus*. In: Rothlisberg PC, Hill BJ, Staples DJ (eds) Second Australian national prawn seminar, pp. 77–83. Cleveland, Australia

Hill BJ, Williams MJ, Dutton P (1982) Distribution of juvenile, subadult and adult *Scylla serrata* (Crustacea: Portunidae) on tidal flats in Australia. Mar Biol 69:117–120

Hindley JPR (1975) Effects of endogenous and some exogenous factors on the activity of the juvenile banana prawn *Penaeus merguiensis*. Mar Biol 29:1–8

Hobson ES (1965) Diurnal-nocturnal activity of some inshore fishes in the Gulf of California. Copeia 3:291–302

Hobson ES (1968) Predatory behavior of some shore fishes in the Gulf of California. US Fish Wildl Serv Res Rep 73:1–92

Hobson ES (1972) Activity of Hawaiian reef fishes during evening and morning transitions between daylight and darkness. Fish Bull US 70:715–740

Hobson ES (1973) Diel feeding migrations in tropical reef fishes. Helgoländ Wiss Meerunters 24:361–370

Hobson ES (1974) Feeding relationships of teleostean fishes on coral reefs in Kona, Hawaii. Fish Bull 72:915–1031

Hobson ES (1975) Feeding patterns among tropical reef fishes. Am Sci 63(4):382–392

Hobson ES, Chess JR (1973) Feeding oriented movements of the atherinid fish *Pranesus pinquis* at Majuro Atoll, Marshall Islands. Fish Bull US 71:777–786

Hoeinghaus DJ, Layman CA, Arrington DA et al (2003) Spatiotemporal variation in fish assemblage structure in tropical floodplain creeks. Environ Biol Fish 67:379–387

Holland KN, Peterson JD, Lowe CG et al (1993) Movements, distribution and growth rates of the white goatfish *Mulloides flavolineatus* in a fisheries conservation zone. Bull Mar Sci 52:982–992

Howard AE (1980) Substrate and tidal limitations on the distribution and behaviour of the lobster and edible crab. Prog Underwar Sci 52:165–169

Howard AE, Nunny RS (1983) Effects of near-bed current speed on the distribution and behaviour of the lobster *Homarus gammarus* (L.). J Exp Mar Biol Ecol 71:27–42

Hyland SJ, Hill BJ, Lee CP (1984) Movement within and between different habitats by the portunid crab *Scylla serrata*. Mar Biol 80:57–61

Irlandi EA, Crawford MK (1997) Habitat linkages: the effect of intertidal salt marshes and adjacent subtidal habitats on abundance, movement, and growth of an estuarine fish. Oecologia 110:222–230

Ishibashi T (1973) The behavioural rhythms of the gobioid fish *Boleophthalmus chinensis* (Osbeck). Fukuoka Univ Sci Rep 2:69–74

Jernakoff P (1987) Foraging patterns of juvenile western rock lobsters *Panulirus cynus* George. J Exp Mar Biol Ecol 113:125–144

Jernakoff P, Phillips BF (1988) Effect of a baited trap on the foraging movements of juvenile western rock lobsters, *Panulirus cygnus* George. Aust J Mar Freshw Res 39:185–192

Jernakoff P, Phillips BF, Fitzpatrick JJ (1993) The diet of post-puerulus western rock lobster, *Panulirus cygnus* George, at Seven Mile Beach, Western Australia. Mar Freshw Res 44:649–655

Joll LM, Phillips BF (1984) Natural diet and growth of juvenile western rock lobsters *Panulirus cygnus* George. J Exp Mar Biol Ecol 75:145–169

Jones RS, Chase JA (1975) Community structure and distribution of fishes in an enclosed high island lagoon in Guam. Micronesia 11:127–148

Kangas MI (2000) Synopsis of the biology and exploitation of the blue swimmer crab, *Portunus pelagicus* Linnaeus, in Western Australia. Fish Res Rep Fish West Aust 121:1–22

Kaunda-Arara B, Rose GA (2004) Homing and site fidelity in the greasy grouper *Epinephelus tauvina* (Serranidae) within a marine protected area in coastal Kenya. Mar Ecol Prog Ser 277:245–251

Kelletat DH (1995) Atlas of coastal geomorphology and zonality. J Coast Res, special issue No. 13, 286 pp.

Kleypas J, Dean JM (1983) Migration and feeding of the predatory fish, *Bairdiella chrysura* Lacépède, in an intertidal creek. J Exp Mar Biol Ecol 72:199–209

Klumpp DW, Nichols PD (1983) A study of food chains in seagrass communities. II. Food of the rock flathead, *Platycephalus laevigatus* Cuvier, a major predator in a *Posidonia autralis* seagrass bed. Aust J Mar Freshw Res 34:745–754

Kneib RT (1995) Behaviour separates potential and realized effects of decapod crustaceans in salt marsh communities. J Exp Mar Biol Ecol 193:239–256

Kneib RT (1997) The role of tidal marshes in the ecology of estuarine nekton. Oceanogr Mar Biol Ann Rev 35:163–220

Kneib RT, Wagner SL (1994) Nekton use of vegetated marsh habitats at different stages of tidal inundation. Mar Ecol Prog Ser 106:227–238

Kochzius M (1999) Interrelation of ichthyofauna from a seagrass meadow and coral reef in the Philippines. Proc 5th Internat Indo-Pacific Fish Conf pp. 517–535

Kopp D, Bouchon-Navaro Y, Louis M et al (2007) Diel differences in the seagrass fish assemblages of a Caribbean island in relation to adjacent habitat types. Aquat Bot 87:31–37

Krumme U (2004) Pattern in the tidal migration of fish in a north Brazilian mangrove channel as revealed by a vertical split-beam echosounder. Fish Res 70:1–15

Krumme U, Brenner M, Saint-Paul U (2008) Spring-neap cycle as a major driver of temporal variations in feeding of intertidal fishes: evidence from the sea catfish *Sciades herzbergii* (Ariidae) of equatorial west Atlantic mangrove creeks. J Exp Mar Biol Ecol 367:91–99

Krumme U, Liang TH (2004) Tidal-induced changes in a copepod-dominated zooplankton community in a macrotidal mangrove channel in northern Brazil. Zool Stud 43:404–414

Krumme U, Saint-Paul U (2003) Observation of fish migration in a macrotidal mangrove channel in Northern Brazil using 200 kHz split-beam sonar. Aquat Living Resour 16:175–184

Krumme U, Saint-Paul U, Rosenthal H (2004) Tidal and diel changes in the structure of a nekton assemblage in small intertidal mangrove creeks in northern Brazil. Aquat Living Resour 17:215–229

Kuipers B (1973) On the tidal migration of young plaice (*Pleuronectes platessa*) in the Wadden Sea. Neth J Sea Res 6:376–388

Kuo SR, Lin HJ, Shao KT (1999) Fish assemblages in the mangrove creek of Northern and Southern Taiwan. Estuaries 22:1004–1015

Kuwamura T (1985) Social and reproductive behaviour of three mouthbrooding cardinalfishes, *Apogon doederleini, A. niger* and *A. notatus*. Environ Biol Fish 13:17–24

Kvale EP (2006) The origin of neap-spring tidal cycles. Mar Geol 235:5–18

Laegdsgaard P, Johnson CR (1995) Mangrove habitats as nurseries: unique assemblages of juvenile fish in subtropical mangroves in eastern Australia. Mar Ecol Prog Ser 126:67–81

Larkum AWD, Orth RJ, Duarte CM (2006) Seagrasses: biology, ecology, and conservation. Springer, Dordrecht

Laroche J, Baran E, Rasoanandrasana NB (1997) Temporal patterns in a fish assemblage of a semiarid mangrove zone in Madagascar. J Fish Biol 51:3–20

Little MC, Reay PJ, Grove SJ (1988) The fish community of an East African mangrove creek. J Fish Biol 32:729–747

Lohmann KS (1985) Geomagnetic field detection by the western Atlantic spiny lobster, *Panulirus argus*. Mar Behav Physiol 12:1–17

Longley WH, Hildebrand SF (1941) Systematic catalogue of the fishes of Tortugas, Florida. Carnegie Institute of Washington Publication 535

Loya Y (1972) Community structure and species diversity of hermatypic corals at Eilat, Red Sea. Mar Biol 13:100–123

Lozano-Alvarez E, Briones-Fourzan P, Ramos-Aguilar ME (2003) Distribution, shelter fidelity, and movements of subadult spiny lobsters (*Panulirus argus*) in areas with artificial shelters (casitas). J Shellfish Res 22:533–540

Lozano-Alvarez E, Carrasco-Zanini G, Briones-Fourzan P (2002) Homing and orientation in the spotted spiny lobster, *Panulirus guttatus* (Decapoda, Palinuridae), towards a subtidal coral reef habitat. Crustaceana 75:859–873

Lugendo BR, Nagelkerken I, Kruitwagen G et al (2007) Relative importance of mangroves as feeding habitats for fishes: a comparison between mangrove habitats with different settings. Bull Mar Sci 80:497–512

Major PF (1977) Predatory-prey interactions in schooling fishes during periods of twilight: a study of the silverside *Pranesus insularum* in Hawaii. Fish Bull US 75:415–426

Marnane MJ (2000) Site fidelity and homing behaviour in coral reef cardinalfishes (family Apogonidae). J Fish Biol 57:1590–1600

Marnane MJ, Bellwood DR (2002) Diet and nocturnal foraging in cardinalfishes (Apogonidae) at One Tree Reef, Great Barrier Reef, Australia. Mar Ecol Prog Ser 231:261–268

Mayo HJ, Dudley DL (1970) Movements of tagged blue crabs in North Carolina waters. Comm Fish Rev 32:29–35

McFarland WN, Hillis Z (1982) Observations on agonistic behavior between members of juvenile French and white grunts - Family Haemulidae. Bull Mar Sci 32:255–268

McFarland WN, Munz FW (1976) The visible spectrum during twilight and its implications to vision. In: Evans GC, Bainbridge R, Rackham O (eds) Light as an ecological factor: II, pp. 249–270. Blackwell, Oxford

McFarland WN, Ogden JC, Lythgoe JN (1979) The influence of light on the twilight migrations of grunts. Environ Biol Fish 4:9–22

Meager JJ, Vance DJ, Williamson I et al (2003) Microhabitat distribution of juvenile *Penaeus merguiensis* de Man and other epibenthic crustaceans within a mangrove forest in subtropical Australia. J Exp Mar Biol Ecol 294:127–144

Ménard A, Turgeon K, Kramer D (2008) Selection of diurnal refuges by the nocturnal squirrelfish, *Holocentrus rufus*. Environ Biol Fish 81:59–70

Metcalfe JD, Hunter E, Buckley AA (2006) The migratory behaviour of North Sea plaice: currents, clocks and clues. Mar Freshw Behav Physiol 39:25–36

Meyer CG, Holland KN, Papastamatiou YP (2007b) Seasonal and diel movements of giant trevally *Caranx ignobilis* at remote Hawaiian atolls: implications for the design of Marine Protected Areas. Mar Ecol Prog Ser 333:13–25

Meyer CG, Holland KN, Wetherbee BM et al (2000) Movement patterns, habitat utilization, home range size and site fidelity of whitesaddle goatfish, *Parupeneus porphyreus*, in a marine reserve. Environ Biol Fish 59:235–242

Meyer CG, Papastamatiou YP, Holland KN (2007a) Seasonal, diel, and tidal movements of green jobfish (*Aprion virescens*, Lutjanidae) at remote Hawaiian atolls: implications for marine protected area design. Mar Biol 151:2133–2143

Meyer JL, Schultz ET (1985) Migrating haemulid fishes as a source of nutrients and organic matter on coral reefs. Limnol Oceanogr 30:146–156

Meyer JL, Schultz ET, Helfman GS (1983) Fish schools: an asset to corals. Science 220:1047–1149

Mohan R, Selvam V, Azariah J (1995) Temporal distribution and abundance of shrimp postlarvae and juveniles in the mangroves of Muthupet, Tamilnadu, India. Hydrobiologia 295:183–191

Morin JG, Harrington A, Nealson K et al (1975) Light for all reasons: versatility in the behavioural repertoire of the flashlight fish. Science 190:74–76

Mumby PJ, Edwards AJ, Arias-Gonzales JE et al (2004) Mangroves enhance the biomass of coral reef fish communities in the Caribbean. Nature 427:533–536

Munro ISR (1975) Biology of the banana prawn (*Penaeus merguiensis*) in the south-east corner of the Gulf of Carpentaria. In: Young PC (ed) National prawn seminar, pp. 60–78. Aust Govt Publ Serv, Canberra

Nagelkerken I, Dorenbosch M, Verberk WCEP et al (2000) Day-night shifts of fishes between shallow-water biotopes of a Caribbean bay, with emphasis on the nocturnal feeding of Haemulidae and Lutjanidae. Mar Ecol Prog Ser 194:55–64

Nagelkerken I, Roberts CM, van der Velde G et al (2002) How important are mangroves and seagrass beds for coral-reef fish? The nursery hypothesis tested on an island scale. Mar Ecol Prog Ser 244:299–305

Nakamura Y, Sano M (2004) Overlaps in habitat use of fishes between a seagrass bed and adjacent coral and sand areas at Amitori Bay, Iriomote Island, Japan: importance of the seagrass bed as juvenile habitat. Fish Sci 70:788–803

Nanami A, Yamada H (2008) Size and spatial arrangement of home range of checkered snapper *Lutjanus decussatus* (Lutjanidae) in an Okinawan coral reef determined using a portable GPS receiver. Mar Biol 153:1103–1111

Natajaran P (1989a) Persistent locomotor rhythmicity in the prawns *Penaeus indicus* and *P. monodon*. Mar Biol 101:339–346

Natajaran P (1989b) External synchronizers of tidal activity rhythms in the prawns *Penaeus indicus* and *P. monodon*. Mar Biol 101:347–354

Nishikawa M, Ishibashi T (1975) Entrainment of the activity rhythm by the cycle of feeding in the mudskipper, *Periophthalmus cantonensis* (Osbeck). Zool Mag Tokyo 84:184–189

Nishimoto RT, Herrnkind WF (1978) Directional orientation in blue crabs, *Callinectes sapidus* Rathbun: escape responses and influence of wave direction. J Exp Mar Ecol Biol 33:93–112

Ogden JC, Buckman NS (1973) Movements, foraging groups, and diurnal migratons of the striped parrotfish *Scarus croicensis* Bloch (Scaridae). Ecology 54:589–596

Ogden JC, Ehrlich PR (1977) The behavior of heterotypic resting schools of juvenile grunts (Pomadasyidae). Mar Biol 42:273–280

Ogden JC, Zieman JC (1977) Ecological aspects of coral reef-seagrass bed contacts in the Caribbean. Proc 3rd Int Coral Reef Symp 1:377–382

Ogden JC (1997) Ecosystem interactions in the tropical seascape. In: Birkeland C (ed) Life and death of coral reefs, pp. 288–297. Chapman and Hall, New York

Okuda N, Yanagisawa Y (1996) Filial cannibalism by mouthbrooding males of the cardinalfish, *Apogon doederleini*, in relation to their physical condition. Environ Biol Fish 45:397–404

Parrish JD (1989) Fish communities of interacting shallow-water habitats in tropical oceanic regions. Mar Ecol Prog Ser 58:143–160

Paul RKG (1981) Natural diet, feeding and predatory activity of the crabs *Callinectes arcuatus* and *C. toxotes* (Decapoda, Brachyura, Portunidae). Mar Ecol Prog Ser 6:91–99

Penn JW (1984) The behaviour and catchability of some commercially exploited penaeids and their relationships to stock and recruitment. In: Gulland JA, Rothschild BJ (eds) Penaeid shrimps: their biology and management, pp. 173–186. Fishing News Books. Surrey, England

Phillips BF, Joll LK, Ramm DC (1984) An electromagnetic tracking system for studying the movements of rock (spiny) lobsters. J Exp Mar Biol Ecol 79:9–18

Pittman SJ, McAlpine CA (2003) Movements of marine fish and decapod crustaceans: process, theory and application. Adv Mar Biol 44:205–294

Popple ID, Hunte W (2005) Movement patterns of *Cephalopholis cruentata* in a marine reserve in St Lucia, W.I., obtained from ultrasonic telemetry. J Fish Biol 67:981–992

Potter MA, Sumpton WD, Smith GS (1991) Movement, fishing sector impact and factors affecting the recapture rate of tagged sand crabs, *Portunus pelagicus* (L.) in Moreton Bay, Queensland. Aust J Mar Freshw Res 42:751–760

Potts GW (1973) The ethology of *Labroides dimidiatus* (Cuv. and Val.) (Labridae: Pisces) on Aldabra. Anim Behav 21:250–291

Primavera JH (1998) Mangroves as nurseries: shrimp populations in mangrove and non-mangrove habitats. Estuar Coast Shelf Sci 46:457–464

Quinn NJ, Koijs BL (1981) The lack of changes in nocturnal estuarine fish assemblages between new and full moon phases in Serpentine Creek, Queensland. Environ Biol Fish 6:213–218

Quinn NJ, Koijs BL (1987) The influence of the diel cycle, tidal direction and trawl alignment on beam trawl catches in an equatorial estuary. Environ Fish Biol 19:297–308

Quinn TP, Ogden JC (1984) Field evidence of compass orientation in migrating juvenile grunts (Haemulidae). J Exp Mar Biol Ecol 81:181–192

Radakov DV, Silva A (1974) Some characteristics of the schooling behavior of *Jenkinsia lamprotaenia*. J Ichthyol 14:283–286

Rahman MJ, Cowx IG (2006) Lunar periodicity in growth increment formation in otoliths of hilsa shad (*Tenualosa ilisha*, Clupeidae) in Bangladesh waters. Fish Res 81:342–344

Randall JE (1963) An analysis of the fish populations of artificial and natural reefs in the Virgin Islands. Caribb J Sci 3:31–47

Randall JE (1967) Food habits of reef fishes of the West Indies. Stud Trop Oceanogr 5:665–847

Robblee MB, Zieman JC (1984) Diel variation in the fish fauna of a tropical seagrass feeding ground. Bull Mar Sci 34:335–345

Robertson AI (1980) The structure and organization of an eelgrass fish fauna. Oecologia 47:76–82

Robertson AI (1988) Abundance, diet and predators of juvenile banana prawns, *Penaeus merguiensis*, in a tropical mangrove estuary. Aust J Mar Freshw Res 39:467–478

Robertson AI, Duke NC (1987) Mangroves as nursery sites: comparisons of the abundance and species composition of fish and crustaceans in mangroves and other nearshore habitats in tropical Australia. Mar Biol 96:193–205

Robertson AI, Duke NC (1990) Mangrove fish communities in tropical Australia: spatial and temporal patterns in densities, biomass and community structure. Mar Biol 104:369–379

Robertson AI, Howard RK (1978) Diel trophic interactions between vertically-migrating zooplankton and their fish predators in an eelgrass community. Mar Biol 48:207–213

Rönnbäck P, Troell M, Kautsky N et al (1999) Distribution pattern of shrimps and fish among *Avicennia* and *Rhizophora* microhabitats in the Pagbilao mangroves, Philippines. Estuar Coast Shelf Sci 48:223–234

Rooker JR, Dennis GD (1991) Diel, lunar and seasonal changes in a mangrove fish assemblage off southwestern Puerto Rico. Bull Mar Sci 49:684–698

Rountree RA, Able KW (1993) Diel variarion in decapod and fish assemblages in New Jersey polyhaline marsh creeks. Estuar Coast Shelf Sci 37:181–201

Rozas LP (1995) Hydroperiod and its influence on nekton use of the salt marsh: a pulsing ecosystem. Estuaries 18:579–590

Ruiz GM, Hines AH, Posey MH (1993) Shallow water as a refuge habitat for fish and crustaceans in non-vegetated estuaries: an example from Chesapeake Bay. Mar Ecol Prog Ser 99:1–16

Ruppert EE, Barnes RD (1994) Invertebrate zoology, 6th ed. Saunders College Publ, Fort Worth

Ryer CH (1987) Temporal patterns of feeding by blue crabs (*Callinectes sapidus*) in a tidal-marsh creek and adjacent seagrass meadow in the lower Chesapeake Bay. Estuaries 10:136–140

Sale P (2006) Coral reef fishes: dynamics and diversity in a complex ecosystem. Academic Press, San Diego

Sasekumar A, Chong VC, Leh MU et al (1992) Mangroves as a habitat for fish and prawns. Hydrobiologia 247:195–207

Schaffmeister BE, Hiddink JG, Wolff WJ (2006) Habitat use of shrimps in the intertidal and shallow subtidal seagrass beds of the tropical Banc d'Arguin, Mauritania. J Sea Res 55:230–243

Schories D, Barletta-Bergan A, Krumme U et al (2003) The keystone role of leaf-removing crabs in mangrove forests of north Brazil. Wetlands Ecol Manag 11:243–255

Schwamborn R, Bonecker ACT (1996) Seasonal changes in the transport and distribution of meroplankton into a Brazilian estuary with emphasis on the importance of floating mangrove leaves. Arch Biol Tec 39:451–462

Shapiro DY (1987) Reproduction in groupers. In: Polovina JJ, Ralston S (eds) Tropical snappers and groupers. Biology and fisheries management. Westview Press, Boulder

Sheaves M (2005) Nature and consequences of biological connectivity in mangrove systems. Mar Ecol Prog Ser 302:293–305

Sheppard CRC, Price ARG, Roberts CM (1992) Marine ecology of the Arabian region: patterns and processes in extreme tropical environments. Academic Press, London

Shimps EL, Rice, JA, Osborne JA (2005) Hypoxia tolerance in two juvenile estuary-dependent fishes. J Exp Mar Biol Ecol 325:146–162

Smith GS, Sumpton WD (1987) Sand crabs a valuable fishery in southeast Queensland. Qld Fisherman 5:13–15

Sogard SM, Powell GVN, Holmquist JG (1989) Utilization by fishes of shallow, seagrass-covered banks in Florida Bay: 2. Diel and tidal patterns. Environ Biol Fish 24:8–92

Spalding M, Blasco F, Field C (1997) World mangrove atlas. The International Society for Mangrove Ecosystems (ISME), Okinawa

Spalding MD, Ravilious C, Green EP (2001) World atlas of coral reefs. University of California Press, Berkeley

Staples DJ, Vance DJ (1979) Effects of changes in catchability on sampling of juvenile and adolescent banana prawns, *Penaeus merguiensis* de Man. Aust J Mar Freshw Res 30:511–519

Starck WA II (1971) Biology of the gray snapper, *Lutjanus griseus* (Linnaeus), in the Florida Keys. Stud Trop Oceanogr (Miami) 10:1–150

Starck WA II, Davis WP (1966) Night habits of fishes of Alligator Reef, Florida. Ichthyologica/The Aquarium Journal 38:313–356

Steele CW (1985) Absence of a tidal component in the diel pattern of locomotory activity of sea catfish, *Arius felis*. Environ Biol Fish 12:69–73

Sudara S, Satumanatpan S, Nateekarnjanalarp S (1992) A study of the interrelationship of fish communities between coral reefs and seagrass beds. In: Chou LM, Wilkinson CR (eds) Proc 3rd ASEAN Science and Technology Week Conference, Vol. 6. Marine science: living coastal resources, pp. 321–326. Dept of Zoology, National University of Singapore and National Science and Technology Board, Singapore

Sumpton WD, Smith GS (1991) The facts about sand crabs. Qld Fisherman, June, pp. 29–31

Szedlmayer ST, Able KW (1993) Ultrasonic telemetry of age-0 summer flounder, *Paralichthys dentatus*, movements in a southern New Jersey estuary. Copeia 1993:728–736

Thompson AA, Mapstone BD (2002) Intra- versus inter-annual variation in counts of reef fishes and interpretations of long-term monitoring studies. Mar Ecol Prog Ser 232:247–257

Tongnunui P, Ikejima K, Yamane T et al (2002) Fish fauna of the Sikao Creek mangrove estuary, Trang, Thailand. Fish Sci 68:10–17

Trendall J, Bell S (1989) Variable patterns of den habitation by the ornate rock lobster, *Panulirus ornatus*, in the Torres Strait. Bull Mar Sci 45:564–573

Tulevech SM, Recksiek CW (1994) Acoustic tracking of adult white grunt, *Haemulon plumierii*, in Puerto Rico and Florida. Fish Res 19:301–319

Tupper M, Juanes F (1999) Effects of a marine reserve on recruitment of grunts (Pisces: Haemulidae) at Barbados, West Indies. Environ Biol Fish 55:53–63

Unsworth RKF, Bell JJ, Smith DJ (2007b) Tidal fish connectivity of reef and seagrass habitats in the Indo-Pacific. J Mar Biol Ass UK 87:1287–1296

Unsworth RKF, Wylie E, Bell JJ et al (2007a) Diel trophic structuring of seagrass bed fish assemblages in the Wakatobi Marine National Park, Indonesia. Estuar Coast Shelf Sci 72:81–88

Unsworth RKF, Salinas de Leon P, Garrard SL et al (2008) High connectivity of Indo–Pacific seagrass fish assemblages with mangrove and coral reef habitats. Mar Ecol Prog Ser 353:213–224

van der Veer HW, Bergman MJN (1987) Development of tidally related behaviour of a newly settled 0–group plaice (*Pleuronectes platessa*) population in the western Wadden Sea. Mar Ecol Prog Ser 31:121–129

van Montfrans J, Ryer CH, Orth RJ (1991) Population dynamics of blue crabs *Callinectes sapidus* Rathbun in a lower Chesapeake Bay tidal marsh creek. J Exp Mar Biol Ecol 153:1–14

Vance DJ (1992) Activity patterns of juvenile penaeid prawns in response to artificial tidal and day-night cycles: a comparison of three species. Mar Ecol Prog Ser 87:215–26

Vance DJ, Haywood MDE, Heales DS et al (1996) How far do prawns and fish move into mangroves? Distribution of juvenile banana prawns *Penaeus merguiensis* and fish in a tropical mangrove forest in northern Australia. Mar Ecol Prog Ser 131:115–124

Vance DJ, Haywood MDE, Heales DS et al (2002) Distribution of juvenile penaeid prawns in mangrove forests in a tropical Australian estuary, with particular reference to *Penaeus merguiensis*. Mar Ecol Prog Ser 228:165–177

Vance DJ, Haywood MDE, Kerr JD (1990) Use of a mangrove estuary as a nursery area by postlarval and juvenile banana prawns, *Penaeus merguiensis* de Man, in northern Australia. Estuar Coast Shelf Sci 31:689–701

Vance DJ, Staples DJ (1992) Catchability and sampling of three species of juvenile penaeid prawns in the Embley River, Gulf of Carpentaria, Australia. Mar Ecol Prog Ser 87:201–213

Vannini M, Cannicci S (1995) Homing behaviour and possible cognitive maps in crustacean decapods. J Exp Mar Biol Ecol 193:67–91

Verweij MC, Nagelkerken I (2007) Short and long-term movement and site fidelity of juvenile Haemulidae in back-reef habitats of a Caribbean embayment. Hydrobiologia 592:257–270

Verweij MC, Nagelkerken I, Hol KEM et al (2007) Space use of *Lutjanus apodus* including movement between a putative nursery and a coral reef. Bull Mar Sci 81:127–138

Verweij MC, Nagelkerken I, Wartenbergh SLJ et al (2006) Caribbean mangroves and seagrass beds as diurnal feeding habitats for juvenile French grunts, *Haemulon flavolineatum*. Mar Biol 149:1291–1299

Vezzosi R, Barbaresi, Anyona D et al (1995) Activity patterns in *Thalamita crenata* (Portunidae, Decapoda): a shaping by the tidal cycles. Mar Freshw Behav Physiol 24:207–214

Vidy G, Darboe FS, Mbye EM (2004) Juvenile fish assemblages in the creeks of the Gambia Estuary. Aquat Living Resour 17:56–64

Vivien ML (1975) Place of apogonid fish in the food webs of a Malagasy coral reef. Micronesica 11:185–196

Vivien ML, Peyrot-Clausade M (1974) A comparative study of the feeding behaviour of three coral reef fishes (Holocentridae), with special reference to the polychaetes of the reef cryptofauna as prey. Proc 2nd Int. Coral Reef Symp 1:179–192

Wassenberg TJ, Hill BJ (1987) Feeding by the sand crab *Portunus pelagicus* on material discarded from prawn trawlers in Moreton Bay, Australia. Mar Biol 95:387–393

Wehrtmann SI, Dittel AI (1990) Utilization of floating mangrove leaves as a transport mechanism of estuarine organisms, with emphasis on decapod Crustacea. Mar Ecol Prog Ser 60:67–73

Weinstein MP, Heck KL (1979) Ichthyofauna of seagrass meadows along the Caribbean coast of Panama and in the Gulf of Mexico: composition, structure and community ecology. Mar Biol 50:97–107

Weisberg SB, Whalen R, Lotrich VA (1981) Tidal and diurnal influence on food consumption of a salt marsh killifish *Fundulus heteroclitus*. Mar Biol 61:243–246

Weiss HM, Lozano-Alvarez E, Briones-Fourzan P (2008) Circadian shelter occupancy patterns and predator–prey interactions of juvenile Caribbean spiny lobsters in a reef lagoon. Mar Biol 153:953–963

Weissburg MJ, James CP, Smee DL et al (2003) Fluid mechanics produces conflicting constraints during olfactory navigation of blue crabs, *Callinectes sapidus*. J Exp Biol 206:171–180

Weissburg MJ, Zimmer-Faust RK (1993) Life and death in moving fluids: hydrodynamic effects on chemosensory-mediated predation. Ecology 74:1428–1443

Weissburg MJ, Zimmer-Faust RK (1994) Odor plumes and how blue crabs use them to find prey. J Exp Biol 197:349–375

Williams JG (1979) Estimation of intertidal harvest of Dungeness crab, *Cancer magister*, on Puget Sound, Washington, beaches. US Natl Mar Fish Serv Fish Bull 77:287–292

Williams MJ (1982) Natural food and feeding in the commercial sand crab *P. pelagicus* Linnaeus, 1766 (Crustacea: Decapoda: Portunidae) in Moreton Bay, Queensland. J Exp Mar Biol Ecol 59:165–176

Wilson JP, Sheaves M (2001) Short-term temporal variations in taxonomic composition and trophic structure of a tropical estuarine fish assemblage. Mar Biol 139:878–796

Wirjoatmodjo S, Pitcher TJ (1984) Flounders follow the tides to feed: evidence from ultrasonic tracking in an estuary. Estuar Coast Shelf Sci 19:231–241

Wolanski E (1994) Physical oceanographic processes of the Great Barrier Reef. CRC Press, Boca Raton

Wolanski E, Mazda Y, Ridd P (1992) Mangrove hydrodynamics. In: Robertson AI, Alongi DM (eds) Tropical mangrove ecosystems, pp. 43–62. Coastal and Estuarine Studies 41 American Geophysical Union, Washington DC

Wolff N, Grober-Dunsmore R, Rogers CS et al (1999) Management implications of fish trap effectiveness in adjacent coral reef and gorgonian habitats. Environ Biol Fish 55:81–90

Wothke A, Greven H (1998) Field observations on four-eyed fishes, *Anableps anableps* (Anablepidae, Cyprinodontiformes), in Trinidad. Z Fischk 5:59–75

Yamahira K, Nojima S, Kikuchi T (1996) Age specific food utilization and spatial distribution of the puffer, *Takifugu niphobles*, over an intertidal sand flat. Environ Biol Fish 45:311–318

Zeller DC (1997) Home range and activity patterns of the coral trout *Plectropomus leopardus* (Serranidae). Mar Ecol Prog Ser 154:65–77

Zeller DC (1998) Spawning aggregations: patterns of movement of the coral trout *Plectropomus leopardus* (Serranidae) as determined by ultrasonic telemetry. Mar Ecol Prog Ser 162:253–263

Zimmer-Faust RK, Finelli CM, Pentcheff ND et al (1995) Odor plumes and animal navigation in turbulent water flow: a field study. Biol Bull 188:11–116

Zimmer-Faust RK, O'Neill PB, Schar DW (1996) The relationship between predator activity state and sensitivity to prey odor. Biol Bull 190:82–87

Chapter 9
Living in Two Worlds: Diadromous Fishes, and Factors Affecting Population Connectivity Between Tropical Rivers and Coasts

David A. Milton

Abstract Among the large range of life history patterns of tropical fishes, about 200 species from 20 fish families undertake diadromous migrations. In most diadromous fish species, only a proportion of the population undertakes migrations and this proportion varies widely between species and families. Three different types of migration are anadromy, catadromy, and amphidromy. Tropical anadromous species are mostly clupeoids, including several shads and herrings. These species spawn in freshwater and migrate to the sea as juveniles, and most of the population matures there before returning to breed in freshwater. Catadromous species have the opposite behavior – they spawn in the sea before migrating to freshwater where they mature. Anguillid eels, mullets, and many centropomids from tropical regions are catadromous. The most common form of diadromy in the tropics is amphidromy. The largest groups of amphidromous fishes are the gobies and gudgeons. Amphidromous fishes spawn in freshwater and the larvae migrate to the sea before migrating back to freshwater, and are common on many islands of all the major oceans. The freshwater and marine components of diadromous fish populations rely on freshwater flows to maintain their connectivity. Most tropical diadromous fishes migrate between habitats during seasonal monsoonal floods. The construction of dams, and shifts in the intensity and reductions in the quantity of rainfall from changing climate are two of the major threats to maintaining connectivity between freshwater and marine populations. The examples of these effects presented here suggest that tropical diadromous fishes will face increasing challenges in maintaining their populations unless greater effort is made to facilitate their migrations.

Keywords Diadromy · Migration · Droughts · Dams · Climate change

D.A. Milton (✉)
Wealth from Oceans Flagship, CSIRO Marine and Atmospheric Research, P.O. Box 120, Cleveland, Queensland 4163, Australia
e-mail: david.milton@csiro.au

I. Nagelkerken (ed.), *Ecological Connectivity among Tropical Coastal Ecosystems*,
DOI 10.1007/978-90-481-2406-0_9, © Springer Science+Business Media B.V. 2009

9.1 Introduction

Fish that migrate regularly between freshwater and coastal marine environments in the tropics must adapt to a wide range of environmental conditions. In freshwater, they will often occur in strong currents many kilometers from the sea. To reach these preferred habitats can require negotiating both artificial barriers, such as dams and weirs, as well as natural barriers like waterfalls. In the estuary and sea, they have to adapt to the physiological stress of increased salinity. Relatively few fish species regularly undertake these migrations between marine and freshwaters (between 250 and 300 species, McDowall 1997, Riede 2004). Over two thirds of these species are found in the tropics (~201 species, Table 9.1). The majority of the fishes that regularly migrate between freshwater and the sea are members of genera that also contain non-migratory species. The life cycles adopted by fish species that utilize both freshwater and adjacent coastal marine habitats form a subset of the range of adaptive strategies used by fishes. In this chapter I discuss the range of types of diadromous migrations made by tropical fish species. I then identify some of the factors that influence or are likely to influence the ability of the marine and freshwater populations of these species to remain connected.

In this chapter, riverine habitats are defined as all freshwater reaches of rivers and smaller streams and are above tidal influence. The intention is to consider the connectivity of fish populations between freshwater and adjacent coastal marine habitats. This will focus on species that live above the estuary during at least part of their lives and spend variable periods in the lower estuary and adjacent coastal marine habitats.

Species that undertake regular and predictable migrations between freshwater and the sea are called diadromous (Myers 1949, McDowall 1988, 2001). Within diadromous fishes, there have been three recognized variations in migratory behavior – anadromy, catadromy, and amphidromy (Myers 1949). Anadromous fishes undertake migrations from the sea as mature adults to breed in freshwater. Temperate salmonids are the best known example of this type of migration. Catadromous fishes such as *Anguilla* eels and several centropomids make spawning migrations from freshwater to the sea as mature adults. The third sub-group within the diadromous fishes are the amphidromous species. These species undertake migrations in both directions between freshwater and the sea usually for trophic reasons (McDowall 2007). McDowall (1988) lists about 227 fish species that are diadromous, but McDowall (1997) suggested that over 250 species may eventually be found to be diadromous, as the ecology of many species remained unknown. More recently, Riede (2004) re-examined the status of diadromous fishes and his summary of the more recent literature has expanded the number of species known to be diadromous to at least 300 species (Froese and Pauly 2003). Diadromy is not as common among tropical fish species, but some forms of diadromy such as amphidromy, mostly occur in tropical species (McDowall 2007).

Table 9.1 Tropical fish species that regularly migrate between freshwater and coastal marine habitats. Migratory status as documented in Froese and Pauly (2003). The number of species in each family with each migratory behavior are shown in brackets. Many of the species identified as diadromous by Froese and Pauly (2003) are based on an uncited report by Riede (2004) and require further verification

Family	Type of migration	Distribution	Species	Preferred adult habitat	Migratory season
Ambassidae	Amphidromous	southern Asia and western Pacific (4 spp.)	*Ambassis gymnocephalus, A. kopsii, A. miops, A. nalua*; other 15 species are mostly estuarine or coastal marine	Estuaries and lower freshwater reaches of rivers	Wet season
	Catadromous	western Pacific (1 spp.)	*Ambassis interrupta*	Coast and estuaries	Wet season
Anguillidae	Catadromous	Indo-West Pacific (11 spp.)	*Anguilla bengalensis, A. bicolor, A. celebesensis, A. interioris, A. malgumora, A. marmorata, A. megastoma, A. mossambica, A. nebulosa, A. obscura, A. rheinhardii*; other six species are temperate	Freshwater	Wet season
Ariidae	Amphidromous	southern Asia (10 spp.)	*Ameiurus melas, Arius jella, Cephalocassia jatia, Cochlefelis burmanica, Hemiarius sona, Hexanematichthys sagor, Nemapteryx caelata, Netuma thalassina, Plicofollis platystomus, P. tenuispinis*; other six species of *Ameiurus*, three species of *Cephalocassia*, three species of *Cochlefelis*, four species of *Hemiarius*, one species of *Hexanematichthys*, five species of *Nemapteryx*, six species of *Plicofollis*, and two species *Netuma* are mostly estuarine or marine	Mostly estuaries and tidal reaches of rivers	Possibly wet season
	Anadromous	Indo-West Pacific (2 spp.)	*Arius madagascariensis, Neoarius graeffei*; other 30 species of *Arius* and eight species of *Neoarius* are either found in marine or freshwater	Freshwater and estuaries	Possibly wet season
Atherinopsidae	Anadromous	eastern Pacific and western Atlantic (2 spp.)	*Atherinella chagresi, A. guatemalensis*; other 33 species of *Atherinella* are estuarine or marine	Freshwater and estuaries	Wet season

Table 9.1 (continued)

Family	Type of migration	Distribution	Species	Preferred adult habitat	Migratory season
Centropomidae	Amphidromous	eastern Pacific and western Atlantic (6 spp.)	*Centropomus ensiferus, C. medius, C. nigrescens, C. parallelus, C. robalito, C. undecimalis*; other four species of *Centropomus* are estuarine and marine	Mostly coastal marine and estuaries	Wet season
	Anadromous	western Atlantic (1 sp.)	*Centropomus poeyi*	Coastal marine and estuaries	Probably wet season
	Catadromous	southern and Southeast Asia to Australia (1 sp.)	*Lates calcarifer*; other eight *Lates* species all live in freshwater	Mostly estuarine and coastal marine	Pre–wet season
		western Atlantic (1 sp.)	*Centropomus pectinatus*	Estuaries and freshwater rivers	Probably wet season
Clupeidae	Amphidromous	Indo-West Pacific (2 spp.)	*Escualosa thoracata, Sardinella melanura*; other 20 species of *Sardinella*, and *E. elongata* are coastal marine	Coastal marine and estuaries	Possibly wet season
	Anadromous	western Pacific and eastern Atlantic (10 spp.)	*Anodontostoma chacunda, A. thailandiae, Herklotsichthys gotoi, Hilsa kelee, Nematalosa galatheae, N. nasus, Pellonula leonensis, P. vorax, Tenualosa ilisha* (not all populations), *T. reevesi*; one species of *Anodontostoma* and other 11 species of *Herklotsichthys* are marine, two species of *Tenualosa* are estuarine or coastal, and one species of *Tenualosa* lives in freshwater	Coastal marine and estuaries	Wet season
	Catadromous	eastern Atlantic (1 sp.)	*Ethmalosa fimbriata*	Mostly estuaries and adjacent coastal marine	Wet season

Table 9.1 (continued)

Family	Type of migration	Distribution	Species	Preferred adult habitat	Migratory season
Eleotridae	Amphidromous	eastern and western Pacific, and western Atlantic (21 spp.)	*Bostrychus africanus, B. sinensis, Bunaka gyrinoides, B. pinguis, Butis amboinensis, B. butis, B. humeralis, B. koilomatodon, B. melanostigma, Dormitator latifrons, D. lebretonis, D. maculatus, Eleotris acanthopoma, E. amblyopsis, E. fusca, E. mauritianus, E. melanosoma, E. sandwicensis, Guavina guavina, Ophieleotris aporos, Ophiocara porocephala*; other species of *Guavina* occurs in estuaries or coastal waters, four species of *Bostrychus*, one species of *Butis*, three species of *Dormitator*, one *Ophieleotris*, and one species of *Ophiocara* are found in freshwater	Mostly freshwater rivers and estuaries	Probably differs among species; both wet and dry season
Elopidae	Catadromous	Pacific and Atlantic (6 spp.)	*Eleotris annobonensis, E. balia, E. pisonis, E. senegalensis, E. vittata, gobiomorus dormitory*, other 20 species of *Eleotris* and two *Gobiomorus* species are either mostly marine or only in freshwater	Mostly estuaries and freshwater rivers	Probably wet season
	Amphidromous	western Atlantic (1 sp.)	*Elops saurus*	Estuaries and coastal marine	Wet season

Table 9.1 (continued)

Family	Type of migration	Distribution	Species	Preferred adult habitat	Migratory season
Engraulidae	Anadromous	western Pacific (1 sp.)	*Elops hawaiensis*; other five species of *Elops* are mostly marine	Coastal marine	Wet season
	Amphidromous	Indo-West Pacific (9 spp.)	*Coilia dussumieri, C. mystus, C. neglecta, C. ramcarati, C. reynaldi, Thryssa dussumieri, T. gautamiensis, T. hamaltonii, T. kammalensoides*; other eight species of *Coilia* are marine, and 19 species of *Thryssa* are confined to marine or freshwater	Coastal marine and estuaries	Probably wet season
	Anadromous	western Atlantic (2 spp.)	*Anchoviella lepidentostole, Lycengraulis grossidens*; other 15 species of *Anchoviella* live mostly in marine or freshwater, and three species of *Lycengraulis* live in freshwater	Coastal marine and estuaries	Unsure, possibly wet season
	Catadromous	Papua New Guinea (1 sp.)	*Thryssa scratchleyi*	Freshwater rivers	Presumably wet season
Gerreidae	Amphidromous	Indo-Pacific, and eastern and western Atlantic (6 spp.)	*Eucinostomus melanopterus, Gerres cinereus, G. filamentosus, G. limbatus, G. longirostris, G. setifer*; other 18 species of *Gerres*, and 10 species of *Eucinostomus* are estuarine or coastal marine	Coastal marine and estuaries	Unknown, possibly wet season

Table 9.1 (continued)

Family	Type of migration	Distribution	Species	Preferred adult habitat	Migratory season
Gobiidae	Amphidromous	Indo-Pacific and eastern Atlantic (55 spp.)	*Awaous grammepomus, A. guamensis, A. ocellaris, Cotylopus acutipinnis, Glossogobius aureus, G. celebius, G. giuris, Gobioides broussonnetii, G. sagitta, Gobionellus occidentalis, G. oceanicus, G. thoropsis, Lentipes armatus, L. concolor, L. whittenorum, Periophthalmodon schlosseri, P. septemradiatus, Periophthalmus argentilineatus, P. barbarus, P. malaccensis, P. modestus, P. novemradiatus, P. weberi, Porogobius schlegelii, Pseudapocryptes elongates, Pseudogobius javanicus, P. melanostictus, P. poicilosoma, Schismatogobius roxasi, Sicydium plumieri, Sicyopus auxilimentus, S. jonklaasi, S. leprurus, S. zosterophorum, Sicyopterus fuliag, S. griseus, S. lacrymosus, S. lagocephalus, S. macrostetholepis, S. micrurus, S. rapa, Stenogobius blokzeyli, S. genivittatus, S. hawaiiensis, Stiphodon aurorostrum, S. elegans, S. stevensoni, S. surrufus, Redigobius balteatus, R. bikolanus, R. dispar, R. macrostoma, R. roemeri, R. sapangus, Zappa confluentus*; other nine species of *Awaous*, one species of *Cotylopus*, nine species of *Lentipes*, 19 species of *Glossogobius*, three species of *Gobioides*, nine species of *Gobionellus*, one *Pseudapocryptes*, 10 species of *Sicyopus*, and 23 species of *Stiphodon* occur mostly in freshwater. Other 19 species of *Glossogobius*, nine species of *Gobionellus*, four species of *Pseudogobius*, seven species of *Redigobius*, 10 species of *Schismatogobius*, 23 species of *Sicyopterus*, and 22 species of *Stenogobius* occur in freshwater or coastal marine waters	Mostly freshwater or estuaries	Wet season

Table 9.1 (continued)

Family	Type of migration	Distribution	Species	Preferred adult habitat	Migratory season
	Anadromous	western and eastern Atlantic (1 sp.)	*Sicydium punctatum*; other 13 species of *Sicydium* are either freshwater or marine	Freshwater	Wet season
	Catadromous	eastern Atlantic (1 sp.)	*Awaous tajasica*	Freshwater	Wet season
Kuhliidae	Amphidromous	Hawaii (1 sp.)	*Kuhlia sandvicensis*	Variable, freshwater and coastal marine	Wet season
	Catadromous	Indo-Pacific (4 spp.)	*Kuhlia boninensis, K. caudivittata, K. marginata, K. rupestris*; other eight *Kuhlia* species mostly marine (6 spp.) or freshwater (2 spp.)	Variable from freshwater to tidal freshwater and estuaries, depending on species	Wet season
Lutjanidae	Catadromous	Papua New Guinea and Southeast Asia (2 spp.)	*Lutjanus goldiei, Lutjanus maxweberi*; other 64 *Lutjanus* species mostly marine or coastal, and occur in all oceans	Freshwater rivers to estuaries	Wet season
Mugilidae	Amphidromous	Indo-West Pacific and eastern Atlantic (9 spp.)	*Liza macrolepis, L. melinoptera, L. parmata, L. subviridus, L. vaigiensis, Valamugil buchanani, V. cunesius, V. seheli, V. speigleri*; other 16 species of *Liza*, and other five *Valamugil* species mostly coastal marine	Coastal marine and estuaries	Unclear, possibly wet season
	Catadromous	Atlantic and Indo-Pacific (10 spp.)	*Agonostomus monticola, A. telfairii, Crenimugil heterocheilus, Joturus pichardi, Liza falcipinnis, L. grandisquamis, L. parmata, L. parsia, Mugil cephalus, M. curema*; other 15 species of *Mugil* are marine	Variable, mostly freshwater and estuaries	Pre-wet season and wet season depending on species

Table 9.1 (continued)

Family	Type of migration	Distribution	Species	Preferred adult habitat	Migratory season
Megalopidae	Amphidromous	eastern and western Atlantic (1 sp.)	*Megalops atlanticus*	Mostly coastal marine	Spring/summer
		Indo-West Pacific (1 sp.)	*M. cyprinoides*	Mostly coastal marine	Wet season
Pangasidae	Anadromous	Southeast Asia (1 sp.)	*Pangasius krempfi*; other 24 species of *Pangasius* mostly confined to freshwater	Coastal marine and estuaries	Dry season
Pristigasteridae	Anadromous	Indo-West Pacific, eastern and western Atlantic (4 spp.)	*Ilisha filigera, I. megaloptera, I. sirishai, Pellona ditchella*; other 10 species of *Ilisha*, and five species of *Pellona* are mostly freshwater or marine	Coastal marine and estuaries	Wet season
	Amphidromous	southern and Southeast Asia (3 spp.)	*Ilisha kampeni, I. melastoma, I. novacula*	Estuaries and freshwater rivers	Wet season
Rhyacichthyidae	Amphidromous	western Pacific (2 spp.)	*Rhyacichthys aspro, R. guilberti*	Freshwater streams	Probably wet season
Syngnathidae	Amphidromous	Indo-Pacific (3 spp.)	*Hippichthys cyanospilos, H. spicifer, Microphis leiaspis*; other three species of *Hippichthys*, and 16 species of *Microphis* live in freshwater or estuaries	Freshwater streams and estuaries	Unknown, possibly dry season
	Anadromous	Indo-Pacific and western Atlantic (1 sp.)	*Microphis brachyurus*	Freshwater streams	Unknown, possibly dry season
Toxotidae	Amphidromous	Indo-Pacific (3 spp.)	*Toxotes blythii, T. chatareus, T. jaculatrix*; other four species of *Toxotes* occur in freshwater	Estuaries and freshwater streams	Wet season

9.2 Types of Connectivity

The classification of fish species into the three forms of migratory behavior can be somewhat arbitrary and are part of a continuum of recognizable fish movement patterns (Myers 1949, McDowall 1988, Elliot et al. 2007). McDowall (1988, 2007) summarizes the types of migrations that are associated with each migratory behavior. Within species showing each type of migratory behavior, some individuals and even populations can also be non-migratory. For example, among the anguillid eels, most individuals of tropical eel species have so far been shown to be obligate migrants, whereas recent studies have shown that some populations of temperate species are facultative migrants and only some migrate from coastal spawning grounds to rivers (Daverat et al. 2006, Edeline 2007, Thibault et al. 2007). Thus for facultative diadromous fishes, some individuals in a population may migrate annually or infrequently, and others remain in the same habitat throughout their lives (Tsukamoto et al. 1998, Milton and Chenery 2005, Thibault et al. 2007).

In tropical regions, there are at least 20 families of fish that contain diadromous species (Table 9.1) that make a variable proportion of the species in each family. Diadromous species occur in all oceans, and this migratory behavior appears to have evolved independently multiple times (McDowall 1997). Four of the 20 families in Table 9.1 have species that make each of the three forms of migration (Centropomidae, Clupeidae, Engraulidae, and Gobiidae). For other families such as the Anguillidae, all diadromous species are catadromous. There are fewer anadromous species among tropical fishes, and few tropical families are entirely anadromous. This form of migration usually only occurs in a subset of species within a family (Table 9.1).

The presence of many amphidromous and catadromous species in freshwater on isolated tropical islands throughout the Indo-Pacific suggests that gene flow and connectivity among populations should be limited. Indeed, McDowall (2004) asked how remote and isolated islands (such as Hawaii or Guam) have any freshwater fish fauna. He found that the distribution patterns of many diadromous species suggest that dispersal between regions was increased among migratory species. Within fish families such as the gobies and eleotrids that have diadromous and non-diadromous species, diadromous species are more widely distributed (McDowall 2001). Chubb et al. (1998) also found limited evidence of inter-island genetic structuring in four species of Hawaiian amphidromous fishes. Similarly, Keith et al. (2005) examined genetic structure among nine species of amphidromous sicydiine gobies, including the widely distributed *Sicyopterus lagocephalus*. They suggested that *S. lagocephalus* had colonized islands throughout the Indo-Pacific relatively recently (~3.5 million yrs ago). Other species of sicydiine goby that were endemic to single island groups had evolved earlier. McDowall (2003b) and Keith et al. (2005) hypothesized that interspecific differences in the duration of the larval and juvenile phase, and historical ocean current patterns had lead to the current distribution patterns. All these studies suggest that diadromy has enhanced population dispersal in fishes during their marine phase by allowing colonization of new habitats.

9.2.1 Amphidromy

This form of diadromy is the most widespread among tropical fishes, with at least 137 species (68% of 201 tropical species identified as diadromous) that undertake amphidromous migrations (Table 9.1). The family with the largest number of amphidromous species is the Gobiidae (55 species). Among the amphidromous gobies, the sicydiine gobies are well represented (Fig. 9.1). These gobies are mostly found on islands in the Pacific, Indian Ocean, and in the Caribbean, where they are an important component of the freshwater fish faunas (McDowall 2004). Other families with several amphidromous species include gudgeons (Eleotridae), silver biddies (Gerreidae), and catfishes (Ariidae) (Table 9.1). The catfish species are found in large river systems around the Indo-Pacific, but the amphidromous gudgeons are found widely in rivers along the margins of the eastern and western Pacific and both sides of the Atlantic Ocean.

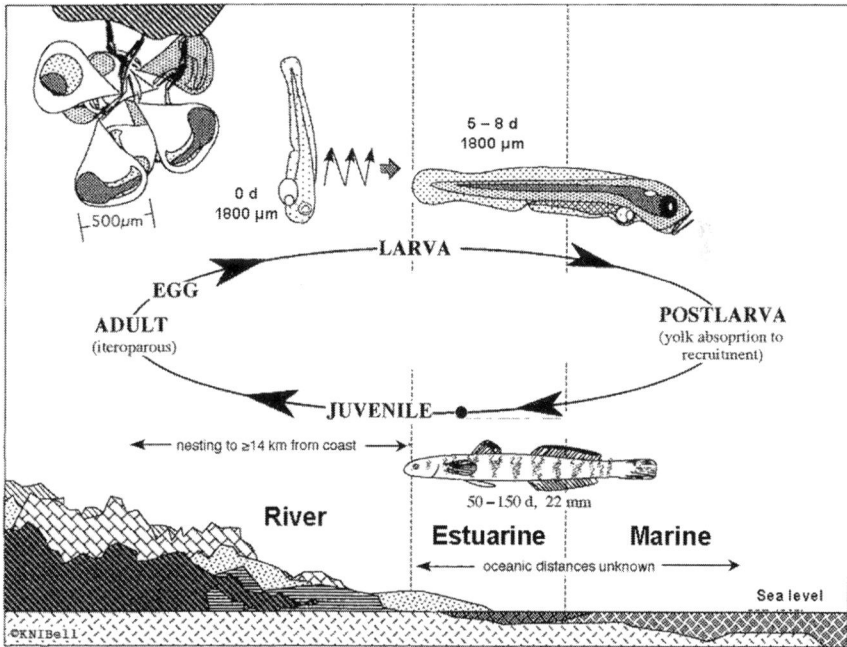

Fig. 9.1 The life cycle of amphidromous sicydiine gobies from Dominica in the Caribbean (redrawn from Bell et al. 1995). The pyriform eggs laid under stones hatch and vertically swim as they drift downstream while absorbing the yolk sac over 5–8 days, by which time they reach the sea. Juveniles then return to freshwater at between 50–150 days (Bell et al. 1995)

9.2.2 Anadromy

This form of migration occurs widely in a small number of species in many of the tropical diadromous fish families. A total of 25 tropical species of fish are facultative or obligate anadromous species (Table 9.1), especially in the Clupeiformes (Clupeidae, Engraulidae, and Pristigasteridae; McDowall 2003a). Gross (1987) has hypothesized that anadromy should have evolved from freshwater species, but as McDowall (1997) argues, there are limited data to support this contention. Among tropical species that are anadromous, most of these families and genera contain other species that are mostly marine. Gross et al. (1988) suggested that anadromy was more common in temperate waters as marine habitats in temperate regions are more productive than nearby freshwater habitats. More recently, McDowall (2003) showed that anadromy was most prevalent in northern latitudes, with few species in tropical or southern temperate regions undertaking these types of movements. Thus, anadromy may also have evolved in northern temperate regions with large rivers that favor this form of migratory behavior.

9.2.3 Catadromy

At least 39 species of migratory tropical fishes have been classified as catadromous (Table 9.1) and they mostly occur in the Indo-Pacific region. In this region, anguillid eels and the mullets (Mugilidae) are two tropical families with several species that live mostly in freshwater and migrate to the sea to spawn (Table 9.1). Most other families and even individual fish show a broad range of variation in their degree of catadromy. An important commercial catadromous species in Indo-Pacific region is barramundi *Lates calcarifer* (Blaber 2000). Other *Lates* species are restricted to freshwater and confined to various African lakes.

Gross et al. (1988) found that while anadromy was relatively common in temperate regions, catadromy was more prevalence among fishes from tropical waters. They hypothesized that this may be due to the higher productivity of tropical freshwater habitats relative to the adjacent marine spawning grounds. Data from studies comparing growth and feeding by freshwater and marine populations of the tropical catadromous fish *Lates calcarifer* (Anas 2008; see Section 9.3.3) appear to support this hypothesis.

9.3 Examples of Tropical Diadromous Fish Life Cycles

9.3.1 Amphidromous Gobies and Gudgeons

In anadromous and amphidromous species, spawning occurs in freshwater, but species vary in the life history strategies that they adopt. Amphidromy is relatively common among the Gobiidae (55 species, Table 9.1). Amphidromous gobies are present in all the world's tropical oceans and are an important component of the freshwater fish faunas on many tropical islands. Some species of tropical amphidromous gobies have some of the smallest eggs and larvae in the Gobiidae (Miller 1984). They lay their eggs on the substrate (Fig. 9.1) and these are guarded by males until hatching (Keith 2003). Spawning occurs during increased river flow in the wet season (Erdman 1961, Fitzsimmons et al. 2002, Lim et al. 2002). After hatching, larvae move downstream with the current into estuarine and coastal marine waters at 1–4 mm in size (Han et al. 1998, Keith et al. 1999). Larvae actively swim with the current (Bell and Brown 1995) as there is little or no feeding during this migration (Iguchi and Mizuno 1999). At this time, rivers are probably turbid and the reduced water clarity should minimize predation (Blaber 2000). Plankton production is also higher in tropical coastal waters during the wet season (Longhurst and Pauly 1987) and this may enhance larval feeding and thus growth and survival.

Larvae of tropical amphidromous gobies remain in coastal waters for variable periods from 30–266 days (Bell et al. 1995, Shen et al. 1998, Radtke et al. 1988, 2001, Hoareau et al. 2007, Yamasaki et al. 2007). They can have among the longest larval phases of any tropical fish species with marine larvae (Radtke et al. 2001). Following their long larval phase, the gobies recruit to their adult freshwater habitats at 13–25 mm depending on the species (Keith 2003). Several species move upstream in mass migrations and rapidly cue in to the direction of the current in order to migrate upstream (Bell and Brown 1995, Keith 2003). This behavior of trying to find and swim into the current probably helps them find their preferred freshwater habitats (Erdmann 1986, Fievet and Le Guennec 1998). Tate (1997) found that the behavioral differences among the Hawaiian amphidromous gobies contribute to their dispersal and settlement patterns within streams. Both the diel timing of recruitment and inter- and intraspecific agonistic behavior were shown to influence the ultimate distance fish migrate upstream. The presence of predatory adult *Kuhlia rupestris* and *Anguilla marmorata* in the lower reaches of many freshwater streams of the islands in the Pacific may also have influenced the in-stream distribution patterns (Nelson et al. 1997).

Once in freshwater, the larvae grow more rapidly than on the coast and metamorphose from their pelagic larval shape to the benthic adult form (Erdman 1961, Tate 1997, Nishimoto and Kuamo'o 1997, Shen et al. 1998, Radtke et al. 2001). Many species of amphidromous goby (e.g., *Sicydium punctatum*) are known to be strong swimmers. This strong swimming ability and their ability to use their pelvic fins to grip and climb allow them to find and negotiate waterfalls and presumably artificial barriers during migration (Smith and Smith 1998). Some species school in the estuary before migrating into freshwater and undertake mass migrations. Other species migrate as larvae, and change their color and metamorphose once they have negotiated the first barrier to large predators such as waterfalls or rapids (Tate 1997, Keith 2003). Once the fishes have negotiated these barriers, they change into their adult form and settle and find their preferred microhabitat.

There are several tropical amphidromous gudgeon (sleepers – Eleotridae) that have similar life cycles to the gobies. *Gobiomorphus, Dormitator,* and *Eleotris* species from the tropical western Atlantic also have adhesive eggs that they lay on the substrate in the lower estuaries or tidal freshwater (Nordlie 1981, Teixeira 1994). Spawning occurs during the wet season and larvae form schools, as in the gobies, and probably disperse upstream as they grow (Nordlie 1981, Winemiller and Ponwith 1998). Unlike the gobies, they tend to grow to a larger size (>200 mm standard length (SL)) and lack the fused pelvic fins that gobies use to adhere to the substrate in strong currents. Instead, they favor off-river water bodies or reaches of rivers with lower flow (Nordlie 1981, 2000).

Among Pacific *Eleotris* species, *E. sandwicensis* in Hawaii and *E. acanthopoma* elsewhere in the northern tropical Pacific live in the lower reaches of freshwater streams and rarely ascend above small (1–2 m) barriers (Fitzsimmons et al. 2002, Maeda and Tachihara 2004). In contrast, *Eleotris fusca* is capable of ascending above falls of up to 10 m, similar to many of the amphidromous gobies (Fitzsimmons et al. 2002, Maeda and Tchihara 2004, Maeda et al. 2007). Larval duration of *Eleotris* can also be quite long (2–4 months), and they remain unpigmented until recruiting into waters with low salinity (Maeda and Tachihara 2005). All species were found to recruit to freshwater at similar sizes to gobies (10–19 mm SL; Shen et al. 1998, Maeda et al. 2007), and *E. fusca* migrated upstream against the flow similar to amphidromous gobies (Maeda Tachihara 2005).

9.3.2 Anadromous Clupeoid Fishes

Anadromy is a relatively rare migratory behavior in tropical fishes and there are few studies of obligate anadromous fishes from the tropics. The majority of tropical anadromous species are clupeiform fishes (Table 9.1). Within these families (Clupeidae, Engraulidae, and Pristigasteridae) most are non-migratory species that spend their lives in estuarine or marine waters (Blaber 2000). Among the pristigasterids, Blaber et al. (1998) documented the biology of six species of *Ilisha* from the northern coast of Borneo, including the anadromous *Ilisha megaloptera* and *I. filigera*. *Ilisha filigera* reach up to almost 1 m in length, making them one of the largest clupeoids in the world (Whitehead 1985). Both species spawn during the wet season (*I. filigera*) or early dry season (*I. megaloptera*) in rivers with high turbidities (34–1000 NTU), strong freshwater flows, and large tidal ranges (Blaber et al. 1997). Fry and small juveniles (<30 mm SL) were found in the upper estuaries of the larger rivers in low salinities (0–5) at the end of the wet season (March) suggesting spawning took place nearby during the period of increased river flow. Larval and juvenile fishes then move downstream to the lower estuary and adjacent coast. Spawning at this time appears to be adapted to reduce predation from visual predators (Blaber 2000). Coastal plankton production is also higher during the wet season (Longhurst and Pauly 1987) and this should enhance larval and juvenile growth and survival.

Both *Ilisha* species are widespread in tropical estuaries and adjacent coasts from the west coast of India to Southeast Asia (Whitehead 1985). A strong seasonal monsoonal rainfall pattern is common throughout their range. Both species are fast-growing, with *I. megaloptera* reaching sexual maturity within nine months and

I. filigera in 18 months, and live for 2–4 yrs. They are both multiple spawners with similar relative fecundities (Blaber et al. 1998). This life history strategy is common among other tropical coastal clupeoids (Milton et al. 1994, 1995) and appears to be an adaptation to the spatial and temporal dynamics of these tropical environments.

Another widespread anadromous clupeoid, *Anadontostoma chacunda*, is also abundant in the lower estuaries and adjacent coastal waters throughout the Indo-West Pacific (Munro et al. 1998). Like *Ilisha, Anodontostoma* also spawn during the wet season (November–February) (Munro et al. 1998), but it appears not to be strongly associated with large, fast-flowing turbid rivers. Instead, they seem to avoid turbid waters in the upper estuary during the wet season (Cyrus and Blaber 1992) and are more abundant in sandy rather than mangrove habitats (Jaafar et al. 2004). In northern Australia, fish were caught only in the lower estuary except during the late dry season when freshwater flows were minimal (Cyrus and Blaber 1992). This preference for less turbid unstructured habitats in low flow areas suggests that the factors influencing connectivity of freshwater and coastal components of the population are quite different from those of *Ilisha*.

One of the best studied tropical anadromous clupeoid fishes is hilsa *Tenualosa ilisha*. Hilsa are an abundant species of significant fisheries importance in southern Asia, from the Arabian Gulf to northeastern Indonesia (Sumatra) (Blaber et al. 2003a, see http://dx.doi.org/10.1007/978-90-481-2406-0). It only occurs in rivers with large flow volumes, and the adjacent coast where salinity is reduced (Blaber et al. 2003a). Until recently, hilsa were thought to be strictly anadromous (Coad et al. 2003, Blaber et al. 2003a). However, a detailed study of the biology and movements of hilsa in Bangladesh by Blaber et al. (2003b) and Milton and Chenery (2003) have shown that fish can spawn in a range of salinities and many may not migrate into freshwater at all (Fig. 9.2). Both studies found that most hilsa do spawn in

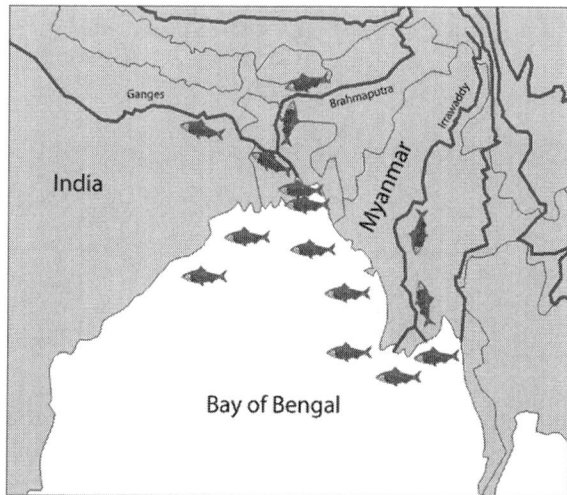

Fig. 9.2 The life cycle of the anadromous clupeid, hilsa *Tenualosa ilisha* in the Bay of Bengal. Fish spawn throughout the region in both coastal marine and freshwater reaches of the major rivers where salinity is low (<5) during the monsoonal wet season (May – September)

freshwater, mainly during the monsoon wet season (March–September). Spawning conditions in rivers and on the coast at this time are similar to those experienced by *Ilisha*. Turbidity is high and river flow is strong and many shallow coastal habitats are inundated and colonized by small juveniles. Juveniles that are spawned in freshwater migrate downstream and reach the lower reaches of the rivers or enter the estuary at about 3–4 months of age. Fish grow rapidly and reach maturity within 12 months, when they migrate back to the spawning grounds (Milton and Chenery 2003). Milton and Chenery (2003) did not find that all fish returned to their natal areas to spawn, although this appears to be the most common migration strategy.

9.3.3 Catadromous Barramundi

Barramundi (*Lates calcarifer*) is a protandrous hermaphrodite and has a complex life history (Grey 1987; Fig. 9.3). Spawning occurs in high salinity in coastal waters adjacent to estuary mouths during the summer wet season, with peak activity linked to spring high tides (Moore 1982, Davis 1982, 1986). Spawning at this time enables the larvae to migrate to, and utilize shallow coastal littoral marine and freshwater wetlands with few predators and an abundance of prey (Moore 1982, Davis 1988, Pender and Griffin 1996, Sheaves et al. 2007). Recent otolith chemistry studies have shown that most larval barramundi do not use freshwater habitats (Milton and Chenery 2005, Anas 2008, Milton et al. 2008). This contrasts with the widely-held belief that coastal freshwater wetlands were the most important nursery areas (Moore 1982, Russell and Garrett 1985). Juvenile barramundi stay in protected tidal

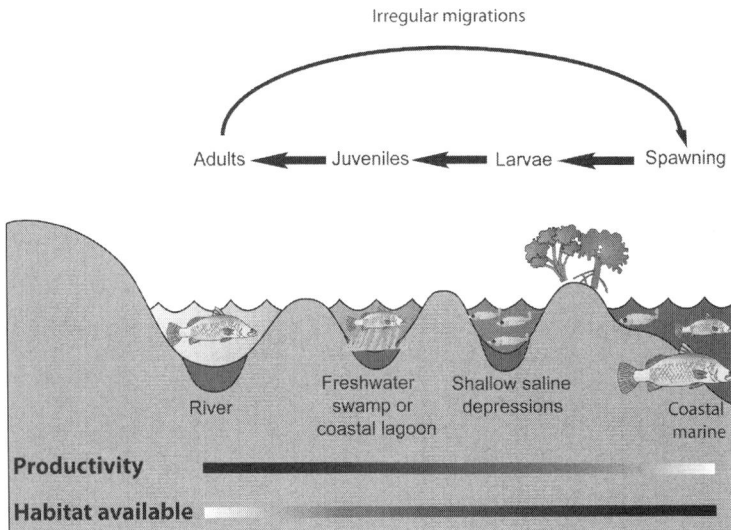

Fig. 9.3 A schematic overview of the life cycle of barramundi *Lates calcarifer* in southern Papua New Guinea

creeks in the lower estuary until 1–2 yrs of age, when they disperse more widely (Moore and Reynolds 1982, Russell and Garrett 1985, 1988). The majority of adult barramundi live in large estuaries or coastal marine waters (Grey 1987).

In contrast, the barramundi larvae that migrate into coastal freshwater wetlands do so at 2–4 months of age (Fig. 9.3). Once in freshwater, fish stay for periods between 3 and 8 months (Papua New Guinea, Anas 2008) and 4 years (northern Australia, Milton et al. 2008). After this time, fish move from these coastal wetlands back to marine waters and disperse. Tagging studies in Papua New Guinea found that immature fish (1–2 yrs old) in the coastal marine waters migrate into freshwater parts of rivers from 2–3 yrs of age (Moore and Reynolds 1982; Fig. 9.4). Many remain in freshwater for several years, apparently not contributing to the spawning population (Milton and Chenery 2005).

In southern Papua New Guinea, the percentage of juvenile barramundi in coastal waters that have grown in freshwater nurseries and the period they spent in freshwater varies between years (Anas 2008; Fig. 9.5). Fish are more likely to use freshwater in years when there is sufficient rainfall following the peak spawning period in October–November (Moore 1982; Fig. 9.6). In other years, the majority of fish remain in coastal marine waters. Anas (2008) found that juvenile barramundi in freshwater swamps grew faster than fish from adjacent coastal marine nursery areas. These freshwater swamps dry each year and the fish need to leave before the swamps drain and they become trapped. Anas (2008), found however, that many fish do get trapped in the swamp and either die or are eaten by scavenging birds or people. How some fish know when to leave remains a mystery?

These studies show, that although connectivity between coastal marine and freshwater habitats is not obligatory for many diadromous species, such as barramundi, those that do migrate into freshwater are more likely to grow faster and thus survive better and produce more progeny by migrating to these more productive tropical habitats (Gross et al. 1988). However, once in freshwater, catadromous fishes need to return to the sea to spawn. They must rely on increased water levels during floods to reconnect their isolated freshwater habitats with the sea.

Fig. 9.4 A young adult male barramundi (3 yrs old) caught in a coastal freshwater waterhole in northern Queensland. The pale silvery color is characteristic of fish that have grown up in freshwater

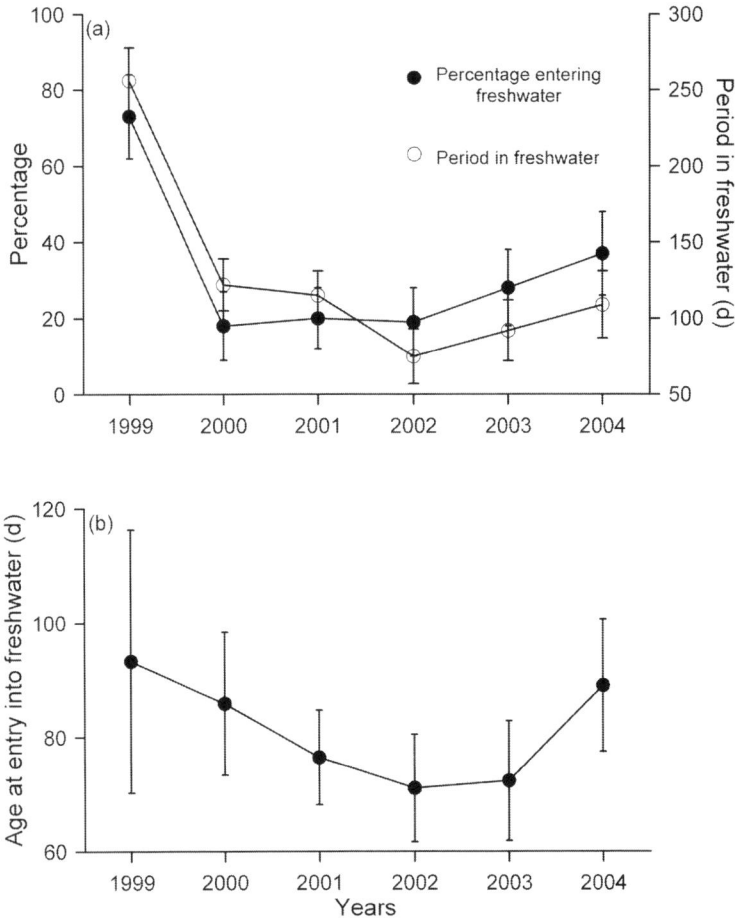

Fig. 9.5 The percentage of juvenile barramundi (± standard error) in southern Papua New Guinea entering freshwater each year from 1999–2004, the length of time they spent in freshwater (**a**), and the age (in days) they entered freshwater (**b**) (data from Anas 2008)

9.4 Factors Affecting the Connectivity Between Freshwater and Marine Populations

9.4.1 Droughts and Floods

Seasonality in tropical areas is primarily associated with variation in rainfall (wet and dry seasons). Most diadromous fishes in the tropics have evolved to take advantage of these seasonal rainfall patterns and associated river flooding to migrate to more productive habitats for breeding (Gross et al. 1988). Increased rainfall and river flooding is believed to bring increased productivity to riverine, estuarine, and

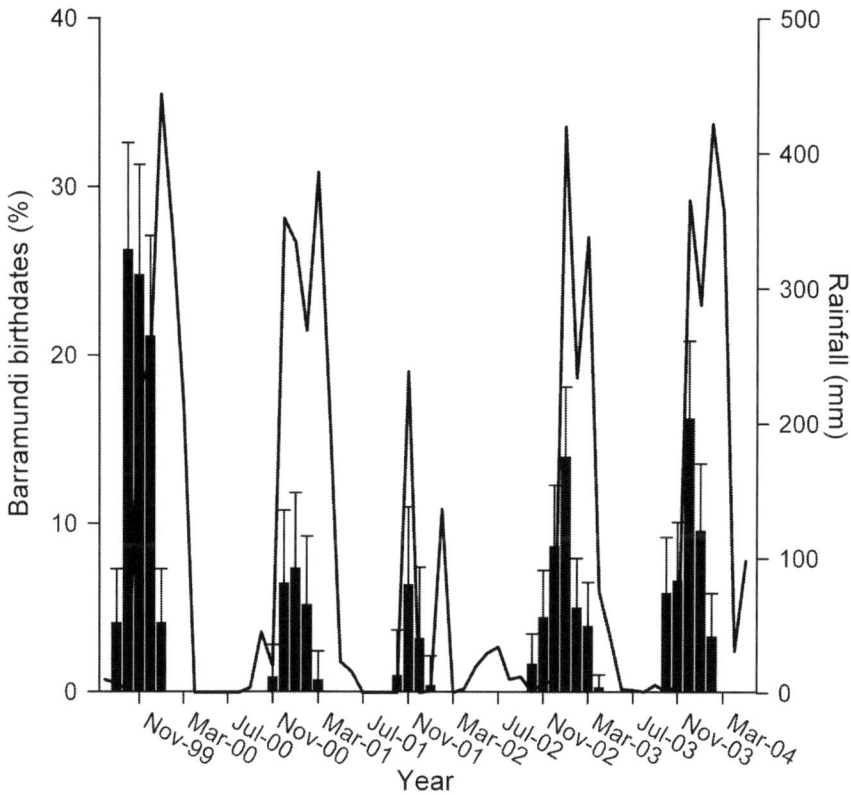

Fig. 9.6 The percentage (+ standard error) of juvenile barramundi (9–12 months old) in southern Papua New Guinea born each month that had spent time in freshwater (data from Anas 2008). Monthly rainfall (in mm) during the period is shown by the solid line. All fish were caught on the coast after returning from their nursery habitats in either the coastal swamps or small mangrove creeks

coastal waters by transporting exogenous nutrients and sediments that stimulate increased primary production (Robins et al. 2006). Many studies have shown that the wet season is the primary growth and reproductive season for fishes in tropical areas (Lowe-McConnell 1987). In the tropics, this annual cycle of flush and renewal is heavily influenced by the large-scale cyclical oceanographic-atmospheric phenomena known as El Niño–Southern Oscillation (ENSO) (Rasmusson and Wallace 1983). One of the main consequences of the ENSO is the strong interannual variation in rainfall throughout the tropics (Ropelewski and Halpert 1996). This interannual variation results from the Southern Oscillation surface pressure differential in the Pacific Ocean that 'flips' irregularly at intervals of 2–10 yrs (Mol et al. 2000). Changes in the direction of this differential have the consequence of causing droughts and floods in different regions of the tropics. Impacts of drought on fish communities have been documented at a range of scales from local population declines, loss of habitat, inducing movements to major changes in community composition (Matthews and Marsh-Matthews 2003). Diadromous fishes need water flow to maintain connectivity between freshwater and estuarine and coastal habitats, and so their populations are often heavily impacted by severe or prolonged drought.

The large catadromous Indo-Pacific centropomid, barramundi (*Lates calcarifer*) is an important commercial and recreational species of estuaries and freshwater wetlands throughout southern and southeast Asia, Papua New Guinea and northern Australia. In northern Australia, it reaches up to 30 kg and occurs widely in most coastal marine, estuarine, and freshwater habitats. However, barrages and dams have limited upstream access to freshwater by barramundi (and other species) in many rivers. In other rivers, fish migrate upstream during floods and then are trapped in freshwater impoundments and thus stopped from moving downstream to spawn. The effect can be extreme in some rivers, to the extent that large predator species such as barramundi become rare or no longer occur in freshwater reaches where they were previously abundant (Fieve et al. 2001, Hogan and Vallance 2005).

In northern Australia, the barramundi that enter freshwater do so between 2 and 6 months of age, with some staying in freshwater until sexual maturity at age 4 (Milton et al. 2008). The freshwater wetlands they use are deeper and more permanent than those in other parts of their range (such as in Papua New Guinea). They are also only intermittently connected to the main river or estuary (Sheaves et al. 2007). Thus, high river flows at critical times can enhance the productivity of such diadromous fish populations by either maintaining connectivity between habitats, increasing nutrient and food supply, or allowing larval and juvenile fishes to access more productive shallow freshwater habitats (Gross et al. 1988) with fewer predators. Once these fish have returned to the estuary, they usually disperse alongshore and many migrate upstream into freshwater habitats (Russell and Garrett 1988). Many of the barramundi that migrate into freshwater remain there until maturity (>4 yrs old), when they move to coastal spawning grounds as mature males (Moore 1982, Davis 1986).

Barramundi of all sizes show extreme plasticity in growth rates among individuals, with fish within a single age cohort varying in size by 40–50% (Staunton-Smith et al. 2004). Faster-growing fish in an age-class are more likely to have used freshwater habitats than those that remained in the estuary (Milton et al. 2008; Fig. 9.7). Staunton-Smith et al. (2004) and Robins et al. (2006) found barramundi year-class strength and growth rates correlated with river flows. They hypothesized three possible mechanisms to explain these patterns: by (1) an increase in the spawning population with migrants from land-locked freshwaters, (2) increased survival of larvae and early juveniles through access to inundated littoral habitats, or (3) increased survival (and growth) of older juveniles and sub-adults by accessing more productive freshwater habitats. The studies of Anas (2008), Milton and Chenery (2005), McCulloch et al. (2005), and Milton et al. (2008) have all found that most barramundi do not access freshwater during their larval and early juvenile stages, but only enter freshwater as older juveniles (>3 months old). This would support the third hypothesis that access to more permanent freshwater lagoons by older juveniles in

Fig. 9.7 The percentage (+ standard error) of fast and slow growing adult barramundi (defined by Staunton-Smith et al. 2004) in the Fitzroy River estuary, Australia, caught after a large flood in February 2003. Fish were separated according to whether they had accessed freshwater or remained in the estuary throughout their life based on their otolith chemistry (data from Milton et al. 2008)

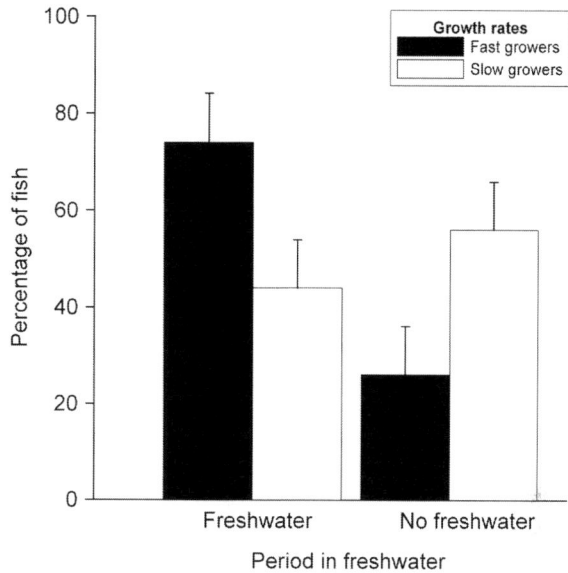

northern Australia is contributing to increased survival and growth of barramundi during years of high freshwater flow (Fig. 9.7).

The advantage of using freshwater habitats appears to be from increased growth due to a greater food supply (Sheaves et al. 2007) and possibly fewer predators of larger fish. River floods would help fish find and access coastal freshwater lagoons. In most years, the majority of barramundi do not use coastal freshwater wetlands as juveniles, as the habitat is not widespread or as available as tidal creeks and estuaries. Yet, for fish that can access freshwater, there appears to be a clear adaptive benefit.

The mass upstream migratory behavior of amphidromous gobies (Erdmann 1986) and cueing to water flow (Fievet and Le Guennec 1998) can cause the potential loss of entire cohorts if they become confused between natural flowing rivers and canalized water courses that outflow from facilities such as power stations. Fievet and Le Guennec (1998) describe how migrating postlarval *Sicyium plumieri* entered the wastewater canals of a hydroelectric power station on Guadeloupe in the Caribbean when natural flowing streams were not accessible.

Clearly, the large numbers of dams on large and small river systems in tropical areas (Nilsson et al. 2005) pose a major risk to the ability of diadromous fishes to maintain the connectivity between freshwater and adjacent marine habitats. Water resource managers need to include appropriately-timed freshwater flows of sufficient magnitude to facilitate migrations. The alternative will be that more fish species will become threatened like the anadromous shad *Tenualosa reevesi* (Weimin et al. 2006). This species has become endangered from a combination of unsustainable fisheries catch (He and Chen 1996) and the construction of terrace dams that have disrupted their spawning migrations (Wang 2003).

9.4.2 Impacts of dams

Water extraction for human activities has a major impact globally on the productivity of rivers, their estuaries, and associated wetlands (Gopal and Sah 1993, Jensen 2001, Hillman and Brierley 2002, Gillanders and Kingsford 2002). To reduce the scale and severity of these impacts, governments worldwide are increasingly moving from engineering solutions to an environmental management approach to water allocation (Finlayson and Brizga 2000). The recognition of the need for environmental flows has now been incorporated in water management policy in over 30 countries (Hillman and Brierley 2002).

In order to maintain connectivity between freshwater and marine populations, managers need to develop effective water management that includes freshwater flows for riverine and estuarine organisms. However, in order to achieve this, we need a greater understanding of the important mechanisms involved in maintaining population connectivity (Staunton-Smith et al. 2004, Robins et al. 2005). Most studies to date have focused on comparing catch rates of commercial fisheries with freshwater flows (Sutcliffe et al. 1977, Lloret et al. 2001, Quiñones and Montes 2001, Robins et al. 2005). These studies have found a significant relationship between flow and catch of many marine and estuarine species. They suggest that freshwater flows can influence fish spawning, survival, and growth during their first year of life (Drinkwater and Frank 1994, Robins et al. 2006).

Mechanisms by which freshwater flow enhances estuarine and coastal marine fish populations will vary between species and depend on their life history (Robins et al. 2005). For diadromous fish species, freshwater flow is required to maintain natural movements between freshwater and coastal marine habitats. In many regulated rivers, this also requires development of structures to enable migratory species to negotiate dams and barrages. Catadromous fish species such as centropomids can have enhanced survival and growth in years when larvae and juveniles from coastal spawning grounds access more productive freshwater habitats (Staunton-Smith et al. 2004). Shea and Peterson (2007) further show that even the stability and quantity of habitats available within regulated rivers are also strongly influenced by the temporal and spatial variations in flow.

Many river systems in the tropics (e.g., Mekong, Indus, and Ganges) have barrages or dams that restrict or even halt flows during the dry season low-flow period (Chang and Naves 1984, Robins et al. 2005, Halliday and Robins 2007). They can even cause physical alteration of the stream bed, and these habitat modifications can have an impact on migration (McDowall 1995, Smith et al. 2003). Disruption of fish migrations as a result can cause major changes in the abundance of species as well as change the community structure above and below these structures (Duque et al. 1998, Greathouse 2006a, b). In order to maintain the connectivity between freshwater and marine populations of diadromous fishes there needs to be a minimum flow in these regulated rivers before fish can migrate (Whitfield 2005, Greathouse et al. 2006b, James et al. 2007). In seasonal tropical habitats, the timing of flows is also important to allow larval and juvenile fishes to enter these more productive freshwater habitats.

In some tropical areas, such as northern Australia, unregulated rivers show high annual variability in the timing and intensity of discharge (Finlayson and McMahon 1988). River flows in these seasonally dry tropical areas are stochastic, resulting from pulse rainfall events (Benke et al. 2000, Gillanders and Kingsford 2002). In many tropical areas, these flood events have high sediment loads. Diadromous clupeoids spawn at this time and the associated high turbidity may reduce the level of predation by visual predators. Regulating river flow and altering the natural pulse intensity will reduce the natural turbidity and potentially increase predation on migrating larvae and juveniles of diadromous fishes. It would also increase the time taken for migrating larvae of amphidromous and anadromous fishes to reach coastal feeding habitats and thus increase mortality due to starvation (Iguchi and Mizuno 1999, March et al. 2003).

9.4.3 Global Climate Change

The IPCC (Intergovernmental Panel on Climate Change 2002, 2007) clearly shows that the world is warming faster now than ever before during recorded human history. It predicts that the effects of this warming on tropical parts of the world will be measured through rising sea levels and increased precipitations in moist tropical areas such as southern Asia. This precipitation is expected to be more intense and less predictable than in the past. This alteration of rainfall patterns is predicted to cause more frequent and intense flooding of tropical coastal wetlands and result in increasing erosion (IPCC 2002). In contrast, dryland rivers in continents such as Africa and Australia are expected to have lower rainfall in their catchments leading to less runoff and lower and more variable flows. These changes in the river flow and inundation of coastal habitats are likely to alter the connectivity between coastal and freshwater populations of many diadromous fishes.

In western Africa, the salinity gradient in the estuary of the Saloum River in Senegal has reversed as rainfall in the catchment has declined since the 1950s (Pages and Citeau 1990). The salinity gradient in the estuary during the dry season now reaches >100 in the upper estuary, 100 km from the coast. Annual precipitation and the number of rainy days have also sharply declined in Ivory Coast further south in the Gulf of Guinea (Servat et al. 1997). Pages and Citeau (1990) further predicted that this salinization would worsen in the Saloum and the adjacent Casamance, Senegal, and Gambia Rivers, as sea levels and temperatures rise.

These changes in the rainfall and subsequent increases in salinity have modified the fish species composition in the Saloum Estuary compared with adjacent estuaries with a 'normal' salinity gradient (Albaret et al. 2004, Simier et al. 2004, Ecoutin et al. 2005). The catadromous clupeid *Ethmalosa fimbriata* and the amphidromous gerreid *Eucinostomus melanopterus* appear to be similarly abundant in the Saloum and the 'normal' Gambia River Estuary. Studies of the effects of these contrasting salinity regimes on fish reproduction showed that the catadromous mullets *Liza falcipinnis* and *L. grandisquamis* as well as *Ethmalosa* had a longer and more intensive reproductive season in the hypersaline Saloum Estuary (Panfili et al. 2004, 2006). These species appear to show broad phenotypic plasticity in their responses to changes in environmental conditions (Guyonnet et al. 2003). *Ethmalosa* grew more slowly due to the poorer conditions. However, they adapted by maturing earlier with increased fecundity in the hypersaline reaches of the Saloum Estuary (Panfili et al. 2004). *Liza falcipinnis* showed the opposite pattern, increasing in size at maturity in the hypersaline Saloum Estuary (Panfili et al. 2006). Thus, although reproduction is strongly linked to the wet season and rainfall in most tropical diadromous fishes, many appear to be capable of adapting their life history to changes in environmental conditions that will be expected as the climate changes.

The amphidromous gobies (subfamily Sicydinae) and eleotrids from high islands in the Pacific Ocean have small and fragmented populations. These species are more vulnerable to changes in the river flows and loss of coastal habitats from erosion and sea level rise as the area of their preferred freshwater habitats are small. Populations of these species would be more vulnerable than species from continental areas as the amount of available habitat on continents is usually much larger.

Many other tropical diadromous fishes also rely on coastal marine and estuarine habitats during their larval and juvenile life stages. Predicted intensification of *El Niño* (ENSO) events (Poloczanska et al. 2007) will be expected to strongly influence recruitment and migration patterns of fishes in tropical estuarine habitats (Garcia et al. 2001, 2003, 2004). Garcia et al. (2001, 2004) found that the catadromous *Mugil curema* were more abundant in the Patos Lagoon Estuary in Brazil during the drier *La Niña* years, whereas anadromous *Lycengraulis grossidens* were more abundant during *El Niño* years when rainfall was high. The impacts of these climate-related phenomena on the fish communities of estuaries and adjacent coastal waters are complex (Garcia et al. 2004), and there are likely to be diadromous species that increase as well as others that decrease as a result of environmental changes.

Increased storm surge from more violent tropical storms (Poloczanska et al. 2007) will result in increased erosion of natural habitats and increasing human restructuring of the coast to protect the large coastal human population in all continents. Both effects will reduce the quality and quantity of nursery habitats of many species, such as mullets, tarpon (*Megalops*), and centropomids. These species tend to take several years to reach maturity and thus are adapted to low adult mortality and stable environments (Blaber 2000). Productivity per female is very high ($>10^6$ eggs per spawning; Moore 1982, Garcia and Solano 1995, Blaber 2000) and most are multiple spawners. Many species rely on tidal inundation to access preferred coastal habitats, and sea level rise is predicted to alter the magnitude of local tidal ranges (Poloczanska et al. 2007). Catadromous and amphidromous species that use mangroves as nurseries will also be at risk as they are considered particularly susceptible to rapidly changing sea levels or prolonged reduction in salinity from altered river flow.

Other groups such as anguillid eels and clupeiform fishes are probably less likely to be affected by rising sea levels and changes to river flow. They tend to use deeper, less-structured habitats during most coastal life-stages. These habitats should be more resilient to changes in the frequency and intensity of storms and changes in river runoff. Nevertheless, prolonged droughts and changes in river flow patterns may impact on the ability of anguillid leptocephala and glass eels to negotiate both natural and artificial barriers and find suitable freshwater habitats. August and Hicks (2008) found that temperate eel leptocephala have a water temperature optimum, and that rising water temperatures (which is expected as a result of global warming) reduce these migrations. Of greater concern is the potential for ocean warming to increase leptocephala starvation, and thus reduce recruitment from reduced spring thermocline mixing and nutrient circulation and productivity in anguillid spawning areas (Poloczanska et al. 2007). Knights (2003) showed that reductions in eel recruitment between 1952 and 1995 were correlated with sea surface temperature anomalies in both the Atlantic and Pacific Oceans.

There is increasing evidence that catadromy is less widespread in temperate species compared to the tropical *Anguilla* (Shiao et al. 2003, Briones et al. 2007, Edeline 2007, Thibault et al. 2007). Although the temperate *Anguilla* eels have evolved from tropical species (Aoyama et al. 2001, Minegishi et al. 2005), the temperate species are much more flexible in their use of freshwater habitats than their tropical counterparts (Edeline 2007, Thibault et al. 2007). Most individuals of all the tropical species of eel studied so far have spent time in freshwater (Shiao et al. 2003, Briones et al. 2007). However, not all individuals showed this migratory behavior. This suggests that maintaining connectivity between freshwater and coastal marine habitats may be less critical for sustaining some tropical eel populations than for other groups such as the sicydiine gobies and diadromous eleotrids. Clearly, there will be tropical diadromous fish species that are advantaged by changes in world climate and others that are disadvantaged. Our task is to take the necessary steps to minimize the additional impacts of anthropogenic factors on natural changes in climate.

Acknowledgment I thank the three anonymous reviewers for their constructive comments.

References

Albaret J-J, Simier M, Darboe FS et al (2004) Fish diversity and distribution in the Gambia estuary, West Africa, in relation to environmental variables. Aquat Living Resour 17:35–46

Anas A (2008) Early life history dynamics and recruitment of the Barramundi, *Lates calcarifer Lates calcarifer* (Bloch) in Western Province, Papua New Guinea. Unpub PhD thesis, University of Queensland, Australia

Aoyama J, Nishida M, Tsukamoto K (2001) Molecular phylogeny and evolution of the freshwater eel genus *Anguilla*. Mol Phylogenet Evol 20:450–459

August SM, Hicks BJ (2008) Water temperature and upstream migration of glass eels in New Zealand: implications of climate change. Environ Biol Fish 81:195–205

Bell KNI, Pepin P, Brown JA (1995) Seasonal, inverse cycling of length- and age-at-recruitment in the diadromous gobies *Sicydium punctatum* and *Sicydium antillarum* in Dominica, West Indies. Can J Fish Aquat Sci 52:1535–1545

Bell KNI, Brown JA (1995) Active salinity choice and enhanced swimming endurance in 0 to 8-d-old larvae of diadromous gobies, with emphasis on *Sicydium punctatum* (Pisces) in Dominica, West Indies. Mar Biol 121:409–417

Benke AC, Chaubey I, Ward GM et al (2000) Flood pulse dynamics of an unregulated floodplain in the southeastern U.S. coastal plain. Ecology 81:2730–2741

Blaber SJM (2000) Tropical estuarine fishes: ecology, exploitation and conservation. Blackwell Science, Oxford

Blaber SJM, Farmer MJ, Milton DA et al (1997) The ichthyoplankton of selected estuaries in Sarawk and Sabah: composition, distribution and habitat affinities. Estuar Coast Shelf Sci 45:197–208

Blaber SJM, Milton DA, Brewer DT et al (2003a) Biology, fisheries, and status of tropical shads *Tenualosa spp.* in south and southeast Asia. Am Fish Soc Symp 35:49–58

Blaber SJM, Milton DA, Fry G et al (2003b) New insights into the life history of *Tenualosa ilisha* and fishery implications. Am Fish Soc Symp 35:223–240

Blaber SJM, Staunton-Smith J, Milton DA et al (1998) The biology and life history strategies of *Ilisha* (Teleostei: Pristigasteridae) in the coastal waters and estuaries of Sarawak. Estuar Coast Shelf Sci 47:499–511

Briones AA, Yambot AV, Shiao JC et al (2007) Migratory patterns and habitat use of tropical eels *Anguilla* spp. (Teleostei: Anguilliformes: Anguillidae) in the Philippines, as revealed by otolith microchemistry. Raff Bull Zool Suppl 14:141–149

Chang BD, Navas W (1984) Seasonal variations in growth, condition and gonads of *Dormitator latifrons* in the Chone River basin, Ecuador. J Fish Biol 24:637–648

Chubb AL, Zink RM, Fitzsimons JM (1998) Patterns of mtDNA in Hawaiian freshwater fishes: the phylogeographic consequences of amphidromy. J Hered 89:9–16

Coad BW, Hussain NA, Ali TS et al (2003) Middle eastern shads. Am Fish Soc Symp 35:59–67

Cyrus DP, Blaber SJM (1992) Turbidity and salinity in a tropical northern Australian estuary and their influence on fish distribution. Estuar Coast Shelf Sci 35:545–563

Daverat F, Limburg KE, Thibault I et al (2006) Phenotypic plasticity of habitat use by three temperatel eel species, *Anguilla anguilla, A. japonica* and *A. rostrata*. Mar Ecol Prog Ser 306: 31–241

Davis TLO (1982) Maturity and sexuality in barramundi, *Lates calcarifer* (Bloch), in the Northern Territory and south-eastern Gulf of Carpentaria. Aust J Mar Freshw Res 33:529–545

Davis TLO (1986) Migration patterns in barramundi, *Lates calcarifer*, in van Diemen Gulf, Australia, with estimates of fishing mortality in specific areas. Fish Res 4:243–258

Davis TLO (1988) Temporal changes in the fish fauna entering a tidal swamp system in tropical Australia. Environ Biol Fishes 21:161–172

Drinkwater KF, Frank KT (1994) Effects of river regulation and diversion on marine fish and invertebrates. Aquat Conserv Freshw Mar Ecosyst 4:135–151

Duque AB, Taphorn DC, Winemiller KO (1998) Ecology of the coporo, *Prochilodus mariae* (Characiformes: Prochilodontidae), and status of annual migrations in western Venezuela. Environ Biol Fishes 53:33–46

Ecoutin J-M, Richard E, Simier M et al (2005) Spatial versus temporal patterns in fish assemblages of a tropical estuarine coastal lake: the Ebrie Lagoon (Ivory Coast). Estuar Coast Shelf Sci 64:623–635

Edeline E (2007) Adaptive phenotypic plasticity of eel diadromy. Mar Ecol Prog Ser 341:229–232

Elliott M, Whitfield AK, Potter IC et al (2007) The guild approach to categorizing estuarine fish assemblages: a global review. Fish Fish 8:241–268

Erdman DS (1961) Notes on the biology of the gobiid fish *Sicydium plumeri* in Puerto Rico. Bull Mar Sci 11:448–456

Erdmann DS (1986) The green stream goby *Sicydium plumeri*, in Puerto Rico. Trop Fish Hobby 2:70–74

Fievet E, de Morais LT, de Morais AT et al (2001) Impacts of an irrigation and hydro-electricity scheme in a stream with a high rate of diadromy (Guadeloupe, Lesser Antilles): can downstream alterations affect upstream faunal assemblages? Archiv Hydrobiol 151:405–425

Fievet E, Le Guennec B (1998) Mass migration of *Sicydium* spp. (Gobiidae)in the streams of Guadeloupe island (French West Indies): implications for the derivation race of small hydro-electric power stations. Cybium 22:293–296

Finlayson BL, Brigza SO (Eds) (2000) River management: the Australian experience. John Wiley and Sons, Chichester

Finlayson BL, McMahon TA (1988) Australia vs the world: a comparative analysis of streamflow characteristics. In: Warner RF (ed) Fluvial geomorphology of Australia, pp. 17–40. Academic Press, Sydney

Fitzsimmons JM, Parham JE, Nishimoto RT (2002) Similarities in behavioural ecology among amphidromous and catadromous fishes on oceanic islands of Hawaii and Guam. Environ Biol Fish 65:123–129

Froese R, Pauly D (eds) (2003) FishBase 2003: concepts, design and data sources. ICLARM, Philippines

Garcia AM, Vieira JP, Winemiller KO (2001) Dynamics of the shallow-water fish assemblage of the Patos Lagoon estuary (Brazil) during cold and warm ENSO episodes. J Fish Biol 59:1218–1238

Garcia AM, Vieira JP, Winemiller KO (2003) Effects of the 1997–1998 El Nino on the dynamics of the shallow water fish assemblages of the Patos Lagoon estuary (Brazil). Estuar Coast Shelf sci 57:489–500

Garcia AM, Vieira JP, Winemiller KO et al (2004) Comparison of the 1982–1983 and 1997–1998 El Niño effects on the shallow-water fish assemblageof the Patos Lagoon estuary (Brazil). Est 27:905–914

Garcia CB, Solano OD (1995) *Tarpon atlanticus* in Colombia: a big fish in trouble. Naga 18:47–49

Gillanders BM, Kingsford MJ (2002) Impact of changes in flow of freshwater on estuarine and open coastal habitats and the associated organisms. Ocean Mar Biol Ann Rev 40:233–309

Gopal B, Sah M (1993) The conservation and management of rivers in India – case study of the river Yamuna. Environ Conserv 20:243–254

Greathouse EA, Pringle CM, Holmquist JG (2006b) Conservation and management of migratory fauna: dams in tropical streams of Puerto Rico. Aquat Conserv Mar Freshw Ecosyst 16:695–712

Greathouse EA, Pringle CM, McDowall WH et al (2006a) Indirect upstream effects of dams: consequences of migratory consumer extirpation in Puerto Rico. Ecol Appl 16:339–352

Grey DL (1987) Introduction. In: Copeland JW, Grey DL (eds) Proceedings of ACIAR international workshop on the management of wild and cultured sea bass – barramundi (*Lates calcarifer*). ACIAR Proc 20

Gross MR (1987) The evolution of diadromy in fishes. Am Fish Soc Symp 1:14–25

Gross MR, Coleman RM, McDowall RM (1988) Aquatic productivity and the evolution of diadromous fish migration. Science 239:1291–1293

Guyonnet B, Aliaume C, Albaret J-J et al (2003) Biology of *Ethmalosa fimbriata* and fish diversity in the Ebrie Lagoon (Ivory Coast), a multipolluted environment. ICES J Mar Sci 60:259–267

Halliday I, Robins J (2007) Environmental flows for subtropical estuaries: understanding the freshwater needs of estuaries for sustainable fisheries production and assessing the impacts of water regulation. Final report FRDC Project No. 2001/022. Coastal zone project FH3/AF

Han KH, Kim YU, Choe KJ (1998) Spawning behaviour and development of eggs and larvae of the Korea freshwater goby *Rhinogobius brunneus* (Gobiidae: Perciformes). J Kor Fish Soc 31:114–120

He SP, Chen YY (1996) The status of the endangered freshwater fishes. In: CCICED (ed) Conserving China's biodiversity. China Environmental Science Press, Wuhan, China

Hillman M, Brierley G (2002) Information needs for environmental-flows allocation: a case study of the Lachlan River, New South Wales, Australia. Ann Assoc Am Geog 92:617–630

Hoareau TB, Lecomte-Finiger R, Grondin H et al (2007) Oceanic larval life of La Reunion 'bichiques', amphidromous gobiid post-larvae. Mar Ecol Prog Ser 333:303–308

Hogan AE, Vallance TD (2005) Rapid assessment of fish biodiversity in southern Gulf of Carpentaria catchments. Project report number QI04074, Qld Depart Prim Indust Fish, Walkamin

Iguchi K, Mizuno N (1999) Early starvation limits survival in amphidromous fishes. J Fish Biol 54:705–712

Intergovernmental Panel on Climate Change (2002) Climate change and biodiversity. IPCC Techn Paper V:1–26

Intergovernmental Panel on Climate Change (2007) Fourth assessment report – climate change 2007: synthesis report. IPCC, http://www.ipcc.ch/pdf/assessment-report/ar4/syr/ar4_syr_spm.pdf. Cited 14 Dec 2007

Jaafar Z, Hajisamae S, Chou LM (2004) Community structure of coastal fishes in relation to heavily impacted human modified habitats. Hydrobiologia 511:113–123

James NC, Cowley PD, Whitfield AK et al (2007) Fish communities in temporarily open/closed estuaries from the warm- and cool-temperate regions of South Africa: a review. Rev Fish Biol Fish 17:565–580

Jensen JG (2001) Managing fish, flood plains and food security in the lower Mekong basin. Wat Sci Tech 43:157–164

Keith P (2003) Biology and ecology of amphidromous Gobiidae of the Indo-Pacific and the Caribbean regions. J Fish Biol 63:831–847

Keith P, Galewski T, Cattaneo-Berrebi G et al (2005) Ubiquity of *Sicyopterus lagocephalus* (Teleostei: Gobioidei) and phylogeography of the genus *Sicyopterus* in the Indo-Pacific area inferred from mitochondrial cytochrome b gene. Mol Phyolog Evol 37:721–732

Keith P, Vigneux E, Bosc P (1999) Atlas des poissons et crustacés d'eau douce de la Réunion. Patrimoines naturels, Vol. 39. Mus Nat d'Hist Nat, Paris

Knights B (2003) a review of the possible impacts of long-term oceanic and climate changes and fishing mortality on recruitment of anguillid eels of the northern hemisphere. Sci Tot Environ 310:237–244

Lim P, Meunier F, Keith P et al (2002) Atlas des poissons d'eau douce de la Martinique. Patrimoines naturels,Vol 51. Mus Nat Hist Natur, Paris

Lloret J, Lleonart J, Solé I et al (2001) Fluctuations of landings and environmental conditions in the north-western Mediterranean Sea. Fish Oceanogr 10:33–50

Longhurst AR, Pauly D (1987) Ecology of tropical oceans. Academic Press, San Diego

Lowe-McConnell RH (1987) Ecological studies in tropical fish communities. Cambridge University Press, Cambridge

Maeda K, Tachihara K (2004) Instream distributions and feeding habits of two species of sleeper, *Eleotris acanthopoma* and *Eleotris fusca*, in the Teima River, Okinawa Island. Ichthyol Res 51:233–240

Maeda K Tachihara K (2005) Recruitment of amphidromous sleepers *Eleotris acanthopoma*, *Eleotris melanosoma* and *Eleotris fusca* in the Teima River, Okinawa Island. Ichthyol Res 52:325–335

Maeda K, Yamasaki N, Tachihara K (2007) Size and age at recruitment and spawning season of sleeper, genus *Eleotris* (Teleostei: Eleotridae) on Okinawa Island, southern Japan. Raff Bull Zool Suppl No 14 199–207

March JG, Benstead JG, Pringle CM et al (2003) Damming tropical island streams: problems. solutions, and alternatives. Biosci 53:1069–1078

Matthews WJ, Marsh-Matthews E (2003) Effects of drought on fish across axes of space, time and ecological complexity. Freshw Biol 48:1232–1253

McCulloch M, Cappo M, Aumend J et al (2005) Tracing the life history of individual barramundi using laser-ablation MC-ICP-MS Sr-isotopic and Sr/Ba ratios in otoliths. Mar Freshw Res 56:637–644

McDowall RM (1988). Diadromy in fishes: migration between freshwater and marine environments. Croom Helm, London

McDowall RM (1995) Seasonal pulses in migration of New Zealand diadromous fish and the potential impacts of river mouth closure. NZ J Mar Freshw Res 29:517–526

McDowall RM (1997) The evolution of diadromy in fishes (revisited) and its place in phyolgenetic analysis. Rev Fish Biol Fish 7:443–462

McDowall RM (2001) Diadromy, diversity and divergence: implications for speciation processes in fishes. Fish Fish 2:278–285

McDowall RM (2003a) Shads and diadromy: implications for ecology, evolution and biogeography. Am Fish Soc Symp 35:11–23

McDowall RM (2003b) Hawaiian biogeography and the islands' freshwater fish fauna. J Biogeogr 30:703–710

McDowall RM (2004) Ancestry and amphidromy in island freshwater fish faunas. Fish Fish 5:75–85

McDowall RM (2007) On amphidromy, a distinct form of diadromy in aquatic organisms. Fish Fish 8:1–13

Miller PJ (1984) The tokology of goboid fishes. In: Potts GW, Wootton JR (eds) Fish reproduction – strategies and tactics. Academic Press, New York

Milton DA, Blaber SJM, Rawlinson NJ (1994) Reproductive biology and egg production of three species of Clupeidae from Kiribati, tropical central Pacific. Fish Bull US 92:102–121

Milton DA, Blaber SJM, Rawlinson NJ (1995) Fecundity and egg production of four species of short-lived clupeoid from Solomon Islands, tropical south Pacific. ICES J Mar Sci 52:111–125

Milton DA, Chenery SR (2003) Movement patterns of the tropical shad (*Tenu-alosa ilisha*) inferred from transects of ^{87}Sr/^{86}Sr isotope ratios in their otoliths. Can J Fish Aquat Sci 60:1376–1385

Milton DA, Chenery SR (2005) Movement patterns of barramundi *Lates calcarifer*, inferred from ^{87}Sr/^{86}Sr and Sr/Ca ratios in otoliths, indicate non-participation in spawning. Mar Ecol Prog Ser 270:279–291

Milton DA, Halliday I, Sellin M et al (2008) The effect of habitat and environmental history on otolith chemistry of barramundi *Lates calcarifer* in a regulated tropical river. Estuar Coast Shelf Sci 78: 301–315

Minegishi Y, Aoyama J, Inoue JG et al (2005) Molecular phylogeny and evolution of the freshwater eels genus *Anguilla* based on the whole mitochondrial genome sequences. Mol Phylogenet Evol 34:134–146

Mol JH, Resida D, Ramlal JS et al (2000) Effects of El-Nino-related drought on freshwater and brackish-water fishes in Suriname, South America. Environ Biol Fishes 59:429–440

Moore R (1982) Spawning and early life history of barramundi, *Lates calcarifer* in Papua New Guinea. Aust J Mar Freshw Res 33:647–661

Moore R, Reynolds LF (1982) Migration patterns of barramundi *Lates calcarifer* in Papua New Guinea. Aust J Mar Freshw Res 33:671–682

Munro TA, Wngratana T, Nizinski MS (1998) Clupeidae. In: Carpenter KE, Niem VH (eds) The living marine resources of the western central Pacific. Vol. 3. Batoid fishes, chimaeras and bony fishes part 1 (Elopidae to Linophyrnidae). FAO, Rome

Myers GS (1949) Usage of anadromous, catadromous and allied terms for migratory fishes. Copeia 1949:89–97

Nelson SG, Parham RB, Tibbatts FA et al (1997) Distribution and microhabitat of the amphidromous gobies in streams of Micronesia. Micronesica 30:83–91

Nilsson C, Reidy CA, Dynesius M et al (2005) Fragmentation and flow regulation of the world's large river systems. Science 308:405–408

Nishimoto RT, Kuamo'o DGK (1997) Recruitment of goby postlarvae into Hakalau stream, Hawaii Island. Micronesica 30:41–49

Nordlie FG (1981) Feeding and reproductive biology of eleotrids fishes in a tropical estuary. J Fish Biol 18:97–110

Nordlie FG (2000) Patterns of reproduction and development of selected resident teleosts of Florida salt marshes. Hydrobiologia 434:165–182

Pages J, Citeau J (1990) Rainfall and salinity of a sehalian estuary between 1927 and 1987. J Hydrol 113:325–341

Panfili J, Durand J-D, Mbow A et al (2004) Influence of salinity on life history traits of the bonga shad *Ethmalosa fimbriata* (Pisces, Clupeidae): comparisons between the Gambia and Saloum estuaries. Mar Ecol Prog Ser 270:241–257

Panfili J, Thior D, Ecoutin J-M et al (2006) Influence of salinity on the size at maturity of fish species reproducing in contrasting West African estuaries. J Fish Biol 69:95–113

Pender PJ, Griffin RK (1996) Habitat history of barramundi *Lates calcarifer* in a north Australian river system based on barium and strontium levels in scales. Trans Am Fish Soc 125:679–689

Poloczanska ES, Babcock RC, Butler A et al (2007) Climate change and Australian marine life. Ocean Mar Biol Ann Rev 45:409–480

Quiñones RA, Montes RM (2001) Relationship between freshwater input to the coastal zone and the historical landings of the benthic/demersal fish *Eleginops maclovinus* in central-south Chile. Fish Oceanogr 10:311–328

Radtke RL, Kinzie III RA, Folsom SD (1988) Age at recruitment of Hawaiian freshwater gobies. Environ Biol Fish 23:205–213

Radtke RL, Kinzie III RA, Shafer DJ (2001) Temporal and spatial variation in length of larval life and size at settlement of the Hawaiian amphidromous goby *Lentipes concolor*. J Fish Biol 59:928–938

Randall JE, Randall HA (2001) Review of the fishes of the genus *Kuhlia* (Perciformes: *Kuhliidae*) of the Central Pacific. Pac Sci 55:227–256

Rasmusson EM, Wallace JM (1983) Meteorological aspects of El Nino/Southern Oscillation. Science 222:1195–1202

Riede K (2004) Global register of migratory species - from global to regional scales. Final report of the R&D-projekt 808 05 081. Fed Agency Nat Conserv, Bonn, Germany

Robins J, Mayer DG, Staunton-Smith J et al (2006) Variable growth rates of a tropical estuarine fish species (barramundi, *Lates calcarifer*) under different freshwater flow conditions. J Fish Biol 69:379–391

Robins JB, Halliday IA, Staunton-Smith J et al (2005) Freshwater-flow requirements of estuarine fisheries in tropical Australia: a review of the state of knowledge and application of a suggested approach. Mar Freshw Res 56:343–360

Ropelewski CF, Halpert MS (1996) Quantifying Southern Oscillation – precipitation relationships. J Climat 9:1043–1059

Russell DJ, Garrett RN (1985) Early life history of barramundi *Lates calcarifer* in north-eastern Queensland. Aust J Mar Freshw Res 36:191–201

Russell DJ, Garrett RN (1988) Movements of juvenile barramundi *Lates calcarifer* (Bloch), in north-eastern Queensland. Aust J Mar Freshw Res 39:117–123

Servat E, Paturel JE, Lubes H et al (1997) Climatic variability in humid Africa along the Gulf of Guinea part I: detailed analysis of the phenomenon in Cote d'Ivoire. J Hydrol 191:1–15

Shea CP, Peterson JT (2007) An evaluation of the relative influence of habitat complexity and habitat stability on fish assemblage structure in unregulated and regulated reaches of a large southeastern warmwater stream. Trans Am Fish Soc 136:943–958

Sheaves M, Johnson R, Abrantes K (2007) Fish fauna of dry tropical and subtropical estuarine floodplain wetlands. Mar Freshw Res 58:931–943

Shen K, Lee Y, Tzeng WN (1998) Use of otolith microchemistry to investigate the life history pattern of gobies in a Taiwanese stream. Zool Stud 37:322–329

Shiao JC, Iizuka Y, Chang CW et al (2003) Disparity in habitat use an migratory behaviour between tropical eel *Anguilla marmorata* and temperate eel *A. japonica* in four Taiwanese rivers. Mar Ecol Prog Ser 261:233–242

Simier M, Blanc L, Aliaume C et al (2004) Spatial and temporal structure of fish assemblages in an "inverse estuary", the Sine Saloum system (Senegal) Estuar Coast Shelf Sci 59:69–86

Smith GC, Covich AR, Brasher AMD (2003) An ecological perspective on the biodiversity of tropical island streams. BioScience 53:1048–1051

Smith RJF, Smith MJ (1998) Rapid acquisition of directional preferences by migratory juveniles of two amphidromous Hawaiian gobies, *Awaous guamensis* and *Sicyopterus stimpsoni*. Environ Biol Fishes 53:275–282

Staunton-Smith J, Robins JB, Mayer DG et al (2004) Does the quantity and timing of fresh water flowing into a dry tropical estuary affect year-class strength of barramundi (*Lates calcarifer*). Mar Freshw Res 55:787–797

Sutcliffe WH, Drinkwater K, Muir BS (1977) Correlations of fish catch and environmental factors in the Gulf of Maine. J Fish Res Bd Can 34:19–30

Tate DC (1997) The role of behavioural interactions of immature Hawaiian stream fishes (Pisces: Gobiodei) in population dispersal and distribution. Micronesica 30:51–70

Teixeira RL (1994) Abundance, reproductive period, and feeding habits of eleotrids fishes in estuarine habitats of north-east Brazil. J Fish Biol 45:749–761

Thibault IJJ, Dodson F, Caron W et al (2007) Facultative catadromy in American eels: testing the conditional strategy hypothesis. Mar Ecol Prog Ser 344:219–229

Tsukamoto K, Nakai I, Tesch WV (1998) Do all freshwater eels migrate? Nature 396:635–636

Wang HP (2003) Biology, population dynamics, and culture of Reeves Shad *Tenualosa reevesii*. Am Fish Soc Symp 35:77–84

Weimin W, Abbas K, Xufa M (2006) Threatened fishes of the world: *Macrura reevesi* Richardson 1846 (Clupeidae). Environ Biol Fish 77:103–104

Whitehead PJP (1985) FAO species catalogue 7, Clupeoid fishes of the world. Part 1 – Chirocentridae, Clupeidae and Pristigasteridae. FAO Fisheries Synposis 125, Vol. 7

Whitfield AK (2005) Preliminary documentation and assessment of fish diversity in sub-Saharan African estuaries. Afr J Mar Sci 27:307–324

Winemiller KO, Ponwith BJ (1998) Comparative ecology of eleotrids fishes in central Amercian coastal streams. Environ Biol Fish 53:373–384

Yamasaki N, Maeda K, Tachihara K (2007) Pelagic larval duration and morphology at recruitment of *Stiphodon pernopterygionus* (Gobiidae: Sicydiinae). Raff Bull Zool Suppl No 14:209–214

Chapter 10
Evaluation of Nursery function of Mangroves and Seagrass beds for Tropical Decapods and Reef fishes: Patterns and Underlying Mechanisms

Ivan Nagelkerken

Abstract Shallow-water tropical coastal habitats, such as mangroves and seagrass beds, have long been associated with high primary and secondary productivity. The ubiquitous presence of juvenile fish and decapods in these systems has led to the hypothesis that they act as nurseries. Earlier studies mainly focused on the faunal community structure of these systems, leaving us with little detailed insight into their potential role as nurseries. Habitats are considered nurseries if their contribution, in terms of production, to the adult population is greater than the average production of all juvenile habitats, measured by the factors density, growth, survival, and/or movement. High food abundance and low predation risk form the most likely factors that contribute to the attractiveness of tropical nursery habitats. Here, the current state of knowledge on nursery function of shallow-water coastal habitats, particularly mangroves and seagrass beds, is reviewed for each of the above-mentioned factors. Most data show that mangroves and/or seagrass beds have high densities of various fish species and some of their food items, and a lower predation risk for fish and decapods due to factors such as low predator abundance, high water turbidity, and complex habitat structure. In contrast, growth rates of fish appear higher on coral reefs. There is increasing evidence that at least part of the fish or decapod population in these putative nurseries eventually moves away to offshore habitats. The current review shows that mangrove and/or seagrass habitats may act as nurseries through higher juvenile densities and survival rates than offshore habitats, but that trade-offs may exist to the detriment of growth rate. With the lack of detailed movement studies, the exact degree to which mangroves and seagrass beds sustain offshore fish and decapod populations remains largely unclear.

Keywords Nursery function · Mangroves · Seagrass beds · Coral reef fish · Ecological connectivity

I. Nagelkerken (✉)
Department of Animal Ecology and Ecophysiology, Institute for Water and Wetland Research, Faculty of Science, Radboud University Nijmegen, Heyendaalseweg 135, P.O. Box 9010, 6500 GL Nijmegen, the Netherlands
e-mail: i.nagelkerken@science.ru.nl

I. Nagelkerken (ed.), *Ecological Connectivity among Tropical Coastal Ecosystems,*   357
DOI 10.1007/978-90-481-2406-0_10, © Springer Science+Business Media B.V. 2009

10.1 Introduction

Mangrove forests and seagrass beds are prominent features of sheltered tropical coastlines. They typically occur in shallow-water environments that are protected from the ocean surge, such as lagoons, embayments, and estuaries. One conspicuous feature of these habitats is their high densities of juvenile fish and decapods, including species for which the adults live on adjacent coral reefs or offshore areas (Ogden and Zieman 1977, Parrish 1989, Robertson and Blaber 1992). Based on this spatial separation between juvenile and adult populations, mangroves and seagrass beds have been assumed to function as important nursery areas that contribute to adult populations (Fig. 10.1; Parrish 1989). Not only was this assumption long-time based on qualitative observations (see Nagelkerken et al. 2000c), also few efforts were made to define clearly what a nursery habitat constitutes. Beck et al. (2001) provided a clear definition of a nursery habitat with testable predictions related to the nursery-role concept: 'A habitat is a nursery for juveniles of a particular species if its contribution per unit area to the production of individuals that recruit to adult populations is greater, on average, than production from other habitats in which juveniles occur'. Beck et al. (2001) further argue that 'The ecological processes operating in nursery habitats, as compared with other habitats, must support greater contributions to adult recruitment from any combination of four factors: (1) density, (2) growth, (3) survival of juveniles, and (4) movement to adult habitats'. Dahlgren et al. (2006) suggested an additional definition asserting that the contribution of nursery habitats could also be calculated on basis of total number of individuals per habitat instead

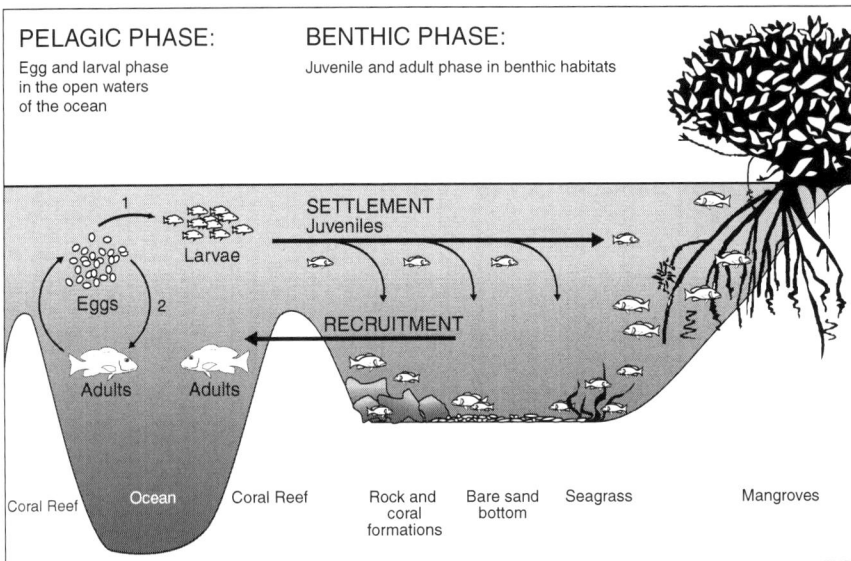

Fig. 10.1 Overview of life cycle of reef fishes defined as ontogenetic shifters (*sensu* Adams et al. 2006)

of individuals per unit surface area. This approach would be more valuable in conserving habitats that provide highest overall contribution to adult populations but that would otherwise not be defined as nursery habitats following the definition by Beck et al. (2001). For example, conserving seagrass beds with large surface areas but lower per unit faunal abundances instead of targeting small coral patches with small surface areas but relatively high densities.

Three hypotheses have been proposed to explain the attractiveness of mangroves and seagrass beds for faunal assemblages: (1) the food availability hypothesis, which suggests that these habitats harbor a high abundance of food, (2) the predation risk hypothesis, which suggests that the lower densities of predators, and the higher water turbidity compared to offshore habitats result in a lower predation pressure, and (3) the structural heterogeneity hypothesis, which suggests that animals are attracted to the complex structure provided by mangrove prop-roots or seagrass shoots (Parrish 1989, Blaber 2000, Laegdsgaard and Johnson 2001, Verweij et al. 2006a, Nagelkerken and Faunce 2008). These hypotheses are not mutually exclusive; for example, increased habitat structural complexity due to presence of vegetation could provide for more living space, higher food abundance due to a larger habitat surface area that functions as living space for small prey organisms and epibionts, and a reduction in predation risk. The above hypotheses are all related to the 'minimize μ/g hypothesis' which states that habitats are selected where the ratio of mortality risk (μ) to growth rate (g) is minimized (Utne et al. 1993, Dahlgren and Eggleston 2000). As such, the factors food abundance, predator abundance, turbidity, and structure likely contribute either directly or indirectly to the underlying mechanisms that regulate the nursery-role measures of density, growth, and survival.

Until recently, little was known about growth, survival, and movement of fish and decapods across multiple tropical coastal habitats (Beck et al. 2001, Heck et al. 1997, 2003, Sheridan and Hays 2003). Fish density has by far received the most attention, but even in this case studies have provided us with relatively little information on potential linkages among habitats (see Section 10.2.1 and Chapter 11). Although faunal density as a factor has been reviewed earlier, there is need for a more detailed review. Earlier reviews were written from either a mangrove (Sheridan and Hays 2003, Manson et al. 2005, Faunce and Serafy 2006, Nagelkerken 2007) or seagrass (Heck et al. 1997, Jackson et al. 2001, Heck et al. 2003, Minello et al. 2003) perspective. In addition, most studies did not provide a comparison with the coral reef habitat, or were largely based on temperate coastal habitats (Beck et al. 2001, Jackson et al. 2001, Heck et al. 2003, Minello et al. 2003). The current review provides a first-time overview of all hypothesized nursery-role factors and potential underlying mechanisms in multiple tropical coastal habitats.

The focus of the current review is on mangrove, seagrass, and coral reef ecosystems as they occur worldwide as extensive areas along (sub)tropical shorelines. I specifically only review studies that compared multiple tropical habitats, and that included juvenile habitats (i.e., various habitats located in embayments, lagoons, or estuaries) as well as adult habitats (i.e., coral reefs or offshore areas) for fish and decapods, so as to evaluate the nursery function of the former for species living as adults in the latter. Only tropical studies are reviewed here (the tropics defined as the area located between latitudes 23°N and 23°S), with the exception of some studies from the Indo-Pacific (Japan, up to 28°N; eastern Australia, down to 27°S) and studies from the western Atlantic (southern Florida, up to 26°N). As there are still relatively few published studies on nursery-role factors, I not only review the results of these studies, but also comment on some of the methodology used and how this has or has not increased our understanding of the juvenile habitat function of mangroves and seagrass beds. I start by reviewing the existing data on the main components of nursery-role determination, i.e., density, growth, survival, and movement. Second, I review potential underlying mechanisms that determine the attractiveness of shallow-water tropical habitats for juvenile fish and decapods, including food abundance and predation risk. Finally, I combine these results and synthesize the current state of knowledge on nursery role of shallow-water tropical coastal habitats for species that live in offshore habitats as adults.

10.2 Existing Evidence for Nursery role of Shallow-Water Tropical Coastal Habitats

10.2.1 Density

Of all fish species found on Indo-Pacific and Caribbean coral reefs only a few dozen co-occur in shallow-water habitats located in back-reef areas, and even fewer (at least 17 species in the Caribbean) occur there abundantly as juveniles (see review by Nagelkerken 2007). Reef fish species of which the juveniles are mainly found in embayments or lagoons, and the adults mainly on the coral reef have been called 'nursery species' in the Caribbean (Nagelkerken and van der Velde 2002). In the Indo-Pacific, this term was first used for species of which the mean juvenile density in mangroves and/or seagrass beds is >70% of the total juvenile density in all habitats, and of which the mean adult density on coral reefs is >70% of the total adult density in all habitats (Dorenbosch et al. 2005a). These definitions do not imply, however, that these species occur in equal densities in all available shallow-water habitat types. In addition, they were formulated for studies done at single geographical locations, and there is no reason to assume beforehand that these patterns hold true for other locations as well. Therefore, this section reviews for the nursery species as identified by Nagelkerken and van der Velde (2002) and Dorenbosch et al. (2005a), whether other studies also found them to be most abundant as juveniles in back-reef areas, and which specific shallow-water habitat is most commonly used by juveniles of individual species. The comparison is restricted to fish, as size-frequency studies for tropical decapods in multiple habitats are practically non-existent.

The majority of studies focusing on juvenile fish densities in multiple shallow-water coastal habitats have been done in the Caribbean (Table 10.1). Ten out of 15 studies in this region show that most or all nursery fish species studied had highest juvenile densities in the mangroves only. Only a few species showed highest

Table 10.1 Overview of comparative daytime studies on juvenile fish densities in different tropical coastal habitats, using a single methodology across all habitats. Studies are ordered alphabetically by location names (separately for Indo-Pacific and Caribbean). Only studies that have included coral reef habitat are listed, and for studies sampling complete fish faunas only results for nursery species (see below) are reported. Area: Ca = Caribbean, IP = Indo-Pacific. Habitat type: AB = algal beds, BCM = branching coral with associated macroalgae, BR = back-reef coral habitat, C = coral, Ch = tidal channels, CR = coral reef, CRb = coral rubble, IF = intertidal flats with patches of coral/sand/seagrass/rubble, LP = limestone pavement, MF = subtidal mud flats with low cover of macroalgae, M = macroalgae, Mg = mangrove, PR = patch reef, S = unvegetated sand areas, Sg = seagrass bed. Species selection: Cp = complete fish fauna surveyed, Sl = selected species surveyed or size class data provided for selected species only. For the Caribbean, nursery species are defined as 'reef fishes of which the juveniles mainly occur in bays and the adults mainly on the coral reef' (*sensu* Nagelkerken et al. 2000b), and includes 17 fish species (see Nagelkerken et al. 2000b). For the Indo-Pacific, nursery species are defined as 'fish species of which the mean juvenile density in mangroves or seagrass beds is >70% of the total juvenile density in all habitats, and of which the mean adult density on coral reefs is >70% of the total adult density in all habitats' (*sensu* Dorenbosch et al. 2005a). All = all nursery species pooled

Reference	Location	Area	Habitat types compared	Methodology and species selection	Taxa	No. of nursery species	Highest fish density
Vagelli (2004)	Indonesia	IP	CR, S, Sg	Visual census (Sl)	Fish	1[2]	Sg[4]
Nakamura and Sano (2004b)	Japan	IP	CR, S, Sg	Visual census (Cp)	Fish	1[2]	Sg
Nakamura and Tsuchiya (2008)	Japan	IP	BR, CR, Sg	Visual census (Cp)	Fish	7[2]	Sg
						1[2]	BR
Shibuno et al. (2008)	Japan	IP	CR, CRb, Mg, S, Sg	Visual census (Sl)	Fish	3[2]	Mg
						3[2]	Sg
Tupper (2007)	Palau	IP	BCM, C, CR, CRb, LP, M, Mg, S, Sg	Visual census (Sl)	Fish	1	BCM[4]
Dorenbosch et al. (2005a)	Tanzania + Comoros	IP	AB, CR, IF, Mg, Sg	Visual census (Cp)	Fish	All	Sg
Dorenbosch et al. (2005b)	Tanzania	IP	CR, Sg	Visual census (Cp)	Fish	All	Sg

Table 10.1 (continued)

Reference	Location	Area	Habitat types compared	Methodology and species selection	Taxa	No. of nursery species	Highest fish density
Dorenbosch et al. (2006)	Tanzania	IP	CR, Sg	Visual census (Sl)	Fish	1	Sg
Dorenbosch et al. (2006)	Aruba	Ca	CR, Mg	Visual census (Sl)	Fish	1	Mg
Dorenbosch et al. (2007)	Aruba	Ca	CR, Mg, Sg	Visual census (Cp)	Fish	All	Mg[5]
Lindquist and Gilligan (1986)	Bahamas	Ca	CR, lagoon with Mg/S/Sg	Visual census (Sl)	Fish	1	Lagoon
Mumby et al. (2004)	Belize	Ca	CR, Mg, Sg	Visual census (Sl)	Fish	1	Mg
Huijbers et al. (2008)	Bermuda	Ca	[1]Mg, PR, Sg	Visual census (Sl)	Fish	2	Mg + Sg
						1	PR
						1	Mg + PR + Sg
Nagelkerken et al. (2000c)	Bonaire	Ca	CR, Mg, Sg	Visual census (Sl)	Fish	4	Mg
						4	Sg
						1	CR
Cocheret de la Morinière et al. (2002)	Curaçao	Ca	CR, Mg, Sg	Visual census (Sl)	Fish	5	Mg
						2	Sg
Nagelkerken and van der Velde (2002)	Curaçao	Ca	[1]Ch, CR, MF, Mg, Sg	Visual census (Cp)	Fish	All	Mg
						8	Mg
						3	Ch
						2	Ch + Mg
Pollux et al. (2007)	Curaçao	Ca	CR, Mg, Sg	Visual census (Sl)	Fish	1	Mg[4]
						1	Mg + Sg[4]

Table 10.1 (continued)

Reference	Location	Area	Habitat types compared	Methodology and species selection	Taxa	No. of nursery species	Highest fish density
Serafy et al. (2003)	Florida, USA	Ca	CR, Mg	Visual census (SI)	Fish	3	Mg
						2	CR + Mg
Eggleston et al. (2004)	Florida, USA	Ca	Ch, CR, Mg, PR, Sg	Visual census (SI)	Fish	4	Mg
Rooker (1995)	Puerto Rico	Ca	CR, Mg	Spearfishing (SI)	Fish	1	Mg
Christensen et al. (2003)	Puerto Rico	Ca	CR, Mg, Sg	Visual census (Cp)	Fish	Haemulidae	Mg
						Lutjanidae	Mg
Aguilar-Perera and Appeldoorn (2007)	Puerto Rico	Ca	BR, CR, Mg, Sg	Visual census (SI)	Fish	2	Sg
						2	CR
						1	BR
						1	BR + Sg
						1	Mg
						1	Mg + Sg
Gratwicke et al. (2006)	Tortola (British Virgin Islands)	Ca	CR, lagoons with AB/Mg/S/Sg	Visual census (SI)	Fish	6	Lagoon
						4[3]	Lagoon

[1] fossil reef boulders and undercut notches excluded here due to their small surface areas
[2] includes additional nursery species – not listed as such by Dorenbosch et al. (2005a)
[3] includes other species showing similar pattern of lagoonal habitat use by their juveniles as nursery species
[4] recruits
[5] except one seagrass bed with equal total density

densities in seagrass or channel habitats (four studies), while some species occurred commonly in multiple shallow-water habitats (seven studies). Three studies found high(est) juvenile densities on the coral reef, but just for one or two species.

In the Indo-Pacific region, comparative habitat utilization studies have only been done in the last few years. Six out of eight studies show that seagrass beds alone harbored highest juvenile fish densities of various nursery species (Table 10.1). One study showed highest densities in mangroves for three species, while two studies found highest juvenile densities for two species on corals habitats located in back-reef areas.

The nursery function of mangroves has long been questioned in the Indo-Pacific (Blaber and Milton 1990, Thollot 1992; Chapter 11). Also the current review found little support for their nursery function, in terms of juvenile fish densities of reef fish. Some studies in the Indo-Pacific have found higher fish densities in mangroves than in seagrass beds (Robertson and Duke 1987, Blaber et al. 1989), but juvenile and adult densities were not separated nor were nursery species distinguished from species that spent most of their life cycle in estuaries. Hence, little information on nursery function can be obtained from such studies. Blaber et al. (1989) showed that a pattern of higher fish densities in mangroves was only present for truly estuarine species, while the strong correlation between mangrove abundance and offshore fisheries in the Indo-Pacific mainly applies to estuarine-dependent species as opposed to coral reef species (see Chapter 15).

Although numerous studies have investigated fish communities of mangroves, seagrass beds, or other shallow-water habitats, few can be used to support the nursery-role hypothesis with respect to density. The main caveats are: (1) most studies focused on a single habitat, (2) in cases where multiple habitats were studied the adult habitat was excluded, (3) no distinction was made between juvenile and adult densities, or (4) different methodologies were used across habitats. In addition, conclusions of studies vary due to differences in definitions, and variation in time, space, and species (Chapter 11). As few studies are currently available that have tackled these caveats, we still know relatively little of the precise ontogenetic habitat shifts of decapods and fish among tropical coastal habitats.

Comparison of species overlap or total fish densities among habitats, as has been done in some studies in the past (e.g., Thollot 1992), does little to increase our understanding of the potential habitat linkages, because of the large differences in ecology among species. These studies correctly identified that the majority of the species living on coral reefs are not associated with mangrove or seagrass habitats. However, for some common and ecologically important species this association (based on density) is highly evident, and these species show significantly smaller adult population sizes when nearby mangrove and seagrass juvenile habitats are lacking (Nagelkerken et al. 2002, Dorenbosch et al. 2004, Mumby et al. 2004). Recently, studies have been done for complete coral reef fish communities to identify the specific species that are commonly associated with mangroves/seagrass beds during their juvenile life stage (Nagelkerken et al. 2000b, Dorenbosch et al. 2005a). For these species, there is growing evidence that of various shallow-water habitats, mangroves or seagrass beds commonly harbor the highest densities of juveniles,

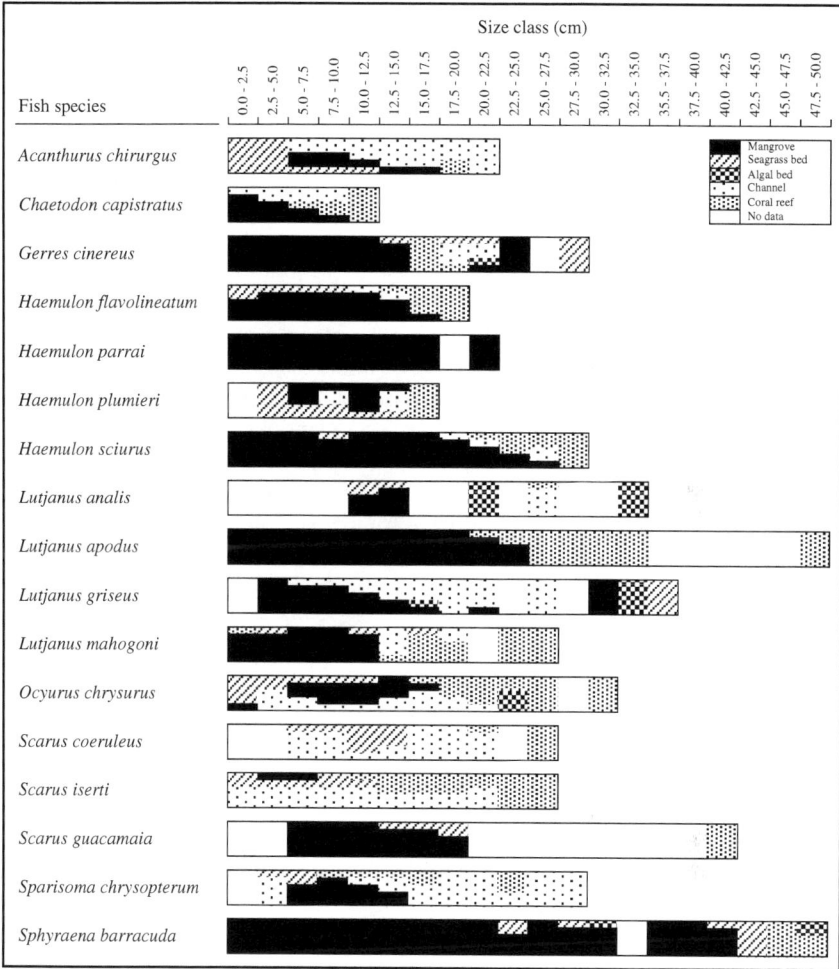

Fig. 10.2 Ontogenetic utilization of tropical coastal habitats by 17 Caribbean nursery fish species in Curaçao. Nursery species are defined as species that largely inhabit back-reef areas as juveniles and coral reefs as adults (*sensu* Nagelkerken and van der Velde 2002). For each size class of each fish species the relative abundance per habitat is shown as the percentage of the total abundance per size class (in all habitats), rounded off to portions of 20%. Figure reproduced from Nagelkerken et al. (2000b), with kind permission from Inter-Research Science Centre

depending on the species and life stage considered (Fig. 10.2). This pattern seems to be consistent across various study locations and geographic regions, and is probably caused by the fact that vegetated habitats harbor higher faunal densities than unvegetated habitats (Orth et al. 1984, Heck et al. 2003). As stated earlier, total habitat surface area is an important factor to consider when comparing habitat value for juveniles.

10.2.2 Growth

It has been hypothesized that growth rates of fauna are higher in vegetated than in unvegetated habitats because the former provide a greater abundance of food sources (seagrass: Heck et al. 1997, mangrove: Laegdsgaard and Johnson 2001). Very few studies have compared growth rates of fish among the coral reef–seagrass–mangrove continuum, and such studies are lacking for decapods. Two studies (Grol et al. 2008; M Grol et al., Radboud University Nijmegen, unpubl. data) showed that for a wide size range of the grunt *Haemulon flavolineatum* growth rates were higher on the coral reef than in mangroves and seagrass beds (Table 10.2, Fig. 10.3), which is the opposite of that expected from the putative nursery role of the latter two habitats. Also lagoonal (i.e., located in shallow marine lagoons) habitats harboring coral colonies showed higher fish growth rates than other lagoonal habitats (Dahlgren and Eggleston 2000, Tupper 2007). Grol et al. (2008) found that differences in fish growth rates (from enclosures) among habitats were consistent with patterns of food densities, providing a potential explanation for the among-habitat differences in fish growth (see Section 10.3.1). Comparison between mangroves and seagrass beds alone shows contrasting results. While a preliminary study showed higher growth for early juveniles of two fish species in mangroves than in seagrass beds (I Mateo, University of Rhode Island, pers. comm.), another study did not find any differences in growth rate between these two habitats (Grol et al. 2008; Fig. 10.3).

Some growth studies were done in microhabitats (i.e., small-sized habitats within ecosystems) within lagoonal environments, and are therefore less useful for testing the nursery-role concept for larger and spatially separated habitats such as mangroves, seagrass beds, and coral reefs. For example, two studies focused on microhabitats located within a relatively small seascape (i.e., microhabitats located at single 'lagoon sites' by Tupper 2007, and microhabitats located within 'tidal creeks' by Dahlgren and Eggleston 2000). With various microhabitats in such close proximity to one another, fish could have shown regular short-distance migrations among microhabitats, making it difficult to separate the effect on individual microhabitats and extrapolating these results to larger-sized habitats. Another important consideration is that habitats may show different growth rates for different size classes of fish. The currently available data mostly apply to the early juvenile stage (<6 cm total length; Table 10.2). Dahlgren and Eggleston (2000) and Grol et al. (unpubl. data) were the only ones to study growth rates for different fish size classes, but found a consistent pattern of habitat differences through ontogeny.

Various methods have been used to study growth of fishes in different habitats, including *in situ* measurements of somatic growth in enclosures (Grol et al. 2008), *in situ* mark and recapture of fish (Tupper 2007), incremental growth in stained otoliths (Reichert et al. 2000), incremental width between otolith growth rings (Levin et al. 1997), and length–age regressions based on otolith ageing (Rypel and Layman 2008). Potential problems of the different techniques include: stress as a result of fish handling (enclosures, mark–recapture, otolith staining), reduced home range

Table 10.2 Overview of comparative growth studies on fishes among different tropical coastal habitats. Studies comparing microhabitats within major habitat types (e.g., various growth forms or sizes of corals) are not included. Habitat type: BC = branching coral, BCM = low branching coral with associated macroalgae, BM = bushy macroalgae, C = coral, CR = coral reef, CRb = coral rubble, FR = fore-reef, Mg = mangrove, PR = patch reef, Sg = seagrass, Sp = sponges. Unit of measurement: B = biomass, L = length, W/L = weight/length ratio. TL = total length

Reference	Location	Habitat types compared	Methodology	Unit of measurement	Taxa	Species	Fish TL (cm)	Highest growth rate
Dahlgren and Eggleston (2000)	Bahamas	BM, postalgal habitat harboring C/CRb/Sg/Sp	*In situ* enclosures	L	Fish	*Epinephelus striatus*	3.5–4.0 5.0–5.5 7.0–7.5	Postalgal > BM
Tupper (2007)	Palau	BC, BCM, BM, CRb, Sg	Mark-recapture	L	Fish	*Cheilinus undulatus*	3.5–5.0	BCM > all others[1] BM > BC + Sg[2]
Grol et al. (2008)	Aruba + Curaçao	CR, Mg, Sg	*In situ* enclosures	B, L, W/L	Fish	*Haemulon flavolineatum*	3.5–4.2	CR > Mg + Sg[3]
Grol et al. (unpubl. data)	Curaçao	CR, Mg, Sg	Otolith ageing	L	Fish	*Haemulon flavolineatum*	4–18[4]	CR > Mg + Sg
I Mateo (pers. comm.)	Puerto Rico + St. Croix (US Virgin Islands)	Mg, Sg	Otolith increment width	L	Fish	*Haemulon flavolineatum*	2–5	Mg > Sg
					Fish	*Lutjanus apodus*	3–6	Mg > Sg

[1] not significant
[2] not statistically tested
[3] growth in weight and length showed only a higher trend (non-significant) for CR in Curaçao
[4] fork length

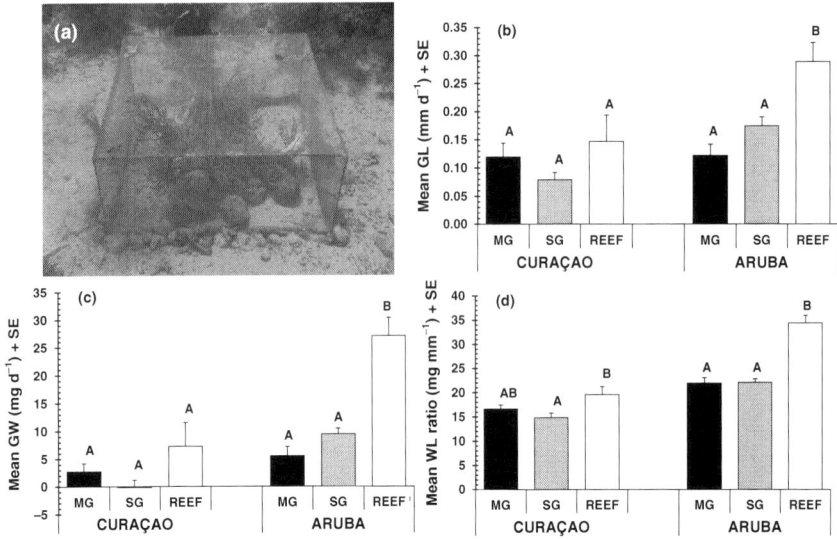

Fig. 10.3 (a) Enclosures used to study somatic growth of *Haemulon flavolineatum* on the coral reef, and results (mean + standard error) of such growth experiments across mangrove (MG), seagrass (SG), and coral reef (REEF) habitats on two Caribbean islands for (**b**) growth in length (GL), (**c**) growth in weight (GW), (**d**) growth in weight–length ratio (WL). Different letters indicate significant differences (p < 0.05) across habitats. Figure reproduced from Grol et al. (2008), with kind permission from Inter-Research Science Centre

(enclosures), movements among habitats which makes the contribution of individual habitats to growth difficult to distinguish (mark–recapture, all otolith studies), and differential growth rates with increasing fish size (mark–recapture when duration of time at liberty of marked fish is not considered).

All five studies from Table 10.2 have measured growth in length, which works well for small fish growing fast lengthwise, but is less suitable for larger fish because at a certain size fishes show a faster increase in biomass than in length (Wootton 1998). This probably makes growth in biomass, or the weight–length ratio, a more appropriate unit of measure for larger fish. Alternative approaches could include various condition indices, body lipid content, tissue RNA:DNA ratios, and glycine uptake rates by scales (Wootton 1998). Furthermore, considering that many fish species make feeding, diel, or tidal movements between habitats (see Chapter 8), measuring growth rates (somatic or from otoliths) for fishes held in enclosures seems to be the best approach, if indeed the enclosures are large enough for the fish to be able to feed normally. Independent of the differences in methodology used among studies, however, there is currently no evidence to support the assumption of higher growth rates of fish or decapods in mangrove/seagrass habitats than in adult (reef) habitats (Table 10.2), although too few studies have been done to accept this as a general conclusion.

10.2.3 Survival

Very few studies have quantified survival of fish and decapods in multiple tropical coastal habitats. Typically, comparative studies among shallow tropical ecosystems have evaluated survival based on predation rates using tethering techniques. However, natural survival depends on more factors than predation alone (e.g., food abundance, competition, interaction between predation and density-dependent survival; Booth and Hixon 1999, Hixon and Jones 2005). Results from tethering studies alone are therefore reviewed in Section 10.3.2.4 which deals with predation risk.

One of the few relevant studies on fish survival found a significantly higher survival over time of juvenile *Cheilinus undulatus* in low branching coral with associated macroalgae compared to seagrass, macroalgal, coral rubble, and coral microhabitats, all of which were situated with a marine lagoon (Tupper 2007). In another relevant study, Risk (1997) found higher survival of juvenile surgeonfish on pavement areas than on sand areas or dead *Acropora palmata* corals in back-reef areas, and attributed this to the presence of abundant shelter holes and food resources in the former. Both studies were done in (micro)habitats located within lagoons, not providing insight into the degree of fish survival at larger spatial scales or between nearshore juvenile and offshore adult habitats.

Probably the most difficult aspect of quantifying survival is accounting for the potential movement of fish or decapods and thus confounding natural mortality with emigration. This is probably why some studies on survival use the term 'persistence' instead of 'survival'. Tupper (2007) addressed this problem by externally tagging fishes in multiple habitats, and found that it was very unlikely that fish had migrated away since tagged individuals showed little movement (i.e., <5 m) from their release sites during the first three months. Early juvenile fish generally show high fidelity to shelter sites in tropical shallow-water habitats (Watson et al. 2002, Verweij and Nagelkerken 2007), making density estimates of small juveniles over time a potential technique to study *in situ* survival. This technique is less reliable or unusable, however, when considering larger-sized animals that show less site fidelity, or when considering animals that show regular feeding, tidal, or diurnal migrations to adjacent habitats (see Chapter 8; Verweij and Nagelkerken 2007, Verweij et al. 2007). In both cases, population size could be affected by mortality in temporary feeding or shelter habitats, and may thus not be representative for the respective habitat studied.

10.2.4 Movement

Movement from juvenile to adult habitats is an important factor underlying the nursery-role concept, because potentially higher production in juvenile habitats such as mangroves and seagrass beds does not transfer to adjacent coral reefs when such linkages are lacking. Tagging studies on fauna that are associated with mangroves or seagrasses during their juvenile life stage have been done as early as the

1960's, and were initially focused on linkages between coastal estuaries and off-shore commercial shrimp stocks (Table 10.3). Fish tagging studies investigating mangrove/seagrass – reef linkages have only been published in the last decade. The majority of the studies used external tags or measurement of stable isotopes in tissue or fish otoliths. Focus has mostly been on penaeid shrimp, and fish species belonging to the families Haemulidae and Lutjanidae. Maximum distances of movement ranged from a few hundred meters to 315 km.

Although the number of studies that shows evidence of nearshore–offshore movement of fauna has been growing, these studies still tell us little about the degree to which offshore populations are replenished by mangrove and seagrass habitats. There are several reasons for this, as listed and discussed below.

First, results from different studies vary considerably. Offshore recaptures of animals that had been artificially tagged in mangroves/seagrass beds are extremely low (range 0.1–6.8%, with all but one study <2%; Table 10.3), and it can be debated to what degree results from tagging studies reflect natural behavior and movement patterns. In contrast, much higher estimates of movement have been obtained on basis of measurement of stable isotopes or microelements in animal tissue or fish otoliths. Such studies suggest contributions ranging from 36–98% of offshore populations that are derived from juveniles living in nearshore mangrove and seagrass areas (Fig. 10.4; Fry et al. 1999, Chittaro et al. 2004, Nakamura et al. 2008, Verweij et al. 2008; M Lara and D Jones, University of Miami, unpubl. data). However, the latter four studies all sampled reefs near mangrove/seagrass areas, while migration distances away from these juvenile habitats can be up to a few hundred kilometers for tropical shrimp as well as fish (Table 10.3; Gillanders et al. 2003). If animals congregate on reefs near mangrove/seagrass areas, then results of studies are a function of distance of sampled reefs to potential nursery areas. Thus, conclusions regarding the contribution of nursery habitats to offshore populations are clearly affected by the study setup and methodology. For more details on advantages and disadvantages of various tagging techniques, see Chapter 13.

Second, none of the studies in Table 10.3 that have found offshore migration of fish and decapods were able to separate the contribution of individual juvenile habitats. Instead, contribution from embayment/estuary vs. offshore was the main variable tested. Therefore, the degree to which, e.g., mangrove vs. seagrass habitats individually contribute to this offshore production remains inconclusive. The question remains whether this is a testable factor in general. Juvenile fish show frequent tidal and diurnal migrations among lagoonal habitats (see Chapter 8), and it will therefore be difficult to separate the contribution of each of these habitats.

Third, there is some ambiguity in what is considered an adult habitat or adult population. For example, gray snapper *Lutjanus griseus* and bluestriped grunt *Haemulon sciurus* use mangroves as juvenile habitats, but inland mangroves in Florida with relatively deep water under their canopy (especially during the dry season) also harbor adult populations (e.g., Faunce and Serafy 2007) even though these species are generally associated with offshore reefs as adults. The stated ambiguity also relates to the variability of the reef habitat. Patch reefs, fringing reefs, and coral patches in back-reef areas are often used as synonyms for 'coral reef' (Nagelkerken et al.

Table 10.3 Overview of tropical studies showing (ontogenetic) movement from juvenile shallow-water areas to adult coral reef or offshore habitats by fishes and decapods. Studies are ordered chronologically. For details on tagging methodology, see Chapter 13. For habitat abbreviations, see Table 10.2. Np = data not provided. N_0 = number of individuals tagged or processed, N_t = number of individuals showing reef-directed or offshore movement. Note that distance moved may depend on study

Reference	Location	Methodology	Taxa	Species	N_0/N_t	Habitat movement	Distance moved (km)
Iversen and Idyll (1960)	Florida, USA	External tags	Shrimp	*Farfantepenaeus duorarum*	1,157/1	Mg estuary to offshore	~96
Costello and Allen (1966)	Florida, USA	Tissue staining	Shrimp	*Farfantepenaeus duorarum*	98,525/350	Mg estuary/Sg to offshore	>278
Lucas (1974)	Australia	External tags	Shrimp	*Penaeus plebejus*	15,143/174	Mg/Sg estuary to offshore	>~25
Somers and Kirkwood (1984)	Australia	External tags	Shrimp	*Penaeus esculentus*	5,011/8	Sg/mud bay to offshore	~70
			Shrimp	*Penaeus semisulcatus*	4,354/46	Sg/mud bay to offshore	<108
Fry et al. (1999)	Florida, USA	Stable isotopes in tissue	Shrimp	*Farfantepenaeus duorarum*	134/~54	Mg estuary/Sg to offshore	<~200
Kanashiro (1998)	Japan	Np	Fish	*Lethrinus nebulosus*	44,092[1]/<154	Sg lagoon to CR	<7
Sumpton et al. (2003)	Australia	External tags	Fish	*Pagrus auratus*	2,700/4	Mg/Sg estuary to offshore	<290
Chittaro et al. (2004)	Belize	Otolith micro-elements	Fish	*Haemulon flavolineatum*	39/14	Mg to CR in back-reef[1]	0.27
Russell and McDougall (2005)	Australia	External tags	Fish	*Lutjanus argentimaculatus*	22,202/35	Mg/Sg estuaries/rivers to offshore CR	<315
Tupper (2007)	Palau	Subcutaneous elastomers	Fish	*Cheilinus undulatus*	250/20	BCM/BM to PR in lagoon[2]	0.054–0.112
Verweij and Nagelkerken (2007), C Huijbers and I Nagelkerken (unpubl. data)	Curaçao		Fish	*Plectropomus areolatus*	73/9	CR to deep PR in lagoon[2]	0.302–0.351
		Internal tags	Fish	*Haemulon flavolineatum*	1,114/3	Rocky bay shoreline to CR	2
Verweij et al. (2007)	Curaçao	External tags	Fish	*Lutjanus apodus*	59/4	Rocky bay shoreline to CR	<0.22
Nakamura et al. (2008)	Japan	Stable isotopes in tissue	Fish	*Lutjanus fulvus*	41/36	Mg to CR	~1

Table 10.3 (continued)

Reference	Location	Methodology	Taxa	Species	N_0/N_t	Habitat movement	Distance moved (km)
Verweij et al. (2008)	Curaçao	Stable isotopes in otoliths	Fish	*Ocyurus chrysurus*	51/50	Mg/Sg embayment to CR	<1.5
Luo et al. (2009)	Florida, USA	Telemetry	Fish	*Lutjanus griseus*	14/7	Mg/Sg embayment to CR	5–15
M Lara and D Jones, University of Miami (pers. comm.)	Florida, USA	Otolith micro-elements	Fish	*Lutjanus griseus*	35/33	Mg/Sg estuary to CR	10–65
C Layman, Florida International University (pers. comm.; http://www.adoptafish.net)	Bahamas	Telemetry	Fish	*Lutjanus apodus*	11/4	All species: Mg lagoon to nearshore rocky outcroppings and PR	>2
			Fish	*Lutjanus cyanopterus*	25/6		>2
			Fish	*Lutjanus griseus*	33/7		>2
			Fish	*Sphyraena barracuda*	8/3		>2

[1] aquarium-reared fish released at on average 10.4 cm fork length
[2] movement to subadult habitat

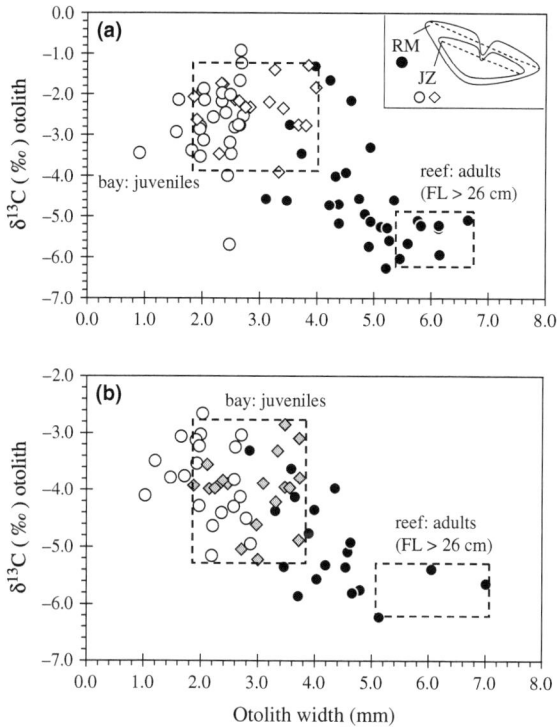

Fig. 10.4 Results of stable carbon isotope ($\delta^{13}C$) analysis in otoliths of juvenile (collected from a mangrove/seagrass embayment) and adult (collected from the adjacent coral reef) *Ocyurus chrysurus*, showing the degree of adult fish on the reef that have lived in seagrass beds during earlier life stages for (**a**) Spanish Water Bay, and (**b**) Piscadera Bay in Curaçao. Inset shows location where carbon samples were taken on cross sections of adult reef fish and juvenile bay fish otolith. Dashed lines represent otolith width (x-axis). JZ = juvenile zone (diamonds = otolith margin of bay juveniles, white circles = juvenile zone of reef fishes), RM = otolith margin of reef fish (black circles), FL = fork length. Figure reprinted from Verweij et al. (2008), p. 1544. Copyright (2008) by the American Society of Limnology and Oceanography, Inc.

2008). Chittaro et al. (2004), for example, calculate the degree to which mangroves contribute to fish populations on 'coral reef habitats'. However, these habitats were located in shallow waters (2 m depth) in the back-reef area while adult grunts are typically found in deeper waters on fore-reefs. Another example is that of Tupper (2007) who showed movement of juvenile fish to deeper patch reefs. These reefs were located within the lagoon where the juvenile habitats were also situated; the ultimate adult habitat for these species is located on the deeper offshore reefs outside the lagoon. To complicate the matter, adult habitat utilization may vary according to the geomorphology of the shelf (see Appeldoorn et al. 1997). If the shelf is narrow, fish are likely to show movement to the adjacent fore-reef. If the shelf is large, however, fish may either take permanent residence on nearby patch reefs on the shelf, or may show a gradual movement towards the shelf break (e.g., Kanashiro 1998,

Bouwmeester 2005) in which case the dispersal distance is related to the degree of isolation of the reefs (Appeldoorn et al. 1997). Bouwmeester (2005), for example, showed that of the 27-recaptured tagged sub-adult *Haemulon flavolineatum*, 12 had moved up to 300 m towards the adjacent reef front within a period at liberty of 3–16 months.

Fourth, when can movement be considered as permanent? Recent studies have suggested that fish may show repetitive movements between juvenile and adjacent adult habitats. For example, Layman (unpubl. data; Table 10.3) observed that ultrasonically tagged fishes moved >2 km from the upper reaches of a mangrove-dominated tidal creek system out of the mouth of the creek and into the marine environment. Residence time in the creek (after the tagging event) ranged from ten days to over a year and some fish moved to the marine environment and then >2 km back into the creek system. Also Verweij et al. (2007) showed regular movement of *Lutjanus apodus* between shallow-water bay habitats and shallow reef habitats, up to a few hundreds meters away from the mouth of the embayment (Fig. 10.5).

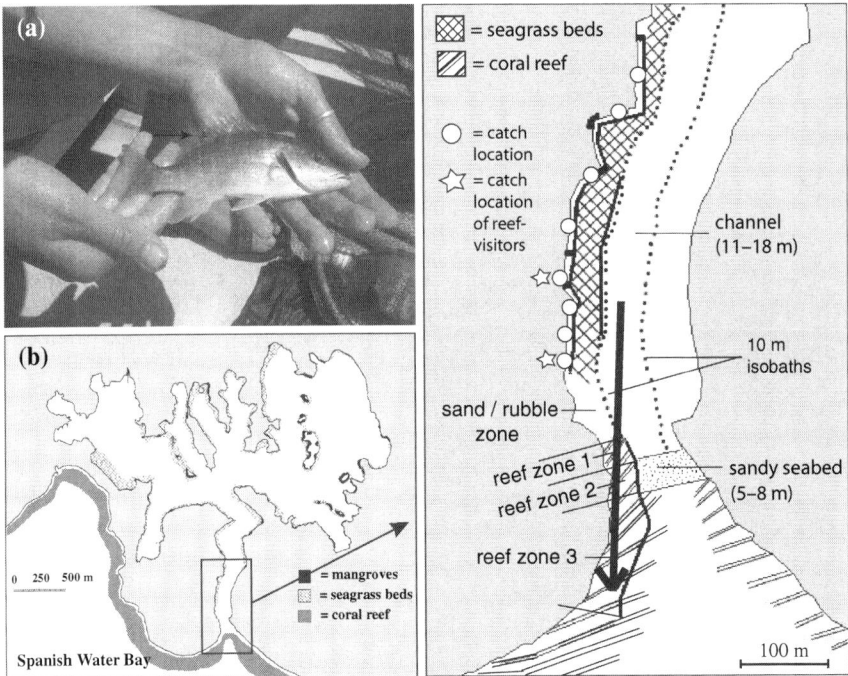

Fig. 10.5 (**a**) *Lutjanus apodus* tagged externally (see arrow) with monofilament line and beads (photo: S Wartenbergh) to study (**b**) movement (indicated by thick arrow) between juvenile back-reef habitats in Spanish Water Bay and the adjacent coral reef in Curaçao. Catch locations of *L. apodus* within embayment are shown for all fish and those that were resighted on the coral reef. Thin lines along coastline (on reef and in seagrass) indicate transect lines searched for tagged fish. Figure modified and reproduced from Verweij et al. (2007), with kind permission from Bulletin of Marine Science

It is unclear whether such movements are a strategy to explore the adult habitat, or if they are perhaps related to spawning migrations on the reef, after which the fish return to their shallow-water shelter sites. If such a mechanism of non-unidirectional movements is more of a rule rather than an occasional event, it will make quantification of juvenile habitat contribution to adult populations more difficult. Two studies have potentially overcome this problem by analyzing stable isotope analysis of fish tissue, their stomach contents, and prey items collected from coral reefs (adult fish) and from mangroves/seagrass areas (juvenile fish) (Nakamura et al. 2008, Verweij et al. 2008). Fish collected from coral reefs that had a mangrove/seagrass stable isotope signature could either live/feed permanently on the reef but have grown up in the mangrove/seagrass areas and recently moved to the reef, or could still live in these nursery areas but temporarily visit the adjacent coral reef (during the time at which they were caught). Stable isotope signatures of stomach contents of the reef fish matched those of prey items collected on the reef, however (Nakamura et al. 2008, Verweij et al. 2008). As these signatures were significantly different from those in mangrove/seagrass areas, it was clear that the collected reef fish were reef residents that fed on the coral reef.

It is clear that several problems still exist in performing movement studies and in interpreting those that have already been published, to support the nursery-role hypothesis. A variety of tagging techniques has been used so far, with different degrees of success (Table 10.3; see Chapter 13). Nevertheless, there is sufficient evidence present that shows the existence of ecological linkages between juvenile mangrove/seagrass habitats and adult offshore areas by fish as well as shrimp, suggesting at least a partial contribution of these shallow-water habitats to adult stocks.

10.3 Underlying Factors Determining Nursery-role Potential

10.3.1 Food Availability Hypothesis

Only a few studies exist that have compared benthic or planktonic food abundance in mangrove, seagrass, as well as coral reef habitats. For most taxa, prey densities or biomass were typically higher in seagrass beds than on coral reefs (Table 10.4). An interesting pattern was observed in a study by Nakamura and Sano (2005): total density of epi- and infauna was higher on seagrass beds than on coral reefs, but the opposite was true for biomass. The higher biomass on the reef was mainly caused by the higher abundance of large-sized benthic invertebrates such as crabs, mollusks, and shrimp, whereas the higher abundance on the seagrass beds mainly consisted of smaller macrofauna such as Annelida, Copepoda, and Tanaidacea.

The usefulness of food density estimations to support nursery-role factors is debatable, however. Several reasons can be mentioned for this. First, interactions exist between presence of predators and food availability. For example, when predators were present, small fish selected suboptimal feeding habitats with lower predation risk which led to lower growth rates than when predators were absent (Werner

Table 10.4 Overview of comparative studies (ordered chronologically) on abundance of potential food items for fish in different tropical coastal habitats. The focus was on studies that included at least the coral reef habitat and mangrove or seagrass habitat. Data are shown at taxonomic levels higher than that of family. Np = data not provided. Habitat type: Ch = tidal channels, CR = coral reef, MF = subtidal mud flats with low cover of macroalgae, Mg = mangrove, S = unvegetated sand areas, Sg = seagrass bed. Unit = unit of measurement: N = numbers, B = biomass. Food taxa: Amp = Amphipoda, Ann = Annelida, Cop = Copepoda, Cum = Cumacea, Iso = Isopoda, Mol = Mollusca, Tan = Tanaidacea

Reference	Location	Collection time (hr)	Habitat types compared	Methodology	Unit	Food taxa (showing habitat differences)	Highest invertebrate abundance
Kitheka et al. (1996)	Kenya	Np	CR, Mg, Sg	Plankton net hauls	B	All invertebrates	Mg/Sg > CR[1,2]
					N	All invertebrates	CR/Sg > Mg[2]
Cocheret de la Morinière (2002)	Curaçao	Plankton: night time	All taxa: CR, Sg, Mg	Plankton net hauls	N	All invertebrates	Sg > CR/Mg
				Plankton net hauls	N	All crustaceans	Sg > CR/Mg
				Plankton net hauls	N	Tanaidacea	Sg > CR/Mg
		Sediment: daytime	All taxa: CR, Sg, Mg	Sediment cores	N	All invertebrates	Sg > CR/Mg
				Sediment cores	N	All crustaceans	Sg > CR/Mg
				Sediment cores	N	Tanaidacea	Sg > CR/Mg
Nagelkerken (2000)	Curaçao	17:00–19:00	CR, MF; Sg	Plankton net hauls	N	All invertebrates	Sg > MF > CR
			CR, MF; Sg	Plankton net hauls	N	All crustaceans	Sg > MF > CR
			Ch, CR, MF, Mg, Sg	Sediment cores	N	All invertebrates	Ch/MF/Mg/Sg > CR
			Ch, CR, MF, Mg, Sg	Sediment cores	N	All crustaceans	Ch/MF/Mg/Sg > CR
Nakamura and Sano (2005)	Japan	10:00–16:00	All taxa: S, Sg, CR	Hand net	B	All invertebrates	CR > Sg > S[3]
				Hand net	N	All invertebrates	Sg > CR > S[3]
				Hand net	N	Ann, Cop, Cum, Iso, Mol, Tan	Sg > CR > S[3]
					N	Crabs[3], Shrimps	CR > Sg/S[3]
				Sediment cores	B	All invertebrates	CR > Sg[3] > S
				Sediment cores	N	All invertebrates	Sg > S/CR[3]
				Sediment cores	N	Amp, Ann	Sg > CR
				Sediment cores	N	Mol	CR > S/Sg
Grol et al. (2008)	Aruba	Daytime	CR, Mg, Sg	Plankton net hauls	N	Cop	CR > Sg/Mg
	Curaçao	Daytime	CR, Mg, Sg	Plankton net hauls	N	Cop	CR > Sg/Mg

[1] only during wet season
[2] not statistically tested
[3] not significantly different

et al. 1983). This left more resources available in optimal, but predator-rich, food habitats for larger size classes of fish that were less vulnerable to predation, and led to higher growth rates than when predators were absent (Werner et al. 1983). In such way, predators alone can determine growth rates of fish to the point that growth is not directly related to food availability. Second, if food is not limiting in a habitat, then more food does not translate to higher growth rates (Sale 1974, Barrett 1999). Third, it is production of prey that is important, and not standing stock of prey. A favored prey item may show a high production rate, but low standing stock due to continuous removal by predators. Fourth, the number of prey available to individual predators should be considered. In this respect, competition among predators for the same prey species is important, and the relative abundance of competitors and prey determines potential prey availability.

Dealing with all of the above issues to determine which topical coastal habitats are optimal for feeding, is probably nearly impossible in most field studies. On top of that, the specific diet of predators needs to be known. Diet is not constant and varies through time and space; fish often show ontogenetic changes in diet (e.g., Eggleston et al. 1998, Cocheret de la Morinière et al. 2003b), and flexibility in food consumption depending on what food sources are available (e.g., Jennings et al. 1997).

Even if productivity of individual prey items, diet of individual predators, and the relative abundance of competitors and prey are known, it will be difficult to establish to degree to which predators are successful in targeting the available prey items. Prey species seek cover and show evasive behavior when attacked (Meager et al. 2005, Scharf et al. 2006). In addition, some prey species may be sheltering in the substratum during daytime when many diurnally active fish species are foraging. It is known that many invertebrate prey species emerge at night (Jacoby and Greenwood 1989, Laprise and Blaber 1992, Ríos-Jara 2005). In addition, size or biomass of prey items are important to consider. Clearly, a large shrimp provides much more food for a fish than a small shrimp. Foraging theory suggests that in order to minimize the energy cost-benefit ratio, predators should select fewer large preys rather than many smaller preys (Schoener 1971, Stephens and Krebs 1986). There are also issues concerning the size of prey that a particular size class of predator is able to handle (Kruitwagen et al. 2007). Additionally, nutritional value may vary among food items (Wilson and Bellwood 1997, Graham 2007). Finally, production of prey items varies, for example, with season (e.g., Day et al. 1989, Edgar and Shaw 1995).

Even though it is difficult to calculate and compare food production and availability among tropical habitats, several studies in the Caribbean have shown that various species of nocturnal zoobenthivorous fishes show daily feeding movements from daytime shelter sites (e.g., mangroves and patch reefs) to nocturnal feeding habitats (typically seagrass beds) (Ogden and Ehrlich 1977, Burke 1995, Nagelkerken et al. 2000a, Nagelkerken and van der Velde 2004, Verweij et al. 2006b, Nagelkerken et al. 2008). This could indicate that production of prey items is higher in seagrass beds, possibly due to the structure provided to prey items by the seagrass blades, as stated earlier. An alternative explanation, however, could be that seagrass beds are preferred feeding habitats because of their more extensive surface areas than other shallow-water habitats, in such way potentially reducing competition for food (Nagelkerken et al. 2000a).

10.3.2 Predation Risk Hypothesis

10.3.2.1 Predator Abundance

Very few comparative studies have been done on predator assemblages in multiple tropical coastal habitats, and they have mainly focused on fish as predators. In the current review, only studies focusing on complete fish predator assemblages were included because different predator species or types (e.g., ambush vs. actively swimming predators; resident vs. transient predators) may have very different effects on juvenile fish communities. All studies listed in Table 10.5 suggest that under most circumstances, in the western Atlantic as well as in the Indo-Pacific, offshore habitats harbor higher abundances or species richness of fish predators than various shallow-water habitats such as coral rubble, sandy areas, seagrass beds, and mangrove channels or creeks. A clear exception is formed by permanently inundated Caribbean mangrove prop-root habitats that can harbor comparable or higher daytime densities of predators than the coral reef (Table 10.5).

One problem with interpreting data on piscivore abundances is related to the question of what constitutes a piscivore, as exemplified by the following three points. First, studies typically focus on strict piscivores and not on species for which fish form a minor component of the diet. Second, various lagoonal/estuarine fish species show ontogenetic shifts in diet from zoobenthivory to piscivory at a size when they are still considered juveniles that reside in back-reef habitats (Fig. 10.6). Third, juvenile densities of large-body predator species are often excluded in quantitative studies. Such juvenile piscivores are much more numerous than their larger conspecifics, and they are more likely to access shallow-water habitats. Therefore, they can exert significant effects on juvenile fish abundances (see review by Sheaves 2001). In fact, various species that utilize shallow-water habitats during their juvenile stage predate on other juvenile fishes in the same juvenile habitats (Dorenbosch et al. 2009). This probably explains why some Caribbean mangroves harbor such high densities of piscivores: larger specimens of various fish species occur in higher densities in the mangroves than in other shallow-water habitats, at a size at which they have switched to from zoobenthivory to piscivory (Nagelkerken et al. 2000b, Cocheret de la Morinière et al. 2003a, b, Nagelkerken and van der Velde 2003).

Spatial and habitat variation clearly cause an additional source of variation with respect to predator densities. Chapter 11 deals in more detail with these kinds of variations for juvenile fish, but the same applies to (juvenile) piscivores. An important observation with respect to predation risk is that piscivore densities seem to decrease with distance away from reefs (Valentine et al. 2007, Vanderklift et al. 2007), potentially causing juvenile populations in areas located further into lagoons/estuaries to experience less predation. In addition, predator densities may decrease from the mangrove fringe towards the shoreline due to the lower

Table 10.5 Overview of comparative studies on predatory fish assemblages across different tropical coastal habitats. The focus is only on studies investigating complete predator assemblages. Np = data not provided. Habitat type: CR = coral reef, CRb = coral rubble, Mg = mangrove, N = undercut notches in fossil reef terrace and associated boulders. OW = open shoreline waters, Sg = seagrass

Reference	Location	Time of survey (hr)	Habitat types compared	Methodology	Unit of measurement	Highest predator density
Blaber (1980)	Australia	Np	Mg creeks, open bay	Fry, gill, seine and stake nets	Relative species richness	Open bay > Mg creeks
Blaber et al. (1985)	Australia	Np	Deep Mg channel, intertidal Mg creek, deep + intertidal OW	Gill nets, small/large beach seines, fry seine nets	Relative abundance Relative species richness Relative biomass Relative abundance Relative species richness	OW > Mg channel[1] OW > Mg channel[1] OW > Mg channel[1] intertidal OW < or > [2] intertidal Mg creek[1] intertidal OW < or > [2] intertidal Mg creek[1]
Nakamura and Sano (2004a)	Japan	10:00–16:00	CR, Sg	Visual census	Density Species richness	CR > Sg CR > Sg
Wilson et al. (2005)	Turks and Caicos	Daytime	CR, Mg, S, Sg	Visual census	Density	CR/Mg > S/Sg
Dorenbosch et al. (2009)	Curaçao	Daytime Daytime Night time Night time	CR, Mg, N, Sg	Visual census	Density Species richness Density Species richness	Mg > CR/N > Sg Mg > N > CR > Sg CR/N > Mg/Sg CR/Mg/N > Sg
M Grol and I Nagelkerken (unpubl. data)	Aruba	Daytime	CR, CRb, Mg, Sg	Visual census	Density Species richness	Mg > CR > CRb/Sg CR > CRb/Mg > Sg

[1] not statistically tested
[2] depending on month of year

Fig. 10.6 Nursery species or predator? (**a**) Juvenile *Lutjanus apodus* mainly as a zoobenthivore in mangroves, and (**b**) as a subadult predator/piscivore on the coral reef

accessibility caused by shallower water and a denser prop-root system (Vance et al. 1996).

Another source of variation is temporal variation. Piscivore densities in mangroves are often higher during daytime than at night (Table 10.5), because piscivores that utilize mangroves as shelter habitats during daytime typically leave at night to forage in adjacent seagrass and mud flat habitats (Rooker and Dennis 1991, Nagelkerken et al. 2000a, Dorenbosch et al. 2009). In addition, densities of roving piscivores may be underestimated because they are transient (e.g., from adjacent coral reefs or deeper waters; Blaber et al. 1992). Even though they are often only temporarily present in estuarine/lagoonal habitats, they can cause high mortality rates among early juveniles in a relatively short time period (Carr and Hixon 1995, Hixon and Carr 1997, I Nagelkerken pers. observ.).

Several recent studies have shown high abundances of predators that forage on fish and shrimp in some estuarine habitats (e.g., Salini et al. 1990, Brewer et al. 1995, Baker and Sheaves 2005, Kulbicki et al. 2005, Baker and Sheaves 2006) and it has been argued that the protective value of these areas to juvenile fishes may be smaller than previously assumed (Sheaves 2001). Nevertheless, when comparing specific shallow-water juvenile habitats to offshore adult habitats, the studies in Table 10.5 show significantly lower piscivore densities in the former than in the latter, supporting a refuge role for fish of various estuarine/lagoonal habitats; mangroves sometimes appear to form an important exception to this pattern, though.

10.3.2.2 Turbidity Hypothesis

The turbidity hypothesis provides another potential explanation for lower predation risk in estuarine or lagoonal habitats: due to the typically higher turbidity in these shallow-water habitats, predation pressure is significantly lowered (Blaber and Blaber 1980). This hypothesis has been heavily debated, however. It has been tested using a variety of field and aquarium studies, but with different outcomes (Blaber and Blaber 1980, Cyrus and Blaber 1987, Benfield and Minello 1996, Macia et al. 2003, Meager et al. 2005, Johnston et al. 2007). The interpretation of these results are heavily affected by, for example, (1) the difference between field studies describing fish distribution in areas with different turbidities vs. aquarium studies testing predation risk at various turbidity levels, (2) the differential effects of light intensity vs. light scattering on predation risk, (3) the different behavioral reactions of prey and predators in situations with different turbidity and habitat types, (4) differences in absolute turbidity levels studied, and (5) differences in species, size classes, and habitats tested.

10.3.2.3 Structural Heterogeneity Hypothesis

The ultimate reason why structure is attractive to fauna is related to provision of food and/or reduction of predation. Increased structural complexity can increase densities of potential prey items (Heck and Orth 1980, Orth et al. 1984, Verweij et al. 2006a), but little evidence is present of how the structure of different shallow-water habitats provides for differences in food availability. Therefore, the structural heterogeneity hypothesis is discussed here only in relation to the factor predation risk. There is a wide body of literature available that shows that structure is effective in reducing predation risk, and which specific elements of habitat structure are preferred by fish and decapods (e.g., see references in Hixon and Beets 1993, and reviews by Orth et al. 1984, Horinouchi 2007). Most studies have been done for single habitats, testing variables such as size and number of shelter holes, habitat rigidity, vegetation cover, substratum type, density of mangrove roots or seagrass shoots, structure formed by epibionts, and degree of shade (e.g., Bell and Westoby 1986, Laprise and Blaber 1992, Hixon and Beets 1993, Cocheret de la Morinière et al. 2004, Nakamura et al. 2007, MacDonald et al. 2008). Studies have shown that fish often prefer dense and/or dark structure (e.g., Cocheret de la Morinière et al. 2004), but the attractiveness of the structure partially depends on fish species, size, life-stage, behavior, coloration, etc. For shrimp, it has been shown that they prefer soft bottom over hard bottom, in which they can burrow to escape predation (e.g., Laprise and Blaber 1992). Less common are studies comparing preference of fish or decapods for structure of different habitat types. An exception is formed by a large number of studies having investigated preference of fish or decapods for vegetated (mostly seagrass) vs. unvegetated (mostly sand or mud) habitats (e.g., review by Heck et al. 2003). Table 10.6 lists some of the few studies that have compared preference by fish or decapods for structure of multiple vegetated shallow-water microhabitat types.

Table 10.6 Overview of studies on visual preference of fishes and decapods for various types of habitat structure. Habitat structure: AC = algal clumps, C = corals, CRb = coral rubble, Mg = mangrove roots, S = sand, Sg = seagrass shoots. TL = total length

Reference	Location	Habitat structures compared	Factors excluded	Type of experiment	Duration of experiment	Taxa	Species	TL (cm)	Habitat preferences
Herrnkind and Butler (1986)	Florida, USA	AC, Sg	Predators/food	Aquarium	24 hrs	Lobster	*Panulirus argus*	Pueruli	AC > Sg
								Juveniles	AC > Sg
Verweij et al. (2006a)	Curaçao	All taxa: Mg[1], Mg+Sg[2], S, Sg[1]	None	*In situ* open habitats	3 days	Fish	*Acanthurus bahianus*, *A. chirurgus*, and *Scarus guacamaia*	Np	Mg+Sg > Mg > S/Sg
						Fish	*Haemulon flavolineatum*, *Lutjanus apodus*[4], *L. mahogoni*[5], and *Ocyurus chrysurus*[4]	Np	Mg+Sg/Sg > Mg/S
						Fish	*Sphyraena barracuda*	Np	Mg+Sg/Mg/Sg > S
						Fish	*Eucinostomus* spp.	Np	No preference
						Fish	*Acanthurus bahianus*, *A. chirurgus*, *Eucinostomus* spp., and *Scarus guacamaia*	Np	No preference
		All taxa: Mg[1], Mg+Sg[2], S, Sg[1]	Food	*In situ* open habitats	3 days	Fish	*Haemulon flavolineatum*, *Lutjanus apodus*[4], *L. mahogoni*[5], and *Ocyurus chrysurus*[4]	Np	Mg+Sg/Sg > Mg/S
						Fish	*Sphyraena barracuda*	Np	Mg+Sg/Mg/Sg > S

Table 10.6 (continued)

Reference	Location	Habitat structures compared	Factors excluded	Type of experiment	Duration of experiment	Taxa	Species	TL (cm)	Habitat preferences
Nakamura et al. (2007)	Japan	All taxa: C, S, Sg	None	In situ open habitats	1 week	Fish	Amblyglyphidodon curacao	0.9–1.7	C > S/Sg
						Fish	Cheiloprion labiatus	1.2–1.7	C > S/Sg
						Fish	Dischistodus prosopotaenia	1.2–2.0	C > S/Sg
M Grol and I Nagelkerken (unpubl. data)	Aruba	C, CRb, Mg[1] Sg	Predators	In situ enclosures	2 days	Fish	Haemulon flavolineatum 3.5		C/Sg > CRb>Mg
M Grol and I Nagelkerken (unpubl. data)	Curaçao	C, CRb, Mg, Sg	Predators	In situ enclosures	2 days	Fish	Haemulon flavolineatum 3.5–4.0		Sg > C/CRb > Mg
C Huijbers and I Nagelkerken (unpubl. data)	Curaçao	C, CRb, Mg, Sg	Predators/food/ habitat smell	In situ enclosures	15 min, daytime	Fish	Haemulon flavolineatum 2–3.5 4–15		No preference Mg > C/CRb/Sg
M Igulu and I Nagelkerken (unpubl. data)	Tanzania	C[1], Mg[3], Sg	Predators/food/ habitat smell	Aquarium	15 min, daytime	Fish	Lutjanus fulviflamma	0–5 5–10	C/Sg > Mg C/Sg > Mg

[1] artificial habitat mimics used
[2] habitats combined
[3] standing mangrove pneumatophores instead of hanging mangrove prop-roots
[4] under shaded conditions: Mg+Sg/Mg/Sg > S
[5] under shaded conditions: no habitat preference

Two of the four studies that have included multiple reef and non-reef habitat structures in choice experiments, showed lowest preference for mangroves (Table 10.6). However, when the structure was not accessible to the fish (i.e., separated by glass compartments), fish of 4 – 15 cm in length preferentially chose for the dark mangrove microhabitat. Coral and/or seagrass microhabitats were most often preferred (Table 10.6), though. This result remained the same whether or not fish species were studied that are associated with mangroves/seagrass beds during their juvenile life stage (Nakamura et al. 2007).

Results from a study that only compared preference by fish for the lagoonal habitats mangrove, seagrass, or sand structure, revealed that habitat preference by fish varied with trophic group and activity pattern (Table 10.6; Verweij et al. 2006a). This study used experimental units with various combinations of artificial mangrove roots and seagrass leaves. Diurnal herbivores showed increasing preference for units with increased structure (mangrove and seagrass combined > only mangrove > only seagrass or sand) because they offered a larger surface area for algal grazing (Figs. 10.7a, c); when the structure was cleaned of epibionts, herbivores no longer showed any preference for a specific habitat structure (Verweij et al. 2006a). In contrast, nocturnally active zoobenthivores only sheltered and did not feed in the experimental units during daytime and thus preferred structure above no structure; however, the specific habitat structure preferred showed a strong interaction with absence/presence of shade (Verweij et al. 2006a; Table 10.6, Fig. 10.7d). For decapods, few habitat choice experiments have been done. Herrnkind and Butler (1986) showed that spiny lobsters of two different life stages preferred algal clumps above seagrasses (Table 10.6), a pattern that was also reflected in the density distribution of lobsters in the field.

It remains largely unclear which specific elements of the different microhabitat types explain habitat preferences by fish and decapods, as most of the above studies did not measure the structural heterogeneity of the habitats offered. This is quite difficult as the various habitat types differ so much in their shape, rigidity, color, overgrowth by epibionts, etc. Nakamura et al. (2007) attempted to overcome this problem by manipulating various aspects of the coral and seagrass structure. They showed preference in species of Pomacentridae for rigid over flexible structure,

---

Fig. 10.7 (continued) (**a**) Schematic drawing of an experimental unit with artificial mangrove roots, artificial seagrass leaves, and shade, to study habitat preference by juvenile fishes. Food was either absent or present as fouling algae and associated macrofauna on roots and leaves, and with or without access to zoobenthos in the sandy substratum. (**b**) Juvenile *Sphyraena barracuda* (arrow) hiding in an experimental unit with mangrove root mimics cleaned off epibionts and with shade (photo: A De Schrijver). Results of habitat preference by fishes are shown for (**c**) diurnal herbivores, and (**d**) nocturnal zoobenthivores. Empty = empty unit, AS = artificial seagrass leaves only, AM = artificial mangrove roots only, AS + AM = artificial seagrass leaves and artificial mangrove roots, N = total number of fish observed in all units. Figure reproduced from Verweij et al. (2006), with kind permission from Inter-Research Science Centre

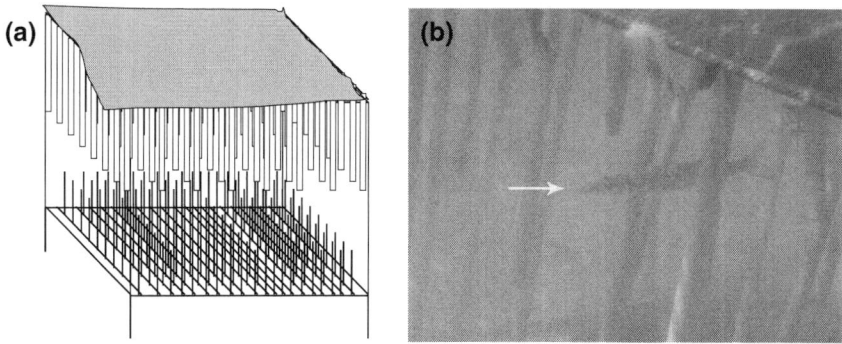

(c) Herbivores (diurnally active)

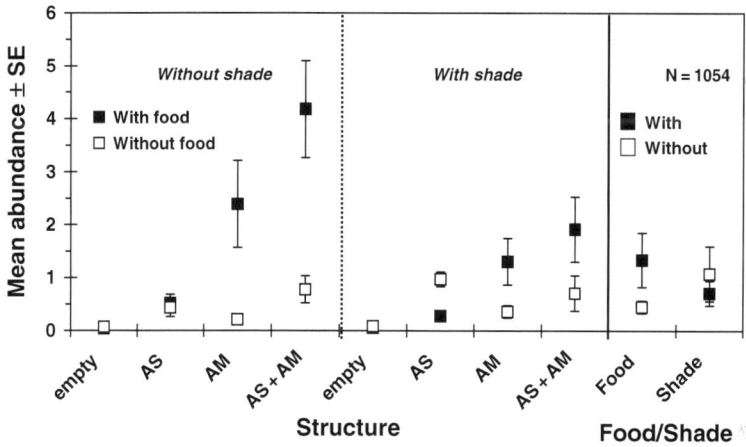

(d) Zoobenthivores (nocturnally active)

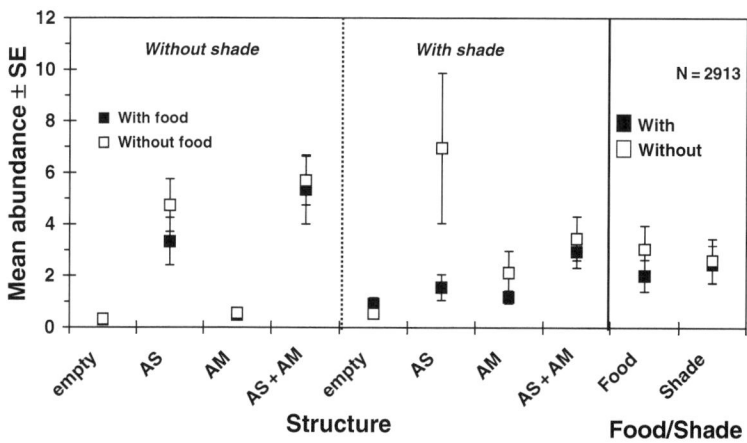

Fig. 10.7   (continued)

independent of shape (coral vs. seagrass mimics), suggesting that rigidity of the habitat structure can play a role in its attractiveness.

Clearly, preference for different habitat structure also varies with body size (Fig. 10.7b; Eggleston and Lipcius 1992, Hixon and Beets 1993); larger fish may outgrow the shelter provided by specific habitat types, large fish are less vulnerable to predation (e.g., Laegdsgaard and Johnson 2001), and dense aquatic vegetation may interfere with their feeding (e.g., Spitzer et al. 2000). Furthermore, preference for degree of habitat heterogeneity and shade varies with species (Cocheret de la Morinière et al. 2004). Finally, some species may be better shaped or colored/camouflaged for one habitat above the other (e.g., Laprise and Blaber 1992).

10.3.2.4 Predation Risk

Predation risk is the resultant of factors such as predator abundance, turbidity, and availability and type of structure. A review of studies on predation risk in multiple habitats (Table 10.7) shows that six out of eight studies found significantly higher predation on fish and decapods on coral reefs, or in coral microhabitats, than in vegetated habitats such as mangroves and seagrass beds. The higher predation was independent of whether the coral habitats were located on the shelf (fore-reefs) or in back-reef areas (patch reefs). Two additional studies show that this effect was only present for the smaller size classes of fish and decapods tested (Table 10.7). Several studies have shown that even at distances as small as 2–20 m away from the edges of patch reefs, predation risk is greatly reduced on seagrass beds compared to that on the patch reefs themselves (Shulman 1985, Sweatman and Robertson 1994, Valentine et al. 2008). A higher predation risk on the coral reef could be expected, as predator density and species richness are also higher there (see Section 10.3.2.1). Whether the higher predation risk on the reef is also influenced by a lower availability of shelter or presence of less suitable shelter on the reef remains unclear.

When only habitats within estuaries/lagoons are considered, vegetated habitats mostly provide more protection from predation than unvegetated habitats (Laprise and Blaber 1992, Heck et al. 2003, Horinouchi 2007; Table 10.7). Within vegetated habitats differences also occur. For example, Dorenbosch et al. (2009) found higher predation rates on small juvenile fish in some mangroves than in seagrass beds, and accounted this to high densities of large nursery species seeking refuge in mangroves. For lobsters, on the other hand, Acosta and Butler (1997) showed higher predation rates in the seagrass beds than in the mangroves.

A detailed comparison of predation risk across studies is difficult to make as predation risk clearly depends on factors such as species and size class considered, methodology used, geomorphology of the habitat, and distance to the coral reef. Predation risk is related to the behavior of prey as well as predator (Main 1987, Primavera 1997), and varies with prey body size relative to shelter size (Eggleston and Lipcius 1992, Bartholomew 2002). Predation risk may also vary according to time of day. Various studies have observed higher predation rates during dusk, dawn, or nighttime compared to daytime (McFarland 1991, Laprise and Blaber 1992,

Table 10.7 Overview of comparative experimental studies on predation risk of juvenile fish and decapods in different tropical coastal habitats. Studies comparing only a single vegetated habitat vs. a sand habitat were excluded. Habitat type: AB = algal beds, AC = algal clumps, BM = bushy macroalgae, C = coral, CR = coral reef, CRb = coral rubble, PR = patch reef, MF = subtidal mud flats, Mg = mangrove, S = unvegetated sand areas, Sg = seagrass, Sp = sponges. Np = data not provided. TL = total length

Reference	Location	Time of survey (hr)	Habitat types compared	Methodology	Taxa	Species	TL (cm)	Highest predation rate
Herrnkind and Butler (1986)	Florida, USA	Day, and day vs. night	AB, AC, S, Sg	Tethering	Lobster	*Panulirus argus*	0.7–1.1[2]	S > AC/Sg > AB
Acosta and Butler (1997)	Belize	Dusk to dusk	Mg, PR, Sg	Tethering	Lobster	*Panulirus argus*	0.8–1.5[2] 2.0–4.5[2]	PR/Sg > Mg Sg > Mg/PR
Acosta and Butler (1999)	Florida, USA	Day	AB, CR, Sg	Tethering	Lobster	*Panulirus argus*	Np; postlarvae	CR > AB/Sg
Shulman (1985)	St. Croix (US Virgin Islands)	Day	PR, S in Sg bed	Tethering	Fish	*Haemulidae*	3.2[3]	PR > S in Sg bed
Sweatman and Robertson (1994)	Panama	09:05–10:40, and 15:20–15:40	PR, Sg	Glass bottle enclosures	Fish	*Acanthurus bahianus* and *A. chirurgus* mixed	2.6–3.8[3]	PR > Sg
Dahlgren and Eggleston (2000)	Bahamas	10:00–19:15	BM, postalgal habitat harboring C/CRb/Sg/Sp	Tethering	Fish	*Epinephelus striatus*	3.5–4.0 5.5–5.5 7.0–7.5	Postalgal > BM Postalgal = BM Postalgal = BM

Table 10.7 (continued)

Reference	Location	Time of survey (hr)	Habitat types compared	Methodology	Taxa	Species	TL (cm)	Highest predation rate
Laegdsgaard and Johnson (2001)	Australia	Np	MF, Mg, Sg	Tethering	Fish	*Sillago* spp.	Np	MF > Mg/Sg
Nakamura and Sano (2004a)	Japan	10:00–16:00	CR, Sg	Tethering	Fish	*Apogon ishigakiensis*	2.9	Sg = CR
					Fish	*Stethojulis strigiventer*	3.0	CR > Sg
Chittaro et al. (2005)	Belize	09:00–11:00, and 14:00–16:30	CR, Mg-Sg[1]	Tethering	Fish	*Haemulon chrysargyreum*	3–6	CR > Mg-Sg[1]
Dorenbosch et al. (2009)	Curaçao	09:00–13:00	CR, Mg, Sg	Tethering	Fish	*Haemulon flavolineatum*	3.1–4.5	[4]CR > Mg > Sg

[1] combined habitat
[2] carapace length
[3] standard length
[4] depending on spatial setting along ocean–bay gradient

Danilowicz and Sale 1999), while Chittaro et al. (2005) found higher predation rates in the afternoon than in the morning at one of two coral reef sites and in the mangroves/seagrass beds. As most studies were done during daytime (Table 10.7), little is known of relative predation risk among habitats during dusk or at night. For lobster, Herrnkind and Butler (1986) found the same pattern for predation rate among habitats at night as during the day, however.

The selected methodology may also affect the results. Tethering is a commonly used technique in predation studies. It has several disadvantages such as unnatural behavior of tethered prey (Zimmer-Faust et al. 1994, Curran and Able 1998), and it may potentially make prey vulnerable to predators that are not normally able to capture them (Haywood et al. 2003). Nevertheless, results from a study using un-tethered juvenile fish in glass bottle enclosures showed higher rates of predator attacks on patch reefs than on seagrass beds (Sweatman and Robertson 1994), just as was the case for tethering studies. Studies have furthermore shown that factors such as distance from reefs (Shulman 1985) and water depth (Rypel et al. 2007) affect predation risk. Although all of the above-mentioned factors cause variability in the results of predation studies, the current evidence shows that fish and decapods utilizing mangroves and/or seagrass beds often benefit from a lower predation risk as compared to coral reef habitats.

10.4 Synthesis: Nursery role of Mangroves and Seagrass beds

The current review does not provide unequivocal evidence that mangroves or seagrass beds act as nurseries for species of fish or decapods that live as adults on coral reefs or in offshore areas. No single study has successfully tested all four nursery-role factors as postulated by Beck et al. (2001) (Table 10.8). The two studies that tested most factors were done in Palau (Tupper 2007) and Curaçao (for *Haemulon flavolineatum*, based on various studies – see Table 10.8), but neither of the two studies showed (significant) movement to adult populations on adjacent coral reefs. Combining separate studies that have investigated different nursery-role factors in the same study area on the same species, shows that in six out of seven cases where mangroves and/or seagrass beds were studied these two habitats (sometimes in combination with another habitat) harbored higher fish densities or showed higher survival rates than coral reefs. Growth rates, on the other hand, were never highest in mangroves or seagrass beds.

All studies listed in Table 10.8 furthermore failed to evaluate their results in terms of production of fish or decapod populations from different habitats, which forms the underlying basis for quantifying nursery contribution (Beck et al. 2001). Most studies also failed to study movement, or show significant movement from putative nurseries to offshore adult populations. It is therefore still unclear to what degree habitats that showed higher fish densities or survival rates contributed to adult populations.

Probably the best evidence of nursery role so far, is from studies that have attempted to quantify which proportion of individuals from adult populations have

Table 10.8 Studies, or combination of studies done at the same location, that have investigated more than one nursery-role factor for the same species. Habitats are shown that had highest density, growth, or survival. + = movement of fauna from shallow-water areas to coral reefs; in same column: − = movement not investigated. Contribution to adult population indicates the percentage of individuals collected from coral reefs that have originated from a specific shallow-water habitat (between brackets). BCM = branching coral with associated macroalgae, BM = bushy macroalgae, Ch = tidal channels, CR = coral reef, CRb = coral rubble, Mg = mangrove, Sg = seagrass. Postalgal = habitat harboring corals, rubble, seagrass, and sponges

Reference	Taxa	Species	Location	Density (juveniles)	Growth	Survival	Movement to coral reef	Contribution to adult population
Dahlgren and Eggleston (2000)	Fish	*Epinephelus striatus*	Bahamas; Great Exuma Island	−	Postalgal	BM (only for 3.5–4.0 cm fish)	−	−
Grol et al. (2008, unpubl. data)	Fish	*Haemulon flavolineatum*	Aruba; coastal lagoon	CRb + Mg	CR	−	−	−
Chittaro et al. (2004, 2005)	Fish	*Haemulon flavolineatum*	Belize; Turneffe Atoll	Mg[1]	−	Mg-Sg	+	36% (Mg)
Nagelkerken and van der Velde (2002), Grol et al. (2009), Dorenbosch et al. (2009), Verweij and Nagelkerken (2007)	Fish	*Haemulon flavolineatum*	Curaçao; Spanish Water Bay	Mg	CR	[3]Mg + Sg	+[4]	−
Nagelkerken and van der Velde (2002), Verweij et al. (2008)	Fish	*Ocyurus chrysurus*	Curaçao; Spanish Water Bay	Ch + Mg	−	−	+	98% (Mg/Sg embayment)
Tupper (2007)	Fish	*Cheilinus undulatus*	Palau; various sites within the lagoon	BCM	BCM[2]	BCM	±[5]	−
Nakamura and Sano (2004a, 2004b)	Fish	*Stethojulis strigiventer*	Japan; Iriomote Island, Amitori Bay	Sg	−	Sg	−	−
Nakamura et al. (2008), Shibuno et al. (2008)	Fish	*Lutjanus fulvus*	Japan; Ishigaki Island, Itona coast	Mg	−	−	+	88% (Mg)

[1] based on relative density; pattern not present for all study sites
[2] not statistically significant
[3] depending on spatial setting along ocean–bay gradient
[4] restricted movement (N = 3)
[5] movement towards adult fore-reef habitat (i.e., coral/algal microhabitats to deeper patch reefs in lagoon)

passed through mangrove/seagrass habitats during their juvenile life stage. Even movement studies that have not incorporated other nursery role factors (see Section 10.2.4 and Table 10.3) are valuable, because they show actual transfer of individuals from juvenile to adult habitats. So far, three studies have shown higher (>50%) contributions to adult populations from areas with mangrove/seagrasses than from coral reefs (Table 10.8). However, these studies were based on number of individuals, and it is therefore unclear whether the higher contribution in number is also reflected by a higher production. Considering the high percentage of contributions (88 – 98%) the latter seems likely, though. It remains unclear whether this contribution from mangrove/seagrass areas was accomplished through a higher density, faster growth, or a higher survival rate compared to coral reefs, as measurements of these factors were not directly linked to the individuals that had moved to the reef.

So where do we go from here? Testing the Beck et al. (2001) nursery hypothesis will be quite a challenge. Measuring the total biomass transferred by individuals that recruit to adult habitats, for example, is not easy. Biomass measurements are in fact only representative when measured at the time of movement from juvenile to adult habitat. Before this transfer, biomass of individuals can still increase during further residency in the juvenile habitats, whereas just after this transfer a potential rapid change in biomass due to arrival in a new environment (e.g., more food), or rapid predation upon recruitment in the new habitat, may obscure the contribution of biomass produced during residency in the juvenile habitats.

Evaluating nursery role on a per-unit-area basis as postulated by Beck et al. (2001) is a specific complicating factor. What is the minimum surface area of a habitat for it to be included in the nursery value calculations (i.e., average for all habitats where juveniles occur)? For habitats with large surface areas there is little doubt, but what about microhabitats? For example, should small sand patches within seagrass beds be considered as separate habitats, or do they form part of the seagrass bed? At what point are coral formations considered as a separate habitat type vs. part of an existing habitat (e.g., seagrass beds)? Seascapes that harbor many small-scale 'patch' habitats may thus form almost impossible cases for the study of nursery-role value. It may therefore be more valuable to use the approach postulated by Dahlgren et al. (2006), based on the overall contribution of each habitat, so that small-sized habitats that contribute little do not need to be considered in detail. The question is furthermore to what degree habitats can be considered separately. Habitats with very large surface areas that have few interlinkages with adjacent habitats are easy to delineate, but this is not the case for complex mosaics of habitats, especially when they are linked by tidal and diurnal feeding migrations (see Chapter 8). For example, shelter is sought in one habitat and food in another, making it almost impossible to separate the individual contribution of each juvenile habitat. In these situations, it is often easier – and more valuable for management purposes – to calculate total contribution per lagoon or estuary to the adult population.

In conclusion, although worldwide many mangroves and/or seagrass beds seem to harbor higher densities and show increased survival of fish and decapods compared to coral reefs or offshore habitats, little direct evidence demonstrates that this translates to a higher production of populations from mangroves or seagrass beds, and there is little empirical evidence of the degree to which this ultimately supports adult population on reefs or offshore habitats. A recent review identified the most urgent studies that need to be undertaken, and provided a research strategy to assess nursery function of back-reef habitats (Adams et al. 2006). Their four-level strategy consists of (1) building conceptual models to guide the research, (2) identifying juvenile habitat use patterns, (3) assessing habitat connectivity, and (4) experimentally examining the underlying mechanisms for patterns observed at levels two and three. Carefully designed experiments to measure nursery-role factors, and use of advanced techniques to study animal movements in detail (see Chapter 13) are critical for a better insight into the nursery role of tropical mangroves and seagrass beds for coral reef populations. This is of utmost importance for the conservation and management of these ecosystems, which are among the most threatened tropical coastal habitats (see Chapter 16; Alongi 2002, Duarte 2002, Hughes et al. 2003)

Acknowledgments I would like to thank Dr. J Serafy and the Cooperative Institute for Marine and Atmospheric Studies at the Rosenstiel School of Marine and Atmospheric Science of the University of Miami for hosting me during my stay in Miami while writing this chapter. This chapter was written while supported by a Vidi fellowship from the Netherlands Organization for Scientific Research (NWO) to I Nagelkerken. I thank Drs. C Layman, S Blaber, C Faunce, and M Haywood for providing critical comments on the manuscript.

References

Acosta CA, Butler IV MJ (1997) Role of mangrove habitat as a nursery for juvenile spiny lobster, *Panulirus argus*, in Belize. Mar Freshw Res 48:721–728

Acosta CA, Butler IV MJ (1999) Adaptive strategies that reduce predation on Caribbean spiny lobster postlarvae during onshore transport. Limnol Oceanogr 44:494–501

Adams AJ, Dahlgren CP, Kellison GT et al (2006) Nursery function of tropical backreef systems. Mar Ecol Prog Ser 318:287–301

Alongi DM (2002) Present state and future of the world's mangrove forests. Environ Conserv 29:331–349

Appeldoorn RS, Recksiek CW, Hill RL et al (1997) Marine protected areas and reef fish movements: the role of habitat in controlling ontogenetic migration. Proc 8th Int Coral Reef Symp 2:1917–1922

Aguilar-Perera A, Appeldoorn RS (2007) Variation in juvenile fish density along the mangrove-seagrass-coral reef continuum in SW Puerto Rico. Mar Ecol Prog Ser 348:139–148

Baker R, Sheaves M (2005) Redefining the piscivore assemblage of shallow estuarine nursery habitats. Mar Ecol Prog Ser 291:197–213

Baker R, Sheaves M (2006) Visual surveys reveal high densities of large piscivores in shallow estuarine nurseries. Mar Ecol Prog Ser 323:75–82

Barrett NS (1999) Food availability is not a limiting factor in the growth of three Australian temperate reef fishes. Environ Biol Fish 56:419–428

Bartholomew A (2002) Faunal colonization of artificial seagrass plots: the importance of surface area versus space size relative to body size. Estuaries 25:1045–1052

Beck MW, Heck KL, Able KW et al (2001) The identification, conservation and management of estuarine and marine nurseries for fish and invertebrates. BioScience 51:633–641

Bell JD, Westoby M (1986) Abundance of macrofauna in dense seagrass is due to habitat preference, not predation. Oecologia 68:205–209

Benfield MC, Minello TJ (1996) Relative effects of turbidity and light intensity on reactive distance and feeding of an estuarine fish. Environ Biol Fish 46:211–216

Blaber SJM (1980) Fish of the Trinity Inlet system of north Queensland with notes on the ecology of fish faunas of tropical Indo-Pacific estuaries. Aust J Mar Freshw Res 31:137–146

Blaber SJM (2000) Tropical estuarine fishes: ecology, exploitation and conservation. Blackwell, Oxford

Blaber SJM, Blaber TG (1980) Factors affecting the distribution of juvenile estuarine and inshore fish. J Fish Biol 17:143–162

Blaber SJM, Milton DA (1990) Species composition, community structure and zoogeography of fishes of mangrove estuaries in the Solomon Islands. Mar Biol 105:259–267

Blaber SJM, Young JW, Dunning MC (1985) Community structure and zoogeographic affinities of the coastal fishes of the Dampier region of north-western Australia. Aust J Mar Freshw Res 36:247–266

Blaber SJM, Brewer DT, Salini JP (1989) Composition and biomasses of fishes in different habitats of a tropical northern Australian estuary – their occurrence in the adjoining sea and estuarine dependence. Estuar Coast Shelf Sci 29:509–531

Blaber SJM, Brewer DT, Salini JP et al (1992) Species composition and biomasses of fishes in tropical seagrasses at Groote-Eylandt, Northern Australia. Estuar Coast Shelf Sci 35:605–620

Booth DJ, Hixon MA (1999) Food ration and condition affect early survival of the coral reef damselfish, *Stegastes partitus*. Oecologia 121:364–368

Bouwmeester BLK (2005) Ontogenetic migration and growth of French grunt (Teleostei: *Haemulon flavolineatum*) as determined by coded wire tags. M.Sc. thesis, University of Puerto Rico, Mayagüez

Brewer DT, Blaber SJM, Salini JP et al (1995) Feeding ecology of predatory fishes from Groote Eylandt in the Gulf of Carpentaria, Australia, with special reference to predation on penaeid prawns. Estuar Coast Shelf Sci 40:577–600

Burke NC (1995) Nocturnal foraging habitats of French and bluestriped grunts *Haemulon flavolineatum* and *H. sciurus*, at Tobacco Caye, Belize. Environ Biol Fish 42:365–374

Carr MH, Hixon MA (1995) Predation effects on early postsettlement survivorship of coral-reef fishes. Mar Ecol Prog Ser 124:31–42

Chittaro PM, Fryer BJ, Sale PF (2004) Discrimination of French grunts (*Haemulon flavolineatum*, Desmarest, 1823) from mangrove and coral reef habitats using otolith microchemistry. J Exp Mar Biol Ecol 308:169–183

Chittaro PM, Usseglio P, Sale PF (2005) Variation in fish density, assemblage composition and relative rates of predation among mangrove, seagrass and coral reef habitats. Environ Biol Fish 72:175–187

Christensen JD, Jeffrey CFG, Caldow C et al (2003) Cross-shelf habitat utilization patterns of reef fishes in southwestern Puerto Rico. Gulf Caribb Res 14:9–27

Cocheret de la Morinière E (2002) Post-settlement life cycle migrations of reef fish in the mangrove-seagrass-coral reef continuum. Ph.D. thesis, University of Nijmegen, Nijmegen

Cocheret de la Morinière E, Pollux BJA, Nagelkerken et al (2002) Post-settlement life cycle migration patterns and habitat preference of coral reef fish that use seagrass and mangrove habitats as nurseries. Estuar Coast Shelf Sci 55:309–321

Cocheret de la Morinière E, Pollux BJA, Nagelkerken I et al (2003a) Diet shifts of Caribbean grunts (Haemulidae) and snappers (Lutjanidae) and the relation with nursery-to-coral reef migrations. Estuar Coast Shelf Sci 57:1079–1089

Cocheret de la Morinière E, Pollux BJA, Nagelkerken I et al (2003b) Ontogenetic dietary changes of coral reef fishes in the mangrove-seagrass-reef continuum: stable isotopes and gut-content analysis. Mar Ecol Prog Ser 246:279–289

Cocheret de la Morinière E, Nagelkerken I, van der Meij H et al (2004) What attracts juvenile coral reef fish to mangroves: habitat complexity or shade? Mar Biol 144:139–145

Costello TJ, Allen DM (1966) Migrations and geographic distribution of pink shrimp, *Penaeus duorarum*, of the Tortugas and Sanibel grounds, Florida. Fish Bull 65:449–459

Curran MC, Able KW (1998) The value of tethering fishes (winter flounder and tautog) as a tool for assessing predation rates. Mar Ecol Prog Ser 163:45–51

Cyrus DP, Blaber SJM (1987) The influence of turbidity on juvenile marine fishes in estuaries. Part 2. Laboratory studies, comparisons with field data and conclusions. J Exp Mar Biol Ecol 109:71–91

Dahlgren CP, Eggleston DB (2000) Ecological processes underlying ontogenetic habitat shifts in a coral reef fish. Ecology 81:2227–2240

Dahlgren CP, Kellison GT, Adams AJ et al (2006) Marine nurseries and effective juvenile habitats: concepts and applications. Mar Ecol Prog Ser 312:291–295

Danilowicz BS, Sale PF (1999) Relative intensity of predation of the French grunt, *Haemulon flavolineatum*, during diurnal, dusk, and nocturnal periods on a coral reef. Mar Biol 133:337–343

Day JW Jr, Hall CAS, Kemp WM et al (1989) Estuarine ecology. Wiley, New York

Dorenbosch M, van Riel MC, Nagelkerken I et al (2004) The relationship of reef fish densities to the proximity of mangrove and seagrass nurseries. Estuar Coast Shelf Sci 60:37–48

Dorenbosch M, Grol MGG, Christianen MJA et al (2005a) Indo-Pacific seagrass beds and mangroves contribute to fish density and diversity on adjacent coral reefs. Mar Ecol Prog Ser 302:63–76

Dorenbosch M, Grol MGG, Nagelkerken I et al (2005b) Distribution of coral reef fishes along a coral reef – seagrass gradient: edge effects and habitat segregation. Mar Ecol Prog Ser 299:277–288

Dorenbosch M, Grol MGG, Nagelkerken I et al (2006) Seagrass beds and mangroves as potential nurseries for the threatened Indo-Pacific humphead wrasse, *Cheilinus undulatus* and Caribbean rainbow parrotfish, *Scarus guacamaia*. Biol Conserv 129:277–282

Dorenbosch M, Verberk WCEP, Nagelkerken I et al (2007) Influence of habitat configuration on connectivity between fish assemblages of Caribbean seagrass beds, mangroves and coral reefs. Mar Ecol Prog Ser 334:103–116

Dorenbosch M, Grol MGG, de Groene A et al (2009) Piscivore assemblages and predation pressure affect relative safety of some back-reef habitats for juvenile fish in a Caribbean bay. Mar Ecol Prog Ser 379:181–196

Duarte CM (2002) The future of seagrass meadows. Environ Conserv 29:192–206

Edgar GJ, Shaw C (1995) The production and trophic ecology of shallow-water fish assemblages in southern Australia. 1. Species richness, size-structure and production of fishes in Western Port, Victoria. J Exp Mar Biol Ecol 194:53–81

Eggleston DB, Lipcius RN (1992) Shelter selection by spiny lobster under variable predation risk, social conditions, and shelter size. Ecology 73:992–1011

Eggleston DB, Grover JJ, Lipcius RN (1998) Ontogenetic diet shifts in Nassau grouper: trophic linkages and predatory impact. Bull Mar Sci 63:111–126

Eggleston DB, Dahlgren CP, Johnson EG (2004) Fish density, diversity, and size-structure within multiple back reef habitats of Key West national wildlife refuge. Bull Mar Sci 75:175–204

Faunce CH, Serafy JE (2006) Mangroves as fish habitat: 50 years of field studies. Mar Ecol Prog Ser 318:1–18

Faunce CH, Serafy JE (2007) Nearshore habitat use by gray snapper (*Lutjanus griseus*) and bluestriped grunt (*Haemulon sciurus*): environmental gradients and ontogenetic shifts. Bull Mar Sci 80:473–495

Fry B, Mumford PL, Robblee MB (1999) Stable isotope studies of pink shrimp (*Farfantepenaeus duorarum* Burkenroad) migrations on the southwestern Florida shelf. Bull Mar Sci 65:419–430

Gillanders BM, Able KW, Brown JA et al (2003) Evidence of connectivity between juvenile and adult habitats for mobile marine fauna: an important component of nurseries. Mar Ecol Prog Ser 247:281–295

Graham NAJ (2007) Ecological versatility and the decline of coral feeding fishes following climate driven coral mortality. Mar Biol 153:119–127

Gratwicke B, Petrovic C, Speight MR (2006) Fish distribution and ontogenetic habitat preferences in non-estuarine lagoons and adjacent reefs. Environ Biol Fish 76:191–210

Grol MGG, Dorenbosch M, Kokkelmans EMG et al (2008) Mangroves and seagrass beds do not enhance growth of early juveniles of a coral reef fish. Mar Ecol Prog Ser 366:137–146

Haywood MDE, Manson FJ, Loneragan NR et al (2003) Investigation of artifacts from chronographic tethering experiments – interactions between tethers and predators. J Exp Mar Biol Ecol 290:271–292

Heck KL, Nadau DA, Thomas R (1997) The nursery role of seagrass beds. Gulf Mexico Sci 15: 50–54

Heck KL, Orth RJ (1980) Structural components of eelgrass (*Zostera marina*) meadows in the lower Chesapeake Bay – decapod crustacea. Estuaries 3:289–295

Heck KL, Hays G, Orth RJ (2003) Critical evaluation of the nursery role hypothesis for seagrass meadows. Mar Ecol Prog Ser 253:123–136

Herrnkind WF, Butler IV MJ (1986) Factors regulating postlarval settlement and juvenile microhabitat use by spiny lobsters *Panulirus argus*. Mar Ecol Prog Ser 34:23–30

Hixon MA, Beets JP (1993) Predation, prey refuges, and the structure of coral-reef fish assemblages. Ecol Monogr 63:77–101

Hixon MA, Carr MH (1997) Synergistic predation, density dependence, and population regulation in marine fish. Science 277:946–949

Hixon MA, Jones GP (2005) Competition, predation, and density-dependent mortality in demersal marine fishes. Ecology 86:2847–2859

Horinouchi M (2007) Review of the effects of within-patch scale structural complexity on seagrass fishes. J Exp Mar Biol Ecol 350:111–129

Hughes TP, Baird AH, Bellwood DR et al (2003) Climate change, human impacts, and the resilience of coral reefs. Science 301:929–933

Huijbers CM, Grol MGG, Nagelkerken I (2008) Shallow patch reefs as alternative habitats for early juveniles of some mangrove/seagrass-associated fish species in Bermuda. Rev Biol Trop 56 (Suppl. 1):161–169

Iversen ES, Idyll CP (1960) Aspects of the biology of the Tortugas pink shrimp, *Penaeus duorarum*. Trans Am Fish Soc 89:1–8

Jackson EL, Rowden AA, Attrill MJ et al (2001) The importance of seagrass beds as a habitat for fishery species. Oceanogr Mar Biol Annu Rev 39:269–303

Jacoby CA, Greenwood JG (1989) Emergent zooplankton in Moreton Bay, Queensland, Australia: seasonal, lunar, and diel patterns in emergence and distribution with respect to substrata. Mar Ecol Prog Ser 51:131–154

Jennings S, Reñones O, Morales-Nin B et al (1997) Spatial variation in the ^{15}N and ^{13}C stable isotope composition of plants, invertebrates and fishes on Mediterranean reefs: implications for the study of trophic pathways. Mar Ecol Prog Ser 146:109–116

Johnston R, Sheaves M, Molony B (2007) Are distributions of fishes in tropical estuaries influenced by turbidity over small spatial scales? J Fish Biol 71:657–671

Kanashiro K (1998) Settlement and migration of early stage spangled emperor, *Lethrinus nebulosus* (Pisces: Lethrinidae), in the coastal waters off Okinawa island, Japan. Nippon Suisan Gakk 64:618–625

Kitheka JU, Ohowa BO, Mwashote BM et al (1996) Water circulation dynamics, water column nutrients and plankton productivity in a well-flushed tropical bay in Kenya. J Sea Res 35:257–268

Kruitwagen G, Nagelkerken I, Lugendo BR et al (2007) Influence of morphology and amphibious life-style on the feeding ecology of the mudskipper *Periophthalmus argentilineatus*. J Fish Biol 71:39–52

Kulbicki M, Bozec YM, Labrosse P et al (2005) Diet composition of carnivorous fishes from coral reef lagoons of New Caledonia. Aquat living Res 18:231–250

Laegdsgaard P, Johnson C (2001) Why do juvenile fish utilise mangrove habitats? J Exp Mar Biol Ecol 257:229–253

Laprise R, Blaber SJM (1992) Predation by Moses perch, *Lutjanus russelli*, and blue-spotted trevally, *Caranx bucculentus*, on juvenile brown tiger prawn, *Penaeus esculentus* – effects of habitat structure and time of day. J Fish Biol 40:627–635

Levin P, Petrik R, Malone J (1997) Interactive effects of habitat selection, food supply and predation on recruitment of an estuarine fish. Oecologia 112:55–63

Lindquist DG, Gilligan MR (1986) Distribution and relative abundance of butterflyfishes and angelfishes across a lagoon and barrier reef, Andros Island, Bahamas. Northeast Gulf Sci 8:23–30

Lucas C (1974) Preliminary estimates of stocks of king prawn, *Penaeus plebejus*, in south-east Queensland. Aust J Mar Freshwat Res 25:35–47

Luo J, Serafy JE, Sponaugle S et al (2009) Movement of gray snapper *Lutjanus griseus* among subtropical seagrass, mangrove, and coral reef habitats. Mar Ecol Prog Ser 380: 255–269

MacDonald JA, Glover T, Weis JS (2008) The impact of mangrove prop-root epibionts on juvenile reef fishes: a field experiment using artificial roots and epifauna. Estuar Coast 31:981–993

Macia A, Abrantes KGS, Paula J (2003) Thorn fish *Terapon jarbua* (Forskål) predation on juvenile white shrimp *Penaeus indicus* H. Milne Edwards and brown shrimp *Metapenaeus monoceros* (Fabricius): the effect of turbidity, prey density, substrate type and pneumatophore density. J Exp Mar Biol Ecol 291:29–56

Main KL (1987) Predator avoidance in seagrass meadows: prey behavior, microhabitat selection, and cryptic coloration. Ecology 68:170–180

Manson FJ, Loneragan NR, Skilleter GA et al (2005) An evaluation of the evidence for linkages between mangroves and fisheries: a synthesis of the literature and identification of research directions. Oceanogr Mar Biol Annu Rev 43:483–513

McFarland WN (1991) The visual world of coral reef fishes. In: Sale PF (ed) The ecology of fishes on coral reefs, pp. 16–38. Academic Press, San Diego

Meager JJ, Williamson I, Loneragan NR et al (2005) Habitat selection of juvenile banana prawns, *Penaeus merguiensis* de Man: testing the roles of habitat structure, predators, light phase and prawn size. J Exp Mar Biol Ecol 324:89–98

Minello TJ, Able KW, Weinstein MP et al (2003) Salt marshes as nurseries for nekton: testing hypotheses on density, growth and survival through meta-analysis. Mar Ecol Prog Ser 246:39–59

Mumby PJ, Edwards AJ, Arias-González JE et al (2004) Mangroves enhance the biomass of coral reef fish communities in the Caribbean. Nature 427:533–536

Nagelkerken I (2000) Importance of shallow-water bay biotopes as nurseries for Caribbean reef fishes. Ph.D. thesis, University of Nijmegen, Nijmegen

Nagelkerken I (2007) Are non-estuarine mangroves connected to coral reefs through fish migration? Bull Mar Sci 80:595–607

Nagelkerken I, van der Velde G (2002) Do non-estuarine mangroves harbour higher densities of juvenile fish than adjacent shallow-water and coral reef habitats in Curaçao (Netherlands Antilles)? Mar Ecol Prog Ser 245:191–204

Nagelkerken I, van der Velde G (2003) Connectivity between coastal habitats of two oceanic Caribbean islands as inferred from ontogenetic shifts by coral reef fishes. Gulf Caribb Res 14:43–59

Nagelkerken I, van der Velde G (2004) Are Caribbean mangroves important feeding grounds for juvenile reef fish from adjacent seagrass beds? Mar Ecol Prog Ser 274:143–151

Nagelkerken I, Faunce CH (2008) What makes mangroves attractive to fish? Use of artificial units to test the influence of water depth, cross-shelf location, and presence of root structure. Estuar Coast Shelf Sci 79:559–565

Nagelkerken I, Dorenbosch M, Verberk WCEP et al (2000a) Day-night shifts of fishes between shallow-water biotopes of a Caribbean bay, with emphasis on the nocturnal feeding of Haemulidae and Lutjanidae. Mar Ecol Prog Ser 194:55–64

Nagelkerken I, Dorenbosch M, Verberk WCEP et al (2000b) Importance of shallow-water biotopes of a Caribbean bay for juvenile coral reef fishes: patterns in biotope association, community structure and spatial distribution. Mar Ecol Prog Ser 202:175–192

Nagelkerken I, van der Velde G, Gorissen MW et al (2000c) Importance of mangroves, seagrass beds and the shallow coral reef as a nursery for important coral reef fishes, using a visual census technique. Estuar Coast Shelf Sci 51:31–44

Nagelkerken I, Roberts CM, van der Velde G et al (2002) How important are mangroves and seagrass beds for coral-reef fish? The nursery hypothesis tested on an island scale. Mar Ecol Prog Ser 244:299–305

Nagelkerken I, Bothwell J, Nemeth RS et al (2008) Interlinkage between Caribbean coral reefs and seagrass beds through feeding migrations by grunts (Haemulidae) depends on habitat accessibility. Mar Ecol Prog Ser

Nakamura Y, Sano M (2004a) Is there really lower predation risk for juvenile fishes in a seagrass bed compared with an adjacent coral area? Bull Mar Sci 74:477–482

Nakamura Y, Sano M (2004b) Overlaps in habitat use of fishes between a seagrass bed and adjacent coral and sand areas at Amitori Bay, Iriomote Island, Japan: importance of the seagrass bed as juvenile habitat. Fish Sci 70:788–803

Nakamura Y, Sano M (2005) Comparison of invertebrate abundance in a seagrass bed and adjacent coral and sand areas at Amitori Bay, Iriomote Island, Japan. Fish Sci 71:543–550

Nakamura Y, Tsuchiya M (2008) Spatial and temporal patterns of seagrass habitat use by fishes at the Ryukyu Islands, Japan. Estuar Coast Shelf Sci 76:345–356

Nakamura Y, Kawasaki H, Sano M (2007) Experimental analysis of recruitment patterns of coral reef fishes in seagrass beds: effects of substrate type, shape, and rigidity. Estuar Coast Shelf Sci 71:559–568

Nakamura Y, Horinouchi M, Shibuno T et al (2008) Evidence of ontogenetic migration from mangroves to coral reefs by black-tail snapper *Lutjanus fulvus*: stable isotope approach. Mar Ecol Prog Ser 355:257–266

Ogden JC, Ehrlich PR (1977) The behavior of heterotypic resting schools of juvenile grunts (Pomadasyidae). Mar Biol 42:273–280

Ogden JC, Zieman JC (1977) Ecological aspects of coral reef–seagrass bed contacts in the Caribbean. Proc 3rd Int Coral Reef Symp 1:377–382

Orth RJ, Heck KL, van Montfrans J (1984) Faunal communities in seagrass beds: a review of the influence of plant structure and prey characteristics on predator prey relationships. Estuaries 7:339–350

Parrish JD (1989) Fish communities of interacting shallow-water habitats in tropical oceanic regions. Mar Ecol Prog Ser 58:143–160

Pollux BJA, Verberk WCEP, Dorenbosch M et al (2007) Habitat selection during settlement of three Caribbean coral reef fishes: indications for directed settlement to seagrass beds and mangroves. Limnol Oceanogr 52:903–907

Primavera JH (1997) Fish predation on mangrove-associated penaeids – the role of structures and substrate. J Exp Mar Biol Ecol 215:205–216

Reichert MJM, Dean JM, Feller RJ et al (2000) Somatic growth and otolith growth in juveniles of a small subtropical flatfish, the fringed flounder, *Etropus crossotus*. J Exp Mar Biol Ecol 254:169–188

Ríos-Jara E (2005) Effects of lunar cycle and substratum preference on zooplankton emergence in a tropical, shallow-water embayment, in southwestern Puerto Rico. Caribb J Sci 41:108–123

Risk A (1997) Effects of habitat on the settlement and post-settlement success of the ocean surgeonfish *Acanthurus bahianus*. Mar Ecol Prog Ser 161:51–59

Robertson AI, Duke NC (1987) Mangroves as nursery sites – comparisons of the abundance and species composition of fish and crustaceans in mangroves and other nearshore habitats in tropical Australia. Mar Biol 96:193–205

Robertson AI, Blaber SJM (1992) Plankton, epibenthos and fish communities. In: Robertson AI, Alongi DM (eds) Tropical mangrove ecosystems. Coastal Estuar Stud 41:173–224

Rooker JR (1995) Feeding ecology of the schoolmaster snapper *Lutjanus apodus* (Walbaum), from southwestern Puerto Rico. Bull Mar Sci 56:881–894

Rooker JR, Dennis GD (1991) Diel, lunar and seasonal changes in a mangrove fish assemblage off southwestern Puerto Rico. Bull Mar Sci 49:684–698

Russell DJ, McDougall AJ (2005) Movement and juvenile recruitment of mangrove jack, *Lutjanus argentimaculatus* (Forsskål), in northern Australia. Mar Freshw Res 56:465–475

Rypel AL, Layman CA (2008) Degree of aquatic ecosystem fragmentation predicts population characteristics of gray snapper (*Lutjanus griseus*) in Caribbean tidal creeks. Can J Fish Aquat Sci 65:335–339

Rypel AL, Layman CA, Arrington DA (2007) Water depth modifies relative predation risk for a motile fish taxon in Bahamian tidal creeks. Estuar Coast 30:518–525

Sale PF (1974) Overlap in resource use, and interspecific competition. Oecologia 17:245–256

Salini JP, Blaber SJM, Brewer DT (1990) Diets of piscivorous fishes in a tropical Australian estuary, with special reference to predation on penaeid prawns. Mar Biol 105:363–374

Scharf FS, Manderson JP, Fabrizio MC (2006) The effects of seafloor habitat complexity on survival of juvenile fishes: species-specific interactions with structural refuge. J Exp Mar Biol Ecol 335:167–176

Schoener TW (1971) Theory of feeding strategies. Annu Rev Ecol Sys 2:369–404

Serafy JE, Faunce CH, Lorenz JJ (2003) Mangrove shoreline fishes of Biscayne Bay, Florida. Bull Mar Sci 72:161–180

Sheaves M (2001) Are there really few piscivorous fishes in shallow estuarine habitats? Mar Ecol Prog Ser 222:279–290

Sheridan P, Hays C (2003) Are mangroves nursery habitat for transient fishes and decapods? Wetlands 23:449–458

Shibuno T, Nakamura Y, Horinouchi M et al (2008) Habitat use patterns of fishes across the mangrove-seagrass-coral reef seascape at Ishigaki Island, southern Japan. Ichthyol Res 55:218–237

Shulman MJ (1985) Recruitment of coral-reef fishes: effects of distribution of predators and shelter. Ecology 66:1056–1066

Somers IF, Kirkwood GP (1984) Movements of tagged tiger prawns, *Penaeus* spp., in the western Gulf of Carpentaria. Mar Freshw Res 35:713–723

Spitzer PM, Mattila J, Heck KL (2000) The effects of vegetation density on the relative growth rates of juvenile pinfish, *Lagodon rhomboides* (Linneaus), in Big Lagoon, Florida. J Exp Mar Biol Ecol 244:67–86

Stephens DW, Krebs JR (1986) Foraging theory. Princeton University Press, Princeton

Sumpton WD, Sawynok B, Carstens N (2003) Localised movement of snapper (*Pagrus auratus*, Sparidae) in a large subtropical marine embayment. Mar Freshw Res 54:923–930

Sweatman H, Robertson DR (1994) Grazing halos and predation on juvenile Caribbean surgeonfishes. Mar Ecol Prog Ser 111:1–6

Thollot P (1992) Importance of mangroves for Pacific reef fish species, myth or reality? Proc 6th Int Coral Reef Symp 2:934–941

Tupper M (2007) Identification of nursery habitats for commercially valuable humphead wrasse *Cheilinus undulatus* and large groupers (Pisces: Serranidae) in Palau. Mar Ecol Prog Ser 332:189–199

Utne ACW, Aksnes DL, Giske J (1993) Food, predation risk and shelter: an experimental study on the distribution of adult two-spotted goby *Gobiusculus flavescens* (Fabricius). J Exp Mar Biol Ecol 166:203–216

Vagelli AA (2004) Ontogenetic shift in habitat preference by *Pterapogon kauderni*, a shallow water coral reef apogonid, with direct development. Copeia 2004: 364–369

Valentine JF, Heck KL, Blackmon D et al (2007) Food web interactions along seagrass-coral reef boundaries: effects of piscivore reductions on cross-habitat energy exchange. Mar Ecol Prog Ser 333:37–50

Valentine JF, Heck KL, Blackmon D et al (2008) Exploited species impacts on trophic linkages along reef-seagrass interfaces in the Florida keys. Ecol Appl 18:1501–1515

Vance DJ, Haywood MDE, Heales DS et al (1996) How far do prawns and fish move into mangroves? Distribution of juvenile banana prawns *Penaeus merguiensis* and fish in a tropical mangrove forest in northern Australia. Mar Ecol Prog Ser 131:115–124

Vanderklift MA, How J, Wernberg T et al (2007) Proximity to reef influences density of small predatory fishes, while type of seagrass influences intensity of their predation on crabs. Mar Ecol Prog Ser 340:235–243

Verweij MC, Nagelkerken I (2007) Short and long-term movement and site fidelity of juvenile Haemulidae in back-reef habitats of a Caribbean embayment. Hydrobiologia 592:257–270

Verweij MC, Nagelkerken I, de Graaff D et al (2006a) Structure, food and shade attract juvenile coral reef fish to mangrove and seagrass habitats: a field experiment. Mar Ecol Prog Ser 306:257–268

Verweij MC, Nagelkerken I, Wartenbergh SLJ et al (2006b) Caribbean mangroves and seagrass beds as daytime feeding habitats for juvenile French grunts, *Haemulon flavolineatum*. Mar Biol 149:1291–1299

Verweij MC, Nagelkerken I, Hol KEM et al (2007) Space use of *Lutjanus apodus* including movement between a putative nursery and a coral reef. Bull Mar Sci 81:127–138

Verweij MC, Nagelkerken I, Hans I et al (2008) Seagrass nurseries contribute to coral reef fish populations. Limnol Oceanogr 53:1540–1547

Watson M, Munro JL, Gell FR (2002) Settlement, movement and early juvenile mortality of the yellowtail snapper *Ocyurus chrysurus*. Mar Ecol Prog Ser 237:247–256

Werner EE, Gilliam JF, Hall DJ et al (1983) An experimental test of the effects of predation risk on habitat use in fish. Ecology 64:1540–1548

Wilson S, Bellwood DR (1997) Cryptic dietary components of territorial damselfishes (Pomacentridae, Labroidei). Mar Ecol Prog Ser 153:299–310

Wilson SK, Street S, Sato T (2005) Discarded queen conch (*Strombus gigas*) shells as shelter sites for fish. Mar Biol 147:179–188

Wootton RJ (1998) Ecology of teleost fishes. 2nd edition. Fish and Fisheries Series 24. Kluwer Academic Publishers, Dordrecht

Zimmer-Faust RK, Fielder DR, Heck KL Jr et al (1994) Effects of tethering on predatory escape by juvenile blue crabs. Mar Ecol Prog Ser 111:299–303

Chapter 11
Sources of Variation that Affect Perceived Nursery Function of Mangroves

Craig H. Faunce and Craig A. Layman

Abstract Mangroves are considered among the most productive ecosystems on the planet. While mangroves provide numerous critical ecosystem services to surrounding environments, there is particular interest in the role of mangroves as nursery habitats for fish and decapods. Despite this interest, scientific consensus regarding the role of mangroves as nurseries remains elusive. In this chapter, we identify four principal sources of variability that underlie conflicting conclusions regarding the function of mangroves as nursery habitat. We provide brief sketches of the reasons why these sources of variability may affect the role of mangroves as nursery habitat, drawing particularly on recent empirical advances in the field, and conclude with a conceptual model summarizing the different levels at which the nursery function of mangroves is evaluated.

'It is time that we biologists accept diversity and variability for what they are, two of the essential features of the biological world. We would be wise to restructure our search for orderly patterns in the natural world. We should stop thinking primarily in terms of central tendencies . . . .Variation among and within species is fundamental to organisms. Analysis of variation can offer insights just as surely as can traditional delineation of central tendencies.' (Bartholomew 1986).

Keywords Variance · Nursery · Biogeography · Hydrology · Conceptual model

11.1 Introduction

Mangroves are considered among the most productive ecosystems on the planet, and provide numerous other critical ecosystem services (Costanza et al. 1997, see Chapter 16). They often are believed to augment fishery production in estuaries and

C.H. Faunce (✉)
National Marine Fisheries Service, Alaska Fisheries Science Center, 7600 Sand Point Way NE, Seattle, Washington, USA
e-mail: Craig.Faunce@noaa.gov

I. Nagelkerken (ed.), *Ecological Connectivity among Tropical Coastal Ecosystems*, DOI 10.1007/978-90-481-2406-0_11, © Springer Science+Business Media B.V. 2009

adjacent areas due to the abundant food resources and protection from predators that they may provide (see papers in Serafy and Araujo 2007, and Chapter 10). Of particular interest is the role that mangroves may play as nursery habitat, and thus the way mangroves are inter-connected with other 'back reef' habitats through the export of fish biomass (Adams et al. 2006). However, the extent to which mangroves serve as nurseries, or serve a mangrove nursery function, remains a subject of much debate in the scientific literature (Blaber 2007). We believe that much of the disagreement regarding their value as nursery habitat stems directly from the underlying variability in mangrove systems (Ewel et al. 1998, Blaber 2007), as well as the way in which the systems are studied. In the present chapter, we will outline some of these sources of variability, and discuss how they explicitly affect the interpretation of mangroves' role as nursery habitat.

Variability in abiotic and biotic variables is an intrinsic property of biological systems (Bartholomew 1986). Yet scientists seek to identify general rules, laws, or theories that would unite scientific inquiry across these sources of variability. The study of mangrove ecology/biology is no exception. For example, there seems to be a pervasive desire to provide a singular answer to the question: 'are mangroves nursery habitats?' Yet there is likely no simple answer to this question. Mangroves are likely critical nursery habitat in some instances, and have no nursery function in others.

In this chapter, we identify four principal sources of variability that underlie the conflicting conclusions regarding the function of mangroves as nursery habitat. The first is related to how researchers define nursery habitat and mangrove ecosystems, and the next three deal with intrinsic sources of biological variability: variation in space (geomorphology, habitat type, and configuration), variation in time (hydrodynamics, time of day), and variation in species (assemblage vs. species-level analyses). This is not intended as another comprehensive review of the important functions mangroves play with respect to nursery function and fishery production (e.g., Sheridan and Hays 2003, Faunce and Serafy 2006, Blaber 2007, Nagelkerken 2007, see Chapter 10). Instead, we endeavor to provide targeted examples of how variability may lead to different conclusions regarding the role of mangroves as nursery habitat. We hope the end result will be that researchers explicitly consider each of these sources of variability (and others) when evaluating the role of mangroves as nursery habitat.

11.2 Variation in Definitions

11.2.1 What Defines a Nursery?

One of the reasons that researchers have reached different conclusions regarding the role of mangroves as nursery habitat relates to the specific definition of 'nursery' that has been employed (Table 11.1). Historically, nurseries were regarded as those areas that supported a higher density or abundance of immature fishes than other

Table 11.1 A summary of different connotations of the term 'nursery'

Definition	Description	Source
Historical connotation	A habitat type which supports a higher density or abundance of juvenile individuals than other habitat types	–
Predation/food-based	A habitat must provide adequate protection from predators or a food source which is both varied and concentrated	Thayer et al. (1978)
Juvenile contribution function	Nursery habitats for a particular species are those that contribute a greater than average number of individuals to the adult population on a per-unit-area basis in comparison to other habitat types used by juveniles	Beck et al. (2001)
Effective juvenile habitat	A habitat type is considered a nursery if, for a particular species, it contributes a greater proportion of individuals to the adult population than the mean level contributed by all habitats used by juveniles	Dahlgren et al. (2006)
Essential fish habitat	Those waters and substrate necessary to fish for spawning, breeding, feeding, or growth to maturity	NOAA (1996)

adjacent habitats. This criterion allows much latitude in attributing the nursery label to a particular habitat type and provides no standardized guide for evaluation. In this context, the methodologies employed to estimate faunal densities are critical to assessment of nursery function, and identifying nursery habitat often may depend as much on the sampling method employed as the underlying biological or ecological drivers. And since methodologies often are not employed in a fashion that allows direct comparisons among habitat types (Faunce and Serafy 2006), it is not surprising that many different conclusions have been reached regarding the role of mangroves as nursery habitat.

The lack of rigor in defining nursery habitat led Beck et al. (2001) to propose a more stringent set of criteria: 'A habitat is a nursery for juveniles of a particular species if its contribution per unit area to the production of individuals that recruit to adult populations is greater, on average, than production from other habitats in which juveniles occur.' In this context, nursery habitats could support greater production through increased density, growth, survival, or export of juveniles. The main limitation of this definition is that area coverage of habitat types is not considered, so one habitat type may support fewer individuals per unit area, but yet still be the most important contributor in absolute numbers to an adult population. To this end, Dahlgren et al. (2006) suggested that in some contexts it may be useful to identify 'effective juvenile habitats': a habitat for a particular species that contributes a greater proportion of individuals to the adult population than the mean level contributed by all habitats used by individuals, regardless of area coverage. Different conclusions can be reached regarding which habitat types are nurseries depending on which of these two approaches is employed (e.g., see the example outlined in Dahlgren et al. 2006).

Despite the suggestion that a standard, quantifiable, framework is essential to adequately determine whether a habitat functions as a nursery (Beck et al. 2001, Dahlgren et al. 2006), the majority of studies rely on the historical connotation of the term. That is, mangroves are nominally alluded to as 'nurseries' simply if they support a relatively high abundance of juvenile individuals. This is likely due to the difficulty in assessing the production (based on abundance, growth, and survival) and export of individuals that utilize habitats of interest. Any one of these factors is challenging to measure alone, and to measure all four simultaneously may be impossible in many situations (although there are some recent attempts toward this end, see Koenig et al. 2007, Valentine-Rose et al. 2007, Faunce and Serafy 2008a). As such, there is a dearth of information that can used to quantitatively infer mangroves' role in supporting secondary production based on the more stringent definitions provided by Beck et al. (2001) and Dahlgren et al. (2006).

For the remainder of this chapter, we will follow the nursery definition of Beck et al. (2001). As such, we endeavor to point out some of the sources of variability that affect the estimation of the density, growth, mortality, or export of juveniles within coastal habitat types, and how these sources of variability may contribute to the lack of an established consensus of whether mangrove habitat functions in a nursery role.

11.2.2 What Defines a Mangrove Forest?

There have been several attempts to provide a framework for the study of mangrove forests. These frameworks were considered necessary after it was recognized that several different forest types, each with their own physical configuration and production properties, could result from identical mangrove communities exposed to different abiotic regimes. Lugo and Snedaker (1974) and later Lugo (1980) described six types of Florida mangrove forest based on topographic location and geomorphologic form. Moving across a landscape in an upland direction, these forests include those: (1) completely inundated by daily tides (overwash; up to 7 m tall), (2) fringing emergent shorelines (fringe; up to 10 m tall), (3) along flowing waters (riverine; up to 18 m tall), (4) located in a depression behind a berm (basin; up to 15 m tall), (5) located in extreme environments, e.g., poor water exchange (dwarf, or 'scrub'; less than 2 m tall; Fig. 11.1), and (6) located on 'peat islands' within the Everglades (hammock; up to 5 m tall). Woodroffe (1992) developed a more general classification system in Australia that included river-dominated, tide-dominated, and interior mangrove forests. Extending these works, Ewel et al. (1998) developed a hybrid classification scheme used (which we use hereafter): tide-dominated systems are termed fringe mangroves, river-dominated mangroves are termed riverine mangroves, and interior mangroves are termed basin mangroves.

Forest-type is rarely defined in studies of mangrove-associated fauna. Yet differences in forest type have important implications on our perceived value of mangroves as nursery habitats because each forest type serves a different ecological function and is utilized by different motile fauna. For example, because they connect

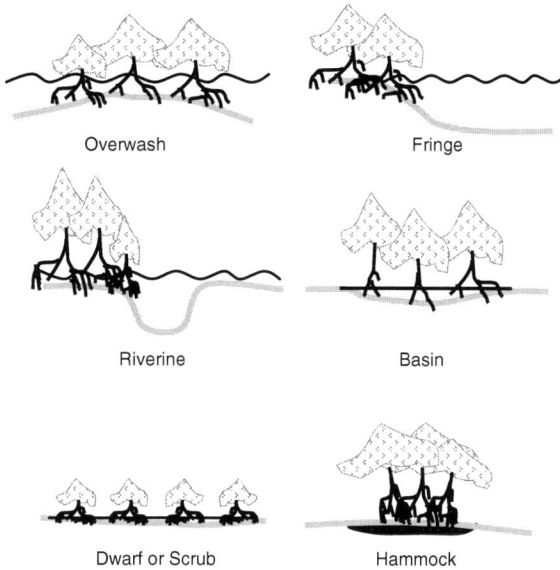

Fig. 11.1 Schematic cross-section view of various forest-type architectures exhibited from a single species of mangrove (*Rhizophora mangle*) in south Florida (following Lugo 1980)

upstream freshwater sources and downstream estuarine waters, riverine mangroves are used opportunistically as a conduit for motile fauna on seasonal, lunar, or daily tidal cycles. However, the same static patch of riverine mangrove may experience a suite of salinities throughout a tidal cycle or season. Consequently, identical locations of habitat may be inhabited by animals from freshwater, estuarine, or marine guilds, and decisions as to the relative importance of the mangle in the lives of fishes becomes an ever-changing target that must be carefully qualified. In contrast, basin-type forests are inundated much less often, but for longer duration than fringing forests (Lewis 2005). These forests are utilized during the flooded period by small-bodied (<100 mm) individuals that include juveniles of estuarine and marine spawning species and resident species that spend their entire lives within the mangle. The dynamics of these fauna are strongly driven by water levels. Within a seasonally flooded Florida mangle, density, biomass, and ultimately secondary production of fishes were positively related to water level and flooding duration (Lorenz 1999). However as water levels decrease, animals must seek deep water refugia at the edges of riverine or fringe forests or be stranded, causing negative correlations between abundance metrics and water levels in these forest types (e.g., Faunce et al. 2004, Serafy et al. 2007). Thus upper basin-type forests function as fish nurseries when flooded, and as important food sources during subsequent dry periods for animals such as birds. Because they are both speciose and abundant, resident fauna are largely responsible for trends in assemblage metrics (e.g., species richness and total abundance). For example, in the Philippines it was found that the density of fishes among stands of different mangrove species and distance to open water were greatest within the upper (shallow) *Avicennia* portions (Rönnbäck et al. 1999). The remainder of our discussion will focus on variation in animal use of well-studied fringing forests.

11.3 Spatial Variation

11.3.1 Geographic Regions

The common ancestry of mangroves has resulted in the global distribution between the 20 °C aquatic isotherms (Alongi 2002), yet individual regions have unique oceanographic and geologic histories. Spalding et al. (1997) identified five regions based on present-day geomorphology and biodiversity of mangrove forests: Australasia (Australia, Papua New Guinea, New Zealand, and the South Pacific islands), South and Southeast Asia (Pakistan to the west, China and Japan to the northeast, including Indonesia), East Africa and the Middle East (Iran to South Africa eastwards, including the islands in the Indian Ocean), West Africa, and the Americas (north, central, and south). Based on mangrove forest composition and richness, West Africa is most similar to the Americas (hereafter Western Atlantic), and East Africa to the Indo-Pacific (hereafter Indo-Pacific), with the latter group roughly three times more speciose than the former (Hogarth 2007).

The divide between the Western Atlantic and the Indo-Pacific is reflected in the current body of literature on mangrove use by motile fauna, with generally all studies within the former accepting the paradigm that mangroves serve a nursery function, and challenges to this paradigm arising from studies conducted in the latter. Such differences in opinion can be explained by the differences in the spatial configuration of shelves, habitat configuration, and/or hydrology between regions.

11.3.2 Shelf Configuration

The availability of mangroves to juveniles determines their nursery value. For species that spawn offshore, availability of mangrove habitats depends directly on the amount of submerged shoreline, the location of reproduction relative to mangroves, the prominent oceanographic conditions during and after the spawn, and the larval duration. These factors are substantially influenced by bathymetry. Obligate-group and pair-group spawning strategies have evolved within functionally and taxonomically related species in the Caribbean region (e.g., Lutjanidae: snappers). Which strategy prevails is related to local differences in shelf slope and resultant mangrove area. In obligate-group spawning, fish aggregate *en masse* at very specific geographic locations to reproduce, and these locations are near local gyres that ideally retain larvae nearshore for a period of time approximating their average larval duration (Heyman et al. 2005, Paris et al. 2005). This spawning strategy is largely documented in locations with limited emergent coastline and steep shelves with limited available hard-bottom promontories, e.g., Belize, southwest Cuba, and the lower Florida Keys.

Because the area of mangrove relative to alternative submerged habitats is relatively small, and the presence and persistence of gyres needed for favorable larval advection are variable, spawning on promontories may be a very risky reproduction strategy if juveniles require mangrove-lined bays to survive. Parrish (1989) proposed that mangrove-lined embayments act as 'waiting rooms' that collect excess larvae from species that spawn offshore, and that the majority of offspring necessary for the maintenance of adult populations are resident to the reef. Parrish (1989) also postulated that mangrove residence may act to mitigate the negative effects of poor juvenile recruitment to adulthood within reef environments. Thus, while juveniles of marine-spawning species on steep slopes do utilize mangroves, their reliance on these systems does not appear to be obligate and they likely do not function as nurseries in this context.

A pair-group spawning strategy is employed by dominant snapper within low-relief continental geomorphologies within the Caribbean (e.g., southeastern Florida, northeastern Cuba, Yucatan peninsula). For example, *Lutjanus griseus* (known locally as mangrove snapper) has evolved a life history strategy to take advantage of the comparatively larger areas of emergent vegetation proximate to broad shelves. This species is capable of spawning in pairs or in groups of <20 individuals; small aggregations can form at numerous, less-specified, locations, and individuals are commonly found in mangroves at most post-larval stages (Serafy et al. 2003, Faunce et al. 2007). Therefore, the reliance on mangroves for the maintenance of healthy adult populations of fish and decapods may be greater within continental low-relief systems with large mangrove area than within steep-sloped insular systems with less mangrove area. Indeed, when data for the same genera (Lujanidae and Haemulidae) residing in mangroves are compared, groupings based on either continental (low relief) or insular (high relief) geomorphologies are evident (Fig. 11.2).

11.3.3 Habitat Configuration

Because they can tolerate a variety of abiotic conditions, mangroves occur in many different areas of coastal and estuarine ecosystems. It is important that the relative position, area, and configuration of the mangrove patch, as well as the developmental stage of the individual, be considered when determining nursery value of mangrove habitat. Because they are located closest to marine source populations, mangroves along oceanic-facing shorelines are more likely to receive marine-derived post-larval recruits than other mangrove locations. Yet, unlike locations within bays, ocean-facing shorelines within much of the Caribbean lack sufficient sediment (due to erosion), and mangrove roots may penetrate into the water only a few centimeters. Thus, along ocean-facing shorelines, their availability (relative area

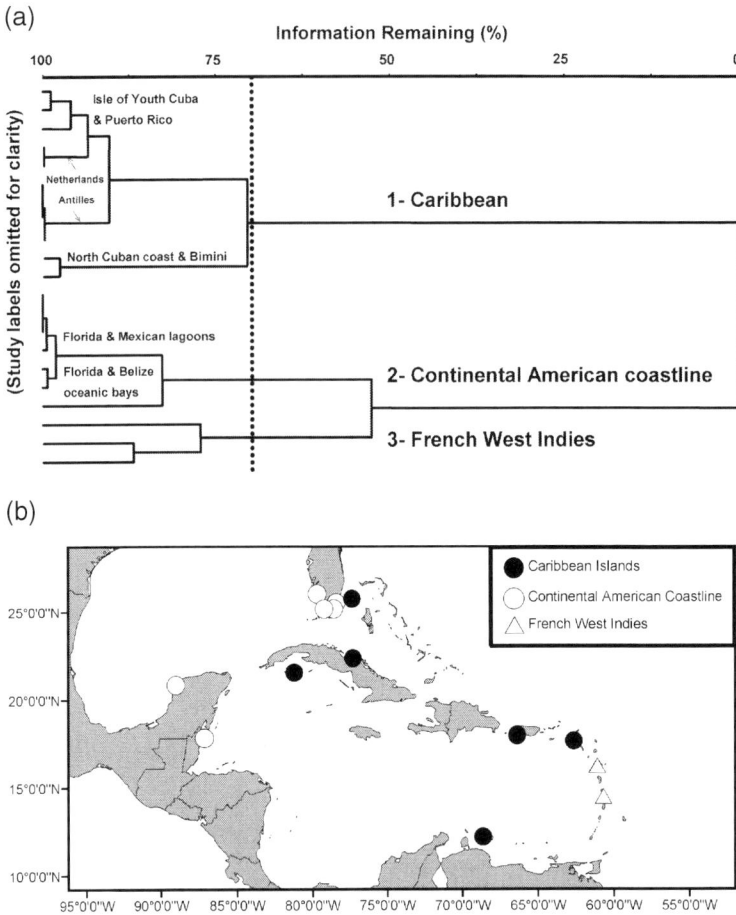

Fig. 11.2 Summary of data for snappers (Lutjanidae) and grunts (Haemulidae) reported from 21 studies conducted within Florida-Caribbean mangroves vetted from the literature (1971–2005) following Faunce and Serafy (2006). Density or biomass were relativized to maximum within each study and entered into agglomerative cluster analysis using Bray-Curtis distances with flexible beta (−0.25) linkage (**a**). Indicator Species Analysis (Dufrêne and Legendre 1997) identified three groups that are related to shelf configuration (**b**). The species *Haemulon flavolineatum* and *Lutjanus apodus* distinguished the Caribbean group, *L. griseus* distinguished the continental American group, and *H. bonariense*, *H. aurolineatum*, and *L. jocu* distinguished the French West Indies group

coverage) may be comparatively low. In cases where mangrove roots are submerged enough to create fish habitat, comparative study has demonstrated that oceanic fringes are utilized much more than their availability would suggest, indicating positive selection for this shoreline type (Faunce and Serafy 2008b).

　　Compared to oceanic fringes, a much greater proportion of our current knowledge of fish and decapod use of mangroves comes from studies conducted within inlets and protected bays. It follows that for species of marine origin, the accessibility of mangroves would not only be influenced by proximity of shoreline to spawning locations (described above), but also by the width and depth of the bay-ocean interface, local currents, and tidal flow. A decline in total abundance and richness of reef-associated demersal fishes has been observed with distance inland from the outer bay mouth in the Caribbean (Nagelkerken et al. 2000a), Brazil (Araujo et al. 2002), Africa (Little et al. 1988), and the Indo-Pacific (Quinn 1980, Blaber et al. 1989, Hajisamae and Chou 2003). Because the pool of available species is likely larger within offshore areas relative to bays, comparisons of total species, total density, and species-specific density will result in a negative relation between these metrics and distance of the mangrove patch from marine source waters.

Given the variation in geomorphology and hydrology described above, it follows that different basins within a single system may vary substantially in their physical and environmental properties, and this will be reflected in animal use patterns. Robertson and Duke (1987) were among the first to propose that each mangrove embayment may be considered its own unit, and that nursery function changes from unit to unit. Ley et al. (1999) first provided evidence consistent with this hypothesis by showing that distinct fish assemblages existed within three connected embayments with varying levels of freshwater flow in Florida. These results compare well with those reported from northeastern Australia, where it has been demonstrated that faunal assemblages can be delimited largely based on characterization of estuaries by catchment hydrology (tide or wave dominated), configuration of estuary mouth, substrate, and mangrove area (Ley 2005). Characterization of a nursery will depend on whether the species under investigation is of freshwater or marine origin, and where the mangle is located relative to fresh and marine water sources.

Mangroves are not the only habitat available to fish and decapods within subtropical and tropical bays, and the relative importance of mangroves compared to other structurally complex habitats is a major focus of current research (Faunce and Serafy 2006). Comparisons of fish size has revealed larger size-class occurs within mangroves than seagrass beds in Florida (Eggleston et al. 2004, Faunce and Serafy 2007) and Curaçao (Nagelkerken et al. 2000a, b, Cocheret de la Morinière et al. 2002). From these observations it has been concluded that mangroves act as secondary habitats for fishes of the region, and it is for this reason that the evaluation of mangroves as nursery habitat need to be carefully considered. For species that undergo ontogenetic habitat shifts, e.g., from seagrass to mangroves to coral reefs, comparisons of relative abundance between habitats are flawed because population dynamics dictate that the smallest and youngest individuals will have the greatest absolute abundance (Ricker 1975). In this example, even for equally-sized patches, total abundance will be lower in mangroves compared to seagrass beds, and yet higher within mangroves compared to coral reefs. This situation, i.e. where seagrass beds comprise the greatest area and contain a greater absolute number of animals relative to mangroves, may explain why comparisons by Sheridan and Hays (2003) failed to find a nursery function role for mangroves.

11.4 Temporal Variation

11.4.1 Hydrology

A temporal perspective also reveals intrinsic differences in the function of man-grove ecosystems in the Western Atlantic and the Indo-Pacific. In the latter region, there is greater influence of freshwater from larger catchments and dramatic changes in water level with tidal flow. These differences at the bay scale translate into great differences in the nature of variation in habitat availability to motile fauna. In the two often studied portions of the Western Atlantic, Southeast Florida, and the Caribbean islands, smaller tidal ranges result in the availability (i.e., inundation) of structurally-heterogeneous habitats (largely fringe mangroves) nearly year-round (Provost 1973). Under this temporal regime, animals are able to reside and select between different microhabitats best suited for their survival, and a positive rela-tionship between depth and body size is apparent (Dahlgren and Eggleston 2000). In contrast, large tidal fluctuations (>2 m) can completely drain and re-flood man-grove forests twice daily in portions of the Indo-Pacific (Wolanski et al. 1992, Blaber 2000). This dynamic forces fishes and shrimps to reside within subtidal riverine forests and adjacent mudflats during ebb periods, and rapidly utilize basin mangroves during flood periods (Wassenberg and Hill 1993, Lugendo et al. 2007). Under such a regime, it becomes apparent why animal assemblages in mangroves are more similar to mud flats than to coral reefs in such areas, and how segrega-tion of prey from predators may be poorly maintained (Thollot and Kulbicki 1988, Sheaves 2001, Baker and Sheaves 2006).

11.4.2 Time of Day

Another source of variation relevant to mangroves' nursery function is the time of day sampling is conducted. Comparisons between day- and night-time use of man-groves have consistently demonstrated that this habitat is predominantly utilized during the former period (Rooker and Dennis 1991, Nagelkerken et al. 2000c). This has major implications, since virtually all observations of mangroves are taken dur-ing the day. Results of multifactorial experiments demonstrate that the relative influ-ence of structure, food, and shade in attracting fishes is dependent upon the diurnal activity of the species; artificial mangrove units with structure and shade were the most attractive to nocturnally active zoobenthivores compared to diurnally active herbivores (Verweij et al. 2006a). For the former taxa, assimilation of energy and resultant growth are the result of foraging in adjacent habitats such as seagrass beds (Loneragan et al. 1997, Cocheret de la Morinière et al. 2003). For this reason, the proximity of a mangrove stand to suitable nocturnal feeding areas (inter-patch dis-tance) may, at least partially, explain why the mangroves support a higher density of fish during the day. It follows that the value of mangroves as nursery habitat may be over-estimated in systems with extensive connectivity among different habitats.

11.5 Species Variation

11.5.1 Variation Among Species

Each mangrove system may be inhabited by different species, and the number of shared species will be conditional on a variety of factors. Nonetheless, the literature is rich in examples of studies that have concluded that mangroves are not nursery habitats because of few shared species between mangroves and adjacent habitats. The most prominent comparison is between mangroves and coral reefs, and this may be one of the primary reasons opinions differ over whether mangroves are nursery habitats (Table 11.2). Blaber et al. (1985) cited that only 22 of over

Table 11.2 Summary of studies (in chronological order, 1971–present) we feel have made important statements counter to the paradigm that mangroves are nursery habitats. Although some of these studies examined both fish and decapods, all statements pertain to mangroves as nursery habitats for fish. Geographic region following Spalding et al. (1997)

Author	Location	Region	Rationale
Blaber and Blaber (1980)	Trinity Inlet system, Australia	Australasia	Fish assemblages result of quiet water, not mangrove presence
Blaber et al. (1985)	Dampier region, Northwest Australia	Australasia	Only 22 of 1,000 species on shelf occur as juveniles within the estuary
Robertson and Duke (1987)	Alligator Creek, Australia	Australasia	Only 3 of top 30 species of commercial importance
Thollot and Kulbicki (1988)	Saint-Vincent Bay, New Caledonia	Australasia	Overlap in species present between mangroves and reef low (13) compared to soft-bottom and coral reef (92) First to state the interaction between mangroves and coral reefs has been overstated
Blaber and Milton (1990)	Solomon Islands, Western Pacific	Australasia	Only 8–9% of snappers of marine origin
Chong et al. (1990)	Selangor, Malaysia	South and Southeast Asia	Fishes found within mangroves ubiquitous within estuary
Weng (1990)	Moreton Bay, Australia	Australasia	Only 5 of 86 species within mangroves were of marine origin, and only two of these of commercial importance
Dennis (1992)	La Parguera, Puerto Rico	Americas	Proposes fundamental difference between mangroves on islands and their counterparts on continental margins First to state value of mangroves as fish nurseries has been overstated in Caribbean

1,000 species that occur on the northeast Australian shelf were found in mangroves. Thollot and Kulbicki (1988) found that there was little overlap in faunal assemblages between mangroves and coral reefs in New Caledonia and concluded that linkages between the two were exaggerated. Similar conclusions have been drawn by Blaber and Milton (1990), Weng (1990), and Lin and Shao (1999). All of these studies have been conducted in the Indo-Pacific. Although not widely acknowledged, similar observations have been made in the Western Atlantic. For example, in southeastern Florida less than ten of over 70 species within mangroves can be considered reef fishes (Ley et al. 1999). Thus, for both the Indo-Pacific and Western Atlantic, when assessed at the level of *entire* fish assemblages, mangroves do not appear to be significant nurseries for coral reef fishes.

How then did such widely different opinions on the function of mangroves as nursery habitats evolve between the Indo-Pacific and the Caribbean? One explanation may be the level at which the majority of studies in the regions are conducted. Whereas studies from the Indo-Pacific have stressed the lack of congruence in faunal composition between mangroves and coral reefs at the assemblage-level, many studies from the Caribbean basin focus on the nursery function of mangroves with respect to particular species. For example, in the Florida Keys, mangroves contain the greatest densities of gray snapper (*Lutjanus griseus*) and barracuda (*Sphyraena barracuda*) relative to other available habitats (Eggleston et al. 2004). The relative abundance of *Haemulon flavolineatum, H. sciurus*, and *Lutjanus apodus* was greater within mangroves that within six other biotopes in Curaçao (Nagelkerken et al. 2000a). Further, the presence of bays containing mangroves has been shown to be positively related to adult fish stocks of certain species. Offshore Curaçao, the densities of grunts (*H. sciurus*), snapper (*L. analis, L. apodus, L. mahogoni, O. chrysurus*), parrotfish (*Scarus coeruleus*), and barracuda (*Sphyraena barracuda*) are greater on coral reefs adjacent to bays containing seagrass beds and mangroves than on coral reefs adjacent to bays without these habitats (Dorenbosch et al. 2004). Similarly, adult biomass of grunts (*H. sciurus, H. flavolineatum, H. plumieri*), and snapper (*O. chrysurus, L. apodus*) have been shown to be substantially greater in proximity to 'rich' mangrove forests than near 'mangrove scarce' areas (Nagelkerken et al. 2002, Mumby et al. 2004). This trend is not simply indicative of mangroves of the Western Atlantic; recent studies using the species-based approach have concluded that mangroves in the Indo-Pacific also contribute to the maintenance of healthy adult populations located in other habitats (Dorenbosch et al. 2005, 2006, Lugendo et al. 2005).

Because positive relationships between fishery yield and mangrove area has been demonstrated for shrimps and fishes (see review by Manson et al. 2005 and Chapter 15), some have used economic importance as a basis for making decisions as to the relevance of mangroves as nursery habitat. This approach is often problematic because ecological factors relevant to individuals of a population (e.g., growth and survival) are different from factors influencing the population maintenance with respect to fishery yield (e.g., catch and effort regulations). In addition, what species are exploited will vary based on location and the type (gear, size, and technology) of the fishery. In Southeast Florida, for example, recreational landings outnumber those from the commercial sector, shifting the label of 'economically important species' towards groupers, snappers, and grunts (i.e., fishes which utilize bays with seagrass beds and mangroves; Ault et al. 1998). In contrast, Robertson and Duke (1987) concluded that Australian mangroves were not important nursery habitats because only three of the top 30 species in total catch were of commercial importance, and Dennis (1992) concluded that the role of mangroves in Puerto Rico may be over-estimated since mangrove dependent species made up a small portion of commercial catches.

11.5.2 Variation Within Species

Another source of variation that confounds attributing nursery function to mangrove habitat is intraspecific variation in habitat utilization, behavior, or diet. Species long have been treated as homogenous units, with intraspecific variation among individuals regarded as non-existent or unimportant. Yet increasing evidence, across a broad range of taxa, suggest that intraspecific variation in niche characteristics may be substantial, and critical to include in ecological models (Bolnick et al. 2003, Bolnick et al. 2007). Ecology of tropical and sub-tropical organisms is no exception, e.g., almost all of the examples in this book seek to identify patterns at the level of 'species' or 'population', tacitly ignoring important aspects of intraspecific variation.

For species that undergo ontogenetic habitat shifts, the life-history stage of the individual must be qualified for adequate comparison of nursery value. For ingressing larvae from marine sources, mangroves may offer an attractive habitat for settlement (compared to bare substrates) because of their structural complexity. However, while modeling exercises indicate that mangroves may create complex currents that act as a hydrodynamic 'trap' for incoming larvae in the case of prawns (Wolanski and Sarenski 1997), relatively few eggs and larval fishes have been collected within mangroves compared to other habitats or life-stages in India (Krishnamurthy and Jeyasslan 1981), Australia (Robertson and Duke 1990), Puerto Rico (Dennis 1992), and Brazil (Barletta-Bergan et al. 2002). Mangroves appear to be utilized much more by individuals after settlement, and as we have discussed, comparison of size-distributions reveals that mangroves likely act as a secondary habitat in the Caribbean after seagrass beds for many species. However, while species-specific comparisons of abundance among different habitats have been extensively conducted, evaluation of how different life-history stages are distributed within patches of the same habitat has been rarely studied. In southeastern Florida, comparison of mangrove shoreline use by two marine fishes revealed that juveniles (age 0), sub-adults (<50% maturity), and adults were physically sorted along a bay-ocean gradient >10 km, with juveniles almost exclusively present near the bay-ocean mouth and adults restricted to inland portions (Faunce and Serafy 2007). Such patterns are likely due to an expansion of home range and mobility with body size, as well as intraspecific variation in habitat utilization among individuals.

11.5.3 Variation Among Individuals

Intraspecific variation can occur at a much finer scale. Even individuals of the same species and size (or age) class, which are 'resident' to the same habitat or area, may develop diverse behavioral (and presumably dietary) patterns (e.g., Verweij et al. 2006b). Such variation rarely is incorporated into the study of purported nursery habitats. We present a simplistic empirical example of individual habitat choice in Fig. 11.3. These data are drawn from an extensive acoustic telemetry monitoring program (Vemco equipment system) on Abaco Island, Bahamas (see http://www.adoptafish.net/). Individual fishes had acoustic transmitters surgically implanted into their body cavity, and stationary receivers recorded each time the 'tagged' fishes passed within their detection range (for more detail on such methodologies see Szedlmayer and Schroepfer (2005)). Such studies provide for remote monitoring of fishes, and a means by which to assess their habitat utilization and presumed foraging excursions. In Fig. 11.3, we depict the proportion of time two *Lutjanus cyanopterus* (cubera snapper) spent at different locations within an intertidal, mangrove-dominated, creek system. Each fish was tagged on the same day, was approximately the same size, and each spent the majority of daylight hours associated with a subtidal sinkhole adjacent to a mangrove stand. At night, each fish

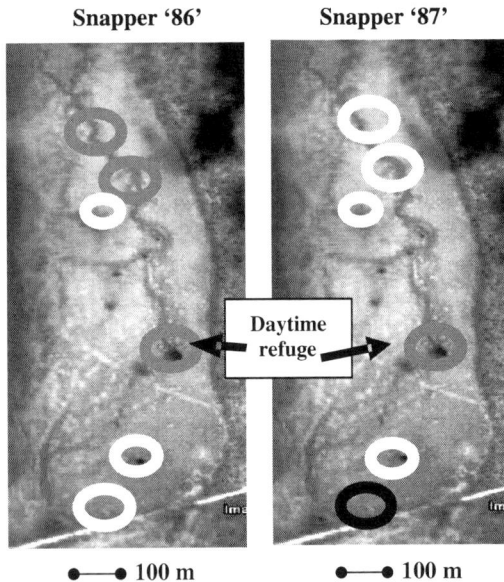

Fig. 11.3 Proportion of time (white <5%, gray 5–50%, black >50%) over a two week period two cubera snapper (*Lutjanus cyanopterus*) spent at different locations in a mangrove-lined intertidal creek system (Abaco Island, Bahamas). Both fishes are daytime residents to the same subtidal habitat, but exhibit distinct nighttime movement patterns. Snapper '86' moves upstream at night and snapper '87' downstream (presumably related to feeding movements). Data from www.adoptafish.net, and based on an acoustic telemetry monitoring system. Size of the symbols represents approximate detection range of telemetry receivers

exhibited a distinct behavioral pattern. Over the same two week period, fish '86' repeatedly moved upstream at night, whereas fish '87' moved downstream. Presumably, this reflects differential utilization of proximate habitats to feed. When scaling up from two fish to an entire population, it is easy to envision how such variation among individuals renders it difficult to assess which habitats may serve as nurseries for an entire species.

In presumed 'generalist' species, intraspecific variation in habitat utilization and foraging behaviors likely increases with increasing heterogeneity of the environment (Layman et al. 2007). That is, the more diverse the habitat mosaic (e.g., seagrass, macroalgal beds, rocky reefs, etc.), and the more diverse the associated food resources, the more likely that intraspecific variation in feeding behaviors may develop. In this context, perhaps *individuals* is the level at which habitat utilization and nursery function should be evaluated. In seeking generalities for an entire species, we may be in danger of oversimplification when attributing a single habitat as *the* 'nursery' for a species. Especially for those species which are characterized by high intraspecific variation in behaviors or dietary patterns, answering the question 'are mangroves nursery habitat?' becomes yet more complicated.

Individual variation can be reflected in the resultant chemical composition of various organs or hard parts such as otoliths. For example, comparison of the signatures deriving from the juvenile portion of the otolith relative to the adult portion can yield information on the relative contribution of individuals from bays containing mangroves to the adult populations located elsewhere. The estimated contribution of nearshore habitats (expressed as a percentage of the total adult population that inhabited bays as juveniles) is estimated at 41% for blue grouper (*Achoerodus viridis*), 32–65% for stone flounder (*Platichthys bicoloratus*), 7–53% for snapper (*Pagrus suratus*), and 40% for *Haemulon flavolineatum* (Gillanders and Kingsford 1996, Yamashita et al. 2000, Gillanders 2002, Chittaro et al. 2004). Because contribution of individuals from mangroves is not 100%, it can be concluded that while bays containing mangroves contribute individuals to the adult population, such contributions are limited. Because mangroves are not the only source of recruits to offshore populations of adult marine fishes, it appears that the export of individuals from mangroves provide *enhancement* (and not maintenance) of offshore populations of certain species, as originally proposed by Bardach (1959) and later Parrish (1989).

11.6 Conclusions

The sources of variability outlined herein are only a partial list of the myriad of factors that affect the role of mangroves as nursery habitat. Two important themes have emerged relevant to mangrove nursery function: (1) attribution of nursery function is influenced by how 'nursery' or 'mangrove ecosystem' is defined, and (2) the importance of mangroves as nursery habitat is dependent on ecological, biological, and hydrological factors that operate at multiple scales. Recent studies have made great strides toward developing more rigorous frameworks for precise quantification and categorization of the most important nursery habitats for various organisms (Beck et al. 2001, Dahlgren et al. 2006). Yet even these frameworks remain limited in many ways, largely because of the difficulty in estimating the production (based on abundance, growth, and survival) and export of individuals that utilize mangroves (or other habitats). We hope this chapter will encourage researchers to state more explicitly their study approach (definitions and analysis focus) and to acknowledge how resultant opinions regarding the importance of mangroves as nurseries stem from real underlying differences among mangrove ecosystems or species of interest.

So where does the future lie? In Fig. 11.4, we outline two parallel frameworks for identifying: (1) 'essential fish habitat' (NOAA 1996), and (2) the levels of study outlined in this chapter at which nursery function can be evaluated. These frameworks are analogous in that the endpoint (highest level) each requires a diverse suite of detailed information (from lower levels). Yet most existing data sets fall far short of these rigorous requirements. As has been emphasized throughout this chapter, much study on the nursery function of mangroves remains at the 'assemblage-' and 'species-' levels, i.e., identifying which species are present and their relative abun-

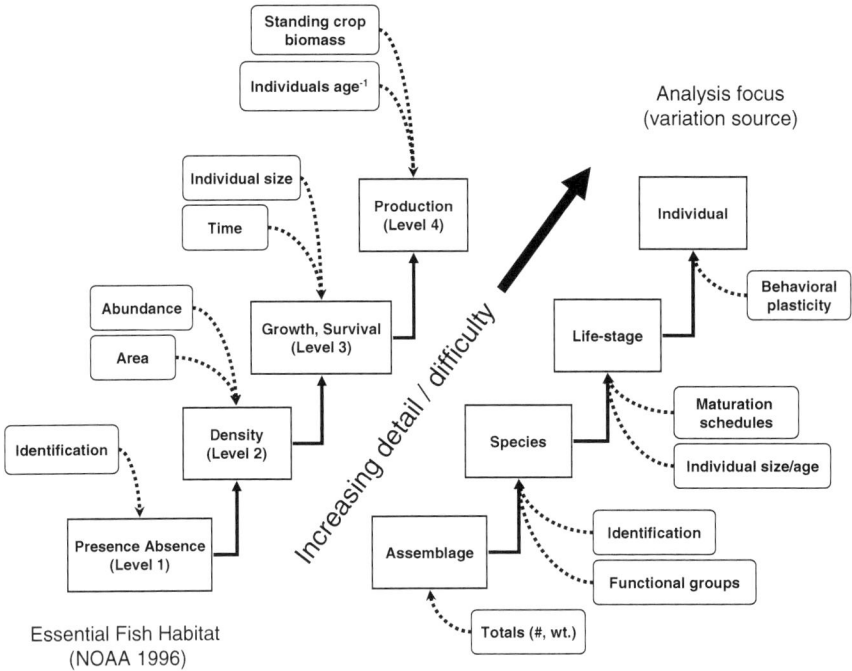

Fig. 11.4 Scheme depicting the parallels between the four levels of information used to assess essential fish habitat by the US Federal Government (left) and the levels of analysis focus described herein to assess the nursery value of mangroves. Each framework is designed so that the analysis of each successive tier (box, solid arrows) poses additional data requirements, some examples of which are illustrated in ovals with dashed arrows. Abbreviations: # = number of fish, wt. = weight of fish

dance among a range of potential habitats. Yet the most robust investigations of the nursery value of mangroves require far more extensive and specific data. As scientists continue to move toward compiling these data sets, we hope that an understanding of the inherent sources of variability in attributing nursery function will remain at the forefront of such efforts.

References

Adams A, Dahlgren C, Kellison GT et al (2006) The juvenile contribution function of tropical backreef systems. Mar Ecol Prog Ser 318:287–301

Alongi DM (2002) Present state and future of the world's mangrove forests. Environ Conserv 29:331–349

Araujo FG, De Azevedo MCC, Silva MA et al (2002) Environmental influences on the demersal fish assemblages in the Sepetiba Bay, Brazil. Estuaries 25:441–450

Ault JS, Bohnsack JA, Meester GA (1998) A retrospective (1979–1996) multispecies assessment of coral reef fish stocks in the Florida Keys. US Fish Bull 96:395–414

Baker R, Sheaves M (2006) Visual surveys reveal high densities of large piscivores in shallow estuarine nurseries. Mar Ecol Prog Ser 323:75–82

Bardach J (1959) The summer standing crop of fish on a shallow Bermuda reef. Limnol Oceanogr 4:77–85

Barletta-Bergan A, Barletta M, Saint-Paul U (2002) Community structure and temporal variability of ichthyoplankton in North Brazilian mangrove creeks. J Fish Biol 61:33–51

Bartholomew GA (1986) The role of natural history in contemporary biology. BioScience 36: 324–329

Beck MW, Heck KL, Able KW et al (2001) The identification, conservation, and management of estuarine and marine nurseries for fish and invertebrates. Bioscience 51:633–641

Blaber SJM, Blaber TG (1980) Factors affecting the distribution of juvenile estuarine and inshore fish. J Fish Biol 17:143–162

Blaber SJM (2000) Tropical estuarine fishes: ecology, exploitation, and conservation. Blackwell Science Ltd., London

Blaber SJM (2007) Mangroves and fishes: issues of diversity, dependence, and dogma. Bull Mar Sci 80:457–472

Blaber SJM, Milton DA (1990) Species composition, community structure and zoogeography of fishes of mangrove estuaries in the Solomon Islands. Mar Biol 105:259–267

Blaber SJM, Young JW, Dunning MC (1985) Community structure and zoogeographic affinities of the coastal fishes of the Dampier Region of north-western Australia. Aust J Mar Freshw Res 36:247–266

Blaber SJM, Brewer DT, Salini JP (1989) Species composition and biomasses of fishes in different habitats of a tropical northern Australian estuary: their occurrence in the adjoining sea and estuarine dependence. Estuar Coast Shelf Sci 29:509–531

Bolnick DI, Svanbäck R, Fordyce JA et al (2003) The ecology of individuals: incidence and implications of individual specialization. Am Nat 161:1–28

Bolnick DI, Svanback R, Araujo M et al (2007) More generalized populations are also more heterogeneous: comparative support for the niche variation hypothesis. Proc Natl Acad Sci USA 104:10075–10079

Chittaro PM, Fryer BJ, Sale PF (2004) Discrimination of French grunts (*Haemulon flavolineatum*, Desmarest, 1823) from mangrove and coral reef habitats using otolith microchemistry. J Exp Mar Biol Ecol 308:169–183

Chong VC, Sasekumar A, Leh MUC et al (1990) The fish and prawn communities of a Malaysian coastal mangrove system, with comparisons to adjacent mud flats and inshore waters. Estuar Coast Shelf Sci 31:703–722

Cocheret de la Morinière E, Pollux BJA, Nagelkerken I et al (2002) Postsettlement life cycle migration patterns and habitat preference of coral reef fish that use seagrass and mangrove habitats as nurseries. Estuar Coast Shelf Sci 55:309–321

Cocheret de la Morinière E, Pollux BJA, Nagelkerken I et al (2003) Ontogenetic dietary changes of coral reef fishes in the mangrove-seagrass-reef continuum: stable isotopes and gut-content analysis. Mar Ecol Prog Ser 246:279–289

Costanza R, dArge R, deGroot R et al (1997) The value of the world's ecosystem services and natural capital. Nature 387:253–260

Dahlgren CP, Eggleston (2000) Ecological processes underlying ontogenetic habitat shifts in a coral reef fish. Ecology 81:2227–2240

Dahlgren CP, Kellison GT, Adams AJ et al (2006) Marine nurseries and effective juvenile habitats: concepts and applications. Mar Ecol Prog Ser 312:291–295

Dennis GD (1992) Island mangrove habitats as spawning and nursery areas for commercially important fishes in the Caribbean. Proc Gulf Caribb Fish Inst 41:205–225

Dorenbosch M, van Riel MC, Nagelkerken I et al (2004) The relationship of reef fish densities to the proximity of mangrove and seagrass nurseries. Estuar Coast Shelf Sci 60:37–48

Dorenbosch M, Grol MGG, Christianen MJA et al (2005) Indo-Pacific seagrass beds and mangroves contribute to fish density and diversity on adjacent coral reefs. Mar Ecol Prog Ser 302:63–76

Dorenbosch M, Grol MGG, Nagelkerken I et al (2006) Seagrass beds and mangroves as potential nurseries for the threatened Indo-Pacific humphead wrasse, *Cheilinus undulatus* and Caribbean rainbow parrotfish, *Scarus guacamaia*. Biol Conserv 129:277–282

Dufrêne M, Legendre P (1997) Species assemblages and indicator species: the need for a flexible asymetrical approach. Ecol Monogr 67:345–366.

Eggleston, DB, Dahlgren C, Johnson EG (2004) Fish density, diversity and size-structure within multiple back-reef habitats of Key West National Wildlife Refuge, USA. Bull Mar Sci 75:175–204

Ewel KC, Twilley RR, Ong JE (1998) Different kinds of mangrove forests provide different goods and services. Glob Ecol Biogeogr 7:83–94

Faunce CH, Serafy JE (2006). Mangroves as fish habitat: 50 years of field studies. Mar Ecol Prog Ser 318:1–18

Faunce CH, Serafy JE (2007) Nearshore habitat use by gray snapper (*Lutjanus griseus*) and bluestriped grunt (*Haemulon sciurus*): environmental gradients and ontogenetic shifts. Bull Mar Sci 80:473–495

Faunce CH, Serafy JE (2008a) Growth and secondary production of an eventual reef fish during mangrove residency. Estuar Coast Shelf Sci 79:93–100

Faunce CH, Serafy JE (2008b) Selective use of mangrove shorelines by snappers, grunts, and great barracuda. Mar Ecol Prog Ser 356:153–162

Faunce CH, Serafy JE, Lorenz JJ (2004) Density-habitat relationships of mangrove creek fishes within the southeastern saline Everglades (USA), with reference to managed freshwater releases. Wetlands Ecol Manage 12:377–394

Faunce CH, Ault E, Ferguson K et al (2007) Reproduction of four Florida snappers with comparisons between an island and continental reef system. In: Barbieri L Colvocoresses J (eds) Southeast Florida reef fish abundance and biology. Five-year performance report to the U.S. Fish and Wildlife Service. Florida Fish and Wildlife Research Institute, St. Petersburg

Gillanders BM (2002) Connectivity between juvenile and adult fish populations: do adults remain near their recruitment estuaries? Mar Ecol Prog Ser 240:215–223

Gillanders BM, Kingsford MJ (1996) Elements in otoliths may elucidate the contribution of estuarine recruitment to sustaining coastal reef populations of a temperate reef fish. Mar Ecol Prog Ser 141:13–20

Hajisamae S, Chou LM (2003) Do shallow water habitats of an impacted coastal strait serve as nursery grounds for fish? Estuar Coast Shelf Sci 56:281–290

Heyman, WD, Kjerfve B, Graham RT et al (2005) Spawning aggregations of *Lutjanus cyanopterus* (Cuvier) on the Belize Barrier Reef over a six year period. J Fish Biol 67:83–101

Hogarth, PJ 2007. The biology of mangroves and seagrasses. Oxford University Press, London

Koenig CC, Coleman FC, Eklund A et al (2007) Mangroves as essential nursery habitat for goliath grouper (*Epinephelus itajara*). Bull Mar Sci 80:567–586

Krishnamurthy K, Jeyasslan MJP (1981) The early life history of fishes from Pichavaram mangrove ecosystem of India. In: Lasker R, Sherman K (eds) The early life history of fish: recent studies. The second ICES symposium. Conseil International Pour L'Exploration de la Mer Palegade 2–4, Copenhagen, Denmark

Layman CA, Quattrochi JP, Peyer CM et al (2007) Niche width collapse in a resilient top predator following ecosystem fragmentation. Ecol Lett 10:937–944

Ley JA (2005) Linking fish assemblages and attributes of mangrove estuaries in tropical Australia: criteria for regional marine reserves. Mar Ecol Prog Ser 305:41–57

Ley JA, McIvor CC, Montague CL (1999) Fishes in mangrove proproot habitats of northeastern Florida Bay: distinct assemblages across an estuarine gradient. Estuar Coast Shelf Sci 48:701–723

Lewis RR (2005) Ecological engineering for successful management and restoration of mangrove forests. Ecol Eng 24:403–418

Lin HJ, Shao KT (1999) Seasonal and diel changes in a subtropical mangrove fish assemblage. Bull Mar Sci 65:775–794

Little MC, Reay PJ, Grove SJ (1988) Distribution gradients of ichthyoplankton in an East African mangrove creek. Estuar Coast Shelf Sci 26:669–677

Loneragan NR, Bunn SE, Kellaway DM (1997) Are mangroves and seagrasses sources of organic carbon for penaeid prawns in a tropical Australian estuary? A multiple stable isotope study. Mar Biol 130:289–300

Lorenz JJ (1999) The response of fishes to physicochemical changes in the mangroves of northeast Florida Bay. Estuaries:500–517

Lugendo BR, Pronker A, Cornelissen I et al (2005) Habitat utilisation by juveniles of commercially important fish species in a marine embayment in Zanzibar, Tanzania. Aquat Living Resour 18:149–158

Lugendo BR, Nagelkerken I, Kruitwagen G et al (2007) Relative importance of mangroves as feeding habitat for fish: a comparison between mangrove habitats with different settings. Bull Mar Sci 80:497–512

Lugo AE, Snedaker SC (1974) The ecology of mangroves. Annu Rev Ecol Syst 5:39–64

Lugo AE (1980) Mangrove ecosystems: sucessional or steady state? Biotropica 12:65–72

Manson FJ, Loneragan NR, Skilleter GA et al (2005) An evaluation of the evidence for linkages between mangroves and fisheries: a synthesis of the literature and identification of research directions. Annu Rev Oceanogr Mar Biol 43:483–513

Mumby PJ, Edwards AJ, Arias-Gonzales JE et al (2004) Mangroves enhance the biomass of coral reef fish communities in the Caribbean. Nature 427:533–536

Nagelkerken I (2007) Are none-stuarine mangroves connected to coral reefs through fish migration? Bull Mar Sci 80:595–607

Nagelkerken I, Dorenbosch M, Verberk WCEP et al (2000a) Importance of shallow-water biotopes of a Caribbean bay for juvenile coral reef fishes: patterns in biotope association, community structure and spatial distribution. Mar Ecol Prog Ser 202:175–192

Nagelkerken I, van der Velde G, Gorissen MW et al (2000b) Importance of mangroves, seagrass beds and the shallow coral reef as a nursery for important coral reef fishes, using a visual census technique. Estuar Coast Shelf Sci 51:31–44

Nagelkerken I, Dorenbosch M, Verberk WCEP et al (2000c) Day-night shifts of fishes between shallow-water biotopes of a Caribbean bay, with emphasis on the nocturnal feeding of Haemulidae and Lutjanidae. Mar Ecol Prog Ser 194:55–64

Nagelkerken I, Roberts CM, van der Velde G et al (2002) How important are mangroves and seagrass beds for coral-reef fish? The nursery hypothesis tested on an island scale. Mar Ecol Prog Ser 244:299–305

NOAA (National Oceanic and Atmospheric Administration, National Marine Fisheries Service) (1996) Magnuson-Stevens Fishery Conservation and Management Act, as amended through October 11, 1996. NOAA technical memorandum NMFS-F/SPO, 121pp.

Paris CB, Cowen RK, Claro R et al. (2005) Larval transport path-ways from Cuban snapper (Lutjanidae) spawning aggregations based on biophysical modeling. Mar Ecol Prog Ser 296:93–106

Parrish JD (1989) Fish communities of interacting shallow-water habitats in tropical oceanic regions. Mar Ecol Prog Ser 58:143–160

Provost MW (1973) Mean high water mark and use of tidelands in Florida. Fla Sci 36:50–66

Quinn NJ (1980) Analysis of temporal changes in fish assemblages in Serpentine Creek, Queensland. Environ Biol Fishes 5:117–133

Ricker WE (1975) Computation and interpretation of biological statistics of fish populations. Bull Canadian Fish Res Board 1–392

Robertson AI, Duke NC (1987) Mangroves as nursery sites: comparisons of the abundance and species composition of fish and crustaceans in mangroves and other nearshore habitats in tropical Australia. Mar Biol 96:193–205

Robertson AI, Duke NC (1990) Mangrove fish-communities in tropical Queensland, Australia: spatial and temporal patterns in densities, biomass and community structure. Mar Biol 104: 369–379

Rönnbäck P, Troell M, Kautsky N et al (1999) Distribution pattern of shrimps and fish among *Avicennia* and *Rhizophora* microhabitats in the Pagbilao Mangroves, Philippines. Estuar Coast Shelf Sci 48:223–234

Rooker JR, Dennis GD (1991) Diel, lunar and seasonal changes in a mangrove fish assemblage off southwestern Puerto Rico. Bull Mar Sci 49:684–698

Serafy JE, Araujo RJ (2007) First international symposium on mangroves as fish habitat – Preface. Bull Mar Sci 80:453–456

Serafy JE, Faunce CH, Lorenz JJ (2003) Mangrove shoreline fishes of Biscayne Bay, Florida. Bull Mar Sci 72:161–180

Serafy JE, Valle M, Faunce CH et al (2007) Species-specific patterns of fish abundance and size along a subtropical mangrove shoreline: an application the delta approach. Bull Mar Sci 80:609–624

Sheaves MJ (2001) Are there really few piscivorous fishes in shallow estuarine habitats? Mar Ecol Prog Ser 222:279–290

Sheridan P, Hays C (2003) Are mangroves nursery habitat for transient fishes and decapods? Wetlands 23:449–458

Spalding M, Blasco F, Field C (1997) World mangrove atlas. The International Society for Mangrove Ecosystems (ISME), Okinawa

Szedlmayer ST, Schroepfer RL (2005) Longterm residence of red snapper on artificial reefs in the northeastern Gulf of Mexico. Trans Am Fish Soc 134:315–325

Thayer GW, Stuart HH, Kenworthy WJ et al (1978) Habitat values of salt marshes, mangroves, and seagrasses for aquatic organisms. In: Greeson PE, Clark JR, Clark JE (eds) Wetlands functions and values: the state of our understanding. American Water Resources Association, Washington DC

Thollot P, Kulbicki M (1988) Overlap between the fish fauna inventories of coral reefs, soft bottoms and mangroves in Saint-Vincent Bay (New Caledonia). In: Choat JH, Barnes D, Borowitzka MA et al (eds) Sixth International Coral Reef Symposium, Townsville

Valentine-Rose L, Layman CA, Arrington DA et al (2007) Habitat fragmentation decreases fish secondary production in Bahamian tidal creeks. Bull Mar Sci 80:863–877

Verweij MC, Nagelkerken I, de Graaff D et al (2006a) Structure, food and shade attract juvenile coral reef fish to mangrove and seagrass habitats: a field experiment. Mar Ecol Prog Ser 306:257–268

Verweij MC, Nagelkerken I, Wartenbergh SLJ et al (2006b) Caribbean mangroves and seagrass beds as daytime feeding habitats for juvenile French grunts, *Haemulon flavolineatum*. Mar Biol 149:1291–1299

Wassenberg TJ, Hill BJ (1993) Diet and feeding behaviour of juvenile and adult banana prawns (*Penaeus merguiensis* de Man) in the Gulf of Carpentaria, Australia. Mar Ecol Prog Ser 94:287–295

Weng HT (1990) Fish in shallow areas in Moreton Bay, Queensland and factors affecting their distribution. Estuar Coast Shelf Sci 30:569–578

Wolanski E, Mazda Y, Ridd P (1992) Mangrove hydrodynamics. In: Robertson AI, Alongi DM (eds) Tropical mangrove ecosystems. Coastal and Estuarine Studies 41. American Geophysical Union, Washington DC

Wolanski E, Sarenski J (1997) Larvae dispersion in coral reefs and mangroves. Am Sci 85:236–243

Woodroffe C (1992) Mangrove sediments and geomorphology. In: Robertson AI, Alongi DM (eds) Tropical mangrove ecosystems. Coastal and Estuarine Studies 41. American Geophysical Union, Washington DC

Yamashita Y, Otake T, Yamada H (2000) Relative contributions from exposed inshore and estuarine nursery grounds to the recruitment of stone flounder, *Platichthys bicoloratus*, estimated using otolith Sr:Ca ratios. Fish Oceanogr 9:316–327

Part III
Tools for Studying Ecological and Biogeochemical Linkages

Chapter 12
Tools for Studying Biogeochemical Connectivity Among Tropical Coastal Ecosystems

Thorsten Dittmar, Boris Koch and Rudolf Jaffé

Abstract To understand ecosystem functioning in coastal zones it is essential to identify the main pathways and magnitude of nutrient and organic matter fluxes. The different methods that have been applied to quantify material fluxes in tropical coastal ecosystems can be categorized into two fundamentally distinct approaches: direct flux measurements (Section 12.2) and chemical tracer techniques (Section 12.3). For direct flux measurements, the bidirectional flow of water is determined and multiplied with concentrations to obtain fluxes of inorganic nutrients or organic constituents. Water discharge can be measured directly with help of current meters and gauges, or indirectly through use of water tracers. The source of nutrients and organic matter can then be identified with help of specific chemical tracers, mainly isotopes or molecular properties. A combination of tracer techniques and direct flux measurements is most powerful to obtain quantitative information on the fluxes of organic matter and nutrients from the various sources in coastal systems but has very rarely been applied. Regarding source assessments of suspended and sedimentary organic matter, a large number of molecular biomarkers are readily available (Section 12.4). For dissolved organic matter, emerging molecular fingerprinting techniques including ultra-high resolution mass spectrometry may lead to major advances in the future.

Keywords Outwelling · Molecular tracers · Isotopes · Organic matter · Nutrients

12.1 Introduction

Estuaries are at the interface where the land meets the sea and are important conduits of nutrients and organic matter to the ocean (e.g., Hedges and Keil 1995, Gordon and Goñi 2004). Commonly, the contributions of terrigenous and marine-derived

T. Dittmar (✉)
Max Planck Research Group for Marine Geochemistry, Carl von Ossietzky University, Institute for Chemistry and Biology of the Marine Environment, 26111 Oldenburg, Germany
e-mail: tdittmar@mpi-bremen.de

I. Nagelkerken (ed.), *Ecological Connectivity among Tropical Coastal Ecosystems*,
DOI 10.1007/978-90-481-2406-0_12, © Springer Science+Business Media B.V. 2009

organic matter in estuaries have been assessed by a two end-member mixing model (e.g., Prahl et al. 1994, Gordon and Goñi 2004) with little attention on the estuarine biomass contributions to the organic matter pool. Fringe mangrove ecosystems in tropical and subtropical coastal zones are one such estuarine organic matter contributor and need to be carefully considered as an important bioreactor for nutrient fluxes to the oceans (e.g., Lee 1995, Dittmar et al. 2006, Bouillon et al. 2007).

Nutrient fluxes in coastal zones have sharply increased over the last decades for multiple reasons. Changes in land use associated with enhanced mineralization of soil organic matter, runoff of excess nutrients from agricultural areas, and sewage discharge from urban areas are the main causes. It has been argued that mangroves may trap excess nutrients and thus alleviate negative impacts of nutrient pollution in tropical coastal regions (e.g., Primavera et al. 2007, Maie et al. 2008). Pristine mangrove systems, however, can also present a significant net-source of nutrients to the coastal ocean. For instance, considerable outwelling of dissolved inorganic nutrients (N, P, and Si compounds) was observed from North Brazilian mangroves that exceeded the local riverine fluxes by orders of magnitude (Dittmar and Lara 2001a, b). Autochthonous nutrients in the mangrove sediments are probably derived from nitrogen fixation and gradual weathering of P- and Si-rich minerals. Other mangrove systems are nutrient limited and their net ecosystem productivity increases with inputs of inorganic N, P, or both (e.g., McKee et al. 2002). Those systems can efficiently absorb nutrients from external sources. Nutrients are assimilated into biomass by a variety of different primary producers in mangroves, mainly trees and phytobenthos that can grow in dense mats on the sediments and roots. A significant fraction of net-primary production is exported as particulate and dissolved organic matter (POM and DOM) to adjacent coastal waters (e.g., Dittmar et al. 2001a). On a global scale, numerous studies indicate that mangrove forests are a significant net-source of detritus and DOM to adjacent coastal water, and it was estimated that mangroves account for >10% of the terrestrially-derived DOM transported to the ocean (Dittmar et al. 2006), while they cover only <0.1% of the continents' surface.

DOM in aquatic environments has been widely studied because of its importance in a variety of physical, geochemical, and biological processes (e.g., Scully and Lean 1994, Alberts and Takacs 1999, Cai et al. 1999, Del Castillo et al. 2000). Mangrove-derived organic matter that is exported to the ocean consists of a complex mixture of many thousands of different organic compounds (Tremblay et al. 2007). Some of these compounds resist rapid degradation by microorganisms, others are labile and comprise a source of energy and nutrients to heterotrophic organisms (Dittmar et al. 2006). Outwelling of organic matter can thus fuel secondary production in coastal areas. Nutrients are released back into the water column and oxygen is consumed. The extent of this process not only depends on the amount of organic matter being released from a mangrove system but also on its nutritional quality. Refractory compounds can be dispersed over large distances on continental shelves and nutrients are slowly released in small quantities into the water column (Dittmar et al. 2006).

Labile compounds, on the other hand, can be quickly consumed in direct vicinity of the mangrove, which can return large quantities of inorganic nutrients into the water column and may reduce oxygen even to hypoxia levels.

Coastal outwelling is different in several aspects from riverine transport. While river mouths often act as a point-source destination for continental materials onto the shelf, intertidal zones generally accommodate highly productive ecosystems covering broad geographic areas. Transport mechanisms and water sources that may facilitate coastal outwelling include precipitation, surface water runoff, overland flow, groundwater discharge, tidal recharge/discharge through sediments, and bidirectional tidal currents linking the mangrove to the ocean (Valiela et al. 1978, Harvey et al. 1987, Childers et al. 1993, Troccaz et al. 1994, Krest et al. 2000, Tobias et al. 2001, Dittmar and Lara 2001b). The complex geomorphology of tidal creek systems resembles a fractal geometry in which a high proportion of the sediment surface is exposed to tidal waters (Morris 1995, Fagherazzi and Sun 2004, Mudd et al. 2004, D'Alapaos et al. 2005). Likewise, a high proportion of the ecosystem is in contact with coastal aquifers (Novakowski et al. 2004, Gardner 2005). Tidal pumping of surface and pore waters, and groundwater discharge can contribute a substantial water input to the coastal ocean (Moore 1999, Taniguchi et al. 2002, Wilson and Gardner 2006).

Although the importance of estuarine and coastal waters in the global DOM cycling has been recognized, sources, transport, and transformation of DOM are not sufficiently understood. A key challenge in this field is tracing DOM from different sources in complex ecosystems such as coastal wetlands and estuaries. Quantitative determinations of dissolved and particulate organic carbon (DOC and POC) are commonly reported, but also the 'quality', source, and the degree of degradation of organic matter need to be determined to better understand organic matter dynamics in these ecosystems. A suite of analytical methods has been developed and applied for such purpose, ranging from simple optical DOM property measurements in bulk water samples (e.g., Jaffé et al. 2004, Maie et al. 2006a) to complex molecular characterizations (e.g., Maie et al. 2005, Tremblay et al. 2007, Xu et al. 2007).

The different methods that have been applied to assess nutrient and organic matter fluxes in mangrove ecosystems can be categorized into two fundamentally distinct approaches: direct flux measurements (Section 12.2) and chemical tracer techniques (Section 12.3). For direct flux measurements, the bidirectional flow of water is determined and multiplied with concentrations to obtain fluxes of inorganic nutrients or organic constituents. Water discharge can be measured directly with help of current meters and gauges, or indirectly through use of water tracers. The source of nutrients and organic matter can then be identified with help of specific chemical tracers, mainly isotopes or molecular properties. A combination of both approaches is most powerful to obtain quantitative information on the fluxes of organic matter and nutrients from the various sources in mangrove-fringed systems and to assess the impact of these fluxes on coastal and marine ecosystem functioning (e.g., Dittmar et al. 2006).

12.2 Direct Flux Measurements

Water flow patterns in estuaries and tidal creeks are highly complex. Water flow is turbulent, continuously changes in magnitude and direction with the tides, and does not take place in well-defined flow channels. This complexity makes direct water-discharge measurements via gauges and current meters a challenging task in tidal environments. A detailed discussion on the hydrodynamics of mangrove swamps and their coastal waters can be found in Wolanski and Ridd (1990) and Wolanski (1992). Numerical models have been developed for a detailed description of fluxes and associated material fluxes in these challenging environments. The 'General Estuarine Transport Model' (GETM) has been specifically designed for reproducing baroclinic, bathymetry-guided flows where the tidal range may exceed the mean water depth in large parts of the domain such that drying and flooding processes are relevant (Burchard and Bolding 2002). This model has been successfully applied, e.g., in a three-dimensional simulation of the Elbe estuary and its turbidity zone in Northern Germany (Burchard et al. 2004) and on the extensive tidal flats in the East Frisian Wadden Sea (Stanev et al. 2003). The use of this or similar numerical models in the context of this chapter is not known, but it has enormous potential for the quantitative analysis of the nutrient and organic matter exchange in tropical coastal environments. In the following, two comparatively simple and common approaches are described to quantify material fluxes on estuarine and tidal creek scales.

A common method to quantify estuarine source and sink terms is to determine river discharge as a proxy for total freshwater input and salinity as a conservative tracer for seawater in the estuary and coastal zone (e.g., Bianchi 2007). The uptake or release of nutrients and organic matter in the estuary can then be assessed via a simple two-source mixing model. This approach has often been applied to identify non-conservative behavior of constituents in the estuarine water column (e.g., Dittmar and Lara 2001a). If the hydrology of the estuary is sufficiently known, the same model can be used in a quantitative way to determine estuarine fluxes (Fig. 12.1). Input parameters are riverwater discharge, nutrient concentrations of the riverine and marine endmembers, and nutrient concentrations along the salinity gradient of the estuary. Output parameter is an estuarine flux term that quantitatively integrates all additional fluxes in and out of the estuary, aside riverine and oceanic fluxes. The input parameters can be determined at relatively high precision. An adequate sampling strategy is crucial for the successful application of this approach. Deviations from conservative mixing are rarely as consistent as suggested in Fig. 12.1. Plankton blooms, accumulation of detritus along fronts, outwelling blooms and other localized features can cause inconsistent distribution patterns of nutrient and organic matter concentrations along the salinity gradient. The sampling density in the estuary has to be sufficient to capture these heterogeneities in space and time.

An important assumption of the two-source mixing model is that all freshwater is riverine. This assumption is reasonable for many, if not most estuaries. In some coastal areas, however, submarine groundwater discharge may impact the freshwater

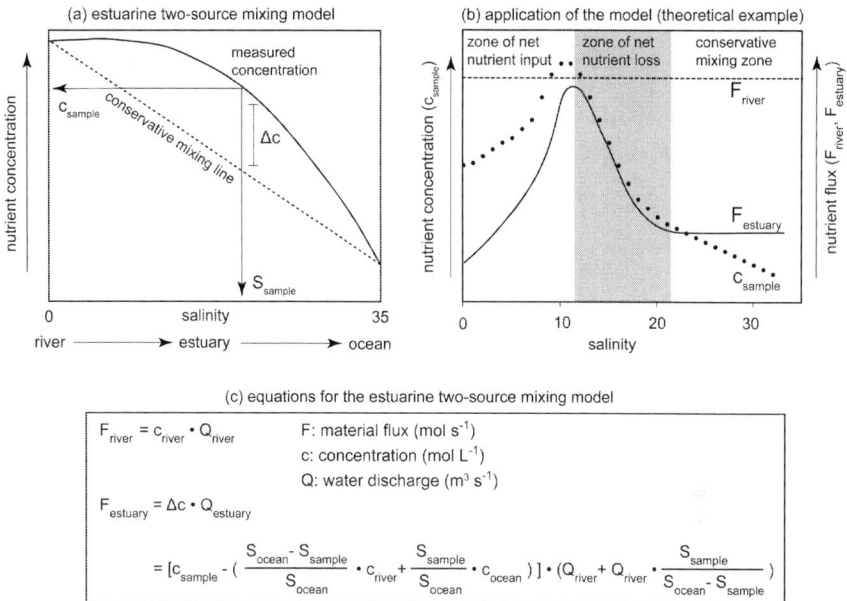

Fig. 12.1 An estuarine two-source mixing model, using salinity as a conservative tracer for freshwater. (**a**) Deviations from conservative mixing can be explored to quantify inputs or losses of nutrients and other constituents in the estuary. (**b, c**) Equations and a theoretical example for estuarine flux calculations are given. Details on the applicability and assumptions are given in the text

balance of estuaries. Submarine groundwater discharge can also be a significant source of nutrients and organic matter to the coastal ocean (e.g., Santos et al. 2008). The submarine discharge of groundwater and associated solutes has received increased attention during recent decades, and its impact on coastal biogeochemistry is widely recognized (Moore 2006). The flux of groundwater into coastal waters can be estimated with help of radioisotopes. In particular ^{226}Ra and ^{228}Ra are widely applied as tracers for submarine groundwater discharge (Moore 2006). Ra is a decay product of Thorium, which is tightly bound to particles, whereas Ra may desorb. As groundwater percolates through sediments, Ra can be transferred to the water column, where it behaves conservatively and can thus be used as a tracer for submarine groundwater discharge. If the nutrient or organic matter concentrations of the groundwater endmember are sufficiently known, the respective fluxes can be estimated by multiplying the water discharge with the concentrations in the groundwater endmember (e.g., Hwang et al. 2005). This approach is complicated by the fact that nutrients and organic matter are subject to extensive biogeochemical transformations in the sediments and that endmember concentrations can often not be established precisely. Nutrient and organic matter concentrations can vary over several orders of magnitude even in a single aquifer (Santos et al. 2008). The discharge of nutrients and organic matter through submarine groundwater has not been studied in mangrove systems yet, but should gain increased attention in future studies.

Although the estuarine mixing model can quantify estuarine source and sink terms of nutrients and organic matter, the sources and sinks can only be localized relative to salinity, but not directly linked to specific processes or environments. For instance, in mangrove-fringed estuaries organic matter or nutrient inputs at mid salinity can be of planktonic or mangrove origin which the estuarine mixing model cannot distinguish. Flux measurements on a smaller scale can yield quantitative information related to a specific source or sink. Many mangroves, in particular in macrotidal regions, exchange water and constituents with the estuary through well-defined tidal creeks (e.g., Dittmar and Lara 2001a). The bidirectional flux of water in these tidal channels can be quantified with current meters and water gauges. The published data for nutrient and organic matter fluxes in mangroves (see below) were all obtained with conventional one-point current meters. The recent advent of shallow-water acoustic Doppler current profilers (ADCP) that continuously monitor two-dimensional current profiles in cross sections has strongly improved our capabilities to precisely monitor currents in tidal creeks. Current measurements combined with water sampling over a tidal cycle yield flux data of solutes or suspended matter (Fig. 12.2). Net fluxes between the mangrove and estuary are obtained by integrating the fluxes over a complete ebb and flood cycle. This method has been applied in most flux studies (e.g., Boto and Bunt 1981, Twilley 1985, Alongi et al. 1998, Dittmar and Lara 2001a). Wattayakorn et al. (1990) and Ayukai et al. (1998) modified this approach by calculating the material flux in mangrove creeks from the tidal diffusion equation assuming a conservative behavior of the material of interest. Dittmar and Lara (2001b) pointed out that fluxes at day and night can be significantly different, because of aquatic primary production and respiration, leading to significant asymmetries in material fluxes between day and night. It is therefore

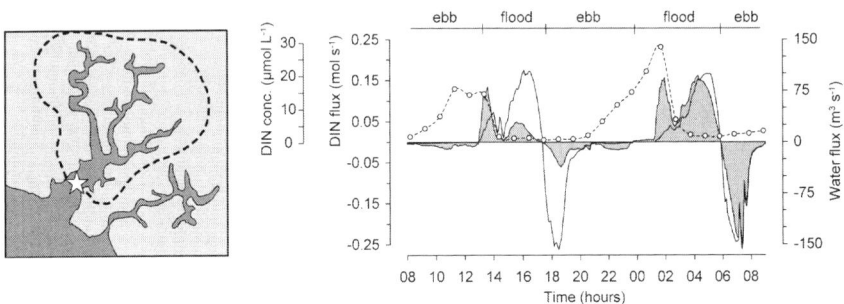

Fig. 12.2 The determination of dissolved inorganic nitrogen (DIN) fluxes in a mangrove tidal creek in Northern Brazil (right-hand panel): DIN concentrations (dotted line) can be multiplied with water discharge (black line) to obtain DIN fluxes (grey areas). The net flux can be calculated as the integral of the DIN fluxes, i.e., the sum of all in- and output over a complete tidal cycle. If the catchment area of the tidal creek is well know, net-fluxes can be normalized to the mangrove area to obtain flux rates (mol. d^{-1}. m^{-2}). The schematic sketch (left-hand panel) illustrates a tidal creek system with a well-defined catchment area (area within dotted line); the star represents the location of a sampling station at the mouth of the tidal creek where the fluxes in and out of the catchment area can be determined. Data from Dittmar and Lara (2001a)

important to monitor tidal fluxes of nutrients and organic matter at least over the course of one day, and not only over one tidal cycle. Vertical stratification can be strong in tidal creeks, which has to be properly addressed. Stratification can be tested with CTD sensors (electric conductivity, temperature, and depth), and an adequate water sampling strategy should be developed accordingly.

If the drainage area of a creek is well-defined, the net material and water fluxes can be normalized to the specific drainage area to obtain area-normalized net-exchange rates (Dittmar and Lara 2001a). It is important that below-ground connections between tidal creeks or creek connections through forest inundation at high tide can be excluded (Fig. 12.2). Some studies applied artificial flumes to monitor fluxes in mangroves. Rivera Monroy et al. (1995) and Romigh et al. (2006) used an artificial 12-m flume to determine material exchange between a small area of mangroves and a tidal creek in the Everglades (Florida). This approach is particularly advantageous if tidal creeks with well-defined flow channels and drainage areas are absent. However, flumes usually drain smaller mangrove areas than natural creeks. Therefore, tidal signals in artificial flumes are usually less pronounced than in creeks, which may cause larger methodological uncertainties.

A general drawback of all direct flux measurements is the fact that net-fluxes have to be calculated as the difference between net-inflow and net-outflow. Potentially large errors are introduced by the inaccuracy of flow and concentration measurements (e.g., Kjerfve et al. 1981). Flow measurements have a relatively large methodological uncertainty and the difference between in- and outflow can be similar or less than the methodological error margin, in which case reliable net-fluxes cannot be determined. Dittmar and Lara (2001a) introduced an approach to correct for systematic uncertainties with help of salinity, assuming that the net balance of salt over a tidal cycle is zero. The overall quality of material net flux estimates can be assessed by the analysis of water balance asymmetries. Ideally, the water flux in a tidal creek with a well-defined drainage area is balanced, i.e., the outflow of water is equal to its inflow. The precision of flux estimates can be further tested during major rain events. For instance, Dittmar and Lara (2001a) detected a net outflow of $36 \ mm \cdot d^{-1}$ from a $2.2 \ km^2$ mangrove area in Brazil, which corresponded to a major rain event of $34 \ mm \cdot d^{-1}$ during flood. The method was therefore sensitive enough to detect this water flux asymmetry caused by a rainwater pulse.

12.3 Tracing the Source of Organic Matter in the Water Column

12.3.1 General Remarks

Direct flux measurements as described in the previous section are essential for any quantitative understanding of element fluxes in the coastal zone. The ultimate source and the processing of organic matter and nutrients in the coastal zone, however, remain often unclear. For instance, photochemical reactions in combination with bacterial remineralization can remove terrigenous DOM from the water column (e.g., Dittmar et al. 2006) and release nutrients at the same time. These nutrients can be taken up by phytoplankton in the estuary and are ultimately returned to the DOM pool. In this case DOC and nutrient concentrations can show conservative behavior, even though terrigenous DOM is replaced by planktonic DOM through complex biogeochemical processes in the estuary. The identification of these processes is essential for a full understanding of ecosystem functioning. Several natural tracer techniques have been developed over the last decades that shed more light on the origin and processing of estuarine organic matter. The different approaches that have been successfully applied in tropical coastal systems will be discussed in the following sub-sections.

An ideal tracer is specific for an organic matter source (species, vegetation type, etc.) and is persistent enough to survive degradation processes on the time scale considered. In addition, the most suitable tracer techniques also provide quantitative information on degradation processes. Obviously, no single tracer technique can provide all this information at the same time. In addition, all available biomarkers are to some extent selectively degraded in comparison to the bulk organic material and therefore are limited in quantitatively characterizing organic matter fluxes. The large variety of different compound classes in fresh organic material is subject to different partitioning and decay mechanisms such as adsorption, photo- or microbial degradation. For each type of organic molecule the reaction dynamics in these processes are different. Some biomolecules are preserved for longer time scales but the majority is degraded quickly, and most of the degradation products escape our analytical window (Hedges et al. 2000). To overcome these difficulties a combination of different approaches can be used, each of which has its strength and drawbacks.

Bulk chemical information, such as optical properties, isotope ratios (e.g., $\delta^{13}C$), or elemental composition (e.g., C/N ratios) can usually be determined on a larger number of samples than molecular biomarkers. However, bulk parameters often lack the specificity of biomarkers, and modifications in the course of degradation often limit their quantitative application. Some of the molecular biomarkers are highly source specific and organic matter from individual ecosystems or even species can be traced in sediments and in the water column. The validation of chemical approaches is often possible through comparison with microfossil analysis (e.g., pollen or foraminifera shells). Since this opportunity is not given in the dissolved phase, the tracing of DOM fluxes is more challenging.

12.3.2 Optical Approaches

Due to the ease of operation, high sample throughput, and high sensitivity, UV-visible and fluorescence spectroscopic techniques have been widely used to characterize sources, degree of degradation, and transformation of DOM in many aquatic environments (e.g., de Souza Sierra et al. 1994, Coble 1996, Lombardi and Jardim 1999, McKnight et al. 2001, Clark et al. 2002). The term CDOM represents the chromophoric or optically active fraction of the DOM pool. The absorbance at 254 nm

(A_{254}) is often used as a quantitative measure for CDOM (e.g., Martin-Mousset et al. 1997, Jaffé et al. 2004), while the DOC-normalized absorbance (or Specific Ultra violet Absorbance-SUVA) at 254 nm reflects the degree of aromaticity of the sample (Weishaar et al. 2003). Other optical parameters used to assess DOM quality are the UV-Visible Index (A_{254}/A_{436}) and a related parameter, the exponential slope (S). S can be estimated after applying a nonlinear exponential regression to the UV-Vis absorbance spectrum (290–700 nm). These parameters have been successfully used to assess the source and transformation history of CDOM (e.g., Zepp and Schlotzhauer 1981, Blough and Green 1995, Jaffé et al. 2004). For example, S values of about 0.012 nm^{-1} were reported for tropical rivers (Battin 1998) while brown coastal waters and blue oligotrophic waters were characterized by values of 0.018 and 0.020 respectively (Blough et al. 1993). Thus, these parameters should be useful in tracing DOM in coastal systems. However, limitations have been reported for aquatic ecosystems with significant inputs of freshly leached, plant-derived DOM (Jaffé et al. 2004) as such materials are susceptible to photobleaching (Scully et al. 2004) causing changes of the UV-Vis based indices.

Fluorescence-based optical properties are also commonly used to trace DOM sources, in particular the maximum fluorescence emission wavelength (λ_{max}) and the Fluorescence Index (F.I.). The λ_{max} value, determined at 313 nm excitation, has been reported to be higher for terrestrially-derived DOM compared with marine-derived DOM. λ_{max} has been used to assess DOM source changes along estuarine salinity gradients and on spatial and temporal scales (e.g., de Souza Sierra et al. 1994, 1997, Jaffé et al. 2004, Maie et al. 2006a). Similarly, the F.I. as initially proposed by McKnight et al. (2001) is a fluorescence emission-based index ($f_{450/500}$) determined at 370 nm excitation. F.I. has been used to differentiate between autochthonous and allochtonous CDOM in tropical rivers (Battin 1998), sub tropical wetlands (Lu et al. 2003) and estuaries (Jaffé et al. 2004, Maie et al. 2006a), and for a variety of different bodies of water (McKnight et al. 2001, Jaffe et al. 2008). The index is based on terrestrial and aquatic/microbial DOM end member values reported as 1.4 and 1.9, respectively (McKnight et al. 2001). It has recently been suggested that more consistent F.I. values can be obtained using $f_{470/520}$ instead of the originally proposed index as fluorescence values at these wavelengths are less affected by analytical and instrument corrections (Cory and McKnight 2005, Maie et al. 2006a, Jaffe et al. 2008).

A detailed study of DOM optical characteristics in estuaries of the Everglades National Park (Jaffé et al. 2004) suggested that DOM was mainly derived from freshwater marshes, mangrove forests, and marine organisms. Spatial data on the F.I. ($f_{450/500}$) and synchronous fluorescence (see below) showed clear differences in DOM optical characteristics between the geomorphologically variable sub-regions. Although bulk DOM concentration behaved conservatively in the estuaries of the southwestern Florida Everglades (Clark et al. 2002), at salinities ≥ 30 there was a clear change in the DOM fluorescence characteristics. This change suggested a switch from allochtonous to autochthonous DOM, i.e., a change in the nature and origin of the dominant organic matter, and not just a simple dilution of terrestrial DOM (Jaffé et al. 2004).

In addition to the optical methods described above, other fluorescence techniques have been successfully applied for DOM source characterizations. Synchronous excitation-emission fluorescence is a two-dimensional fluorescence method that has been used in a variety of DOM studies (e.g., de Souza Sierra et al. 1994, Kalbitz et al. 2000, Lu and Jaffé 2001, Lu et al. 2003). Synchronous fluorescence spectra of DOM usually show the presence of four distinct, but broad peaks (Ferrari and Mingazzini 1995, Lu et al. 2003) that can be assigned to: Peak-I polyphenolic and/or proteinaceous materials, Peak-II compounds of two condensed ring systems, Peak-III fulvic acids, and Peak-IV humic acids and other humic-like substances. As such, synchronous fluorescence has the potential to discriminate between marine and terrestrial DOM in estuaries. While it has been suggested that Peak I is indicative of fresh, marine-derived, possibly protein-like DOM components in estuaries (de Souza Sierra et al. 1994, Jaffé et al. 2004), it is important to keep in mind that polyphenols (such as mangrove-derived tannins) may also produce a fluorescent signal in this spectral region (Ferrari and Mingazzini 1995, Maie et al. 2007).

A higher spectral resolution compared with synchronous fluorescence can be obtained using three-dimensional fluorescence, or excitation-emission-matrix (EEM) fluorescence techniques (Fig. 12.3). These have been widely applied for DOM source assessments, particularly in marine systems (e.g., Coble 1996, Del Castillo et al. 2000, Marhaba et al. 2000, Kowalczuk et al. 2003, Yamashita and Tanoue 2003). The presence of different EEM fluorescence maxima allows for the characterization of DOM into humic-like, marine humic-like, and protein-like contributions (e.g., Coble 1996, Coble et al. 1998, Parlanti et al. 2000). More recently, this EEM 'peak picking' based technique was further refined for DOM tracing purposes by incorporating parallel factor analysis (PARAFAC) for the statistical processing of EEM data (Stedmon et al. 2003). The PARAFAC application can decon-

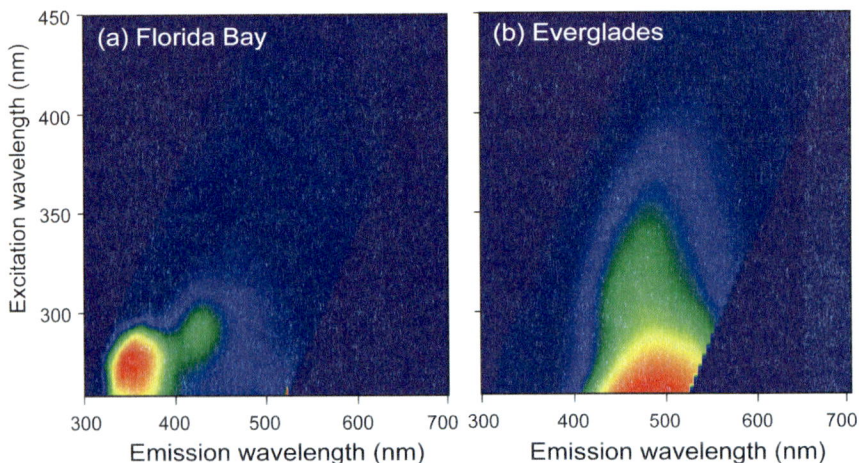

Fig. 12.3 Excitation emission florescence spectra (EEM) of two water samples from (**a**) open water in Florida Bay and (**b**) within the Everglades mangrove swamp (R Jaffé et al. unpubl. data)

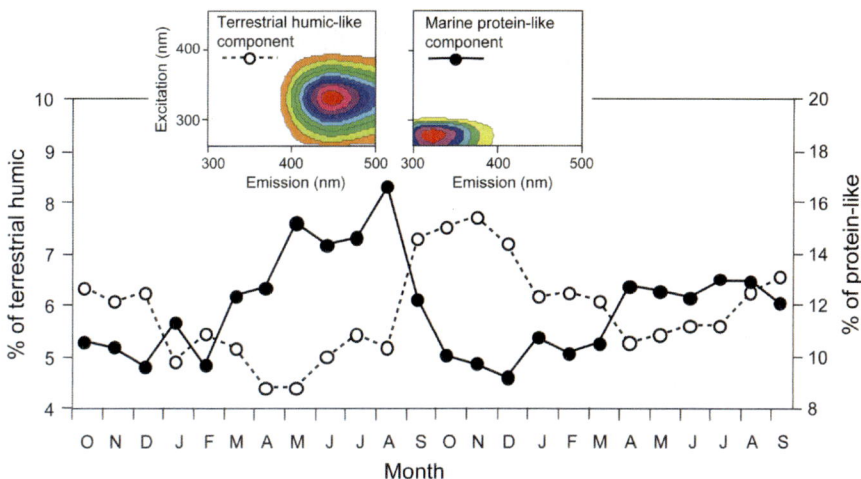

Fig. 12.4 PARAFAC analysis of excitation emission matrix florescence spectra in the Florida Everglades and adjacent coastal water distinguished between eleven DOM components, of which two were most suited to distinguish allochthonous (terrestrial) DOM (dotted line, open circles) from autochthonous (marine) DOM (black line, closed circles). The input of terrigenous DOM during high flow conditions and autochthonous DOM during low flow conditions could be clearly identified (modified from Jaffé et al. 2008)

volute the EEM spectra into individual DOM components which can be modeled and quantified individually (Fig. 12.4). This statistical analysis avoids problems with peak overlap and other analytical interferences, and thus leads to increased resolution.

EEM-PARAFAC has now been applied to several DOM tracing studies including watersheds and estuaries (Stedmon et al. 2003, Stedmon and Markager 2005, Hall et al. 2005, Hall and Kenny 2007, Yamashita et al. 2008), in the assessment of environmental redox conditions based on DOM characteristics (Fulton et al. 2004, Cory and McKnight 2005), water quality assessment (Wang et al. 2007), DOM in soils (Ohno and Bro 2006, Ohno et al. 2007), and laboratory simulation and mesocosm experiments (Muller et al. 2005, Stedmon et al. 2007). For most reported applications DOM components of terrestrial, microbial/marine, and of protein-like origin were reported (Stedmon et al. 2003, Stedmon and Markager 2005). As such, EEM-PARAFAC has an enormous potential to become the leading DOM tracer method, once DOM endmembers are better characterized and a larger user base is established.

Recent monitoring data for a sampling grid across Florida Bay using EEM-PARAFAC resulted in the identification of eleven DOM components ranging in origin from terrestrial, to marine and protein-like. Some of these components showed seasonal trends (Fig. 12.4; Jaffé et al. 2008), which were clearly controlled by a combination of hydrological drivers, such as terrestrial CDOM inputs from the Everglades freshwater marshes and fringe mangrove communities during the high water discharge period (September–December), and by primary productivity, likely from the seagrass community, during the peak summer season (May–August). As such, the application of CDOM optical properties in tropical and sub tropical ecosystems with the objective of tracing organic matter sources has advanced our understanding of DOM dynamics in these environmental settings, but many more such studies are required to make significant strides in this field of biogeosciences.

12.3.3 Isotopic Approaches

Stable isotope measurements, in particular carbon and nitrogen, have often been used to trace the fate of organic matter in aquatic environments and food webs. Like molecular or optical tracer techniques, the stable isotope approach to trace element transfers within the environment relies on different producers having distinct isotopic ratios. The lighter isotope has slightly higher reactivity in metabolic reactions (Fry and Sherr 1984). ^{12}C is preferentially incorporated into biomass during photosynthesis, leading to lower $^{13}C/^{12}C$-ratios in plant tissue than in the initial inorganic carbon. The exact isotope ratio in plant material depends mainly on the photosynthetic mechanism and the source of inorganic carbon. Several primary producers therefore exhibit different isotope ratios. C3-plants generally show lower $^{13}C/^{12}C$-ratios than C4-plants, and algae which assimilate dissolved inorganic carbon from the water show different isotope fractionations during photosynthesis than terrestrial plants (Fry and Sherr 1984). Decomposition often leads to isotope fractionation. Isotopic fractionations caused by degradation are usually small compared with the sharp isotopic differences between terrigenous and algal-derived organic matter in coastal zones. Stable carbon isotope measurements were applied to trace mangrove-derived detritus in coastal food webs or to study the dynamics of suspended and dissolved organic carbon in mangrove environments (e.g., Rodelli et al. 1984, Zieman et al. 1984, Lin et al. 1991, Hemminga et al. 1994, Primavera 1996, Dittmar et al. 2006, Bouillon et al. 2007; see Chapter 3).

Stable carbon isotope ratios are usually expressed as $\delta^{13}C$ relative to the Pee Dee Belemnite (PDB) standard:

$$\delta^{13}C = \left(\frac{(^{12}C/^{13}C)_{\text{sample}}}{(^{12}C/^{13}C)_{\text{PDB}}} - 1 \right) \times 1000\,‰$$

Typical $\delta^{13}C$ in mangrove ecosystems are, for instance: detritus from mangrove trees $-28‰$, seagrass: $-11‰$, and phytoplankton: $-21‰$. The actual $\delta^{13}C$ values in mangroves can deviate substantially from these approximate average values. For instance, heterotrophic respiration cause significant isotopic fractionation, and phytoplankton in estuaries can be isotopically very light, because it often grows on recycled carbon from terrestrial or aquatic sources (e.g., Peterson and Fry 1989).

The determination of $\delta^{13}C$ in particulate samples is routinely performed with isotope-ratio mass spectrometers (ir-MS) that are coupled to high-temperature combustion units. The determination of $\delta^{13}C$ in DOC is technically more challenging, because a coupling between a liquid-phase combustion unit and ir-MS is commercially not yet available. Dittmar et al. (2006, 2008) isolated DOM from seawater via solid-phase extraction (SPE). These DOM extracts are salt-free and can be freeze-dried to obtain a powder that can be analyzed via conventional ir-MS. The disadvantage of this approach is that approximately half of DOM cannot be isolated via SPE and thus escapes the analytical window. Recently, Bouillon et al. (2006) and Beaupré et al. (2007) introduced new methods where conventional DOC oxidation units were coupled to ir-MS to obtained $\delta^{13}C$ on bulk DOC. Examples for the application of this new technique are discussed in Chapter 3.

The parameter $\delta^{13}C$ changes linearly with respect to ^{12}C concentrations, but not linearly with respect to total C ($^{13}C + {}^{12}C$) concentrations (Perdue and Koprivnjak 2007). However, because ^{13}C contributes only about 1.1% to total carbon, it can be assumed that $\delta^{13}C$ represents total C ($^{13}C + {}^{12}C$), which facilitates the establishment of linear mixing models. Because of the relatively large differences, most marine sources can be distinguished from mangrove detritus with help of stable carbon isotopes. For instance, based on a simple two-source mixing model Dittmar et al. (2006) identified mangrove-derived DOM on the North Brazilian shelf (Fig. 12.5). Rezende et al. (1990) proposed a high contribution of marine-derived organic matter to total outwelling of POM from mangroves in Sepetiba Bay (Rio de Janeiro, Brazil) and suggested that outwelling may be much less significant than expected by simple mass balance studies. In complex mangrove systems with multiple terrigenous and marine sources, a single parameter model ($\delta^{13}C$) is not sufficient to distinguish between all possible sources. In addition, only a broad classification of organic matter sources is possible on the basis of $\delta^{13}C$, e.g., different terrestrial C3-plants can usually not be distinguished.

Multi-isotope approaches, in particular the combination of stable carbon ($\delta^{13}C$) and stable nitrogen ($\delta^{15}N$) isotopes can help to differentiate multiple sources and organic matter processing in tropical estuaries and their trophic systems (e.g., Primavera 1996, Marguillier et al. 1997). If the nitrogen concentration in a sample is sufficiently high, $\delta^{13}C$ and $\delta^{15}N$ are simultaneously obtained through conventional ir-MS. Techniques are also available for $\delta^{15}N$ determination on inorganic nitrogen species at natural-abundance levels (Sigman et al. 1997). $\delta^{15}N$ is normalized to atmospheric N_2, thus freshly fixed nitrogen and synthetic fertilizers usually

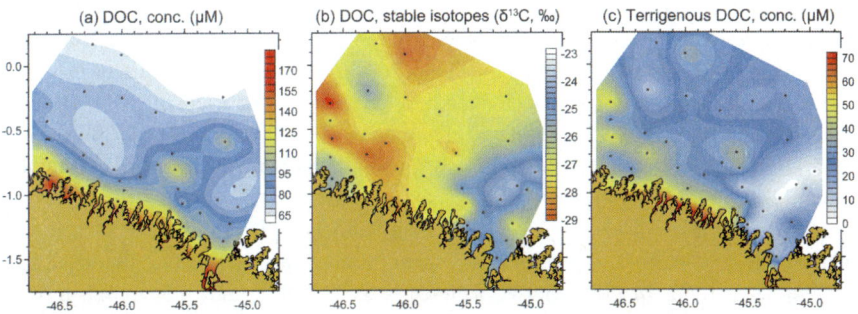

Fig. 12.5 On the North Brazilian shelf, a two-source mixing model for marine and mangrove-derived DOC could be established by using $\delta^{13}C$ (modified after Dittmar et al. 2006)

have $\delta^{15}N$ close to 0‰. During the process of denitrification, $\delta^{15}N$ of the remaining organic matter increases sometimes to values beyond 30‰ in manure or septic systems (e.g., Chang et al. 2002). $\delta^{15}N$ can therefore be a good tracer for nitrogen fixation/denitrification or anthropogenic impacts. However, in many tropical estuaries these processes occur simultaneously which blurs the isotopic signal and can make the identification of a single biogeochemical process or the identification of sources ambiguous.

There are relatively few estuarine isotope studies that have used other elements than carbon and nitrogen. Stable sulfur isotopes ($\delta^{34}S$) have been used to further clarify trophic relationships in tropical coastal systems (e.g., Hsieh et al. 2002, Connolly et al. 2004). $\delta^{34}S$ can also be a powerful tracer for organic matter from coastal wetlands, because redox processes strongly fractionate sulfur isotopes. Dissimilatory sulfate reduction causes enrichment of ^{32}S in sulfide relative to the residual sulfate, and the uptake of sulfide by marsh plants results in isotopically depleted $\delta^{34}S$ plant tissue. This fractionation caused well-defined sedimentary $\delta^{34}S$ profiles of organic matter and inorganic sulfur species in the Everglades, Florida (Bates et al. 1989).

Combinations of molecular biomarker distributions and compound-specific isotope ($\delta^{13}C$) measurements using GC-ir-MS have also been successfully applied to trace organic matter sources in fringe mangrove estuaries (Mead et al. 2005, Hernandez et al. 2001). This technique offers enhanced resolving power compared with either bulk $\delta^{13}C$ determinations or biomarker analyses alone.

12.3.4 Molecular Approaches and Multiple-Source Mixing Models

Molecular tracer approaches can be analytically challenging, but some of them are highly specific, and organic matter from different terrestrial ecosystems or even species can be distinguished. It is possible to establish multiple-source mixing models through the combination of several tracers. For DOM, few molecular tracers are established, and lignin is the only molecular tracer used thus far to trace DOM in mangrove systems (Dittmar et al. 2001b). Lignin is unique for vascular plant material and it is possible to distinguish vegetation types, e.g., woody angiosperms, gymnosperms, or non-woody vascular plants (e.g., Hedges et al. 1986). Therefore, it has been used to trace the fate and transport of terrestrial organic matter in rivers and marine environments (e.g., Hedges and Ertel 1982, Ertel et al. 1984, Hedges et al. 1986, Moran and Hodson 1994, Kattner et al. 1999, Maie et al. 2005). The most commonly used analytical approach for lignin is the quantification of phenolic subunits (Fig. 12.6) via gas or liquid chromatography after sample oxidation with CuO. Some studies used thermochemolytical techniques (Maie et al. 2005). Either method yields an array of several phenolic subunits, whose relative abundance can be indicative for a specific source. Benner et al. (1990) found that lignin-derived phenols are leached in considerable amount from mangrove leaves (*Rhizophora mangle*). Mangrove-derived DOM is therefore rich in lignin.

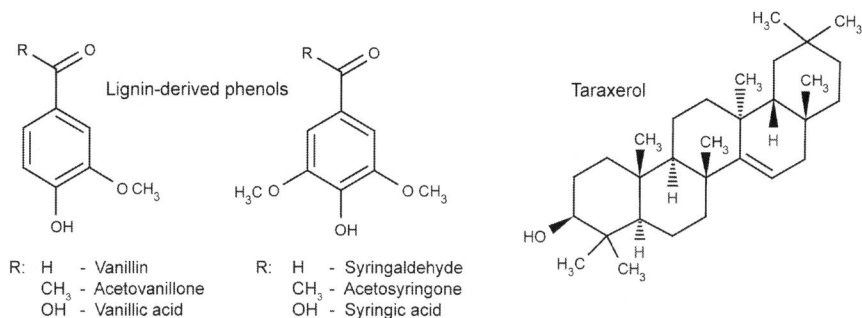

Fig. 12.6 The molecular structure of some lignin-derived phenols which were used to distinguish mangrove-derived DOM from other sources in a North Brazilian estuary (see Fig. 12.7), and taraxerol which is indicative for mangrove detritus in sediments

Moran et al. (1991) traced DOM from a mangrove swamp ecosystem at the Berry Islands (Bahamas) by analysis of dissolved lignin-derived phenols and naturally fluorescing compounds. Dittmar et al. (2001b) established a three-source mixing model for dissolved and suspended organic carbon in a North Brazilian mangrove-fringed estuary (Fig. 12.7). This model uses different lignin-derived phenols and stable carbon isotopes to quantitatively distinguish between mangrove, algae, and riverine organic matter (Fig. 12.8). It is important to note that in contrast to δ^{13}C the use of phenolic ratios requires the establishment of non-linear mixing equations. Dittmar et al. (2001b) also proposed a method for error propagation in three-source models.

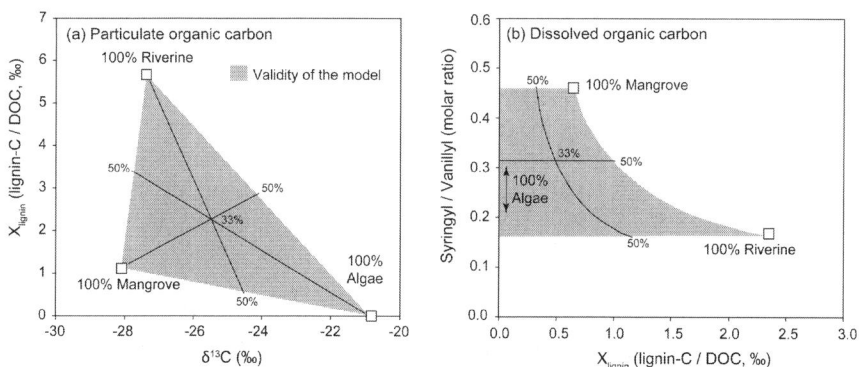

Fig. 12.7 Three-source mixing models to distinguish mangrove, algal-derived, and riverine particular and dissolved organic matter in a North Brazilian estuary (Dittmar et al. 2001b). For these models, carbon-normalized lignin yields (X_{lignin}), δ^{13}C, and the ratio of syringyl and vanillyl phenols were used. Algal-derived organic matter does not contain lignin, and mangrove detritus has a higher content of syringyl phenols than riverine organic matter (modified after Dittmar et al. 2001b)

Fig. 12.8 The application of the three-source mixing model (see Fig. 12.7) revealed a major contribution of mangrove-derived DOM in the mouth of a North Brazilian estuary (modified after Dittmar et al. 2001b)

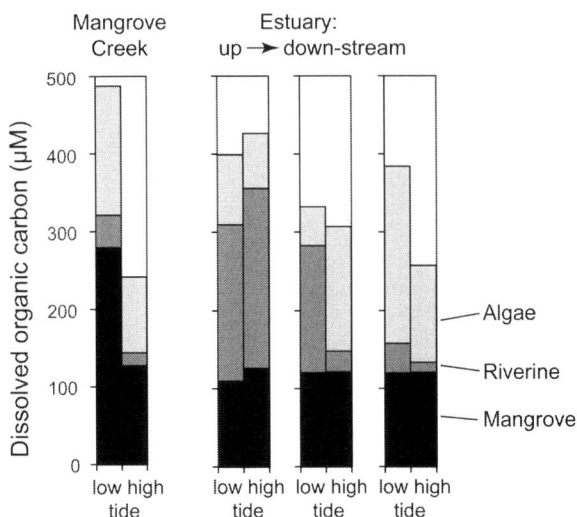

In addition to lignin, amino acid enantiomers, neutral sugars, and amino sugars can be analyzed on the molecular level. These tracers have been used mainly in open ocean environments to trace the fate of planktonic DOM in the water column (e.g., Dittmar et al. 2001a) but have also recently been applied to assess organic matter sources in coastal wetlands and estuaries (Jones et al. 2005, Maie et al. 2005, 2006b). These tracers could be included in multiple-source mixing models in mangrove systems to further improve current models. Most mangrove trees, in particular *Rhizophora mangle*, are rich in tannins (Hernes et al. 2001). Tannins are very soluble in seawater and easily degrade in the water column. Tannins can strongly sorb to sediments and react with organic nitrogen (Maie et al. 2007). Tannins thus have the potential to be used as a highly specific tracer for labile mangrove-derived DOM in the coastal zone. Recently, an analytical method for the molecular-level determination of combustion-derived organic matter ('black carbon') in DOM was presented (Dittmar 2008). Above-ground biomass frequently burns in the tropics, whereas submerged vegetation (including vascular seagrasses) cannot burn. Black carbon may therefore be a powerful molecular tool for source identification of DOM and POM in coastal systems.

12.3.5 Molecular Fingerprinting Techniques

Molecular analyses of lignin phenols and other individual components can yield valuable information on a molecular level, suitable to distinguish sources or processes involved in the flux of organic matter. However, these analytical parameters only represent structural subunits of the original molecules. Extensive research efforts have been undertaken to unravel the molecular structure of the original organic matter in order to identify new biomarkers and to improve our understanding of bioavailability and preservation of organic matter.

Liquid chromatography (LC) and capillary electrophoresis have been used for the separation and property determination of organic matter (e.g., Caron et al. 1996, Frimmel 1998, Wu et al. 2003), in some cases in combination with mass spectrometry (Reemtsma and These 2003, Schmitt-Kopplin and Kettrup 2003, Reemtsma et al. 2006a, Dittmar et al. 2007) or nuclear magnetic resonance (Piccolo et al. 2002, Simpson et al. 2004). Each of these techniques resulted in important chemical and physico-chemical information on organic matter fractions such as polarity, size, or functional environments. Size exclusion chromatography (SEC) coupled to online carbon detection, absorbance, or fluorescence detection can yield information on the molecular size and bulk functional properties of DOM from different sources (Dittmar and Kattner 2003, Scully et al. 2004, Maie et al. 2007, Tzortziou et al. 2008).

The combination of chromatography and mass spectrometry takes advantage of an additional layer of information. Although the resolution of conventional mass spectrometers is not sufficient to separate all molecules in the organic matter samples, the spectra still inherently reflect the overall composition of the samples. Multivariate statistical methods such as cluster analysis, multi-dimensional scaling, and principal component and discriminant analyses are suitable to evaluate and compare mass spectra. Mass spectra in combination with appropriate statistics can yield fingerprints for sources and transformation processes for organic matter in tropical and temperate coastal systems (e.g., Minor et al. 2002, Dittmar et al. 2007). Photodegradation of pore water from a mangrove area in Northern Brazil, for example, caused significant chemical alteration on a molecular level. The exported DOM from the mangroves carried an unequivocal terrestrial $\delta^{13}C$ signature and molecular fingerprinting techniques revealed considerable photodegradation during transport across the shelf (Dittmar et al. 2006, Dittmar et al. 2007).

Even by applying the most advanced separation techniques, the complexity of DOM prevents unequivocal molecular and structural identification with conventional spectroscopic detections or fragmentation experiments (Reemtsma 2001, Mopper et al. 2007). Ultrahigh-resolution Fourier transform ion cyclotron resonance mass spectrometry (FT-ICR-MS; e.g., Marshall et al. 1998) is an analytical technique which opened a new analytical window and yielded extensive molecular information on the otherwise uncharacterizable fraction of organic matter. FT-ICR-MS resolves thousands of individual molecules and provides molecular formulas for most of these molecules in complex organic mixtures such as oils (Schaub et al. 2005), terrestrial humic substances (e.g., Kujawinski et al. 2002, Stenson et al. 2003), aerosols (Reemtsma et al. 2006b), marine organic matter (Koch et al. 2005b, Hertkorn et al. 2006), and groundwater (Einsiedl et al. 2007). It allows for the differentiation between the molecular composition of different organic matter sources and specific processes (Kujawinski et al. 2004).

Figure 12.9 shows the effect of photodegradation on the molecular composition of mangrove pore water determined by FT-ICR-MS. The sample extracts (isolated through solid phase extraction) were directly injected by electrospray ionization

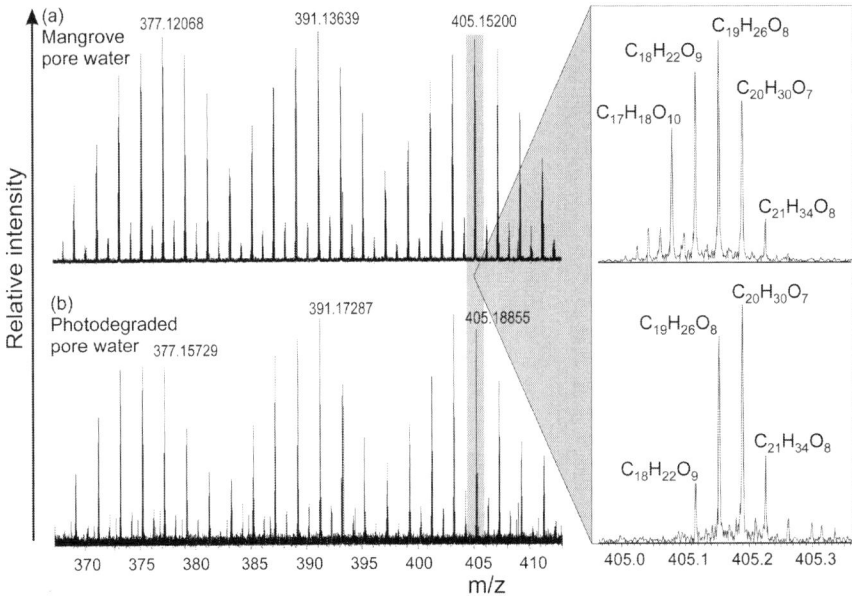

Fig. 12.9 Positive mode electrospray ionization FT-ICR mass spectra of (**a**) mangrove pore water, and (**b**) the same pore water after 14 days of sterile photodegradation. On the left side the m/z range from 370 to 410 is displayed. The right part shows a close-up for the nominal mass 405 m/z. From the exact masses molecular formulas can be calculated (B Koch et al. unpubl. data, Dittmar et al. 2007)

into the FT-ICR mass spectrometer (Fig. 12.9). Ten thousand and more individual masses can be resolved with this technique at a mass accuracy of <1 ppm. Molecular formulas can then be calculated from the exact molecular masses. In order to visually represent the resulting numerous molecular formulas two types of diagrams are commonly used (Fig. 12.10): (1) in the Kendrick plot the exact mass of a molecule is normalized to the exact mass of a functional group such as CH_2 (the 'Kendrik mass defect', KMD). Therefore molecules which belong to the same homologous series plot on horizontal lines (Kendrick 1963, Hughey et al. 2001, Stenson et al. 2003; Fig. 12.10a). (2) In the molecular van Krevelen diagram, molecular formulas are plotted by their H/C versus O/C ratios (Kim et al. 2003; Fig. 12.10b). Each dot in the van Krevelen diagram displays at least one molecular formula represented by its molecular O/C and H/C ratio. Photodegradation of mangrove pore water resulted in a general shift of the average KMD and in decreased O/C ratios and increased H/C ratios.

While providing molecular information in unsurpassed detail, FT-ICR-MS has so far provided little structural information on DOM (Stenson et al. 2003, Koch and Dittmar 2006). In some cases, however, the molecular formula by itself (as provided by FT-ICR-MS) can provide structural information. For instance, hydrogen-deficient molecules ('black carbon') that can only be derived from thermogenic processes in Earth's crust or from biomass burning were recently discovered in

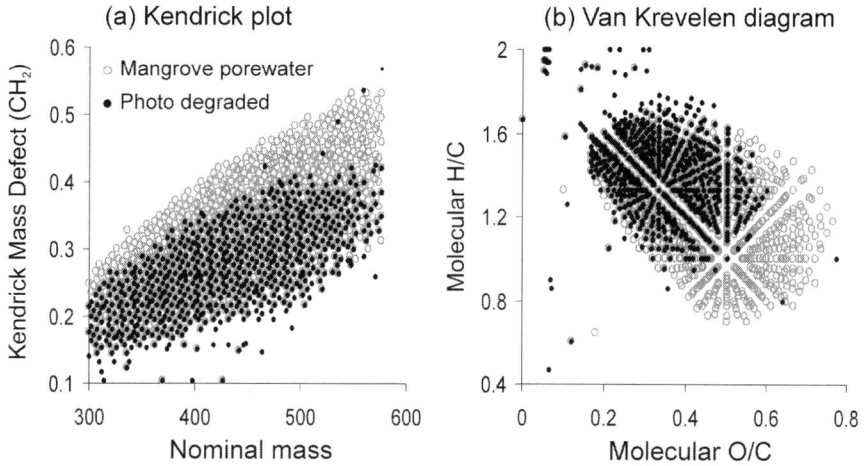

Fig. 12.10 Kendrick plot (**a**) and van Krevelen diagram (**b**) for a mangrove pore water and a photodegraded pore water (B Koch et al. unpubl. data, Dittmar et al. 2007). Photodegradation leads to a relative decrease in oxygen and an increase in hydrogen representing decarboxylation and dearomatization

riverine and marine DOM (Kim et al. 2004, Dittmar and Koch 2006). The aromaticity index (AI) which can be calculated from a molecular formula allows the identification of combustion-derived aromatic or polyaromatic structures (Koch and Dittmar 2006). A combination of reversed-phase chromatography with FT-ICR-MS seems promising for a more detailed structural characterization and adds an additional dimension of information for the molecular-level characterization (Koch et al. 2008).

12.4 Tracing the Source of Organic Matter in Mangrove and Coastal Sediments

The previous sections focused on dissolved and suspended organic matter in the water column. Mangrove-derived DOM can be transported further offshore where it is biogeochemically modified or mineralized. Although it has been observed that suspended mangrove detritus can be transported as POC in mangrove-fringed rivers (Jaffé et al. 2001) and across estuaries (Xu et al. 2006), most of it settles within or in direct vicinity of the mangrove forests. In sediments, a wide array of different biomarkers has been established. Contemporary research on the export of organic matter from tropical coastal ecosystems to estuarine and coastal sediments applies two general techniques: organic matter source identification by (1) microfossils and (2) organic or inorganic chemical markers. Microfossils such as pollen and planktonic shells yield species-specific information about primary producers and are often applied for the identification of the sources of organic matter (e.g., Behling and da Costa 2000, 2001, Cohen et al. 2005a, Cohen et al. 2005b, Scourse et al. 2005).

However, both microfossils and molecular biomarkers are subject to selective degradation and therefore are often not suitable to quantify the bulk organic-matter flux into the sediments. Therefore, the quantification of the burial rate of organic matter to coastal sediments is still an important aspect of contemporary research.

Molecular biomarkers in sediments allow the identification of sources and degradation of sedimentary organic matter, especially when microfossils are not preserved. Most compounds used as molecular biomarkers are assigned to the compound class of lipids. Lipids are defined as compounds that are soluble in apolar solvents and comprise a large variety of chemical classes. Their apolar characteristics hinder microbial decay, an important reason for efficient organic matter preservation in sediments. Lipids occur either in a free (e.g., in cuticular waxes of mangrove leaves) or a bound form in the plant matrix. They are analytically easily accessible by solvent extraction (after saponification for bound lipids) and are usually analyzed by gas chromatography and flame ionization or mass spectrometry detection. After a rapid initial decay, the relative biomarker distributions can be preserved during diagenesis on time scales of several thousand years and are suitable to identify mangrove remains in ancient sedimentary organic matter (Koch et al. 2003, Xu et al. 2007). Various isoprenoids are often used as chemotaxonomic tracers to characterize sources of terrestrial, estuarine, and marine organic matter (e.g., Volkman 1986, Johns et al. 1994, Killops and Frewin 1994, Munoz et al. 1997, Bianchi and Canuel 2001, Jaffé et al. 2001). Sterols or triterpenols occur in cuticular waxes of terrestrial plants and can deliver species-specific information (e.g., Pant and Rastogi 1979, Das and Mahato 1983, Dodd et al. 1995, Wollenweber et al. 1999). Hopanoids are indicative of bacterial/cyanobacterial contributions to the sedimentary organic matter pool (e.g., Mfilinge et al. 2005, Volkman et al. 2007). Various sterols can be used for the identification of organic matter from diatoms, green algae, and dinoflagellates (Volkman et al. 1998, Volkman 2005).

In the subtropical region of South Florida, biomarkers have been used to selectively distinguish mangrove-derived organic matter from other local sources such as seagrasses and other higher plants from coastal wetlands (Jaffé et al. 2006, Xu et al. 2006). For example, the n-alkane based proxy Paq ($C_{23} + C_{25}/C_{23} + C_{25} + C_{29} + C_{30}$) differentiated between mangrove and seagrass inputs. Mead et al. (2005) reported that Paq values in emergent terrestrial plants (0.13–0.51), including mangroves, were generally lower than those in marine submerged vegetation such as seagrass (ca. 1.0). In addition, compound-specific $\delta^{13}C$ values for the n-alkanes (C_{23} to C_{31}) were distinctively different for terrestrial emergent and freshwater/marine submerged plants when compared with seagrass. While the Paq proxy has the potential to assess organic matter inputs in complex ecosystems, it is not directly specific to mangroves. The 3-hydroxy-triterpenoid taraxerol (see Fig. 12.6) was used as a proxy especially for *Rhizophora* species (red mangrove) in coastal zones of northern Brazil, Southwest Africa, Florida, and Japan (Koch et al. 2003, Versteegh et al. 2004, Koch et al. 2005a, Scourse et al. 2005, Jaffé et al. 2006, Xu et al. 2006, Basyuni et al. 2007, Xu et al. 2007) in the absence of other major vascular plant organic matter inputs. Although taraxerol is susceptible to microbial degradation (Koch et al. 2005a) and reactive to UV radiation by generating secondary derivatives under simulated sunlight exposure (Simoneit et al. 2009), it showed an overall stability in sediments. This stability makes it a suitable tracer for *Rhizophora* biomass on times scales of more than one million years (Versteegh et al. 2004). However, many other plant species are known to contain taraxerol and to contribute to coastal organic matter (e.g., Volkman et al. 2000). Therefore, a combination of several triterpenoids increases their reliability as chemotaxonomic tracers especially in relatively young sediments (Koch et al. 2005a). A multi-proxy biomarker approach along a surface sediment transect in the shallow Florida Bay estuary showed that the Paq values significantly increased from nearshore to offshore (i.e., increase in seagrass-derived organic matter). At the same time taraxerol substantially decreased (decrease in mangrove-derived organic matter), suggesting a clear spatial variation of organic matter sources throughout this ecosystem (Fig. 12.11; Xu et al. 2006). Generally, the sites in the northeast contained mixed organic matter sources of both mangrove- and seagrass-derived organic matter, while the sites in Central and Southwest Florida Bay were strongly dominated by seagrass-derived organic matter.

Similarly, temporal and spatial variations in the composition of POM from Florida Bay were examined (Xu and Jaffé 2007). While plankton sources to POM were suggested as a bay-wide phenomenon, several biomarker proxies including Paq and taraxerol indicated a spatial shift in sources: mangrove-derived organic matter rapidly decreased, while seagrass and microbial organic matter markedly increased along a northeastern to southwestern transect. These patterns followed the vegetation distribution and transport patterns in this region. On a temporal scale, POM collected during the dry season was enriched in terrestrial mangrove-derived constituents relative to the wet season. This enrichment is likely a result of reduced primary productivity of planktonic species and seagrasses during the dry season.

Taraxerol has also been applied to assess recent historical changes (ca. 160 yrs) in the environmental conditions of Florida Bay (Xu et al. 2007). The most significant environmental changes were recorded as oscillations in the amplitude and frequency of biomarkers during the 20th century. A substantial rise in abundance of taraxerol in the 1980s was likely a result of increased mangrove primary productivity along the shore of Northeast Florida Bay, stemming from hydrological alterations in South Florida. In a recent study (Fig. 12.12) two sediment cores from Florida Bay were examined for biomarker distributions covering a time period of approximately the last 4,000 years. Taraxerol combined with Paq revealed the sedimentary environmental changes from freshwater marshes to mangrove swamps and then to a

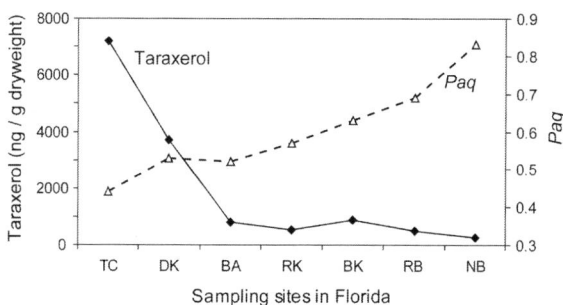

Fig. 12.11 Distribution of the Paq and taraxerol concentrations along a NE—SW transect for surface sediments in Florida Bay, USA (adapted from Xu et al. 2006)

Fig. 12.12 Depth profile of
Paq distribution and taraxerol
concentration for a ca.
4,000 yr sediment core from
Florida Bay, USA. Bottom of
the core (>150 cm) is peat
(Y Xu and R Jaffé unpubl.
data)

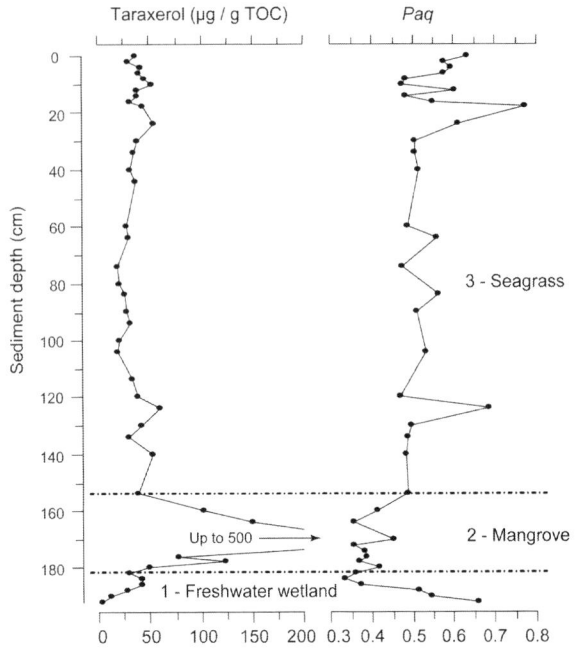

seagrass-dominated marine ecosystem, likely as a result of sea level rise in Florida
Bay since the Holocene. Thus, taraxerol is an excellent tracer of mangrove-derived
organic matter in coastal environments over a wide range of time scales.

12.5 Conclusions

To understand ecosystem functioning in coastal zones it is essential to identify the
main pathways and the magnitude of nutrient and organic matter fluxes. Probably
the most straightforward approach that requires least technological efforts is the
establishment of estuarine mixing models with help of salinity (Section 12.2). Estuarine source and sink terms can be quantified with these models, and most input
parameters can be obtained through routine analytical techniques that are part of
most estuarine monitoring programs. For successful application of these models it is
essential to have sufficient spatial and temporal sampling density to distinguish systematic changes in concentration (as a function of salinity) from localized features
in the estuary. Estuarine-scale studies provide information on a relatively large scale.
Processes within individual ecosystems, however, remain often unknown. Flux studies in tidal creeks or artificial flumes yield data on the exchange of nutrients and
other aquatic constituents on the scale of small tidal drainage areas, ranging from
few m^2 to several km^2. The precise determination of water discharge can be a challenging task in complex tidal environments. The recent advent of modern acoustic

Doppler current profilers (ADCP) for shallow water will make this approach more straightforward and precise in future studies.

To assess the potential impact of organic matter fluxes on estuarine and coastal productivity it is important to identify the source and bioavailability of organic matter (Section 12.3). Plankton-derived organic matter is often more bioavailable than the more refractory humic substances produced in mangrove sediments. The sources of organic matter can be distinguished with help of stable isotopes (Section 12.3.3). Several techniques have also been developed to trace the source of DOM in coastal zones through optical analyses of DOM, in particular fluorescence and absorbance spectroscopy (Section 12.3.2). Optical and isotopic analyses can be routinely performed on a relatively large number of samples, but often lack the specificity to identify the contribution of individual ecosystem components. For instance, mangrove-derived DOM is similar to most riverine DOM in terms of optical and isotopic properties.

Molecular-level analyses can provide more detailed information on source and turnover of organic matter, but most techniques require large analytical efforts and are often not practical for routine studies on a large number of samples. Through the combination of molecular lignin and stable carbon isotope analyses, three-source mixing models can be established to distinguish mangrove from aquatic and riverine organic matter sources (Section 12.3.4). Aside from lignin, a number of additional molecular tracers are available which may prove helpful in future studies. Recent advances in ultra-high resolution mass spectrometry now allow the separation of thousands of individual molecules in any complex organic matter sample. This emerging technique has already provided very detailed molecular fingerprints of DOM in mangrove-fringed estuaries and may lead to major advances in coastal biogeochemistry (Section 12.3.5). In sediments, a number of very specific lipid molecular biomarkers are available. In particular in combination with microfossil analysis (e.g., pollens), the analysis of lipid biomarkers in sediment cores is a powerful tool that allows the reconstruction of coastal vegetation on geological and historic time-scales (Section 12.4).

Acknowledgments The authors thank Ivan Nagelkerken for the invitation to participate in the preparation of this book, and Nagamitsu Maie, Yunping Xu, and Matthias Witt (Bruker Daltonics) for providing previously unpublished data for inclusion into this chapter. The work presented was financially supported by the National Oceanic and Atmospheric Administration (NOAA), the National Science Foundation (NSF), the South Florida Water Management District (SFWMD), and the Deutsche Forschungsgesellschaft (DFG). This is an FCE-LTER publication and Contribution number 429 from the Southeast Environmental Research Center.

References

Alberts JJ, Takacs M (1999) Importance of humic substances for carbon and nitrogen transport into southeastern United States estuaries. Org Geochem 30:385–395

Alongi DM, Ayukai T, Brunskill GJ et al (1998) Sources, sinks, and export of organic carbon through a tropical, semi-enclosed delta (Hinchinbrook Channel, Australia). Mangroves and Salt Marshes 2:237–242

Ayukai T, Miller D, Wolanksi E, Spagnol S (1998) Fluxes of nutrients and dissolved and particulate organic matter in two mangrove creeks in north-eastern Australia. Mangroves and Salt Marshes 2: 223–230

Basyuni M, Oku H, Baba S et al (2007) Isoprenoids of Okinawan mangroves as lipid input into estuarine ecosystem. J Oceanogr 63:601–608

Bates AL, Spiker EC, Holmes CW (1989) Speciation and isotopic composition of sedimentary sulfur in the Everglades, Florida, USA. Chem Geol 146:155–170

Battin TJ (1998) Dissolved organic materials and its optical properties in a blackwater tributary of the upper Orinoco River, Venezuela. Org Geochem 28:561–569

Beaupré SR, Druffel ERM, Griffin S (2007) A low-blank photochemical extraction system for concentration and isotopic analyses of marine dissolved organic carbon. Limnol Oceanogr Methods 5:174–184

Behling H, da Costa ML (2000) Holocene environmental changes from the Rio Curuá record in the Caxiuana region, Eastern Amazon Basin. Quart Res 53:369–377

Behling H, da Costa ML (2001) Holocene vegetational and coastal environmental changes from the Lago Crispim record in northeastern Pará State, eastern Amazonia. Rev Palaeobot Palyno 114:145–155

Benner R, Weliky K, Hedges JI (1990) Early diagenesis of mangroves leaves in a tropical estuary: molecular-level analyses of neutral sugars and lignin-derived phenols. Geochim Cosmochim Acta 54:1991–2001

Bianchi TS, Canuel EA (2001) Organic geochemical tracers in estuaries – Introduction. Org Geochem 32:451–451

Bianchi TS (2007) Biogeochemistry of estuaries. Oxford University Press, Oxford

Blough NV, Green SA (1995) Spectroscopic characterization and remote sensing of non-living organic matter. In: Zepp RG, Sonntag C (eds) The role of non-living organic matter in the Earth's carbon cycle, pp. 23–45. Proc Dahlem Conf Wiley, New York

Blough NV, Zafiriou OC, Bonilla J (1993) Optical absorption spectra of waters from the Orinoco River outflow: terrestrial input of colored organic matter to the Caribbean. J Geophys Res 98:2271–2278

Boto KG, Bunt JS (1981) Tidal export of particulate organic matter from a northern Australian mangrove system. Estuar Coast Shelf Sci 13:247–255

Bouillon S, Dehairs F, Velimirov B et al (2007) Dynamics of organic and inorganic carbon across contiguous mangrove and seagrass systems (Gazi Bay, Kenya). J Geophys Res 112:G02018

Bouillon S, Korntheuer M, Baeyens W et al (2006) A new automated setup for stable isotope analysis of dissolved organic carbon. Limnol Oceanogr Methods 4:216–226

Burchard H, Bolding K (2002) GETM, a general estuarine transport model. Scientific documentation. Technical report, European Commsission, Ispra

Burchard H, Bolding K, Villarreal MR (2004) Three-dimensional modelling of estuarine turbidity maxima in a tidal estuary. Ocean Dyn 54:250–265

Cai Y, Jaffé R, Jones R (1999) Interaction between dissolved organic carbon and mercury species in surface waters of the Florida Everglades. Appl Geochem 14:395–407

Caron F, Elchuk S, Walker ZH (1996) High–performance liquid chromatographic characterization of dissolved organic matter from low-level radioactive waste leachates. J Chromatogr A 739:281–294

Chang CCY, Kendall C, Silva SR et al (2002) Nitrate stable isotopes: tools for determining nitrate source among different land uses in the Mississippi River basin. Can J Fish Sci 59:1874–1885

Childers D, Cofershabica S, Nakashima L (1993) Spatial and temporal variability in marsh water column interactions in a southeastern USA salt-marsh estuary. Mar Ecol Prog Ser 95:25–38

Clark CD, Jimenez-Morais J, Jones G et al (2002) A time-resolved fluorescence study of dissolved organic matter in a riverine to marine transition zone. Mar Chem 78:121–135

Coble PG, del Castillo CE, Avril B (1998) Distribution and optical properties of CDOM in the Arabian Sea during the 1995 southwest Monsoon. Deep Sea Res II 45:2195–2223

Coble PG (1996) Characterization of marine and terrestrial DOM in seawater using excitation-emission matrix spectroscopy. Mar Chem 51:325–346

Cohen MCL, Behling H, Lara RJ (2005a) Amazonian mangrove dynamics during the last millennium: the relative sea-level and the Little Ice Age. Rev Palaeobot Palynol 136:93–108

Cohen MCL, Souza Filho PWM, Lara RJ et al (2005b) A model of Holocene mangrove development and relative sea-level changes on the Bragança Peninsula (northern Brazil). Wetlands Ecol Manage 13:433–443

Connolly RM, Guest MA, Melville AJ et al (2004) Sulfur stable isotopes separate producers in marine food-web analysis. Oecologia 138:161–167

Cory RM, DM Mcknight (2005) Fluorescence spectroscopy reveals ubiquitous presence of oxidized and reduced quinines in dissolved organic matter. Environ Sci Technol 39:8142–8149

D'Alapaos A, Lanzoni S, Marani M (2005) Tidal network ontogeny: channel initiation and early development. J Geophys Res 110:F02001

Das MC, Mahato SB (1983) Triterpenoids. Phytochem 22:1071–1095

De Souza Sierra MM, Donard OFX, Lamotte M (1997) Spectral identification and behaviour of dissolved organic fluorescent material during estuarine mixing processes. Mar Chem 58:51–58

De Souza Sierra MM, Donard OFX, Lamotte M et al (1994) Fluorescence spectroscopy of coastal and marine waters. Mar Chem 47:127–144

Del Castillo CE, Gilbes F, Coble PG et al (2000) On the dispersal of colored dissolved organic matter over the West Florida Shelf. Limnol Oceanogr 45:1425–1432

Dittmar T, Fitznar HP, Kattner G (2001a) Origin and biogeochemical cycling of organic nitrogen in the eastern Arctic Ocean as evident from D- and L-amino acids. Geochim Cosmochim Acta 65:4103–4114

Dittmar T, Hertkorn N, Kattner G et al (2006) Mangroves, a major source of dissolved organic carbon to the oceans. Global Biogeochem Cycles 20:1–7

Dittmar T, Kattner G (2003) Recalcitrant dissolved organic matter in the ocean: major contribution of small amphiphilics. Mar Chem 82:115–123

Dittmar T, Koch BP (2006) Thermogenic organic matter dissolved in the abyssal ocean. Mar Chem 102:208–217

Dittmar T, Koch BP, Hertkon N et al (2008) A simple and efficient method for the solid-phase extraction of dissolved organic matter (SPE-DOM) from seawater. Limnol Oceanogr Methods 6:230–235

Dittmar T, Lara RJ (2001a) Do mangroves rather than rivers provide nutrients to coastal environments south of the Amazon River? Evidence from long-term flux measurements. Mar Ecol Prog Ser 213:67–77

Dittmar T, Lara RJ (2001b) Driving forces behind nutrient and organic matter dynamics in a mangrove tidal creek in North Brazil. Estuar Coast Shelf Sci 52:249–259

Dittmar T, Lara RJ, Kattner G (2001b) River or mangrove? Tracing major organic matter sources in tropical Brazilian coastal waters. Mar Chem 73:253–271

Dittmar T, Whitehead K, Minor EC et al (2007) Tracing terrigenous dissolved organic matter and its photochemical decay in the ocean by using liquid chromatography/mass spectrometry. Mar Chem 107:378–387

Dittmar T (2008) The molecular-level determination of black carbon in marine dissolved organic matter. Org Geochem 39:396–407

Dodd RS, Fromard F, Rafii ZA et al (1995) Biodiversity among West African *Rhizophora*: foliar wax chemistry. Biochem Syst Ecol 23:859–868

Einsiedl F, Hertkorn N, Wolf M et al (2007) Rapid biotic molecular transformation of fulvic acids in a karst aquifer. Geochim Cosmochim Acta 71:5474–5482

Ertel JR, Hedges, JI, Perdue EM (1984) Lignin signature of aquatic humic substances. Science, 223:485–487

Fagherazzi S, Sun T (2004) A stochastic model for the formation of channel networks in tidal marshes. J Geophys Res Lett 31:L21503

Ferrari GM, Mingazzini M (1995) Synchronous fluorescence spectra of dissolved organic matter (DOM) of algal origin in marine coastal waters. Mar Ecol Progr Ser 125:305–315

Frimmel FH (1998) Characterization of natural organic matter as major constituents in aquatic systems. J Contam Hydrol 35:201–216

Fry B, Sherr EB (1984) $\delta^{13}C$ measurements as indicators of carbon flow in marine and freshwater ecosystems. Contrib Mar Sci 27:13–47

Fulton JR, McKnight DM, Foreman CM et al (2004) Changes in fulvic acid redox state through the oxyline of a permanently ice-covered Antarctic lake. Aquat Sci 66:27–46

Gardner LR (2005) Role of geomorphic and hydraulic parameters governing porewater seepage from salt marsh sediments. Water Resour Res 41:W07010

Gordon ES, Goñi MA (2004) Controls on the distribution and accumulation of terrigenous organic matter in sediments from the Mississippi and Atchafalaya river margin. Mar Chem 92:331–352

Hall GJ, Kenny JE (2007) Estuarine water classification using EEM spectroscopy and PARAFAC–SIMCA. Anal Chim Acta 581:118–124

Hall GJ, Clow KE, Kenny JE (2005) Estuarial fingerprinting through multidimensional fluorescence and multivariate analysis. Environ Sci Technol 39:7560–7567

Harvey J, Germann P, Odum W (1987) Geomorphological control of subsurface hydrology in the creekbank zone of tidal marshes. Estuar Coast Shelf Sci 25:677–691

Hedges JI, Eglinton G, Hatcher PG et al (2000) The molecularly-uncharacterized component of nonliving organic matter in natural environments. Org Geochem 31:945–958

Hedges JI, Keil RG (1995) Sedimentary organic matter preservation: an assessment and speculative synthesis. Mar Chem 49:81–115

Hedges JI, Clark WA, Quay PD et al (1986) Composition and fluxes of particulate organic matter in the Amazon River. Limnol Oceanogr 31:717–738

Hedges JI, Ertel JR (1982) Characterization of lignin by gas capillary chromatography of cupric oxide oxidation products. Anal Chem 54:174–178

Hemminga MA, Slim FJ, Kazungu J et al (1994) Carbon outwelling from a mangrove forest with adjacent seagrass beds and coral reefs (Gazi Bay, Kenya). Mar Ecol Prog Ser 106:291–301

Hernandez ME, Mead R, Peralba MC et al. (2001) Origin and transport of n-alkane-2-ones in a subtropical estuary: potential biomarkers for seagrass-derived organic matter. Org Geochem 32:21–32 2001

Hernes PJ, Benner R, Cowie GL et al (2001) Tannin diagenesis in mangrove leaves from a tropical estuary: a novel molecular approach. Geochim Cosmochim Acta 65:3109–3122

Hertkorn N, Benner R, Frommberger M et al (2006) Characterization of a major refractory component of marine dissolved organic matter. Geochim Cosmochim Acta 70:2990–3010

Hsieh HL, Chen CP, Chen YG et al (2002) Diversity of benthic organic matter flows through polychaetes and crabs in a mangrove estuary: delta C-13 and delta S-34 signals. Mar Ecol Prog Ser 227:125–155

Hughey CA, Hendrickson CL, Rodgers RP et al (2001) Kendrick mass defect spectrum: a compact visual analysis for ultrahigh-resolution broadband mass spectra. Anal Chem 73:4676–4681

Hwang DW, Kim GB, Lee YW et al (2005) Estimating submarine inputs of groundwater and nutrients to a coastal bay using radium isotopes. Mar Chem 96:61–71

Jaffé R, Boyer JN, Lu X et al (2004) Sources characterization of dissolved organic matter in a mangrove-dominated estuary by fluorescence analysis. Mar Chem 84:195–210

Jaffé R, Mead R, Hernandez ME et al (2001) Origin and transport of sedimentary organic matter in two subtropical estuaries: a comparative, biomarker-based study. Org Geochem 32:507–526

Jaffé R, McKnight D, Maie N et al (2008) Spatial and temporal variations in DOM composition in ecosystems: the importance of long-term monitoring of optical properties. J Geophys Res 113:G04032

Jaffé R, Rushdi AI, Medeiros PM et al (2006) Natural product biomarkers as indicators of sources and transport of sedimentary organic matter in a subtropical river. Chemosphere 64:1870–1884

Johns RB, Brady BA, Butler MS et al (1994) Organic geochemical and geochemical studies of inner Great Barrier Reef sediments – IV. Identification of terrigenous and marine sourced inputs. Org Geochem 21:1027–1035

Jones V, Collins MJ, Penkman KEH et al (2005) An assessment of the microbial contribution to aquatic dissolved organic nitrogen using amino acid enantiomeric ratios. Org Geochem 36:1099–1107

Kalbitz K, Geyer S, Geyer W (2000) A comparative characterization of dissolved organic matter by means of original aqueous samples and isolated humic substances. Chemosphere 40: 1305–1312

Kattner G, Lobbes JM, Fitznar HP et al (1999) Tracing dissolved organic substances and nutrients from the Lena River through Laptev Sea (Arctic). Mar Chem 65:25–39

Kendrick E (1963) A mass scale based on $CH_2 = 14.0000$ for high resolution mass spectrometry of organic compounds. Anal Chem 35:2146–2154

Killops SD, Frewin NL (1994) Triterpenoid diagenesis and cuticular preservation. Org Geochem 21:1193–1209

Kim S, Kramer RW, Hatcher PG (2003) Graphical method for analysis of ultrahigh-resolution broadband mass spectra of natural organic matter, the Van Krevelen Diagram. Anal Chem 75:5336–5344

Kim SW, Kaplan LA, Benner R et al (2004) Hydrogen-deficient molecules in natural riverine water samples – evidence for the existence of black carbon in DOM. Mar Chem 92:225–234

Kjerfve B, Stevenson LH, Proehl JA (1981) Estimation of material fluxes in an estuarine cross section: a critical analysis of spatial measurement density and errors. Limnol Oceanogr 26: 325–335

Koch BP, Dittmar T (2006) From mass to structure: an aromaticity index for high-resolution mass data of natural organic matter. Rapid Commun Mass Spectrom 20:926–932

Koch BP, Harder J, Lara RJ et al (2005a) The effect of selective microbial degradation on the composition of mangrove derived pentacyclic triterpenols in surface sediments. Org Geochem 36:273–285

Koch BP, Ludwichowski K-U, Kattner G et al (2008) Advanced characterization of marine dissolved organic matter by combining reversed-phase liquid chromatography and FT-ICR-MS. Mar Chem 111:233–241

Koch BP, Rullkötter J, Lara RJ (2003) Evaluation of triterpenols and sterols as organic matter biomarkers in a mangrove ecosystem in northern Brazil. Wetlands Ecol Manage 11:257–263

Koch BP, Witt M, Engbrodt R et al (2005b) Molecular formulae of marine and terrigenous dissolved organic matter detected by electrospray ionizations Fourier transform ion cyclotron resonance mass spectrometry. Geochim Cosmochim Acta 69:3299–3308

Kowalczuk P, Cooper WJ, Whitehead RF et al (2003) Characterization of CDOM in an organic rich river and surrounding coastal ocean in the South Atlantic Bight. Aquat Sci 65:381–398

Krest J, Moore W, Gardner L et al (2000) Marsh nutrient export supplied by groundwater discharge: evidence from radium measurements. Global Biogeochem Cycles 14:167–176

Kujawinski EB, Del Vecchio R, Blough NV et al (2004) Probing molecular-level transformations of dissolved organic matter: insights on photo-chemical degradation and protozoan modification of DOM from electrospray ionization Fourier transform ion cyclotron resonance mass spectrometry. Mar Chem 92:23–37

Kujawinski EB, Freitas MA, Zang X et al (2002) The application of electrospray ionization mass spectrometry (ESI MS) to the structural characterization of natural organic matter. Org Geochem 33:171–180

Lee SY (1995) Mangrove outwelling: a review. Hydrobiologia 295:203–212

Lin G, Banks T, Sternberg LS (1991) Variation in delta ^{13}C values for the seagrass *Thalassia testudinum* and its relations to mangrove carbon. Aquat Bot 40:333–341

Lombardi AT, Jardim WF (1999) Fluorescence spectroscopy of high performance liquid chromatography fractionated marine and terrestrial organic materials. Water Res 33:512–520

Lu XQ, Jaffé R (2001) Interaction between Hg(II) and natural dissolved organic matter: a fluorescence spectroscopy based study. Water Res 35:1793–1803

Lu XQ, Maie N, Hanna JV et al (2003) Molecular characterization of dissolved organic matter in freshwater wetlands of the Florida Everglades. Water Res 37:2599–2606

Marguillier S, van der Velde G, Dehairs F et al (1997) Trophic relationships in an interlinked mangrove-seagrass ecosystem as traced by delta ^{13}C and delta ^{15}N. Mar Ecol Prog Ser 151: 115–121

Maie N, Boyer JN, Yang CY et al (2006a) Spatial, geomorphological, and seasonal variability of CDOM in estuaries of the Florida coastal Everglades. Hydrobiologia 569:135–150

Maie N, Parish KJ, Watanabe A et al (2006b) Chemical characteristics of dissolved organic nitrogen in an oligotrophic subtropical coastal ecosystem. Geochim Cosmochim Acta 70: 4491–4506

Maie N, Pisani O, Jaffé R (2008) Mangrove tannins in aquatic ecosystems: their fate and possible role in dissolved organic nitrogen cycling. Limnol Oceanogr 53:160–171

Maie N, Scully NM, Pisani O et al (2007) Composition of a protein-like fluorophore of dissolved organic matter in coastal wetland and estuarine ecosystems. Water Res 41: 563–570

Maie N, Yang CY, Miyoshi T et al (2005) Chemical characteristics of dissolved organic matter in an oligotrophic subtropical wetland/estuarine ecosystem. Limnol Oceanogr 50:23–35

Marhaba TF, Van D, Lippincott RL (2000) Rapid identification of dissolved organic matter fractions in water by spectral fluorescent signatures. Water Res 34:3543–3550

Marshall AG, Hendrickson CL, Jackson GS (1998) Fourier transform ion cyclotron resonance mass spectrometry: a primer. Mass Spectrom Rev 17:1–35

Martin-Mousset B, Croue JP, Lefebvre E et al (1997) Distribution and characterization of dissolved organic matter of surface waters. Water Res 31:541–553

McKee KL, Feller IC, Popp M et al (2002) Mangrove isotopic (delta N-15 and delta C-13) fractionation across a nitrogen vs. phosphorus limitation gradient. Ecology 83:1065–1075

McKnight DM, Boyer EW, Westerhoff PK et al (2001) Spectrofluorometric characterization of dissolved organic matter for indication of precursor organic material and aromaticity. Limnol Oceanogr 46:38–48

Mead R, Xu YP, Chong J et al (2005) Sediment and soil organic matter source assessment as revealed by the molecular distribution and carbon isotopic composition of nalkanes. Org Geochem 36:363–370

Mfilinge PL, Meziane T, Bachok Z et al (2005) Total lipid and fatty acid classes in decomposing mangrove leaves of *Bruguiera gymnorrhiza* and *Kandelia candel*: significance with respect to lipid input. J Oceanogr 61:613–622

Minor EC, Simjouw J-P, Boon JJ et al (2002) Estuarine/marine UDOM as characterized by size-exclusion chromatography and organic mass spectrometry. Mar Chem 78:75–102

Moore WS (1999) The subterranean estuary: a reaction zone of ground water and sea water. Mar Chem 65:111–125

Moore WS (2006) The role of submarine groundwater discharge in coastal biogeochemistry. J Geochem Explor 88:89–393

Mopper K, Stubbins A, Ritchie JD et al (2007) Advanced instrumental approaches for characterization of marine dissolved organic matter: extraction techniques, mass spectrometry, and nuclear magnetic resonance spectroscopy. Chem Rev 107:419–442

Moran MA, Wicks RJ, Hodson RE (1991) Export of dissolved organic matter from a mangrove swamp ecosystem: evidence from natural fluorescence, dissolved lignin phenols, and bacterial secondary production. Mar Ecol Prog Ser 76:75–184

Moran MA, Hodson RE (1994) Dissolved humic substances of vascular plant origin in a coastal marine environment. Limnol Oceanogr 39:762–771

Morris J (1995) The mass balance of salt and water in intertidal sediments: results from North Inlet, South Carolina. Estuaries 18:556–567

Mudd S, Fagherazzi S, Morris J et al (2004) Flow, sedimentation, and biomass production on a vegetated salt marsh in South Carolina: toward a predictive model of marsh morphologic and ecologic evolution. In: Fagherazzi S, Marani M, Blum L (eds) The ecogeomorphology of tidal marshes, Vol. 59. American Geophysical Union, Coastal and Estuarine Studies, Washington DC

Muller FLL, Larsen A, Stedmon CA et al (2005) Interactions between algal–bacterial populations and trace metals in fjord surface waters during a nutrient-stimulated summer bloom. Linmol Oceanogr 50:1855–1871

Munoz D, Guiliano M, Doumenq P et al (1997) Long term evolution of petroleum biomarkers in mangrove soil (Guadeloupe). Mar Pollut Bull 34:868–874

Novakowski K, Torres R, Gardner L et al (2004) Geomorphic analysis of tidal creek networks. Water Resour Res 40:W05401

Ohno T, Bro R (2006) Dissolved organic matter characterization using multiway spectral decomposition of fluorescence landscapes. Soil Sci Soc Am J 70:2028–2037

Ohno T, Fernandez IJ, Hiradate S et al (2007) Effects of soil acidification and forest type on water soluble soil organic matter properties. Geoderma 140:176–187

Pant P, Rastogi RP (1979) The triterpenoids. Phytochem 18:1095–1108

Parlanti E, Worz K, Geoffroy L et al (2000) Dissolved organic matter fluorescence spectroscopy as a tool to estimate biological activity in a coastal zone submitted to anthropogenic inputs. Org Geochem 31:1765–1781

Perdue EM, Koprivnjak J-F (2007) Using the C/N ratio to estimate terrigenous inputs of organic matter to aquatic environments. Estuar Coast Shelf Sci 73:65–72

Peterson BJ, Fry B (1989) Stable isotopes in ecosystem studies. Annu Rev Ecol Syst 18:293–320

Piccolo A, Conte P, Trivellone E et al (2002) Reduced heterogeneity of a lignite humic acid by preparative HPSEC following interaction with an organic acid. Characterization of size-separates by Pyr-GC-MS and ^{1}H-NMR spectroscopy. Environ Sci Technol 36: 76–84

Prahl FG, Ertel JR, Goñi MA et al (1994) Terrestrial organic carbon contributions to sediments on the Washington margin. Geochim Cosmochim Acta 58:3035–3048

Primavera JH (1996) Stable carbon and nitrogen isotope ratios of penaeid juveniles and primary producers in a riverine mangrove in Guimaras, Philippines. Bull Mar Sci 58:675–683

Primavera JH, Altamirano JP, Lebata M et al (2007) Mangroves and shrimp pond culture effluents in Aklan, Panay Is., Central Philippines. Bull Mar Sci 80:795–804

Reemtsma T (2001) The use of liquid chromatography-atmospheric pressure ionization mass spectrometry in water analysis – Part II: obstacles. Trac-Trend Anal Chem 20:533–542

Reemtsma T, These A (2003) On-line coupling of size exclusion chromatography with electrospray ionization-tandem mass spectrometry for the analysis of aquatic fulvic and humic acids. Anal Chem 75:1500–1507

Reemtsma T, These A, Springer A et al (2006a) Fulvic acids as transition state of organic matter: indications from high resolution mass spectrometry. Environ Sci Technol 40:5839–5845

Reemtsma T, These A, Venkatachari P et al (2006b) Identification of fulvic acids and sulfated and nitrated analogues in atmospheric aerosol by electrospray ionization Fourier transform ion cyclotron resonance mass spectrometry. Anal Chem 78:8299–8304

Rezende CE, Lacerda LD, Ovalle ARC et al (1990) Nature of POC transport in a mangrove ecosystem: a carbon stable isotopic study. Estuar Coast Shelf Sci 30:641–645

Rivera Monroy VH, Day JW, Twilley RR et al (1995) Flux of nitrogen and sediment in a fringe mangrove forest in Terminos Lagoon, Mexico. Estuar Coast Shelf Sci 40:139–160

Rodelli MR, Gearing JN, Gearing PJ et al (1984) Stable isotope ratio as a tracer of mangrove carbon in Malaysian ecosystems. Oecologia 61:326–333

Romigh MM, Davis SE, Rivera-Monroy VH et al (2006) Flux of organic carbon in a riverine mangrove wetland in the Florida coastal Everglades. Hydrobiologia 569:505–516

Santos IR, Burnett WC, Chanton J et al (2008) Nutrient biogeo-chemistry in a Gulf of Mexico subterranean estuary and groundwater-derived fluxes to the coastal ocean. Limnol Oceanogr 53:705–718

Schaub TM, Rodgers RP, Marshall AG et al (2005) Speciation of aromatic compounds in petroleum refinery streams by continuous flow field desorption ionization FT-ICR mass spectrometry. Energy Fuels 19:1566–1573

Schmitt-Kopplin P, Kettrup A (2003) Capillary electrophoresis – electrospray spray ionization-mass spectrometry for the characterization of natural organic matter: an evaluation with free flow electrophoresis-off-line flow injection electrospray ionization-mass spectrometry. Electrophoresis 24:3057–3066

Scourse J, Marret F, Versteegh GJM et al (2005) High-resolution last deglaciation record from the Congo fan reveals significance of mangrove pollen and biomarkers as indicators of shelf transgression. Quaternary Res 64:57–69

Scully NM, Maie N, Dailey S et al (2004) Photochemical and microbial transformation of plant derived dissolved organic matter in the Florida Everglades. Limnol Oceanogr 49:1667–1678

Scully NM, Lean DRS (1994) The attenuation of ultraviolet light in temperate lakes. Arch Hydrobiol 43:135–144

Sigman DM, Altabet MA, Michener R et al (1997) Natural-abundance level measurements of the nitrogen isotopic composition of oceanic nitrate: an adaptation of the ammonium diffusion method. Mar Chem 57:227–242

Simpson AJ, Tseng LH, Simpson MJ et al (2004) The application of LC-NMR and LC-SPE-NMR to compositional studies of natural organic matter. Analyst 129:1216–1222

Simoneit BRT, Xu Y, Neto RR et al (2009) Photochemical alteration of 3-oxy-triterpenoids: implications for the origin of des-A-triterpenoids in aquatic sediments and soils. Chemosphere 74:543–550

Stanev EV, Wolff JO, Burchard H et al (2003) On the circulation in the East Frisian Wadden Sea: numerical modeling and data analysis. Ocean Dyn 53:27–51

Stedmon CA, Markager S (2005) Resolving the variability in dissolved organic matter fluorescence in a temperate estuary and its catchment using PRAFAC analysis. Limnol Oceanogr 50: 686–697

Stedmon CA, Markager S, Bro R (2003) Tracing dissolved organic matter in aquatic environments using a new approach to fluorescence spectroscopy. Mar Chem 82:239–254

Stedmon CA, Markgaer S, Tranvik L et al (2007) Photochemical production of ammonium and transformation of dissolved organic matter in the Baltic Sea. Mar Chem 104:227–240

Stenson AC, Marshall AG, Cooper WT (2003) Exact masses and chemical formulas of individual Suwannee River fulvic acids from ultrahigh resolution electrospray ionization Fourier transform ion cyclotron resonance mass spectra. Anal Chem 75:1275–1284

Taniguchi M, Burnett W, Cable J et al (2002) Investigation of submarine groundwater discharge. Hydrol Process 16:2115–2129

Tobias CR, Harvey JW, Anderson IC (2001) Quantifying groundwater discharge through fringing wetlands to estuaries: seasonal variability, methods comparison, and implications for wetland-estuary exchange. Limnol Oceanogr 46:604–615

Tremblay LB, Dittmar T, Marshall AG et al. (2007) Molecular characterization of dissolved organic matter in a North Brazilian mangrove porewater and mangrove-fringed estuaries by ultrahigh resolution Fourier transform ion cyclotron resonance mass spectrometry and excitation/emission spectroscopy. Mar Chem 105:15–29

Troccaz O, Giraud F, Bertru G et al (1994) Methodology for studying exchanges between salt marshes and coastal marine waters. Wetlands Ecol Manage 3:37–48

Twilley RR (1985) The exchange of organic carbon in basin mangrove forest in a southwestern Florida estuary. Estuar Coast Shelf Sci 20:543–557

Tzortziou M, Neale PJ, Osburn CL et al (2008). Tidal marshes as a source of colored dissolved organic matter in the Chesapeake Bay. Limnol Oceanogr 53:148–159

Valiela I, Teal J, Volkmann S et al (1978) Nutrient and particulate fluxes in a salt marsh ecosystem: tidal exchanges and inputs by precipitation and groundwater. Limnol Oceanogr 23: 798–812

Versteegh GJM, Schefuß E, Dupont L et al (2004) Taraxerol and *Rhizophora* pollen as proxies for tracking past mangrove ecosystems. Geochim Cosmochim Acta 68:411–422

Volkman JK (1986) A review of sterol markers for marine and terrigenous organic matter. Org Geochem 9:83–99

Volkman JK (2005) Sterols and other triterpenoids: source specificity and evolution of biosynthetic pathways. Org Geochem 36:139–159

Volkman JK, Barrett SM, Blackburn SI et al (1998) Microalgal biomarkers: a review of recent research developments. Org Geochem 29:1163–1179

Volkman JK, Revill AT, Bonham PI et al (2007) Sources of organic matter in sediments from the Ord River in tropical northern Australia. Org Geochem 38:1039–1060

Volkman JK, Rohjans D, Rullkotter J et al (2000) Sources and diagenesis of organic matter in tidal flat sediments from the German Wadden Sea. Cont Shelf Res 20:1139–1158

Wang Z, Liu W, Zhao N et al (2007) Composition analysis of colored dissolved organic matter in Taihu Lake based on three dimension excitation-emission fluorescence matrix and PARAFAC model, and the potential application in water quality monitoring, J Environ Sci 19:787–791

Wattayakorn G, Wolanski E, Kjerfve B (1990) Mixing, trapping and outwelling in the Klong Ngao mangrove swamp, Thailand. Estuar Coast Shelf Sci 31:667–688

Weishaar JL, Aiken GR, Bergamashi BA et al (2003) Evaluation of specific ultraviolet absorbance as an indicator of the chemical composition and reactivity of dissolved organic carbon. Environ Sci Technol 37:4702–4708

Wilson A, Gardner L (2006) Tidally driven groundwater flow and solute exchange in a marsh: numerical simulations. Water Resour Res 42:W01405

Wolanski E, Ridd P (1990) Mixing and trapping in Australian tropical coastal waters. In: Cheng RT (ed), Residual currents and long-term transport, pp. 165–183. Springer, New York

Wolanski E (1992) Hydrodynamics of mangrove swamps and their coastal waters. Hydrobiologia 247:141–161

Wollenweber E, Doerr M, Siems K et al (1999) Triterpenoids in lipophilic leaf and stem coatings. Biochem Syst Ecol 27:103–105

Wu FC, Evans RD, Dillon PJ (2003) Separation and characterization of NOM by high-performance liquid chromatography and on-line three-dimensional excitation emission matrix fluorescence detection. Environ Sci Technol 37:3687–3693

Xu Y, Holmes C, Jaffé R (2007) A lipid biomarker record of environmental change in Florida Bay over the past 150 years. Estuar Coast Shelf Sci 73:201–210

Xu Y, Jaffé R (2007) Lipid biomarkers in suspended particulates from a subtropical estuary: assessment of seasonal changes in sources and transport of organic matter. Mar Environ Res 64:666–678

Xu Y, Mead R, Jaffé R (2006) A molecular marker based assessment of sedimentary organic matter sources and distributions in Florida Bay. Hydrobiologia 569:179–192

Yamashita Y, Jaffé R, Maie N et al (2008) Assessment of the dynamics of fluorescent dissolved organic matter in coastal environments by EEM-PARAFAC. Limnol Oceanogr 53:1900–1908

Yamashita Y, Tanoue E (2003) Chemical characterization of protein-like fluorophores in DOM in relation to aromatic amino acids. Mar Chem 82:255–271

Zepp RG, Schlotzhauer PF (1981) Comparison of photochemical behavior of various humic substances in water: 3. Spectroscopic properties of humic substances. Chemosphere 10:479–486

Zieman JC, Macko SA, Mills AL (1984) Role of seagrass and mangroves in estuarine food webs: temporal and spatial changes in stable isotope composition and amino acid content during decomposition. Bull Mar Sc 35:380–392

Chapter 13
Tools for Studying Biological Marine Ecosystem Interactions—Natural and Artificial Tags

Bronwyn M. Gillanders

Abstract   Determining connectivity of organisms is difficult especially for early life history stages (larvae and juveniles). Fortunately, a variety of natural and artificial tags, some of which date back to the 1600's, have been developed to help address the issues of movement. Over the years a vast literature on tagging has emerged, of which I provide an updated review. In this chapter, I discuss five broad areas of tagging (external tags, external marks, internal tags, telemetry, and natural tags) and provide additional information on genetic and chemical methods. For each method I highlight their advantages and disadvantages, and provide examples, where possible, of connectivity among tropical coastal ecosystems. Advances in many of the methodologies are expected to continue, and future studies should consider combining more than one approach especially where natural tags are utilized.

Keywords Acoustic tagging · Otolith chemistry · Stable isotopes · Natural tag · Genetics

13.1 Introduction

The degree of connectivity among aquatic populations in tropical systems is largely unknown for many species. Connectivity of organisms is determined by the dispersal abilities of adults, as well as their eggs and larvae. Almost all coral reef fish have a two phase life cycle where larvae are planktonic (Leis 1991) and therefore are potentially capable of considerable dispersal, whereas adults are relatively sedentary and demersal. In addition, many species also show spatially segregated juvenile and adult habitats separated by distances of a few meters to hundreds of kilometers (Gillanders et al. 2003). Populations can therefore be linked by exchange of larvae, recruits, juveniles, and/or adults (Palumbi 2004). Although knowledge of the extent to which fish move, the habitats they utilize, and the connectedness of populations

B.M. Gillanders (✉)
Southern Seas Ecology Laboratories, DX 650 418, School of Earth and Environmental Sciences, University of Adelaide, SA 5005, Australia
e-mail: bronwyn.gillanders@adelaide.edu.au

I. Nagelkerken (ed.), *Ecological Connectivity among Tropical Coastal Ecosystems*,   457
DOI 10.1007/978-90-481-2406-0_13, © Springer Science+Business Media B.V. 2009

is critical for effective fisheries management and conservation, it has been difficult to quantify.

For a few species direct observations of larval or juvenile movement have been made (e.g., Olson and McPherson 1987). Generally, though, it is difficult to obtain information on population connectivity from direct observation (because the time scale is too short), and therefore it is necessary to employ a mark or tag. A mark is generally considered anything used for recognition purposes that is external, internal, or incorporated into the integument of the organism, whereas a tag generally contains specific identification information and can be attached externally or internally (Guy et al. 1996). The two terms have however been used interchangeably in the literature.

Thorrold et al. (2002) lists four key elements for a successful tagging method in marine environments. First, marked individuals must be able to be unambiguously identified some time after tagging which requires that the tag be retained for a length of time appropriate for the temporal scale of the study. For example, Levin et al. (1993) were interested in tracking dispersal of invertebrates over a tidal cycle (6–24 h) and so a short-term tag was suitable. By comparison, Jones et al. (2005) were interested in determining settlement of coral reef fish after the marking of embryos and therefore the minimum time that the tag needed to be retained for was 15–19 days (i.e., the length of the larval period plus the period prior to hatching). Second, the behavior, growth, and survival of tagged individuals should not differ from those of the untagged population. Tagged organisms should not be more susceptible to predation (in latter sections, I highlight studies testing whether tagging affects behavior, growth, and survival). Malone et al. (1999) showed that juvenile fish tagged with fluorescent visual implant elastomer tags were no more susceptible to predation than non tagged fish. Third, the tagging method needs to be able to mark large numbers of the study organism in a cost-effective manner due to dilution effects resulting from mortality of early life history stages. To date, most studies have focused on directly marking either the larval or juvenile stage, and only recently has the potential for transgenerational marking been investigated (see Thorrold et al. 2006, Almany et al. 2007). Transgenerational marking offers the ability to mark tens of thousands of offspring from a single maternal injection (see chemical marks). Finally, because of dilution effects in the population with tagged individuals generally forming a small proportion of all individuals, methods of detection need to be quick and inexpensive. Many of the individuals being screened for a tag may be unmarked individuals. Simple methods that alert researchers to a tagged individual (e.g., removal of adipose fin as in salmonid tagging) are needed. This then allows fewer individuals to be examined in greater detail (e.g., for a coded wire tag that provides additional information once the code is deciphered) and may be more cost effective than a single tag.

The earliest tagging records are from brown trout in European streams, and date back to the 1600's (Nielsen 1992). At this time wool threads were tied around the tails of fish (Nielsen 1992). Since then the methods available for studying movement and migration of organisms have expanded greatly and become more sophisticated. An enormous literature on tagging (>120,000 peer-reviewed journal articles listed on CSA Illumina up to 2008 for a search combining tag and invertebrate or fish) exists including several recent reviews (e.g., Thorrold et al. 2002, Elsdon and Gillanders 2003a, Semmens et al. 2007, Elsdon et al. 2008). Broadly speaking, these methods can be grouped into five broad categories encompassing external tags, external marks, internal tags, telemetry, and natural marks. However, both genetic identification and marking, and chemical marking may encompass both natural marks and internal tags and therefore I have devoted separate sections to each of these approaches. Below, I briefly explain each of the methods and then provide examples of studies which have used the various methods to show population connectivity. My focus is generally on examples from tropical coastal ecosystems, but temperate examples are provided when little published literature exists for tropical systems.

13.2 External Tags

External tags have been used for group and individual identification, and are generally attached to the animal in one of three ways: transbody tags, dart-style tags, and internal-anchor tags (Nielsen 1992). Briefly, transbody tags protrude through both sides of the animal's body, whereas dart-style and internal-anchor tags only protrude through one side of the animal's body. Transbody tags generally include a shaft that passes through the animal's body and is enlarged on both ends to prevent the tag from coming off (Nielsen 1992). Dart-style tags include a protruding shaft and embedded within the animal's body is an anchor that prevents removal of the tag (e.g., dart or T-bar tags, Nielsen 1992) (Fig. 13.1). Internal-anchor tags are similar to dart-style tags, but the anchoring device is usually a flat disc lying against the inside wall of the fish's body cavity. Identifying information is then found on the protruding part of the tag.

External tags are inexpensive, easily visible, and widely used. However, they are generally restricted to large juvenile and often adult organisms and a critical assumption is that the tag and tagging process do not influence the animal in any way (Nielsen 1992) (see Table 13.1). They can enable individual recognition of organisms, but also require individual handling.

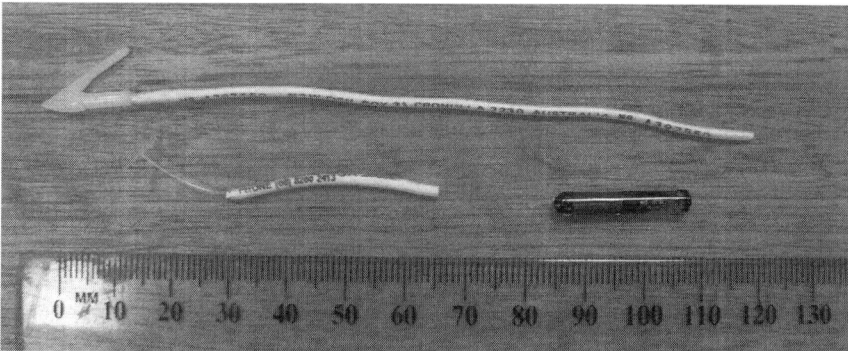

Fig. 13.1 Photo of dart and T-bar tags (external tags), and PIT tag (internal tag)

Table 13.1 Different methods of tagging showing advantages and disadvantages as well as a brief description of each method. Also indicated are published examples from tropical literature

Tagging method	Brief description of method	Advantages	Disadvantages	Tropical examples
EXTERNAL TAGS		Individuals can be recognized; tags inexpensive	Requires individual handling of organisms; restricted to juvenile and adult organisms	Burke (1995), Dorenbosch et al. (2004), Verweij and Nagelkerken (2007), Verweij et al. (2007)
Transbody tags	Protrudes through both sides of animal's body; shaft and enlargements on ends (e.g., discs, beads, bubbles)	Easy to tell if animal properly tagged; can be used on organisms of different sizes; easily detected	Difficult and extensive tagging process; may become entangled; may accumulate algae; may affect growth and behavior	Dart and T-bar anchor tags: Sumpton et al. (2003), Verweij et al. (2007)
Dart-style tags	One end with anchor embedded in animal; has protruding shaft holding vinyl tube, disc, or bead	Can be used on range of organisms of larger size; applied easily and rapidly including by volunteers; easily detected	Not useful on small animals; quality of attachment important and difficult to control; loss of tagging information from abrasion or separation of legend from shaft; tagging process can produce abrasion or enlarged tagging wound	
Internal-anchor tags	Protruding part similar to dart tag; anchor is flat disc on inside wall of fish's body cavity	Generally only used on fish	Not useful on small animals; difficult and extensive tagging procedures; danger to internal organs; loss of tagging information from abrasion or separation of legend from shaft; less easily detected	

Table 13.1 (continued)

Tagging method	Brief description of method	Advantages	Disadvantages	Tropical examples
EXTERNAL MARKS				
Fin clipping	Part or all of one or more fins from a fish's body are removed	Absence of protruding tag Easy and rapid process but requires individual handling; suitable for many sizes and types of organism; can be used for short-term and long-term studies	Generally requires individual handling of organisms Individual marks not possible; limited number of group marks available; effects on mortality variable; may be problems recognizing and interpreting clips	van Rooij et al. 1995
Branding	Hot or cold object applied to skin to produce recognizable mark	Easy and rapid process; minor effects on growth, mortality, and behavior	Suitable primarily for scaleless or fine-scaled fishes; individual marks not available and limited number of group marks possible; generally limited to short-term studies as long-term retention is variable; may cause stress to individual	Zeller and Russ (1998, 2000)
Pigment marking (e.g., tattoos, latex injection)	Pigment injected or embedded in skin sometimes via spraying	Can be used for individual or batch marking; highly visible marks	Requires large amount of handling; mortality and mark retention variable over long-term; may require organism to be anaesthetized	
	Grit marking	Individual handling of fish not required	Individual marks not possible; limited number of group marks available	

Table 13.1 (continued)

Tagging method	Brief description of method	Advantages	Disadvantages	Tropical examples
INTERNAL TAGS				
Coded wire tags	Very small tag (1.1 mm long x 0.25 mm diameter) injected into animal	Absence of protruding tag; much smaller than external tags / Little effect on growth, behavior, and mortality; suitable for long-term studies; suitable for very small organisms through to large individuals; high tag retention; large scale marking possible	Individual handling of fish required for tagging / Reading of tag information involves sacrificing animal or dissection of tag; extensive tag detection process; capital intensive, especially for automated tagging and detection systems	Beukers et al. (1995), Verweij and Nagelkerken (2007)
Passive integrated transponder tags	Miniature signal-relay station, tag is about 12 mm long x 2.1 mm diameter with components enclosed in glass or similar tube	Suitable for many species and sizes of organisms; can be read without sacrificing animal; individuals can be recognized; suitable for long-term studies; no handling needed to identify animal, therefore repeated observations possible	Short signal detection range—animal has to pass close by reader to be detected; tags still relatively large therefore not suitable for very small fish	McCormick and Smith (2004)
Visible implant tags	Internal tag that is readable from outside; small, flat rectangle coded with colors and multidigit, alphanumeric code	Can be read without sacrificing animal; individuals can be identified; tags very cheap compared to other internal tags	Not suitable for very small fish; variable in retention and readability, particularly over long-term	
Visible implant elastomer tags	Made of two part silicon which is mixed immediately prior to use; liquid injected into organism which then cures into solid	Can be used on range of species and very small individuals; tags very cheap compared to other internal tags	Variable in retention and readability, particularly over long-term	Beukers et al. (1995), Frederick (1997a, b)

Table 13.1 (continued)

Tagging method	Brief description of method	Advantages	Disadvantages	Tropical examples
TELEMETRY		More detailed information on movement provided than just location of capture and recapture	Limited to large animals due to size of tags	
Acoustic telemetry	Tag is a high frequency transmitter that can be heard by a hydrophone	No need to recapture individuals to gain movement information; near real-time information possible	Tags typically quite large (20–100 mm) and expensive; manual following may be necessary if moored automatic tracking systems are not available	Zeller (1998), (1999), Zeller and Russ (1998), see also Pacific Ocean Shelf Tracking (POST) project (www.post.coml.org)
Archival tags	Tag records and stores a range of information	Suitable for obtaining environmental and movement information for large organisms	Need to recover tag to download information, although satellite recovery now possible (e.g., pop-up archival tags); tags typically quite large (30–80 mm) and relatively expensive	Gunn et al. (2003), Prince et al. (2005)
NATURAL MARKS		No need to mark fish thereby eliminating handling; marking process is free; every animal is marked reducing number of animals that must be recaptured to require a reliable sample for analysis	Need to be present and stable through the time period under study; can be more difficult to identify on older individuals; identification method frequently observational and statistical	
Morphometric marks	Based on differences in relative measurements of body or body structures (e.g., scale and otolith measurements) or coloration	Morphometric features frequently overlap across groups; may require extensive sample preparation in case of scales and otoliths		Gaines and Bertness (1992) (temperate), Swearer et al. (1999), Wilson et al. (2006)

Table 13.1 (continued)

Tagging method	Brief description of method	Advantages	Disadvantages	Tropical examples
Meristic marks	Based on differences in number of skeletal features (e.g., counts of number of vertebrae)		Meristic features frequently overlap across groups; unlikely to be reliable markers	
Parasitic marks	Parasite should be present in every individual from one group and no individuals from all other groups, or should be differences in infection rates or genetics of parasites from different groups		Difficult to find parasites that are common in one group and rare in another group for same host; parasite needs to remain on host even after host has moved; requires extensive evaluation before use	Grutter (1998), Cribb et al. (2000)
GENETIC MARKS		See generic comments under 'natural marks'	See generic comments under 'natural marks'	
Estimates of migration	Molecular markers used to indirectly estimate levels of dispersal based on models of gene flow or assign individuals to source populations	Information passes between generations	Samples can be degraded if not handled properly	Brazeau et al. (2005)
Parentage analysis	Marker of choice is usually microsatellites		All potential parents should be genotyped; sometimes difficult to exclude all but one pair of parents as likely parents for offspring; development costs may be high; genotype errors may affect accuracy of parentage assignment	Jones et al. (2005)

Table 13.1 (continued)

Tagging method	Brief description of method	Advantages	Disadvantages	Tropical examples
CHEMICAL MARKS		See generic comments under 'natural marks'	See generic comments under 'natural marks'	
Artificial marks	Fluorescent compounds that are bone seeking are incorporated into calcified structures, and when viewed under appropriate filters the mark fluoresces	Can mark all life history stages including eggs with limited handling of organisms; usually no affect on growth or mortality; marks typically long-lasting; high mark efficiency	Individual marks not possible, although can get batches of marks; marks of some fluorescent compounds degrade over time	Jones et al. (1999), (2005)
	Elements and isotopes can be applied by immersing fish in solution containing elements/isotopes; injecting larger individuals or by transgenerational marking	Marks can be obtained that are clearly artificial	For immersion may need to hold organism in solution for extended period of time for good mark; ability to detect mark may decrease over time	Almany et al. (2007)
	Temperature fluctuations used to induce marks which alter spacing of increments in otoliths	Chemical free method of marking; permanent mark applied with high efficiency; can mark large numbers of fish		
Natural marks	Elemental and/or isotopic signatures used as natural tag		Natural environmental fluctuations may produce similar marks; interpretation of band pattern may be problematic	Swearer et al. (1999), Patterson et al. (2004, 2005), Chittaro et al. (2004), Verweij et al. (2008)
			Information generally more difficult to interpret than artificial tag; need to have distinct signatures among areas of interest	

Several studies have investigated the degree of movement between different habitat types using external tags. Transbody tags (comprising monofilament line and colored beads) were used as short-term tags on juveniles and subadults of three species of Caribbean reef fish (*Haemulon flavolineatum, Haemulon sciurus*, and *Lutjanus apodus*) (Verweij and Nagelkerken 2007, Verweij et al. 2007) and two species (*Lutjanus fulviflamma* and *Lutjanus ehrenbergii*) of African reef fish (Dorenbosch et al. 2004). Dart style tags (Floy tags) were also used on *Lutjanus apodus* (Verweij et al. 2007). Tagging mortality was low [11/12 (92%) fish survived]; however, within two weeks 17% of tagged fish (2 out of 12) had lost their external tag (Verweij and Nagelkerken 2007). After six weeks most fish had either lost their tags or algal fouling prevented recognition of the tags (Verweij et al. 2007). These studies did, however, provide direct evidence of connectivity between mangrove/seagrass habitats (juvenile areas) and the adult coral reef habitat, or between mangrove/rocky shoreline and seagrass beds for feeding. Overall, high fidelity to small spatial areas was found for all species. Burke (1995) used small, glowing Cyalume light sticks sutured to the dorsal musculature of fish to investigate their nocturnal foraging habitats.

Sumpton et al. (2003) also investigated localized movement of snapper (*Pagrus auratus*) in a large subtropical embayment (Queensland, Australia) and the contribution of juvenile to offshore populations by tagging fish with anchor and dart tags. They found that most snapper movements were localized (about 1% of movements exceeded 100 km) and few fish (4 out of 2,500 tagged) moved out of the bay suggesting that the bay was not a significant contributor to the offshore fishery.

13.3 External Marks

An external mark alters an animal's appearance so that it is identifiable (Nielsen 1992). External marks comprise three major techniques: fin clipping, branding, and pigment marking. Fin clipping involves removing all or part of one or more fins from a fish's body, whereas branding involves scarring the skin tissue in a distinctive pattern, and pigment marking involves embedding inert colored material beneath the animal's skin or exoskeleton (e.g., tattooing, latex injection, fluorescent marking; Nielsen 1992). External marking methods are usually used to identify a few groups of organisms (e.g., distinguish hatchery-raised fish from wild fish) rather than lots of individuals (Table 13.1).

Several studies have used freeze branding (where cold temperatures are used to scar the skin tissue in a distinctive pattern) to mark adult coral reef fish largely to determine movement across boundaries of marine protected areas or to estimate population size (Zeller and Russ 1998, 2000; Fig. 13.2). In general, hot and freeze branding methods are considered harmful to juveniles (but see Saura 1996). Marks can last several years, devices for marking are relatively simple and inexpensive to construct and large numbers can be marked (Hargreaves 1992, Saura 1996).

Fin clipping has been widely practiced on salmonids (adipose fin is removed) often in conjunction with the insertion of a coded wire tag (see below) (van der

Fig. 13.2 Photos of freeze branding a coral trout (*Plectropomus leopardus*) (**a**) and a freeze-branded fish *in situ* (**b**). Photo credits: Dirk Zeller (a) and Roger Grace (b)

Haegen et al. 2005). Some studies report lower survival from fin clips (Wertheimer et al. 2002) while others found no differences in survival or growth (Thompson et al. 2005) suggesting an evaluation on a species by species basis may be required. A study of protogynous parrotfish *Sparisoma viride* on a fringing reef on Bonaire (Netherlands Antilles) found no affect of fin clipping on growth (van Rooij et al. 1995). This study focused on growth of different life phases rather than connectivity. I am not aware of any coral reef studies that have used this methodology for marking individuals to determine connectivity.

Various pigment marking techniques have been employed on fish. Ogden and Ehrlich (1977) used an air blast gun to mass-mark 100–200 French grunts by forcing fluorescent pigment granules into the epidermis. Although the pigment was not detectable to the naked eye because most of the pigment on the surface sloughed off, it was easily seen using ultraviolet light in the dark. Pigment was retained for up to two months, and little mortality was found. Experiments showed that juvenile grunts were capable of moving kilometers to return to home reefs (Ogden and Ehrlich 1977). More recently, Hayes et al. (2000) investigated the use of high-pressure injection of photonic (polymethylacrylate paint fluorescent pigment encapsulated in latex microspheres) paint to mark adult salmon. Marks to the pectoral girdle had no additional affect on mortality, were easy to apply and identify, and were retained for up to 45 days. In addition, tattoos are another form of pigment marking and have been used for individual recognition.

13.4 Internal Tags

Internal tags are tags completely embedded in the animal's body (Table 13.1). Over the years internal tags have become smaller and smaller (Nielsen 1992). The most commonly used is the coded wire tag (CWT), which is a tiny piece of magnetized, thin wire injected into the animal. Tags are detected using magnetic detectors (either manual or automatic) and the etched code read under a microscope. Several other types of internal tags are also available, namely passive integrated transponder (PIT) and visible implant (VI) tags. More recently, Northwest Marine Technology (http://www.nmt.us/index.htm) has manufactured the visual implant elastomer (VIE) tags. VIE tags are injected internally but are visible externally, and are generally used for batch marking.

Coded wire tags are widely used on salmonids in the USA. Many hatchery-reared salmonids are tagged with CWT on release to rivers, enabling their origins to be determined when they are recaptured as either juveniles or adults at sea (Courtney et al. 2000). Several studies have used CWT's on coral reef fish, although the primary purpose of the study was not always an investigation of movement. Beukers et al. (1995) showed that retention rates of CWT (and VIE tags) were high for two size classes of coral reef fish (100% for 10–20 mm juveniles and 80–100% for 30–40 mm adults), and that survivorship and growth were not influenced by tagging. These tags (along with VI and VI fluorescent tags) also showed good retention rates (>90%) for several species of temperate reef fish (Buckley et al. 1994). CWT's have also been used for tagging juveniles of invertebrates (e.g., clam, Lim and Sakurai 1999; lobster, Sharp et al. 2000; shrimp, Kneib and Huggler 2001; crabs, Davis et al. 2004; mussels, Layzer and Heinricher 2004; holothurians, Purcell et al. 2006b). In a coral reef application, Verweij and Nagelkerken (2007) tagged 1,114 *Haemulon flavolineatum* with CWT's, but only 4.6% were recaptured after 163–425 days at liberty. Reef-directed movements were found for two individuals suggesting that some *H. flavolineatum* may move from bay nurseries to coral reefs at latter life history stages.

PIT tags are electromagnetically coded tags injected under the skin of various animals (Figs. 13.1, 13.3). They are relatively small (12 mm long x 2 mm diameter), inexpensive, and allow individual identification of organisms. PIT detection systems containing a magnetic field energizer, radio receiver, and processor are required for detection of the tag. These systems require the tagged organism to pass close by the detector for the animal to be identified. A range of different readers are possible, including portable tag readers, electronic gates (e.g., fishways), or flat-bed antennae (Semmens et al. 2007). PIT tags have been used in a wide range of studies of fish habitat use and movement primarily in freshwater systems (e.g., Ombredane et al. 1998, Das Mahapatra et al. 2001, Knaepkens et al. 2007). On coral reefs, McCormick and Smith (2004) used PIT tags to quantify small-scale space use of *Pomacentrus amboinensis*. They found that females showed strong periodicity in visits to male nests.

Visible implant tags are small (2–4 mm long, 0.5–2 mm diameter, 0.1 mm thick) with a printed alphanumeric code allowing for individual identification of animals. They are often implanted into sites on the head of fish (e.g., adipose tissue of salmonids) and a key advantage is that they can be read *in situ*. These tags are generally not suitable for very small fish and the readability of the tag may decrease over time as pigmentation of the fish occurs. Although they have been implanted into fish, these have not to my knowledge included coral reef species.

Fig. 13.3 Photos showing (**a**) tagging of a common snook (*Centropomus undecimalis*) with a PIT tag (arrow), and (**b**) an electronic gate (white arrows) in a mangrove creek for monitoring passing PIT-tagged fish. Note also a hand-held receiver to the right of the fish being tagged (**a**). Photo credit: Ivan Nagelkerken

Visible implant elastomer tags are made from a two part silicon based material that is injected as a liquid and cures into a solid. It is available in both fluorescent and non fluorescent colors and has been successfully used to tag small (8–56 mm) coral reef fish (Beukers et al. 1995, Frederick 1997a, Tupper 2007). A number of studies have investigated tag retention and effects on growth and survival, but few studies have used VIE tagged fish to address ecological questions. Recently-settled coral reef fish tagged with VIE tags were observed to move, over bare sand, a distance of up to 100 m (Frederick 1997b). Tupper (2007) found no movement of *Cheilinus undulatus* for three months after settlement, but then found that fish moved an average of 90–106 m, depending on the site they were tagged at, over the next three months. *Plectropomus areolatus* moved to deeper water, generally covering over 300 m, soon after settlement (Tupper 2007). Several non coral reef studies have used VIE tagged individuals to investigate site fidelity (Willis et al. 2001, Skinner et al. 2005).

13.5 Telemetry

Several types of electronic devices with either a unique code or frequency can be implanted or attached to the animal. As the animal swims around, the code or frequency is transmitted to a receiver, along with environmental information in some cases (Voegeli et al. 2001, Heupel et al. 2006, Semmens et al. 2007). Two types of telemetry methods (acoustic telemetry and archival tags) are referred to below, but at present are only suitable for larger individuals (see Table 13.1).

Acoustic telemetry involves tagging an organism with a high-frequency transmitter that can be heard by a mobile or fixed hydrophone linked to a receiver. Either a presence/absence record, a geographical position, or the relative direction of the animal is obtained. Tags are typically quite large (15 mm and 0.5 g in water, Vemco, Halifax, Canada, www.vemco.com) potentially limiting their applicability, but the technology is improving at such a rate that tags are continually getting smaller and smaller. Additional issues include the cost of tags and the receiving equipment, which are expensive. Advantages include that individuals do not need to be recaptured to obtain movement information, and considerably more information is obtained than just capture and recapture position (Table 13.1). For example, local movement patterns including territory and home range size can be obtained.

Szedlmayer and Able (1993) used acoustic transmitters (37 x 16 mm cylinders) to estimate residence time and movements of juvenile summer flounder *Paralichthys dentatus* in a subtidal creek. All tagged fish eventually moved out of the creek and it was suggested that this was part of a seasonal migration to the adult habitat. Several coral reef studies have also used acoustic telemetry to address questions related to movement and activity patterns largely in relation to movement in and out of marine protected areas (Tulevech and Recksiek 1994, Zeller 1998, Zeller and Russ 1998, Zeller 1999). These coral reef studies all focused on larger fish (>30 cm fork length). More recently, Beets et al. (2003) tagged two species of reef fish with sonic tags and noted migrations into seagrass habitats from reef habitats (87–767 m).

Archival tags record and store information including light intensity (to estimate position), and pressure (to estimate depth), as well as water and body temperature (Arnold and Dewar 2001). The major disadvantage of archival tags is that they have to be recovered to download the data, although the recovery issue is continually improving with pop-up archival transmitting (PAT) tags (Wildlife computers, Washington, USA, www.wildlifecomputers.com) and communicating history acoustic transponder (CHAT) tags (Vemco); the latter download their data, and in the case of 'business card' tags the data from other tags encountered along the way, to moored or mobile acoustic receivers (see Table 13.1, Fig. 13.4). Unfortunately, the tags are currently quite large (30–80 mm) and relatively expensive, limiting their widespread use in connectivity studies. Tag sizes should weigh less than 1.5–2% of the tagged animal's body weight in water (Nielsen 1992). To date, archival tags have primarily been used to investigate behavior and migration of sharks and large fish (e.g., marlin, swordfish, tuna; Gunn et al. 2003, Takahashi et al. 2003, Prince et al. 2005, Schaefer et al. 2007). With continued miniaturization of electronic components, telemetry may offer a viable alternative to other artificial tagging methods for very small organisms (Sibert and Nielsen 2001).

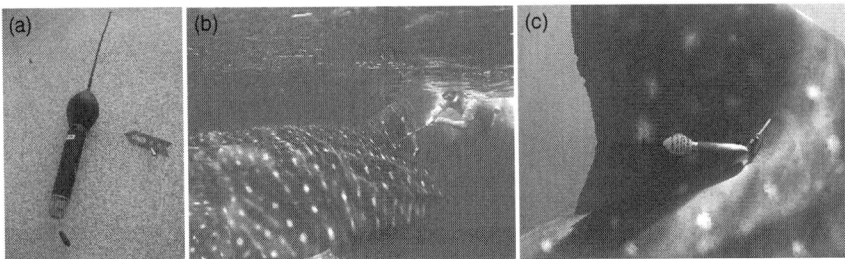

Fig. 13.4 Photos of pop-up archival transmitting tag (**a**), tagging a whale shark (**b**), and a PAT tag on a whale shark *in situ* (**c**). Photo credits: Steve Wilson (a) and Cary McLean (b and c)

13.6 Natural Marks

Rather than use of artificial marks and tags, researchers have investigated whether natural marks can be used to identify animals. Natural marks can be broadly grouped into five key areas: morphometric marks (shape, color, or markings of body features), meristic marks (intraspecific differences in numbers of repeated tissue features such as gill rakers or fin rays), parasitic marks (presence/absence of parasites in animals from different areas or in genetics of parasites among areas), chemical marks (differences in chemical composition of animal tissue), and genetic marking. Chemical marks and genetic marks will not be discussed in this section as they are discussed latter in this chapter (see Sections 13.7 and 13.8). Several studies have used natural marks to distinguish individuals (Grimes et al. 1986, Connell and Jones 1991, Wilson et al. 2006; Fig. 13.5), and at least one study has used the size of individuals to determine origins of larvae (Gaines and Bertness 1992). This latter application was possible because bay larvae were substantially larger than larvae that developed over the continental shelf such that larvae flushed from the bay could be distinguished at coastal sites (Gaines and Bertness 1992).

These methods all require similar basic steps. First, it is necessary to obtain structures or information from organisms of known origin (e.g., from fish collected from different locations) and ensure that differences exist between groups. Second, validation of the mark's reliability is required. Frequently, this is done using the same group of organisms as used to determine whether differences among groups occur; however, in an ideal situation additional organisms should be collected for validation and assigned to groups to determine potential error rates of assignment. Third, structures or information from organisms of unknown origin can now be used to assign fish to different groups and thereby determine potential connectivity. However, it is important to remember that an assumption of these and some other tagging techniques is that all potential source populations have been characterized (see Gillanders 2005b).

Fig. 13.5 Photo of barracuda (*Sphyraena barracuda*) showing natural markings. Photo credit: Shaun Wilson

In reality, morphometric and meristic marks are unlikely to be useful for investigating connectivity between tropical coastal ecosystems except in unusual situations. Parasitic marks are also likely to have limited utility, but have been used to detect movement from estuary to adult habitat and determine relative stock composition (Olson and Pratt 1973, Moles et al. 1990). Frequencies of infection of the brain parasite *Myxobolus neurobius* were used to determine stock composition of sockeye salmon since individuals from southeast Alaska showed high infection rates ($>85\%$) and Canadian stocks showed low infection rates ($<10\%$) (Moles et al. 1990). In a separate study, Olson and Pratt (1973) found that certain parasites (e.g., the acanthocephalan *Echinorhynchus lageniformis*) were acquired by English sole *Pleuronectes vetulus* only while in the estuary and not whilst offshore. The incidence of infection in estuarine fish before emigration was similar to the incidence in 0-group fish collected offshore after emigration, suggesting that there was little or no influx of young from potential non-estuarine habitats (Olson and Pratt 1973).

Several coral reef studies have suggested that parasites may also be useful for indicating limited small-scale movement (Grutter 1998, Cribb et al. 2000). For example, monogeneans (*Benedenia* sp.) were in significantly greater abundance on *Hemigymnus melapterus* (Labridae) from the reef flat than the reef slope at Heron Island (Great Barrier Reef, Australia) suggesting that these fish do not move between the two habitats, which are separated by only a few hundred meters (Grutter 1998).

For parasites to be used to determine connectivity among tropical coastal ecosystems the parasite would need to affect young fish in the juvenile habitat (e.g., seagrass or mangroves), but the adults in the coral reef habitat should not be susceptible to further infection. The parasite infection would also need to be maintained for a sufficient period once the fish has moved to the adult habitat such that fish which have moved could be detected (see Williams et al. 1992, and MacKenzie and Abaunza 1998, for more details). Another possibility is to investigate whether there are differences in the genetics of the parasites of individuals in different areas (see also Criscione et al. 2006, Nieberding and Olivieri 2007).

13.7 Genetic Identification and Marking

Molecular genetic methods can be used to estimate migration rates of individuals between and among populations. Genetic approaches tend to focus on longer time scales (10^2–10^3 generations rather than a single generation as in otolith chemistry (see below) and can be biased toward rare mixing events (Becker et al. 2007, Craig et al. 2007, but see Palumbi 2004). However, genetic methods are the only method that can measure effective dispersal between populations since it can determine those individuals that survive and breed after dispersal to a new population (Purcell et al. 2006a). Marine population genetics needs to be able to distinguish between evolutionarily significant gene flow and demographically relevant migration (Marko et al. 2007).

Early genetic methods using allozymes and mitochondrial DNA showed mixed success in resolving genetic structure. More sensitive genetic markers (e.g., nuclear DNA) and more powerful analytical tools are now being utilized. Microsatellites, introns, randomly amplified polymorphic DNA, and restriction fragment length polymorphisms (RFLPs) often reveal high levels of diversity when allozymes and mtDNA show relatively little variation (Davies et al. 1999). Early statistical procedures were developed for fisheries management and involved determining the composition of the catch via mixed stock analysis (MSA) (Davies et al. 1999). MSA uses a maximum likelihood approach, focuses at the population level and determines the combination of potential source populations in a sample based on populations being defined *a priori*. Such analyses are likely to miss individuals with an unusual origin (Davies et al. 1999). Assignment tests use genetic information to assign individuals or populations to different sources and assess dispersal among populations (Waser and Strobeck 1998, Davies et al. 1999, Manel et al. 2005). Two basic approaches have been taken: one involving classification where individuals are assigned to *a priori* groups and the other involving clustering where categories are not predefined (Manel et al. 2005).

Microsatellite markers have been used to estimate connectivity among populations using three main methods: (1) indirect estimates of migration based on the level of differentiation between populations; (2) direct estimates of migration based on assigning individuals to source populations; (3) using genes as natural tags (e.g., parentage analysis) (Carmen and Ablan 2006).

The majority of genetic studies report indirect estimates of connectivity among populations (see Hellberg et al. 2002 for a review), for example, F_{ST} statistics or analogs (e.g., Fauvelot and Planes 2002, Planes and Fauvelot 2002, Lessios et al. 2003, Dorenbosch et al. 2006, Costantini et al. 2007, Fauvelot et al. 2007, Haney et al. 2007). F_{ST} values show the extent to which populations differ from one another, or the level of inbreeding within populations, and can be used to estimate the number of migrants per generation (N_m) (Hartl and Clark 1997). Levels of exchange among populations are used to indicate the extent of connectivity between populations such that greater differentiation reflects low levels of exchange and little differentiation reflects high levels of exchange (Fig. 13.6). Indirect estimates are often based on simplified, unrealistic population models and there can be large variances associated with the estimates (Paetkau et al. 2004). In addition, estimates of N_m reflect long-term dispersal rates and focus at the population rather than individual level (Paetkau et al. 2004).

Direct estimates of gene flow can be obtained from individual-based information on multi-locus genotypes (Carmen and Ablan 2006). Estimates of migration are obtained by assigning individuals to possible source populations (Manel et al. 2005). Methods have also been developed to detect migrants within a population (e.g., Rannala and Mountain 1997). Although assignment tests have been applied to a number of fisheries questions including in tropical waters (e.g., population structure, van Herwerden et al. 2003), few have addressed issues of connectivity (but see Brazeau et al. 2005). Brazeau et al. (2005) used amplified fragment length polymorphism (AFLP) primers to genotype adults from colonies of the coral *Agaricia*

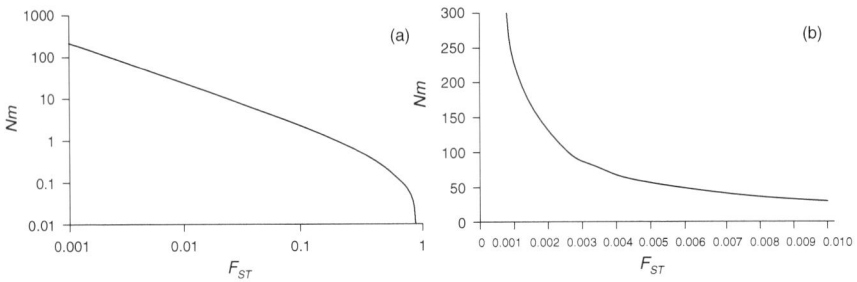

Fig. 13.6 Relationship between genetic differentiation (F_{ST}) and estimates of gene flow (Nm). Small changes in F_{ST} can lead to large differences in Nm because of the reciprocal relationship between the two (**a**), especially when F_{ST} is small (**b**). From Palumbi (2003), with kind permission of Ecological Society of America

agaricites at three locations (Bahamas, Florida, and Gulf of Mexico). They also genotyped recruits from one of the sites and then assigned them to source populations. Recruits were assigned to the reef from which they were collected suggesting one of three possibilities, namely self-seeding of larvae, selective post-settlement mortality of larvae derived from more distant sites, or larvae coming from distant populations which show similar genotypes to the local population (Brazeau et al. 2005).

Genes may also be used as natural tags similar to that of chemical and parasite tags. If genetic tags are to be effective then individuals from a cohort need to be able to be distinguished from untagged individuals when they are recaptured in the wild (this is a similar requirement to other methods) (Carmen and Ablan 2006). Three approaches to genetic tags are possible: (1) use of individuals from a limited number of parents whose genotypes are known, (2) introducing rare genes into a population such that they occur at greater frequencies than in wild populations, and (3) introducing novel sequences into the DNA, which would not be found in wild individuals (e.g., polyploidy and transgenic technology) (Thorrold et al. 2002, Carmen and Ablan 2006). Both these latter approaches are of some concern in terms of environmental consequences, and are not discussed further.

Parentage analysis, where the parents of a particular individual are identified is a specific kind of assignment method (Manel et al. 2005). Ideally, all potential parents have been genotyped, and all but one pair can be excluded as the likely parents for the offspring (Manel et al. 2005). Much work has focused on the use of these methods in aquaculture or hatchery-based stocking programs (Wilson and Ferguson 2002), although the approach has recently been applied to coral reef fishes. Jones et al. (2005) sampled all reproductive anemone fish from a population ($n = 85$) on Schumann Island (Papua New Guinea), as well as all new recruits (offspring, $n = 73$), and screened them for 11 microsatellite DNA markers. Paternity was then assessed using a likelihood approach. Results showed that 23 of the newly settled individuals were spawned by local pairs, although the various subareas did not contribute equally (Jones et al. 2005). This finding of 31.5% returning to the natal area was identical to that obtained with tetracycline-marked embryos.

13.8 Chemical Marks

13.8.1 Artificial Marks

Chemical marks may be induced and therefore artificial (e.g., organism immersed in chemicals that are then incorporated into body tissues) or occur naturally. Artificial environmental markers have been used to label tissues or calcified structures of a variety of organisms. Fluorescent compounds, such as tetracycline, calcein, alizarin complexone, and alizarin red S (Fig. 13.7), have frequently been used primarily for age validation rather than connectivity studies. Marking is generally via immersion, but several studies have used food via the diet or injection for marking. Studies utilizing fluorescent compounds have encompassed a wide range of organisms and life history stages (see Levin 1990, and Table 1 in Thorrold et al. 2002).

Several studies have used tetracycline to mass mark large numbers of damselfish eggs to determine whether larvae return to their natal reef (Jones et al. 1999, Jones et al. 2005). Jones et al. (1999) marked over 10 million *Pomacentrus amboinensis* embryos on Lizard Island (Great Barrier Reef, Australia) with tetracycline. They found 15 marked juveniles out of 5,000 examined and inferred that 15–60% of juveniles returned to the natal population since they estimated they had marked 0.5–2% of embryos. This was the first study to show that juveniles of a coral reef fish can return to their natal reef, thus challenging the concept that long distance dispersal and open populations are the norm (Jones et al. 1999). In a subsequent study, Jones et al. (2005) marked all *Amphiprion polymnus* embryos at Schumann Island (Kimbe Bay, northern coast of New Britain, Papua New Guinea) with tetracycline, and then examined all juveniles collected over a three month period ensuring they corresponded to within 9–12 days after embryo marking, thus allowing for the length of larval duration. Again, they were able to show that many individuals (15.9% in 2002 and 31.5% in 2003) settled close to their natal sites (Jones et al. 2005). Paternity analysis also predicted 31.5% self-recruitment in 2003 supporting the tetracycline results (see Section 13.7).

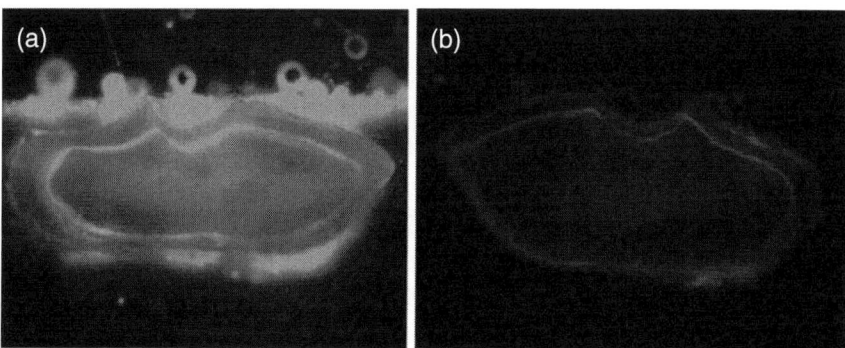

Fig. 13.7 Photos showing sectioned otoliths marked with fluorescent compounds, namely calcein (**a**) and alizarin red S (**b**). Photo credit: Dave Crook

    Besides fluorescent compounds, a range of elements (e.g., Sr, rare earth elements) and isotopes (e.g., enriched isotopes of Ba and Sr) have also been used to successfully mark invertebrates and fish via immersion (see Levin 1990, and Table 1 in Thorrold et al. 2002). Strontium is the element most commonly used to mark a variety of calcified structures of fish largely because it replaces Ca in calcified structures (e.g., Behrens Yamada and Mulligan 1987, Schroder et al. 1995, Pollard et al. 1999). One of the main limitations of using Sr is that it occurs naturally in fish. In addition, whilst many researchers assume that Sr is in low concentrations in freshwater relative to marine waters and therefore can indicate movement between the two environments, Sr:Ca ratios of freshwater can be as high or higher than marine waters (Kraus and Secor 2004). These two scenarios may potentially lead to problems distinguishing between artificial and natural signatures. Thus, to ensure that the Sr:Ca signature is distinct from natural signatures, knowledge of the natural variation in the Sr:Ca of the waters the fish may encounter is required and large amounts of Sr may be required to ensure that the mark produced is indeed artificial.

    Elements that occur in naturally low concentrations in invertebrates and fish may offer more reliable artificial marks than Sr. Several studies have investigated use of rare earth elements, or lanthanides, primarily for marking juvenile salmonids. Giles and Attas (1993) injected fingerling rainbow trout either singly or in combination with rare earth elements (dysprosium, europium, or samarium) and showed that over time marks were retained and were primarily associated with the gut. Other studies have marked salmon fry by immersing them in lanthanum, cerium, or samarium for up to six weeks (Ennevor and Beames 1993, Ennevor 1994). More recently, rare earth elements (La, Gd, Er) have shown variable success in producing marks in fish especially over short time periods (e.g., 1 day–7 weeks) (Munro et al. unpubl. data). Campana and Gillanders (unpubl. data) also had little success in marking two species of fish with lanthanide elements (Ce, Pr, Nd, Sm, Eu, Gd, Tb, Dy, Ho, Er, Ho, Er, Tm, Yb, and Lu) after immersion for 24 hrs, despite indications that the lanthanide elements were present in the water.

    Rare earth elements have also been investigated as a tagging method for larval invertebrates (Levin 1990, Levin et al. 1993, Anastasia et al. 1998). A variety of experiments on clam, barnacle, and polychaete larvae used rare earth elements that occurred in naturally low concentrations (Lu, Eu, and Sm) (Levin et al. 1993). Few larvae contained Eu and Sm, and Lu led to decreased survivorship, but showed adequate retention on time scales of hours to days (Levin et al. 1993). Experiments tracking rare earth element-marked larvae in the field have not been undertaken.

    Other elements have also been assessed primarily for marking invertebrates. For example, Anastasia et al. (1998) found that selenium was readily taken up by larval crabs from their food and was retained for weeks making it a potential tag for monitoring larval dispersal. Mn has also been investigated as a marker of abalone shells (Hawkes et al. 1996). If elements are to be good tags, they need to be inexpensive, easy to apply, long-lasting, non-toxic, not cause changes in behavior or metabolism of the organism, and be measured using standard instrumentation (Crook et al. 2005).

Stable isotopes may offer an alternative to fluorescent and simple elemental marking of invertebrates and fish. To date, most studies have relied on natural variation in isotope ratios (see Section 13.8.2), but several studies have investigated use of enriched stable isotopes to create marks that are unequivocally artificial and distinct from the natural signature (Thorrold et al. 2006, Walther and Thorrold 2006, Almany et al. 2007, Munro et al. 2008). Munro et al. (2008) reared small juvenile fish in water enriched in ^{137}Ba and ^{86}Sr for various lengths of time. Fish exposed to increased levels of ^{137}Ba (either 5 µg/L for 8 days or 15 µg/L for 4 days) had significantly lower ^{138}Ba/^{137}Ba ratios in their otoliths relative to the natural ratio of control fish. In addition, using combinations of ^{137}Ba (0–5 µg/L) and ^{86}Sr (0–100 µg/L) eight unique signatures could be produced, although marking was done over 24 days. Larvae have also been marked using different concentrations of ^{137}Ba (0–90 µg/L) by immersing them for 1–5 days prior to first feeding (S Woodcock et al., University of Adelaide, unpubl. data).

Mass marking via immersion would only be possible in the field under special circumstances. Recently, two studies have shown the potential to mark large numbers of larvae via marking the mother (Thorrold et al. 2006, Almany et al. 2007). Referred to as transgenerational marking, the mother is injected with the enriched isotope and then passes this to the egg material where it is eventually incorporated into the otolith (Thorrold et al. 2006). The technique worked on both benthic spawning and pelagic spawning species (Thorrold et al. 2006). For the benthic spawning species, marked larvae were produced for at least 90 days after a single injection to the mother and across multiple clutches (Thorrold et al. 2006; Fig. 13.8). This method has also been used in the field (Almany et al. 2007). Almany et al. (2007) tagged 176 clownfish females and 123 butterflyfish from the reef around Kimbe Island (northern coast of New Britain, Papua New Guinea) with enriched ^{137}Ba in December 2004. They then returned in February 2005 and collected 15 clownfish and 77 butterflyfish that had recently settled into their benthic habitat. After ageing to confirm that they were born after injection of the adults, the Ba isotope ratios in their otolith cores were quantified. Otoliths of nine clownfish and eight butterflyfish were classified as tagged fish suggesting natal homing. Since all clownfish larvae produced at Kimbe Island were assumed tagged and 17.3% of butterflyfish were injected with enriched Ba, both species had around 60% of larvae return to their natal reef (Almany et al. 2007).

Several radioactive isotopes have also been used for marking organisms. Only a few isotopes are considered environmentally acceptable, and the reader is referred to Thorrold et al. (2002) for further details.

Short-term temperature fluctuations, while not a chemical method, have also been used to induce patterns onto otoliths as a means of mass marking (commonly referred to as thermal marking) (Volk et al. 1999). The method is typically used on salmonids and generally involves changing exposure to different water temperatures to alter the width and contrast between dark and light zones of daily growth increments (see Thorrold et al. 2002 for picture). The geographic origins of chum salmon from the high seas have been determined using thermal marks in otoliths (Urawa et al. 2000). This method is likely only suitable for species that are hatchery-raised

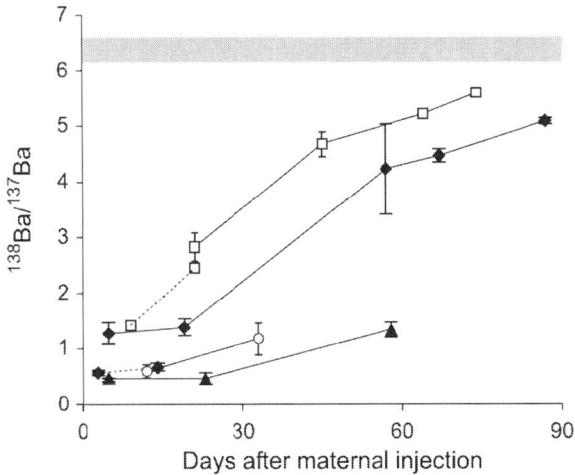

Fig. 13.8 ^{138}Ba/^{137}Ba isotope ratios in cores of larval otoliths (mean $\pm$ standard deviation) from clutches of female clownfish (*Amphiprion melanopus*) injected with ^{137}BaCl$_2$ at four dosage rates (open squares 0.45 μg ^{137}Ba per g female, solid diamonds 2.3 μg ^{137}Ba per g female, open circles 4.5 μg ^{137}Ba per g female, solid triangles 23 μg ^{137}Ba per g female) and then left to spawn naturally. The shaded horizontal bar indicates ^{138}Ba/^{137}Ba isotope ratios in control larval otoliths (mean $\pm$ standard deviation, $N = 20$). From Thorrold et al. (2006), with kind permission of NRC Research Press

and in areas where water temperatures can be raised (since the expense of cooling water is likely to be too great). The species being marked will also need to be able to tolerate a temperature increase.

13.8.2 Natural Elemental and Isotopic Signatures

Mass marking of large numbers of offspring to determine connectivity is often not feasible due to the large numbers that need marking, high mortality rates at early life history stages, and the dilution of marked offspring among non-marked offspring meaning that a large number of individuals need to be examined to get any marked individuals. Researchers have therefore investigated the possibility of using natural elemental and isotopic marks to investigate connectivity (for reviews see Campana 1999, Hobson 1999, Thorrold et al. 2002, Elsdon and Gillanders 2003a, Gillanders et al. 2003, Gillanders 2005a, 2005b, Elsdon et al. 2008). Elemental signatures in calcified structures of a variety of organisms including statoliths of gastropods (Zacherl 2005) and squid (Ikeda et al. 2003), coral skeletons (Fallon et al. 2002), bivalve shells (Becker et al. 2005), walrus teeth (Evans et al. 1995), and fish otoliths, scales, and vertebrae (Gillanders 2001) have been used to distinguish groups of organisms. Natural tags obviously have a number of advantages over artificial marking methods since they eliminate the need to handle the organisms, and essentially every organism contains a tag, meaning information is available from every organism that is captured. Information obtained from natural tags is generally more equivocal to interpret than that obtained from artificial tags (Thorrold et al. 2002).

Natural elemental and isotopic marks rely on there being differences in signatures between organisms collected from different areas since the physical and chemical environment in which the organisms are found differs. Fish otoliths are well suited to this application since they grow throughout the life of the fish, are not resorbed or altered through time, and chemical information can be related to the age of the fish. It is beyond the scope of this chapter to review factors contributing to differences in elemental/isotopic chemistry (e.g., temperature, salinity, growth rate) of otoliths (or other structures), but the reader is referred to Elsdon et al. (2008) for a recent review.

As a first step towards assessing connectivity, it is important to determine whether organisms residing in different areas do in fact have different elemental or isotopic signatures. Many studies have suggested that spatial differences in otolith elemental or isotopic signatures (Fig. 13.9) do occur in coral reef fishes (e.g., Dufour et al. 1998, Swearer et al. 1999, Chittaro et al. 2004, Patterson et al. 2004a, Lo-Yat et al. 2005, Patterson and Kingsford 2005, Patterson et al. 2005, Chittaro et al. 2006b, Ruttenberg and Warner 2006, Lara et al. 2008). However, while these studies demonstrate that spatial variation in otolith chemistry is found, the variation occurs at a range of scales (e.g., between reefs, between northern and southern groups of reefs) and these scales may differ across studies; therefore, it is important to determine the scale of variation before assessing connectivity. For example, Ruttenberg and Warner (2006) found significant small-scale variation (among islands, 10s of km, and among clutches of eggs within an island) in natal otolith chemistry of near-term benthic eggs of a damselfish (*Stegastes beebei*), but not at the larger scale (among regions separated by 100–150 km). In contrast, Patterson and Kingsford (2005) found that the greatest spatial variation in otolith chemistry of another damselfish (*Acanthochromis polyacanthus*) occurred at the largest scale (between northern and southern Great Barrier Reef (GBR), separated by 1,000's km), but they also detected significant variation at smaller scales (e.g., broods within a site (1–10 m) and to a lesser extent sites within a reef (100 m–100 km)). Determining spatial variation is, however, only the first step in assessing connectivity between habitats and ideally the scale of variation in otolith chemistry should match the dispersal scale of the organism.

Use of natural elemental chemistry relies on the ability of the otolith elemental composition to reflect the environment in which the fish resides. Experimental studies suggest that scale or otolith chemistry of several elements (e.g., Ba, Mg, Sr) reflects that of water chemistry (e.g., Bath et al. 2000, Wells et al. 2000, Elsdon and Gillanders 2003b). Several studies have now reported elevated levels of several elements in the eggs or larvae of fish (e.g., Mn, Brophy et al. 2004; Mn, Mg, and Ba, Ruttenberg et al. 2005; Mn, Zn, Sn, Ba, Ce, and Pb, Chittaro et al. 2006a), and thus the environment may not be the primary determinant of embryo otolith chemistry. Several reasons for elevated concentrations in embryo otoliths relative to juveniles have been suggested (e.g., differences in crystal structure,

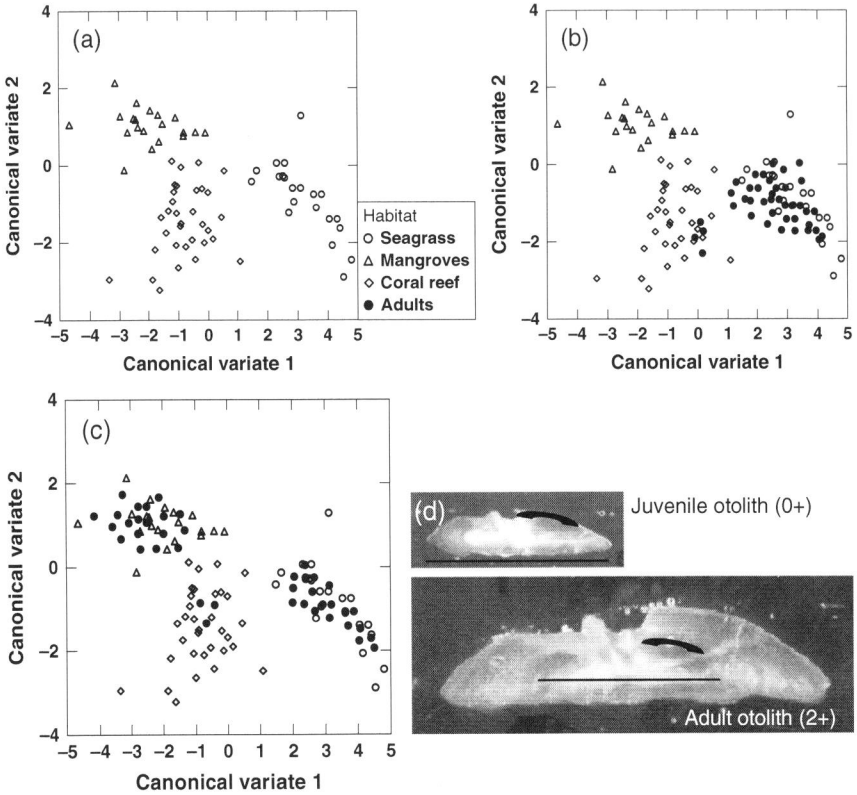

Fig. 13.9 Hypothetical diagrams showing (**a**) differences in otolith chemistry among habitats where either larvae or juveniles are found (open circles: seagrass, open triangles: mangrove habitats, open diamonds: coral reef habitat), and (**b**) otolith chemistry for the larval or juvenile region of adult otoliths showing a situation where most fish originate from seagrass, or (**c**) where similar numbers originate from seagrass and mangrove habitats compared to coral reef habitats. Adults are indicated by closed circles. Note that ideally all possible larval/juvenile habitats should be sampled. Also shown are a juvenile and an adult otolith (**d**), with the two black curved lines showing the area analyzed for otolith chemistry

embryonic development/yolk sac contribution, Ca-binding proteins, and otolith protein content); however, to interpret natal elemental concentrations of otoliths further research is required (Chittaro et al. 2006a). At present, it would seem that non-core areas of the otolith (e.g., edge) can not be used as a proxy for the core region, but the spike in Mn (or other elements) may be a good indicator for the core region of the otolith (Patterson et al. 2004a, Ruttenberg et al. 2005).

Similarly, researchers have investigated whether other proxies (e.g., edge of otoliths of juveniles or adults, resin-based elemental accumulators, samples of seawater, otoliths of other species) can be used to predict geographic differences in natal signatures (i.e., represent the otolith core). Warner et al. (2005) found that although three proxies (edge of adult otoliths, diffusive gradients in thin films deployed for around 14 days, seawater samples) all showed significant regional differences in concentrations of some elements, there was little congruency in spatial patterns seen with those found in the larval otoliths. Most studies also show significant variation in otolith chemistry between species suggesting that elemental signatures for each species of interest need to be determined (e.g., Chittaro et al. 2006b).

The temporal stability of elemental signatures is particularly important for retrospective determination of origins of fish. It is important to either evaluate the temporal stability of elemental signatures or to match the juvenile (or adult) signatures to the same larval (or juvenile) cohort (see for example Almany et al. 2007). In the latter case, ageing errors need to be minimized. For coral reef fish, temporal variation across years (e.g., between consecutive years) has been found (Chittaro et al. 2004, Bergenius et al. 2005, Patterson and Kingsford 2005). For example, Bergenius et al. (2005) found that otolith chemistry of coral trout *Plectropomus leopardus* varied between years, and two other serranid species (*Cephalopholis cyanostigma* and *Epinephelus fasciatus*) varied between two samples collected 4 yrs apart. Smaller scale temporal variation (among months) was not detected in one study (Patterson and Kingsford 2005), but few coral reef studies have investigated variation at this scale.

A number of otolith chemistry papers from tropical waters have their ultimate goal as determining connectivity among populations, however, to date most have simply documented spatial variation in otolith chemistry (see above) which is a precursor to tracking fish movement. Two studies have focused on connectivity between juvenile and sub-adult/adult fish. Patterson et al. (2004b) provided a preliminary attempt to detect philopatry in red drum (*Sciaenops ocellatus*) by using the discriminant functions from juvenile otoliths to classify otolith cores of adults. Most of the adults collected from nearshore waters off Tampa Bay (Florida, USA) were assigned as having recruited from Tampa Bay (17 out of 20 fish), one came from nearby Cedar Key, and two others came from the Atlantic coast. Another study aimed to determine connectivity between mangroves and reefs (across a distance of 0.25–7.1 km) in oceanic locations (Chittaro et al. 2004). They found no overall mangrove or reef signatures evident, although for specific regions (e.g., Bahamas, Belize) there was variation between the two broad habitats. The juvenile portion of the otoliths of fish collected from reefs in Belize was then analyzed, and using discriminant functions developed from fish collected the previous year the juvenile portions were then classified to either reef or mangrove habitat. 36% of fish had a signature more representative of mangroves than reef habitats suggesting connectivity between mangrove and reef habitats (Chittaro et al. 2004).

A hypothetical example of determining connectivity is provided in Fig. 13.9. Here juveniles were sampled from three habitats (seagrass, mangroves, coral reef) and their otolith chemistry analyzed. Significant differences were found between the habitats (Fig. 13.9a) which then meant that adults (for example 2+ fish collected two years after the juveniles) could be examined to determine which habitat they originated from. In this hypothetical example most fish came from seagrass (Fig. 13.9b), or equal contributions from mangroves and seagrass were found (Fig. 13.9c). In both cases, few fish originated from coral reef habitats. The region of the otolith sampled in juvenile fish and adult fish can also be seen (Fig. 13.9d). The edge of juvenile fish would represent the habitat in which the fish was collected, and the region sampled in the adult then represents where it spent its juvenile life.

Two recent papers (Beck et al. 2001, Dahlgren et al. 2006) have suggested how the contribution of juvenile habitats to the adult population can be determined. The key distinction between the two approaches is that for Beck et al. (2001) the contribution to adult populations is based on the area that juvenile habitats comprise, whereas for Dahlgren et al. (2006) it is based on absolute production. Both approaches require all potential juvenile habitats to be assessed. Several papers based in temperate systems have applied these concepts to fish. Kraus and Secor (2005) calculated nursery value of tidal freshwater and brackish littoral habitats of white perch (*Morone americana*) in the Patuxent River estuary (USA). They found that for dominant year classes brackish habitats were more important, whereas in all other age classes freshwater habitats contributed more per unit area to the adult population. Fodrie and Levin (2008) investigated the nursery role of four coastal ecosystems (exposed coasts, bays, lagoons, and estuaries) to California halibut (*Paralichthys californicus*) populations based on elemental analysis of otoliths. In both years (2003 and 2004) exposed coasts and bays contributed most halibut to adult populations, which was a similar result to that obtained from juvenile distribution data. Otolith chemistry also suggested that individuals migrated only small distances (<10 km) from their nursery habitats (Fodrie and Levin 2008). Several other temperate reef studies have also investigated the contribution of juvenile habitats to adult populations based on otolith chemistry (e.g., Gillanders and Kingsford 1996, Hamer et al. 2005, Brown 2006).

Several other otolith chemistry studies have focused on connectivity of populations via larval dispersal. Swearer et al. (1999) compared larval growth history and otolith trace element composition of newly settled individuals of the bluehead wrasse (*Thalassoma bifasciatum*) from St Croix (US Virgin Islands) to determine the source of recruits. They predicted that larvae that had been retained in coastal waters could grow faster and settle at larger sizes and would have significantly higher levels of trace elements than larvae dispersing throughout oceanic waters. Their results suggest that most recruits to two leeward reefs were from locally retained larval populations, whereas the single windward reef was replenished by dispersing larvae. Given that recruitment is greater to leeward reefs their findings suggest local retention of larvae is important for maintenance of populations and suggest that connectivity among distant populations may not be as great as previously thought (Swearer et al. 1999). Patterson et al. (2005) also examined the early life history of a coral reef fish via their otolith elemental signatures. They suggested that fish from reefs in the southern GBR had multiple larval sources because their pre-settlement otolith chemistry differed indicating that they had occupied different water masses, whereas those from the northern GBR had a single larval source based on similar otolith chemistry (Patterson et al. 2005). While the focus for natural chemical marks has been on otoliths, elemental signatures of invertebrates may also be used to determine connectivity (see for example DiBacco and Levin 2000, Becker et al. 2007).

Connectivity among different environments can also be assessed by analyzing elemental profiles along the otolith (e.g., Elsdon and Gillanders 2005, 2006, Hamer et al. 2006). For such applications knowledge of spatial and temporal variation in water chemistry is required. Researchers also need to realize that a change in otolith chemistry may also represent a change in environment surrounding a stationary fish. Ontogenetic effects on otolith chemistry can confound profile analyses particularly if profiles cover different life history stages (e.g., larval, juvenile, subadult, and adult growth). McCulloch et al. (2005) used Sr isotopes and elemental abundances to characterize habitats occupied by barramundi (*Lates calcarifer*). They showed that barramundi have flexible life histories occupying marine and freshwater nurseries with some individuals spending their entire life history in freshwater, some entirely in marine and others moving between the two habitats (see also Milton and Chenery 2003 for similar applications, Milton and Chenery 2005, see Chapter 9).

Stable isotopes (other than Sr, see above) can also be used to trace the origin or movement of organisms because isotopic signatures in animal tissues reflect those of local food webs or of the aquatic habitat in which they have grown. Isotopic signatures of food webs or water masses vary spatially depending on biogeochemical processes (Hobson 1999, Kennedy et al. 2000). The contribution of diet versus water to the isotopic signal is likely to depend on the isotope. Several studies have used a variety of stable isotopes (e.g., δ^{13}C, δ^{15}N, δ^{34}S) to investigate movement largely using tissue samples (e.g., Fry 1981, 1983, Fry et al. 1999), although several studies have investigated the use of stable isotopes in otoliths (Dufour et al. 1998, Kennedy et al. 2002, Augley et al. 2007, Huxham et al. 2007, Verweij et al. 2008). Fry (1981) examined δ^{13}C values in tissues of brown shrimp *Penaeus aztecus* as they moved from inshore seagrass beds to offshore areas. Offshore habitats with a phytoplankton-based food web are depleted in ^{13}C relative to a seagrass-based food web. Sub-adult individuals collected offshore had δ^{13}C values typical of individuals in seagrass meadows, suggesting that they had moved from seagrass to offshore regions (Fry 1981). Similarly, Cocheret de la Morinière et al. (2003) used δ^{13}C signatures to discriminate between nursery habitats and coral reef since δ^{13}C of food items between these habitats varied. Because adult individuals collected from the reef and juveniles collected from the nursery habitats varied it was assumed that regular diurnal feeding migrations did not occur between the two areas. In general, most studies of stable isotopes in fish tissue have used isotopes to track energy and nutrition sources (e.g., relative importance of primary producers in an animal's diet, see Chapter 3). There have been recent calls for more experimental work to determine causes of variation in stable isotopes, fractionation effects, and turnover rate of tissues, as this would aid interpretations of field data (e.g., Logan et al. 2006, Barnes and Jennings 2007, Guelinckx et al. 2007).

13.9 Summary and Future Directions

A variety of natural and artificial tags exist for determining connectivity among populations. The majority are applicable for larger individuals (e.g., external tags and marks, many of the internal tagging methods) and some at present are really only suitable for adults (e.g., acoustic telemetry, archival tags). Some of the internal tags have been used on fish as small as 7 mm (e.g., visible implant elastomer), but methods involving individual tagging of organisms are not really suitable for large scale studies of connectivity since large numbers need to be tagged to overcome dilution effects associated with natural mortality and dispersal. Natural elemental signatures have demonstrated connectivity among populations (e.g., mangroves and coral reefs, estuaries and open coast, among coral reefs) but many of these studies have been hampered by variation at different spatial scales. Several studies have now used artificial marks in otoliths to demonstrate that larvae recruit to natal reefs, however, a definitive demonstration that larvae disperse to other reefs has not been made since larvae tagged on one reef have not been recovered at distant reefs. Thus, knowledge of the proportion of larvae that go to different reefs is largely unknown. Artificial chemical marks are an exciting area for future research but may be limited in their field application to a few species and/or places where these methods are tractable. To date, they have typically been used on small demersal species or fish in breeding aggregations. In addition, knowledge of the proportion of the population tagged is essential.

Several studies have used a couple of approaches to provide independent estimates of connectivity. For example, Jones et al. (2005) used both artificial marking of embryo's with tetracycline as well as paternity analysis to estimate self-recruitment of *Amphiprion polymnus*. Both approaches showed that 31.5% of recruits were from marked embryos or resident parents. Comparative approaches to estimating connectivity are encouraged especially when using indirect methods since they help confirm and verify results. Such approaches are likely to be easier for larger organisms where conventional and acoustic tagging can be combined with say natural elemental signatures. Besides multiple approaches to determining connectivity, a two-stage approach may also be useful where, for example, an external tag or mark could be used to indicate that this organism warrants further attention. Additional attention would then be focused on the few organisms that are detected in the first round of screening rather than examining all organisms. Thus, there would be significant cost savings. Currently, many salmonids are fin clipped (adipose fin removed as an external mark) and coded wire tagged (internal tag) such that the fin clip indicates that the CWT is present which can then be examined for further information. A similar approach might incorporate a fluorescent compound as an external mark (Crook et al. 2007), and either natural or artificial chemical marks could provide additional information (e.g., reef, etc). In a slightly different variant, molecular approaches could be utilized to focus in on a smaller spatial scale such that only those organisms from a certain set of reefs are examined further for say chemical information.

In conclusion, a range of methods exist which are likely to be applicable to different life history stages, organisms, and locations. Determining connectivity of organisms is clearly difficult especially for early life history stages, but information on connectivity is critical for a number of ecological as well as management-related questions.

References

Almany GR, Berumen ML, Thorrold SR et al (2007) Local replenishment of coral reef fish populations in a marine reserve. Science 316:742–744

Anastasia JR, Morgan SG, Fisher NS (1998) Tagging crustacean larvae: assimilation and retention of trace elements. Limnol Oceanogr 43:362–368

Arnold G, Dewar H (2001) Archival and pop-up satellite tagging of Atlantic bluefin tuna. In: Sibert JR, Nielsen J (eds) Electronic tagging and tracking in marine fisheries. Kluwer Academic Publishers, Dordrecht, The Netherlands

Augley J, Huxham M, Fernandes TF et al (2007) Carbon stable isotopes in estuarine sediments and their utility as migration markers for nursery studies in the Firth of Forth and Forth Estuary, Scotland. Estuar Coast Shelf Sci 72:648–656

Barnes C, Jennings S (2007) Effect of temperature, ration, body size and age on sulphur isotope fractionation in fish. Rapid Commun Mass Spectrom 21:1461–1467

Bath GE, Thorrold SR, Jones CM et al (2000) Strontium and barium uptake in aragonitic otoliths of marine fish. Geochim Cosmochim Acta 64:1705–1714

Beck MW, Heck KL, Able KW et al (2001) The identification, conservation, and management of estuarine and marine nurseries for fish and invertebrates. BioScience 51:633–641

Becker BJ, Fodrie FJ, McMillan PA et al (2005) Spatial and temporal variation in trace elemental fingerprints of mytilid mussel shells: a precursor to invertebrate larval tracking. Limnol Oceanogr 50:48–61

Becker BJ, Levin LA, Fodrie FJ et al (2007) Complex larval connectivity patterns among marine invertebrate populations. Proc Natl Acad Sci USA 104:3267–3272

Beets J, Muehlstein L, Haught K et al (2003) Habitat connectivity in coastal environments: patterns and movements of Caribbean coral reef fishes with emphasis on bluestriped grunt, *Haemulon sciurus*. Gulf Caribb Res 14:29–42

Bergenius MAJ, Mapstone BD, Begg GA et al (2005) The use of otolith chemistry to determine stock structure of three epinepheline serranid coral reef fishes on the Great Barrier Reef, Australia. Fish Res 72:253–270

Beukers JS, Jones GP, Buckley RM (1995) Use of implant microtags for studies on populations of small reef fish. Mar Ecol Prog Ser 125:61–66

Brazeau DA, Sammarco PW, Gleason DF (2005) A multi-locus genetic assignment technique to assess sources of *Agaricia agaricites* larvae on coral reefs. Mar Biol 147:1141–1148

Brophy D, Jeffries TE, Danilowicz BS (2004) Elevated manganese concentrations at the cores of clupeid otoliths: possible environmental, physiological, or structural origins. Mar Biol 144:779–786

Brown JA (2006) Using the chemical composition of otoliths to evaluate the nursery role of estuaries for English sole *Pleuronectes vetulus* populations. Mar Ecol Prog Ser 306:269–281

Buckley RM, West JE, Doty DC (1994) Internal microtag systems for marking juvenile reef fishes. Bull Mar Sci 55:848–857

Burke NC (1995) Nocturnal foraging habitats of French and bluestriped grunts, *Haemulon flavolineatum* and *H. sciurus*, at Tobacco Caye, Belize. Environ Biol Fish 42:365–374

Campana SE (1999) Chemistry and composition of fish otoliths: pathways, mechanisms and applications. Mar Ecol Prog Ser 188:263–297

Carmen MA, Ablan A (2006) Genetics and the study of fisheries connectivity in Asian developing countries. Fish Res 78:158–168

Chittaro PM, Fryer BJ, Sale R (2004) Discrimination of French grunts (*Haemulon flavolineatum* Desmarest, 1823) from mangrove and coral reef habitats using otolith microchemistry. J Exp Mar Biol Ecol 308:169–183

Chittaro PM, Hogan JD, Gagnon J et al (2006a) In situ experiment of ontogenetic variability in the otolith chemistry of *Stegastes partitus*. Mar Biol 149:1227–1235

Chittaro PM, Usseglio P, Fryer BJ et al (2006b) Spatial variation in otolith chemistry of *Lutjanus apodus* at Turneffe Atoll, Belize. Estuar Coast Shelf Sci 67:673–680

Cocheret de la Morinière E, Pollux BJA, Nagelkerken I et al (2003) Ontogenetic dietary changes of coral reef fishes in the mangrove-seagrass-reef continuum: stable isotopes and gut-content analysis. Mar Ecol Prog Ser 246:279–289

Connell SD, Jones GP (1991) The influence of habitat complexity on postrecruitment processes in a temperate reef fish population. J Exp Mar Biol Ecol 151:271–294

Costantini F, Fauvelot C, Abbiati M (2007) Fine-scale genetic structuring in *Corallium rubrum*: evidence of inbreeding and limited effective larval dispersal. Mar Ecol Prog Ser 340:109–119

Courtney DL, Mortensen DG, Orsi JA et al (2000) Origin of juvenile Pacific salmon recovered from coastal southeastern Alaska identified by otolith thermal marks and coded wire tags. Fish Res 46:267–278

Craig MT, Eble JA, Bowen BW et al (2007) High genetic connectivity across the Indian and Pacific Oceans in the reef fish *Myripristis berndti* (Holocentridae). Mar Ecol Prog Ser 334:245–254

Cribb TH, Anderson GR, Dove ADM (2000) *Pomphorhynchus heronensis* and restricted movement of *Lutjanus carponotatus* on the Great Barrier Reef. J Helminthol 74:53–56

Criscione CD, Cooper B, Blouin MS (2006) Parasite genotypes identify source populations of migratory fish more accurately than fish genotypes. Ecology 87:823–828

Crook DA, Munro AR, Gillanders BM et al (2005) Review of existing and proposed methodologies for discriminating hatchery and wild-bred fish. Murray Darling Basin Commission, Native fish strategy project R5003

Crook DA, O'Mahony D, Gillanders BM et al (2007) Production of external fluorescent marks on golden perch fingerlings through osmotic induction marking with alizarin red S. N Am J Fish Manage 27:670–675

Dahlgren CP, Kellison GT, Adams AJ et al (2006) Marine nurseries and effective juvenile habitats: concepts and applications. Mar Ecol Prog Ser 312:291–295

Das Mahapatra K, Gjerde B, Reddy P et al (2001) Tagging: on the use of passive integrated transponder (PIT) tags for the identification of fish. Aquac Res 32:47–50

Davies N, Villablanca FX, Roderick GK (1999) Determining the source of individuals: multilocus genotyping in nonequilibrium population genetics. Trends Ecol Evol 14:17–21

Davis JLD, Young-Williams AC, Hines AH et al (2004) Comparing two types of internal tags in juvenile blue crabs. Fish Res 67:265–274

DiBacco C, Levin LA (2000) Development and application of elemental fingerprinting to track the dispersal of marine invertebrate larvae. Limnol Oceanogr 45:871–880

Dorenbosch M, Pollux BJA, Pustjens AZ et al (2006) Population structure of the Dory snapper, *Lutjanus fulviflamma*, in the western Indian Ocean revealed by means of AFLP fingerprinting. Hydrobiologia 568:43–53

Dorenbosch M, Verweij MC, Nagelkerken I et al (2004) Homing and daytime tidal movements of juvenile snappers (Lutjanidae) between shallow-water nursery habitats in Zanzibar, western Indian Ocean. Environ Biol Fish 70:203–209

Dufour V, Pierre C, Rancher J (1998) Stable isotopes in fish otoliths discriminate between lagoonal and oceanic residents of Taiaro Atoll (Tuamotu Archipelago, French Polynesia). Coral Reefs 17:23–28

Elsdon TE, Wells BK, Campana SE et al (2008) Otolith chemistry to describe movements and life-history parameters of fishes: hypotheses, assumptions, limitations, and inferences. Oceanogr Mar Biol: Annu Rev 46:297–330

Elsdon TS, Gillanders BM (2003a) Reconstructing migratory patterns of fish based on environmental influences on otolith chemistry. Rev Fish Biol Fish 13:219–235

Elsdon TS, Gillanders BM (2003b) Relationship between water and otolith elemental concentrations in juvenile black bream *Acanthopagrus butcheri*. Rev Fish Biol Fish 260:263–272

Elsdon TS, Gillanders BM (2005) Alternative life-history patterns of estuarine fish: barium in otoliths elucidates freshwater residency. Can J Fish Aquat Sci 62:1143–1152

Elsdon TS, Gillanders BM (2006) Identifying migratory contingents of fish by combining otolith Sr:Ca with temporal collections of ambient Sr:Ca concentrations. J Fish Biol 69:643–657

Ennevor BC (1994) Mass marking coho salmon, *Oncorhynchus kisutch*, fry with lanthanum and cerium. Fish Bull 92:471–473

Ennevor BC, Beames RM (1993) Use of lanthanide elements to mass mark juvenile salmonids. Can J Fish Aquat Sci 50:1039–1044

Evans RD, Richner P, Outridge PM (1995) Micro-spatial variations in heavy metals in the teeth of walrus as determined by laser ablation ICP-MS: the potential for reconstructing a history of metal exposure. Arch Environ Contam Toxicol 28:55–60

Fallon SJ, White JC, McCulloch MT (2002) *Porites* corals as recorders of mining and environmental impacts: Misima Island, Papua New Guinea. Geochim Cosmochim Acta 66:45–62

Fauvelot C, Lemaire C, Planes S et al (2007) Inferring gene flow in coral reef fishes from different molecular markers: which loci to trust? Heredity 99:331–339

Fauvelot C, Planes S (2002) Understanding origins of present day genetic structure in marine fish: biologically or historically driven patterns? Mar Biol 141:773–788

Fodrie FJ, Levin LA (2008) Linking juvenile habitat utilization to population dynamics of California halibut. Limnol Oceanogr 53:799–812

Frederick JL (1997a) Evaluation of fluorescent elastomer injection as a method for marking small fish. Bull Mar Sci 61:399–408

Frederick JL (1997b) Post-settlement movement of coral reef fishes and bias in survival estimates. Mar Ecol Prog Ser 150:65–74

Fry B (1981) Natural stable carbon isotope tag traces Texas shrimp migrations. Fish Bull 79:337–345

Fry B (1983) Fish and shrimp migrations in the northern Gulf of Mexico analyzed using stable C, N and S isotope ratios. Fish Bull 81:789–801

Fry B, Mumford PL, Robblee MB (1999) Stable isotope studies of pink shrimp (*Farfantepenaeus duorarum* Burkenroad) migrations on the southwestern Florida shelf. Bull Mar Sci 65:419–430

Gaines SD, Bertness MD (1992) Dispersal of juveniles and variable recruitment in sessile marine species. Nature 360:579–580

Giles MA, Attas EM (1993) Rare earth elements in internal batch marks for rainbow trout: retention, distribution, and effects on growth of injected dysprosium, europium, and samarium. Trans Am Fish Soc 122:289–297

Gillanders BM (2001) Trace metals in four structures of fish and their use for estimates of stock structure. Fish Bull 99:410–419

Gillanders BM (2005a) Otolith chemistry to determine movements of diadromous and freshwater fish. Aquat Living Resour 18:291–300

Gillanders BM (2005b) Using elemental chemistry of fish otoliths to determine connectivity between estuarine and coastal habitats. Estuar Coast Shelf Sci 64:47–57

Gillanders BM, Able KW, Brown JA et al (2003) Evidence of connectivity between juvenile and adult habitats for mobile marine fauna: an important component of nurseries. Mar Ecol Prog Ser 247:281–295

Gillanders BM, Kingsford MJ (1996) Elements in otoliths may elucidate the contribution of estuarine recruitment to sustaining coastal reef populations of a temperate reef fish. Mar Ecol Prog Ser 141:13–20

Grimes CB, Able KW, Jones RS (1986) Tilefish, *Lopholatilus chamaeleonticeps*, habitat, behaviour and community structure in Mid-Atlantic and southern New England waters. Environ Biol Fish 15:273–292

Grutter AS (1998) Habitat-related differences in the abundance of parasites from a coral reef fish: an indication of the movement patterns of *Hemigymnus melapterus*. J Fish Biol 53:49–57

Guelinckx J, Maes J, Van Den Driessche P et al (2007) Changes in $\delta_{13}C$ and $\delta_{15}N$ in different tissues of juvenile sand goby *Pomatoschistus minutus*: a laboratory diet-switch experiment. Mar Ecol Prog Ser 341:205–215

Gunn JS, Patterson TA, Pepperell JG (2003) Short-term movement and behaviour of black marlin *Makaira indica* in the Coral Sea as determined through a pop-up satellite archival tagging experiment. Mar Freshw Res 54:515–525

Guy CS, Blankenship HL, Nielsen LA (1996) Tagging and marking. In: Murphy BR, Willis DW (eds) Fisheries techniques. American Fisheries Society, Bethesda, Maryland

Hamer PA, Jenkins GP, Coutin P (2006) Barium variation in *Pagrus auratus* (Sparidae) otoliths: a potential indicator of migration between an embayment and ocean waters in south-eastern Australia. Estuar Coast Shelf Sci 68:686–702

Hamer PA, Jenkins GP, Gillanders BM (2005) Chemical tags in otoliths indicate the importance of local and distant settlement areas to populations of a temperate sparid, *Pagrus auratus*. Can J Fish Aquat Sci 62:623–630

Haney RA, Silliman BR, Rand DM (2007) A multilocus assessment of connectivity and historical demography in the bluehead wrasse (*Thalassoma bifasciatum*). Heredity 98:294–302

Hargreaves NB (1992) An electronic hot-branding device for marking fish. Progressive Fish-Culturist 54:99–104

Hartl DL, Clark AG (1997) Principles of population genetics Sinauer Associates Inc, Sunderland, Maryland

Hawkes GP, Day RW, Wallace MW et al (1996) Analyzing the growth and form of mollusc shell layers, in situ, by cathodoluminescence microscopy and Raman spectroscopy. J Shell Res 15:659–666

Hayes MC, Focher SM, Contor CR (2000) High-pressure injection of photonic paint to mark adult Chinook salmon. N Am J Aquac 62:319–322

Hellberg ME, Burton RS, Neigel JE et al (2002) Genetic assessment of connectivity among marine populations. Bull Mar Sci 70:273–290

Heupel MR, Semmens JM, Hobday AJ (2006) Automated acoustic tracking of aquatic animals: scales, design and deployment of listening station arrays. Mar Freshw Res 57:1–13

Hobson KA (1999) Tracing origins and migration of wildlife using stable isotopes: a review. Oecologia 120:314–326

Huxham M, Kimani E, Newton J et al (2007) Stable isotope records from otoliths as tracers of fish migration in a mangrove system. J Fish Biol 70:1554–1567

Ikeda Y, Arai N, Kidokoro H et al (2003) Strontium:calcium ratios in statoliths of Japanese common squid *Todarodes pacificus* (Cephalopoda: Ommastrephidae) as indicators of migratory behavior. Mar Ecol Prog Ser 251:169–179

Jones GP, Milicich MJ, Emslie MJ et al (1999) Self recruitment in a coral reef fish population. Nature 402:802–804

Jones GP, Planes S, Thorrold SR (2005) Coral reef fish larvae settle close to home. Curr Biol 15:1314–1318

Kennedy BP, Blum JD, Folt CL et al (2000) Using natural strontium isotopic signatures as fish markers: methodology and application. Can J Fish Aquat Sci 57:2280–2292

Kennedy BP, Klaue A, Blum JD et al (2002) Reconstructing the lives of fish using Sr isotopes in otoliths. Can J Fish Aquat Sci 59:925–929

Knaepkens G, Maerten E, Tudorache C et al (2007) Evaluation of passive integrated transponder tags for marking the bullhead (*Cottzis gobio*), a small benthic freshwater fish: effects on survival, growth and swimming capacity. Ecol Freshw Fish 16:404–409

Kneib RT, Huggler MC (2001) Tag placement, mark retention, survival and growth of juvenile white shrimp (*Litopenaeus setiferus* Perez Farfante, 1969) injected with coded wire tags. J Exp Mar Biol Ecol 266:109–120

Kraus RT, Secor DH (2004) Incorporation of strontium into otoliths of an estuarine fish. J Exp Mar Biol Ecol 302:85–106

Kraus RT, Secor DH (2005) Application of the nursery role hypothesis to an estuarine fish. Mar Ecol Prog Ser 291:301–305

Lara MR, Jones DL, Chen Z et al (2008) Spatial variation of otolith elemental signatures among juvenile gray snapper (*Lutjanus griseus*) inhabiting southern Florida waters. Mar Biol 153:235–248

Layzer JB, Heinricher JR (2004) Coded wire tag retention in ebony shell mussels *Fusconaia ebena*. N Am J Fish Manage 24:228–230

Leis JM (1991) The pelagic stage of reef fishes: the larval biology of coral reef fishes. In: Sale PF (ed) The ecology of fishes on coral reefs. Academic Press, San Diego

Lessios HA, Kane J, Robertson DR (2003) Phylogeography of the pantropical sea urchin *Tripneustes*: contrasting patterns of population structure between oceans. Evolution 57:2026–2036

Levin LA (1990) A review of methods for labeling and tracking marine invertebrate larvae. Ophelia 32:115–144

Levin LA, Huggett D, Myers P et al (1993) Rare-earth tagging methods for the study of larval dispersal by marine invertebrates. Limnol Oceanogr 38:346–360

Lim BK, Sakurai N (1999) Coded wire tagging of the short necked clam *Ruditapes philippinarum*. Fish Sci 65:163–164

Lo-Yat A, Meekan M, Munksgaard N et al (2005) Small-scale spatial variation in the elemental composition of otoliths of *Stegastes nigricans* (Pomacentridae) in French Polynesia. Coral Reefs 24:646–653

Logan J, Haas H, Deegan L et al (2006) Turnover rates of nitrogen stable isotopes in the salt marsh mummichog, *Fundulus heteroclitus*, following a laboratory diet switch. Oecologia 147:391–395

MacKenzie K, Abaunza P (1998) Parasites as biological tags for stock discrimination of marine fish: a guide to procedures and methods. Fish Res 38:45–56

Malone JC, Forrester GE, Steele MA (1999) Effects of subcutaneous microtags on the growth, survival and vulnerability to predation of small reef fishes. J Exp Mar Biol Ecol 237:243–253

Manel S, Gaggiotti OE, Waples RS (2005) Assignment methods: matching biological questions with appropriate techniques. Trends Ecol Evol 20:136–142

Marko PB, Rogers-Bennett L, Dennis AB (2007) MtDNA population structure and gene flow in lingcod (*Ophiodon elongatus*): limited connectivity despite long-lived pelagic larvae. Mar Biol 150:1301–1311

McCormick MI, Smith S (2004) Efficacy of passive integrated transponder tags to determine spawning-site visitations by a tropical fish. Coral Reefs 23:570–577

McCulloch M, Cappo M, Aumend J et al (2005) Tracing the life history of individual barramundi using laser ablation MC-ICP-MS Sr-isotopic and Sr/Ba ratios in otoliths. Mar Freshw Res 56:637–644

Milton DA, Chenery SR (2003) Movement patterns of the tropical shad hilsa (*Tenualosa ilisha*) inferred from transects of ^{87}Sr/^{86}Sr isotope ratios in their otoliths. Can J Fish Aquat Sci 60:1376–1385

Milton DA, Chenery SR (2005) Movement patterns of barramundi *Lates calcarifer*, inferred from ^{87}Sr/^{86}Sr and Sr/Ca ratios in otoliths, indicate non-participation in spawning. Mar Ecol Prog Ser 301:279–291

Moles A, Rounds P, Kondzela C (1990) Use of the brain parasite *Myxobolus neurobius* in separating mixed stocks of sockeye salmon. Am Fish Soc Symp 7:224–231

Munro AR, Gillanders BM, Elsdon TS et al (2008) Enriched stable isotope marking of juvenile golden perch *Macquaria ambigua* otoliths. Can J Fish Aquat Sci 65:276–285

Nieberding CM, Olivieri I (2007) Parasites: proxies for host genealogy and ecology? Trends Ecol Evol 22:156–165

Nielsen LA (1992) Methods of marking fish and shellfish. Special publication 23. American Fisheries Society, Bethesda, Maryland

Ogden JC, Ehrlich PR (1977) The behavior of heterotypic resting schools of juvenile grunts (Pomadasyidae). Mar Biol 42:273–280

Olson RE, Pratt I (1973) Parasites as indicators of English sole (*Parophrys vetulus*) nursery grounds. Trans Am Fish Soc 102:405–411

Olson RR, McPherson R (1987) Potential vs realized larval dispersal – fish predation on larvae of the ascidian *Lissiclinium patella* (Gottschaldt). J Exp Mar Biol Ecol 110:245–256

Ombredane D, Bagliniere JL, Marchand F (1998) The effects of Passive Integrated Transponder tags on survival and growth of juvenile brown trout (*Salmo trutta* L.) and their use for studying movement in a small river. Hydrobiologia 372:99–106

Paetkau D, Slade R, Burden M et al (2004) Genetic assignment methods for the direct, real-time estimation of migration rate: a simulation-based exploration of accuracy and power. Mol Ecol 13:55–65

Palumbi SR (2003) Population genetics, demographic connectivity, and the design of marine reserves. Ecol Appl 13:S146–S158

Palumbi SR (2004) Marine reserves and ocean neighborhoods: the spatial scale of marine populations and their management. Annu Rev Environ Resour 29:31–68

Patterson HM, Kingsford MJ (2005) Elemental signatures of *Acanthochromis polyacanthus* otoliths from the Great Barrier Reef have significant temporal, spatial, and between-brood variation. Coral Reefs 24:360–369

Patterson HM, Kingsford MJ, McCulloch MT (2004a) Elemental signatures of *Pomacentrus coelestis* otoliths at multiple spatial scales on the Great Barrier Reef, Australia. Mar Ecol Prog Ser 270:229–239

Patterson HM, Kingsford MJ, McCulloch MT (2005) Resolution of the early life history of a reef fish using otolith chemistry. Coral Reefs 24:222–229

Patterson HM, McBride RS, Julien N (2004b) Population structure of red drum (*Sciaenops ocellatus*) as determined by otolith chemistry. Mar Biol 144:855–862

Planes S, Fauvelot C (2002) Isolation by distance and vicariance drive genetic structure of a coral reef fish in the Pacific Ocean. Evolution 56:378–399

Pollard MJ, Kingsford MJ, Battaglene SC (1999) Chemical marking of juvenile snapper, *Pagrus auratus* (Sparidae), by incorporation of strontium into dorsal spines. Fish Bull 97:118–131

Prince ED, Cowen RK, Orbesen ES et al (2005) Movements and spawning of white marlin (*Tetrapturus albidus*) and blue marlin (*Makaira nigricans*) off Punta Cana, Dominican Republic. Fish Bull 103:659–669

Purcell JFH, Cowen RK, Hughes CR et al (2006a) Weak genetic structure indicates strong dispersal limits: a tale of two coral reef fish. Proc R Soc B-Biol Sci 273:1483–1490

Purcell SW, Blockmans BF, Nash WJ (2006b) Efficacy of chemical markers and physical tags for large-scale release of an exploited holothurian. J Exp Mar Biol Ecol 334:283–293

Rannala B, Mountain JL (1997) Detecting immigration by using multilocus genotypes. Proc. Natl. Acad Sci USA 94:9197–9201

Ruttenberg BI, Hamilton SL, Hickford MJH et al (2005) Elevated levels of trace elements in cores of otoliths and their potential for use as natural tags. Mar Ecol Prog Ser 297:273–281

Ruttenberg BI, Warner RR (2006) Spatial variation in the chemical composition of natal otoliths from a reef fish in the Galapagos Islands. Mar Ecol Prog Ser 328:225–236

Saura A (1996) Use of hot branding in marking juvenile pikeperch (*Stizostedion lucioperca*). Annu Zool Fenn 33:617–620

Szedlmayer ST, Able KW (1993) Ultrasonic telemetry of age-0 summer flounder, *Paralichythys dentatus*, movements in a southern New Jersey estuary. Copeia 1993:728–736

Schaefer KM, Fuller DW, Block BA (2007) Movements, behavior, and habitat utilization of yellowfin tuna (*Thunnus albacares*) in the northeastern Pacific Ocean, ascertained through archival tag data. Mar Biol 152:503–525

Schroder SL, Knudsen CM, Volk EC (1995) Marking salmon fry with strontium chloride solutions. Can J Fish Aquat Sci 52:1141–1149

Semmens JM, Pecl GT, Gillanders BM et al (2007) Approaches to resolving cephalopod movement and migration patterns. Rev. Fish Biol. Fish 17:401–423

Sharp WC, Lellis WA, Butler MJ et al (2000) The use of coded microwire tags in mark-recapture studies of juvenile Caribbean spiny lobster, *Panulirus argus*. J Crust Biol 20:510–521

Sibert JR, Nielsen J (2001) Electronic tagging and tracking in marine fisheries. Kluwer Academic Publishers, Dordrecht, The Netherlands

Skinner MA, Courtenay SC, Parker WR et al (2005) Site fidelity of mummichogs (*Fundulus heteroclitus*) in an Atlantic Canadian estuary. Water Qual Res J Canada 40:288–298

Sumpton WD, Sawynok B, Carstens N (2003) Localised movement of snapper (*Pagrus auratus*, Sparidae) in a large subtropical marine embayment. Mar Freshw Res 54:923–930

Swearer SE, Caselle JE, Lea DW et al (1999) Larval retention and recruitment in an island population of a coral-reef fish. Nature 402:799–802

Takahashi M, Okamura H, Yokawa K et al (2003) Swimming behaviour and migration of a swordfish recorded by an archival tag. Mar Freshw Res 54:527–534

Thompson JM, Hirethota PS, Eggold BT (2005) A comparison of elastomer marks and fin clips as marking techniques for walleye. N Am J Fish Manage 25:308–315

Thorrold SR, Jones GP, Hellberg ME et al (2002) Quantifying larval retention and connectivity in marine populations with artificial and natural markers. Bull Mar Sci 70:291–308

Thorrold SR, Jones GP, Planes S et al (2006) Transgenerational marking of embryonic otoliths in marine fishes using barium stable isotopes. Can J Fish Aquat Sci 63:1193–1197

Tulevech SM, Recksiek CW (1994) Acoustic tracking of adult white grunt, *Haemulon plumieri*, in Puerto Rico and Florida. Fish Res 19:301–319

Tupper M (2007) Identification of nursery habitats for commercially valuable humphead wrasse *Cheilinus undulatus* and large groupers (Pisces: Serranidae) in Palau. Mar Ecol Prog Ser 332:189–199

Urawa S, Kawana M, Anma G et al (2000) Geographic origin of high seas chum salmon determined by genetic and thermal otolith markers. N Pac Anad Fish Comm Bull 2:283–290

van der Haegen GE, Blankenship HL, Hoffmann A et al (2005) The effects of adipose fin clipping and coded wire tagging on the survival and growth of spring Chinook salmon. N Am J Fish Manage 25:1161–1170

van Herwerden L, Benzie J, Davies C (2003) Microsatellite variation and population genetic structure of the red throat emperor on the Great Barrier Reef. J Fish Biol 62:987–999

van Rooij JM, Bruggemann JH, Videler JJ et al (1995) Plastic growth of the herbivorous reef fish *Sparisoma viride* – field evidence for a trade-off between growth and reproduction. Mar Ecol Prog Ser 122:93–105

Verweij MC, Nagelkerken I (2007) Short and long-term movement and site fidelity of juvenile Haemulidae in back-reef habitats of a Caribbean embayment. Hydrobiologia 592: 257–270

Verweij MC, Nagelkerken I, Hans I et al (2008) Seagrass nurseries contribute to coral reef fish populations. Limnol Oceanogr 53:1540–1547

Verweij MC, Nagelkerken I, Hol KEM et al (2007) Space use of *Lutjanus apodus* including movement between a putative nursery and a coral reef. Bull Mar Sci 81:127–138

Voegeli FA, Smale MJ, Webber DM et al (2001) Ultrasonic telemetry, tracking and automated monitoring technology for sharks. Environ Biol Fish 60:267–281

Volk EC, Schroder SL, Grimm JJ (1999) Otolith thermal marking. Fish Res 43:205–219

Walther BD, Thorrold SR (2006) Water, not food, contributes the majority of strontium and barium deposited in the otoliths of a marine fish. Mar Ecol Prog Ser 311:125–130

Warner RR, Swearer SE, Caselle JE et al (2005) Natal trace-elemental signatures in the otoliths of an open-coast fish. Limnol Oceanogr 50:1529–1542

Waser PM, Strobeck C (1998) Genetic signatures of interpopulation dispersal. Trends Ecol Evol 13:43–44

Wells BK, Bath GE, Thorrold SR et al (2000) Incorporation of strontium, cadmium, and barium in juvenile spot (*Leiostomus xanthurus*) scales reflects water chemistry. Can J Fish Aquat Sci 57:2122–2129

Wertheimer AC, Thedinga JF, Heintz RA et al (2002) Comparative effects of half-length coded wire tagging and ventral fin removal on survival and size of pink salmon fry. N Am J Aquac 64:150–157

Williams HH, MacKenzie K, McCarthy AM (1992) Parasites as biological indicators of the population biology, migrations, diet, and phylogenetics of fish. Rev Fish Biol Fish 2:144–176

Willis TJ, Parsons DM, Babcock RC (2001) Evidence for long-term site fidelity of snapper (*Pagrus auratus*) within a marine reserve. NZ J Mar Freshw Res 35:581–590

Wilson AJ, Ferguson MM (2002) Molecular pedigree analysis in natural populations of fishes: approaches, applications, and practical considerations. Can J Fish Aquat Sci 59:1696–1707

Wilson SK, Wilson DT, Lamont C et al (2006) Identifying individual great barracuda *Sphyraena barracuda* using natural body marks. J Fish Biol 69:928–932

Yamada SB, Mulligan TJ (1987) Marking nonfeeding salmonid fry with dissolved strontium. Can J Fish Aquat Sci 44:1502–1506

Zacherl DC (2005) Spatial and temporal variation in statolith and protoconch trace elements as natural tags to track larval dispersal. Mar Ecol Prog Ser 290:145–163

Zeller DC (1998) Spawning aggregations: patterns of movement of the coral trout *Plectropomus leopardus* (Serranidae) as determined by ultrasonic telemetry. Mar Ecol Prog Ser 162:253–263

Zeller DC (1999) Ultrasonic telemetry: its application to coral reef fisheries research. Fish Bull 97:1058–1065

Zeller DC, Russ GR (1998) Marine reserves: patterns of adult movement of the coral trout (*Plectropomus leopardus* (Serranidae)). Can J Fish Aquat Sci 55:917–924

Zeller DC, Russ GR (2000) Population estimates and size structure of *Plectropomus leopardus* (Pisces : Serranidae) in relation to no-fishing zones: mark-release-resighting and underwater visual census. Mar Freshw Res 51:221–228

Chapter 14
A Landscape Ecology Approach for the Study of Ecological Connectivity Across Tropical Marine Seascapes

Rikki Grober-Dunsmore, Simon J. Pittman, Chris Caldow, Matthew S. Kendall and Thomas K. Frazer

Abstract Connectivity across the seascape is expected to have profound consequences for the behavior, growth, survival, and spatial distribution of marine species. A landscape ecology approach offers great utility for studying ecological connectivity in tropical marine seascapes. Landscape ecology provides a well developed conceptual and operational framework for addressing complex multi-scale questions regarding the influence of spatial patterning on ecological processes. Landscape ecology can provide quantitative and spatially explicit information at scales relevant to resource management decision making. It will allow us to begin asking key questions such as 'how much habitat to protect?', 'What type of habitat to protect?', and 'Which seascape patterns provide optimal, suboptimal, or dysfunctional connectivity for mobile marine organisms?'. While landscape ecology is increasingly being applied to tropical marine seascapes, few studies have dealt explicitly with the issue of connectivity. Herein, we examine the application of landscape ecology to better understand ecological connectivity in tropical marine ecosystems by: (1) reviewing landscape ecology concepts, (2) discussing the landscape ecology methods and tools available for evaluating connectivity, (3) examining data needs and obstacles, (4) reviewing lessons learned from terrestrial landscape ecology and from coral reef ecology studies, and (5) discussing the implications of ecological connectivity for resource management. Several recent studies conducted in coral reef ecosystems demonstrate the powerful utility of landscape ecology approaches for improving our understanding of ecological connectivity and applying results to make more informed decisions for conservation planning.

Keywords Seascape ecology · Landscape ecology · Connectivity · Spatial scale · Pattern metrics · Fish

R. Grober-Dunsmore (✉)
Institute of Applied Sciences, Private Bag, Laucala Campus,
University of South Pacific, Suva, Fiji Islands
e-mail: rikkidunsmore@gmail.com; dunsmore_l@usp.ac.fj

I. Nagelkerken (ed.), *Ecological Connectivity among Tropical Coastal Ecosystems*,
DOI 10.1007/978-90-481-2406-0_14, © Springer Science+Business Media B.V. 2009

14.1 Conceptual Framework

Tropical marine ecosystems often exist as dynamic and spatially heterogeneous seascapes in which different habitat types (e.g., coral reef, seagrass, open water, mangrove, sand) are connected to one another by a variety of biological, physical, and chemical processes (Fig. 14.1). Water movements, including tides and currents, facilitate the exchange of nutrients, chemical pollutants, pathogens, sediments, and organisms among components of the seascape. The active movement of organisms also connects habitat patches across the seascape (Sale 2002, Gillanders et al. 2003). For example, many tropical marine species exhibit complex life histories that utilize resources from spatially and compositionally discrete habitat patches (Parrish 1989, Pittman and McAlpine 2003). Highly mobile species can connect patches through daily foraging movements, including tidal and diel migrations, as well as, broader scale excursions for spawning and seasonal migrations (Zeller 1998, Kramer and Chapman 1999; see Chapters 4 and 8). Furthermore, many species of fish and crustaceans exhibit distinct shifts in habitat through ontogeny (Dahlgren and Eggleston 2000, Nagelkerken and van der Velde 2002). The ability of an organism to successfully navigate among several (often critical) 'ontogenetic stepping stones' or to move successfully to spawning locations will likely be influenced by both the composition of the seascape (i.e., the patch type and the abundance and richness of patch types) and the spatial configuration or spatial arrangement of patches (e.g., distance

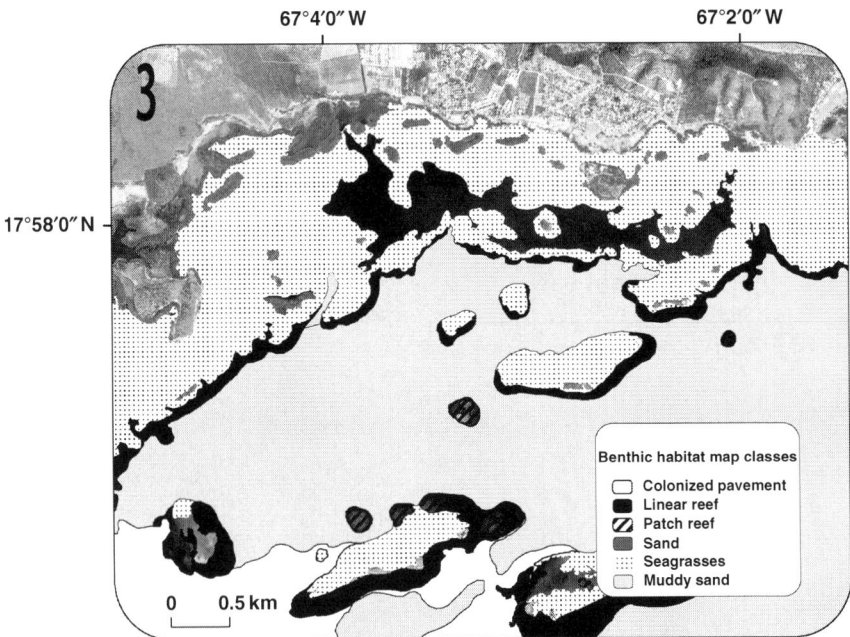

Fig. 14.1 IKONOS image with marine portions classified into six benthic habitat types for the La Parguera coast of Puerto Rico

to suitable patches, juxtaposition of complementary resources). The composition and configuration of the seascape encompasses many quantifiable structural features that are likely to influence ecological connectivity, with some configurations providing better connectivity for a species (or assemblage) than others (Mumby 2006, Grober-Dunsmore et al. 2007, Pittman et al. 2007b).

Improving our understanding of ecological connectivity in tropical marine ecosystems is one of the most pressing needs of resource managers and decision-makers today. For example, optimally-connected seascapes for specific species can be identified and mapped providing valuable spatially explicit information in support of resource management activities such as the design of Marine Protected Areas (MPAs). In addition, such information can also contribute to the design of optimally-connected habitat-restoration projects. At present, we have little knowledge of the behavior of tropical marine organisms at spatial and temporal scales relevant to their key life-cycle movements. Consequently, we remain largely ignorant of the spatial and temporal patterns of ecological connectivity that are likely to exist in marine environments. This greatly inhibits our ability to understand the influence of seascape patterning on connectivity.

Several important research questions emerge from this conceptual framework which can and must be asked of marine environments: (1) 'How does the spatial patterning of the seascape influence connectivity?', (2) 'What are the factors that inhibit or facilitate exchange or flows of materials and energy among spatial elements of the seascape?', (3) 'How does loss of habitat or change in habitat configuration alter connectivity and thus change the functioning of the seascape?', and (4) 'What and where are the optimally connected seascapes'?

Recent technological advances in remote sensing, acoustic telemetry, Geographical Information Systems (GIS), and spatial statistics now allow us to capture, manage, and analyze the data needed for connectivity studies in a spatially explicit way and at appropriately broad spatial scales (Crooks and Sanjayan 2006). Integrating spatial technologies with the discipline of landscape ecology provides both the operational and conceptual frameworks necessary to tackle these complex ecological problems at multiple spatial scales. Landscape ecology is a discipline that deals with environmental complexity including spatial heterogeneity and the importance of scale (Wu 2006) and has demonstrated great utility in the examination of ecological connectivity in terrestrial environments (With et al. 1997, Crooks and Sanjayan 2006). An extensive suite of concepts, terminology, and analytical tools for understanding the linkages between spatial patterning of land surfaces and ecological processes have recently been developed (Turner 2005). Landscape ecologists have shown that a spatially explicit and quantitative examination of fluxes in heterogeneous systems (e.g., Turner 1989) is the key to improving our understanding of how the physical structure and temporal dynamics of spatial mosaics influence ecological connectivity (Crooks and Sanjayan 2006).

Many tropical marine organisms, particularly fish and crustaceans, exhibit a strong linkage with benthic structure. For this reason, landscape ecology has been strongly and increasingly advocated as an ecologically meaningful approach for examining species-environment relationships in a wide range of structured shallow-water marine habitat types (Robbins and Bell 1994, Irlandi et al. 1995, Pittman et al. 2004, Grober-Dunsmore 2005, Grober-Dunsmore et al. 2008). We argue here that the application of a landscape ecology approach to the highly heterogeneous structures that typify coral reef ecosystems will assist in understanding the interactions between movement behavior and the spatial patterning of the seascape. Ultimately, this should lead to more ecologically-meaningful decision making in resource management.

This chapter examines the value of applying a landscape ecology approach when attempting to examine ecological connectivity in tropical marine ecosystems by: (1) presenting a landscape ecology conceptual framework for understanding connectivity, (2) reviewing existing landscape ecology methods and tools for evaluating connectivity, (3) discussing data needs and limitations, (4) reviewing lessons learned from terrestrial landscape and coral reef ecology studies, and (5) discussing the many implications for resource management. The focus here is on highly mobile species, with particular emphasis on marine fish, but we also draw on examples from terrestrial systems to highlight some similarities and differences in applying such an approach in marine systems. We do not address marine connectivity from a metapopulation (Hanski 1998) or genetic perspective (Cowen et al. 2006), although these approaches can also be spatially explicit and overlap in techniques and terminology does sometimes occur. Furthermore, we will not deal with the many approaches for studying larval connectivity. Instead, emphasis is placed on connectivity associated with active movement of individuals across the benthic seascape. These techniques, however, may also be applicable to studies of nutrient fluxes or other exchanges of materials across the seascape.

14.1.1 Definitions and Concepts

14.1.1.1 Some Commonly Used Landscape Ecology and GIS Terms

Various concepts, terminology, structural relationships, and analytical techniques used in terrestrial landscape ecology are also appropriate for studying ecological patterns and processes in shallow-water benthic seascapes (Carleton Ray 1991, Robbins and Bell 1994, Table 14.1). In landscape ecology terminology, patches are the basic spatial element in the landscape, and have been defined simply as a relatively homogeneous nonlinear area that differs from its surroundings (Forman and Godron 1986). In coral reef ecosystems, a wide variety of patch types have been classified (e.g., linear reef, patch reef, seagrass, sand, or some more biologically specific class such as gorgonian-dominated hard-bottom; Mumby and Harborne 1999, Kendall et al. 2002). Patch types differ from habitat types, in that habitat types do not infer any structural boundaries, although these terms are often used interchangeably. The structural boundaries of patches are referred to as edges or ecotones, each of which can be represented as a sharp boundary or a gradual transition from one structural type or community to another. A patch or an aggregation of patches can form a corridor, which is a linear feature that differs from its surroundings and connects

Table 14.1 Definitions for major concepts of landscape ecology, with examples of application to coral reef ecosystems (adapted from Forman 1995)

Concept	Definition	Coral reef example
Matrix	The dominant element in a landscape	Sand or seagrass
Patches	Patches are the basic spatial element in the landscape	Reef patch
Mosaic	A combination of different patch types that are usually interspersed amongst one another	Patches of seagrass, reef, sand, mangrove
Seascape	A heterogeneous marine area that can exist at a wide range of scales and may be described as a mosaic pattern or spatial gradient	The home range of a fish is an ecologically meaningful seascape
Seascape structure	The composition and spatial arrangement of patches, but may also include bathymetric complexity or structure in the water column	The distribution, diversity, and spatial geometry of structure at relevant spatial scales
Patch context	The position of a patch relative to surrounding seascape elements	A patch can be surrounded by seagrass or sand habitat
Heterogeneity	The uneven, non-random distribution of objects	Distribution of habitat patches that comprise a reef
Seascape connectivity	The degree to which the seascape facilitates or impedes movement among resource patches	Cross-shelf movement of grunts through ontogeny
Structural connectivity	Physical linkages within a seascape	A map of a reef area portrays structural connectivity
Potential connectivity	Measure of connectivity that incorporates indirect and limited information on mobility of organism	Extrapolating vagility of all jacks based on information on one species
Actual connectivity	Measure of connectivity that quantifies the movement of individuals through a habitat or landscape	Spatial information from acoustic tracking of fish, conchs, or lobsters
Functional connectivity	How the structure of the seascape interacts with the properties of the organisms, disturbances, or materials to influence how they move	How the spatial arrangement of seagrass beds influences the movement of grunts
Stepping stone connectivity	A row of small patches (stepping stones) can connect an otherwise disconnected set of patches	Seagrass patch connecting reef patches in sand matrix
Spatial scale/ Temporal scale	A measure of the resolution or extent perceived or considered	Depends on question asked and may be selected using species home range or other ecological processes
Extent	The size of the study area or the duration of time under consideration	The area of interest
Grain	The finest level of spatial or temporal resolution possible within a given data set	The smallest unit or minimum mapping unit (e.g., a 1 m^2 patch reef)

patches (Forman and Godron 1986). More broadly, an aggregation of patches may form a mosaic or habitat mosaic. The most abundant and well-connected component of the landscape or seascape is sometimes referred to as the matrix (Forman and Godron 1986). These elements of landscapes are usually arbitrarily defined and typically determined by the observer and depend upon the perspective, scale, and question of interest (Wiens et al. 1993).

The seascape can also be considered as a spatial unit or sampling unit (i.e., seascape unit) within a GIS, within which seascape structure can be quantified and characterized into two categories of structure: (i) composition, and (ii) configuration. Essentially, landscape composition, also referred to as marine landscape composition (Grober-Dunsmore et al. 2004, Pittman et al. 2004) and subsequently seascape composition (*sensu* Pittman et al. 2007b), encompasses the variety and abundance of patch types, whereas landscape or seascape configuration (also referred to as spatial arrangement) is the physical distribution of patches in space (Dunning et al. 1992, Pittman et al. 2004).

GIS are routinely used in quantitative landscape ecology. A GIS is an organized collection of specific computer hardware, software, geographic data, and personnel designed to efficiently capture, store, update, manipulate, analyze, and display all forms of geographically referenced information. The two major types of internal data organization used in GIS are raster (grid) and vector. Raster systems superimpose a regular grid over the area of interest and associate each cell or pixel with one or more data records (Malczewski 1999). Vector systems are based primarily on coordinate geometry and take advantage of the convenient division of spatial data into point, line, and polygon types. GIS are frequently used to investigate questions regarding ecological connectivity including the application of algorithms or methods for quantifying spatial pattern from habitat maps.

14.1.1.2 What is Connectivity?

The term *connectivity* appears in diverse contexts in the ecological science literature, sometimes with much ambiguity, in both terrestrial and marine research. To landscape ecologists, it often refers to the interactive pathways that link organisms and ecological processes with landscape elements (Crooks and Sanjayan 2006). Changes in composition and configuration are capable of altering the physical connectivity of landscapes (see Section 14.2). Each species' unique biological and behavioral characteristics interact with the physical landscape structure to determine the functional connectivity of a particular landscape. To understand this complexity, connectivity is often described and quantified in three ways: (1) structural, (2) potential, and (3) actual connectivity (Calabrese and Fagan 2004, Fagan and Calabrese 2006). Structural connectivity usually refers to spatial characteristics of the physical structure of the environment. It is the type of connectivity that one envisions when examining a map, and is typically measured by quantifying the configuration of a landscape with limited reference to the movement of organisms, materials, or energy (Crooks and Sanjayan 2006; Fig. 14.2). Potential connectivity considers some limited, albeit, indirect information on the dispersal or movement ability of the organism or process of interest (Fagan and Calabrese 2006). Actual connectivity

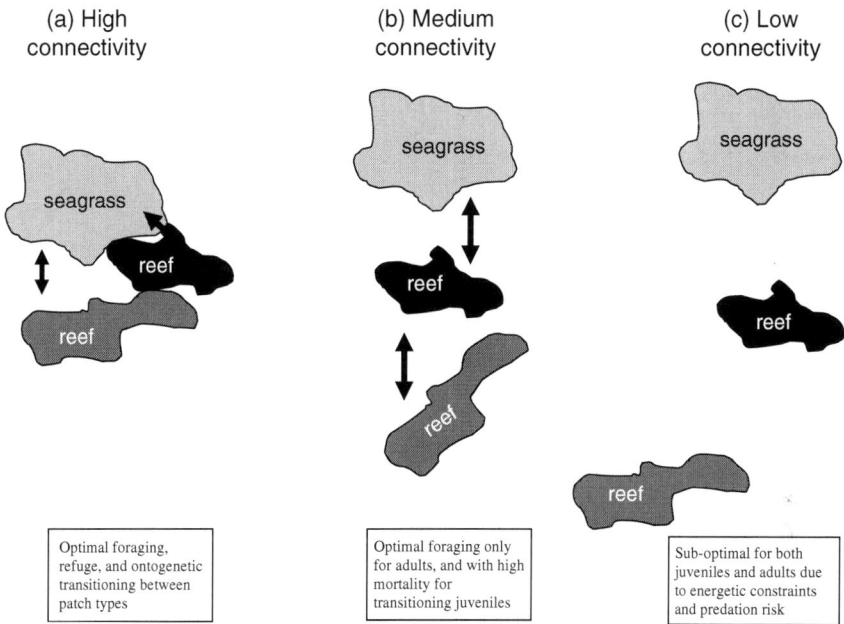

Fig. 14.2 Schematic representation of different seascape structure for a hypothetical species that requires multiple patch types in close proximity. (**a**) Highest or optimal connectivity occurs where all three essential resources exist in close proximity and even juveniles are able to easily traverse between patches. Seascapes (**b**) and (**c**) represent seascapes with sub-optimal configuration for our species with only adults able to traverse between seagrass and coral reefs due to greater distances of travel required over unsuitable low structure patches (i.e., sand) to reach essential resources. Dispersal ability and movement patterns of organisms will influence the degree to which these seascapes are connected

quantifies the movement of individuals through a habitat or seascape (e.g., acoustic tracking), thereby providing a direct measure of the potential linkages that may exist among habitat patches or seascape elements. The latter two types of connectivity are synonymous with functional connectivity, i.e., measures which examine how the structure of the landscape interacts with ecological processes such as disturbances or the movement of organisms and other materials across the seascape (Wiens 2006).

14.1.1.3 The Importance of Spatial Scale

The relationship between scale and pattern is considered one of the most important issues in ecology (Levin 1992, Schneider 2001). Our perception and measurement of pattern is determined by scale selection, and species respond to pattern individualistically at a range of spatial scales (Wiens and Milne 1989; Fig. 14.3). In landscape ecology, the concept of scale has two subcomponents: grain and extent (Forman and Godron 1986, Turner 1989). Spatial grain refers to the size of the sample unit area. Spatial extent is the overall area encompassed by an investigation. Therefore, extent

	m²	10 m²	km²	km² × 10	km² × 10²	km² × 10³
goby	▬					
damselfish	▬▬▬					
butterflyfish		▬▬▬▬▬				
reef crab		▬▬▬▬▬				
conch			▬▬▬▬			
parrotfish			▬▬▬▬▬▬			
lobster				▬▬▬▬▬▬▬		
jack				▬▬▬▬▬▬▬▬▬▬▬		
experimental ecologist		▬▬▬▬				
MPA manager			▬▬▬▬▬▬			
fishery manager				▬▬▬▬▬▬▬▬▬▬▬▬▬		

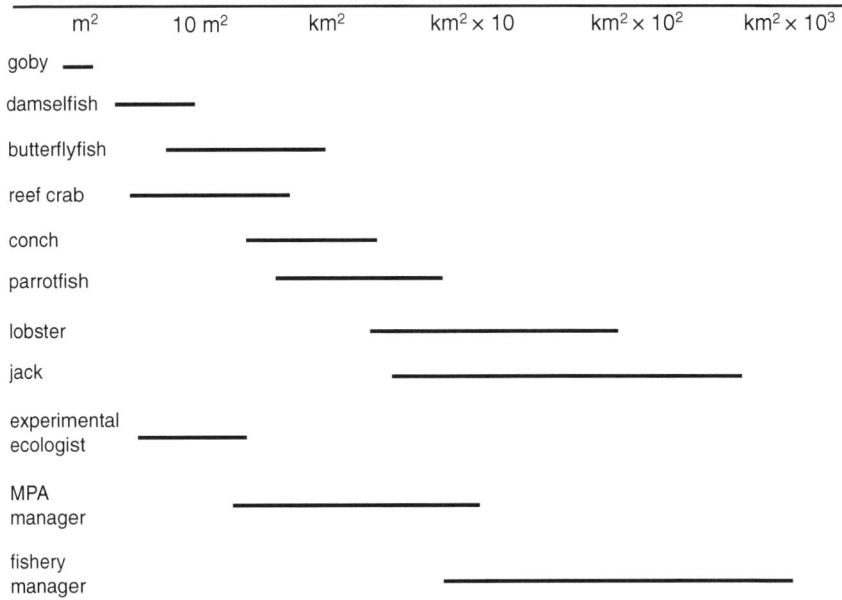

Fig. 14.3 Diagrammatic representation of the scaling windows or domains of various organisms occupying tropical marine ecosystems, and of the humans who manage or conduct research on these resources across spatial scales (adapted from Wiens et al. 2002)

and grain define the upper and lower limits of resolution of a study and constrain any inferences about the scale-dependency of ecological phenomena (Wiens 1989). Spatial resolution is an alternative term for spatial grain that is usually indicated by pixel or cell size or the minimum mapping unit (MMU). The thematic resolution, or the level of habitat classification, can also differ with various maps. Maps delineate habitat classes or patch types, but the level or detail of information can vary. Often a hierarchical habitat classification scheme is created to define and delineate habitat maps. A hierarchical scheme allows users to collapse or expand the level of detail depending upon their specific needs. For instance, at the lowest level of thematic resolution, a map may indicate a patch as soft bottom. At a finer resolution, the same patch would be classified as seagrass, and at an even finer thematic resolution, the same patch may have information on the species, relative height and density of seagrasses. Varying spatial resolution and thematic resolution of data will have important consequences for studies of the influence of patterns in the seascape (Kendall and Miller 2008).

14.1.2 The Emergence of Landscape Ecology in Tropical Marine Ecology

The importance of seascape composition and configuration has long been recognized as an important structuring mechanism for coral reef fishes. Early studies highlighted the importance of ecological interactions between adjacent habitat types in a mosaic of patch types (i.e., seagrasses, coral reefs, mangroves) (Gladfelter et al. 1980, Ogden and Gladfelter 1983, Birkeland 1985, Parrish 1989). More recently, the importance of the location of various patch types such as mangroves, seagrasses, and coral reefs to fish species and communities was demonstrated (Nagelkerken et al. 2000a, b, Nagelkerken and van der Velde 2002, Dorenbosch et al. 2005, 2006a). Furthermore, Dorenbosch et al. (2004a) highlighted the importance of proximity between adjacent habitat types (seagrasses and coral reefs) on faunal abundance and diversity. Increasingly, the movements of fishes are being correlated with the presence of specific habitats (e.g., Verweij et al. 2007).

While these studies highlighted the ecological importance of spatial patterning of patches and linkages between patches, they are not considered landscape ecology studies. Spatial patterning was not explicitly quantified and incorporated as an explanatory variable in these studies (i.e., surveys were conducted without a quantitative spatial context). In addition, they were conducted without the adoption of a landscape ecology conceptual and operational framework. Several early studies adopted an island biogeography perspective (MacArthur and Wilson 1967) to examine the importance of patch size and the spatial arrangement of patches (i.e., patch isolation) using artificial units simulating seagrasses or patch reefs (Molles 1978, Bohnsack et al. 1994) at relatively fine spatial scales ($1-10$ m^2), though they also lacked a consideration of the seascape context.

Further recognition of the importance of scale together with spatial technologies such as GIS and an increased availability of benthic habitat maps allows us to quantify seascape patterning at multiple spatial scales. Investigators have used this information to examine the influence of seascape patterning on species distributions and assemblage richness, biomass, and abundance (Turner et al. 1999, Kendall et al. 2003, Pittman et al. 2004, 2007a, b, Grober-Dunsmore et al. 2007, 2008; Table 14.2). These studies have successfully used spatial information (features and habitat classes) that exist in digital benthic habitat maps as explanatory variables. Yet, very few marine applications of landscape ecology have targeted the subject of connectivity directly. Instead, connectivity is increasingly being investigated through examination of the spatial patterns of species abundance, size class, and movement data (i.e., acoustic telemetry). While these approaches are useful, such studies often neglect to incorporate and quantify patterns in the seascape structure. Without spatially explicit information on seascape structure, studies will be missing a suite of potentially important explanatory variables and results will have limited application to resource management.

14.2 Operational Framework: Designing a Landscape Ecology Study

14.2.1 Scale Selection

It is essential to define connectivity from the perspective of the organism, species, or process of interest (e.g., terrestrial studies: Wiens and Milne 1989, With et al. 1997; marine studies: Pittman and McAlpine 2003, Pittman et al. 2004, 2007b,

Table 14.2 Examples from coral reef ecosystems of studies that may provide insights into the importance of specific seascape features on connectivity for mobile marine organisms. Major findings by seascape feature and potential general principles that may be useful for understanding ecological connectivity in coral reef ecosystems are offered

Landscape feature	Major findings	Reference	General principle for coral reef management
COMPOSITION AND ARRANGEMENT			
Patch size and distribution	Patch size and shape influences reef fish distribution	Grober-Dunsmore et al. (2004), Lugendo et al. (2007a), Ault and Johnson (1998)	Habitat configuration influences reef fish distribution, colonization
	Contribution of mangrove habitat as a feeding source may depend on habitat configuration		Areas with similar configurations may function similarly
	Species diversity differed in isolated and continuous reef patches		
Habitat composition	Presence of juvenile grunts related to area of soft bottom	Kendall et al. (2003), Grober-Dunsmore et al. (2007, 2008), Pittman et al. (2004), Chittaro et al. (2004), Appeldoorn et al. (2003)	Landscape composition influences presence of reef fishes
	Total abundance, species richness, and species presence correlated to particular habitats		Specific habitat associations for groups of reef fishes, though dependence may not be obligate
	Certain species present only where mangrove/seagrass occurs		
Presence of specific habitats	Assemblage structure differed with presence of mangrove/seagrass	Nagelkerken et al. (2001, 2002), Dorenbosch et al. (2006b), Nagelkerken and van der Velde (2002)	Presence/absence of specific habitats may influence fish assemblage structure and fish density
	Higher densities of reef fishes on reefs with seagrass		
	Densities of nursery species higher where nursery habitats present		

Table 14.2 (continued)

Landscape feature	Major findings	Reference	General principle for coral reef management
Movement	Few movements when reefs separated by 20 m of sand/rubble Conch density reduced with occurrence of sand Movement of reef fishes reduced with presence of specific seascape features	Chapman and Kramer (2000), Tewfik and Bene (2003), Grober-Dunsmore and Bonito (2009)	Certain features/habitats may serve as barriers to movement for some species; spillover may be reduced by configuration of habitats
Edge	Reef-associated species almost exclusively on reef edge Piscivore abundance increased with perimeter:area ratio Density predicted to be higher in reserves with high area:edge ratio Species richness and density of fishes affected by edge habitat	Dorenbosch et al. (2005), Grober-Dunsmore et al. (2004), Kramer and Chapman (1999), Jelbart et al. (2006)	Edges can affect ecological processes including movement of species; high area:edge ratio may help retain mobile marine organisms
Proximity	Species richness increased with proximity to topographically complex areas Density of nursery species higher adjacent to seagrass/mangrove Lutjanid and haemulid biomass greater near mangrove/seagrass Adult densities of nursery species higher on reefs adjacent to seagrass/mangrove Density and colonization related to distance from reef High species overlap in adjoining habitats compared to those that were spatially separated	Pittman et al. (2007b), Dorenbosch et al. (2004a, 2005, 2007), Appeldoorn et al. (2003), Nagelkerken and Faunce (2007), Lugendo et al. (2007b)	Proximity to specific habitats may be important for certain taxa; habitat proximity may interact with degree of connectivity; distance between juvenile and adult habitats may influence colonization Similarity in community composition may be a function of distance between habitats

Table 14.2 (continued)

Landscape feature	Major findings	Reference	General principle for coral reef management
Fragmentation	Juvenile blue crab survival significantly lower in patches separated by expanses of unvegetated sediment than patches separated by <1 m of unvegetated sediment (connected patches)	Hovel and Lipscius (2002)	Fragmentation and connectivity influences population dynamics of blue crabs
Thresholds of habitat availability	Abrupt decline in species richness and density (fish and decapods) occurred at 20% seagrass cover. Above 30%, reef fish diversity and density not controlled by seagrass	Pittman et al. (2004), Grober-Dunsmore et al. (2008)	Habitat loss important driver of population decline. At certain threshold, the relative influence of particular habitats may change
Interactions of seascape features	Fish assemblages differed between day and night, likely due to nocturnal migrations of juvenile snappers from low-tide shelter habitat to high-tide shelter habitat (notches)	Nagelkerken et al. (2000b), Dorenbosch et al. (2004b)	Importance of habitats may differ with proximity/composition of other habitats. Movement of fishes between habitats may be influenced by surrounding habitats
CONNECTIVITY			
In mangrove, coral reef, and seagrass	Biomass of some fishes doubled when adult habitat connected to mangrove. Temporal availability of mangroves influenced by tidal fluctuations. Reefs with greater connectivity to mangroves may have increased immigration and productivity. *Scarus guacamaia* largely absent on reefs isolated from mangroves. Habitat use suggests movement to offshore reefs with increased size	Mumby et al. (2004), Sheaves (2005), Mumby (2006), Nagelkerken et al. (2002), Chittaro et al. (2004), Lugendo et al. (2006)	Connectedness between mangrove and reef increased biomass of particular species. Position of patches relative to tidal flux can influence connectivity. Algorithms can generate a connectivity matrix to identify connected corridors of habitat. Connectivity of reef and mangrove habitat beneficial to some species

Table 14. 2 (continued)

Landscape feature	Major findings	Reference	General principle for coral reef management
In back-reef	Otolith microchemistry suggests grunts pass through mangrove nursery before moving to reef Different habitats used as feeding grounds (based on isotopes) by different fish species High site fidelity to small spatial scales: <171 m linear distribution range for haemulids	Verweij and Nagelkerken (2007)	Connectivity between back-reef habitats through fish movement
In reef habitat	Individuals of highly vagile species are able to move among isolated patches in response to habitat preferences or resource availability. The continuous shelter provided by contiguous reef may allow sedentary species to migrate to more favorable areas	Ault and Johnson (1998)	Patterns in distribution and abundance established at recruitment are modified by post-settlement migration. Migration may differ for isolated reef patches and connected continuous reefs
In seagrass, reefs, and saltmarsh	Movement (crabs) increased when seagrass connected to reefs/marsh Fishes (snappers) tagged in an embayment resighted offshore, crossing a 115 m open sand zone	Micheli and Peterson (1999), Verweij and Nagelkerken (2007)	Predation and foraging movements influenced by connectivity of matrix Direct evidence of connectivity through movement

Grober-Dunsmore et al. 2007, 2008). The same seascape will likely differ in functional connectivity for different processes, species, and even for different life stages of the same species. Natural history attributes of specific organisms (e.g., life history strategy, mobility, dispersal, resource requirements, habitat generalist or specialist, behavioral attributes, etc.) must be considered as an important determinant of the potential response to seascape structure (Pittman et al. 2004, Grober-Dunsmore et al. 2008), though some generality may be expected to occur among taxa (Stamps et al. 1987, Sisk et al. 1997, Mitchell et al. 2001). For example, a seagrass patch boundary may function as a constraint to the scale and direction of movements for a resident seagrass specialist, but may be relatively inconsequential to a more generalist species (Pittman et al. 2004; Fig. 14.3). In addition, investigations of seascape connectivity must recognize that for an individual species, connectivity may be an integrated function of responses to structural characteristics or ecological processes existing at multiple scales in time and space (Crooks and Sanjayan 2006).

When adopting an organism-based or organism-centered approach, the spatial resolution of the maps or sample units and the extent of the study should be appropriate to the scales at which the organisms responds and utilizes its environment. For instance, studies focused on understanding seascape connectivity throughout the daily home range (i.e., routine foraging or territorial movements) would probably select different scales than studies focused on connectivity throughout the life cycle (Pittman and McAlpine 2003). Identifying the appropriate scale(s) is essential for assessing connectivity, though it remains a significant challenge. Without data at the appropriate scale(s), interpretation of results can be incorrect and misleading due to scale dependencies. Ultimately, scales selected should always be relevant to the questions being asked (Wiens and Milne 1989, Li and Wu 2004). To obtain interpretable results, it is highly advisable to conduct pilot studies to establish a reliable estimate of the relevant spatial and temporal scales. For faunal studies, the use of acoustic tracking, tagging, and appropriately designed extractive sampling or visual census surveys can provide data suitable for selecting the temporal and spatial extents of habitat use (Pittman and McAlpine 2003). Predictable patterns of behavior, such as migrations (diel, tidal, seasonal, and spawning), residence times within patches, and home range sizes, provide ecologically meaningful, organism-based scales for connectivity investigations (e.g., Meyer et al. 2007).

Efforts to define the most influential spatial scales for fish–environment relationships have demonstrated complex, scale-dependent, and species-specific relationships (Kendall et al. 2003, Pittman et al. 2004, 2007b, Grober-Dunsmore et al. 2007, 2008; Table 14.2). These exploratory studies have linked fish and crustacean distributions to variability in seascape structure quantified at multiple spatial scales from classified benthic habitat maps. Scale issues have importance for the construction and application of benthic habitat maps. Both the spatial (cell size or minimum mapping unit) and thematic resolution (level of detail in patch composition) can affect results and the types of questions that can be asked (Kendall and Miller 2008). Practical considerations such as data availability and the type of research methods employed will also affect the scale selected.

Selecting the grain size and spatial extent is a crucial first consideration in any investigation of ecological connectivity. In addition, the spatial extent (size of the study area) and the temporal extent (maximum duration of time for a study or process) are important attributes of scale to examine early in the planning stage since they define the level of detail and set the time and space bounds for the study. These decisions will impact all stages of an ecological study from budgeting, to data collection and interpretation of results.

14.2.2 Use of Spatially Referenced Faunal Distribution Data

Several types of faunal distribution data can be used to apply landscape ecological principles to the study of connectivity in coral reef ecosystems. Three approaches are commonly used in marine ecology and are discussed here in order of increasing strength of inference for examining seascape connectivity: (1) non-extractive survey (e.g., visual observation) or extractive sampling (e.g., traps and enclosures, netting), (2) tag recapture/resighting, and (3) hydro-acoustic telemetry. To be useful these data types must be spatially explicit, that is they must have geographic coordinates or other positional information that can be used to understand an organism's location with respect to the surrounding seascape.

The type of data selected for study will be largely driven by what aspects of connectivity are of interest. Marine resource managers are increasingly interested in understanding: (1) which combinations of habitat types, and more specifically, which spatial configurations provide functional connectivity or even maximum connectivity for a species or community metric (i.e., species richness)?, (2) the consequences of habitat loss on functional connectivity; and (3) the actual pathways of movement across the seascape for marine organisms in relation to protected area boundaries.

It is important to be aware of the tradeoffs when choosing among data sources, since certain questions can only be addressed with specific types of data and the availability of information varies, as does the cost of acquiring the information. Typically, structural connectivity metrics require less data input and are relatively inexpensive compared to actual connectivity metrics (Fig. 14.4). Metrics of potential and actual connectivity require considerable information specific to the organism of study, which often limits the available data for study, when compared to structural connectivity. Such tradeoffs must be considered when selecting the type of spatially referenced distribution data for your study.

14.2.2.1 Observational Studies

In coral reef ecosystems, underwater visual counts of fish species (i.e., abundance, body size, and species composition) (e.g., Brock 1954, Bohnsack and Bannerot 1986) are the most common data collected. Similar data types can also be provided using extractive sampling via traps or nets, although both passive and active fishing gears can be highly selective (Recksiek et al. 1991, Rozas and Minello 1998). These

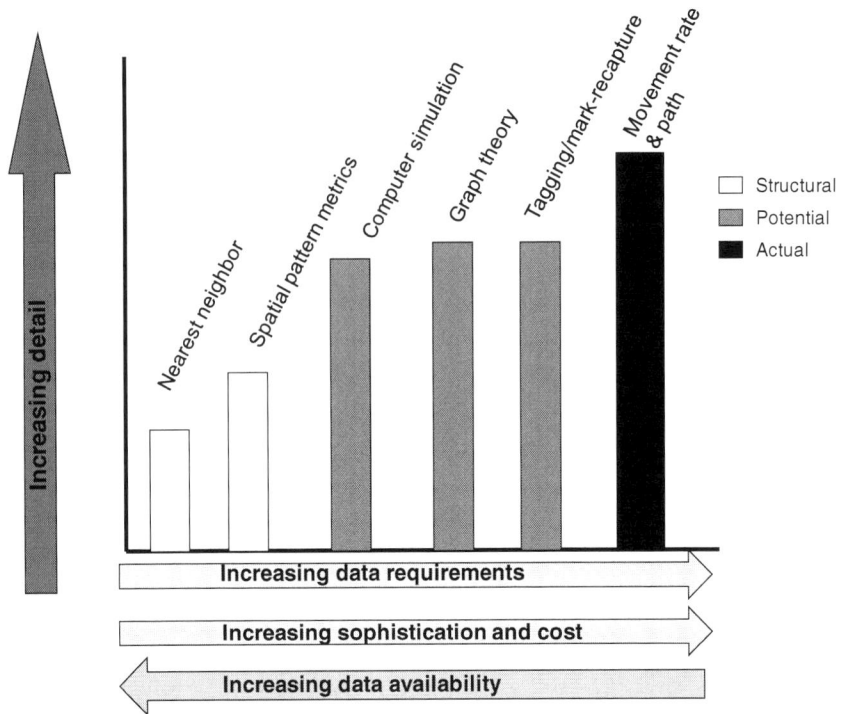

Fig. 14.4 Schematic representation of the tradeoff between information content and data requirements among connectivity metrics and approaches (structural, potential, and actual). Both information content and data requirements increase going from nearest neighbor to actual movement rates, as does the level of detail for connectivity increases on the y-axis. Factors such as technological sophistication, cost, and availability of information also influence the decision regarding which approach and metric to use for studying connectivity (modified from Calabrese and Fagan 2004)

techniques are fieldwork-intensive and provide only a snap shot of the fish community at a given time and place, yet with an appropriate sampling design, driven by specific questions regarding connectivity, spatial patterns in species abundance patterns and size distributions can be examined in relation to seascape structure.

This linking of faunal pattern to environmental pattern often serves as an effective and exploratory precursor to more detailed species–environment studies (Underwood et al. 2000). In coral reef ecosystems, several studies document correlative relationships between underwater visual census data and proximity of adjacent habitat types or the juxtapositioning of seascape elements (Nagelkerken et al. 2002, Grober-Dunsmore et al. 2004, 2007, 2008, Dorenbosch et al. 2007, Jelbart et al. 2007, Pittman et al. 2007b, Vanderklift et al. 2007; Table 14.2). Slightly stronger inference can be made regarding connectivity in the case of observational data on species that undergo habitat-dependent ontogenetic shifts (i.e., different size classes associated with different habitat types; Nagelkerken et al. 2000a, Christensen et al. 2003, Mumby et al. 2004; Table 14.2). For example, if juveniles of a species only occur in seagrasses and adults only occur in coral reefs then it is likely, but not certain, that juveniles shift from seagrass to coral reefs at some point in time.

Census effort along ecotones or habitat boundaries, if properly timed, in some cases may provide strong inference of connectivity among seascape elements. Observing the daily migrations of grunts into adjacent soft-bottom habitats (Ogden and Ehrlich 1977, Helfman et al. 1982) provides direct evidence of the functional connectivity of these habitat types, but does not clarify the spatial extent of the connection. Similarly, nets and traps may be deployed along habitat boundaries and used to infer movement between seascape features (and in the case of nets, direction of the movement) (Clark et al. 2005). This approach provides some important information and in general these census techniques have the advantage of being relatively low cost with few technological requirements and have contributed to the identification of spatial variables that could potentially influence connectivity.

Yet, visual census and extractive sampling alone is rarely sufficient to piece together the details of connectivity, such as pathways and responses to boundaries across the seascape. Since the unit of study for these observational methods is at the species level, rather than the individual level, scientists can only identify species–habitat associations, which provide only indirect evidence of connectivity. Yet studies without spatially explicit data can still address important questions on habitat-use patterns. If information on size and age class has been collected, insights into why a particular habitat is important can also be inferred. If juveniles are observed in one habitat and adults in another, one can infer that the two habitat types are connected; however, direct observation of movement is necessary to confirm that these specific habitat patches are connected. With careful experimental design (by selecting varying patches sizes), observational studies can also address questions such as: 'How much habitat is needed for a species to occur?' and 'How close should certain patch types or resources be to optimize habitat use?' These observational studies do not confirm connectivity, but can provide a valuable initial springboard for further studies on connectivity and lead to construction of more specific and testable hypotheses. Furthermore, sufficient information can be gathered to develop map products and empirical models of connectivity that support decision-making in resource management (Mumby 2006).

14.2.2.2 Tagging Studies

Mark–recapture/resighting techniques (e.g., subcutaneous dyes, plastic wire, fin clips) have also been used effectively to examine habitat use in coral reef ecosystems (e.g., Zeller and Russ 1998; Chapter 13). When analyzed using a seascape context, these techniques can provide direct evidence of connectivity (Fig. 14.5), while remaining relatively inexpensive and requiring materials that are widely available. Although potentially more powerful than census data in determining connectivity between landscape elements, the approach still lacks the capability to answer many key questions regarding functional connectivity. For example, a marked or tagged fish released in habitat A and subsequently resighted in habitat B is obvious evidence of a connection, yet as with census methods, the timing of movements, exact route

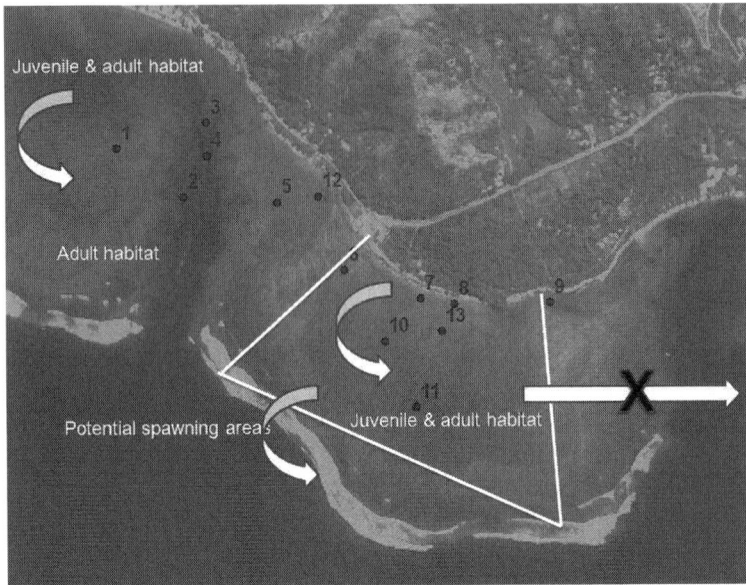

Fig. 14.5 Map of potential connectivity of inshore and offshore reef habitat based on mark–recapture and visual census studies of Lethrinidae in Vitu Levu, Fiji. Key areas for juveniles, adults, and spawning aggregations were identified. Small-sized numbers indicate the locations of acoustic receiving stations distributed inside and outside the MPA (MPA boundaries are depicted by white lines). Arrows indicate likely areas of movement and connectivity for different life history stages. Adult fishes move across the forereef to spawn, from the continuous back reef areas. Connectivity within the continuous back reef habitat occurs for juveniles and adults, however, deep water channels may serve as a barrier to movement, limiting connectivity among alongshore MPAs

followed, and any intermediary stops along the way are usually not known. Knowing the trajectory of movement and the response of organisms to particular habitat features will be particularly important for considering the flow of energy and materials among elements of the seascape.

14.2.2.3 Telemetry Studies

The most revealing and sophisticated techniques for addressing connectivity across the seascape are those that provide spatially continuous, temporally referenced movement data for individual animals or other mobile components. Hydroacoustic telemetry (Holland et al. 1993, Meyer et al. 2000, Starr et al. 2007) and other tracking techniques, such as close observation of tagged fish (Burke 1995), collect continuous positional data as fish move across the seascape. Tracking can either be conducted manually (e.g., Beets et al. 2003), in real time with directional hydrophones (i.e., following the fish by boat with an acoustic receiver), or automated with an array of fixed receivers (see Chapter 13). Mapping these movements and their timing provides a wealth of information at spatial and temporal resolution

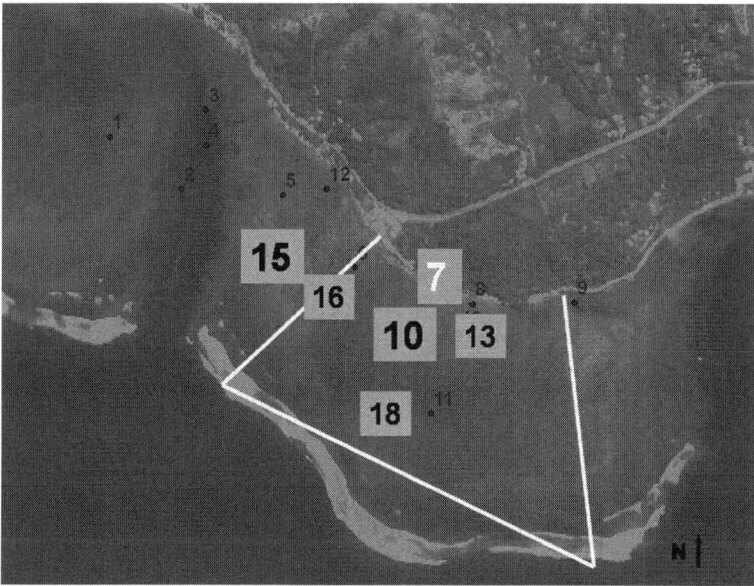

Fig. 14.6 Map of actual connectivity of one individual lethrinid inside and outside marine protected areas in Vitu Levu, Fiji. MPA boundaries are depicted by white lines. Numbers represent acoustic receiving stations (those not shaded are locations where the fish was not detected). Shaded numbers indicate locations where the fish was detected. The size of the number is scaled to represent the number of detections: a larger number indicates a greater number of detections. This fish (tagged at station seven, indicated in white shading) moved freely across the MPA boundary within the continuous back reef habitat patch, suggesting that minor modifications in the boundaries of the MPA may include the daily home range movements of this fish

not available using non-tracking approaches (e.g., Chateau and Wantiez 2007). Connections can be plotted among habitat patches and the timing of transitions can be identified. Exact pathways of travel and obstacles to dispersal can be identified by overlaying animal tracks onto benthic maps (Fig. 14.6). The principle disadvantage of these techniques is that they require an initially high investment in cost, technology, and field set-up (Fig. 14.4). Although many telemetry studies have now been conducted (e.g., Meyer et al. 2007, Starr et al. 2007), few researchers are linking movement data to spatially explicit information on seascape features (e.g., channels, patch edges) (Grober-Dunsmore and Bonito 2009). Much could be learned through reinterpretation of existing tracking data with seascape structure that is typically represented in benthic habitat maps (Pittman and McAlpine 2003).

14.2.3 Analytical Tools for Examining Seascape Connectivity

Many analytical tools are available to assist our understanding of the structural and functional connectivity of seascapes and the consequences on species distribution and behavior. Often adapted from engineering and systems analysis, these methods have been successfully applied to examine ecological connectivity in both terrestrial and marine landscapes. Here we describe three types of spatial analytical tools of particular relevance to seascape analyses: (1) spatial pattern metrics, (2) graph-theoretic approaches, and (3) computer simulation models. These tools address three types of connectivity: (1) structural, (2) potential, and (3) actual connectivity (Calabrese and Fagan 2004). Data requirements and complexity increase from type 1–3 often requiring more site- and species-specific information to address actual connectivity, but so too does ecological realism and therefore explanatory performance (Calabrese and Fagan 2004; Fig. 14.4). In general, structural connectivity is more easily visualized and measured than functional connectivity; however, it generally ignores the behavioral response of organisms to the landscape. Most spatial pattern metrics are useful for quantifying seascape structure and can be used to help explain the influence of seascape composition and configuration (Table 14.2) including structural connectivity. Structural connectivity metrics can be used to examine the relationship between seascape structure and species distributions and to determine whether differences in seascape structure matter. For instance, 'Does the proximity or juxtaposition of complementary resources in two discrete patch types influence species distributions, growth, and movement?' (e.g., Irlandi and Crawford 1997). However, before such information could be used to design movement corridors or predict dispersal pathways, more information would be required to provide information on the spatial processes relevant to functional connectivity. It is important to realize that seascapes that are structurally connected may not necessarily be functionally connected for all species.

In contrast, functional connectivity metrics incorporate various levels of movement information; therefore their use broadens the types of questions that can be answered to include ecological processes. Potential and actual connectivity metrics define seascape structure using indirect and direct knowledge of an organism's dispersal ability or behavior. Potential connectivity metrics can be parameterized using estimates of mobility derived from body size or trophic guild (Kramer and Chapman 1999), or measurements with limited spatial detail, such as mean or maximum recapture distances from mark–capture studies. Potential connectivity can be used to address questions such as, 'Is there a threshold of habitat below which a landscape is fragmented?' (Table 14.2) or 'How will dispersal pathways be affected by degradation or removal of certain habitat patches?' Potential connectivity metrics are capable of addressing many more resource management questions than structural metrics and are relatively cost-effective compared to the more data-intensive actual connectivity metrics.

Metrics of actual connectivity directly link individual movement data to spatially explicit patterns of landscape structure and are useful for modeling population dynamics (immigration, colonization, dispersal) in response to landscape features (Rothley and Rae 2005). They can also predict dispersal pathways, be used to design networks of reserves, and assess the flexibility in habitat requirements of certain organisms. Numerous methods exist to provide the most direct estimate of actual connectivity, though these are generally costly and labor-intensive (Fagan and Calabrese 2006). Acoustic tracking of the precise movement pathways of individual animals is the most direct measure (Fagan and Calabrese 2006). However, radiotracking has been used to provide critical long-distance dispersal information (Gillis and Krebs 2000), mark–recapture can be used to compare dispersal abilities in different landscapes (Pither and Taylor 1998) and genetic methods to explore the genetic consequences of connectivity (Andreassen and Ims 2001).

14.2.3.1 Spatial Pattern Metrics

Spatial pattern metrics measure structural connectivity and are usually in the form of mathematical equations or algorithms designed to quantify the composition and spatial arrangement of landscapes (Table 14.3). Structural metrics are measured on maps or GIS images (though they can be calculated from paper habitat maps or hand delineated polygons from aerial photography). The computer-based approach, however, provides higher spatial accuracy and greater flexibility in data processing. Software packages such as Fragstats v3.3. (McGarigal et al. 2002) offer a wide selection of structural connectivity metrics such as contagion (the aggregation of patches; Li and Reynolds 1993), proximity index (the isolation of patches; Gustafson and Parker 1992), patch cohesion (area-weighted mean perimeter–area ratio divided by area-weighted mean patch shape index; Schumaker 1996), connectance index (number of functional joinings between all patches of the same type divided by total number of possible joinings; McGarigal et al. 2002), and lacunarity (a measure of the distribution of gap sizes; Plotnick et al. 1993). Several of these metrics quantify similar geometric properties of spatial pattern, and therefore are often collinear (Riitters et al. 1995). In addition, patch area and patch quality interact with spatial patterning to determine connectivity, such that a range of metrics and additional information may also be required to quantify ecologically meaningful structural connectivity.

Meta-analyses have shown that structural connectivity metrics such as nearest-neighbor distance or inter-patch distance are often more sensitive to sample size and less likely to detect a significant effect than functional metrics (Moilanen and Nieminen 2002; Table 14.3); since they are often applied without any knowledge of species resource requirements and space-use patterns. In a terrestrial forest system, Schumaker (1996) found only weak correlations between nine commonly used pattern metrics and the results from simulation models of dispersal indicating that pattern metrics may not always be appropriate for predicting connectivity. In contrast, Tischendorf (2001) found strong correlations, but with some highly variable results using similar comparisons between pattern metrics and simulated dispersal processes.

Some spatial pattern metrics allow functional information on how species use the landscape to be taken into consideration. For example, connectance, also referred to as CONNECT (Fragstats v3.3), can be defined on the number of functional joinings between patches of a specified patch type. The metric allows the user to input a threshold distance for a particular species to determine if a pair of patches is connected or not. Then FRAGSTATS computes connectance as a percentage of the maximum possible connectance given the number of patches. The threshold distance

Table 14.3 A summary of the data-dependent classification framework for connectivity metrics (from Calabrese and Fagan 2004)

Connectivity metrics	Type of connectivity	Habitat-level data	Species-level data	Method
Nearest neighbor distance	Structural	Nearest neighbor distance	Patch occupancy	Patch-specific field surveys
Spatial pattern indices	Structural	Spatially explicit	None	GIS/remote sensing
Scale–area slope	Structural	None	Point or grid based occurrences	Occurrences databases, presence/absence sampling
Graph theoretic	Potential	Spatially explicit	Dispersal ability	GIS/remote sensing and dispersal studies
Buffer, radius, incidence function metapopulation model	Potential	Spatially explicit, including patch area	Patch occupancy and dispersal ability	Multi-year, patch-specific field surveys or single year, patch occupancy study with dispersal study
Movement distance (emigration, immigration, dispersal, spawning)	Actual	Variable, depends on method	Movement pathways or location specific dispersal ability	Track movement pathways, mark–recapture studies

can be based on either Euclidean distance or functional distance (McGarigal et al. 2002).

Very few marine examples exist where spatial pattern metrics have been applied to seascapes (but see Garrabou et al. 1998, Turner et al. 1999, Andrefouet et al. 2003, Pittman et al. 2004, 2007a, b, Grober-Dunsmore et al. 2007, 2008, Kendall and Miller 2008) (Table 14.2), with no examples of studies that have focused specifically on connectivity. Further studies are needed to determine the ecological relevance of seascape structure as quantified by structural connectivity metrics for marine species and marine processes. In time, such studies should also provide the necessary information for evaluating the suitability of pattern metrics to investigate seascape connectivity.

While exploratory studies that include a wide range of metrics may be fruitful, marine ecologists should choose pattern metrics judiciously, with some understanding of the approach that will be adopted and the intended purpose for the data. In addition, several of these metrics quantify similar geometric properties of spatial pattern and are often collinear (Riitters et al. 1995). However, exploratory statistical techniques may also be useful in selecting the best predictors amongst similar data structures or in simplifying complex and collinear multivariate data to a more parsimonious set of orthogonal variables. Techniques such as Principal Components

Analysis (PCA) have been used for such a purpose (McGarigal and McComb 1995). Any novel application of these methods (e.g., in marine settings) should also involve exploration of the behavior of spatial pattern metrics when applied to data of varying spatial and thematic resolution (Hargis et al. 1998, Saura and Martinez-Millan 2001). Furthermore, some caution is necessary when interpreting results from species-metric studies since existing metrics are unlikely to capture all of the relevant spatial information in marine environments and new metrics specific to aquatic systems may be required. Ultimately, of course, the metric must be relevant to the questions being asked, and researchers must recognize that some metrics will not be appropriate for practical applications (Crooks and Sanjayan 2006).

14.2.3.2 Graph Theory

In landscape ecology, graph theoretic approaches, which integrate habitat maps with information on the movement and behavior of fauna or any other mobile component of the ecosystem, offer several advantages for connectivity analyses. Graph theoretic approaches are usually considered a potential connectivity technique (Calabrese and Fagan 2004), since graphs link structural seascape pattern to estimates of dispersal ability and thereby offer the ability to go beyond structural connectivity and closer to an understanding of functional connectivity. A graph includes a set of 'nodes' usually indicating the center of a habitat patch, and lines termed 'edges' that link the nodes of two connected patches (reviewed by Urban and Keitt 2001). If the distance between a given pair of patches is less than or equal to the distance the organism can move, then the patches are considered to be connected or potentially connectable.

Connectivity can be weighted using variables such as patch type, size, isolation, and other measures of patch quality to create least-cost movement pathways (Bunn et al. 2000, Urban 2005). The pair-wise connections are then scaled-up to measure connectivity across the entire seascape or area of interest and graph theory provides a set of metrics to summarize various attributes of the connections. For example, patterns of connectivity can be evaluated and models can be constructed that examine the dysfunction that may result from changes to the spatial arrangement of the seascape, i.e., the loss of a node or patch (Urban and Keitt 2001). Graph theory has provided valuable insight into the spread of terrestrial invasive species (Urban and Keitt 2001), and could prove similarly valuable for predicting the rate of spread and spatial pathways of marine invasive species or disease. For species that undergo distinct ontogenetic habitat shifts, graphs could identify or rank seascapes based on their potential connectivity, and these models could then be evaluated using abundance data or fish telemetry or tag–recapture/resight data. Furthermore, the technique allows one to calculate the area of connected habitat that falls within and outside existing MPA boundaries and also provides connectivity surfaces to inform the design of MPAs and MPA networks.

While relatively novel to marine systems, graph theory is well developed in terrestrial urban planning, computer science, and protected area design (Urban and Keitt 2001, Rothley and Rae 2005). Treml et al. (2008) was the first to develop a

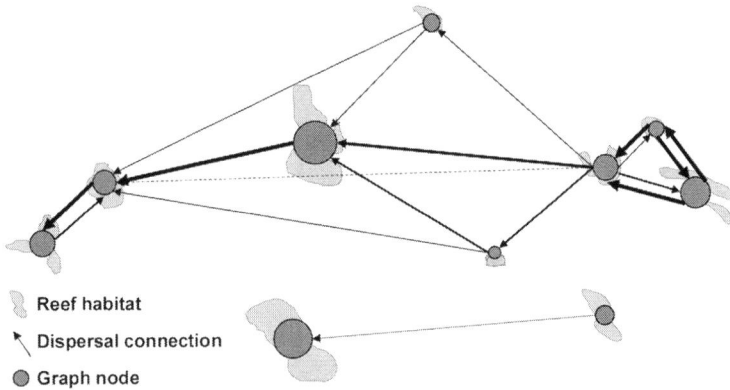

Fig. 14.7 A graph-theoretic illustration of marine connectivity. Coral reef habitat is represented by nodes within the graph framework. When larvae from a source reef reach a downstream reef site, a dispersal connection is made. This dispersal connection and direction is represented by an arrow, or 'edge' within the graph. The thickness of the arrow reflects the strength of connection (from Treml et al. 2008, with kind permission of Springer Science+Business Media)

marine application of graph theory. The authors applied a metapopulation conceptual framework and utilized an advection–diffusion biophysical model to develop connectivity estimates between islands for dispersing coral larvae in the tropical Pacific Ocean (Fig. 14.7). When combined with benthic habitat maps, the technique offers great promise for mapping connectivity and for the identification of optimally-connected seascapes to support marine protected area designation efforts and efficacy. Further studies are required to evaluate the utility of this technique for marine systems.

14.2.3.3 Computer Simulation Models

Given constraints associated with broad-scale field manipulations and data collection, simulation models are valuable tools for examining the potential influence of seascape structure on species distributions and individual movements. When employed as an exploratory tool, model results can be used to construct testable hypotheses to reveal ecological mechanisms underlying spatial patterns. In terrestrial systems, a special suite of spatially-explicit landscape simulation models, termed neutral models, have proven very effective in examining connectivity (Gardner and O'Neill 1991, With 1997). Neutral models use a set of decision rules to create random structural patterns independent of ecological processes. In these models, landscape structure is typically binary (suitable and unsuitable patches), although more complex models can incorporate more of the natural variability in ecological systems such as hierarchical random landscapes and gradients using fractal algorithms (techniques) to generate complex clustered spatial patterns (With 1997). Patch-mosaic neutral models have also been developed. These models simulate mosaic structure rather than the configuration of pixels in a raster grid and focus on incorporating aspects of composition and spatial arrangement (Gaucherel et al. 2006).

Ecological thresholds can be important phenomena in nature, and knowing when and where a threshold will occur is extremely important information for resource management. Neutral models have been used to identify thresholds in connectivity particularly in relation to the loss of habitat that occurs along a fragmentation gradient (With and Crist 1995, Pearson et al. 1996). Percolation theory proposes that beyond a predictable threshold (approx. 60%) an abrupt change will occur in system behavior (Plotnick et al. 1993, With and Crist 1995). In terrestrial landscape ecology, studies of fragmentation effects have detected thresholds at approximately 30% of remaining suitable habitat (70% loss) for a wide range of fauna (Andrén 1994), although the exact effect will be both species-specific and scale-dependent. Such rough guidelines can be useful for predicting the response of populations to degradation or loss of habitat (Taylor et al. 2007), or determining how a reduction in connectivity will influence the population dynamics of a species (With and Crist 1995).

14.2.3.4 Ecological Thresholds in Seascape Structure

In marine systems, less is known about critical ecological thresholds and how they vary between species, though recent evidence suggests that they also occur in shallow-water marine ecosystems (Table 14.2). In seagrass beds of Moreton Bay (Australia) a gradual decline in resident fish abundance was detected, along spatial gradients in seagrass cover, until approximately 15–20% seagrass cover, beyond which many abundant species were absent (Pittman 2002, Pittman et al. 2004). In the Caribbean (Virgin Islands, Florida Keys, and Turks and Caicos), an examination of fish communities on coral reefs along a spatial gradient in seagrass cover revealed that fish diversity and abundance increased from 0 to 20–30% seagrass coverage then plateaued out at 40% indicating a threshold-like response (Grober-Dunsmore 2005). Thresholds for the amount of habitat available to support fish appear to occur at lower percent cover values for seagrasses than those detected for mammals in terrestrial systems (i.e., threshold at 15–30% for fish vs. 30% for mammals) although much variability exists. Differences in thresholds for fish using seagrasses may result from the highly dynamic patchiness of seagrass beds with some patches relatively ephemeral due to die-offs and storms combined with the highly mobile nature of many fish which enables them to traverse relatively large distances to inhabit even small patches of seagrass, albeit at low abundance. Critical thresholds imply that absence, loss, or degradation of habitat can have deleterious effects on population dynamics, and are therefore crucially important to studies of connectivity in tropical marine systems. Models for marine organisms can easily be developed (Fig. 14.8) as data on the responses to varying levels of habitat availability and movement becomes accessible.

Coupling neutral models with individual-based models of dispersal or models of gene flow and population dynamics should also be developed for marine species (Butler at al. 2005). Individual-based correlated random walk models (Schippers

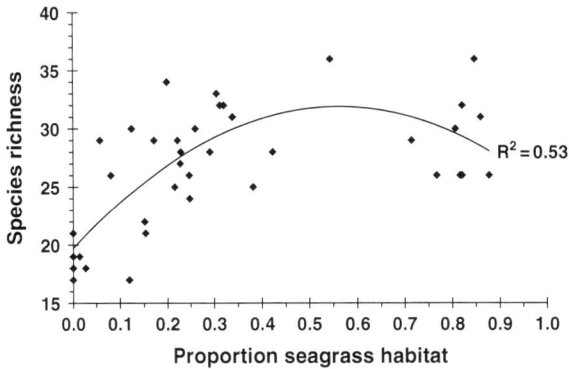

Fig. 14.8 Relationship of reef fish species richness (mobile invertebrate feeders) and seagrass cover (proportion of habitat). Species richness declines considerably below 30% seagrass coverage. Above 40% seagrass coverage, increases in seagrass do not result in additional species

et al. 1996) and other types of movement simulations reveal which spatial patterns facilitate or impede movements across a landscape and the relative cost–benefits associated with certain pathways (Tischendorf and Fahrig 2000). An individual-based model was helpful to test how change in habitat area and spatial configuration of seagrass beds influenced predator–prey interactions and cohort size for a group of settling juvenile blue crabs (*Callinectes sapidus*) (Hovel and Regan 2008). Prey cohort size was maximized in patchy seagrasses, which corresponded to results from field experiments, whereas mobile prey able to detect and avoid predators had higher survival in continuous seagrass beds (Hovel and Regan 2008). If individual-based spatially explicit seascape models are to be useful in studying connectivity, considerable effort will be required to identify and quantify behavioral responses to seascape structure. Then, simulations can be parameterized with a meaningful behavioral response or threshold effect.

14.3 Important Considerations

14.3.1 Data Needs

To investigate connectivity in terrestrial environments, landscape ecologists typically use a wide variety of spatial datasets often requiring integration and additional digital processing within a GIS (e.g., vegetation maps, digital elevation models, tracking data). In marine environments, studies of connectivity may also require multiple spatial data sets including: (1) benthic habitat maps to capture information on the distribution of patch types that can determine connectivity in a region, (2) oceanographic characteristics (e.g., sea surface temperature, frontal boundaries, upwelling zones, prevailing current patterns), (3) bathymetric surfaces (e.g., linear features, canyons, continental shelf, banks, seamounts, promontories), and (4) ecological factors (e.g., distribution of predators, prey, competitors) and human uses (e.g., point or non-point source pollution, ship traffic, fishing areas) that may act as facilitators, barriers, or modifiers to movement for marine organisms. For species that are strongly linked to the benthos, a benthic habitat map alone may be sufficient environmental data to begin studies of seascape connectivity.

Many of these environmental data are freely available via online data portals or digital archives, however, for many regions of the earth additional data may need to be collected or acquired. Several obstacles may inhibit the use of existing data when applying landscape ecology approaches to connectivity including inadequate spatial coverage, a mismatch of temporal coincidence, and most commonly, inappropriate or mismatched scales of data layers. An important first step in any broad scale study of connectivity is to evaluate the availability and quality of data for the study region. One of the most significant obstacles faced by marine spatial ecologists is the absence of appropriate spatial data (i.e., benthic habitat maps). Even for data-rich areas, often the available data were acquired for an entirely different purpose, and therefore may not necessarily represent the environmental reality from the organism-based perspective. Given the paucity of suitable data, researchers frequently proceed with environmental data with unknown accuracy and data that often do not match the spatial and temporal resolution of the ecological processes under investigation. This problem is made more challenging by that fact that very little scientific information is available to guide scale selection such as identifying the appropriate spatial grain and extent for a study. Also, very little is known about the relative importance to marine species of different seascape features or variables.

As a general rule, the environmental data must be available at a finer scale (finer spatial grain) than the scale of the process under investigation. This then provides an opportunity to coarsen the resolution of patterning so that the linkage between organism and environment can be explored at multiple spatial scales.

Even when the necessary data are readily available or easily collected, the value of subsequent seascape analyses is compromised if researchers fail to assess the accuracy of spatial data (Turner et al. 2001). Error associated with the remote sensing and GIS data acquisition, processing, analysis, conversion, and final product presentation can have a significant impact on the confidence of decisions made using the data (Lunetta et al. 1991). Potential sources of error include the age of data, completeness of aerial coverage, and map scale (Burrough 1986). Those that occur with natural variation in original measurements include positional and content accuracy, while other sources of error occur during processing (i.e., numerical computation, classification) (Burrough 1986). Errors in spatial data can obscure or distort species–habitat relationships and may even result in spurious correlations (Karl et al. 2000). The saying 'garbage in, garbage out' applied to any analytical process, and typically landscape ecology analyses are highly susceptible to data quality. Nevertheless, if a strong ecological signal exists then even with relatively crude data (with some minor errors) it may still be possible to detect the influence of pattern on process. Consequently, a validation step will be essential when developing derived products such as modeled outputs to enable appropriate statements of accuracy. In addition to testing the accuracy of the original data, the use of multiple techniques to analyze relationships or create models can verify the robustness of research results by providing information on potential bias in the data, techniques, and interpretation of the final model results.

14.3.2 Not all Habitat Patches are Created Equal

Many ecological processes operate to influence connectivity. Yet, relevant or appropriately-scaled data often do not exist or are impractical to obtain at sufficiently broad spatial scales. For instance, species interactions (e.g., predation, competition) are known to significantly affect organism distribution and habitat utilization patterns, but are difficult to incorporate, since spatial data that represent their multidimensional complexity typically do not exist. The fact that predation and competition can decouple or obscure species–environment data is rarely considered in landscape ecology analyses, where habitat patterns are the primary focus. For example, within a life stage, the presence of a large number of predators or prey may have a greater impact on determining the connectivity of a given species to its environment, either by influencing the species attraction or avoidance of certain patches, or by introducing different mortality rates among patches. Similarly, anthropogenic factors such as fishing pressure or pollution may influence distribution patterns and energy fluxes. Studies are now required that are capable of partitioning the influence of species interactions from the influence of benthic seascape structure.

14.4 Lessons Learned

14.4.1 Lessons from Terrestrial Landscape Ecology

Connectivity is a central theme in terrestrial landscape ecology (Turner et al. 2001, Turner 2005, Crooks and Sanjayan 2006), providing specific concepts, analytical tools, and unique insights in the linkage between ecological patterns and processes. The general principles that have emerged from landscape ecology appear to be applicable across a range of ecological subjects and natural resource management topics (see Turner et al. 2001, Gutzwiller 2002, Taylor et al. 2006, Wiens 2006). The generality of these principles will now urgently require further testing in marine systems and these efforts will set the stage for a deepening of the knowledge base of seascape ecology. Here we list eight principles or ecological statements, some of which can be formulated as testable hypotheses for future work in coral reef ecosystems and the broader marine environment.

- Connectivity is a key feature of landscape structure. Actual connectivity of a landscape is more complicated than simple corridors or proximity between two patch types (Crooks and Sanjayan 2006). Understanding how the fabric of a landscape is woven together to facilitate or impede movement of organisms, materials, or energy will be critical for conservation efforts.

- Landscape connectivity is species-specific. Different organisms will respond to landscape structure in different ways (Taylor et al. 2006). The same landscape will differ in connectivity for various processes, species, and life-history stages. Species-specific responses must be recognized when managing entire ecosystems across a range of spatial scales and taxa.
- Resource managers should manage the entire landscape mosaic. Managing the landscape mosaic offers an effective means for preserving connectivity (Taylor et al. 2006). Since single species-level approaches are difficult, and managing for individual habitat patches creates challenges, managers must consider not only the focal patch type but also the surrounding area (Turner et al. 2001).
- Real landscapes are not random because ecological patterns and processes are not random (Forman 1995, Taylor et al. 2006). Landscapes contain barriers to movement, detrimental habitat, areas of high predation risk, and areas that contain patches with higher and lower quality habitat, which result from a variety of causes including biotic and abiotic interactions, natural disturbances, and patterns of human activities and stressors, and such heterogeneity will have profound consequences on species distributions and ecological processes.
- Connectivity is a necessary, but not sufficient condition for species conservation (Taylor et al. 2006). Landscape connectivity influences reproduction, mortality, fitness, and access to resources; however, other landscape characteristics are also critical, and must be considered for effective management.
- Connectivity is a dynamic concept. Landscapes are ever-changing and are being modified by physical forces (e.g., hurricanes, climate change) and biological processes (e.g., competition) over short and long time scales (Taylor et al. 2006). Landscape connectivity will be related to behavioral characteristics, the degree of natural and anthropogenic disturbance, and interactions among landscape elements. Variation in connectivity due to such effects must be recognized when designing and interpreting results of landscape studies.
- Scale is crucial. There is no 'right' scale for studying landscapes (Wiens et al. 2002), but scale effects must be carefully considered when designing and evaluating landscape connectivity.
- Consider potentially confounding effects. Because of the large spatial scale over which landscape studies occur and the possible influence of non-measured explanatory variables, the potential effects of fishing, predation, and other non-measured attributes should be considered. Carefully designed experiments that control or account for these potentially confounding factors are recommended.

14.4.2 Insights for Seascape Ecology

Though few landscape ecology studies have been conducted in coral reef ecosystems, a wealth of research provides insight into how various spatial elements of the seascape may influence connectivity (Table 14.2). Such studies, though not always designed with a landscape perspective, can provide a foundation for understanding the consequences of connectivity (Table 14.2). The general principles derived from major findings from examples of recent coral reef studies that address some aspect of landscape structure (Table 14.2) are presented as a starting point for designing seascape studies and interpreting results for resource managers.

- Patch size, distribution, and configuration are perhaps the most basic aspects of seascape pattern, influencing the distribution of marine organisms. Patch size may also affect a number of important ecological processes such as colonization, reproduction, mortality, predator-prey interactions, and the transport of materials, energy, and marine organisms across seascapes. The spatial configuration of habitat patches (i.e., shape, clustering, edge:perimeter ratio, contiguity) may also control the movement patterns of marine organisms.
- Habitat composition also influences the distribution of marine organisms. The composition of habitat patches within an area and in surrounding areas can determine the presence, diversity, and abundance of marine organisms.
- Specific habitats can be crucial for certain marine organisms. While some species are habitat generalists and have flexibility in their habitat requirements, other species are habitat specialists that depend upon particular habitat types. These dependencies may vary with life history stage.
- Movements interact with and are likely modified by seascape structure. Certain features or configurations may function as facilitators or inhibitors of movement. While some habitat types or features (e.g., extremely deep or shallow areas) may facilitate movement, other features may impede it. If dispersal capacity of a marine organism is low, connectivity may depend more heavily on the seascape structure of features immediately surrounding them. If dispersal capacity is high, seascape structure across larger spatial scales may have greater influence on connectivity.
- Edge effects result from a combination of biotic and abiotic factors that alter environmental conditions along patch edges compared to patch interiors, and the presence and type of edges within a seascape may influence connectivity.
- Proximity to specific patch types or seascape features determines species richness, abundance, and density and therefore may influence colonization, survivorship, or mobility across seascapes.
- Fragmentation (either through loss or degradation of particular habitats) can influence important ecological processes, and may have possible consequences on survivorship, or dispersal of marine organisms.
- Thresholds of habitat availability appear to occur in tropical seascapes, with habitat becoming either connected or disconnected at some unknown threshold of habitat abundance. As little is known about thresholds of habitat in marine systems, it should be considered as a potential factor in structuring marine communities.
- Interactions between seascape features may confound individual effects. Connectivity may be determined by a combination of multiple seascape features, and discerning the contribution of a single feature may prove challenging.

- Connectivity is increasingly being identified as a vital element of seascape structure, and has been shown to influence biomass, habitat use, site fidelity, and movement of marine organisms. A few habitat types have been investigated, but the importance of connectivity of other untested habitats and seascape features should be considered.

With additional research, general principles useful for managing mobile marine organisms in coral reef ecosystems may be further developed. Field researchers are testing these concepts in coral reef ecosystems, and data obtained can be applied to modeling efforts to solve complex resource management questions.

14.5 Implications for Resource Management

The ability to identify functionally well-connected seascapes and evaluate their relative importance to species and communities is of great value to resource managers faced with the challenge of protecting an optimal subset of seascapes. The knowledge of how spatial patterns in the environment influence connectivity will also facilitate the design of habitat restoration or habitat creation plans that maximize organism survival, growth, productivity, and the species diversity of communities. This is a major knowledge-gap in applied marine ecology that requires urgent attention since many millions of dollars are spent on selecting and implementing MPAs and on restoration projects that do not have spatially explicit information on connectivity.

By focusing specifically on ecological connectivity and carefully scaling investigations of seascape patterns to specific ecological processes, a landscape ecology approach will allow us to determine the amount, type, configuration, and location of patch types required to maintain ecological connectivity. These are the central questions that must be addressed in tropical marine systems to successfully identify essential fish habitat, predict effects of habitat alteration, and prioritize among management options. A landscape ecology approach also offers great potential for the study of the spread of marine invasive species, with some seascapes being less rapidly colonized than others seascapes. In terrestrial systems, landscape ecology has made substantial contributions to our understanding of the direction and rate of spread of invasive plants, wildfires, and climate-induced shifts in species distributions (With 2002).

Clearly, identifying optimal seascape composition and arrangement for marine protected areas and networks of marine protected areas require the consideration of interactions between structural and functional connectivity across multiple spatial and temporal scales (Ward et al. 1999). Increasingly, resource managers will need to manage mosaics of coral reef habitat within and among protected areas, rather than focusing on individual habitat types or patches. Using landscape ecology principles, concepts, and tools, connectivity for various species can be identified and evaluated in relation to existing or planned jurisdictional boundaries to optimize conservation efforts across broad spatial scales.

In terrestrial landscape ecology, several decision support tools have been successfully applied to design reserves, create corridors of habitat, reduce the effects of forest fragmentation, and optimize connectivity of important landscape features for targeted organisms (Crooks and Sanjayan 2006). As ecological connectivity in marine ecosystems is further investigated, similar applications can be developed and incorporated into spatial tools to support resource management and conservation, especially marine spatial planning (Possingham et al. 2000, Mumby 2006). The benefits of selecting one habitat type or patch over another or choosing among alternate combinations of habitat patches can be evaluated using optimization algorithms such as those used in software programs MARXAN (Possingham et al. 2000) and C-Plan (Pressey 1999, Margules and Pressey 2000). These approaches are ideal for: (1) evaluating the costs and benefits of alternate protected area designs, (2) predicting the impacts of degrading or excluding specific seascape features when designating essential fish habitat, and (3) assessing the consequences of reducing or increasing ecological connectivity for a wide spectrum of organisms. Such decision support tools are now urgently required for enhanced management of the heavily used and highly valued tropical marine seascapes worldwide. These approaches can facilitate the selection and comparison of multiple candidate protected area networks allowing resource managers to prioritize areas based on their ecological connectivity (Mumby 2006). As landscape ecology approaches and tools are increasingly applied in tropical marine ecosystems, the utility of such concepts for improving our understanding of ecological connectivity and applying results to make more informed decisions for conservation planning will be realized.

Acknowledgments The support of the National Marine Fisheries Service, Fisheries Ecology Division at the Southwest Fisheries Science Center in Santa Cruz, CA, is greatly appreciated. In particular, Churchill Grimes was instrumental in providing support for Dr. Rikki Dunsmore. In addition, insights from Chris Jeffrey and Mark Monaco of NOAA's Center for Coastal Monitoring and Assessment were valuable. The Biogeography Branch coral reef monitoring and mapping activities are supported by funding from the Coral Reef Conservation Program (CRCP).

References

Andreassen HP, Ims RA (2001) Dispersal in patchy vole populations: role of patch configuration, density dependence, and demography. Ecology 82:2911–2926

Andrefouet S, Kramer P, Torres-Pulliza D et al (2003) Multi-site evaluation of IKONOS data for classification of tropical coral reef environments. Remote Sens Environ 88:128–143

Andrén H (1994) Effect of habitat fragmentation on birds and mammals in landscapes with different proportions of suitable habitat: a review. Oikos 71:355–366

Appeldoorn RS, Friedlander A, Sladek Nowlis J et al (2003) Habitat connectivity in reef fish communities and marine reserve design in Old Providence-Santa Catalina, Colombia. Gulf Caribb Res 14:61–77

Ault TR, Johnson CR (1998) Spatially and temporally predictable fish communities on coral reefs. Ecol Monogr 68:25–50

Beets J, Muehlstein L, Haught K et al (2003) Habitat connectivity in coastal environments: patterns and movements of Caribbean coral reef fishes with emphasis on bluestriped grunt, *Haemulon sciurus*. Gulf Caribb Res 14:29–42

Birkeland C (1985) Ecological interactions between mangroves, seagrass beds and coral reefs. In: Birke-land C (ed) Ecological interactions between tropical coastal ecosystems. UNEP Regional Seas Reports 73. Earth-print, Stevenage, UK

Bohnsack JA, Bannerot SP (1986) A stationary visual census technique for quantitatively assessing community structure of coral reef fishes. NOAA Technical Report NMFS 41

Bohnsack JA, Harper DE, McClellan DB et al (1994) Effects of reef size on colonization and assemblage structure of fishes at artificial reefs off southeastern Florida, USA. Bull Mar Sci 55:796–823

Brock VE (1954) A method of estimating reef fish populations. J Wildl Manage 18:297–308

Bunn AG, Urban DL, Keitt TH (2000) Landscape connectivity: a conservation application of graph theory. J Environ Manage 59(SI 4):265–278

Burke N (1995) Nocturnal foraging habitats of French and bluestriped grunts, *Haemulon flavolineatum* and *H. sciurus* at Tobacco Caye, Belize. Environ Biol Fish 42:365–374

Burrough PA (1986) Principles of Geographic Information Systems for land resources assessment. Oxford University Press, Oxford, UK

Butler MJ, Dolan TW, Hunt JH et al (2005) Recruitment in degraded marine habitats: a spatially explicit, individual-based model for spiny lobster. Ecol Appl 15:902–918

Calabrese JM, Fagan WF (2004) A comparison-shopper's guide to connectivity metrics. Front Ecol Environ 2:529–536

Carleton Ray G (1991) Coastal-zone biodiversity patterns. BioScience 41:490–498

Chapman MR, Kramer DL (2000) Movements of fishes within and among fringing coral reefs in Barbados. Environ Biol Fish 57:11–24

Chateau O, Wantiez L (2007) Site fidelity and activity patterns of a humphead wrasse, *Cheilinus undulates* (Labridae), as determined by acoustic telemetry. Environ Biol Fish 80:503–508

Chittaro PM, Fryer BJ, Sale R (2004) Discrimination of French grunts (*Haemulon flavolineatum* Desmarest, 1823) from mangrove and coral reef habitats using otolith microchemistry. J Exp Mar Biol Ecol 308:169–183

Christensen JD, Jeffrey CFG, Caldow C et al (2003) Cross-shelf habitat utilization patterns of reef fishes in southwestern Puerto Rico. Gulf Caribb Res 14:9–27

Clark R, Monaco ME, Appeldoorn RS, Roque B (2005) Fish habitat utilization in a Puerto Rico coral reef ecosystem. Proc Gulf Caribb Fish Inst 56:467–485

Cowen RK, Paris CB, Srinivasan A (2006) Scaling of connectivity in marine populations. Science 311:522–527

Crooks KR, Sanjayan M (2006) Connectivity conservation. Cambridge University Press, Cambridge

Dahlgren CP, Eggleston DB (2000) Ecological processes underlying ontogenetic habitat shifts in a coral reef fish. Ecology 81:2227–2240

Dorenbosch M, Grol MGG, Christianen MJA et al (2005) Indo-Pacific seagrass beds and mangroves contribute to fish density and diversity on adjacent coral reefs. Mar Ecol Prog Ser 302:63–76

Dorenbosch M, Grol MGG, Nagelkerken I et al (2006a) Different surrounding landscapes may result in different fish assemblages in East African seagrass beds. Hydrobiologia 563:45–60

Dorenbosch M, Grol MGG, Nagelkerken I et al (2006b) Seagrass beds and mangroves as potential nurseries for the threatened Indo-Pacific humphead wrasse, *Cheilinus undulatus* and Caribbean rainbow parrotfish, *Scarus guacamaia*. Biol Conserv 129:277–282

Dorenbosch M, van riel MC, Nagelkerken I et al (2004a) The relationship of reef fish densities to the proximity of mangrove and seagrass nurseries. Estuar Coast Shelf Sci 60:37–48

Dorenbosch M, Verberk WCEP, Nagelkerken I et al (2007) Influence of habitat configuration on connectivity between fish assemblages of Caribbean seagrass beds, mangroves and coral reefs. Mar Ecol Prog Ser 334:103–116

Dorenbosch M, Verweij MC, Nagelkerken I et al (2004b) Homing and daytime tidal movements of juvenile snappers (Lutjanidae) between shallow-water nursery habitats in Zanzibar, western Indian Ocean. Environ Biol Fish 70:203–209

Dunning JB, Danielson BJ, Pulliam HR (1992) Ecological processes that affect populations in complex landscapes. Oikos 65:169–175

Fagan WF, Calabreses JM (2006) Quantifying connectivity: balancing metric performance with data requirements. In: Crooks KR, Sanjayan M (eds) Connectivity conservation. Conservation biology 14, Cambridge University Press, Cambridge

Forman RTT (1995) Land mosaics: the ecology of landscapes and regions. Cambridge University Press, Cambridge

Forman RTT, Godron M (1986) Landscape ecology. John Wiley and Sons, New York

Gardner RH, O'Neill RV (1991) Pattern, process and predictability: the use of neutral models for landscape analysis. In: Turner MG, Gardner RH (eds) Quantitative methods in landscape ecology: the analysis and interpretation of landscape heterogeneity. Springer-Verlag, New York

Garrabou J, Riera J, Zabala M (1998) Landscape pattern indices applied to Mediterranean subtidal rocky benthic communities. Landsc Ecol 13:225–247

Gaucherel C, Fleury D, Auclair D (2006) Neutral models for patchy landscapes. Ecol Modell 197:159–170

Gillanders BM, Able KW, Brown JA et al (2003) Evidence of connectivity between juvenile and adult habitats for mobile marine fauna: an important component of nurseries. Mar Ecol Prog Ser 247:281–295

Gillis EA, Krebs CJ (2000) Survival of dispersing versus philopatric juvenile snowshoe hares: do dispersers die? Oikos 90:343–346

Gladfelter WB, Ogden JC, Gladfelter EH (1980) Similarity and diversity among coral reef fish communities: a comparison between tropical western Atlantic (Virgin Islands) and tropical central pacific (Marshall Islands) patch reefs. Ecology 61:1156–1168

Grober-Dunsmore R (2005) The application of terrestrial landscape ecology principles to the design and management of marine protected areas in coral reef ecosystems. Ph.D. dissertation submitted to University of Florida, Department of Fisheries and Aquatic Sciences, Florida, 219 pp.

Grober-Dunsmore R, Beets J, Frazer T et al (2008) Influence of landscape structure on reef fish assemblages. Landsc Ecol 23(SI):37–53

Grober-Dunsmore R, Bonito V (2009) Movement of reef fishes inside and outside of Votua MPA, Fiji Islands. Report to NOAA Coral Reef International 2009 Coral Reef library, 24pp.

Grober-Dunsmore R, Frazer T, Beets J et al (2004) The significance of adjacent habitats on reef fish assemblage structure: are relationships detectable and quantifiable at a landscape scale? Proc Gulf Caribb Fish Inst 55:713–734

Grober-Dunsmore R, Frazer TK, Lindberg WJ et al (2007) Reef fish and habitat relationships in a Caribbean seascape: the importance of reef context. Coral Reefs 26:201–216

Gustafson EJ, Parker GR (1992) Relationships between landcover proportion and indexes of landscape spatial pattern. Landsc Ecol 7:101–110

Gutzwiller KJ (2002) Applying landscape ecology in biological conservation. Springer-Verlag, New York

Hanski I (1998) Metapopulation dynamics. Nature 396:41–49

Hargis CD, Bissonette JA, David JL (1998) The behavior of landscape metrics commonly used in the study of habitat fragmentation. Landsc Ecol 13:167–186

Helfman GS, Meyer JL, McFarland WN (1982) The ontogeny of twilight migration patterns in grunts (Pisces, Haemulidae). Anim Behav 30:317–326

Holland KN, Peterson JD, Lowe CG et al (1993) Movements, distribution and growth rates of the white goatfish *Mulloides flavolineatus* in a fisheries conservation zone. Bull Mar Sci 52:982–992

Hovel KA, Lipcius RN (2002) Effects of seagrass habitat fragmentation on juvenile blue crab survival and abundance. J Exp Mar Biol Ecol 271:75–98

Hovel KA, Regan HM (2008) Using an individual-based model to examine the roles of habitat fragmentation and behavior on predator–prey relationships in seagrass landscapes. Landsc Ecol 23(S1):75–89

Irlandi EA, Ambrose WG, Orlando BA (1995) Landscape ecology and the marine environment-how spatial configuration of seagrass habitat influences growth and survival of the Bay scallop. Oikos 72:307–313

Irlandi EA, Crawford MK (1997) Habitat linkages: the effect of intertidal saltmarshes and adjacent sub-tidal habitats on abundance, movement, and growth of an estuarine fish. Oecologia 110:222–230

Jelbart JE, Ross PM, Connolly RM (2006) Edge effects and patch size in seagrass landscapes: an experimental test using fish. Mar Ecol Prog Ser 319:93–102

Jelbart JE, Ross PM, Connolly RM (2007) Fish assemblages in seagrass beds are influenced by the proximity of mangrove forests. Mar Biol 150:993–1002

Karl JW, Heglund PJ, Garton EO et al (2000) Sensitivity of species habitat-relationship model performance to factors of scale. Ecol Appl 10:1690–1705

Kendall MS, Christensen JD, Hillis-Starr Z (2003) Multi-scale data used to analyze the spatial distribution of French grunts, *Haemulon flavolineatum*, relative to hard and soft bottom in a benthic landscape. Environ Biol Fish 66:19–26

Kendall MS, Kruer CR, Buja KR, Christensen JD, Finkbeiner M, Monaco ME (2002) Methods used to map the benthic habitats of Puerto Rico and the U.S. Virgin Islands. NOAA/NOS Biogeography Program Technical Re-port. Silver Spring, MD, p 45

Kendall MS, Miller T (2008) The influence of thematic and spatial resolution on maps of a coral reef ecosystem. Marine Geodesy 31:75–102

Kramer DL, Chapman MR (1999) Implications of fish home range size and relocation for marine reserve function. Environ Biol Fish 55:65–79

Levin SA (1992) The problem of pattern and scale in ecology. Ecology 73:1943–1967

Li HB, Reynolds JF (1993) A new contagion index to quantify spatial patterns of landscapes. Landsc Ecol 8:155–162

Li HB, Wu JG (2004) Use and misuse of landscape indices. Landsc Ecol 19:389–399

Lugendo BR, Nagelkerken I, Jiddawi N et al (2007b) Fish community composition of a tropical non-estuarine embayment in Zanzibar (Tanzania). Fish Sci 73:1213–1223

Lugendo BR, Nagelkerken I, Kruitwagen G et al (2007a) Relative importance of mangroves as feeding habitat for fish: a comparison between mangrove habitats with different settings. Bull Mar Sci 80:497–512

Lugendo BR, Nagelkerken I, van der Velde G et al (2006) The importance of mangroves, mud and sand flats, and seagrass beds as feeding areas for juvenile fishes in Chwaka Bay, Zanzibar: gut content and stable isotope analyses. J Fish Biol 69:1639–1661

Lunetta RS, Congalton RG, Fenstermaker LK et al (1991) Photogramm Eng Remote Sens 57: 677–687

MacArthur RH, Wilson EO (1967) The theory of island biogeography. Princeton University Press, Princeton, New Jersey

Malczewski J (1999) GIS and multicriteria decision analysis. John Whiley & Sons, New York

Margules CR, Pressey RL (2000) Systematic conservation planning. Nature 405: 243–253

McGarigal K, Cushman SA, Neel MC (2002) FRAGSTATS: spatial pattern analysis program for categorical maps. University of Massachusetts Amherst, Massachusetts. http://www.umass.edu/landeco/research/ fragstats/fragstats.html/

McGarigal K, McComb WC (1995) Relationships between landscape structure and breeding birds in the Oregon Coast Range. Ecol Monogr 65:235–260

Meyer CG, Holland KN, Papastamatiou YP (2007) Seasonal and diel movements of giant trevally *Caranx ignobilis* at remote Hawaiian atolls: implications for the design of marine protected areas. Mar Ecol Prog Ser 333:13–25

Meyer CG, Holland KN, Wetherbee BM et al (2000) Movement patterns, habitat utilization, home range size and site fidelity of whitesaddle goatfish, *Parupeneus porphyreus*, in a marine reserve. Environ Biol Fish 59:235–242

Micheli F, Peterson CH (1999) Estuarine vegetated habitats as corridors for predator movements. Conserv Biol 13:869–881

Mitchell MS, Lancia RA, Gerwin JA (2001) Using landscape-level data to predict the distribution of birds on a managed forest: effects of scale. Ecol Appl 11:1692–1708

Moilanen A, Nieminen M (2002) Simple connectivity measures in spatial ecology. Ecology 83:1131–1145

Molles MC (1978) Fish species-diversity on model and natural reef patches-experimental insular bio-geography. Ecol Monogr 48:289–305

Mumby PJ (2006) Connectivity of reef fish between mangroves and coral reefs: algorithms for the design of marine reserves at seascape scales. Biol Conserv 128:215–222

Mumby PJ, Edwards AJ, Arias-Gonzalez JE et al (2004) Mangroves enhance the biomass of coral reef fish communities in the Caribbean. Nature 427:533–536

Mumby PJ, Harborne AR (1999) Development of a systematic classification scheme of marine habitats to facilitate regional management of Caribbean coral reefs. Biol Conserv 88:155–163

Nagelkerken I (2007) Are non-estuarine mangroves connected to coral reefs through fish migration? Bull Mar Sci 80:595–607

Nagelkerken I, Dorenbosch M, Verberk WCEP et al (2000a) Importance of shallow-water biotopes of a Caribbean bay for juvenile coral reef fishes: patterns in biotope association, community structure and spatial distribution. Mar Ecol Prog Ser 202:219–230

Nagelkerken I, Dorenbosch M, Verberk WCEP et al (2000b) Day-night shifts of fishes between shallow-water biotopes of a Caribbean bay, with emphasis on the nocturnal feeding of Haemulidae and Lutjanidae. Mar Ecol Prog Ser 194:55–64

Nagelkerken I, Faunce CH (2007) Colonisation of artificial mangroves by reef fishes in a marine seascape. Estuar Coast Shelf Sci 75:417–422

Nagelkerken I, Kleijnen S, Klop T et al (2001) Dependence of Caribbean reef fishes on mangroves and seagrass beds as nursery habitats: a comparison of fish faunas between bays with and without mangroves/seagrass beds. Mar Ecol Prog Ser 214:225–235

Nagelkerken I, Roberts CM, van der Velde G et al (2002) How important are mangroves and seagrass beds for coral-reef fish? The nursery hypothesis tested on an island scale. Mar Ecol Prog Ser 244:299–305

Nagelkerken I, van der Velde G (2002) Do non-estuarine mangroves harbour higher densities of juvenile fish than adjacent shallow-water and coral reef habitats in Curacao (Netherlands Antilles)? Mar Ecol Prog Ser 245:191–204

Ogden JC, Ehrlich PR (1977) Behavior of heterotypic resting schools of juvenile grunts (Pomadasyidae). Mar Biol 42:273–280

Ogden JC, Gladfelter EH (1983) Coral reefs, seagrass beds and mangroves: their interaction in the coastal zones of the Caribbean. UNESCO Rep Mar Sci 23:1–133

Parrish JD (1989) Fish communities of interacting shallow-water habitats in tropical oceanic regions. Mar Ecol Prog Ser 58:143–160

Pearson SM, Turner MG, Gardner RH et al (1996) An organism-based perspective of habitat fragmentation. In: Szaro RC (ed) Biodiversity in managed landscapes: theory and practice. Oxford University Press, California

Pither J, Taylor PD (1998) An experimental assessment of landscape connectivity. Oikos 83:166–174

Pittman SJ (2002) Linking fish and prawns to their environment in shallow-water marine landscapes. Ph thesis, Geographical Sciences Department and The Ecology Centre, University of Queensland, Brisbane, Australia

Pittman SJ, Caldow C, Hile SD et al (2007b) Using seascape types to explain the spatial patterns of fish in the mangroves of SW Puerto Rico. Mar Ecol Prog Ser 348:273–284

Pittman SJ, Christensen JD, Caldow C et al (2007a) Predictive mapping of fish species richness across shallow-water seascapes in the Caribbean. Ecol Modell 204:9–21

Pittman SJ, McAlpine CA (2003) Movement of marine fish and decapod crustaceans: process, theory and application. Adv Mar Biol 44:205–294

Pittman SJ, McAlpine CA, Pittman KM (2004) Linking fish and prawns to their environment: a hierarchical landscape approach. Mar Ecol Prog Ser 283:233–254

Plotnick RE, Gardner RH, O'Neill RV (1993) Lacunarity indexes as measures of landscape texture. Landsc Ecol 8:201–211

Possingham H, Ball I, Andelman S (2000) Mathematical models for identifying representative reserve networks. In: Ferson S, Burgman M (eds) Quantitative methods for conservation biology. Springer-Verlag, New York

Pressey RL (1999) Applications of irreplaceability analysis to planning and management problems. Parks 9:42–51

Recksiek CW, Appeldoorn RS, Turningan RG (1991) Studies of fish traps as stock assessment devices on a shallow reef in south-western Puerto Rico. Fish Res 10:177–197

Riitters KH, O'Neill RV, Hunsaker CT et al (1995) A factor analysis of landscape pattern and structure metrics. Landsc Ecol 10:23–39

Robbins BD, Bell SS (1994) Seagrass landscapes: a terrestrial approach to the marine subtidal environment. Trends Ecol Evol 9:301–304

Rothley KD, Rae C (2005) Working backwards to move forwards: graph-based connectivity metrics for reserve network selection. Environ Modell Assess 10:107–113

Rozas LP, Minello TJ (1998) Nekton use of salt marsh, seagrass, and nonvegetated habitats in a south Texas (USA) estuary. Bull Mar Sci 63:481–501

Sale PF (2002) The science we need to develop for more effective management. In: Sale PF (ed) Coral reef fishes: dynamics and diversity in a complex ecosystem. Academic Press, London

Saura S, Martinez-Millan J (2001) Sensitivity of landscape pattern metrics to map spatial extent. Photogramm Eng Remote Sens 67:1027–1036

Schippers P, Verboom J, Knaapen JP et al (1996) Dispersal and habitat connectivity in complex heterogeneous landscapes: an analysis with a GIS-based random walk model. Ecogeography 19:97–106

Schneider MF (2001) Habitat loss, fragmentation and predator impact: spatial implications for prey conservation. J Appl Ecol 38:720–735

Schumaker NH (1996) Using landscape indices to predict habitat connectivity. Ecology 77: 1210–1225

Sheaves M (2005) Nature and consequences of biological connectivity in mangrove sytems. Mar Ecol Prog Ser 302:293–305

Sisk TD, Haddad NM, Ehrlich PR (1997) Bird assemblages in patchy woodlands: modeling the effects of edge and matrix habitats. Ecol Appl 7:1170–1180

Stamps JA, Buechner M, Krishnan VV (1987) The effects of edge permeability and habitat geometry on emigration from patches of habitat. Am Nat 129:533–552

Starr RM, Sala E, Ballesteros E et al (2007) Spatial dynamics of the Nassau grouper *Epi-nephelus striatus* in a Caribbean atoll. Mar Ecol Prog Ser 343:239–249

Taylor DS, Reyier EA, Davis WP et al (2007) Mangrove removal in the Belize cays: effects on mangrove-associated fish assemblages in the intertidal and subtidal. Bull Mar Sci 80:879–890

Taylor PD, Fahrig L, With KA (2006) Landscape connectivity: a return to the basics. In: Crooks KR, Sanjayan M (eds) Connectivity conservation. Cambridge University Press, Cambridge

Tewfik A, Bene C (2003) Effects of natural barriers on the spillover of a marine mollusc: implications for fisheries reserves. Aquat Conserv 13:473–488

Tischendorf L (2001) Can landscape indices predict ecological processes consistently? Landsc Ecol 16:235–254

Tischendorf L, Fahrig L (2000) How should we measure landscape connectivity? Landsc Ecol 15:633–641

Treml E, Halpin P, Urban D et al (2008) Modeling population connectivity by ocean currents, a graph-theoretic approach for marine conservation. Landsc Ecol 23(S1):19–36

Turner MG (1989) Landscape ecology the effect of pattern on process. Annu Rev Ecol Syst 20:171–197

Turner MG (2005) Landscape ecology: what is the state of the science? Annu Rev Ecol Evol Syst 36:319–344

Turner MG, Gardner RH, O'Neill RV (2001) Landscape ecology in theory and practice: pattern and process. Springer-Verlag, New York

Turner SJ, Hewitt JE, Wilkinson MR et al (1999) Seagrass patches and landscapes: the influence of wind-wave dynamics and hierarchical arrangements of spatial structure on macrofaunal seagrass communities. Estuaries 22:1016–1032

Underwood AJ, Chapman MG, Connell SD (2000) Observations in ecology: you can't make progress on processes without understanding the patterns. J Exp Mar Biol Ecol 250:97–115

Urban D, Keitt T (2001) Landscape connectivity: a graph theoretic perspective. Ecology 82:1205–1218

Urban DL (2005) Modeling ecological processes across scales. Ecology 86:1996–2006

Vanderklift MC, How J, Wernberg T et al (2007) Proximity to reef influences density of small predatory fishes, while type of seagrass influences intensity of their predation on crabs. Mar Ecol Prog Ser 340:235–243

Verweij MC, Nagelkerken I (2007) Short and long-term movement and site fidelity of juvenile Haemulidae in back-reef habitats of a Caribbean embayment. Hydrobiologia 592:257–270

Vierweij MC, Nagelkerken I, Hol KEM et al (2007) Space use of *Lutjanus apodus* including movement between a putative nursery and a coral reef. Bull Mar Sci 81:127–138

Ward TJ, Vanderklift MA, Nicholls AO et al (1999) Selecting marine reserves using habitats and species assemblages as surrogates for biological diversity. Ecol Appl 9:691–698

Wiens J (1989) Spatial scaling in ecology. Funct Ecol 3:385–39

Wiens JA (2006) Connectivity research – what are the issues? In: Crooks KR, Sanjayan M (eds) Connectivity conservation. Cambridge University Press, Cambridge

Wiens JA, Milne BT (1989) Scaling of landscapes in landscape ecology, or landscape ecology from a beetle's perspective. Landsc Ecol 3:87–96

Wiens JA, Stenseth NC, Vanhorne B et al (1993) Ecological mechanisms and landscape ecology. Oikos 66:369–380

Wiens JA, Van Horne B, Noon BR (2002) Landscape structure and multi-scale management. In: Liu J, Taylor WW (eds) Integrating landscape ecology into natural resource management. Cambridge University Press, Cambridge

With KA (1997) The application of neutral landscape models in conservation biology. Conserv Biol 11:1069–1080

With KA (2002) The landscape ecology of invasive spread. Conserv Biol 16:1192–1203

With KA, Crist TO (1995) Critical thresholds in species responses to landscape structure. Ecology 76:2446–2459

With KA, Gardner RH, Turner MG (1997) Landscape connectivity and population distributions in heterogeneous environments. Oikos 78:151–169

Wu JG (2006) Landscape ecology, cross-disciplinarity, and sustainability science. Landsc Ecol 21:1–4

Zeller DC (1998) Spawning aggregations: patterns of movement of the coral trout *Plectropomus leopardus* (Serranidae) as determined by ultrasonic telemetry. Mar Ecol Prog Ser 162:253–263

Zeller DC, Russ GR (1998) Marine reserves: patterns of adult movement of the coral trout *(Plectropomus leopardus* (Serranidae)). Can J Fish Aquat Sci 55:917–924

Part IV
Management and Socio-economic Implications

Chapter 15
Relationships Between Tropical Coastal Habitats and (offshore) Fisheries

Stephen J.M. Blaber

Abstract The economic welfare and productivity of many tropical fisheries, inshore or offshore, depends on the integrity of coastal habitats, particularly mangroves and coral reefs. Fisheries within coastal systems in developing countries are usually artisanal or subsistence in nature, whereas offshore fisheries are usually commercial or industrial. Relationships between fisheries production and areas of mangrove have been quantified, notably for penaeid prawns, but in most cases the causal links have not been established. Nevertheless, evidence of the value of mangroves to fisheries continues to mount and the importance of the relationship has gained widespread acceptance. Coral reef fisheries are largely the domain of small-scale fishers, but their relative importance is very great with global catches in excess of 2×10^6 tonnes. Their fisheries productivity is less than that of estuarine and coastal waters. The connectivity between reef fisheries and mangroves and seagrasses, and connectivity between reefs are relevant to offshore fisheries. Coral reefs support some pelagic fisheries, such the pole-and-line tuna fleets in the Pacific and Indian Oceans. For most tropical fisheries, the key issue is the depleted state of the resources, e.g., for most tropical Asian countries biomass has declined to <10% of baseline estimates. The major contributor to this is overfishing linked to poverty among fishing communities—symptoms of lack of effective management. Strategies to address the situation relate to ecological connectivity and dependence on mangroves or coral reefs, the balance between small-scale and industrial fisheries, and scales of management as well as use of Marine Protected Areas (MPAs).

Keywords Fisheries · Mangroves · Coral reefs · Baitfish · Management

S.J.M. Blaber (✉)
CSIRO Marine and Atmospheric Research, P.O. Box 120, Cleveland, Queensland 4163, Australia
e-mail: steve.blaber@csiro.au

I. Nagelkerken (ed.), *Ecological Connectivity among Tropical Coastal Ecosystems*,
DOI 10.1007/978-90-481-2406-0_15, © Springer Science+Business Media B.V. 2009

15.1 Introduction

Tropical coastal habitats are usually dominated by turbid mangrove-lined shores and estuaries, or by clear-water coral reef systems, both of which have been significantly impacted by human activities. The ongoing and increasing concern for the maintenance of the health of mangroves and coral reefs is driven not only by conservation and aesthetic considerations, but also in relation to their economic importance. Fisheries in the tropics consist of very diverse industries whose economic welfare and productivity depends most on the integrity of mangrove and coral reef ecosystems.

Costanza et al. (1997) calculated that the economic value of estuaries in terms of services and natural capital per hectare was the highest of all ecosystems. Tropical mangrove systems in particular, are zones of high productivity (Blaber 2000) and assuming that most of the fisheries productivity is closely linked to mangroves, a number of recent studies have emphasized the economic value of mangroves, especially in the developing world (Barbier and Strand 1998, Hamilton et al. 1989, Nickerson 1999, Barbier 2000). Rönnbäck (2001) stated that 'one major driving force behind the loss of more than 50% of the world's mangroves during the last decades, and its continuation (Duke et al. 2007), is the inability among economists to recognize and value all goods and services produced by this ecosystem'. Barbier et al. (2002) in estimating the welfare effects of mangrove–fisheries linkages in Thailand commented that the fisheries most likely to be affected by habitat losses and impacts are those containing a high proportion of artisanal fishers.

Coral reef fisheries are also very valuable and recent estimates (Agardy et al. 2005) show that those in Southeast Asia generate US$2.5 billion per year and worldwide contribute about a quarter of the annual fish catch in developing countries providing food to about 1 billion people in Asia alone. The marginalization of fishers in many developing countries is largely responsible for increasing rates of overfishing (Pauly 1997), and once coral reefs are destroyed restoration is extremely difficult (Moberg and Rönnbäck 2003).

However, it is important to distinguish between *fisheries within coastal systems*, usually of an artisanal or subsistence nature in developing countries, and *offshore fisheries* that are usually commercial or industrial concerns, although not always, as in the case of the artisanal tuna fisheries of the Maldives. In the former case, the activities by traditional or artisanal fishermen may be long-established and are totally dependent on the existence of the mangrove or reef system. Many of these fisheries have been studied in detail and examples are given in Blaber (2000), Jhingran (2002), Islam and Haque (2004), Kathiresan and Qasim (2005), Munro (1996), Cheung et al. (2007), and Wilkinson et al. (2006). Although the major threats to mangrove and reef fishes are usually linked to environmental degradation, such as removal of mangroves and coral destruction, there is also evidence to suggest that many fish species in coastal regions, particularly in South and Southeast Asia, are declining in abundance primarily as a result of overfishing. Such overfishing is strongly linked to issues of food security, but unlike many other human activities, fisheries are almost completely dependent on the maintenance of ecosystem integrity (Blaber 2007).

Key issues that are explored in this chapter relate to the connectivity between mangroves and/or coral reefs and offshore fisheries. Of major importance is an understanding of how the productivity of these fisheries may be influenced by coastal processes and particularly by human-induced perturbations to mangrove and coral reef habitats.

15.2 Fisheries

The types of fisheries can be divided into three main sectors (Harden Jones 1994, Rawlinson et al. 1995):

(1) Subsistence fisheries, where the fishers predominantly consume all of their catch or give it away, but do not sell it.
(2) Artisanal fisheries, where the fishers sell part of their catch, but also retain part for their own consumption.
(3) Commercial fisheries, where the fishers sell all of their catch.

Within the subsistence and artisanal sectors are included 'traditional fisheries'. Many of these have a very long history and form part of the culture of human coastal communities. They may also have a longstanding and complex interrelationship with the environment, and are increasingly coming to be regarded as part of the overall ecology of the tropical environment (Agardy et al. 2005).

An additional category exists in developed subtropical and tropical countries, such as Australia, South Africa, and the USA.

(4) Recreational fisheries, where fishing is carried out as a sport or leisure pastime and not primarily for producing food or income. Nevertheless, the service infrastructure associated with recreational fishing usually encompasses economically important income-generating activities.

In developing countries there are often resource conflicts between subsistence or artisanal fisheries and commercial fisheries, and in developed countries between recreational and commercial fisheries. Both situations require important resource allocation decisions from government entities and NGOs that may manage the resources.

The complexity and relative importance of the different fisheries sectors, particularly in relation to the value of the recreational component, is demonstrated by the situation in South Africa (Griffiths and Lamberth 2002). Here the marine line fishery targets almost 200 species, of which 31 contribute substantially to catches. Fishers comprise recreational, commercial, and subsistence components. The commercial component consists of about 18,600 participants (2,600 vessels of 5–15 m long), which target both pelagic and demersal species beyond the surf-zone. The recreational component may be divided into estuarine anglers (72,000), shore anglers (412,000), spearfishers (7,000), and a recreational boat-based sector (12,000 participants). Subsistence line fishing is largely limited to estuarine and shore-based activities in the Transkei and KwaZulu-Natal. The line fishery, excluding the estuarine component, is estimated to provide employment for approximately 132,000 people, and to contribute about R2.2 billion (US$0.3 billion) to the South African GDP. Although the commercial component is responsible for 79% of the estimated total catch, the recreational component provides 81% of the employment and generates 82% of the revenue. Species targeted by the line fishery display diverse life-history strategies, including long life-spans (>20 years), estuarine-dependence, sex change, and aggregating behavior, that cause populations to be particularly vulnerable to overfishing.

15.3 Mangrove/Estuary—Fisheries Connectivity

Links between offshore fisheries production and mangroves, and in particular the effects of mangrove loss on these links, are much harder to quantify, mainly because the large-scale removal of mangroves for aquaculture and development purposes has coincided with increased fishing pressure and more efficient fishing technologies. Over the last four decades, many studies have demonstrated a strong relationship between mangrove presence and fish catch (Turner 1977, Yáñez-Arancibia et al. 1985, Pauly and Ingles 1986, Lee 2004, Manson et al. 2005, Meynecke et al. 2007), with fishery catch being influenced by the relative abundance of mangroves in a region (Table 15.1). Correlations have also been found between the extent (area or linear extent) of mangroves and the catches of prawns (particularly banana prawns) in the fisheries adjacent to the mangroves (Turner 1977, Staples et al. 1985, Pauly and Ingles 1986, reviewed in Baran 1999; Fig. 15.1). Such studies provided important information on the mangrove—fisheries relationship. This observed relationship mainly derives from a group of economically important species classified as estuarine-dependent (Cappo et al. 1998) or (non-estuarine) bay-habitat dependent (Nagelkerken and van der Velde 2002). Mangroves, or similar environments, are the principal habitat for at least one part of their life cycle (Blaber et al. 1989, Nagelkerken et al. 2000). Typically, the adults spawn offshore, producing eggs that disperse in the water column for varying lengths of time. The eggs then develop into planktonic larvae which move, or are carried by currents, into inshore and estuarine waters. The subadult or adults migrate out of the estuary or lagoon, and back towards the offshore areas or adjacent coral reefs. Therefore, mangroves could function as an important link in the chain of habitats that provide complementary resources and benefits, e.g., as nursery areas for fish, prawns, and crabs (Sheridan and Hays 2003, Crona and Rönnbäck 2005; see Chapter 10), with spatial complexity at a scale that provides refuge to small prey, and abundant food for commercial species at certain stages in their life cycle (Chong et al. 1990).

Depending on species, location and time scale, however, the significance of relationships between commercial catch and mangroves is highly variable, indicating that the link is more complex than a linear function (Baran 1999). The predictors used in most regression analyses are themselves strongly correlated, and catch

Table 15.1 Relationships between mangroves and offshore fisheries production that have been quantified (modified from Blaber 2007)

Source and area	Formula	X function	Y function	r^2	n
Martosubroto and Naamin (1977), Indonesia	$Y = 0.1128X + 5.473$	Mangrove area ($*10,000$ ha)	Penaeid production ($*1,000$ tonnes)	0.79	–
Turner (1977), USA	$Y = 1.96X - 4.39$	% of saline vegetation in a hydrological unit	Percentage of *Penaeus aztecus*	0.92	7
Staples et al. (1985), Australia	$Y = 1.074X + 218.3$	Mangrove shoreline in km	Catch of *Penaeus merguiensis* (tonnes)	0.58	6
Yáñez-Arancibia et al. (1985), Mexico	$LnY = 0.496LnX + 6.07$	Coastal marshes in km^2	Fish capture (tonnes)	0.48	10
Pauly and Ingles (1986), tropics	$Log_{10}MSY = 0.4875log_{10}\ AM - 0.0212L + 2.41$	MSY = maximum sustainable yield of penaeids; AM = area of mangroves; L = degrees of latitude	–	–	–
Paw and Chua (1989), Southeast Asia	$Y = 0.8648X + 0.0991$	Log_{10} of mangrove area	$Log_{10}$of penaeid catch (tonnes)	0.66	17

Table 15.1 (continued)

Source and area	Formula	X function	Y function	r^2	n
Nickerson (1999), Philippines, leiognathids	$K(t+1) = K(t) - [M(t) - M(t+1)]D$	K = metric tons of fish and invertebrates if unexploited in the total mangrove area; M = hectare of fringing mangrove area; and D = metric tonnes of fish and invertebrates per hectare of mangrove area	—	—	—
Barbier et al. (2002), Gulf of Thailand	$HD_{it} = b_0 + b_1 EDLM_{it} + b_2 ED^2_{it} + \mu_{it}$	Pooled time series—cross sectional analysis of the relationship between harvest, effort, and mangrove area for demersal fisheries, where: $i = 1, \ldots, 5$ zones; $t = 1, \ldots, 11$ years (1983–1993); HD_{it} = demersal fish harvest (kg) for zone i at time t; $EDLM_{it}$ = demersal fisheries effort (hrs) × log mangrove area (km^2) for zone i at time t; ED^2_{it} = demersal fisheries effort (hrs) squared for zone i at time t	—	—	—
Manson et al. (2005), Queensland, Australia	Relationships between mangrove environment and CPUE data. PCA and multiple regression analyses	For mangrove-related species (two shrimps, one crab, and one fish) mangrove area and perimeter accounted for most variation; for non-mangrove estuarine species (two shrimps, one crab, and one fish) mangrove perimeter was significant, but latitude was dominant variable	—	—	—

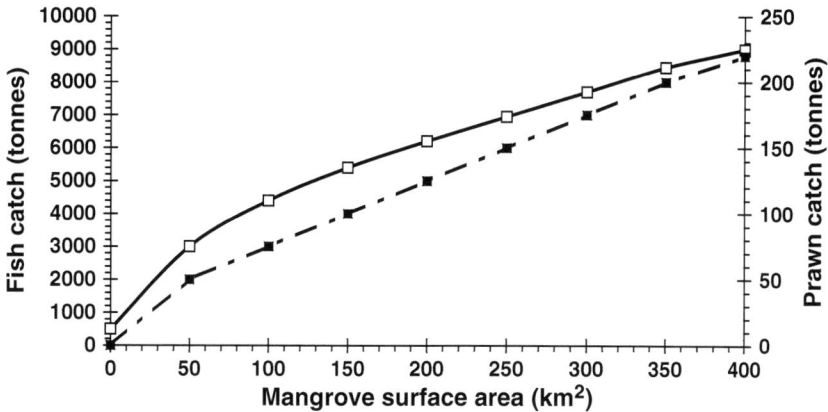

Fig. 15.1 Direct relationships between surface area of mangroves and fish (solid line) and prawn (dashed line) production simulated by Baran (1999) from data in Yáñez-Arancibia et al. (1985) and Paw and Chua (1989). Modified from Baran (1999)

statistics are often not well delineated. There is high variation within the data sets (mangrove forest distribution, commercial records, effect of stock size, and fishing pressure), and difficulty in distinguishing links against a background of highly variable temperature, rainfall, ocean currents, and fishing effort. In this chapter, the relationships between coastal waters and fisheries involving both penaeid prawns and fishes are discussed.

15.3.1 Penaeid Prawns

Much of the evidence about the importance of mangroves to fisheries production has come from studies of penaeid prawns. Research in the Gulf of Mexico (Turner 1977), Indonesia (Martosubroto and Naamin 1977), India (Kathiresan and Rajendran (2002), Australia (Staples et al. 1985, Vance et al. 1996), and the Philippines (Paw and Chua 1989, Primavera 1998), provides good evidence that there is a correlation between commercial offshore prawn catches and the total area of adjacent mangroves. Baran (1999) re-plotted the relationship demonstrated by Paw and Chua (1989) on an ordinary scale and showed that the relationship is not linear and exhibits an inflexion at a certain abscissa (Fig. 15.1), below which small reduction in the surface area of mangroves implies a drastic reduction in production of prawns. Pauly and Ingles (1986) concluded that most of the variance in the catches of penaeids could be explained by a combination of mangrove area and latitude (Table 15.1).

15.3.1.1 Malaysia

There is considerable evidence for a relationship between the extent of mangroves and penaeid prawn landings in Peninsular Malaysia. However, Loneragan et al.

(2005) suggested that landings may have been maintained or increased despite large losses of mangroves, and the results of their study are particularly interesting and illustrate the difficulty of understanding the relationships. They document changes in catches of all prawns, white prawns (mainly *Penaeus merguiensis*) in relation to mangrove extent, rainfall, and the area of shallow water. Although there was a significant linear relationship between the landings of total prawns and mangrove area in both the 1980s and 1990s, this was not the case for the mangrove-dependent white prawns where a significant relationship was found only for the 1990s. The area of shallow water accounted for the greatest proportion of variation in landings of both all prawns and white prawns, and was the most significant variable fitted to multiple regressions of landings and coastal attributes (area of shallow water, mangrove area, and length of coastline). Landings of all prawns and white prawns in the states of Selangor and Johor, where large areas of mangrove have been lost, appear to have been maintained or increased in the 1990s. The lack of a clear relationship between mangrove loss and prawn landings may be due to the migration of prawns from adjacent areas, or because other attributes of mangroves, such as the length of mangrove-water interface, may be more important for the growth and survival of prawn populations than total area of mangroves. This is further reinforced by a recent review by Chong (2006), which showed that landings of penaeids for the entire west coast of Peninsular Malaysia actually declined from 60,967 tonnes in 1989 to 39,296 tonnes in 2003 (35% reduction) despite a reduction in fishing effort, while their main nursery habitat (mangroves) shrunk in area (23% loss).

15.3.1.2 Gulf of Carpentaria, Australia

The relationship between the offshore catches of *Penaeus merguiensis* in the Gulf of Carpentaria and a suite of factors influencing the life-cycle of the species has been studied for over 20 years. A strong positive relationship was established between rainfall and subsequent offshore commercial catches (Staples et al. 1985, Vance et al. 1985) where over a 26 yrs period in the Southeast Gulf, annual rainfall accounted for 81% of annual variation in catch. This relationship was, however, not as strong in other areas, particularly the Northeast Gulf, where there was no significant relationship between rainfall and catch. Further work by Vance et al. (1996) showed that the lack of a strong relationship between rainfall and offshore catch in the Northeast Gulf, is not due to a difference in the response of prawns to rainfall, but is rather due to differences in rainfall levels and the physical characteristics of the different river systems.

15.3.1.3 India

Strong correlations between penaeid prawn catches, mainly of *Penaeus indicus, P. merguiensis*, and *Metapenaeus dobsoni*, and the extent of adjacent mangrove areas have been reported from a number of areas of India (Kathiresan et al. 1994, Mohan et al. 1997, Kathiresan and Bingham 2001, Rönnbäck et al. 2002). An analysis of the value of the Godavari mangroves by Rönnbäck et al. (2002) showed that mangroves subsidize both total fisheries catch by prawn trawlers and the aquaculture industry's dependence on inputs of lime, seed, spawners, and feed. The results show that 32,600 ha of mangroves support an annual fisheries catch over 100,000 tonnes, much of it penaeids. This means that each hectare of mangrove generates 3.1 tonnes of catch, which corresponds to a gross financial value of US$3,900 annually. Subsistence fisheries comprised more than one-third of the total catch by weight, and together with penaeid prawns, trawl catches were the most important resource by value.

15.3.1.4 Gulf of Mexico

Turner (1977) found a positive correlation between penaeid shrimp catches and the vegetated surface area of estuaries in the Gulf of Mexico and in Louisiana; the percentage of brown shrimps (*Penaeus aztecus*) in total shrimp catches was correlated with the area of adjacent mangrove habitats. Pink shrimp (*Farfantepenaeus duorarum*) is an important commercial species in the eastern Gulf of Mexico, with annual landings between 2,300 and 4,500 metric tonnes during the past decade (Ehrhardt et al. 2001). The adults are exploited by a trawl fishery in the Dry Tortugas region, while juveniles inhabit Florida Bay. The onshore–offshore ontogenetic migrations create many opportunities for disruption of cohorts. Potential linkage between the abundance of juveniles in nursery areas and of recruits to the fishery was examined by relating recruitment success to juvenile density spanning 123 months of data. The fitted recruitment success model predicts well the general trend in the ratio of recruitment to juvenile density. However, as in other parts of the world, the relationships are complex and the full magnitude of the expected trends is not fully explained by the model, and environmental variables on recruitment may have a sizeable effect. However, there is a significant relationship between recruitment to the fishery and juvenile density in the nursery grounds, emphasizing the importance of the juvenile habitat to fisheries production.

Barbier and Strand (1998) determined the effects of changes in mangrove area in the Laguna de Terminos on the production and value of prawn harvests in Campeche from 1980 to 1990. They showed that mangroves have an important and essential input into the Campeche prawn fishery, but that the low levels of deforestation between 1980 and 1990 of about 2 km^2 annually (2.3%) caused a loss of 28.8 tonnes of catch equivalent to only 0.4% loss of harvest and revenue. Of greater significance in terms of impact on the commercial catch was the increase in fishing effort with the number of boats increasing from 4,500 in 1980 to over 7,200 in 1990. Barbier and Strand (1998) go on to state that the management implications from these results are clear, and that although protection of mangroves may be important for preventing losses in the fishery, control of overfishing is more critical. As long as fishing effort continues to increase, catches will fall, even if mangrove areas are fully protected.

15.3.2 Fishes

The significance of the relationships between fishes and mangroves is rather more equivocal than that for penaeid prawns, and Robertson and Blaber (1992) concluded that in spite of the correlations between mangrove area and commercial fish catches, a causal link has not been established experimentally. Nevertheless, there is little doubt that many tropical estuaries and coastal waters are zones of high productivity due to a combination of shallowness and high nutrient input from rivers. In addition, the vegetation in and adjacent to tropical estuaries, particularly the mangroves, contributes to this productivity, and the fish yields in terms of tonnes landed per km^2 per year of a selection of subtropical and tropical estuaries and non-coral reef coastal waters are shown in Table 15.2. The reported values are mainly from larger systems, albeit still estuarine *sensu* Blaber (2000), because these are the ones that support significant fisheries and for which data exist. The tonnages are based on total catches in relation to estuarine area and in most cases do not reflect sustainable yields. The values range from 1–38 tonnes per km^2 per year and are generally higher than those for tropical rivers and lakes, but the range is similar to that reported for tropical continental shelves and coral reefs (Lowe-McConnell 1975, Marten and Polovina 1982). The key features of a selection of the larger fisheries are given below.

Table 15.2 Production of fish in selected tropical estuaries and non-coral reef coastal systems (modified partly from Marten and Polovina 1982, and Blaber 2000)

Country	Fishery area	Tonnes per km^2 per year	References
Colombia	Cienaga Grande	12.0	Rueda and Defeo (2001)
El Salvador	Jiquilisco	1.7	Hernandez and Calderon (1974), Phillips (1981)
Ghana	Sakumo lagoon	15.0	Pauly (1976)
India	Lake Chilka	3.7	Jhingran and Natarajan (1969)
	Lake Pulicat	2.6	Jhingran and Gopalakrishnan (1973)
	Mandapam lagoon	5.6	Tampi (1959)
	Hooghly-Matlah	11.4	Jhingran (1991)
	Vellar-Coloroon	11.1	Venkatesan (1969)
Ivory Coast	Ébrié lagoon	16.0	Durand et al. (1978)
Malagasy	Pangalanes lagoon	3.7	Laserre (1979)
Malaysia	Larut-Matang	38.64*	Choy (1993)
Mexico	Caimanero lagoon	34.5	Warburton (1979)
	Terminos lagoon	20.0	Yáñez-Arancibia and Lara Dominguez (1983)
	Tamiahua lagoon	4.7	Garcia (1975)
Philippines	San Miguel Bay	23.8**	Mines et al. (1986)
South Africa	Kosi system	1.0	Kyle (1988, 1999)
USA	Texas bays	12.1	Jones et al. (1963)
Venezuela	Lake Maracaibo	1.9	Nemoto (1971)
	Tacarigua lagoon	11.0	Gamboa et al. (1971)

* Includes penaeid prawn catches and non-coastal waters
** Probably an overestimate as trawlable biomass is only 2.13 tonnes per km^2

15.3.2.1 The Hilsa Fishery, South Asia

Various aspects of the biology of the Hilsa and its fisheries (*Tenualosa ilisha*) have been summarized by Blaber (2000) and Blaber et al. (2003). This anadromous clupeid is the basis of the world's largest tropical estuarine fishery. The species extends from the Arabian Gulf to at least Burma, but the largest fisheries are in the Bay of Bengal and its estuaries in India and Bangladesh. The popularity, socio-religious significance, and traditional public knowledge of Hilsa are reflected in the proverbs and historical records of the Bengal area, and no other fish is as highly prized (Raja 1985). The largest catches of Hilsa come from the Ganges Delta and upper Bay of Bengal region, with Bangladesh probably taking the largest share (well over 100,000 tonnes per year), followed by India (about 25,000 tonnes) and Burma (about 5,000 tonnes). Unfortunately, the nature of the fishery has precluded very accurate records of annual catches (Dunn 1982), as is the case with many of the fisheries of the developing world, and estimates of total yields vary widely. However, there seems little doubt that the overall catch in the Bay of Bengal region is now at least of the order of 200,000 tonnes. It is the most important fishery of the region and for example, currently makes up about 25% of the total fish landings in Bangladesh.

It has subsistence, artisanal, and commercial sectors although there is considerable overlap between all three, and very large numbers of people are dependent on the fishery. The subsistence sector comprises mainly the fishing activities of women and children who catch juveniles in the estuaries and rivers, the artisanal consists of smaller, mainly non-mechanized boats working the estuaries, while the commercial sector consists of larger mechanized vessels working in the Bay of Bengal. There are conflicts of interest between the sectors, with the commercial fishers believing that the catching of juveniles by the subsistence sector adversely impacts adult stocks, and the artisanal riverine fishers contending that the expansion in the marine fishery has reduced the number of Hilsa available in the estuaries.

15.3.2.2 The Larut-Matang Fishery, Malaysia

The mangrove estuaries and adjacent coastal waters of the Matang area of Peninsular Malaysia are very heavily fished by both commercial and artisanal sectors and as such, are similar to many such areas in Southeast Asia. The forests of the Matang Mangrove Forest Reserve have been managed on the basis of sustainable yields, with a 30 yrs timber harvesting rotation, since the early part of this century. In 1992, 450,000 tonnes of timber products (including charcoal) were produced (Gopinath and Gabriel 1997). The forest industry provides direct employment for about 1,400 and indirect employment for 1,000, and the total annual value of the forest products are about US$9 million. In contrast to this, the fishing industry provides direct employment for 2,500 and indirect employment for 7,500, with an annual value of about US$30 million. The fisheries thus provide four times more employment and economic returns than the forestry, clearly illustrating the value of the links between mangroves and coastal waters to the fisheries (Ong 1982, Khoo 1989).

In 1990 there were 2,540 fishers, of which 1,250 were involved in trawling, and the remainder operated traditional gears, 970 of whom used mainly gill nets. Most of the boats are small, locally constructed and suitable for shallow seas, narrow river mouths and creeks, and 79% are less than 20 gross tones. The above figures exclude about 200 unlicensed mechanized push net operators that fish the very shallow coastal areas and river mouths.

In 1990 a total of 58,300 tonnes were landed (73% from trawling, 7% from gill nets) plus approximately another 2,300 tonnes from the push net operators. The main species landed are sciaenids, *Rastrelliger* spp., ariids, *Sardinella* spp., *Lates calcarifer, Megalops cyprinoides*, and large numbers of juveniles of a wide variety of inshore and estuarine species. The fishery also catches large numbers of commercially important penaeid prawns (Blaber 2000).

15.3.2.3 The Ébrié Lagoon Fishery, Ivory Coast

The lagoon network of the Ivory Coast consists of three different lagoon complexes stretching along the coast for about 300 km. The nature of the fish fauna of the Ébrié Lagoon (a coastal lake of 566 km^2) has been extensively studied (Albaret and Écoutin 1989, 1990) and has well-developed traditional and commercial fisheries.

The commercial fisheries using mainly ring nets and beach seine nets (up to 2 km in length), operated by a salaried workforce, began in about 1960 and increased fourfold between 1964 and 1975, to become responsible for at least 70% of the catch (5,000 tonnes per year). The permanent opening of the lagoon to the sea in 1950 brought a more marine influence and increased the catches of marine migrant species. The traditional fishery takes about 2,000 tonnes per year.

Six species make up the bulk of landings in both fisheries: *Ethmalosa fimbriata* (61%), *Tilapia* spp. (6%), *Elops lacerta* (6%), *Chrysichthys* spp. (5%), *Tylochromis jentinki* (4%), and *Sardinella maderensis* (4%).

15.3.2.4 The Gulf of Nicoya Fishery, Costa Rica

This large mangrove-lined estuarine embayment on the Pacific coast of Costa Rica is the most important fishing ground of Costa Rica. The main Pacific ports of the country are located within the estuary and the estuary supports a substantial artisanal fishery on various stocks in the inner Gulf, as well as a commercial prawn (shrimp) trawl fishery in the outer Gulf.

The artisanal fishery lands about 6,300 tonnes per year of which 43% consists of sciaenids (known locally as 'corvina'), such as *Cynoscion albus, C. squamipinnis, Stellifer* spp., and *Bairdiella* spp. (Szelistowski and Garita 1989, Herrera and Charles 1994). These corvina populations are comprised of high-valued species that are caught in drift gillnet fisheries that are barely managed, and the fishery is regarded as a common property resource. Little is known of the state of the stocks, but the stocks have suffered severe declines under the high fishing pressure. These artisanal fisheries are characterized by the absence of any long-term series of catch records and the only available data are single length-frequency distributions of catch, collected sparsely over time. Hence, any search for alternative harvesting strategies to gill netting and the development of management plans is dependent on the implementation of research and catch monitoring to provide the appropriate information.

15.3.2.5 The Cienaga Grande de Santa Marta Fishery, Colombia

This large (480 km²) and shallow coastal lake on the Caribbean coast of Colombia supports some of the most extensive and varied artisanal fisheries of the region as well as some commercial fishing. Gill nets, cast nets, longlines, and an encircling net known as a 'bolicheo', are the primary techniques used by the fishermen in the lake (Rueda 2007). Species of the families Mugilidae, Gerreidae, Ariidae, Centropomidae, Sciaenidae, and Lutjanidae form the major part of the artisanal catches. The actual species caught vary according to the type of fishing gear, with mojarras, particularly *Eugerres plumieri*, the dominant species taken by the 'bolicheo' nets, and mullets abundant in gillnet catches. Commercial catches are dominated by carangids and haemulids (León and Racedo 1985). Much of the system is now included with conservation areas and there is an urgent need for integrated management plans (Rueda and Defeo 2001).

15.3.3 Mangrove—Fishery Relationships: Conclusions

The few studies that have quantified relationships between mangroves and coastal resources were summarized by Baran (1999) and Manson et al. (2005) and these and others are listed in Table 15.1, but it must be reiterated that these are correlations and that causal links have not been established experimentally. In the Gulf of Mexico, Yáñez-Arancibia et al. (1985) showed a logarithmic relationship (positive correlation) between commercial fish catches and mangrove area. However, the same authors showed that in Terminos Lagoon and the adjacent coastal waters of Campeche Sound in the Gulf of Mexico, fish yields are largely controlled by climatic and meteorological conditions, the amount of river discharge and tidal amplitude—factors that play a major role in affecting the movement patterns of fish between the lagoon and the sea (Yáñez-Arancibia et al. 1985).

Manson et al. (2005) showed that links between mangrove area and coastal fisheries production could be detected for some species at a broad regional scale (1,000s of kilometers) on the east coast of Queensland, Australia. The relationships between catch-per-unit-effort for different commercially caught species in four fisheries—trawl (Fig. 15.2), line, net, and pot—and mangrove characteristics estimated from Landsat images, were analyzed. The species were categorized into three groups based on life history characteristics, namely mangrove-related species (banana prawns *Penaeus merguiensis*, mud crabs *Scylla serrata*, and barramundi *Lates calcarifer*), estuarine species (tiger prawns *Penaeus esculentus* and *Penaeus semisulcatus*, blue swimmer crabs *Portunus pelagicus*, and blue threadfin *Eleutheronema tetradactylum*), and offshore species (coral trout

Fig. 15.2 Commercial penaeid prawn trawler on the east coast of Queensland, Australia

Plectropomus spp.). For the mangrove-related species, mangrove characteristics such as area and perimeter accounted for most of the variation in the model; for the non-mangrove estuarine species, latitude was the dominant parameter, but some mangrove characteristics (e.g., mangrove perimeter) also made significant contributions to the models. In contrast, for the offshore species, latitude was the dominant variable, with no contribution from mangrove characteristics.

As with the Paw and Chua (1989) relationship between mangroves and prawn catch, Baran (1999) re-plotted this relationship on an ordinary scale and showed that the relationship is not linear and exhibits an inflexion at a certain abscissa below which a small reduction in the surface area of mangroves implies a drastic reduction in fish catch (Fig. 15.1). In Vietnam, De Graaf and Xuan (1998) demonstrated a similar relationship and indicated that one hectare of mangrove forest supports a marine catch of 450 kg per year.

The dependence on the mangrove habitat by juveniles of a number of commercially important fishes, including mugilids, polynemids, ariids, clupeids and engraulids is well documented (Blaber 2000). For example, the life history of the Tarpon, *Megalops atlanticus,* of the tropical Atlantic has been summarized by Garcia and Solano (1995) for the Colombian coast. The leptocephali of *M. atlanticus* migrate into mangrove estuaries (Zerbi et al. 1999) where they grow to about 28 mm, they then enter a second larval stage when their length decreases to about 13 mm. After this, growth increases again and the larvae become juveniles. At a length of about 100 mm they move to freshwaters. Prior to sexual maturity the fish leave the estuary (>400 mm) and reach sexual maturity at a length of about 1 m. Spawning aggregations of 25–250 *M. atlanticus* have been recorded up to 25 km from the coast. Fecundity is very high with a 2-m fish containing about 12 million eggs. The life cycle of *Megalops cyprinoides* of the Indo-West Pacific appears to be similar in most respects to that of *M. atlanticus* (Pandian 1969, Coates 1987).

Baran (1999) and Baran and Hambrey (1999), in recent important reviews of the fishery–mangrove relationship, showed that all the studies to date suffer from problems of auto-correlation—with many factors other than just mangrove area, such as river discharge, area of shallow coastal water, intertidal area, and food availability, contributing to the relationships. These reviews also show that finding a relationship between mangrove area and fish production is not straightforward, because:

(1) Closely related fish species can have very different ecological requirements, which blurs possible global relationships.
(2) Results drawn from site-specific studies cannot be generalized to large areas in different geomorphic and climatic settings.
(3) Fishery statistics, in most cases, can not be disaggregated enough to link catches to specific mangrove zones.

More recently, in a review of the role of mangroves as nursery habitats for transient fishes and decapods, Sheridan and Hays (2003) concluded that the case for identifying flooded mangrove forests as critical nursery habitat remains equivocal until sufficient further experimental and quantitative studies have been carried out. Furthermore, in Queensland, Australia, Manson et al. (2005) showed an empirical link between the extent of mangrove habitat and fishery production (mainly Crustacea) for three mangrove-related species, but such links for four non-mangrove estuarine species were less significant, and for these species latitude was the dominant variable.

As indicated earlier, not all mangroves or areas in mangrove systems, have the same relationships with fishes. For example, Vance et al. (1996) showed that in northern Australia, the deeper waters at the fringes of mangrove forests contain much of the functionality compared with the more inland shallower, intertidal areas. There are other studies which show that the inshore shallower area of the mangrove forests are preferred areas by shrimps and small fishes (Rönnbäck et al. 1999, Affendy and Chong 2007) presumably to avoid predation. Hence, if much of the loss of mangroves could be confined to the inland side, and deeper fringing areas left intact, perhaps much of the functional value could be retained. Kapetsky (1985) suggested that much of the functional value of mangroves might be retained from a smaller area of mangroves, for example 75% of nursery function from 50% of original area. However, it is the deeper fringing areas of mangroves adjacent to the sea that have been most attractive and suitable for aquaculture pond development throughout the tropics and have hence suffered greater proportional losses than more inland mangroves. This is not to say that shallower inland mangroves may not be important, particularly with regard to small fish species diversity and conservation, as demonstrated by Taylor et al. (1995) in Florida and Rönnbäck et al. (1999) in the Philippines.

Evidence of the value of mangroves to fish (see Chapter 10) and fisheries is rapidly increasing and appears almost overwhelming, but the case at the moment is not proven because much of the evidence is circumstantial. Therefore, despite the fact that the relationship has gained widespread scientific and public acceptance, there is still an urgent need for experimental and quantitative studies (such as that of Verweij et al. (2006)) to lend weight to the economic arguments that the value of retaining mangroves far exceeds the value of their destruction for whatever purpose. Mangroves are still being lost at an unacceptable rate from all points of view: ecological, economic, conservation, and human safety (e.g., Tsunami protection) (Blaber 2007).

15.4 Coral Reef—Offshore Fisheries Connectivity

Coral reef fisheries are largely the domain of small-scale fishers (Fig. 15.3), but their relative importance is very great with global catches in excess of 2×10^6 tonnes (Munro 1996). Their fisheries productivity is, however, usually much less than that for tropical estuarine and coastal (non-coral reef) waters (Tables 15.2, 15.3). Although most of the fisheries productivity of coral reefs is harvested within

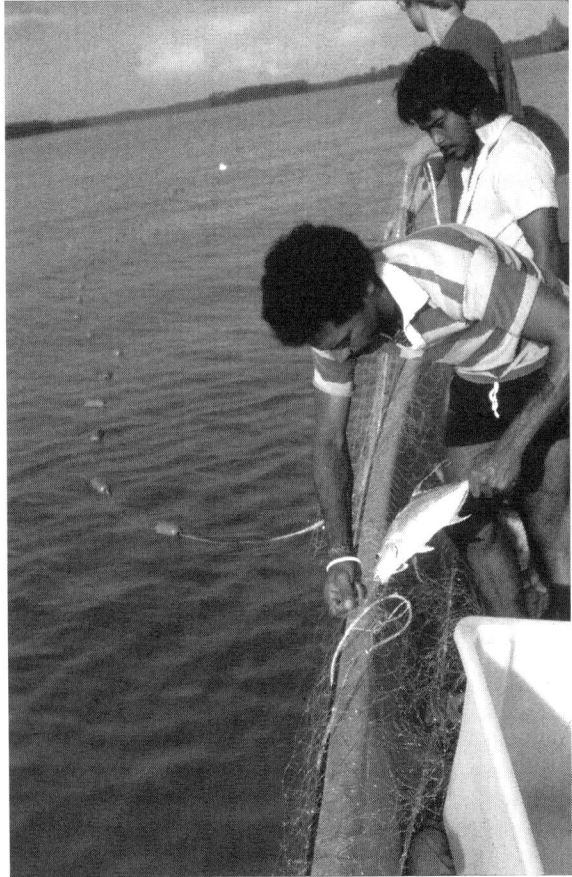

Fig. 15.3 Gill netting is one of the most common types of fishery in coral reef areas, such as here in the Solomon Islands

Table 15.3 Fish yields from various coral reef fisheries (modified partly from Marten and Polovina 1982, and Munro 1996)

Location	Tonnes per km² per year	References
Bahamas	2.4	Gulland (1971)
Caribbean, total mean value	0.4	Munro (1977)
Fiji	3.4	Jennings and Polunin (1995)
Ifaluk (Pacific)	5.1	Stevenson and Marshall (1974)
Jamaica, north coast	3.7	Munro (1977)
Kenya	2.0–4.0	Kaunda-Arara et al. (2003)
Madagascar, Tulear	12.0	Laroche and Ramananarivo (1995)
Mauritius	4.7	Wheeler and Ommaney (1953)
Philippines, Bolinao	12.0	McManus (1992)
Philippines, Luzon	7.0	Christie and White (1994)
Philippines, reef slope	2.7	McManus (1992)
Red Sea (Eritrea)	0.15–0.6	Tsehaye (2007)

the reef system, there are several ways in which their connectivity with other systems is relevant to 'offshore fisheries'. These include the relationship between reef fisheries and mangroves and seagrasses, connectivity between reefs, and ways in which reefs support some pelagic fisheries.

15.4.1 Connectivity Between Reef Fisheries and Mangroves/Seagrasses

This subject is dealt with in detail in Chapters 10 and 11, and only those points related to reef fisheries are discussed here.

Worldwide, most studies on mangrove fish communities and their linkages with offshore fisheries have been done in estuarine mangrove systems (Nagelkerken 2007). However, there are 100–1,000 of small islands in the Caribbean and Indo-Pacific which only harbor non-estuarine mangroves located in marine embayments and lagoons. Although their surface area is mostly much smaller than that of large estuarine mangrove forests, they may be important on an island scale for coral-reef associated fisheries. Only in this millennium have studies started focusing in more detail on the connectivity between non-estuarine mangroves (and seagrass beds) and adjacent coral reefs with regard to fish movement (Nagelkerken 2007), mostly based on multiple habitat density comparisons using a single census technique and distinguishing between fish size classes. This has resulted in the identification of several (commercial) reef fish species which appear to depend on mangroves while juvenile (e.g., Nagelkerken et al. 2000, Cocheret de la Morinière et al. 2002, Christensen et al. 2003, Serafy et al. 2003, Dorenbosch et al. 2007). Studies comparing reef fish communities near and far from mangrove habitats, and with the presence or absence of island mangroves, have shown that the dependence on mangroves is species-specific, but appears to be high for various reef species (Nagelkerken et al.

2000, 2002, Mumby et al. 2004, Dorenbosch et al. 2004, 2005, 2007). Otolith micro-chemistry studies have also suggested a linkage between mangroves and coral reefs (Chittaro et al. 2004).

15.4.2 Connectivity Between Reefs

In coral reef environments it is widely accepted that pelagic larval dispersal is responsible for most of the linkages between reef fish populations (Boehlert 1996), and most coral reef fishes are thought to be highly sedentary with movements limited to a few kilometers (Kaunda-Arara and Rose 2004). This is exemplified by the fishery situation with regard to the Nassau grouper (*Epinephelus striatus*), which is the second most abundant fish landed in the Bahamas (422 tonnes in 2003), and is becoming scarce as the species is mainly sedentary and forms regular and conspicuous spawning aggregations where it is heavily fished (Ehrhardt and Deleveaux 2007).

However, in addition to pelagic reef-associated species, such as some carangids, long distance movements of adult reef fish have been documented and may complement larval dispersal in maintaining connectivity between populations (Kaunda-Arara and Rose 2004). This is further supported by evidence from the Virgin Islands where Beets et al. (2003) suggest that habitat connectivity may be greatly dependent on movement of large organisms, particularly haemulids. The question of scale is important here and results from French Polynesia show that relatively little connectivity is likely among reef systems separated by >300 km, but substantial exchange occurs at smaller spatial scales (Lo-Yat et al. 2006).

15.4.3 Coral Reefs and Pelagic Fisheries

The pelagic fisheries of most significance in relation to coral reefs are some of the scombrid fisheries, primarily for tuna, in the Pacific and Indian Oceans (Dalzell 1996; Table 15.4). The tuna pole-and-line industry relies upon supplies of baitfish taken from coral reef areas (Blaber and Copland 1990). About 20 species are commonly used as live bait (Table 15.5), but large numbers of other small species and juveniles of other larger species, mainly reef-associated, are captured during the baitfishing operations (Rawlinson and Sharma 1993). Although declining in importance relative to purse seining, pole-and-line fleets are still important in the Pacific in Solomon Islands, Fiji, and eastern Indonesia. In the Indian Ocean the very large artisanal fishery for tuna in the Maldives uses pole-and-line.

In the Pacific, commercial pole-and-line boats (Fig. 15.4) collect large quantities of live bait from coral reef areas using bouki-ami or lift nets. The species used are mainly stolephorid anchovies and *Spratelloides* spp. It was estimated that 1.9 million kg of baitfish were captured by the pole-and-line fleet in Solomon Islands in 1992, mainly from 78 bait grounds in the Western Province (Tiroba 1993).

Table 15.4 Summary of catch rates and catch composition for troll fishing on reef stocks and adjacent pelagic stocks in the Caribbean, Indian Ocean, and South Pacific (modified from Dalzell 1996)

Location	CPUE range (kg line per hour)	Mean CPUE	Principal pelagic species in catch	Principal reef components	References
Fiji	1.6–16.2	5.7	*Scomberomorus commerson, Euthynnus affinis, Gymnosarda unicolor, Grammatorcynus bicarinatus*	*Sphyraena qenie, Sphyraena barracuda, Plectropomus areolatus, Caranx ignobilis*	Chapman and Lewis (1982) Lewis et al. (1983)
Maldives	0–7.5	0.93	*Katsuwonus pelamis, Gymnosarda unicolor, Euthynnus affinis, Auxis thazard, Acanthocybium solandri*	*Sphyraena barracuda, Aprion virescens*	Blaber et al. (1990)
New Caledonia	0–7.3	4.0	*Scomberomorus commerson, Euthynnus affinis, Acanthocybium solandri, Thunnus albacares*	*Caranx ignobilis, Sphyraena barracuda, Aprion virescens, Lutjanus bohar, Epinephelus* spp.	Chapman and Cusack (1989)
Palau	3.7–12.6	8.2	*Katsuwonus pelamis, Thunnus albacares, Acanthocybium solandri*	*Caranx ignobilis, Sphyraena* spp.	Anon (1990, 1991)
Papua New Guinea	3.7–6.9	4.9	*Euthynnus affinis, Auxis thazard, Scomberomorus commerson*	*Caranx ignobilis, Sphyraena barracuda, Epinephelus fuscoguttatus, Symphorus nematophorus, Lethrinus* spp.	Anon (1984), Wright and Richards (1985), Dalzell and Wright (1986), Lock (1986)
Tuvalu	0.5–7.0	2.7	*Acanthocybium solandri, Katsuwonus pelamis, Thunnus albacares, Istiophorus platypterus*	*Aprion virescens, Aphareus furcatus, Caranx* spp., *Sphyraena barracuda*	Chapman and Cusack (1988)

Table 15.5 Species commonly captured in coral reef areas by tuna pole-and-line fisheries for use as live bait. PNG = Papua New Guinea

Family	Scientific name	English name	Fisheries
Apogonidae	*Rhabdamia cypselurus*		Solomons, PNG, Fiji, Indonesia, Maldives
	Rhabdamia gracilis		Solomons, PNG, Fiji, Indonesia, Maldives
Atherinidae	*Atherinomorus lacunosus*		Solomons, PNG, Fiji, Indonesia
	Hypoatherina ovalaua		Solomons, PNG, Fiji, Indonesia
Caesionidae	*Caesio* spp.	Fusilier	Solomons, PNG, Fiji, Indonesia, Maldives
	Dipterygonotus balteatus	Whitelined fusilier	Solomons, PNG, Fiji, Indonesia, Maldives
	Pterocaesio spp.	Fusilier	Solomons, PNG, Fiji, Indonesia, Maldives
Carangidae	*Decapterus* spp.	Round scad	Solomons, PNG, Fiji, Indonesia
	Selar spp.	Big eye scad, atule	Solomons, PNG, Fiji, Indonesia
Scombridae	*Rastrelliger kanagurta*	Indian mackerel	Solomons, PNG, Fiji, Indonesia
Clupeidae	*Amblygaster sirm*	Spotted pilchard	Solomons, PNG, Fiji
	Herklotsichthys quadrimaculatus	Gold spot herring	Solomons, PNG, Fiji
	Sardinella fimbriata	Fringescale sardine	Solomons, PNG, Fiji
	Sardinella spp.	Sardine	Solomons, PNG, Fiji
	Spratelloides delicatulus	Blue sprat	Solomons, PNG, Fiji, Indonesia, Maldives
	Spratelloides gracilis	Silver sprat	Solomons, PNG, Fiji, Indonesia, Maldives
	Spratelloides lewisi	Lewis's sprat	Solomons, PNG, Fiji
Engraulididae	*Stolephorus devisi*	Gold anchovy	Solomons, PNG, Fiji, Indonesia
	Stolephorus heterolobus	Blue anchovy	Solomons, PNG, Fiji, Indonesia, Maldives
	Stolephorus indicus		Solomons, PNG, Fiji
	Stolephorus punctifer	Ocean anchovy	Indonesia
	Thryssa baelama	Little priest	Solomons, PNG, Fiji, Indonesia

The Maldives in the Indian Ocean has a long history of tuna fishing going back several centuries. Tuna fishing is one of the main economic activities of the country, with almost 10% of the population involved directly in the fishery (Anon 2003). The main catch is skipjack tuna (115,300 tonnes in 2002), but significant quantities of yellowfin tuna (21,700 tonnes in 2002) are caught as well. During the last few years the average size and fishing power of local pole and line vessels has increased significantly. This has contributed to a major recent increase in tuna catches and catch

Fig. 15.4 Commercial tuna pole-and-line boat—entirely dependent on baitfish caught over coral reefs

rates (Adam et al. 2003). The primarily artisanal fleet (Fig. 15.5) relies on daily supplies of baitfish from reef areas. A detailed description of the baitfishing method is given by Maniku et al. (1990). Catches are made using lift nets deployed over coral 'bommies' where suitable concentrations of fish have been located. Over 20 species of baitfish are in regular use, falling into three categories: fusiliers (Caesionidae), sprats of the genus *Spratelloides*, and cardinal fishes (Apogonidae). It has been estimated that the catch rates of tuna in the Maldives lie within the range of 7−13 kg of tuna per kg of baitfish.

The pole-and-line fisheries for tuna in eastern Indonesia, mainly in Maluku and Sulawesi, which caught 106,677 tonnes of skipjack tuna in 1997, also rely on large quantities of baitfish taken in coral reef areas. Most of the baitfish is captured using fixed fishing platforms (bagans) set up over coral reefs. Lift nets are deployed from the bagans at night using lights to attract the fish over the net (Naamin and Gafa 1998). About 16 species from five families, namely, Engraulididae, Clupeidae, Caesionidae, Scombridae, and Carangidae are commonly used as baitfish, but usually about 70% of the catch consists of *Stolephorus* spp. Naamin and Gafa (1998) report a decreasing trend in the ratio of baitfish utilization to tuna catch from 9−10 kg tuna per kg of baitfish (1968–1971) to 3.5 kg of tuna per kg of baitfish (1986–1995).

15.5 Management and Governance Issues

For most tropical fisheries the key issue is the depleted state of the resources. In a recent analysis of the situation in Asia, Stobutzki et al. (2006) state that for most

Fig. 15.5 Traditional Maldivian pole-and-line tuna 'dhoni' (**a**) and typical catch (**b**), entirely dependent on baitfish caught over coral reefs

Southeast Asian countries for which time series data are available, total biomass has declined to <10% of baseline estimates. The major contributor to the declines is overfishing linked also to poverty among fishing communities—both a symptom of lack of effective fisheries management. A variety of strategies to address the situation is available and is outlined in Stobutzki et al. (2006). Of particular relevance to the connectivity question and the dependence of offshore fisheries on mangroves or coral reefs, are the balance between small-scale and industrial fisheries, the scales of management taking into account, degrees of connectivity, and the use of marine protected areas (MPAs) (see also Chapter 16).

15.5.1 Balance Between Small-Scale and Industrial Fisheries

In most developing countries the offshore resources are harvested by an industrial sector using relatively large boats and sophisticated technologies, whereas the inshore resources in and around mangroves and coral reefs are taken by small-scale artisanal and subsistence fishers often using traditional low technology methods. Depletion by either sector of fisheries resources that are connected inshore–offshore through life-cycles or migrations will obviously impact both sectors. Management measures to address this problem such as delineation of fishing zones, based on depth, distance from shore, and vessel size have been successful in reducing conflict and separating the fishers in a number of Asian countries, but they are still competing for the same connected resources (Garces et al. 2006). Overfishing will continue unless total levels of fishing effort are reduced and rights to specific catch levels allocated to the fisheries sectors (Stobutzki et al. 2006).

15.5.2 Connectivity

Ablan (2006) states that there is an urgent need for habitat connectivity-related information with regard to fisheries in developing countries for four main reasons, and her findings are summarized below:

Firstly, the main strategies to manage the multi-gear and multi-species fisheries in many tropical developing countries are almost exclusively area-based. Asian countries have fisheries management zones, which restrict gear or vessel types in particular areas (Garces et al. 2006). Marine protected areas are now receiving significant attention as a potential fisheries management and conservation tools (Pollnac et al. 2001). The success of these management strategies is dependent on how well management units align with the structure and dynamics of the ecological systems and their connectivity and how well they serve the biological requirements of commercial fish species.

Secondly, management units are becoming more focused at local scales (see Chapter 16). The Local Government Code of 1991 in the Philippines and the Local Autonomous Code of 1998 in Indonesia are some examples of legislation where the authority to develop and regulate coastal resources, including the fisheries, has been formally devolved to local governments. Community-based coastal management initiatives, whereby coastal communities assume many of the responsibilities for implementing, monitoring, and enforcing regulations are advocated (Ferrer et al. 1996, Crawford et al. 2000, Christie 2005). With management parcels shrinking in size, the success of initiatives within these management areas will depend on the extent to which they are independent or dependent on other areas. As discussed in Chapter 16, spatial management to achieve large-scale conservation and sustainable use at a regional scale has good potential; however, at the local scale without planning and coordination taking into account ecosystem connectivity, most community-based management initiatives are likely to fail in relation to fisheries with onshore–offshore dependencies.

Thirdly, demersal species are distributed in patches which are associated with specific habitats. The vulnerability of fish populations to fishing pressure and local extinction is much greater when populations occur in aggregations. Their abundance may decrease substantially when the habitats they require are destroyed. Areas where habitats are healthy and fish are abundant may be important to communities and countries other than those with jurisdiction over them. The Spratly Islands area, in the South China Sea, is a case in point. This rich fishing area is located within the Exclusive Economic Zones (EEZs) of four countries (McManus 1994, Morton and Blackmore 2001). The area is also a source of fish to other areas on the boundary of the South China Sea, and as a result is heavily overfished, and has one of the lowest biomasses of fish per square kilometer (McManus and Menez 1998). Within a country, fishing areas may be within the jurisdiction of two or more local administrative units. Balicasag Island in the Bohol Sea is another example. The area is recognized by local fishers as seasonal spawning grounds for high value species, such as groupers and snappers. Groupers can travel long distances to breed (Pittman et al. (2004), Kaunda-Arara and Rose 2004) and the fish in the area come from localities other than the reef in which spawning occurs. Healthy mangroves and sea-grass beds are nursery grounds for fish that settle elsewhere (Mumby et al. 2004), making them essential habitats for the survival of the fish. Managing these critical areas requires information on fish movements and trophic linkages.

Finally, manpower and financial resources to enforce fisheries management regulations are limited in many tropical countries. Local governments, now held responsible for managing their fisheries, may not have sufficient resources and may be severely hampered without accurate information about the 'connectedness' of their fisheries.

15.5.3 Marine Protected Areas (MPAs)

The designation of areas as 'marine protected areas' (often known as marine sanctuaries or marine parks) has become a widespread practice in many tropical seas primarily as a conservation tool and to attract and encourage ecotourism. They are described in detail in Chapter 16 and only those points relevant to fisheries production are discussed here.

In terms of a more holistic approach to ecosystem management MPAs are a useful fisheries management tool. Fishing is generally prohibited or fishing effort is greatly reduced in MPAs thus providing a number of possible benefits to fisheries management in relation to mangroves and coral reefs (Cabanban 2000):

(1) Protection of spawning stocks
(2) Protection of nursery grounds
(3) Enhancement of catch in adjacent zones through emigration
(4) Provision of larvae and recruits to downstream reef or other nearshore habitats

Defining systems of marine protected areas, and strategies to manage shared stocks, were recommended as priority actions to ensure sustainability of fisheries within countries in South and Southeast Asia (Silvestre et al. 2003). There is relatively little quantitative data about the connectivity and spatial structure of coastal fisheries resources in developing countries, but resource managers in many developing countries are keenly aware of the concept (Ablan 2006). However, there is strong evidence about the positive effects of protecting spawning biomass through the use of MPAs in Australia (Robertson 1999), the Philippines (Alcala and Russ 1990), and the Caribbean (Roberts 1997). Generally, fish catches from adjacent areas increased after protection and declined when the protection was lifted (Cabanban 2000). However, while conservationists, resource managers, scientists, and coastal planners have recognized the broad applicability of MPAs, they are often implemented without a firm understanding of the conservation science—both ecological and socio-economic—underlying the rationale for marine protection (Agardy et al. 2003). Despite considerable investment in monitoring of coral reefs and other coastal habitats, for most MPAs in the tropics, the data available do not show clearly whether biodiversity, socio-economic, or fishery objectives are being met (Wells et al. 2007).

The size of MPAs has received considerable attention and there may be important differences between what works in the tropics and what works in temperate areas (Laurel and Bradbury 2006). For East African coral reef areas, research suggests that tropical fisheries dominated by rabbitfish, emperors, and surgeonfish should be enhanced by closed areas of 10–15% of the total fishery area (McClanahan and Mangi 2000).

In assessing the value of MPAs to downstream fisheries in the tropics it has often been assumed that there are no regulations limiting fishing effort and that MPAs, by themselves, can be used to maintain both sustainable fish stocks and sustainable harvests (Hilborn et al. 2006). However, modeling by Hilborn et al. (2006) showed that the situation can be complex and they found that when a stock is managed at maximum sustainable yield, or is overfished, an MPA is only effective if there is a reduction in total catch in order to avoid increased fishing pressure on the stock outside the MPA. Hilborn et al. (2006) further suggest that catches will be lower as a result of overlaying an MPA on existing fisheries management and only when the stock is so overfished that it is close to extinction does an MPA not lead to lower catches. In a catch-regulated fishery, even if the stock is overfished, MPA implementation may not improve overall stock abundance or increase harvest unless catch is simultaneously reduced in the areas outside the MPA.

References

Ablan MACA (2006) Genetics and the study of fisheries connectivity in Asian developing countries. Fish Res 78:158–168

Adam MS, Anderson RC, Hafiz A (2003) The Maldivian tuna fishery. IOTC Proc 6:202–220

Affendy N, Chong VC (2007) Shrimp ingress into mangrove forests of different age stands, Matang Mangrove Forest Reserve, Malaysia. Bull Mar Sci 80:915

Agardy T, Alder J, Birkeland C et al. (2005) Coastal systems. In: U.N. millenium assessment: condition and trends assessment, pp. 513–550. Island Press, Washington DC

Agardy T, Bridgewater P, Crosby MP et al (2003) Dangerous targets? Unresolved issues and ideological clashes around marine protected areas. Aquat Conserv Mar Freshw Ecosyst 13:353–367

Albaret JJ, Ecoutin JM (1989) Communication mer-lagune: impact d'une reouverture sur l'ictyofaune de la lagune Ébrié (Côte d'Ivoire). Rev d'Hydrobiol Trop 22:71–81

Albaret JJ, Ecoutin JM (1990) Influence des saisons et des variations climatiques sur les peuplements de poissons d'une lagune tropicale en Afrique de l'Ouest. Acta Oecol 11:557–583

Alcala AC, Russ CR (1990) A direct test of the effects of protective management on abundance and yield of tropical marine resources. J Cons Explor Mer 46:40–47

Anon (1984) Annual report 1983, Research and Surveys Branch, Fisheries Division, Department of Primary Industry, Port Moresby, Papua New Guinea

Anon (1990) Annual report 1990, Division of Marine Resources, Ministry of Resource Development, Palau

Anon (1991) Annual report 1991, Division of Marine Resources, Ministry of Resource Development, Palau

Anon (2003) Proposal for a small-scale tagging project in the Republic of Maldives. IOTC Proc 6:1–2

Baran E (1999) A review of quantified relationships between mangroves and coastal resources. Phuket Mar Biol Centre, Res Bull 62:57–64

Baran E, Hambrey J (1999) Mangrove conservation and coastal management in Southeast Asia: what impact on fisheries resources. Mar Pollut Bull 37:431–440

Barbier EB (2000) Valuing the environment as input: review of applications to mangrove fishery linkages. Ecol Econ 35:45–61

Barbier EB, Strand I (1998) Valuing mangrove-fishery linkages. Environ Res Econ 12:151–166

Barbier EB, Strand I, Sathirathai S (2002) Do open access conditions affect the valuation of an externality? Estimating the welfare effects of mangrove-fishery linkages in Thailand. Environ Res Econ 21:343–367

Beets J, Muehlstein L, Haughht K et al (2003) Habitat connectivity in coastal environments: patterns and movements of Caribbean coral reef fishes with emphasis on bluestriped grunt, *Haemulon sciurus*. Gulf Caribb Res 14:29–42

Blaber SJM (2007) Mangroves and fishes: issues of diversity, dependence and dogma. Bull Mar Sci 80:457–472

Blaber SJM (2000) Tropical estuarine fishes: ecology, exploitation and conservation. Blackwell, Oxford

Blaber SJM, Brewer DT, Salini JP (1989) Species composition and biomasses of fishes in different habitats of a tropical northern Australian estuary: their occurrence in the adjoining sea and estuarine dependence. Estuar Coast Shelf Sci 29:509–531

Blaber SJM, Copland J (eds.) (1990) The biology of tuna baitfish. ACIAR Proc 30:211

Blaber SJM, Milton DA, Chenery SR et al (2003) New insights into the life history of *Tenualosa ilisha* and fishery implications. Am Fish Soc Symp 35:223–240

Blaber SJM, Milton DA, Rawlinson NJF (1990) Reef fish and fisheries in Solomon Islands and Maldives and their interactions with tuna baitfisheries ACIAR Proc 30:159–168

Boehlert GW (1996) Larval dispersal and survival in tropical reef fishes. In: Polunin NVC, Roberts CM (eds) Reef fisheries. Chapman and Hall, London

Cabanban AS (2000) Quantifying and showing benefits from marine protected areas for fisheries management. Report of the regional symposium on marine protected areas and their management, 1–4 November 1999, BOBP, Chennai, India. BOBP Report 86:40–48

Cappo M, Alongi DM, Williams DM et al (1998) A review and synthesis of Australian fisheries habitat research. Major threats, issues and gaps in knowledge of coastal and marine fisheries habitats – a prospectus of opportunities for the FRDC "Ecosystem protection program". Australian Institute of Marine Science, Townsville

Chapman LB, Cusack P (1988) Deep sea fisheries development project. Report on second visit to Tuvalu (30 August–7 December 1983). Unpublished report. South Pacific Commission, New Caledonia, 51 pp.

Chapman LB, Cusack P (1989) Deep sea fisheries development project. Report on fourth visit to the territory of New Caledonia at the Belep Islands (18 August–15 September 1986). Unpublished report. South Pacific Commission, New Caledonia, 30 pp.

Chapman LB, Lewis AD (1982) UNDP/MAF survey of Walu and other large coastal pelagics in Fiji waters. Unpublished report. Ministry of Agriculture and Fisheries, Suva, Fiji, 36 pp.

Cheung WL, Watson R, Morato T et al (2007) Intrinsic vulnerability in the global fish catch. Mar Ecol Prog Ser 333:1–12

Chittaro PM, Fryer BJ, Sale PF (2004) Discrimination of French grunts (*Haemulon flavolineatum*, Desmarest, 1823) from mangrove and coral reef habitats using otolith microchemistry. J Exp Mar Biol Ecol 308:169–183

Chong VC (2006) Importance of coastal habitats in sustaining the fisheries industry. In: National fisheries symposium on 'Advancing R and D towards fisheries business opportunities', 26–28 June 2006, Crown Plaza Riverside Hotel, Kuching, Sarawak, Malaysia

Chong VC, Sasekumar A, Leh MUC et al (1990) The fish and prawn communities of a Malaysian coastal mangrove system, with comparisons to adjacent mud flats and inshore waters. Estuar Coast Shelf Sci 31:703–722

Choy SK (1993) The commercial and artisanal fisheries of the Larut Matang district of Perak. In: Sasekumar A (ed) Proceedings of a workshop on mangrove fisheries and connections, August 26–30, 1991, Ipoh, Malaysia, pp 27–40. Ministry of Science, Technology & Environment, Kuala Lumpur, Malaysia

Christensen JD, Jeffrey CFG, Caldow C et al (2003) Cross-shelf habitat utilization patterns of reef fishes in southwestern Puerto Rico. Gulf Caribb Res 14:9–27

Christie P (2005) Is integrated coastal management sustainable? Ocean Coast Manage 48:208–238

Christie P, White AT (1994) Reef fish yield and reef condition for San Salvador Island, Luzon, Philippines. Asian Fish Sci 7:135–148

Coates D (1987) Observations on the biology of Tarpon, *Megalops cyprinoides* (Broussonet) (Pisces: Megalopidae), in the Sepik River, northern Papua New Guinea. Aus J Mar Freshw Res 38:529–535

Cocheret de la Morinière E, Pollux BJA, Nagelkerken I et al (2002) Post-settlement life cycle migration patterns and habitat preference of coral reef fish that use seagrass and mangrove habitats as nurseries. Estuar Coast Shelf Sci 55:309–321

Costanza R, d'Arge R, de Groot S et al (1997) The value of the world's ecosystem services and natural capital. Nature 387:253–260

Crawford B, Balgos M, Pagdilao C (2000) Community-based marine sanctuaries in the Philippines: a report on focus group discussions. Coastal management report # 2224. PCAMRD book series no. 30. Coastal Resource Center, University of Rhode Island, Narragansett, RI, USA, and Philippine Council for Aquatic and Marine Research and Development, Los Baños, Laguna, Philippines

Crona BI, Rönnbäck P (2005) Use of replanted mangroves as nursery grounds by shrimp communities in Gazi Bay, Kenya. Estuar Coast Shelf Sci 65:535–544

Dalzell P (1996) Catch rates, selectivity and yields of reef fishing. In: Polunin NVC, Roberts CM (eds) Reef fisheries. Chapman and Hall, London

Dalzell P, Wright A (1986) An assessment of the exploitation of coral reef fishery resources in Papua New Guinea. In: Maclean JL, Dizon LB, Hosillos LV (eds) The first Asian fisheries forum, Vol. 1. Asian Fisheries Society, Manila, Philippines

De Graaf GJ, Xuan TT (1998) Extensive shrimp farming, mangrove clearance and marine fisheries in the southern provinces of Vietnam. Mangroves and Saltmarshes 2:159–166

Dorenbosch M, Grol MGG, Christianen JA et al (2005) Indo-Pacific seagrass beds and mangroves contribute to fish density and diversity on adjacent coral reefs. Mar Ecol Prog Ser 302:63–76

Dorenbosch M, van Riel MC, Nagelkerken I et al (2004) The relationship of fish densities to the proximity of mangrove and fish nurseries. Estuar Coast Shelf Sci 60:37–48

Dorenbosch M, Verberk WCEP, Nagelkerken I et al (2007) Influence of habitat configuration on connectivity between fish assemblages of Caribbean seagrass beds, mangroves and coral reefs. Mar Ecol Prog Ser 334:103–116

Duke NC, Meynecke JO, Dittmann S et al (2007) A world without mangroves? Science 317:41–42

Dunn IG (1982) The Hilsa fishery of Bangladesh, 1982: an investigation of its present status with an evaluation of current data. A report prepared for the Fisheries Advisory Service, Planning, Processing and Appraisal Project, Field Document 2, pp. 1–70. FAO, Rome

Durand JR, Amon Kothias JB, Ecoutin JM et al (1978) Statistiques de pêche en Lagune Ébrié (Côte d'Ivoire): 1976 et 1977. Centre de Recherches Oceanographiques Abidjan, Documents Scientifique 67:114

Ehrhardt NM, Deleveaux VKW (2007) The Bahamas' Nassau grouper (*Epinephelus striatus*) fishery – two assessment methods applied to data – deficient coastal population. Fish Res 87: 17–27

Ehrhardt NM, Legault CM, Restrepo VR (2001) Density-dependent linkage between juveniles and recruitment for pink shrimp (*Farfantepenaeus duorarum*) in southern Florida. ICES J Mar Sci 58:1100–1105

Ferrer EM, De Cruz LP, Domingo MA (eds) (1996) Seeds of hope. College of Social Work and Community Development, University of the Philippines, Quezon City, Philippines

Garces LR, Silvestre GT, Stobutzki I et al (2006) A regional database management system – the fisheries resource information system and tools (FiRST): its design, utility and future directions. Fish Res 78:119–129

Garcia S (1975) Los recursos pesqueros regionales de Tuxpan, Veracruz a Tampico, Tamps, y su posible industrialización. Instituto Nacional de Pesca (Mexico) Boletin Informativo

Gamboa BR, Garcia AG, Benitey JA et al (1971) Estudio de las condiciones hidrográficas y quimicas en el aqua de la Laguna de Lacarigua. Boletin del Instituto Oceanografico de Venezuela, Universidad de Oriente 10:55–72

García CB, Solano OD (1995) *Tarpon atlanticus* in Colombia: a big fish in trouble. Naga, The ICLARM Q 18:47–49

Gopinath N, Gabriel P (1997) Management of living resources in the Matang Mangrove Reserve, Perak, Malaysia. Intercoastal Netw 1:23

Griffiths MH, Lamberth SJ (2002) Evaluating the marine recreational fishery in South Africa. In: Recreational fisheries: ecological, economic, and social evaluation pp 227–251. Fish Aquat Res Ser 8

Gulland JA (1971) The fish resources of the ocean. Fishing News Books Ltd., Byfleet, England

Hamilton L, Dixon J, Miller G (1989) Mangroves: an undervalued resource of the land and sea. Ocean Yearbook 8:254–288

Harden Jones FR (1994) Fisheries ecologically sustainable development: terms and concepts. IASOS, University of Tasmania, Hobart

Hernandez RRA, Calderon MG (1974) Inventario preliminar de la flora y fauna acuática de la Bahía de Jiquilisco. Ministerio de Agricultura y Granadería, Dirección General de Recursos Naturales Renovables, Servicio de Recursos Pesqueros, El Salvador

Herrera A, Charles AT (1994) Costa Rican coastlines: mangroves, reefs, fisheries and people. In: Wells PG, Ricketts PJ (eds) Coastal zone Canada - 94, cooperation in the coastal zone. Conference Proceedings, Vol 2. Coastal Zone Canada Association, Dartmouth, Canada

Hilborn R, Micheli F, De Leo GA (2006) Integrating marine protected areas with catch regulation. Can J Fish Aquat Sci 63:642–649

Islam MS, Haque M (2004) The mangrove-based coastal and nearshore fisheries of Bangladesh: ecology, exploitation and management. Rev Fish Biol Fish 14:153–180

Jennings S, Polunin NVC (1995) Comparative size and composition of yield from six Fijian reef fisheries. J Fish Biol 46:28–46

Jhingran VG (1991) Fish and fisheries of India, 3rd edn. Hindustan Publishing Corporation, New Delhi

Jhingran VG (2002) Fish and fisheries of India, 3rd edn. Hindustan Publishing Corporation, New Delhi

Jhingran VG, Gopalakrishnan V (1973) Estuarine fisheries resources of India in relation to adjacent seas. J Mar Biol Assoc India 15:323–334

Jhingran VG, Natarajan AV (1969) A study of the fisheries and fish populations of the Chilika Lake during the period 1957–1965. J Inland Fish Soc India 1:49–126

Kapetsky JM (1985) Mangroves, fisheries and aquaculture. FAO Fish Rep 338:17–36

Kathiresan K, Bingham BL (2001) Biology of mangroves and mangrove ecosystems. Adv Mar Biol 40:81–251

Kathiresan K, Moorthy P, Rajendran N (1994) Forest structure and prawn seeds in Pichavaram mangroves. Environ Ecol 12:465–468

Kathiresan K, Qasim SZ (2005) Biodiversity of mangrove ecosystems. Hindustan Publishing Corporation, New Delhi

Kathiresan K, Rajendran N (2002) Fishery resources and economic gain in three mangrove areas on the south-east coast of India. Fish Manage Ecol 9:277–283

Kaunda-Arara B, Rose GA (2004) Effects of marine reef National Parks on fishery CPUE in coastal Kenya. Biol Conserv 118:1–13

Kaunda-Arara B, Rose GA, Muchiri MS et al (2003) Long-term trends in coral reef fish yields and exploitation rates of commercial species from coastal Kenya. Western Indian Ocean J Mar Sci 2:105–116

Khoo HK (1989) The fisheries in the Matang and Merbok mangrove ecosystems. Proc 12th Ann Seminar Malaysian Soc of Mar Sci, pp. 147–169

Kyle R (1999) Gillnetting in nature reserves: a case study from the Kosi Lakes, South Africa. Biol Conserv 88:183–192

Jones RL, Kelley DW, Owen LW (1963) Delta fish and wildlife protection study. Resources and Agriculture, Sacramento, California. Report number 2, pp 73

Laroche J, Ramananarivo N (1995) A preliminary survey of the artisanal fishery on the coral reefs of the Tulear region (southwest Madagascar). Coral Reefs 14:193–200

Laserre G (1979) Bilan de la situation de peches: aux Pangalanes Est. (Zone Tamatave-Andevovanto) au Lac Anony (region Fort Dauphin). Perspective et Amenagement. Consultant's Report to MAG/76/002

Laurel BJ, Bradbury IR (2006) 'Big' concerns with high latitude marine protected areas (MPAs): trends in connectivity and MPA size. Can J Fish Aquat Sci 63:2603–2607

Lee SY (2004) Relationship between mangrove abundance and tropical prawn production: a reevaluation. Mar Biol 145:943–949

León RA, Racedo JB (1985) Composition of fish communities in the lagoon and estuarine complex of Cartagena Bay, Cienaga de Tesca and Cienaga Grande de Santa Marta, Colombian Caribbean. In: Yáñez-Arancibia A (ed) Fish community ecology in estuaries and coastal lagoons: towards an ecosystem integration. UNAM Press, Mexico

Lewis AD, Chapman LB, Sesewa A (1983) Biological notes on coastal pelagic fishes in Fiji. Ministry of Agriculture and Fisheries, Suva, Fiji, Fisheries Division Technical Report 4:1–68

Lock JM (1986) Study of the Port Moresby artisanal reef fishery. Technical Report, Fisheries Division, Department of Primary Industry, Papua New Guinea 86:1–56

Loneragan NR, Ahmad Adnan N, Connolly RM et al (2005) Prawn landings and their relationship with the extent of mangroves and shallow waters in western penin-sula Malaysia. Estuar Coast Shelf Sci 63:187–200

Lowe-McConnell RH (1975) Fish communities in tropical freshwaters. Their distribution, ecology and evolution. Longman Inc., New York, pp 337

Lo-Yat A, Meekan MG, Carleton JH et al (2006) Largescale dispersal of the larvae of nearshore and pelagic fishes in the tropical oceanic waters of French Polynesia. Mar Ecol Prog Ser 325: 195–203

Maniku H, Anderson RC, Hafiz A (1990) Tuna baitfishing in Maldives. In: Blaber SJM, Copland JW (eds) Tuna baitfish in the Indo-Pacific region. ACIAR Proc 30:22–29

Manson FJ, Loneragan NR, Harch B et al (2005) A broad-scale analysis of links between coastal fisheries production and mangrove extent: a case study for northeastern Australia. Fish Res 74:79–86

Marten GG, Polovina JJ (1982) A comparative study of fish yields from various tropical ecosystems. In: Pauly D, Murphy GI (eds) Theory and management of tropical fisheries. ICLARM/CSIRO, Manila. pp 255–285

Martosubroto P, Naamin N (1977) Relationship between tidal forests (mangroves) and commercial shrimp production in Indonesia. Mar Res Indonesia 18:81–86

McClanahan TR, Mangi S (2000) Spillover of exploitable fishes from a marine park and its effect on the adjacent fishery. Ecol Appl 10:1792–1805

McManus JW (1992) How much harvest should there be? In: Resource ecology of the Bolinao coral reef system. ICLARM studies and reviews 22:52–56

McManus JW (1994) The Spratly Islands: a marine park? Ambio 23:181–186

McManus JW, Menez LAB (1998) The proposed international Spratly Island marine park: ecological considerations. In: Lessions H (ed) Proc 8th Int Coral Reef Symp 2: 1943–1948

Meynecke J-O, Lee SY, Duke NC et al (2007) Relationship between estuarine habitats and coastal fisheries in Queensland, Australia. Bull Mar Sci 80:773–793

Mines AN, Smith IR, Pauly D (1986) An overview of the fisheries of San Miguel Bay, Philippines. In: Maclean JL, Dizon LB, Hosillos LV (eds) The first Asian fisheries forum. Asian Fisheries Society, Manila, Philippines. pp 385–388

Moberg F, Rönnbäck P (2003) Ecosystem services of the tropical seascape: interactions, substitutions and restoration. Ocean Coast Manage 46:27–46

Mohan PC, Rao RG, Dehairs F (1997) Role of Godavari mangroves (India) in the production and survival of prawn larvae. Hydrobiologia 358:317–320

Morton B, Blackmore G (2001) South China Sea. Mar Ecol Prog Ser 42:1236–1263

Mumby PJ, Edwards AJ, Arias-González JE et al (2004) Mangroves enhance the biomass of coral reef fish communities in the Caribbean. Nature 427:533–536

Munro JL (1977) Actual and potential production from the coralline shelves of the Caribbean Sea. FAO Fish Rep 200:301–321

Munro JL (1996) The scope of tropical reef fisheries and their management. In: Polunin NVC, Roberts CM (eds) Reef fisheries. Chapman and Hall, London

Naamin N, Gafa B (1998) Tuna baitfish and the pole-and-line fishery in eastern Indonesia - an overview. Indon Fish Res J 4:16–24

Nagelkerken I (2007). Are non-estuarine mangroves connected to coral reefs through fish migration? A review. Bull Mar Sci 80:595–608

Nagelkerken I, Roberts CM, van der Velde G et al (2002) How important are mangroves and seagrass beds for coral-reef fish? The nursery hypothesis tested on an island scale. Mar Ecol Prog Ser 244:299–305

Nagelkerken I, van der Velde G (2002) Do non-estuarine mangroves harbour higher densities of juvenile fish than adjacent shallow-water and coral reef habitats in Curacao (Netherlands Antilles)? Mar Ecol Prog Ser 245:191–204

Nagelkerken I, van der Velde G (2004a) Relative importance of interlinked mangroves and sea-grass beds as feeding habitats for juvenile reef fish on a Caribbean island. Mar Ecol Prog Ser 274:153–159

Nagelkerken I, van der Velde G (2004b) Are Caribbean mangroves important feeding grounds for juvenile reef fish from adjacent seagrass beds? Mar Ecol Prog Ser 274:143–151

Nagelkerken I, van der Velde G, Gorissen MW et al (2000) Importance of mangroves, seagrass beds and the shallow coral ref. as a nursery for important coral reef fishes, using a visual census technique. Estuar Coast Shelf Sci 51: 31–44

Nemoto T (1971) La pesca en el lago de Maracaibo. Projecto de investigacion y desarollo pesquero MAC-PNUD. FAO Technical Paper 24:1–56

Nickerson DJ (1999) Trade-offs of mangrove area development in the Philippines. Ecol Econ 28:279–298

Ong JE (1982) Aquaculture, forestry and conservation of Malaysian mangroves. Ambio 11: 252–257

Pandian TJ (1969) Feeding habits of the fish *Megalops cyprinoides* Broussonet, in the Cooum backwaters, Madras. J Bombay Nat Hist Soc 65:569–580

Pauly D (1976) The biology, fishery and potential for aquaculture of *Tilapia melanotheron* in a small West African lagoon. Aquaculture 7:33–49

Pauly D (1997) Small-scale fisheries in the tropics: marginality, marginalization,and some implications for fisheries management. In: Pikitich EK, Huppert DD, Sissenwine M (eds) Proc 20th Am Fish Soc Symp: Global Trends-Fisheries Management, 14–16 June 1994, Seattle, WA (USA). American Fisheries Society, Bethesda, MD, USA

Pauly D, Ingles J (1986) The relationship between shrimp yields and intertidal vegetation (man-grove) areas: a reassessment. In: IOC/FAO workshop on recruitment in tropical coastal demersal communities, Ciudad de Carmen, Mexico. IOC, UNESCO, Paris

Paw JN, Chua TE (1989) An assessment of the ecological and economic impact of mangrove conversion in Southeast Asia. Mar Pollut Bull 20:335–243

Phillips PC (1981) Diversity and fish community structure in a Central American mangrove embay-ment. Rev Biol Trop 29:227–236

Pittman SJ, McAlpine CA, Pittman KM (2004) Linking fish and prawns to their environment: a hierarchical landscape approach. Mar Ecol Prog Ser 283:233–254

Pollnac R, Crawford BR, Gorospe MLG (2001) Discovering factors that influence the success of community based marine protected areas in the Visayas, Philippines. Ocean Coast Manage 44:683–710

Primavera JH (1998) Mangroves as nurseries: shrimp populations in mangrove and non-mangrove habitats. Estuar Coast Shelf Sci 46:457–464

Raja BTA (1985) Current knowledge of the biology and fishery of Hilsa Shad, *Hilsa ilisha* (Ham. Buch.) of upper Bay of Bengal. Internal report, Bay of Bengal project document, Colombo, Sri Lanka

Rawlinson NJF, Milton DA, Blaber SJM et al (1995) The subsistence fishery of Fiji. ACIAR Monogr 35:1–138

Rawlinson NJF, Sharma SP (1993) Analysis of historical tuna baitfish catch and effort data from Fiji with an assessment of the current status of the stocks. ACIAR Proc 52:26–48

Roberts CM (1997) Connectivity and management of Caribbean coral reefs. Science 278: 1454–1457

Robertson AI, Blaber SJM (1992) Plankton, epibenthos and fish communities. In: Robertson AI, Alongi D (eds) Tropical mangrove ecosystems. Springer-Verlag, New York

Robertson J (1999) Reef closures - do they really protect reef communities? In: Proceedings of the APEC workshop on impacts of destructive fishing practices on the marine environment, 16–18 December 1997, Hong Kong, 315 pp.

Rönnbäck P (2001) Ecological economics of fisheries supported by mangrove ecosystems: economic efficiency, mangrove dependence, sustainability and benefit transfer. 2nd Western Indian Ocean Science Association Scientific Symposium – Book of abstracts, 41 pp.

Rönnbäck P, Macia A, Almqvist G, Schultz L, Troell M (2002) Do penaeid shrimps have a preference for mangrove habitats? Distribution pattern analysis on Inhaca Island, Mozambique. Estuar Coast Shelf Sci 55:427–436

Rönnbäck P, Troell M, Kautsky N et al (1999) Distribution pattern of shrimps and fish among *Avicennia* and *Rhizophora* microhabitats in the Pagbilao mangroves, Philippines. Estuar Coast Shelf Sci 48:223–234

Rueda M (2007) Evaluating the selective performance of the encircling gillnet used in tropical fisheries from Colombia. Fish Res 87:28–34

Rueda M, Defeo O (2001) Survey abundance indices in a tropical estuarine lagoon and their management implications: a spatially explicit approach. ICES J Mar Sci 58:1219–1231

Serafy JE, Faunce CH, Lorenz JJ (2003) Mangrove shoreline fishes of Biscayne Bay, Florida. Bull Mar Sci 72:161–180

Sheridan P, Hays C (2003) Are mangroves nursery habitat for transient fishes and decapods? Wetlands 23:449–458

Silvestre GT, Garces LR, Stobutzki I et al (2003) South and South-East Asian coastal fisheries: their status and directions for improved management: conference synopsis and recommendations. In: Silvestre G, Garces L, Stobutzki I (eds) Assessment, management and future directions for coastal fisheries in Asian countries. World Fish Center Conference Proceedings 67, World Fish Center, Penang

Staples DJ, Vance DJ, Heales D (1985) Habitat requirements of juvenile penaeid prawns and their relationship to offshore fisheries. In: Rothlisberg PC, Hill BJ, Staples DJ (eds) Second Australian national prawn seminar, Kooralbyn, Australia. CSIRO, Cleveland, Queensland, Australia

Stevenson DK, Marshall N (1974) Generalisations on the fisheries potential of coral reefs and adjacent shallow-water environments. Proc 2nd Int Coral Reef Symp 1:147–158

Stobutzki IC, Silvestre GT, Garces LR (2006) Key issues in coastal fisheries in South and Southeast Asia, outcomes of a regional initiative. Fish Res 78:109–118

Szelistowski WA, Garita J (1989) Mass mortality of sciaenid fishes in the Gulf of Nicoya, Costa Rica. Fish Bull US 87:363–365

Tampi PRS (1959) The ecological and fisheries characteristics of a salt water lagoon near Mandapam. J Mar Biol Assoc India 1:113–130

Taylor DS, Davis WP, Turner BJ (1995) *Rivulus marmoratus*: ecology of distributional patterns in Florida and the central Indian River Lagoon. Bull Mar Sci 57:202–207

Tiroba G (1993) Current status of commercial baitfishing in Solomon Islands. ACIAR Proc 52:113–116

Tsehaye I (2007) Monitoring fisheries in data limited situations. A case study of the artisanal reef fisheries of Eritrea. Ph.D. thesis, Wageningen University, the Netherlands, 229 pp.

Turner RE (1977) Intertidal vegetation and commercial yields of penaeid shrimp. Trans Am Fish Soc 106:411–416

Vance DJ, Haywood MDE, Heales DS et al (1996) How far do prawns and fish move into mangroves? Distribution of juvenile banana prawns *Penaeus merguiensis* and fish in a tropical mangrove forest in northern Australia. Mar Ecol Prog Ser 131:115–124

Vance DJ, Staples DJ, Kerr JD (1985) Factors affecting year-to-year variation in the catch of banana prawns (*Penaeus merguiensis*) in the Gulf of Carpentaria, Australia. J Cons Explor Mer 42:83–97

Venkatesan V (1969) A preliminary study of the estuaries and backwaters in south Arcot district, Tamil Nadu (South India). Part II: Fisheries. First All-India symposium on estuarine biology, Tambaram, Madras

Verweij MC, Nagelkerken I, de Graaf D et al (2006) Structure, food and shade attract juvenile coral reef fish to mangrove and seagrass habitats: a field experiment Mar Ecol Prog Ser 306:257–268

Yáñez-Arancibia A, Lara-Domínguez AL (1983) Dinámica ambiental de la Boca de Estero Pargo y estructura de sus comunidades de peces en cambios estacionales y ciclos 24 horas (Laguna de Términos, Sur del Golfo de Mexico). Annales del Instituto de Ciências del Mar y Limnologia del Universidad Nacional Autónoma de México 10:85–116

Wells S, Burgess N, Ngusaru A (2007) Towards the 2012 marine protected area targets in Eastern Africa. Ocean Coast Manage 50:67–83

Wheeler JFG, Ommaney FD (1953) Report on the Mauritius-Seychelles fisheries survey, 1948–49. Colonial Office Fishery Publications, HMSO, London

Wilkinson C, Caillaud A, DeVantier L (2006) Strategies to reverse the decline in valuable and diverse coral reefs, mangroves and fisheries: the bottom of the J curve in Southeast Asia. Ocean Coast Manage 49:764–778

Wright A, Richards AH (1985) A multispecies fishery associated with coral reefs in the Tigak Islands, Papua New Guinea. Asian Mar Biol 2:69–84

Yáñez-Arancibia A, Lara-Dominguez AL, Sanchez-Gil P et al (1985) Ecology and evaluation of fish community in coastal ecosystems: estuary-shelf interrelation-ships in the southern Gulf of Mexico. In: Yáñez-Arancibia A (ed) Fish community ecology in estuaries and coastal lagoons: towards an ecosystem integration. UNAM Press, Mexico

Zerbi A, Aliaume C, Miller JM (1999) A comparison between two tagging techniques with notes on juvenile tarpon ecology in Puerto Rico. Bull Mar Sci 64:9–19

Chapter 16
Conservation and Management of Tropical Coastal Ecosystems

William Gladstone

Abstract All major coastal ecosystems in the tropics are being degraded. The problems include losses of biodiversity, reduced ecosystem functions, and costs to coastal human societies. Declines in species' abundances, and habitat loss and modification are the result of the demands for aquaculture, port construction, trawling, excessive nutrient loads, overfishing and collecting, sedimentation from catchment activities, invasive species, and climate change. A global response to these changes has been conservation and management approaches that aim to reduce, reverse, and prevent unnatural changes and address their underlying causes. Successes in conservation and management are likely when actions are designed to achieve the fundamental ecological goals of ensuring resilience, maintaining ecosystem connectivity, protecting water quality, conserving species-at-risk, conserving representative samples of species and assemblages, and managing at the appropriate spatial scale. Achieving societal aspirations for coastal ecosystems requires that management approaches address the socio-economic aspects of issues and include stakeholder consultation, participation, and education. Achieving long-term success in conservation and management requires coastal nations to address fundamental issues such as lack of information for management decision-making, population growth and poverty, limited technical and management capacity, poor governance, lack of stakeholder participation, the mismatches between the issue and the geographic scale of management, lack of an ecosystem perspective, ineffective governance and management, and a lack of awareness of the effects of human activities.

Keywords Coastal zone management · Marine protected area · Socio-economics · Stakeholder participation · Sustainability

W. Gladstone (✉)
School of Environmental and Life Sciences, University of Newcastle Central Coast, P.O. Box 127, Ourimbah NSW 2258, Australia
e-mail: william.gladstone@newcastle.edu.au

I. Nagelkerken (ed.), *Ecological Connectivity among Tropical Coastal Ecosystems*,
DOI 10.1007/978-90-481-2406-0_16, © Springer Science+Business Media B.V. 2009

16.1 Introduction

Great changes have occurred in many tropical countries in recent decades and these have led to problems in their coastal and marine ecosystems and their dependent human societies. For example, in the Red Sea 'In the late 1960s, probably 98% of the total Red Sea coast was in practically virgin condition . . . ' (Ormond 1987). The rapid development that occurred in parts of the Red Sea since the 1960s (as a direct result of the expansion of petroleum-based economies) had profound consequences for its ecosystems, with the loss of this 'virgin' status in many places. Coral reefs near urban and industrial centers were degraded by land-filling and dredging, port activities, sewage, and tourism. Three-quarters of the Red Sea's mangrove stands were negatively affected by camel grazing, felling, cutting, solid wastes, sewage, burial by mobilized sand dunes, or obstruction to tidal flows. Sharks were over-fished and overfishing by industrial trawlers in the Gulf of Aden depleted cuttlefish and deep-sea lobsters (Gladstone 2008).

A global response to the problems occurring in all tropical coastal ecosystems (including the Red Sea) has been the design and development of a range of con-servation and management tools, approaches, and principles and these will be the focus of this chapter. I begin by justifying the need for conservation and manage-ment from the perspectives of the benefits human societies derive from them, and the ecological, social, and economic costs flowing from their degradation. I then review nine major goals for conservation and management. Each goal is described and jus-tified, and some practical case studies of the ways each is being implemented are provided. There is a rich vocabulary in the disciplines of coastal conservation and management (Kay and Alder 1999) but I have selected 'goals' to illustrate the point that achieving these goals will help achieve the conservation and sustainable use of coastal ecosystems. Readers interested in additional related topics (e.g., financing, legal aspects) will find many relevant references herein. I have deliberately focused on the practical ways of addressing current issues, rather than a detailed review of the issues, and readers interested in the latter can consult several excellent recent reviews (Connell 2007, Fine and Franklin 2007, Glasby and Creese 2007). Exam-ples of the practical actions that can be applied are described in case studies in boxes and many more are listed in Appendix 16.1 at the end of this chapter. The references cited in Appendix 16.1 provide the starting point for further exploration of a diverse and exciting literature.

16.2 The Values of Coastal Ecosystems

Coastal ecosystems in the tropics include coral reefs, mangroves, and seagrass. Coral reefs, described as 'the largest durable bioconstruction projects on Earth' (Knowlton and Jackson 2001), are the major centers of marine diversity. More phyla inhabit coral reefs than tropical rainforests, and coral reefs probably contain close to one million species, although only about 100,000 have been described (Harrison and Booth 2007). The presence of coral reefs influences the physical structure of the coastline and adjacent ecosystems, and they protect mangroves and seagrasses against the sea.

Seagrasses are the only marine representatives of the flowering plants and the habitats they form ('seagrass beds') contain diverse assemblages of other organisms. There are more than 70 seagrass species, with centers of diversity occurring in southwestern Australia, Southeast Asia, and Japan/Republic of Korea (Gillanders 2007). Mangrove forests are the other plant-based coastal habitat that occurs in the high intertidal areas of soft sediment shorelines. Mangroves and seagrass beds contribute to other habitats through export of detritus (see Chapter 3) and the movements of juvenile and adult organisms (see Chapters 8 and 10), and both habitats trap sediments and thereby protect coral reefs (Connolly and Lee 2007).

The conservation and management of coastal ecosystems can be justified by the need to maintain the benefits they provide to human society (Duarte 2000, Turner 2000, UNEP 2006). Ecosystem services, including provisioning, regulating, and cultural services, are the benefits humans derive from ecosystems and their supply is dependent on supporting services (Table 16.1). Provisioning services provide the products used by humans for subsistence, enjoyment, and enterprise, and include pharmaceuticals, curios, building materials, and food from fisheries and aquaculture. Regulating services include shoreline protection and stabilization from waves and storm surges (provided by coral reefs, mangroves, and seagrass), and sediment trapping and pollutant filtering (by mangroves and seagrass).

Cultural and amenity services are the non-material benefits obtained from ecosystems. These include the attributes of ecosystems that are appreciated and used for tourism, recreation, cultural, and spiritual reasons. These services also include the traditional knowledge that forms the basis of much fisheries management, tourism, alternative food sources and medicinals, education, and research (UNEP 2006). Humans use beaches, cliffs, estuaries, open coasts, and coral reefs for recreation and their aesthetic values. Coastal recreational activities such as boating, fishing, swimming, walking, beachcombing, SCUBA diving, and sunbathing produce substantial economic and social returns to coastal nations and communities. The rapid growth of coastal tourism and the associated economic and social benefits means that it is now an essential component of the economies of many small island states (Spurgeon 2006, UNEP 2006).

These ecosystem services depend on the availability of habitats and nurseries, primary productivity, and nutrient cycling. The associated benefits of habitats and nurseries include their usage by a diverse range of species and communities, support for ecologically, recreationally, and commercially significant species, and opportunities for life cycle completion (by providing pathways of connectivity between different habitats) (UNEP 2006).

16.3 Issues for Coastal Ecosystems in the Tropics

All major coastal ecosystems are experiencing degradation throughout tropical regions of the world (summarized in Table 16.2). Coral cover is a case in point. Overall, 30% of global coral reefs are already severely damaged and 60% may be

Table 16.1 The range of services provided by tropical ecosystems and examples of the benefits provided to human society from these services (X indicates the ecosystem provides a significant amount of the service) (adapted from UNEP 2006)

Ecosystem Service	Estuaries	Mangroves	Lagoons and salt ponds	Intertidal	Rock and shelf reefs	Seagrass	Coral reefs	Inner shelf
Cultural services								
Aesthetics	X		X	X			X	
Cultural and amenity	X	X	X	X	X	X	X	X
Education and research	X	X	X	X	X	X	X	X
Recreational	X	X	X	X	X		X	
Provisioning services								
Fiber, timber, fuel	X	X	X					X
Food	X	X	X	X	X	X	X	X
Medicines, other resources	X	X	X				X	
Regulating services								
Atmospheric and climate regulation	X	X	X	X	X	X	X	X
Biological regulation	X	X	X	X	X		X	
Erosion control	X	X	X			X	X	
Flood/storm protection	X	X	X	X	X	X	X	
Freshwater storage and retention	X		X					
Human disease control	X	X	X	X	X	X	X	
Hydrological balance	X		X					
Waste processing	X	X	X			X	X	
Supporting services								
Biochemical	X	X					X	
Nutrient cycling and fertility	X	X	X	X	X	X	X	X

Table 16.2 Synthesis of issues negatively affecting tropical coastal ecosystems

Issue	Mangroves	Seagrass	Intertidal rock and mud	Beaches and dunes	Coral reef	Soft bottoms
Alterations to natural hydrology	X	X				
Climate change	X	X	X	X	X	X
Collection for timber, fodder, fuel	X					
Destructive fishing practices		X			X	X
Disease	X				X	
Excessive nutrients, other pollutants, solid wastes	X	X	X	X	X	X
Grazing by domestic animals	X					
Habitat loss, modification, destruction	X	X	X	X	X	X
Invasive species	X	X	X	X	X	X
Lack of information (and assessment methods) for component biodiversity, distribution, natural patterns of variation	X	X	X	X	X	X
Limited technical and management capacity	X	X	X	X	X	X
Oil spills	X	X	X	X	X	
Overfishing/collecting	X		X		X	X
Poor governance and corrupt politics	X	X	X		X	X
Population growth, poverty	X	X	X	X	X	X
Recreation		X	X	X	X	X
Sedimentation	X	X			X	X
Tourism	X				X	
Tropical storms	X	X		X	X	
Upstream agricultural practices	X	X			X	X

Sources: Gladstone (2006), Wilkinson (2006), UNEP (2006)

lost by 2030 (Wilkinson 2006). Recovery of coral reefs will be slow or not occur at all when they experience multiple stressors (Connell 1997). There has been a region-wide decline in coral cover in the Caribbean from 50 to 10% between 1977 and 2001 (Gardner et al. 2003). This loss of Caribbean coral reefs has been greater than any time in the last 100,000 years (Precht and Aronson 2006). The Indo-Pacific region contains 75% of the world's coral reefs and has experienced substantial declines in coral cover: average coral cover was only 22.1% in 2003 and cover declined at the annual rate of 1% in the past 20 years and 2% between 1997 and 2003 (equivalent to an annual loss of 3,186 km^2) (Bruno and Selig 2007).

Habitat loss and modification are being driven by the demands for aquaculture, port construction, trawling, road construction, and the building industry (UNEP 2006). Approximately 75% of sheltered tropical coasts worldwide were once occupied by mangroves, but this figure is nowadays probably closer to 25% (Dahdouh-Guebas 2002). The use of mangroves and seagrass as nursery habitats by many coastal species, including commercially important species, highlights the more widespread costs that are felt from loss of these habitats.

Invasive species are likely to be an increasing cause of change in coastal ecosystems (UNEP 2006). Invasive species influence fisheries, local ecological interaction, and coastal infrastructure, and their effects will be difficult to reverse. The major route of transfer of invasive species is in ship's ballast water. Ships began using water to control their draught, trim, and heel in the last nineteenth century in place of solid materials. However, it is only in recent years with the advent of larger tankers traveling at faster speeds that the chance of successful transfer of organisms around the world increased substantially. Currently, global shipping annually transfers 12 billion tonnes of ballast water around the world (Facey 2006).

Climate change will be one of the dominant causes of change in coastal ecosystems, especially mangroves, coral reefs, and beaches, through its potential influence on sea level, storm frequency, sea temperatures, and oceanographic processes such as upwellings and surface currents. Changes arising from climate change will be difficult to reverse and are likely to manifest as coral bleaching, coastal erosion, alterations in plankton delivery to coastal zones, and altered calcification processes arising from changes in ocean chemistry (Fine and Franklin 2007).

The food delivered by fisheries is one of the most important services derived from coastal ecosystems (see Chapter 15), e.g., fisheries based around coral reefs in developing countries provide food to about 1 billion people in Asia. After a period of intense growth in catch beginning in the mid-twentieth century, catches began to stagnate and decline at the end of the 1980s due to overfishing (UNEP 2006). The percentage of under-exploited stocks has declined and the percentage of stocks exploited at or beyond their maximum sustainable yield has increased. At the same time increases in per capita consumption of fish stimulated the rapid growth of aquaculture to fill the gap between production and demand, and aquaculture is the fastest growing primary industry globally. Many wild capture fisheries and aquaculture practices are leading to: physical damage to habitats and associated changes in community structure (e.g., from trawling) or complete habitat loss (e.g., due to conversion from mangroves to aquaculture), pollution, over-exploitation of species for fish meal, changes in trophic ecology manifested as reduced numbers of top predators ('fishing down the food web'), effects on by-catch species (especially turtles, seabirds, sharks), and the spread of infectious diseases (UNEP 2006).

Major losses of seagrass have occurred in Florida and Australia and degradation is expected to accelerate in Caribbean and Southeast Asia (UNEP 2006). The major causes of seagrass loss are nutrient loading, sedimentation, dredging, and loss from algae farming. Coral reefs are highly degraded throughout the world: 20% are severely damaged and unlikely to recover, with the areas of most concern being the Caribbean and Southeast Asia (UNEP 2006). Major activities degrading coral reefs include: destructive fishing, collection for construction, overfishing, nutrient loading, bleaching, and sedimentation from catchment activities.

Underlying causes of many issues for coastal ecosystems (Table 16.2) include lack of information for management decision-making, population growth and poverty, limited technical and management capacity, poor governance and corruption, lack of institutional collaboration, a focus on solving single issues, lack of stakeholder participation, mismatch between the issue and the geographic scale of management, lack of an ecosystem perspective, ineffective governance and management, and a lack of awareness of the consequences of human activities (Duda and Sherman 2002). The remainder of this chapter considers goals for conservation and management of coastal ecosystems and the practical steps needed to address the immediate and underlying causes of issues.

16.4 Goals for Conservation and Management of Tropical Ecosystems

The protection of coastal tropical ecosystems and the maintenance of ecosystem services is a highly desirable though complex aim. Conservation and management are more likely to succeed when they are planned with reference to goals or guiding principles that are based on ecological and socio-economic understanding. The remainder of this chapter is a synthesis of nine goals for conservation and management. These goals acknowledge that successful conservation and management requires consideration of species and ecosystems and the people who use and manage them. Five goals relate to the need to conserve biodiversity and associated ecological processes at the scale of whole ecosystems and include: maintenance of resilience, connectivity, and water quality, the recovery of species at-risk of extinction, and conservation of representative samples of biodiversity. Four goals relate to the people and institutions who use and manage coastal ecosystems and include: understanding of the socio-economic context, stakeholder participation, education (which includes capacity building), and management at the appropriate spatial scale. Each goal is supported by examples of the management actions and interventions and many of these (e.g., establishment and management of marine reserves, environmental assessment) are relevant to several goals, which reinforces their general power for conservation and management. Appendix 16.1 is an overview of the practical actions that can be utilized to achieve each goal.

16.4.1 Providing for Resilience

Tropical ecosystems are affected by anthropogenic and natural disturbances such as storms, coral bleaching, crown-of-thorns starfish, invasive species, shipwrecks, pollution events, disease, and fishing. Resilience is the ability of an ecosystem to recover from a disturbance and maintain its production of goods and services (Carpenter et al. 2001). A large number of coral reefs were affected by the 1998 bleaching event and the resilience of reefs to continued bleaching events is a major concern. Resilience requires ecosystems to possess biological and functional diversity including herbivores (especially grazing parrotfish and sea urchins; Mumby et al. 2006, 2007), mobile species that move between ecosystems (such as fishes moving between mangroves, seagrass, and coral reefs), a reef framework consisting of scleractinian corals and coralline algae, predators (that maintain a high diversity of herbivores and control bioeroders), corallivores, and settlement facilitators (such as bacteria, diatoms, coralline algae) (Nyström and Folke 2001, Grimsditch and Salm 2005). An ecosystem's resilience will be facilitated by its connections with source areas that provide large numbers of recruits that maintain populations in sink areas. Resilience will be naturally greater in dense reef networks where individual reefs are highly connected but resilience is likely to be less for isolated reefs (Roberts et al. 2006). Appropriate environmental conditions for successful recruitment are required and these may relate to water quality, light availability, limited sedimentation, and availability of suitable substratum (Grimsditch and Salm 2005).

Resilience can be maintained by a range of management actions (Appendix 16.1). Key functional groups can be conserved through: fisheries management, species-specific action plans (Gladstone 2006), protection of spawning aggregation sites (Gladstone 1986, 1996), and Marine Protected Areas (MPAs). Fisheries management (e.g., banning fish traps) can maintain both functional diversity and abundant populations (Mumby et al. 2007). Population rehabilitation (e.g., via transplantation of urchins) may be necessary to return the resilience of specific sites (Jaap et al. 2006). Populations of targeted species recover in no-take MPAs (Edgar et al. 2007). The grazing intensity of parrotfish in MPAs can be double that occurring in non-reserve areas (Mumby et al. 2006) and is associated with significant increases in the density of coral recruits (Mumby et al. 2007). Coral reefs within MPAs are more resilient to a major natural disturbance and the effects of increasing human usage (see Box 16.1). Populations of a diverse range of species that are protected within MPAs act as 'source' areas by producing large numbers of genetically diverse propagules that will be available for settlement in downstream 'sink' areas. Additional protection of highly important source areas (e.g., spawning aggregation sites) is likely to be necessary because many have been decimated by targeted fishing (Sadovy 1993).

Box 16.1 Maintaining resilience through MPAs

MPAs are one of several management tools that are necessary to maintain the resilience of coral reef ecosystems. The result of unmanaged use is clearly illustrated by the change in Jamaican coral reefs. The resilience of Jamaica's coral reefs to disturbance had been compromised by the loss (through over-fishing) of the predators (triggerfish) and competitors (parrotfish) of the grazing sea urchin *Diadema antillarum* (which controls growth of macroalgae and therefore facilitates coral recruitment and growth). Grazing by the sea urchin was the main control of the growth of algae and necessary to the recovery of Jamaica's coral reefs from the devastating loss of coral caused by Hurricane Allen in 1981. However, pathogen-induced mortality of *D. antillarum* in 1983–1984 led to an explosion in growth of algae and a phase shift of the entire ecosystem from being coral-dominated to an algae-dominated system. Recent surveys indicate recovery of urchin populations is occurring in some areas of the Caribbean with associated increases in coral cover. However, the change in Jamaica's reefs had a significant effect on the local economy. Two MPA experiences illustrate the alternative scenarios that may arise when reefs are managed to maintain resilience. The Bahamas' Exuma Cays Land and Sea Park (ECLSP) has been protected from fishing since 1986 and this has resulted in an increased survival of large-bodied parrotfish (despite the increased density of parrotfish predators). As a result the grazing intensity by parrotfish in the ECSLP is double that of non-reserve areas, which has led to a fourfold decrease in cover of macroalgae and a twofold increase in density of coral recruits. In contrast to the regional-wide decline in coral cover that followed the mass mortality of *D. antillarum*, the reefs of Bonaire did not experience overgrowth of macroalgae and no decline in coral cover. Spearfishing was banned on the reefs of Bonaire in 1971 and the Bonaire Marine Park was established in 1979. The lack of an effect from the loss of sea urchins in Bonaire is attributed to the abundant grazing fish that remained there.
Sources: Hughes (1994), Carpenter and Edmunds (2006), Mumby and Harborne (2006), UNEP (2006), Mumby et al. (2007)

On a larger scale, Integrated Coastal Management (ICM) that includes spatially coordinated protection of connected ecosystems (such as seagrass, mangroves, coral reefs) will sustain adult populations of important functional groups. ICM also provides for management of human activities in associated terrestrial ecosystems (such as catchments) to limit changes in water quality and thereby maintain the environmental conditions required for resilience, e.g., suitable water quality for coral settlement and survival (McCook et al. 2001).

16.4.2 Maintain/Restore Connectivity

Connectivity is the linkage of spatially disjunct populations and systems via dispersal of eggs and larvae, the movements of juvenile and adult organisms, and the passage of water masses. Tropical ecosystems are connected at a range of spatial and temporal scales:

(1) across environments, e.g., via the flow of water and its constituents from catchments to estuaries and then to coral reefs (Torres et al. 2001; see Chapter 2),
(2) across ecosystems, e.g., seagrass, mangroves, coral reefs by the ontogenetic and diurnal migrations of fishes (Ogden and Ehrlich 1977, Mumby et al. 2004, Mumby and Harborne 2006; see Chapters 8 and 10),
(3) between examples of a single system, e.g., between coral reefs by between-reef movement of larvae or adult fishes migrating to spawning aggregation sites (see Chapter 4), and
(4) within a single habitat, e.g., return of larval fishes to their natal reef, the diurnal movements of fishes between reef habitats, or the ontogenetic movements by coral reef fishes among reef habitats (Nagelkerken et al. 2000).

The ecological processes that are supported by connectivity include population replenishment, primary productivity (Meyer and Schultz 1985, Ogden 1997), and habitat formation (Bellwood 1995). Mixing of freshwater runoff and coastal waters adjacent to rivers and estuaries creates a different environment that is occupied by distinct species assemblages (Veron 1995). The connectivity between catchments and coasts that creates these unique coastal environments thereby supports the great biological diversity of tropical coasts.

Ecosystems are resilient when they remain connected to sources of replenishment. Conversely, resilience may be diminished by population declines in source areas (Roberts et al. 2006) and loss or degradation of the habitats required by different ontogenetic stages (Mumby et al. 2004). Management actions to maintain/restore connectivity include the protection or rebuilding of viable populations in areas that are well-connected to downstream areas, the protection of corridors of connected habitats (such as mangroves, seagrass, and coral reefs), and the rehabilitation of degraded habitats (Appendix 16.1). For example, reductions in populations of the rainbow parrotfish (*Scarus guacamaia*) in the western Caribbean are related to loss of nursery habitat (mangroves) and overfishing. However, despite fishing restrictions recovery is non-existent in areas where mangroves are absent (Mumby et al. 2004). *S. guacamaia* is listed as vulnerable on the IUCN Red List. The size, location and number of MPAs needed to maintain connectivity will vary with the density of habitats, the reproductive strategy and habitat requirements of the species of concern, the degree of self-replenishment, and the risks of future loss (Roberts et al. 2006).

A negative consequence of connectivity between terrestrial and coastal ecosystems is the degradation of coastal ecosystems from unmanaged land uses. For example, substantial areas of seagrass have been lost due to declines in water quality associated with poor land use practices in catchments. The unique nearshore coral assemblages adjacent to the Great Barrier Reef have been degraded by declines in water quality arising from extensive land clearing for agriculture in catchments (Furnas 2003). Maintenance of connectivity at this landscape–seascape scale requires integrated actions that address land use in catchments and human uses of each connected habitat (Appendix 16.1).

16.4.3 Protect Water Quality

Nutrient and sediment loads to the coastal zone increase following catchment alterations for agriculture and grazing, urbanization, and industrialization. Increases in these loads can have extreme effects such as the creation of coastal 'dead zones' or zones of hypoxia (Joyce 2000). The effects of elevated nutrients and sediment loads on coastal ecosystems will depend on input levels, historical ambient loads, the natural dispersal processes, and the extent of other simultaneous stresses (Furnas 2003). There is still considerable debate about the relative importance of declining water quality or reductions in herbivores as the cause of change of many coral reef systems from coral to algae-dominated (Precht and Aronson 2006).

Of longer-term significance for coastal ecosystems may be the additional reduction in resilience to natural and anthropogenic disturbances caused by declines in water quality. The most extreme examples of the ecosystem-wide effects of eutrophication have occurred in semi-enclosed bays following point-source discharges of sewage, e.g., Kaneohe Bay, Hawaii (Grigg 1995) or urban-industrial effluent, e.g., Barbados (Tomascik 1990). Excessive nutrients released into Kaneohe Bay caused persistent plankton blooms. Corals suffered extensive mortality due to freshwater runoff and sedimentation. Proliferation of filter-feeders (which fed on the plankton) and macroalgae on the dead coral substratum inhibited coral settlement and prevented reef recovery. Reefs became more unstable for settlement due to crumbling caused by boring organisms. Major improvements in the coral reef ecosystems of Kaneohe Bay followed infrastructure developments including the redirection of sewage offshore (Appendix 16.1). Seagrass beds adjacent to developed catchments are likely to be influenced by increased sediments, nutrients, and the addition of herbicides. Particular concerns relate to the transfer of land-derived herbicides from seagrass to herbivores such as dugong (Furnas 2003), and overgrowth of corals (Miller and Sluka 1999).

Larger-scale effects may follow degradation of adjacent terrestrial systems. On the mainland adjacent to the Great Barrier Reef, land-use practices have increased the quantity of sediment and nutrients in run-off seven-fold since 1850. Although seemingly an enormous increase, it has been difficult to link directly these increases with the degradation that has been observed in some coastal and island fringing coral reefs (in part because of the lack of long-term monitoring). However, most of the disturbed reefs are located adjacent to catchments where there have been significant amounts of land clearing and fertilizer usage. In addition, given the long time periods required for degraded ecosystems to recover and the likelihood of additional anthropogenic and natural stresses compromising this recovery, a precautionary approach to management that includes changing current land-use practices is recommended (Appendix 16.1) (Furnas 2003). For example, the goal of the Reef Water Quality Protection Plan is halting and reversing the decline in water quality entering the Great Barrier Reef within 10 yrs. The two objectives to achieve this goal are to (i) reduce the load of pollutants from diffuse sources in the water entering the Reef, and (ii) rehabilitate and conserve areas of the Reef catchment that have a role in removing water borne pollutants. Some of the practical steps that are being implemented to achieve these objectives include: self management approaches, public education, economic incentives, planning for natural resource management and land use, regulatory frameworks, research and information sharing, government-private partnerships, the setting of priorities and targets, and monitoring and evaluation (The State of Queensland and Commonwealth of Australia 2003). Additional practical steps are provided in Appendix 16.1.

16.4.4 Conservation and Recovery of Species-at-Risk

Some species are especially vulnerable to over-exploitation and habitat loss due to features of their life history (such as slow growth, late maturity, low fecundity), specialized habitat requirements, restricted breeding season or their habit of aggregating in a limited number of localized areas at predictable times to reproduce (Dulvy et al. 2003, Claydon 2004; Box 16.2). The current status of some of these species is that 37% of sharks, rays, and chimaeras are threatened/near-threatened, three species of sea turtles are critically endangered and three are endangered (from a total of seven species), and shorebirds are declining globally (UNEP 2006). The IUCN's Red List of Threatened Species includes 1,530 marine species of which 80 are threatened with extinction and 31 have a high risk of extinction. A particular concern is the rate of new additions to the list of threatened marine species. Fish species that have declined recently include the giant humphead wrasse *Cheilinus undulatus* (Sadovy et al. 2003), humphead parrotfish *Bolbometopon muricatum* (Donaldson and Dulvy 2004), and the Banggai cardinalfish *Pterapogon kauderni* (Allen 2000; Fig.16.1).

Box 16.2 Conservation and recovery of species at-risk: the Banggai cardinalfish

The Banggai cardinalfish *Pterapogon kauderni* is naturally vulnerable because it is endemic to the Banggai Islands, in central-eastern Sulawesi, Indonesia, over an estimated area of 34 km^2. Like other species in the family Apogonidae males of the Banggai cardinalfish incubate the fertilized eggs within their mouth; however, it is unique in the very small number of eggs (12–40) which are large (2.5–3.0 mm) and lack a pelagic larval phase following hatching. Males incubate the eggs for 2–3 weeks and continue to brood the newly hatched juveniles within their mouth for another 6–10 days. Emergent juveniles are independent, but mortality is high. Due to the lack of a pelagic dispersive phase there is no prospect of recovery of locally depleted populations by recruitment from outside sources. The species is highly prized in the aquarium trade because of its beautiful appearance, unique biology, and ease of capture, and large numbers (700,000–900,000) are collected annually. The population declined by 89% between 1995 and 2007 following the start of the aquarium fishery. Further problems include habitat destruction from dynamite fishing and net damage to corals. *P. kauderni* was listed on the IUCN Red List as endangered in 2007. The most promising conservation measure would appear to be replacement of the wild capture industry by captive breeding, including at the community-level; however, there has been little take up of this so far.

Sources: Allen (2000), IUCN (2007)

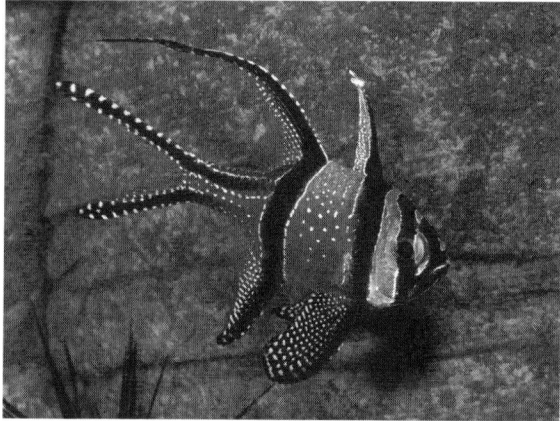

Fig. 16.1 The Banggai cardinalfish (*Pterapogon kauderni*), listed as endangered on the IUCN Red List in 2007 (photo: David Harasti)

Conservation measures are designed to prevent or arrest declines and facilitate recovery of depleted populations. The necessary practical steps include the development of recovery plans, critical habitat protection, captive breeding, trade restrictions, provision of alternative livelihoods for coastal communities that utilize these species, national legislation, international treaties, and community education (Appendix 16.1).

16.4.5 Conservation of Representative Samples of Species and Assemblages

The global biodiversity crisis stresses the need for samples of the variety of species and assemblages to be conserved in perpetuity so that future generations can share the same experiences as us and to fulfill human society's moral responsibilities towards biodiversity. A further rationale is that different ecosystems have different functional values (Mumby and Harborne 2006) and therefore conservation of representative examples of each will ensure maintenance of a suite of ecological functions and processes. Properly managed MPAs are the most appropriate practical action tool to achieve this aim (Appendix 16.1) and they can vary from large multiple-use MPAs (within which areas are zoned for different levels of use with no-take reserves buffered by a zone of less restrictive usage) to networks of smaller no-take marine reserves.

Selection of candidate protected areas requires the clear delineation of a set of MPA selection criteria that fulfill a society's vision for biodiversity conservation. Selection of MPAs within Australia, for example, is guided by the criteria of comprehensiveness, adequacy, and representativeness (Australian and New Zealand Environment and Conservation Council Task Force on Marine Protected Areas 1999). When these criteria are combined with criteria for connectivity, population replenishment, and resilience, conservation planning can achieve multiple objectives for biodiversity conservation and maintenance of ecological functions (see Box 16.3). Deficiencies resulting from a history of *ad hoc* selection of MPAs (Pressey and McNeill 1996) are nowadays addressed by the use of automated and objective reserve selection software (Possingham et al. 2000). Reserve selection programs aim to achieve the selection criteria for the minimum cost and select sites that are complementary (Box 16.3). Conservation planning to represent samples of the variety of biodiversity should ideally be based on accurate spatial data, such as maps for the planning area of the distribution boundaries of species, assemblages, and habitats, as well as ecological understanding of the factors and process (e.g., depth, wave exposure, oceanography) that underlie variation in species and assemblages. However, these data are rarely available because of patchiness in sampling records, access problems, uncertain taxonomy, financial constraints, and limited research (Gladstone 2007). Surrogates are a potential solution to this issue when they can be shown to represent other unmeasured species and assemblages (Gladstone and Owen 2007), and their distribution is already mapped within the planning area or data on their distribution is more easily and cheaply obtained (Appendix 16.1). Recent advances in remote sensing and habitat mapping show great promise for economically and rapidly providing spatial data suitable for conservation planning (Mumby and Harborne 2006). The selection of MPAs must also include socio-economic considerations and these are discussed in the following section on socio-economic assessment.

Box 16.3 Approaches to conservation planning for MPAs

The Seaflower Biosphere Reserve (San Andrés Archipelago, Colombia) was declared a UNESCO international biosphere reserve in recognition of its great significance in the Caribbean for its biodiversity and endemism, and covers an area of 255 km^2. Along with other island groups in the Archipelago its biodiversity values are to be conserved by establishing a multiple-use MPA that includes no-take reserves. The process of designating the boundaries of potential reserves began with the confirmation (via extensive field surveys) that habitats classified *a priori* represented distinct assemblages of species. General criteria for designating reserve boundaries included the requirement that individual reserves should cover at least 10 km^2 to ensure population viability, be placed on every coastal shelf and include representatives of each habitat present, and have straight line boundaries to facilitate compliance enforcement in the field. Additional specific criteria included the need to include within reserves spawning aggregation sites, rare and ecologically significant habitats (e.g., mangroves, seagrass), and corridors of ecologically connected habitats (seagrass, mangroves, coral reef). Stakeholders (local fishers) were consulted about these criteria and provided their own preferences for reserve boundaries. The reserve boundaries nominated by the fishers covered 27–32% of the area, the scientist's boundaries (based on the above general and specific criteria) covered 38–41% of the area (with an average of 30% coverage of each habitat type; Friedlander et al. 2003). The Seaflower MPA was declared in 2005.

Working at a much larger spatial scale, the Great Barrier Reef Marine Park covers 344,400 km^2 (85% of the area of California) and is a World Heritage site. Increasing pressure from a range of different uses, and a recognition that the levels of protection afforded to the Park's biodiversity in no-take areas was inadequate (only 4.5% of the Park was no-take and 80% of this was coral reefs) led to a re-zoning process called the Representative Areas program. The planning units were 70 bioregions, and decisions on candidate locations for no-take areas were guided by scientific operational principles (e.g., no-take areas should have a minimum length of 20 km to maintain population viability) and social, cultural, economic, and management feasibility operational principles. The process involved the identification of alternative sets of no-take areas that achieved the biological objectives (determined by reserve selection software) and the integration of the social, cultural, economic, and management factors (based on a high degree of stakeholder consultation). The final outcome was a re-zoning with more than 33% of the Park's area designated as no-take reserves which represented a five-fold increase in the total global area of no-take reserves (Fernandes et al. 2005).

16.4.6 Understanding the Socio-Economic Context

Twelve percent of the total global population (equivalent to 31% of the global coastal population) lives within 50 km of a coral reef (UNEP 2006). People have favored coastal locations for settlement because, among other benefits, these areas tend to contain the greatest biological productivity. Sixty-five percent of cities with populations above 2.5 million inhabitants are located along the world's coasts. One billion people depend on fish catches from shallow coastal waters dominated by coral reefs (Whittingham et al. 2003). Coastal populations in many countries are growing at double the national rate (Turner et al. 1996). Small island states typically have experienced population growth rates of around 3% per annum, although the rate of emigration is also high in some cases. Many islands are also densely populated with the capital island of the Maldives, Malé, providing an extreme example. It is home to 56,000 people despite being only 1,700 m long and 700 m wide (Pernetta 1992).

Economic valuation of ecosystem services quantifies their contribution to human welfare and provides further support for conservation and management (Costanza et al. 1997, Costanza 1999, Balmford et al. 2002). Methods of economic valuation have been reviewed (Ahmed 2004) and the estimates of economic benefits are impressive. Earth's oceans contribute about US$21 trillion per year to human societies from their provision of food, materials, and services (e.g., atmospheric gas and climate regulation, cycling of water, nutrients, and wastes; Costanza 1999). The net economic benefits of coral reefs are estimated to be US$30 billion per annum, including US$100 million annually from recreational fisheries (UNEP 2006). Fisheries on small island states in the Caribbean provide full-time and part-time direct employment for more than 200,000 people and indirect employment for an additional 100,000 (UNEP 2006). A synthesis of recent economic analyses of the value of tropical coastal ecosystems is provided in Table 16.3.

The human costs of ecosystem degradation can be measured in terms of loss of revenue, opportunities, and social costs such as reduced income and loss of a preferred lifestyle (Table 16.4). The costs of ecosystem degradation are experienced more deeply in the coastal communities of developing countries, which have a greater dependence on coastal ecosystems (Turner et al. 1996, Dahdouh-Guebas 2002). Declines in the production values of ecosystems will lead to social and economic hardships, loss of tourism potential because of declining attractiveness, and loss of option values such as the potential for pharmaceutically active compounds or future tourism ventures (Bruno and Selig 2007). The reliance of human society on coastal ecosystems means that conservation has to be balanced with sustainable use.

An understanding of the importance of cultural factors will increase the likelihood of success of conservation and management. Cultural significance can relate both to places and activities. Culturally significant places are areas that are important to a community because of some attribute of the natural environment or its association with a spiritual activity. MPAs have been used to protect culturally significant sites (Kelleher and Kenchington 1992, Gladstone 2000, Salm et al. 2000). For example, in the Farasan Islands (Red Sea) the local community organizes an annual festival to coincide with the mass spawning of the parrotfish *Hipposcarus harid* in a single bay (Fig. 16.2) (Gladstone 1996). The bay was given the highest level of protection in the zoning scheme for the multiple use Farasan Islands MPA to simultaneously protect the spawning ground and thereby ensure the sustainability

Table 16.3 Summary of economic values derived from coastal and marine tropical ecosystems (all values are in US$)

Ecosystem	Economic value	Basis of economic value	Source
Coral reef	$30 billion per year or $100,000–$600,000 per km^2	Net potential benefits in goods and services, including tourism, fisheries, and coastal protection worldwide	Cesar et al. (2003)
	$1 billion per year (1991/92)	Great Barrier Reef, Australia	Driml (1994)
	$128 million (1991/92)	Commercial fishing sales from the Great Barrier Reef	Driml (1994)
	$94 million (1991/92)	Expenditure on recreational fishing and boating on the Great Barrier Reef	Driml (1994)
	$628 million (1991/1992)	Expenditure by tourists on accommodation and commercial passenger vessels on the Great Barrier Reef	Driml (1994)
	$3.1 billion–$4.6 billion per year (2000)	Fisheries, dive tourism, and shoreline protection services of Caribbean coral reefs	Burke and Maidens (2004)
	$310 million per year (2000)	Coral reef-associated fisheries in the Caribbean region	Burke and Maidens (2004)
	$2.1 billion per year (2000)	Net benefits from dive tourism in the Caribbean region	Burke and Maidens (2004)
	$700 million–$2.2 billion per year	Shoreline protection services provided by Caribbean reefs	Burke and Maidens (2004)
	$2.4 billion per year	Value of coral-reef based fisheries in Southeast Asia	UNEP (2006)
	$1.2 billion per year	Reef-based tourism in Florida Keys, USA	UNEP (2006)
	>$1.0 billion per year	Direct revenue from 1.6 million visitors to Great Barrier Reef	UNEP (2006)
	$300,000–$35 million per year	Recreational value per reef in marine management areas in Hawaiian Islands	UNEP (2006)

Table 16.3 (continued)

Ecosystem	Economic value	Basis of economic value	Source
Mangrove	$9,990 per ha per year	Disturbance regulation, waste treatment, habitat, food production, raw materials, recreation	Costanza et al. (1997)
	$1,500 per km^2	Potential net benefit for medicinal plants	Ruitenbeek (1994)
	$30,000 per km^2 per year (totaling $10 million per year)	Forestry products (timber and charcoal) of the Matang mangroves in Malaysia	Talbot and Wilkinson (2001)
	$750–$16,750 per ha	Seafood supported by mangroves	UNEP (2006)
	$600 per ha	Fisheries yields adjacent to mangroves	UNEP (2006)
	$15 and $61 per ha	Medicinal plants and medicinal values, respectively	UNEP (2006)
	$6,200 per km^2 in the United States to $60,000 per km^2 in Indonesia per year	Annual commercial fish harvests from mangroves	UNEP-WCMC (2006)
	$2.7 million–$3.5 million per km^2	Mangroves in Thailand	UNEP-WCMC (2006)
Seagrass	$19,004 per ha per year	Nutrient cycling raw materials	Costanza et al. (1997)

Table 16.4 Economic losses from issues negatively affecting tropical coastal ecosystems (all values are in US$)

Issue	Economic loss	Source
Coral degradation and death could lead to loss of shoreline protection services in the Caribbean region	Totaling $140–$420 million annually (within the next 50 years)	Burke and Maidens (2004)
Cost of replacing the coastal protection provided by now degraded reefs Sri Lanka	$246,000–836,000 per km	Berg et al. (1998)
Decrease in tourism-generated income, employment, fish productivity, and shoreline protection in Indian Ocean resulting from declines in coral reef quality from 1998 bleaching event	$608 million–$8 billion (over 20 years)	UNEP (2006)
Degradation of Caribbean coral reefs could reduce net annual revenues from coral reef-associated fisheries	Estimated $95–$140 million per year by 2015	Burke and Maidens (2004)
Degraded Great Barrier Reef in Australia as the result of the predicted effects of global warming	$2.5 billion–US$6 billion over 19 years	Hoegh-Guldberg and Hoegh-Guldberg (2004)
Economic loss associated with coral bleaching over a 50 years time horizon with a 3% discount rate	$28.4 billion in Australia and $38.3 billion in Southeast Asia	Cesar et al. (2003)
Increased sea-surface temperatures, sea level rise, and loss of species due to climate change in the Caribbean	$109.9 million	Cesar et al. (2003)
Net loss after 20 years of blast fishing of coral reefs in Indonesia	$300,000 per km^2 in areas with a high potential value of tourism and coastal protection, and $33,900 per km^2 where there is low potential value	Pet-Soede et al. (1999)
Predicted net economic loss from blast fishing, overfishing, and sedimentation	$2.6 billion for Indonesia and $2.5 billion for the Philippines over a 20 years period	Burke et al. (2002)
Reef degradation that is projected to occur in the Caribbean by 2015 may reduce fisheries production by 30–45% resulting in revenue reduction from $310 million to only $140 million	$170 million in the year 2015	Burke and Maidens (2004)
Restoring a 250 meter-long beach following erosion as a result of offshore coral mining in Indonesia	$125,000 annually (over 7 years)	UNEP (2006)

Fig. 16.2 The festival of the harid parrotfish is a culturally significant event in the Farasan Islands, Saudi Arabian Red Sea, which has been incorporated into a management plan for the Farasan Islands Marine Protected Area. (**a**) Locals capturing spawning harid parrotfish, (**b**) Lower guards of the local *emir* maintain orderly conduct during the collection of the parrotfish (photos: William Gladstone)

of the cultural festival (Gladstone 2000). As a culturally significant activity, fishing fulfils many needs in fishers that are unrelated to economic returns. This can make it difficult to implement alternative livelihood schemes, even when catches are declining (Pollnac et al. 2001, Momtaz and Gladstone 2008). Understanding the personal significance of an activity such as fishing will increase the likelihood of more acceptable alternatives being developed.

Coastal communities are not homogeneous entities, consisting of groups of individuals who differ in the ways they perceive and use their environment. Perceptions and uses, in turn, depend upon a host of social, cultural, and economic factors such as age, occupation, income, ethnicity, gender, level of education, and migration status (Cinner and Pollnac 2004). Changing people's behaviors, as the fundamental means of addressing conservation and management problems, is therefore a complex undertaking. A socio-economic assessment provides the framework for comprehending the socio-economic context in which management and conservation have to operate and demonstrates the underlying causes (e.g., poverty, lack of education) of many issues for coastal ecosystems. Management can then be directed at addressing both the underlying causes of issues (which is likely to be a long-term undertaking) and the immediate effects. A socio-economic assessment covers the social, cultural, economic, and political conditions of stakeholders (Bunce et al. 2000, Browman and Stergiou 2005). Specific areas that may be assessed include stakeholder characteristics, resource use patterns, gender issues, stakeholder perceptions of problems and management, organization and resource governance, traditional knowledge, community services and facilities, the local business environment, the incomes of stakeholders, and the economic values of resources (Bunce et al. 2000). Appendix 16.1 provides specific examples for each step of the socio-economic assessment and Box 16.4 provides three case studies that illustrate the ways in which socio-economic understanding has been used to develop conservation and management actions.

Box 16.4 Case studies in the assessment of the socio-economic context for management

Case study 1: Great Barrier Reef Marine Park Representative Areas Program (GBRMPRAP)

The aim of the GBRMPRAP was to comprehensively conserve examples of the Marine Park's biodiversity, which was likely to lead to a substantial increase in the number and total area of no-take areas. There was a high risk of conflict with the Park's existing users: tourism, commercial fishing, and cultural and recreational activities employ 44,000 people and contribute A$3.7 billion annually to the economy of the Park and its catchment (Access Economics 2007). Managers of the GBRMPRAP established a social, economic, cultural steering committee (comprised of representatives of management and stakeholder groups) that developed social, cultural, economic, and management feasibility operational principles that would guide decision-making about the location of no-take areas in partnership with bio-physical operational principles. A key operational principle was to 'maximize complementarity of no-take areas with human values, activities and opportunities' by placing no-take areas where conflicts with indigenous users' aspirations, non-commercial and commercial extractive users, and all non-extractive users, would be minimized. Federal government financial support for displaced fishers enhanced the community's acceptance, and reduced the economic costs, of the greatly expanded network of no-take areas (Fernandes et al. 2005).

Case study 2: The economics of blast fishing

Declines in fish catches, ease of use, and demands of creditors, forced many fishers into blast fishing in Indonesia. Blast fishing targets schooling reef fishes but also kills other fishes and invertebrates that are not collected and damages reef habitat. The latter has opportunity costs such as the foregone benefits of tourism. A lack of political will (arising from lack of awareness of the economic costs of blast fishing) is the main reason for the lack of enforcement of this illegal activity. Pet-Soede et al. (1999) quantified the economic costs of blast fishing in Spermonde Archipelago, Southwest Sulawesi, Indonesia, from observations at sea (numbers of bombs, fish catch biomass), interviews with fishers and middlemen (for data on number of trips, costs, and profits), and from the logbook records of fishers of their daily catches. The authors estimated the blast impacts on corals from surveys done while diving. The projected cost of 20 years of blast fishing in areas of high value coral reef was a net loss of US$306,800 per km^2 through loss of coastal protection and foregone benefits of tourism and non-destructive fisheries, which are four times greater than the net private benefits. Management options suggested by this socio-economic assessment include an awareness program (to inform blast fishers of the links between blast fishing and their own livelihood and the general status of Indonesian coral reefs, the latter to counter the blast fishers' perceptions that catches can be improved by traveling to other reefs), provision of alternative livelihoods (e.g., pelagic fisheries, mariculture, tourism), greater enforcement, and a locally managed credit system (Pet-Soede et al. 1999).

Case study 3: Using understanding of the influence of socio-economics on the perceptions of coastal resource issues to address underlying causes

The coastal resources adjacent to the small fishing village of Mahahaul (Mexico) support fishing and tourism and are therefore socially and economically significant to residents. A socio-economic assessment and interviews revealed that most residents believed the reef and fishery were in poor condition. The migration status, wealth, and education of residents influenced their perception of the causes of the issues with wealth being the most influential: poorer residents only attributed declines in fisheries to fishing whereas wealthier residents understood the issue to be the result of a host of inter-related factors (e.g., fishing, increasing tourism, land-based activities). Wealthier residents are therefore more likely to support ecosystem-based management approaches. Developing an understanding of the need for holistic management throughout the coastal community (and thereby improving its chances of success) requires management approaches that include increasing the wealth of poorer residents, e.g., by providing alternative livelihoods or supporting other income generating activities (Cinner and Pollnac 2004).

16.4.7 Stakeholder Participation

Stakeholders are the 'people, groups, communities and organizations who use and depend on the reef, whose activities affect the reef or who have an interest in these activities...' (Bunce et al. 2000). Stakeholders include government agencies mandated with responsibilities for conservation and management, local and indigenous communities, international and local non-government organizations (NGOs), the international donor and lending organizations, the private sector, educators, and researchers. The number of stakeholder groups participating in a conservation and management issue will vary with the issue (e.g., MPA planning vs. fisheries regulations), its scale (e.g., planning for a local MPA vs. an international network of MPAs), the development status of the planning area, and any mandated requirements. Management actions are usually led by governments, but not always. The development of the Ecoregion Plan for the

Sulu-Sulawesi Marine Ecoregion has been led and sustained by an NGO (see Box 16.7).

The participation of stakeholders in management is an acknowledgement of stakeholders' material and personal interests in the outcomes of conservation and management and the practical benefits of this participation. The active inclusion of stakeholders can overcome the limitations of management that arise from insufficient funding, e.g., the participation of stakeholders in monitoring. Communities may also have a strong motivation to initiate and participate in management because an issue directly affects them, e.g., declining fish catches (Pollnac et al. 2001). Stakeholders' desires to participate also reflect a broader need by individuals to make a personal contribution to the sustainability of the marine environment.

Stakeholder participation leads to improved compliance with management regulations and so is more likely to provide successful management outcomes (Bunce et al. 2000). For example, a comparison of MPAs that had been established and managed by central governments with little/no community participation with MPAs that had been established, planned, and managed with a high level of community involvement found that although there were few differences between the two groups in the bio-physical benefits, stakeholder conflicts were more successfully resolved in the community-based MPAs (Alcala et al. 2006). The likely outcome of this difference is a greater chance of long-term sustainability for the community-based MPAs. Limited involvement by stakeholders has been a significant factor in the failure of MPA management in the Caribbean (Mascia 1999) and elsewhere (Beger et al. 2004).

Additional benefits of stakeholder participation include development of more acceptable management practices (Gladstone 2000, Friedlander et al. 2003, Fernandes et al. 2005), improved relationships between management agencies and stakeholders (Bunce et al. 1999, Fernandes et al. 2005), and reduced stakeholder conflicts due to ease of communication and closer links between stakeholders and community management teams (Mefalopulos and Grenna 2004, Alcala et al. 2006). Agencies benefit from the increased awareness by stakeholders of the management process and the support this engenders for an organization and its aims. The increased awareness of coastal issues amongst participating stakeholders is likely to produce positive flow-on effects to other aspects of people's interaction with the environment.

The opportunities for stakeholder participation include participation in the planning process (e.g., by representing the interests of stakeholder groups), the use of local knowledge in the planning process, the incorporation of traditional management practices in management plans, stakeholder-led planning and management (see Box 16.7), assistance with management implementation (e.g., as volunteer community rangers), public review of draft management plans and environmental assessments, and opportunities for volunteerism (Figs. 16.3, 16.4, Box 16.5). Specific examples of these different opportunities for participation are given in Appendix 16.1.

Fig. 16.3 Recreational
SCUBA divers from the
Solitary Islands Underwater
Research group assisting with
volunteer coral reef
monitoring in the Solitary
Islands Marine Park,
Australia (photo: Ian
Shaw©)

Fig. 16.4 Reef HQ
volunteers are a valuable
source of information for the
general community about the
Reef HQ Aquarium and the
Great Barrier Reef Marine
Park (photo: Great Barrier
Reef Marine Park Authority)

Box 16.5 Community volunteers working for coral reef conservation

The Reef HQ Volunteers Association grew out of community enthusiasm for
active involvement with the work of the Reef HQ Aquarium (Townsville, Australia). The goal of the Reef HQ Aquarium is to 'Inspire all to care for the
Great Barrier Reef'. The Volunteers Association involves the community in
achieving this goal through the ongoing education of the general public about
the Great Barrier Reef and the Marine Park. Volunteers are engaged in nearly
all facets of the Reef HQ Aquarium's operations, including interpretation,
education, curatorial, exhibits, administration, promotions, and marketing.
Interpretive volunteers assist with visitor information, conduct talks/tours and
visitor surveys, craft activities, and provide one-to-one interpretation about
the Reef. Education volunteers assist with school groups during the day and

sleepovers at night. Administrative volunteers assist with databases, mail-outs, and photocopying. Exhibits volunteers help with displays. Curatorial volunteers help maintain tanks, prepare feeds, water changes, and other duties. Marketing and promotions volunteers conduct sales calls to hotels/motels and other tourism outlets and promote the Reef HQ Aquarium at special community events. Since its inception in 1987, the Volunteers Association has trained over 975 volunteers, who have collectively contributed over 290,000 hrs of voluntary service, which is valued in excess of A$4 million dollars. New volunteers undergo an initial 18-h basic training course over a seven-week period, followed by an additional 8 hrs of specialized team training over a four-week period. This induction is designed to equip volunteers with information, skills, and confidence needed to provide visitors with an informative and enjoyable experience. The training course covers diverse topics including: operating structure of the Reef HQ Aquarium and an orientation of the facility, marine biology, coral reef ecology, and communication skills and presentation techniques.

Source: staff and volunteers at the Reef HQ Aquarium

Ensuring the success and sustainability of stakeholder participation is challenging but essential, given most issues usually require long-term solutions. Many stakeholders volunteer in their spare time and often with little support and so avoiding volunteer and stakeholder 'burnout' is a major concern, especially when conservation activities come to depend almost entirely on volunteer workers. Practical steps to ensure success and sustainability include: maintaining the mechanisms for ongoing stakeholder input, demonstrating the links between participation and positive conservation and material outcomes for participants, combining community development plans with conservation, providing information, education, and community activities, mandating continued stakeholder participation in legislation, public recognition of successful partnerships, and partnerships with local rather than central governments (Appendix 16.1).

16.4.8 Education

It is essential to develop people's understanding of the relationship between their actions and the environmental problems these may create. It is also essential for people to appreciate the costs arising from not properly managing human uses of ecosystems. Both are needed as a step in getting people to accept the need to change their behavior. A central means of achieving this understanding is communication, education, and public awareness: 'Without communication, education and public awareness, biodiversity experts, policy makers and managers risk continuing conflicts over biodiversity management, ongoing degradation and loss of ecosystems, their functions and services. Communication, education and public awareness provide the link from science and ecology to people's social and economic reality' (Van Boven and Hesselink 2002). This section reviews the recognition given to the importance of education, the range of potential benefits, and specific examples of successful education approaches (summarized in Appendix 16.1).

The central role of education in conservation and management is recognized in international conventions such as the United Nation's Agenda 21 (1992) and Johannesburg Plan of Implementation (2002). It is also incorporated into regional conventions such as the Protocol Concerning Specially Protected Areas and Wildlife that is part of the Convention for the Protection and Development of the Marine Environment of the Wider Caribbean Region. Education is a key management action also in issues-based programs (e.g., Great Barrier Reef Water Quality Protection Plan). In recognition of the significance of education the United Nations declared 2005–2014 the Decade of Education for Sustainable Development.

Education is a powerful tool to develop and increase political will for the need for management (Pet-Soede et al. 1999) and to demonstrate to stakeholders the consequences of their activities, e.g., overfishing (Bunce et al. 1999). Education can support the development of new management actions by explaining the need for management to stakeholders (Fernandes et al. 2005) and demonstrating the potential benefits from examples of similar management (Rodriguez-Martinez and Ortiz 1999, Alcala et al. 2006). Demonstrations of the positive benefits of establishing community-based MPAs have been effective at stimulating fishing communities in the Philippines to create new MPAs (Alcala et al. 2006). Enhancing users' experiences (e.g., by providing information to tourists) will increase the support amongst the broader community for management. Education is as effective as enforcement at ensuring users' compliance with MPA regulations (Alder 1996). Stakeholders' personal ecological effects can be significantly reduced by brief educational interventions that provide information and skills instruction, and education to reduce the impacts of divers on corals has been particularly effective (Medio et al. 1997, Rouphael and Inglis 2001, Hawkins et al. 2005).

On the other hand, lack of awareness (e.g., MPA boundaries) will constrain management success (Alcock 1991, Kelleher and Kenchington 1992, Bunce et al. 1999). Inadequate education is one of the main causes of unsuccessful MPAs (Browning et al. 2006). However, not all stakeholders and groups respond positively to education and so education has to be seen as one part of a mix of management activities. Although education programs can be more expensive than other forms of management, e.g., compliance enforcement (Alder 1996), the potential for flow-on effects to other individuals (e.g., friends, family) and the lifelong changes in behavior (Browning et al. 2006) greatly amplifies the benefits of education.

Each conservation and management issue that requires education has a specific communication issue. For example, conventional environmental education may not be sufficient to convey the complexity of the connections among coastal ecosystems. Connectivity presents issues in technical understanding, complex concepts, and large-scale thinking. Research in the field of environmental education has shown that it is not sufficient to provide information (e.g., about the state of the coast, value of biodiversity) because there is no cause-and-effect from providing information to changes in attitude and behavior. A more effective approach is to create relevance in the lives of individuals (Denisov and Christoffersen 2000, NSW National Parks and Wildlife Service 2002, Gladstone et al. 2006). Major contributors to developing a commitment to conservation include childhood contact with nature (Box 16.6) and the influence of a significant adult, rather than formal education (Palmer 1995). Approaches to successful environmental education have been reviewed recently (Rickinson 2001, Browning et al. 2006). Examples of specific education activities are provided in Appendix 16.1.

Box 16.6 Educating children about marine conservation: SeaWeek on Lord Howe Island

Today's children play a vital role in the future conservation of the marine environment. It is their actions both now and in adult life that will directly affect coral reef communities. Educating children about marine environments can result in positive attitudes and behaviors that can also influence family and friends. Effective education will make children responsible managers of the world's vulnerable coral reefs. One community that takes marine education seriously is World Heritage listed Lord Howe Island. Residents actively support SeaWeek (the annual national public awareness campaign organized by the Marine Education Society of Australasia—MESA) through planned activities and events for the Central School and local community. SeaWeek on Lord Howe is timed to coincide with coral spawning. Night dives and guided snorkeling enable both children and adults to witness this spectacular phenomenon and to learn about the reproductive cycle of corals. Ecological reef walks are guided by a resident naturalist who educates island visitors about coral reefs and their unique organisms and habitats. School activities include snorkel trips conducted by one of the local tour operators and presentations by the Marine Park Manager and other guests such as visiting scientists and marine educators (Fig. 16.5).

Source: Christine Preston, Faculty of Education and Social Work, University of Sydney

---

The stakeholder groups providing education will vary with the context of the education program and the intended audience. Education can be the responsibility of government agencies responsible for coastal management, NGOs, universities, secondary and primary schools, the private sector (e.g., tourist resorts, private aquaria), and informal education providers (e.g., marine discovery centers, visitor centers) (Fig. 16.5).

Most tropical ecosystems occur in developing countries (Fig. 16.6) where management and technical capacity or experience can be constrained by the limited opportunities for formal education and personal experience with successful management interventions. A further constraint on the capacity of many countries to

Fig. 16.5 Children testing
model fish they made to
demonstrate how fish swim
through the water during a
SeaWeek lesson (photo: John
Johnstone)

Fig. 16.6 Development
status (in 2007) of countries
with coral reefs. Sources:
GCRMN (Global Coral Reef
Monitoring Network) (2004)
Status of coral reefs of the
world; 2004. vol 1. Australian
Institute of Marine Sciences,
Townsville

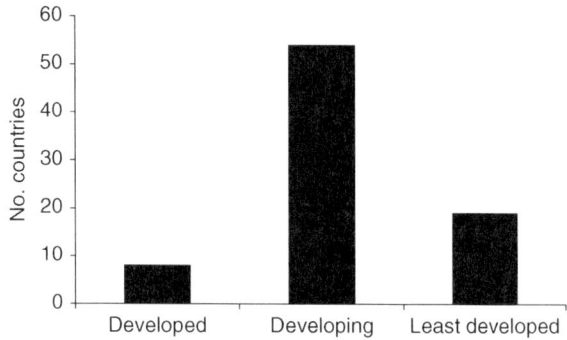

conserve and manage coastal ecosystems is the migration of talented individuals to countries with greater opportunities, further weakening national capacity (Gladstone 2008). In these situations, the continuing education of conservation professionals (i.e., capacity building) (Appendix 16.1) and the development of effective and sustainable conservation and management institutions (i.e., institutional strengthening) must precede or go hand-in-hand with the implementation of more visible forms of management such as MPAs.

16.4.9 Manage at the Most Appropriate Spatial Scale

Planning by agencies and nations for sustainable use and conservation of individual ecosystems, species, catchment-coast linkages, and bioregions may be sufficient in situations where there are few external influences (e.g., isolated oceanic atolls) or no overlapping boundaries with other nations. However, there are compelling reasons for managing at much greater spatial scales. Ecosystem boundaries can be very large when they include the boundaries of relevant catchments and the extent of seaward influence of coastally-derived water masses. The semi-enclosed nature of many seas (e.g., the Caribbean Sea, Red Sea) and the large-scale, trans-boundary nature of some issues (e.g., pollution, climate change, invasive species, coral diseases) means that countries may be affected by issues occurring in neighboring countries. Pathways of connectivity resulting from the trans-boundary movements of pelagic larvae and eggs from spawning sites can cross national boundaries (Domeier 2004). Migratory species such as whale sharks, pelagic fishes, turtles, and cetaceans move between feeding and mating grounds in different countries (Eckert and Stewart 2001). Many countries also share biogeographic regions and catchments. Effective and sustainable solutions to problems are therefore more likely to follow from cooperative approaches rather than single-country actions.

In addition to these bio-physical planning considerations, variations in the development status of countries within a region may mean that the required financial, technical and management expertise for conservation and sustainable use is beyond the capacity of some countries. In these situations support from neighboring countries is required to achieve regional conservation goals (Gladstone et al. 2003). Looking even further afield, the benefits that developed countries gain from the ecosystem goods and services of developing countries places some responsibility on these developed countries to assist, where necessary, with conservation and sustainable use in these developing countries.

The practical options for large-scale management include legally binding global treaties (e.g., United Nations Convention on Law of the Sea), sectoral agreements (e.g., International Convention for the Prevention of Pollution from Ships, International Convention for the Regulation of Whaling), and species-based agreements (e.g., Memorandum of Understanding on the Conservation and Management of Marine Turtles and their Habitats of the Indian Ocean and South-East Asia) (Agardy 2005, UNEP 2006). In the remainder of this section I will focus on the use of spatial management to achieve large-scale conservation and sustainable use.

The Regional Seas Programme was established by the United Nations Environment Programme (UNEP) in 1974 to assist and engage countries sharing a common body of water in cooperative conservation and management activities to address regional issues. The Programme covers 18 regions (and more than 140 countries): the Antarctic, Arctic, Baltic, Black Sea, Caspian, Eastern Africa, East Asian Seas, Mediterranean, North-East Atlantic, North-East Pacific, North-West Pacific, Pacific, Red Sea and the Gulf of Aden, ROPME Sea Area, South Asian Seas, South-East Pacific, Western Africa, and the Wider Caribbean. Countries within each region commit to addressing these issues through their development of a regional plan of action and their ratification of a legally binding convention and associated issue-specific protocols.

The Wider Caribbean Region includes 28 island and continental countries with coasts on the Caribbean Sea and Gulf of Mexico and the adjacent waters of the Atlantic Ocean. Governments of these countries identified the main issues as: land-based sources of wastes and run-off, over-exploitation of marine resources, increasing urbanization and coastal development, unsustainable agricultural and forestry practices, and a lack of government and institutional capacity to address environmental problems. The Convention for the Protection and Development of the Marine Environment in the Wider Caribbean Region (the Cartagena Convention) requires countries to 'protect, develop, and manage their common coastal and marine resources individually and jointly'. The Convention has three associated protocols dealing with oil spills, specially protected areas and wildlife, and pollution from land-based sources and activities. The Protocol on Specially Protected Areas and Wildlife, which became international law in 2000, contains a list of protected species and guidelines for the establishment of protected areas, national and regionally cooperative measures for species' protection, environmental impact assessment, research, and education. Regionally cooperative measures include common measures to protect listed protected species, and common guidelines and criteria for the identification, selection, establishment, and management of protected areas (Appendix 16.1).

Large Marine Ecosystems (LMEs) are regions that cover more than 200,000 km^2 of ocean and extend from the landward boundary of catchments and include estuaries and the coastal zones out to the seaward boundaries of continental shelves or a major current system and also possess a distinct bathymetry, hydrography, and productivity. There are 64 recognized LMEs and many occur within the boundaries of UNEP's Regional Seas. The Global Environment Facility has developed an approach for nations sharing an LME to address jointly their coastal and marine issues (Duda and Sherman 2002). A trans-boundary diagnostic analysis provides a synthesis of trans-boundary concerns and their root causes. Based on this, a strategic action program (SAP) plots the reforms needed regionally and nationally. Box 16.7 illustrates the specific steps undertaken to develop a SAP.

Box 16.7 Case studies in regional approaches to conservation and sustainable use

Case study 1: Strategic Action Programme (SAP) for the Red Sea and Gulf of Aden

The Red Sea and Gulf of Aden SAP's global objective was to safeguard the coastal and marine environments and ensure sustainable use of its resources. The SAP's activities involved: institutional strengthening, reducing navigation risks and maritime pollution, sustainable fisheries, habitat and biodiversity conservation, development of a regional network of marine protected areas, support for integrated coastal zone management, enhancement of public awareness and participation, and monitoring and evaluation of the SAP's outcomes. Implementing regionally-agreed objectives at the national-level has been a major challenge. In the Red Sea and Gulf of Aden regional status assessments were used to develop regional action plans for the conservation of turtles and breeding seabirds. National action plans were developed to facilitate national implementation of the regional needs. Given the discrepancy in capacity among countries, the action plans were adapted to suit each particular country. National implementation is occurring through integrated networks of national and local working groups, government departments, agencies and personnel, non-governmental organizations, and other stakeholders (Gladstone et al. 1999, Gladstone 2006).

Case Study 2: Sulu-Sulawesi Sea Marine Ecoregion

The Sulu-Sulawesi Sea Marine Ecoregion (SSME) (shared among Indonesia, Malaysia, the Philippines) was listed by World Wide Fund for Nature (WWF) as one of its Global 200 ecoregions because of its global significance for coral and fish diversity (see Fig. 16.7), rare and endangered species (including the coelacanth), habitat richness, productive coastal ecosystems, and the large coastal population (35 million people) dependent on its resources. Degradation resulted from unsustainable levels of resource use, poverty, and increasing populations. The SSME Conservation Program involved regional conservation planning combined with specific actions for sites and species. Implementing the Program required a Biodiversity Vision (based on inputs of 70 stakeholders) that guided development of an Ecoregion Conservation Plan (requiring 12 stakeholder workshops across the three nations). The Plan identified 10 objectives with associated actions to be implemented nationally or regionally over 10 years and was formally adopted as national policy by the three nations in a memorandum of understanding in 2004. National actions included enforcement at key sites, an integrated conservation and development plan for key sites, local community education, and a GIS data base. Actions to be implemented collaboratively across the ecoregion include protection of sea turtles, improved fisheries management, and a network of MPAs (Miclat et al. 2006).

Fig. 16.7 The Sulu-Sulawesi Marine Ecoregion is one of the World Wide Fund for Nature's Global 200 ecoregions because of its significance for marine life (photo: David Harasti)

Coordinated networks of MPAs have the potential to achieve local, national and regional conservation goals simultaneously. Regional networks of MPAs have been designed to represent in a complementary manner examples of major regional ecosystems (Gladstone et al. 2003, Agardy 2005), source areas highly connected to areas outside a network, and areas important for species of special concern (Miclat et al. 2006; Box 16.7). To be effective, the functioning of a regional network of MPAs requires a legally binding protocol ratified by all countries within the region, a mechanism for regional coordination of the network, and active on-ground management within each of the participating MPAs. The objective of the regional coordination role should be to support individual MPAs in achieving their management objectives so that regional objectives for sustainable use and conservation are being met. The regional support role can include: development of regionally-agreed guidelines for the selection, establishment and management of MPAs, support for revenue generation, capacity building (e.g., through training in management), and monitoring (Gladstone et al. 2003). The impediments to regional approaches include lack of political will for cooperation with neighboring countries, varying capacity among participating countries, the diversity of stakeholders with varying and conflicting priorities (Agardy 2005), varying opportunities for participation by stakeholders in different countries, and incomplete ratification of legally binding agreements.

16.5 Conclusion

Addressing the issues confronting tropical coastal ecosystems requires management approaches that restore and conserve the natural patterns and functions of ecosystems. The recent advances in our understanding of the dynamics of marine and coastal ecosystems, the scales at which they operate, and the development of new technologies (reviewed in Part 3 of this book) have supported simultaneous advances in conservation and management. These advances include the selection and design of MPAs, habitat rehabilitation, environmental assessment, and the developing field of seascape approaches to management. Achieving society's aspirations for coastal ecosystems requires, on the one hand, changes to people's behaviors and greater awareness about the effects of their actions. This cannot be successful without deep understanding of human coastal societies and economies and their interaction with local coastal ecosystems. The greatest challenges to conservation and management will continue to come from larger factors such as global climate change, poverty, population growth and coastal migration, and low development status.

Acknowledgments I wish to thank the following people for their assistance with this chapter: Steven Lindfield (University of Newcastle), Christine Preston (University of Sydney), David Harasti (for photos), Julie Jones (Great Barrier Reef Marine Park Authority), Julie Spencer (Reef HQ Townsville), Evangeline Miclat (Conservation International–Philippines), Ian Shaw (for photo), and Ivan Nagelkerken for the invitation to participate as an author. I thank Ivan Nagelkerken and two anonymous reviewers for helpful comments on an earlier version of this chapter.

Appendix

Appendix 16.1 Examples of management tools, actions, and interventions to implement the conservation and management goals reviewed in this chapter

Goal	Tools, actions, interventions
Resilience	Education about ecosystems and functional groups
	Establish MPAs to manage/minimize usage
	Fisheries management to maintain functional diversity (e.g., banning fish traps) and abundant populations
	Global action to address causes of coral bleaching
	Identify and protect source reefs (e.g., via MPAs)
	Identify and protect spawning aggregation sites and other sources of replenishment
	Integrated coastal zone management to protect water quality (e.g., catchment management to maintain water quality)
	Minimize disturbance (e.g., oil spills) to currents connecting source reefs with sink reefs
	Protect connected habitats used by all life history stages of key functional groups (Mumby et al. 2004)
	Rehabilitate habitat and species to increase populations and return keystone species (Jaap et al. 2006)
	Status assessments of key functional groups followed by development of action plans for their conservation (Gladstone 2006)
	Trade restrictions on key functional groups and species
Connectivity	Establish MPAs at well-connected reefs to maintain replenishment of downstream reefs (Roberts et al. 2006)
	Integrated coastal zone management (Furnas 2003) to manage external influences on connected habitats, e.g., land use management, development controls, habitat protection plans, environmental assessment
	Protect connected habitats used by all life history stages of key functional groups (Mumby et al. 2004), threatened species, fisheries species in MPA networks
	Rehabilitate connected habitats (Keller and Causey 2005)
Conservation of representative samples of the variety of species and assemblages	Develop maps of spatial distribution of biodiversity or suitable surrogates such as indicator groups (Gladstone 2002), habitats (Friedlander et al. 2003), environmental gradients
	Reserve selection algorithms to identify cost-effective options for MPA networks (Possingham et al. 2000)
	Selection and design guidelines for MPAs (Salm et al. 2000, Fernandes et al. 2005)
Conservation and recovery of species-at-risk	Codes of conduct for industries (e.g., eco-tourism) that interact with these species
	Community education and participation (e.g., in monitoring)
	Conservation and recovery plans
	Identification, protection and management of critical habitats, e.g., nesting sites (Gladstone 2000), spawning aggregations

Appendix 16.1 (continued)

Goal	Tools, actions, interventions
	International treaties
	National conservation and environmental assessment legislation
	Provide alternative livelihoods to communities dependent on threatened species
	Status assessment
	Substitute captive breeding for wild harvest (IUCN 2007)
	Trade restrictions on threatened species, e.g., syngnathids, *Cheilinus undulatus*
Protect water quality	Comprehensive monitoring of water quality, algae, and other ecosystem components
	Economic incentives (e.g., to encourage landholders to implement sustainable management practices and property level planning)
	Environmental assessment for new developments (UNEP 2006)
	Establish targets and priorities (e.g., identify catchments in good condition for protection)
	Planning for natural resource and land management (e.g., habitat rehabilitation, reduce soil and nutrient loss, minimal use of herbicides and pesticides, protect wetlands, riparian zones, and native vegetation important to maintain and improve water quality) (Furnas 2003, The State of Queensland and Commonwealth of Australia 2003)
	Public education and awareness (e.g., increasing stakeholders' awareness of the value for water quality of wetlands and riparian habitat) (The State of Queensland and Commonwealth of Australia 2003)
	Reduce and control point-source inputs via licensing (UNEP 2006), infrastructure improvements
	Regulatory frameworks (e.g., legislation, guidelines, compliance)
	Research and information sharing (e.g., distribute research findings to stakeholder groups)
	Self-management by stakeholders (e.g., industry-led development of best management practice for land, natural resources, and chemical use) (The State of Queensland and Commonwealth of Australia 2003)
	Stakeholder partnerships (e.g., industry, all levels of government, research)
Assessing the socio-economic context for management	Assess effects of management on stakeholders (Elliott et al. 2001) including the need for provision of alternative livelihoods (Pet-Soede et al. 1999, Elliott et al. 2001), or structural adjustments for displaced stakeholders (Fernandes et al. 2005)
	Community services and facilities (Hariri 2006)
	Explore alternative management options with stakeholders (Friedlander et al. 2003, Fernandes et al. 2005)
	Gender issues (Bunce et al. 2000)
	Identification of stakeholders (Bunce et al. 1999)
	Identify and quantify values (market, non-market, non-use) (Ahmed 2004) including costs of current uses (Pet-Soede et al. 1999)

Appendix 16.1 (continued)

Goal	Tools, actions, interventions
	Market attributes (Pet-Soede et al. 1999)
	Organization and resource governance (Elliott et al. 2001)
	Patterns of stakeholders' resource use (Bunce et al. 1999, Gladstone 2000, Cinner and Pollnac 2004)
	Stakeholder characteristics (Gladstone 2000)
	Stakeholder perceptions of the resources and issues (Bunce et al. 1999, Gladstone 2000, Cinner and Pollnac 2004)
	Traditional knowledge (Bunce et al. 2000)
Stakeholder participation	Assistance with management implementation, e.g., installation of mooring buoys (Bunce et al. 1999)
	Combine community development plans with conservation (Pollnac et al. 2001, UNEP 2006)
	Community-based management of MPAs (Pollnac et al. 2001, Beger et al. 2004)
	Demonstrated links between participation and positive conservation and material outcomes for participants, e.g., improvements in fish stocks, living standards (Alcala et al. 2006)
	Incorporation of traditional management practices (Gladstone 2000, Johannes 2002)
	Maintenance of a mechanism that provides for ongoing stakeholder input, e.g., consultative committees (Pollnac et al. 2001)
	Participation in MPA planning by representing the interests of stakeholder groups (Gladstone 2000, Friedlander et al. 2003, Fernandes et al. 2005), providing local knowledge, e.g., spawning aggregations, important species, status of fish stocks, resource-use conflicts (Gladstone 2000, Johannes 2002)
	Partnerships with local rather than central governments (Alcala et al. 2006)
	Provision of information, education, community activities (Pollnac et al. 2001, Alcala et al. 2006)
	Public review and commentary on draft management plans (Fernandes et al. 2005) and environmental assessments
	Stakeholder participation mandated in legislation (Alcala et al. 2006)
	Volunteerism, e.g., community rangers (Alcala et al. 2006), clean-up events (Bunce et al. 1999), education, habitat restoration (Jaap et al. 2006), monitoring (Hodgson 2000), animal rescue
Education	Capacity building and institutional strengthening
	Community outreach (Rodríguez-Martrínez and Ortiz 1999, Browning et al. 2006)
	Diver briefings (Medio et al. 1997)
	Eco-tourism interpretation (Andersen and Miller 2006)
	Education facilities, e.g., aquaria, marine discovery centers, visitor centers, displays, interpretative signage, nature trails (Evans 1997, Browning et al. 2006)
	Educational materials, e.g., posters, stickers, brochures, CDs, DVDs
	Mass media, e.g., radio, newspaper, TV
	Modules in school curricula (Browning et al. 2006)
	On-the-job training and workplace exchanges (Crawford et al. 1993)
	Partnerships with international donors/developed countries (Gladstone et al. 2003)
	Regional training centers (Gladstone 2006)
	Stakeholder education

Appendix 16.1 (continued)

Goal	Tools, actions, interventions
Manage at the appropriate spatial scale	Tertiary education (Smith 2000)
	Training workshops and short courses (Smith 2002)
	Virtual education, e.g., internet discussion forums, list servers, e-mail exchanges, International Coral Reef Information Network's–Coral Reef Education Library
	Analysis of regional issues and root causes (Gladstone et al. 1999)
	Formal adoption by national governments of regional/national actions (Miclat et al. 2006)
	Identification and participation of national and international stakeholders (Gladstone et al. 2003, Miclat et al. 2006)
	Legally binding regional conventions and issue-specific protocols (e.g., protected areas, pollution, biodiversity conservation) (Gladstone 2006, Miclat et al. 2006)
	Programs of complementary regional and national actions to address issues, e.g., MPA networks (Gladstone et al. 2003, Miclat et al. 2006), education, pollution reduction, capacity building, monitoring, regional database and GIS (Gladstone 2006)
	Regional coordinating mechanism (e.g., secretariat), advisory body of representatives of participating governments (Gladstone et al. 1999)

References

Access Economics (2007) Measuring the economic and financial value of the Great Barrier Reef Marine Park, 2005–06. Access Economics Pty Limited, Canberra

Agardy T (2005) Global marine conservation policy versus site-level implementation: the mismatch of scale and its implications. Mar Ecol Prog Ser 300:242–248

Ahmed M (2004) An overview of problems and issues of coral reef management. In: Ahmed M, Chong CK, Cesar H (eds) Economic valuation and policy priorities for sustainable management of coral reefs. WorldFish Center, Penang

Alcala AC, Russ GR, Nillos P (2006) Collaborative and community-based conservation of coral reefs, with reference to marine reserves in the Philippines. In: Cote IM, Reynolds JD (eds) Coral reef conservation. Cambridge University Press, Cambridge

Alcock D (1991) Education and extension: management's best strategy. Aust Park Recreation 27:15–17

Alder J (1996) Costs and effectiveness of education and enforcement, Cairns section of the Great Barrier Reef marine park. Environ Manage 20:541–551

Allen GR (2000) Threatened fishes of the world: *Pterapogon kauderni* Koumans, 1933 (Apogonidae). Environ Biol Fishes 57:142

Andersen MS, Miller ML (2006) Onboard marine environmental education: whale watching in the San Juan Islands, Washington. Tour Mar Environ 2:111–118

Australian and New Zealand Environment and Conservation Council Task Force on Marine Protected Areas (1999) Strategic plan of action for the national representative system of marine protected areas: a guide for action. Environment Australia, Canberra

Balmford A, Bruner A, Cooper P et al (2002) Economic reasons for conserving wild nature. Science 297:950–953

Beger M, Harborne AR, Dacles TP et al (2004) A framework of lessons learned from community-based marine reserves and its effectiveness in guiding a new coastal management initiative in the Philippines. Environ Manage 34:786–801

Bellwood DR (1995) Carbonate transport and within reef patterns of bioerosion and sediment re-lease by parrotfishes (family Scaridae) on the Great Barrier Reef. Mar Ecol Prog Ser 117:127–136

Berg H, Ohman MC, Troeng S et al (1998) Environmental economics of coral reef destruc-tion in Sri Lanka. Ambio 27:627–634

Browman HI, Stergiou KI (2005) Introduction to 'Politics and socio-economics of ecosystem-based management of marine resources' theme section. Mar Ecol Prog Ser 300:241–242

Browning LJ, Finlay RAO, Fox LRE (2006) Education as a tool for coral reef conservation: lessons from marine protected areas. In: Cote IM, Reynolds JD (eds) Coral reef conservation. Cambridge University Press, Cambridge

Bruno JF, Selig ER (2007) Regional decline of coral cover in the Indo-Pacific: timing, extent, and subregional comparisons. PLoS ONE 2:e711

Bunce L, Gustavson K, Williams J et al (1999) The human side of reef management: a case study analysis of the socioeconomic framework of Montego Bay Marine Park. Coral Reefs 18:339–380

Bunce L, Townsley P, Pomeroy R et al (2000) Socioeconomic manual for coral reef man-agement. Australian Institute of Marine Science, Townsville

Burke L, Maidens J (2004) Reefs at risk in the Caribbean. World Resources Institute, Washington DC

Burke L, Selig E, Spalding M (2002) Reefs at risk in Southeast Asia. World Resources Institute, Washington DC

Carpenter RC, Edmunds PJ (2006) Local and regional scale recovery of Diadema promotes recruitment of scleractinian corals. Ecol Lett 9:271–280

Carpenter S, Walker B, Anderies JM et al (2001) From metaphor to measurement: resilience of what to what? Ecosystems 4:765–781

Cesar H, Burke L, Pet-Soede C (2003) The economics of worldwide coral reef degradation. Cesar Environmental Economics Consulting, Arnhem

Cinner JE, Pollnac RB (2004) Poverty, perceptions and planning: why socioeconomics matter in the management of Mexican reefs. Ocean Coast Manage 47:479–493

Claydon J (2004) Spawning aggregations of coral reef fishes: characteristics, hypotheses, threats and management. Oceanogr Mar Biol Annu Rev 42:265–202

Connell JH (1997) Disturbance and recovery of coral assemblages. Coral Reefs 16(suppl):S101–S114

Connell SJ (2007) Water quality and the loss of coral reefs and kelp forests: alternative states and the influence of fishing. In: Connell SJ, Gillanders BM (eds) Marin ecology. Oxford University Press, Melbourne

Connolly RM, Lee SY (2007) Mangroves and saltmarsh. In: Connell SJ, Gillanders BM (eds) Marine ecology. Oxford University Press, Melbourne

Costanza R (1999) The ecological, economic, and social importance of the oceans. Ecol Econ 31:199–213

Costanza R, d'Arge R, de Groot R et al (1997) The value of the world's ecosystem services and natural capital. Nature 387:253–260

Crawford BR, Stanley Cobb J, Friedman A (1993) Building capacity for integrated coastal man-agement in developing countries. Ocean Coast Manage 21:311–337

Dahdouh-Guebas F (2002) The use of remote sensing and GIS in the sustainable management of tropical coastal ecosystems. Environ Dev Sustain 4:93–112

Denisov N, Christoffersen L (2000) Impact of environmental information on decision making processes and the environment. UN Environmental Programme–Global Resources Information Database, Arendal, Norway

Domeier ML (2004) A potential larval recruitment pathway originating from a Florida marine protected area. Fish Oceanogr 13:287–294

Donaldson TJ, Dulvy NK (2004) Threatened fishes of the world: *Bolbometopon muricatum* (Valenciennes 1840) (Scaridae). Environ Biol Fishes 70:373

Driml S (1994) Protection for profit: economic and financial values of the Great Barrier Reef world heritage area and other protected areas. Great Barrier Reef Marine Park Authority, Townsville

Duarte CM (2000) Marine biodiversity and ecosystem services: an elusive link. J Exp Mar Biol Ecol 250:117–131

Duda AM, Sherman KS (2002) A new imperative for improving management of large marine ecosystems. Ocean Coast Manage 45:797–783

Dulvy NK, Sadovy Y, Reynolds JD (2003) Extinction vulnerability in marine populations. Fish Fish 4:25–64

Eckert S, Stewart B (2001) Telemetry and satellite tracking of whale sharks, *Rhyncodon typus*, in the Sea of Cortez, Mexico, and north Pacific Ocean. Environ Biol Fishes 60:299–308

Edgar GJ, Russ GR, Babcock RC (2007) Marine protected areas. In: Connell SJ, Gillanders BM (eds) Marine ecology. Oxford University Press, Melbourne

Elliott G, Mitchell B, Wiltshire B et al (2001) Community participation in marine protected area management: Wakatobi National Park, Sulawesi, Indonesia. Coast Manage 29:295–316

Evans KL (1997) Aquaria and marine environmental education. Aquar Sci Conserv 1:239–250

Facey R (2006) Sea-based activities and sources of pollution. In: Gladstone W (ed) The state of the marine environment report for the Red Sea and Gulf of Aden. Regional Organization for the Conservation of the Environment of the Red Sea and Gulf of Aden, Jeddah

Fernandes L, Day J, Lewis A et al (2005) Establishing representative no-take areas in the Great Barrier Reef: large-scale implementation of theory on marine protected areas. Conserv Biol 19: 1733–1744

Fine M, Franklin LA (2007) Climate change in marine ecosystems. In: Connell SJ, Gillanders BM (eds) Marine ecology. Oxford University Press, Melbourne

Friedlander A, Nowlis JS, Sanchez JA et al (2003) Designing effective marine protected areas in seaflower biosphere reserve, Colombia, based on biological and sociological information. Conserv Biol 17:1769–1784

Furnas M (2003) Catchments and corals: terrestrial runoff to the Great Barrier Reef. Australian Institute of Marine Science, Townsville

Gardner TA, Cote IM, Gill JA et al (2003) Long-term region-wide declines in Caribbean corals. Science 301:958–960

Gillanders BM (2007) Seagrass. In: Connell SJ, Gillanders BM (eds) Marine ecology. Oxford University Press, Melbourne

Gladstone W (1986) Spawning behavior of the bumphead parrotfish, *Bolbometopon muricatum*, at Yonge Reef, Great Barrier Reef. Jpn J Ichthyol 33:326–328

Gladstone W (1996) Unique annual aggregation of longnose parrotfish (*Hipposcarus harid*) at Farasan Island (Saudi Arabia, Red Sea). Copeia 1996:483–485

Gladstone W (2000) The ecological and social basis for management of a Red Sea marine protected area. Ocean Coast Manage 43:1015–1032

Gladstone W (2002) The potential value of indicator groups in the selection of marine reserves. Biol Conserv 104:211–220

Gladstone W (2006) Coastal and marine resources In: Gladstone W (ed) The state of the marine environment report for the Red Sea and Gulf of Aden. Regional Organization for the Conservation of the Environment of the Red Sea and Gulf of Aden, Jeddah

Gladstone W (2007) Requirements for marine protected areas to conserve the biodiversity of rocky reef fishes. Aquat Conserv Mar Freshw Ecosyst 17:71–87

Gladstone W (2008) Towards conservation of a globally significant ecosystem: the Red Sea and Gulf of Aden. Aquat Conserv Mar Freshw Ecosyst 18:1–5

Gladstone W, Krupp F, Younis M (2003) Development and management of a network of marine protected areas in the Red Sea and Gulf of Aden region. Ocean Coast Manage 46: 741–761

Gladstone W, Owen V (2007) The potential value of surrogates for the selection and design of marine reserves for biodiversity and fisheries. In: Day JC, Senior J, Monk S et al (eds) First international marine protected areas congress, 23–27 October 2005, pp. 224–226. IMPAC1, Geelong

Gladstone W, Stanger R, Phelps L (2006) A participatory approach to university teaching about partnerships for biodiversity conservation. Aust J Environ Educ 22:21–32

Gladstone W, Tawfiq N, Nasr D et al (1999) Sustainable use of renewable resources and conservation in the Red Sea and Gulf of Aden: issues, needs and strategic actions. Ocean Coast Manage 42:671–697

Glasby TM, Creese RG (2007) Invasive marine species management and research. In: Connell SJ, Gillanders BM (eds) Marine ecology. Oxford University Press, Melbourne

Grigg RW (1995) Coral reefs in an urban embayment in Hawaii: a complex case history controlled by natural and anthropogenic stress. Coral Reefs 14:253–266

Grimsditch GD, Salm RV (2005) Coral reef resilience and resistance to bleaching. The World Conservation Union (IUCN), Gland, Switzerland

Hariri K (2006) Living marine resources In: Gladstone W (ed) The state of the marine environment report for the Red Sea and Gulf of Aden. Regional Organization for the Conservation of the Environment of the Red Sea and Gulf of Aden, Jeddah

Harrison PR, Booth DJ (2007) Coral reefs: naturally dynamic and increasingly disturbed ecosystems. In: Connell SJ, Gillanders BM (eds) Marine ecology. Oxford University Press, Melbourne

Hawkins JP, Roberts CM, Kooistra D et al (2005) Sustainability of SCUBA diving tourism on coral reefs of Saba. Coast Manage 33:373–387

Hodgson G (2000) Coral reef monitoring and management using Reef Check. Integr Coast Zone Manage 1:169–176

Hoegh-Guldberg H, Hoegh-Guldberg O (2004) Great Barrier Reef 2050: implications of climate change for Australia's Great Barrier Reef. World Wildlife Fund Australia, Sydney

Hughes TP (1994) Catastrophes, phase shifts and large-scale degradation of a Caribbean coral reef. Science 265:1547–1551

IUCN (2007) Banggai Cardinalfish (*Pterapogon kauderni*) fact sheet. World Conservation Union, Gland

Jaap WC, Hudson JH, Dodge RE et al (2006) Coral reef restoration with case studies from Florida. In: Cote IM, Reynolds JD (eds) Coral reef conservation. Cambridge University Press, Cambridge

Johannes RE (2002) The renaissance of community-based resource management in Oceania. Annu Rev Ecol Syst 33:317–340

Joyce S (2000) The dead zones: oxygen-starved coastal waters. Environ Health Perspect 108:A120–A125

Kay R, Alder J (1999) Coastal planning and management. EF & N Spoon, London

Kelleher G, Kenchington R (1992) Guidelines for establishing marine protected areas. IUCN, Gland

Keller BD, Causey BD (2005) Linkages between the Florida Keys National Marine Sanctuary and the South Florida Ecosystem Restoration Initiative. Ocean Coast Manage 48:869–900

Knowlton N, Jackson JBC (2001) The ecology of coral reefs. In: Bertness MD, Gaines SD, Hay ME (eds) Marine community ecology. Sinauer Associates, Sunderland

Mascia MB (1999) Governance of marine protected areas in the Wider Caribbean: preliminary results of an international mail survey. Coast Manage 27:391–402

McCook LJ, Jompa J, Diaz-Pulido G (2001) Competition between corals and algae on coral reefs: a review of evidence and mechanisms. Coral Reefs 19:400–417

Medio D, Ormond RFG, Pearson M (1997) Effects of briefings on rates of damage to corals by SCUBA divers. Biol Conserv 79:91–95

Mefalopulos P, Grenna L (2004) Promoting sustainable development through strategic communication. In: Hamu D, Auchincloss E, Goldstein W (eds) Communicating protected areas. IUCN, Gland

Meyer JL, Schultz ET (1985) Migrating haemulid fishes as a source of nutrients and organic matter on coral reefs. Limnol Oceanogr 30:146–156

Miclat EFB, Ingles JA, Dumaup JNB (2006) Planning across boundaries for the conservation of the Sulu-Sulawesi Marine Ecoregion. Ocean Coast Manage 49:597–609

Miller MW, Sluka R (1999) Coral-seagrass interaction in an anthropogenically enriched lagoon. Coral Reefs 18:368

Momtaz S, Gladstone W (2008) Ban on commercial fishing in the estuarine waters of New South Wales, Australia: community consultation and social impacts. Environ Impact Assess Rev 28:214–225

Mumby PJ, Dahlgren CP, Harborne AR et al (2006) Fishing, trophic cascades, and the process of grazing on coral reefs. Science 311:98–101

Mumby PJ, Edwards AJ, Ernesto Arias-Gonzalez J et al (2004) Mangroves enhance the biomass of coral reef fish communities in the Caribbean. Nature 427:533–536

Mumby PJ, Harborne AR (2006) A seascape-level perspective of coral reef ecosystems. In: Cote IM, Reynolds JD (eds) Coral reef conservation. Cambridge University Press, Cambridge

Mumby PJ, Hastings A, Edwards HJ (2007) Thresholds and the resilience of Caribbean coral reefs. Nature 450:98–101

Nagelkerken I, van der Velde G, Gorissen MW et al (2000) Importance of mangroves, seagrass beds and the shallow coral reef as a nursery for important coral reef fishes, using a visual census technique. Estuar Coast Shelf Sci 51:31–44

NSW National Parks and Wildlife Service (2002) Urban wildlife renewal growing conservation in urban communities research report. NSW National Parks and Wildlife Service, Sydney

Nyström M, Folke C (2001) Spatial resilience of coral reefs. Ecosystems 4:406–417

Ogden JC (1997) Ecosystem interactions in the tropical coastal seascape. In: Birkeland C (ed) Life and Death of Coral Reefs. Chapman and Hall, New York

Ogden JC, Ehrlich PR (1977) The behaviour of heterotypic resting schools of juvenile grunts (Pomdasyidae). Mar Biol 42:273–280

Ormond R (1987) Conservation and management. In: Edwards A, Head S (eds) Key environments: Red Sea. Pergamon Press, Oxford

Palmer JA (1995) Influences on pro-environmental practices: planning education to care for the Earth. IUCN Commission on Education and Communication, Gland

Pernetta JC (1992) Impacts of climate change and sea-level rise on small island states. National and international responses. Global Environ Change 2:19–31

Pet-Soede C, Cesar HSJ, Pet JS (1999) An economic analysis of blast fishing on Indonesian coral reefs. Environ Conserv 26:83–93

Pollnac RB, Crawford BR, Gorospe MLG (2001) Discovering factors that influence the success of community-based marine protected areas in the Visayas, Philippines. Ocean Coast Manage 44:683–710

Possingham H, Ball I, Andelman S (2000) Mathematical models for identifying representative reserve networks. In: Ferson S, Burgman M (eds) Quantitative methods for conservation biology. Springer-Verlag, New York

Precht WF, Aronson RB (2006) Death and resurrection of Caribbean coral reefs: a palaeoecological perspective. In: Cote IM, Reynolds JD (eds) Coral reef conservation. Cambridge University Press, Cambridge

Pressey R, McNeill S (1996) Some current ideas and applications in the selection of terrestrial protected areas: are there any lessons for the marine environment. In: Thackway R (ed) Developing Australia's representative system of marine protected areas: criteria and guidelines for identification and selection. Department of the Environment, Sport and Territories, Canberra

Rickinson M (2001) Learners and learning in environmental education: a critical review of the evidence. Environ Educ Res 7:207–320

Roberts CM, Reynolds JD, Cote IM et al (2006) Redesigning coral reef conservation. In: Cote IM, Reynolds JD (eds) Coral reef conservation. Cambridge University Press, Cambridge

Rodriguez-Martinez R, Ortiz LM (1999) Coral reef education in schools of Quintana Roo, Mexico. Ocean Coast Manage 42:1061–1068

Rouphael AB, Inglis GJ (2001) 'Take only photographs and leave only footprints'?: an experimental study of the impacts of underwater photographers on coral reef dive sites. Biol Conserv 100:281–287

Ruitenbeek HJ (1994) Modelling economy-ecology linkages in mangroves: economic evidence for promoting conservation in Bintuni Bay, Indonesia. Ecol Econ 10:233–247

Sadovy Y (1993) The Nassau grouper, endangered or just unlucky? Reef Encount 13:10–12

Sadovy Y, Kulbicki M, Labrosse P et al (2003) The humphead wrasse, *Cheilinus undulatus*: synopsis of a threatened and poorly known giant coral reef fish. Rev Fish Biol Fish 13:327–364

Salm RV, Clark J, Siirila E (2000) Marine and coastal protected areas: a guide for planners and managers. IUCN, Washington DC

Smith HD (2000) Education and training for integrated coastal area management: the role of the university system. Ocean Coast Manage 43:379–387

Smith HD (2002) The role of the social sciences in capacity building in ocean and coastal management. Ocean Coast Manage 45:379–582

Spurgeon J (2006) Time for a third-generation economics-based approach to coral management. In: Cote IM, Reynolds JD (eds) Coral reef conservation. Cambridge University Press, Cambridge

Talbot F, Wilkinson C (2001) Coral reefs, mangroves and seagrasses: a sourcebook for managers. Australian Institute of Marine Science, Townsville

The State of Queensland and Commonwealth of Australia (2003) Reef water quality protection plan: for catchments adjacent to the Great Barrier Reef world heritage area. Queensland Department of Premier and Cabinet, Brisbane

Tomascik T (1990) Growth rates of two morphotypes of *Montastrea annularis* along a eutrophication gradient, Barbados, W.I. Mar Pollut Bull 21:376–381

Torres R, Chiappone M, Geraldes F et al (2001) Sedimentation as an important environmental influence on Dominican Republic reefs. Bull Mar Sci 69:805–818

Turner RK (2000) Integrating natural and socio-economic science in coastal management. J Mar Syst 25:447–460

Turner RK, Subak S, Adger WN (1996) Pressures, trends, and impacts in coastal zones: interactions between socioeconomic and natural systems. Environ Manage 20:159–173

UNEP-WCMC (2006) In the front line: shoreline protection and other ecosystem services from mangroves and coral reefs. UNEP-WCMC, Cambridge

UNEP (2006) Marine and coastal ecosystems and human well-being: a synthesis report based on the findings of the millennium ecosystem assessment. United Nations Environment Programme, Nairobi

Van Boven G, Hesselink F (2002) Mainstreaming biological diversity: the role of communication, education and public awareness [online]. Available at http://www.iucn.org/webfiles/doc/CEC/Public/ Electronic/CEC/Brochures/CECMainstreaming_anglais.pdf.

Veron JEN (1995) Corals in space and time: the biogeography and evolution of the scleractinia. University of New South Wales Press, Sydney

Whittingham E, Campbell J, Townsley P (2003) Poverty and reefs - a global overview. IMM Ltd Innovation Centre, Exeter University, Exeter

Wilkinson C (2006) Status of coral reefs of the world: summary of threats and remedial actions. In: Cote IM, Reynolds JD (eds) Coral reef conservation. Cambridge University Press, Cambridge

Index

A

Acanthopagrus schlegelii, 147
Accuracy of spatial data, 519
Acoustic telemetry, 287, 414, 463, 469, 470, 484, 495, 501, 507, 510
Active settlement site selection, 192
Actual connectivity, 497, 498, 507, 511, 512, 520
Adult-associated chemical cues, 164
Adult habitat, 83, 90, 95, 163, 164, 171, 189, 192, 211, 216, 239–247, 256, 257, 258, 327, 328, 329, 330, 331, 332, 333, 358, 359, 364, 369, 370, 372, 373, 374, 375, 380, 391, 457, 470, 472, 503, 504
Aesthetascs, 147, 148
Alteration of rainfall patterns, 347
Ambassis jacksoniensis, 56, 57
Amphidromy, 326, 335, 336
Anadontostoma chacunda, 339
Anadromy, 326, 336, 338
Analogies and differences between diel and tidal migrations, 301–304
Anguilla, 144, 326, 327, 337, 349
Ao Nam Bor, 23, 24
Archival tags, 463, 469, 470, 484
Artificial environmental markers, 475
Artificial mangrove roots, 384
Artificial tags on decapods, 257
Artisanal fisheries, 535, 544, 545
Assignment tests, 473
Auditory abilities, 166
Auditory cues, 165–167, 171
Auditory senses, 136, 149–151

B

Baitfish, 550, 553, 554
Barramundi, 51, 79, 86, 88, 336, 340–342, 343, 344, 345, 483, 545
Bay of Bengal, 279, 339, 543

Benthic microalgae, 58, 60
Biomarkers, 432, 438, 441, 443, 444, 445, 447
Biosound, 150, 161, 166
Birgus latro, 238, 246
Black carbon, 440, 442
Bromeliad crabs, 238
Burying behavior, 243

C

Caeté Estuary, 23, 49
Callinectes sapidus, 83, 86, 89, 92, 133, 148, 163, 249, 252, 258, 295, 308, 309, 518
Canalized water, 345
Cannibalism, 251
Capacity building, 571, 592, 596, 599, 600
Catadromous, 82, 326, 327, 328, 329, 330, 332, 334, 336, 340–342, 344, 346, 347, 348
Catadromy, 326, 336, 349
Catchment vegetation, 48
CDOM, 432, 433, 435, 436
Changes in habitat requirements, 206, 207
Chemical conspecific cue, 161, 165, 172
Chemical cues emitted from seagrass bed, 161
Chemical habitat imprinting hypothesis, 160
Chemotaxonomic tracers, 444, 445
Climate change, 3, 10, 46, 52, 347–349, 521, 569, 570, 583, 593, 596
Clupeidae, 78, 282, 290, 328, 334, 336, 338, 552, 553
Coastal populations, 88, 579, 580, 595
Coastal waters, 51, 52, 84, 89, 97, 103, 167, 169, 240, 243, 306, 329, 337, 339, 340, 341, 343, 348, 426, 427, 428, 429, 433, 435, 482, 539, 542, 543, 545, 547, 574, 580
Coded wire tag, 258, 458, 462, 466, 467, 468, 484
Color changes, 288

I. Nagelkerken (ed.) *Ecological Connectivity among Tropical Coastal Ecosystems*,
DOI 10.1007/978-90-481-2406-0_BM2 © Springer Science+Business Media B.V. 2009